10th Annual International Conference "Schumpeterian Readings" (ICSR 2021)

SHS Web of Conferences Volume 116 (2021)

Online and Perm, Russia
7 April and 15 - 16 April 2021

Editors:

Nikolay Akatov
Konstantin Antipyev

ISBN: 978-1-7138-3474-8

Printed from e-media with permission by:

Curran Associates, Inc.
57 Morehouse Lane
Red Hook, NY 12571

Some format issues inherent in the e-media version may also appear in this print version.

For additional information, please contact EDP Sciences – Web of Conferences
at the address below.

EDP Sciences – Web of Conferences
17, Avenue du Hoggar
Parc d'Activité de Courtabœuf
BP 112
F-91944 Les Ulis Cedex A
France

Phone: +33 (0) 1 69 18 75 75
Fax: +33 (0) 1 69 28 84 91

contact-edps@webofconferences.org

Additional copies of this publication are available from:

Curran Associates, Inc.
57 Morehouse Lane
Red Hook, NY 12571 USA
Phone: 845-758-0400
Fax: 845-758-2633
Email: curran@proceedings.com
Web: www.proceedings.com

TABLE OF CONTENTS

International Conference «Schumpeterian Readings»

April 07, 15-16, 2021, Perm, Russia

On April 07, 2021, the Plenary meeting of the International Scientific and Practical conference "Schumpeterian Readings - X" was held. The conference sections were held from 15 to 16 of April.

The conference was attended by more than 100 participants from Russia, CIS states, Europe and Asia, including scientists and practitioners from Kazakhstan, Latvia, Lithuania, the Republic of Belarus, Bulgaria, Malaysia, and Germany. The conference reports were presented in two languages (Russian and English) with simultaneous translation. For the online translation Zoom was used.

This volume comprised of the selected reviewed papers dedicated ideas in the field of management, business education and enterprise development. Keynote papers come from renowned scientists and heads of the companies, and dedicated to the questions of the theory and practice of innovation management.

Organizing Institutions

- Perm National Research Polytechnic University
- Government of Perm Region
- International Society of J.A. Schumpeter
- Russian Association of Business Education
- Institute of Economy the Ural Branch of Russian Academy of Sciences
- Retraining Institute Regional Management Centre – RMC

Co-organizing institutions:

- Development Corporation of Perm Krai
- "New Stellar" science and technology park
- Ministry of Industry and Trade of Perm Region
- MBA Centre of South-Ural State Technical University
- Centre of Business Education of Ulyanovsk State Technical University
- MBA Centre of Ural Federal University

About the forum

Goal

To organize a comprehensive discussion of the theory and practice of innovation management.

Objectives

- discussing the innovation management theory
- presenting the ways to managing innovations in industries
- presenting schemes of social innovation management
- discussing the ecosystems innovation management
- presenting advancing innovations in public-sector
- discussing the innovations in human capital development
- analyzing financial mechanism of innovation investment

Official Languages

The official working languages are Russian and English. Simultaneous Russian-English and English-Russian interpretation will be provided at the Conference's plenary, business sessions, and for keynotes reports.

Expected Outcomes

- adopting a general resolution on the Conference's main theme;
- publishing the materials of some research and practical events of the Conference in the indexed journal;
- establishing international interregional cooperation and exchange of information, best practices, and experiences in the theory and practice of innovation management.

Expected Participants

- employees and representatives of public authorities in local and federal levels;
- employees and representatives of international organizations;
- employees and representatives of non-commercial and public organizations;
- representatives of diplomatic missions and trade representatives of foreign countries;
- academics and educators;
- professionals, employees of national and international companies.

Program Committee

Organizing Committee

Chairman

Akatov N.B	*Dr., prof., PNRPU (Perm, Russia)*

Cochairman

Tyulkina S.Y.	*PhD, PNRPU (Perm, Russia)*

Committee members

Antipyev K.A.	*PhD, PNRPU (Perm, Russia)*
Bryukhanov D.Y.	*PhD, department chair of Yaroslavl State University named after P.G.Demidova, Yaroslavl*
Volkova E.A.	*Dr., Faculty of Economics and Mathematics, Ulyanovsk State Technical University*
Gusakov B.I.	*Dr., prof., Belarusian National Technical University, Minsk (Republic of Belarus)*
Dimitrakiyev D.Y.	*Dr., Director of the Higher International School of the Varna Research Institute (Bulgaria)*
Zinatullin R.R.	*PhD, Head of Department, Department of Science and Innovation, PNIPU*
Isaev S.Y.	*PhD, Director of the STI Competence Center in the direction of Photonika at PGNIU, Assistant General Director for the project to create the Photonika consortium of PJSC PNPPK*
Komarov S.V.	*Dr., prof., PNRPU (Perm, Russia)*
Kustova M.M.	*PhD, Head of Personnel Management of Medisorb JSC*
Menshikov V.	*Dr., prof., Head of the Center for Social Research of the Institute of Humanities and Social Sciences of Daugavpils University, Daugavpils (Lithuania)*
Mingaleva Z.A.	*Dr., prof., PNRPU (Perm, Russia)*
Scott K.E.	*Prof. Dr., Vice-Dean Department of Economics and Business Management, Programme Director – MBA International Trade &Master Business Leadership, Professor for International Management and International Finance Anhalt University of Applied Sciences (Germany)*
Salamzade Yashar	*PhD of Economics, University of Sainz, Gelugor, Pinang (Malaysia)*
Stegniy V.N.	*Dr., editor of the publication "Bulletin of PNIPU: Socio-Economic Sciences"*
Tolchin S.V.	*PhD, project manager "Technopolis" New Star"*
Popov V.L.	*Dr., prof., PNRPU (Perm, Russia)*

edp sciences

SHS Web of Conferences

Statement of Peer review

In submitting conference proceedings to SHS Web of Conferences, I certify to the Publisher that I adhere to the **Policy on Publishing Integrity** of the journal in order to safeguard good scientific practice in publishing.

1. All articles have been subjected to peer review administered by the proceedings editors.
2. Reviews have been conducted by expert referees, who have been requested to provide unbiased and constructive comments aimed, whenever possible, at improving the work.
3. Proceedings editors have taken all reasonable steps to ensure the quality of the materials they publish and their decision to accept or reject a paper for publication has been based only on the merits of the work and the relevance to the journal.

Title, date and place of the conference

10th annual International Conference "Schumpeterian Readings"

April 7, 2021 (virtual sessions)
April 15-16, 2021, Perm, Russia (regular sessions)

Proceedings editor(s):

Akatov Nikolay. Perm National Research Polytechnic University, Perm, Russia
Antipyev Konstantin Perm National Research Polytechnic University, Perm, Russia

Date and editor's signature

14/07/2021

Development of innovative technologies as a reserve of competitiveness of reclamation enterprises

Irina Yurchenko[1*], *Mikhail* Bandurin[2], and *Inna* Bandurina[2]

[1]All-Russia Research Institute of Hydraulic Engineering and Reclamation by the name of AN Kostyakov, 44, Bolshaya Akademicheskaya st. Moscow, 127550, Russia
[2]Kuban State Agrarian University named after I.T. Trubilin, 13, Kalinina st., Krasnodar, 350044, Russia

Abstract. Information support of production processes and production is becoming an unchangeable condition for their efficiency, in terms of importance comparable to the provision of material, technical, financial, labor and other resources. The paper is based on the methods of a systematic approach, analysis and synthesis of world trends in national and international theory and practice of interaction between digital technologies and business. It was found that innovations in the field of digital technologies for land reclamation do not meet the requirements of management automation implemented in the developed sectors of national and global economy. The factors of competitiveness of business processes, which are successfully integrated into management decisions using digital systems, are identified. The influence of managerial innovations on the strategic development of the business structures of an enterprise is characterized. In addition, the authors revealed the reasons for the unsatisfactory implementation of high-tech innovations by enterprises of traditional industries. The authors proposed the transfer of digital technologies from a high-tech sector of production to a low-tech one as a priority measure, representing the important condition for the growth of competitiveness of both traditional and high-tech sectors of economy.

1 Introduction

Recently, the informatization of production is considered as a priority approach to ensure the competitiveness of Russian economy, as a guarantor of the effectiveness of business processes of economic activity subjects (organizations, integrated structures, etc.) [1]. The program "Digital Economy of the Russian Federation", approved by the order of the Government on July 28, 2017 N 1632-r, sets the tasks and timeframes for the implementation of measures to support the development of efficient production in Russia by the state on the basis of complex application of highly integrated computer technologies, automation, "Digital twins", digital modeling and information processing at all stages of planning, development, quality assurance of production, operation up to disposal i.e. the entire life period of a product. The use of these approaches is planned for greater availability of public and commercial services, reduction of the cost of goods and products promotion, optimization of the time of payments and creation of alternative sources of income [2]. The purpose of this paper is to substantiate the principles of formation of digital information systems and mechanisms of functioning of digital information technologies in business processes of economic units of water reclamation complex in order to increase its competitiveness [3]. The relevance of the research is explained by the lack of a proper theoretical and methodological basis in the field of land reclamation to solve these issues, despite their exceptional importance, as well as the contradictory assessments of the effectiveness of the relationship between business strategy and digital methods.

2 Materials and methods

The purpose of the study is to form the principles of the creation of innovative systems and conditions for successful functioning of information technologies in the business processes of enterprises of reclamation water complex, contributing to its competitiveness and, as a consequence, viability [4]. During this study the following basic methods are used: a systematic approach, analysis, synthesis, as well as generalization of global trends, national and international theory and practice of interaction between digital technologies and business, contributing to the competitiveness of enterprises, especially those belonging to traditional sectors of economy, including agricultural production and reclamation. The sources of information were presented by the works of scientific and practical conferences, meetings and seminars, periodicals, the

* Corresponding author: chepura@mail.ru

Internet [5] and data from state statistics bodies. In the modern realities of the evolution of market economy, the most important reserve of enterprise competitiveness, along with material-technical, financial, labor and other resources, is information, the timing, completeness and reliability of which determines the effectiveness of control actions at the horizontal and vertical levels of the functioning of production units. The formation of information for the organization of production of high-quality and profitable products (services) [6] is performed by the digital system of an enterprise, which integrates the members of the production process in a single information field and provides controlled access to information about products, materials and production [7]. The increasing market requirements for the properties of supplied products, the introduction of the Internet into the infrastructure of digital systems leads to the transformation of the methods of production, which makes it necessary to automate not isolated functions, but to carry out comprehensive and integrated automation of processes that determine the competitiveness of an enterprise and products supplied to the market. It is necessary to note that the digital infrastructure in the field of reclamation does not correspond to the level of automation of management and the growth of the field of application of digital systems achieved in the developed sectors of the economy of a country [8]. Initially, the development of digital methods was focused on the automation of investment projects for reclamation and was aimed at the formation of an exclusively digital fund, implemented in the format of models of design objects and special applications that ensure the solution of the problems of creation and operation of a designed object. Design automation in the reclamation sector of economy began in the 60s of the 20th century [9]. First of all, the mechanization of sampling processes was carried out, requiring numerous arithmetic calculations and a small number of analytical dependencies and algorithms. The next step is the automated solution of target tasks that establish the design differences and parameters of the designed water reclamation management complex.

In the late 70s and early 80s of the 20th century, the creation of software products for assessing the filtration and water-salt conditions of irrigated lands, calculations of drainage, predicting the evolution of agricultural production and the consumption of the most important natural resources (water and land) increased [10].

Since the mid-70s, the implementation of computer-aided design systems based on design production lines for the functional structures of the designed reclamation facility has prevailed in the amelioration sector of economy. By the beginning of the reforms of Russian economy in the 90s of the 20th century, more than 350 CAD systems and (according to the estimates of experts) and more than 200 specialized programs were operating in design institutes, intended mainly to automate the design process of reclamation facilities.

3 Results and discussion

The changes in the domestic economic mechanism such as cost accounting, self-financing, self-sufficiency and the emergence of the market - determined both a decrease in design volumes and a lack of demand for CAD. Designers, providing savings in the salary fund, stopped performing specific automation work. In addition to the decline in the quality of the project due to the refusal of the required consideration of possible options and the disruption of the production design process, the lack of demand led to an almost complete loss of the "digital heritage" of reclamation [11]. The modern time, characterized by new "ecological thinking" of society and the transformation of the political and economic infrastructure of production, requires the creation of innovative digital technologies with the involvement of creative directions in methodology, mathematical models and methods of modeling, effective technologies, equipment and means of communication [12].

Today in reclamation enterprises there is a need for a class of modern management systems that provide decision support – Decision Support Systems (DSSs). There is a positive trend in the creation, implementation and use of digital technologies in reclamation production, which is proved by [13]:

- the periodic rotation of computers, equipment and means of information and communication links;

- the application of perfect commercial "digital office" programs [14];

- the organization of joint functioning of heterogeneous software products, technical platforms, communication networks, etc .;

- the priority of using control digital production systems, rather than digital information management systems;

- the adaptation of unified commercial digital systems of the enterprises of advanced economy sectors in the practice of reclamation [15].

However, the few digital technologies operating in reclamation enterprises are separated and implemented, for the most part, on the basis of various (analytical, statistical and technical) platforms for the automation of production subsystems, which reduces the effectiveness of managerial impact and increases production costs, which negatively affect competitiveness of both products and an enterprise as a whole.

The rapid development of digital technologies all over the world indicates the need for them by enterprises, but the ease of replication of digital technologies by competitors and the speed of the subsequent loss of the uniqueness of production create duality in the assessment of digital technologies as a factor in the competitiveness of an enterprise [16].

At the same time, the importance of the role of new information and knowledge of intra-production processes is obvious along with the policies of competitors and suppliers of material and technical resources, consumer requests, state tasks and actions of international organizations and other economic conditions in the competitiveness of enterprises [17]. In the case of a loss of the obtained advantages due to the corresponding reaction of competitors, an enterprise does not decrease the level of efficiency of business processes and it remains possible to operate within the framework of the occupied sphere and the established positioning.

In order to get an information resource, the following are used: software for control, accounting and analysis of production and management operations; document management systems, databases of industry organizations, government and corporate portals, websites and other sources of competitors that predetermine the separation and fragmentation of the initial data. The level of efficiency of business process directly depends on the automation of the processes of systematization of separated and fragmented information of various formats into a single information space and the management of this space, which gives reasons to attribute digital technologies to the factors of enterprise competitiveness.

According to the analysis, the effective factors of the competitiveness of an economic entity that are successfully implemented with the support of management decisions by digital systems formed on the basis of digital technologies include [18]:

- the reduction of the cost of production (services) release processes;

- the improvement of consumer properties of products based on the transformation of design and technological solutions;

- the expansion of sales markets;

- the optimization of the period of renewal of products supplied to the market;

- the efficiency of strategic, tactical and day-to-day planning;

- the strict accounting and consistent actions to transform consumer demand;

- the improvement of logistics services;

- the regulation of the composition and structure of workflow and information flows [19];

- the increase in the effectiveness of intra-production coordination of impacts and regulation of production;

- the increase in labor productivity with its automation and the absence of routine.

It is obvious that the capabilities of digital systems that guarantee the competitiveness of a company largely depend on the quality and availability of information, the qualifications of users and the motivation of personnel to improve business processes that affect the efficiency of the use of analytical information. As the strategic importance of technology for market competition decreases, the role of skills in its daily use can become decisive in the success of an enterprise [20].

The competitiveness of the digital systems themselves is provided by:

- the integrity, due to the scientific and methodological substantiation of the principles of creation and evolution;

- the interrelation of digital funds and their regulation operations, which increase the effectiveness of the organization's coordination;

- the use of unified digital technologies for data transformation and transfer;

- the use of uniform standards for data accumulation, complied with the global ones, contributing to the unification of information flows;

- the modular nature of development, openness and flexibility of architecture, realizing the possibility of constant growth and evolution;

- the high level of integration of specialized programs and automation of information procedures and operations of the functional structures of a company, taking into account the balance of production processes and operations;

- the large volumes of processed data, the speed of their transformation and exchange;

- the degree of accuracy, relevance and security of information about the production operations and products of a company in technological lines;

- the effectiveness of communication, both between structural formations and members of the technological line.

- the information security, authorization and control of access to information;

- the presence of professional specialists interconnected with the main activities of a company, specially trained for the operation of high-tech digital systems.

The unsatisfactory solution to these issues leads to the deviations from the timing of the implementation of planned activities, increase in production costs, loss of consumer properties of products and negative phenomena resulting in the loss of the effectiveness of business processes.

One of the possible ways to improve the processability of digital systems in traditional production sectors can be their transfer from high-tech sectors [21]. At the same time, the increase in the level of competitiveness in the traditional production sector is ensured and caused by innovative transformations. A main condition for the feasibility of such a transfer is the readiness of an enterprise in the traditional sector of economy to introduce innovations. For high-tech sectors of economy, the impact of the degree of interconnections with traditional industries on competitiveness is explained by the level of demand for innovative technologies in enterprises of low-tech industries.

It is obvious that each individual company makes its own choice of innovative development, which is based on a thorough analysis of its needs, taking into account real situations and individual preferences. However, it seems correct not to forget about the "classical" approach to the solution of this problem, formulated by the researcher of the impact of digital technologies on the competitiveness of business processes, Nicholas J. Carr: "spend less; follow the leader, do not rush forward; innovate if the risks are negligible; think about shortcomings, not about opportunities "[5].

In the domestic practice of reclamation, the existing volumes of investment in high-tech automation systems, in fact, deprive the organization of the possibility of full-scale implementation of digital methods. The existing high degree of wear and obsolescence of equipment and structures of the reclamation water management complex impedes the implementation of innovative digital methods of production regulation. Consequently, each digital technology is seen as an addition to traditional control.

In the modern state of the economy of Russia, the evolution of innovation processes in the dominant sector of traditional production can be carried out by the supply from high-tech sectors, which will become a source of dynamic growth of Russian economy as a whole. This aspect transfers the problem of the formation of digital economy in traditional production sectors to the state level and does not contribute to the growth of optimism about its early and favorable solution. However, the experience of overcoming difficulties through the examples of the development of informatization of the country, in general, and the reclamation sector of production, in particular, gives confidence in the reality of a successful solution of this problem.

3 Conclusion

The use of information digital system in the practice of regulation of production reduces the cost of management processes and improves consumer properties of products, thereby increasing the competitiveness of a company.

Digital technologies, performing the functions of a link between separate fragmentary information, knowledge, data within a company, industry and government structures, represent the basis for innovations that combine high technologies with traditional goods and services. This provides a supplied product with new properties that are potentially inaccessible to enterprises in the traditional industry without transfer of innovations from high-tech sectors of economy.

References

1. H. Hirsch-Kreinsen *Chelteham* (UK, Edward Elgar Publishing Limited, 2014)
2. I. V. Bogachev *Forum Cnews Transcript of the speech on November 14, 2013 Moscow* (2013)
3. M. A. Bandurin, I. F. Yurchenko, I. P. Bandurina *International Multi-Conference on Industrial Engineering and Modern Technologies, Far East Conference 2019*, 8933970 (2019)
4. M. Jeffrey *Data driven marketing. 15 indicators that everyone should know* (Moscow, Mann, Ivanov and Ferber, 2013)
5. J. Carr *Publishing house Secret of the firm* (2005)
6. D. A. Markov *Izvestiya of the Russian State Pedagogical University. A.I. Herzen* **12(90)** *(*2009)
7. F. K. Abdrazakov, S. S. Orlova, T. A. Pankova, E. N. Mirkina, O. V. Mikheeva *Journal of Interdisciplinary Research* **8(1)** 154-161 (2018)
8. S. Vladimirov, I. Prikhodko, T. Safronova, E. Chebanova *E3S Web of Conferences* **175** 12010 (2020)
9. L. V. Kireicheva, N. P. Karpenko *Eurasian soil science* **48(5)** 524-532 (2015)
10. *Report "Digital Russia: New Reality"* (2017)
11. M. Porter Competition (M: Ed. House "Williams", 2003)
12. V. I. Olgarenko, G. V. Olgarenko, I. V. Olgarenko *Melioration and water management* **6** 8-11 (2013)
13. V.I. Olgarenko, I.V. Olgarenko, V.I. Olgarenko *IOP Conference Series: Materials Science and Engineering* **698(2)** (2019)
14. B. P. Chesnokov, O. V. Naumova, V. A. Strelnikov, F. K. Abdrazakov, B. A. Tronin *International Journal of Applied Engineering Research*, 11 2140-2144 (2016)
15. L. V. Kireicheva, O. A. Zakharova *Eurasian Soil Science*, **35(9)** 990–995 (2002)
16. M. A. Bandurin, V. A. Volosukhin, V. V. Vanzha,; A. V. Mikheyev, Y. V. Volosukhin *Aer-advances in engineering research* **157** 70-75 (2018)

17. T. Jaegers *Eurostat Statistics in focus* (2013)
18. Vater Hendrick, Bail Elena. *Haufe-Lexware mbH&Co* (2013)
19. P. Watson, N. Chileshe, D. Maslow A *Construction Industry Development 2nd Postgraduate Conference -Cape Town South Africa* 129-141 (2004)
20. M. A. Bandurin, V. A. Volosukhin, A. V. Mikheev, Y. V. Volosukhin, I. P. Bandurina *International conference on mechanical engineering, automation and control systems* **327** (2017)
21. D.V. Kozlov *Water Resources,* **28(2)** 215-219 (2001)

The digital platform features for managing acceleration programs and project activities research

Maria Kamenskikh[1*], *Vladimir* Postnikov[1]

[1]Economy and Management of Industrial Production, Associate professor, Perm National Research Polytechnic University, 29, Komsomolsky av., Perm, 614990, Russia

Abstract. The article is devoted to the study of the essence of digital platforms and their main advantages. Digital platforms are now perceived as one of the most effective tools for conducting acceleration and project management programs. These digital platforms include such platforms that support projects, support work within teams, remind of important events and inform about other opportunities. The following advantages are highlighted: possible participants, availability of accelerators, availability of project showcases, project support, support for work within the team, event reminders, information about support systems, a platform engagement strategy. The main digital platforms for carrying out acceleration and project management programs, which are used in Russia, analyzed their main advantages for startups, investors, experts, ecosystem partners, infrastructure institutions and other stakeholders. There are digital platforms that provide the maximum number of services for program acceleration and project management. Opportunities for the development of digital platforms for conducting acceleration and project management programs have been identified. The research was carried out with the financial support of the Perm Government as part of a research project by the regional research group of agreement SED-26-12-10-5.

1 Introduction

Currently, the digital economy is actively developing. The introduction of information technologies allows more active development of interaction between inventors, investors, infrastructure institutions, small innovative enterprises, experts and other stakeholders in the framework of acceleration programs and project management. It is digital platforms that allow you to most efficiently implement these events.

Digital platforms are the most important objects of the information infrastructure. They appeared in the late 1990s, when networks, various platforms and ICTs began to form. Now the digital platforms increasingly perceived as a qualitatively new basis for the market structure, which contributes to the elimination of intermediaries and the expansion of modern business plans and models.

The growing competition in industries in both domestic and foreign markets, as well as the policies of countries aimed at creating legal and infrastructure conditions for building a digital economy, strengthens the position of companies. Today, one of the leaders in implementing digitalization strategies is the countries of the Asia-Pacific region. One of the main indicators is the number of companies (19%) that launched digital transformation at the fundamental process levels. These include management, manufacturing, logistics, and marketing processes. Then the organizations of the countries of the Americas follow [2]. The integration of processes allows you to calculate the economic effect and efficiency both for the entire chain and for each of its participants separately. In addition, digital platforms are becoming a source for launching innovative processes.

2 Research Background

The digital platform (DP) is one of the significant tools that contribute to the transformation of traditional sectors of the economy and the transition to a digital economy.
Digital platforms are "systems of mutually beneficial relations of a certain number of participants in the economic sphere of activity on the basis of algorithm, carried out in a single information environment. The use of the DP leads to a decrease in transaction expenses due to the use of a package of digital technologies for working with data and a change in the division of labor system" [7]. The definitions of digital platforms by different authors differ from each other. This is largely due to the relevance of the issue and the structured information. So, it is necessary to consider several approaches proposed by different sources (Table 1).

Among all the different definitions presented above, there is a common link in the understanding of digital platforms. The platform is nothing more than a business model. In another way, digital platforms are called multilateral, describing the concept of a bilateral market. The idea behind this concept is that the platform is an intermediary on which participating parties meet, offering each other value. In the basic example of a buyer seller, the first provides a solution to the consumer's problem in

* Corresponding author: permmak13@gmail.com

a commodity/service form, and the second transfers the form that closes his need [3].

Table 1. Approaches to the definition of "digital platform"

No.	Definition	Author
1	The digital platform is a high-tech business model that creates value by facilitating exchanges between two or more interdependent groups of participants.	Aalst, W., Hinz, O. & Weinhardt, C. Big Digital Platforms.
2	The digital platform is a system of mutually beneficial relationships of a significant number of independent industry participants, carried out in a single information environment, leading to a reduction in transaction expenses due to the use of a package of digital data technologies and a change in the division of labor system on the basis of algorithm.	Hein, A., Schreieck, M., Riasanow, T. et al. Digital platform ecosystems.
3	The digital platform is a high-tech business model that creates profits by exchanging between two or more independent groups of participants, the platform directly brings together manufacturers and consumers who are able to interact without intermediaries.	Nadkarni, S., Prügl, R. Digital transformation: a review, synthesis and opportunities for future research.
4	The digital platform is an ecosystem of solutions for customers or a level of business model and customers consider value. The necessary level of products and services provided is created through personalization, customization, functionality, logistics, integrated into the ecosystem.	Veisdal, J. The dynamics of entry for digital platforms in two-sided markets: a multi-case study.

The study of digital platforms is a fairly new phenomenon in economics and management. The technical or functional elements that are most actively used both by project support platforms and in general by the digital market (Russian, world) include the following:

1. Project support.
2. Support for work within the team.
3. Event reminders.
4. Information about support systems.
5. Strategy for engaging participants on the platform.

It is on these elements that it is planned to conduct a study of existing platforms in the field of support for project management and acceleration programs.

The emergence and development of new generation digital technologies (in a different way, Industry 4.0) has become a defining trend in the formation of the modern world economy. Artificial intelligence, robotics, IoT, wireless/ 5G technologies, virtual and augmented reality, beacon beacons and other "end-to-end" technologies create tools that gradually enter the global business and provide performance. For example, according to technology experts, end-to-end technologies can increase productivity in enterprises by 40% [5].

Understanding the economic potential of technology for business, enterprises choose a focus on digitalization. We can say that a certain share of organizations has already entered the initial stage of digitalization: according to the research company - 34.4% of respondents (sample - 300 companies) have developed a digital transformation strategy and are implementing it, and 26.6% plan to develop this direction in the future [4].

Each strategy can be based on different technologies. For example, as the mobile application market grows, multisectoral companies pick up this trend, develop or improve existing applications to optimize customer interaction processes. The technology of 5G can ensure the operation of mobile applications at a qualitatively new level. The business sees potential in this and is ready to invest: according to a Gartner study, 66% of companies plan to use 5G technology in working with their own applications [4].

Through the development of digital technologies, the market for online education, electronic commerce has formed, the market for online transport services is developing: "according to the official forecast of UBS by 2022, the market capacity should grow by more than 8 times and amount to 1 trillion rubles" [10]. In turn, digital platforms act as intermediaries in new markets between customers and service providers, thanks to which the value chain (production, storage, transportation and exchange) is integrated into a single space directly with its participants. Building a platform-based business enables companies to reduce expenses, increase productivity, accelerate decision-making, and ensure a competitive market position.

3 Materials and Methods

Currently, a number of platforms operate in Russia and the world in the field of support for project management and acceleration programs. The entire set of platforms can be conditionally divided into two groups - multifunctional and representing only one function. The latter include such platforms as:

- Information platforms for investors and startups (ICS - Information and communication system for supporting young innovators, PIPELINE, Unified Innovation Portal of the Samara Region RosInfra, Business Platform, Innovation Tatarstan, United Investors, SPARK);
- events of platforms (Calendarius, 2035.events, Leader ID 4ScienceTrello, Miro, Flowly);
- rating platforms (Yarating);

We consider the representatives of each group and describe their solutions in terms of basic parameters. Table 6 provides an analysis of existing platforms for project management support. A comparison of the basic parameters that are necessary for the successful operation of the platform shows that only one platform - the "Innovation Guide" - has implemented more than four functions.

The platforms GenerationS and BusinessChain implement four main parameters, while compared to the leader there is no support for work within the team. Other platforms are iGeneration and AXEL.online. We review existing platforms in project management and support and acceleration programs (Table 2).

https://doi.org/10.1051/shsconf/202111600002

Table 2. Analysis of existing platforms in the field of project management and support and acceleration programs

Platform	Availability of accelerator	Availability of project showcase	Project Support	Support for work within the team	Event Reminders	Information about support systems	Platform Engagement Strategy	Total
Russian								
1	+	-	+	+/-	+	-	+	**4,5**
2	+	-	+	-	+	-	+	**4**
3	+	+	+	-	-	-	+	**4**
4	+	-	+	+/-	-	-	-	**3,5**
5	+	-	+	+/-	+	-	-	**3,5**
6	+	-	+	+/-	-	-	+	**3,5**
7	+/-	-	+	+	-	-	-	**2,5**
Foreign								
8	+	+	+	+/-	-	+	+	5,5
9	+	+	+	+/-	-	+	+	5,5
10	+	+	+	-	-	+	+	5
11	+	-	+	+/-	-	-	+/-	4

In the table 2 there are following platforms:

1. Electronic Acceleration Program Management System, Innovation Guide LLC;
2. GenerationS, RBC JSC;
3. BusinessChain, LLC IPI-Lab;
4. iGeneration, FSAEU HE NNSU;
5. AXEL.online based on the Venture Accelerator; of Tyumen State University;
6. Venture accelerator, Skolkovo;
7. CosmOdis, Education - Society Foundation;
8. BoomStartup, BoomStartup (USA);
9. Gust Accelerator Platform, RocketX (Great Britain);
10. AcceleratorApp, Lickability (USA);
11. Online platform for startups in Central Asia (Uzbekistan).

4 Research Results

CosmOdis has a sufficient number of functions, but it is aimed at school students and students, and not at startups. All other platforms show concentration on two functions.
In addition to Russian, English-language platforms in the field of project management support were analyzed, such as BoomStartup, Gust Accelerator, AcceleratorApp, DealRoom, Extra Founder.
Their main functions are the presentation of startups for investors, as well as online training or information in the form of publication of training materials. All platforms are focused mainly on informing investors about startups, some have coverage of modern trends and platforms for discussing issues. A number of foreign platforms do not provide for regular maintenance, expertise or information about support systems. In addition, on all the platforms under consideration, with the exception of the Online Platform for Startups in Central Asia (Uzbekistan), there is no Russian language (mainly English is used, Spanish, Chinese, French, Arabic are also possible), which makes them inaccessible to the Russian wide audience.
Next, let's take a look at the benefits of using each platform to support project management from the perspective of the target audience and core functionality.
Electronic Acceleration Program Management System, Innovation Guide LLC include:

a. development and implementation of acceleration programs;
b. development and implementation of educational programs;
c. organizing and conducting educational events together with the necessary educational materials: master classes, conferences;
d. organization and conduct of project examination;
e. assistance in preparing projects for participation in third-party competitions, obtaining grant support;
f. investing in innovative startups with high growth potential.

GenerationS, RBC JSC includes:

1. Startups:
   a. start of acceleration programs;
   b. access to business infrastructure for piloting and launch
   c. promotion of the project in the Russian and international markets.
2. Companies:
   a. organization and implementation of acceleration programs;
   b. developing an innovation strategy and conduct an innovative audit.
3. Experts:
   a. access to the project base for examination.
4. Investors:
   a. access to the project base of potentially commercialized projects for subsequent investment.

BusinessChain, LLC IPI-Lab has:

1. Startups:
   a. start of acceleration programs.
2. Experts:
   a. access to the project base for examination.
3. Investors:
   a. access to the project base for investment. Projects are participants who have undergone acceleration.
4. Companies:
   a. organization and implementation of corporate programs based on the platform;
   b. access to projects that have been accelerated for later integration into the business infrastructure.
5. Accelerators:
   a. organization and implementation of platform-based programs.

iGeneration, FSAEU HE NNSU involves:

1. Students:
   a. development of the youth project environment of the university through organizing and conducting acceleration programs.
2. Business:
   a. access to a project base for potentially interested companies.
3. Mentors and experts:
   a. increased competence for mentors and the expert group.

AXEL.online based on the Venture Accelerator of Tyumen State University includes:

1. Startups:
   a. access to acceleration and education programmes.
2. Investors:

a. access to innovative development.
3. Experts:
a. access to the project base for examination.

CosmOdis, Education - Society Foundation involves:
a. access to an educational program on technological access to acceleration and education programmes.
b. entrepreneurship, engineering creativity;
c. access to the project presentation after passing project training, where representatives of schools, business and science from universities will meet.

Venture accelerator, Skolkovo includes:
1. Startups:
a. start of acceleration programs;
b. assistance in finding a team.
2. Companies:
a. organization and implementation of acceleration programs;
b. development of an innovation strategy and conduct an innovative audit.
3. Experts:
a. access to the project base for examination.
4. Investors:
a. access to the project base of potentially commercialized projects for subsequent investment.

BoomStartup, BoomStartup (USA) involves:
1. Startups:
a. provision of an individual acceleration plan;
b. initial evaluation of the project during the registration phase;
c. selection of mentors from the scientific and industrial sphere;
d. no competitive selection
2. Investors:
a. selection of five potentially successful projects.
3. Experts:
a. access to evaluation of various startup projects.

Gust Accelerator Platform, RocketX (Great Britain) has:
1. Startups:
a. access to the world's largest startup network;
b. possibility of using a set of tools and applications for the development of startups;
c. strong team in finance and jurisprudence.
2. Investors:
a. access to the base of startups and innovative projects.
3. Experts:
a. opportunity to work with projects from different countries.

AcceleratorApp, Lickability (USA) includes:
1. Startups:
a. access to an incubation/acceleration application;
b. constant communication with experts and mentors.
2. Investors:
a. access to the base of startups and innovative projects.
3. Experts:
a. access to the project base for examination.

Online platform for startups in Central Asia (Uzbekistan) involves:
1. Startups:
a. programs for different groups of inventors and entrepreneurs.
2. Investors:
a. access to the project base for investment. Projects are participants who have undergone acceleration.
3. Experts:
a. access to the project base for examination.

So, according to the results of the analysis, Russian leaders - platforms in the field of project management - are: electronic acceleration program management system, Innovation Guide LLC has 4.5 out of 7 points; GenerationS, JSC RBC and BusinessChain, LLC IPI-Lab has 4 out of 7 points.

5 Discussions

The comparison in the main parameters reflects that only four platforms - the Innovation Guide, GenerationS and Business Chain have three to five of the declared functions, one of them in part. All platforms, with the exception of Business Chain, have integrated the Support and Reminders function, as well as an activated strategy to attract participating parties to the site. The "Support" function implies the maintenance of participating parties and their support (organizational, technical, information) during acceleration programs.

In GenerationS, partial escort is practiced after the completion of the accelerator. The Reminders function takes the task of alerting: each party receives targeted notifications (directed exclusively to them). The presence of a strategy to attract participants to the platform was determined using the media center, content on social networks, as well as specialized programs for collecting advertisements on Internet resources.

The "Informations" feature is not available not only in the above platforms, but also in other participants. Support for working with participants within the team and additional functionality are partially present. A more detailed analysis of the project management platforms under consideration is provided in table 2.

Based on the analysis, it can be concluded that the most quantitative and qualitative value propositions are formulated on two platforms - GenerationS and Business Chain ("IPI-Lab"). Quantification refers to the number of benefits described for each platform audience. Quality is interpreted in terms of meaning. However, to summarize, we will gradually move along the elements of the table. Positioning is used on each of the three platforms.

1. The Innovation Guide took an auditing position and defined it as a platform for innovative projects and entrepreneurs.
2. The positioning of GenerationS clearly reflects the niche in which they work - corporate innovation. They chose an audience-oriented approach. In this regard, all their offered services are built around the development of innovation in large companies.
3. Business Chain chose a different way of positioning: the platform combined the target benefit and focus on the audience: for whom this platform, as well as what result at the exit its participants will be able to get.

First of all, the platform is designed to accelerate the growth of their capitalization. Startups receive investment and direct it to one of the "pain" areas, experts monetize competencies and form demand in the long term, accelerators launch the program with the support of platform owners, and investors expand investment portfolio and profit. In another way, the owners focused on profitability in the long run.

The next comparison parameter is services - what basic and additional services the platforms offer. The maximum number of characteristics is shown on the platform GenerationS, the smallest on the last. For startups on all platforms, 3 services are offered (taking into account open data): an educational module within the framework of the acceleration program, expertise and mentoring, consulting support. Company-based startup testing and access to innovative infrastructure are offered at the first two sites.

For projects, the Innovation Guide is more focused on the training component, as well as in general, putting it forward as a basis for business representatives. A broader proposal is the RVK platform: post-acceleration tracking in a closed alumni community, as well as a pilot launch in the company and PR support. The second audience is the company. For business, all platforms offer a service for organizing corporate programs, as well as narrower services related to training or scouting projects.

The "Guide to Innovation" focused on education through proposals for organizing master classes, conferences, video training, etc. GenerationS is focused on the practice and integration of startups into the company's activities, therefore it offers services in scouting, conducting strategic sessions, innovative audit and strategy development.

The third group is experts and mentors. For them, the main task is expert evaluation and interaction with startups and investors/companies, since sometimes they can receive a request for assessment, bypassing the course of acceleration. In the first platform, experts can increase competence with participation in educational programs (including international ones), as well as join the project team and continue to work with them.

At the second site, experts are involved in both the acceleration program and the development / implementation of educational modules. On the Business Chain services are exclusively basic. For investors, an open access service is integrated to the base of potentially commercialized/scalable projects at different stages of development. However, if access is provided through accelerators on the first two sites, then a separate base is used on the Business Chain. Projects are submitted to it, but they can be included by the owners of the platform, subject to prior scouting. In addition, the capabilities of the third platform involve investor cooperation and joint investment.

But the only unreached audience is ecosystem partners. The exception is GenerationS. Starting with the platform, they receive services to promote their projects to the markets, the participation of experts GenerationS at partner events, as well as the promotion of the partner and its events among the platform community.

On the list of services, the value proposition does not end, for a complete sheaf, the benefits are described - the advantages that the audience receives when "buying," using these services. The only platform on which the least number of benefits is indicated is the Innovation Guide. They are focused on startup projects. These are precisely the advantages on which each of the participants emphasizes: receiving investments, as well as "packaging the project".

From the point of view of marketing, the value propositions formulated by Business Chain more accurately reflect the need of the audience - faster obtaining investments, reducing the cost of finding independent solutions (including "packaging" the project). For companies, the value proposition focuses on increasing returns through access to high-potential projects. Third-party accelerators receive a funnel of projects and earnings.

GenerationS and Business Chain for experts with the help of developed services provide an influx of new customers. But then the first platform focuses on increasing competence, and the second on increasing demand through increasing recognition. Accordingly, the investor in the long run makes a profit and nurtures the investment portfolio. However, none of the platforms presented when promoting targeted services for companies and a supporter of accelerators have not put forward as a distinctive advantage for companies and third-party accelerators - reducing the cost of organizing and implementing acceleration programs.

The question arises: why is not the emphasis on startup projects, investors or experts? The fact is that companies and third-party accelerators, and other representatives of the innovative infrastructure are customers of platform services. Without their request, one or another program will not be formed - corporate accelerator, intra-corporate, educational module, etc. Since, if there is an event, promotion campaigns will be launched in the future to involve the next participating platform groups - startups, experts and investors. Therefore, the primary audience for platform owners is representatives of business and innovative infrastructure.

The next comparison parameter for innovative project support platforms is the source of monetization. In GenerationS for teams, participation is free, business, acting as a customer of corporate programs, project scouting or innovative audit, is a source of funding. Interaction with ecosystem partners takes place at the level of barter, mutual exchange of services - assistance in organization, marketing, joint launch of events. For experts, the platform becomes a source of earnings, but participation in educational programs is paid. Participation of investors in the platform programs can take place at the invitation of the organizers or by preliminary selection (the investor applies for participation). In general, the interaction process on the platform is free for him. The next platform is the Innovation Guide. Similar to the previous experience on this platform, funding is provided by service customers - representatives of infrastructure and business. Other sources of monetization on the platform are not involved, as for the Business Chain platform.

6 Conclusions and Further Research

So, the study showed that digital platforms have a number of advantages in managing acceleration programs and projects. Next, it is rational to consider the features of the development of a digital platform. Development of a digital project support platform (acceleration, tracking, packaging, incubation), management of acceleration programs (information campaign, acceptance of applications, expertise, acceleration), information about support programs (grants, subsidies, competitions). The goal is to improve the system of attraction to technological entrepreneurship, the development of the information space of innovation support systems, the development of tools for managing acceleration programs and project activities. In the initial phase, the platform will assume the task of organizing and conducting acceleration programs and supporting innovative projects within its framework. The developed platform will allow you to manage and accompany innovative and technological projects as part of acceleration programs and competitions, including registration, filling out a personal account, filling out a project profile, correspondence examination of the project, examination of projects by video link, conducting educational programs, tracking projects, consulting support, communication chains, analytics. Subsequently, as the platform develops, other services can be added that satisfy the requests of the involved groups. The analytical part of the research made it possible to form the competitive advantages of the platform - services that will be directed to each of the involved parties:

- availability of personal rooms for all platform participants;
- possibility of submitting projects for events;
- maintenance of work within the project team;
- project moderation and examination;
- carrying out the acceleration program;
- familiarization with the accelerator community, as well as access to useful contacts through experts, coaches and mentors;
- information on support systems;
- system involvement of participants on the platform;
- convenient interface.

The platform begins with registration. When performing registration actions (filling in the email field and password), the participant enters personal account, where it reflects one's role: administrator; moderator; user (startup); Customer; investor; expert; tracker/mentor.
In the personal account, the participating parties will have opportunities - functions that they can use to perform certain actions.

References

1. W. Aalst, O. Hinz, C. Weinhardt Bus Inf Syst Eng **61**, 645-648 (2019). https://doi.org/10.1007/s12599-019-00618-y
2. F. Aulkemeier, M.E. Iacob, J. van Hillegersberg Electron Markets **29**, 597-608 (2019). https://doi.org/10.1007
3. E.R. Banalieva, C. Dhanaraj J Int Bus Stud **50**, 1372-1387 (2019). https://doi.org/10.1057/s41267-019-00243-7
4. Z. Deng, P.W. Liesch, Z. Wang J Int Bus Stud (to be published). https://doi.org/10.1057/s41267-021-00401-w
5. M. Drewel, L. Özcan, J. Gausemeier, et al. J Knowl Econ (to be published). https://doi.org/10.1007/s13132-021-00772-3
6. A. Hein, M. Schreieck, T. Riasanow, et al. Electron Markets **30**, 87-98 (2020). https://doi.org/10.1007/s12525-019-00377-4
7. A. Hevner, O. Malgonde, Electron Markets **29**, 407-421 (2019). https://doi.org/10.1007/s12525-019-00334-1
8. S. Nadkarni, R. Prügl, Manag Rev Q **19**, 230-237 (2020). https://doi.org/10.1007/s11301-020-00185-7
9. M. Reuver, C. Sørensen, R.C. Basole, J Inf Technol **33**, 124-135 (2018). https://doi.org/10.1057/s41265-016-0033-3
10. A.K. Song Small Bus Econ **53**, 569-590 (2019). https://doi.org/10.1007/s11187-019-00232-y
11. J. Stayton, V. Mangematin, J Technol Transf **44**, 1163-1187 (2019). https://doi.org/10.1007/s10961-017-9646-0
12. F. Sussan, Z.J. Acs, Small Bus Econ **49**, 55-73 (2017). https://doi.org/10.1007/s11187-017-9867-5
13. D.J. Teece, G. Linden J Org Design **6**, 8 (2017). https://doi.org/10.1186/s41469-017-0018-x
14. J. Veisdal Electron Markets **30**, 539-556 (2020).

https://doi.org/10.1007/s12525-020-00409-4

15. F. Zhang, Z. Gao, Q. Ye, Aut. Control Comp. Sci. **49**, 373-379 (2015). https://doi.org/10.3103/S0146411615060127

https://doi.org/10.1051/shsconf/202111600003

Innovations in ensuring information accessibility of state and municipal services in the Russian Federation

Polina Kuznetsova*

PNIPU, State Management and History Department, Russia

Abstract. The most important direction of innovative development for modern state and municipal administration is to improve the quality and accessibility of state and municipal services. Availability of services as a set of different criteria provides opportunities for non-discriminatory access to them. The most important indicator of the services availability is information accessibility - an unhindered and timely opportunity to receive complete and up-to-date information about state and municipal services. In the context of digitalization of public administration, the provision of information is implemented mainly in the electronic form. In this regard, the article analyzes the problems of ensuring information accessibility of state and municipal services through a qualitative study of the portals of government bodies and organizations providing services. The study was conducted on the basis of analysis of two groups of criteria: compliance with regulatory requirements of the legislation to the provided information and analysis of visual content of portals. The results of the study allowed us to identify both successful practices and problems of information accessibility of state and municipal services in the Russian Federation. A comparative analysis of foreign experience in ensuring information accessibility made it possible to develop recommendations for improving the legal regulation in the field of information communications between authorities and organizations with citizens in order to increase the efficiency of providing state and municipal services.

1 Introduction

Improving the efficiency of the state and municipal services organization is the most important direction for developing the system of state and municipal administration at the present stage. The key parameters that make it possible to assess the effectiveness of the services provision by state and municipal authorities are ensuring their high quality and accessibility for consumers-individuals and legal entities. It is in order to improve the quality and accessibility of services that innovative tools and technologies are being introduced into the activities of government bodies, including the transition to electronic service delivery, the development of a system of multifunctional centers for providing state and municipal services (hereinafter referred to as MFC) standardization and regulation of the activities of state and municipal service providers. In the strategic planning documents, the development of electronic systems and technologies for the provision of state and municipal services is identified as the most important direction for improving the provision of services. "The forecast of scientific and technological development of the Russian Federation until 2030" provides for the improvement of electronic state and socially significant services as the priority areas for the development of science and technology [1]. In the "Strategy for the development of the information technology industry in the Russian Federation for 2014-2020 and for the future until 2025", it is noted that information technologies make it possible to improve the quality of public services. As the directions of implementation of the Strategy, the independent implementation of informatization of their activities by state authorities and state organizations is fixed. It is noted that along with solving the problems of informatization of state authorities, the state should take care of creating the infrastructure of socially oriented industries [2]. One of the objectives of the state program of the Russian Federation "Information society" is to expand the use of information and telecommunications technologies in the process of implementing in electronic form the powers of state (municipal) authorities to provide state (municipal) and other socially significant services [3]. The main task of the information state subprogram is defined as: establishing and developing common quality standards and bringing interdepartmental electronic interaction of federal, regional and municipal authorities into compliance with these standards. The expected results of the subprogram implementation include provision of the most widespread and socially significant public services mainly in electronic form; ensuring high quality provision of state and municipal services (implementation of functions), other services and information in electronic form. In the process of implementing the program, it is planned to increase the efficiency of public administration, interaction between the state and citizens, the state and business by expanding the use of modern information technologies.

* Corresponding author: pkuznecova@list.ru

Thus, improving the availability and quality of state and municipal services, organizing and monitoring the quality and availability of their provision by both authorities and institutions, providing services mainly in the electronic form, improving the system of interdepartmental electronic interaction, and introducing an open data system are the most important components of innovative development of state and municipal administration. At the same time, it should be noted that at the present stage, the subject composition of service providers and the process of providing state and municipal services are changing. First, state and municipal authorities are expanding electronic services for the provision of state and municipal services, fully or partially automating them, thereby minimizing direct interaction with citizens who are consumers of services. Second, the MFC's functionality is being expanded as centers for providing state and municipal services, including in electronic form. In addition, MFCs are included in the system of providing other public services of public significance: notaries, banking services, etc. Thirdly, the process of providing services includes not only state and municipal institutions, but also organizations of the non-profit sector. Changing the process of providing services and expanding the list of their suppliers requires a qualitative review of the criteria for the availability and quality of state and municipal services.

2 Methods

The most important type of accessibility to state and municipal services is information accessibility. O.V. Simagina and S.S. Tsukar understand information accessibility as the availability and quality of information, informing citizens about the service and the possibilities of obtaining it. The authors highlight the most important quality characteristic of information accessibility of state and municipal services, which is called personalization of information, which is understood as the adaptation of state information to individual needs of a citizen [4, 163]. L.A. Vidyasova, E.J. Vidasov, J.D. Tensina as the most important characteristics of the use of information technology, including, in the process of obtaining public services, define the concept of "trust information technology", which is manifested in the fact of the use of the services of public services and the estimates of such experience as positive [5, 48]. Ronan McIvor, Marie McHugh, Christine Cadden note that Internet technologies can improve transparency, teamwork and the ability to respond to the needs of citizens in the public sector [6].

We suggest that information accessibility is understood as an unhindered and timely opportunity to receive up-to-date, complete, and up-to-date information about state and municipal services provided by state and municipal government bodies, as well as institutions and organizations (within the meaning of Federal law No. 210 "On the organization of the provision of state and municipal services"). It should be noted that the most important characteristic of information accessibility is precisely the unhindered receipt of information, since we are talking about different groups of service consumers, including those with different psychophysical characteristics. The relevance of information is defined as its significance at a specific time. One of the key characteristics of information is a sign of sufficiency or materiality. It is especially clearly manifested in the conditions of growth of information flows, complication of schemes of their interaction and interrelation, heterogeneity of information. The information should be presented in an easy-to-understand form and not contain redundant data [7, 215]. The efficiency of information reflects the speed of obtaining comprehensive necessary information in real time. Thus, the requirements for information availability are sufficiency and convenience of obtaining information about services. Information can be considered available only if the consumer is highly aware of the nature of the services and the possibilities of obtaining them.

This approach is fully consistent with the norms of the Federal law dated 09.02.2009 n 8-FZ "On ensuring access to information about the activities of state bodies and bodies of local self-government", according to which, the basic principles of access to information about the activities of state bodies and bodies of local self-government are openness, accessibility and accuracy of the provided information, including, placed on the Internet [8].

3 Results and discussion

The development of electronic services for the provision of state and municipal services is implemented both through the functioning of the Unified portal of state and municipal services (hereinafter referred to as the UPSS), and through the creation by government bodies and organizations of their own information resources. As part of the functioning of the UPSS, ensuring information accessibility from the point of view of simplicity and high speed of obtaining information is carried out by concentrating information on the site on the so-called "life circumstances". This approach is an element of applying a personalized, individualized approach to the consumer of services, since it allows you to identify specific needs of citizens at a given time.

It should be noted that in the total number of public services provided in the Russian Federation, a significant place is occupied by the services of the Federal service for state registration, cadastre and cartography (Rosreestr). Website of this organization provides an opportunity to get information about the Federal service's services, download templates of documents required for receiving the service, submit documents in electronic form, and track the degree of readiness of documents. At the same time, the results of surveys of citizens on the assessment of the work quality of the portal rosreestr.gov.ru, posted on the portal, showed an insufficiently high level of satisfaction of citizens. Of the 269 respondents surveyed, assessing the completeness of information, only 24.5% noted that the portal Rosreestr contains all the necessary information, 75.5% of respondents are not fully satisfied with the completeness of the information provided [9]. At the same time, in the course of a survey about the sources of obtaining information about the company's activities in Rosreestr (390 respondents), 76.4% indicated that the portal is the channel for receiving information in Rosreestr [10].

Multifunctional centers for the provision of state and municipal services (MFC PSMU) are the most important participants in the service delivery system. In accordance with Federal law No. 210, the organization of MFC activities is carried out by regional state authorities. In the territory of a subject of the Russian Federation, one MFC is defined, authorized to conclude agreements with the authorities, to coordinate and interact with other MFC's. Resolution of Government regulation No. 1376 sets out the requirements for posting information on the Internet by an authorized MFC [11]. It should be noted that the list of information provided should both contain information about the number of state and municipal services, the procedure for their provision, and reflect compliance with the requirements for the convenience of receiving services, their territorial and temporary availability. This information is posted on the MFC websites created in each subject of the Russian Federation. Maintenance of the MFC's official websites, posting information and uploading data from official Internet resources, is under the MFC's own jurisdiction. In this regard, MFC sites differ significantly both in terms of compliance with regulatory requirements, and in terms of content design, information visualization, and the use of widgets. We analyzed the websites of MFC's in a number of Russian regions, including, Perm region (http://mfc.permkrai.ru/), Sverdlovsk region (https://mfc66.ru/), the city of Moscow (https://md.mos.ru/), the Republic of Dagestan (http://mfcrd.ru/), Jewish Autonomous region (http://xn--79-7lc6ak.xn--p1ai/). The analysis was carried out according to two groups of criteria: compliance with regulatory requirements for information placement and the quality of visual content. The human brain processes visual information 70% faster than it does text. This is due to the fact that it is much easier to perceive such information. Visual content is a great way to convey information to users in a more pleasant and understandable way. Using images makes the site's content not only more attractive, but also more effective.

The analysis of compliance with regulatory criteria showed that almost none of the sites contains complete and exhaustive information, the requirements for which are established by regulatory legal acts. Thus, the website of the Jewish Autonomous region does not have a graphic image of the map of the subject of the Russian Federation indicating the location of the MFC. In the information about the MFC offices, complete information (data of the Manager, area, number of Windows and services provided) is available only on the website of the Republic of Dagestan and the Jewish Autonomous region. However, the same sites do not provide information about the procedure for providing exit services to citizens. Information on the number of services provided during the reporting period is available on the only website – the MFC of the Jewish Autonomous region. On the websites of Moscow MFCs, MFCs of Perm region and Sverdlovsk region, there is no information on how to appeal against actions (inaction), the decisions of the authorities, providing services and workers multifunctional centers; information on the responsibility of MFC's officials; information on the procedure for compensation of harm caused to the claimant as a result of improper performance or non-performance multifunctional center or its employees, as well as the companies involved or their employees duties. Thus, the websites of regional MFCs do not contain complete, up-to-date, and sufficient information about the organization and provision of state and municipal services.

As noted by Vishanth Weerakkody, ZahirIrani, Kawal Kapoor, Uthayasankar Sivarajah, Yogesh K.Dwived, the growth of website usage can be evaluated from the point of view of optimizing user interaction: interface design, the time required to search for the necessary information, and the comprehensibility of various functions/tabs offered on the website [12]. Visual content analysis was performed according to the following criteria: quality of information provision and ease of navigation and search for the necessary information. By the quality of information, we mean its visual component, the form and methods of providing information on the site. The presence of interactive services, photos, infographics, widgets, banners plays an important role, they allow you to divide information into problem blocks, makes information more understandable, attractive and accessible to the consumer, regardless of their psychophysical characteristics. Here we should highlight the MFC websites Moscow region and the Republic of Dagestan, which contain a sufficient amount of visual information. An important parameter of the quality of information delivery is the availability of a version for the visually impaired. It is not available on the website of the MFC of the Jewish Autonomous region, which is a significant violation of the principle of accessibility of services for people with disabilities. We evaluated the navigation and search certificates for the necessary information by the number of clicks to get information and by the presence of a search engine. The most convenient navigation and search system is available on the website of the MFC of the Sverdlovsk region and Moscow city. The website of the MFC of the Jewish Autonomous region has the ability to search for information, but the navigation system requires a significant number of clicks. On the website of the Perm region MFC it is not possible to search for the necessary information with a fairly convenient navigation system. There is no search bar on the website of the MFC of the Republic of Dagestan, information about services is limited, and it is not possible to get it in the list of services. Information about the terms of service provision and the amount of the fee is contained in a separate file.

Thus, the analysis of MFC sites in a number of regions of the Russian Federation revealed a number of significant problems: non-compliance with regulatory requirements for posted information, inconvenience of using sites for consumers, including those with disabilities. We believe that the reasons for these problems are: the lack of a uniform approach to the design of sites, regulation of information visualization methods due to the fact that the MFC of each subject independently determines the content of portals.

For example, the portal of an organization similar to Russian MFCs in Australia – Services of Australia (https://www.servicesaustralia.gov.au/) is based on the centralized principle of organizing the provision of information across the country. The portal contains all regulatory documents defining the procedure for establishing rights to receive services. The visual content of the portal is characterized by high-quality information, a convenient navigation and search system. The portal's services make it possible to clearly define the necessary services for receiving them, based on the principle of distinguishing them not by life situations, but by target groups. So, the portal has sections: elderly people; people looking for work; students and interns; parents; people with disabilities, etc. At the same time, in addition to the standard set of state services for establishing rights to social security and providing social services, the portal's services allow you to organize receiving psychological support for representatives of target groups in various life situations. Also, as a successful example of website design, we should highlight the portal of the Sao Paulo state service center of Brazil Poupatempo (www.poupatempo.sp.gov.br). It is based on the principle of dividing services according to public areas (education, public transport, labor, employment, social security, citizenship, military service, etc.). This approach is quite convenient, as it allows you to quickly navigate the portal and find out the necessary information about the relevant services, as well as get them, since the portal contains a significant number of online services. At the same time, the service center provides not only state and municipal services, but also non-state sector services, information about which is provided on the portal. The site contains comprehensive information about the procedure, methods of providing services, and the level of customer satisfaction. The site's visual content is informative, attractive, and easy to navigate and search for information. The portal of the Canada services center is based on a similar principle of dividing services into public areas (education, business, health, transport), Services of Canada (https://www.canada.ca/en.html). The portal is based on a centralized approach to the provision of services on the principle of "one window" across the country. The program is managed by the Minister of human resources and social development. As an absolute advantage of the portal, the problematic principle of providing information about public services should be noted. Each service contains not only information about the procedure and conditions for obtaining it, but also predicts possible problems that the applicant may encounter with explanations of the procedures for solving them. This approach ensures completeness, efficiency, and accessibility of information about services.

4 Conclusions

Thus, we believe that the most important way to increase the information availability of state and municipal services should be development on the level of Federal legislation standards issuing authority websites and organizations that provide public and municipal services, which includes not only the requirements for completeness, timeliness, adequacy of information provided in accordance with the requirements of normative legal acts, but also the characteristics of the visual content based on the principles of quality provision of information, provided it is received by all categories of consumers of state and municipal services. Currently, the Ministry of economic development of the Russian Federation is preparing a draft Concept for improving the provision of public services and developing the MFC system [13]. The concept provides for the development of MFCs as innovation centers, the prospective expansion of MFC functions due to the closure of government front offices, and the expansion of electronic services for providing services [14]. In this regard, one of the Central ideas of the concept is the development and the application of uniform standards of MFC activity in Russia, including uniformity of the procedure for organizing and financing activities, conditions for providing services and services, etc. However, the list of events does not include the introduction of standardized principles for organizing the operation of MFC sites [15]. Therefore, we consider it necessary to include this area of improving the MFC's activities in the tasks of the Concept, since it is at this stage that, due to fundamental changes in the system of state and municipal services, increasing the information availability of their provision is the most important direction of innovative development of state and municipal management in the Russian Federation.

References

1. *Forecast of scientific and technological development of the Russian Federation for the period up to 2030* (approved by the Government of the Russian Federation on January 3, 2014). Retrieved from: https://www.garant.ru/products/ipo/prime/doc/70484380.
2. *Development strategy of the information technology industry in the Russian Federation for 2014 - 2020 and for the future until 2025.* Retrieved from: https://rg.ru/2013/11/08/texnologii-site-dok.html.
3. *Decree of the Government of the Russian Federation of 15.04.2014 N 313* (as amended on 16.12.2020) «On approval of the state program of the Russian Federation" Information Society» (as amended and supplemented, entered into force on 26.12.2020). Retrieved from: http://www.consultant.ru/document/cons_doc_LA

W_162184/f09179663f3f8d147faddc7e0a531e915284e98b.

4. O.V. Simagina, S.S. Tsukar. *Questions of state and municipal management (VGMU)* **1**, 160-171, (2016)
5. L.A. Vidiasova, E.Iu. Vidiasov, Ia.D. *Tensina. Monitoring.* **5 (153),** 43 – 57 (2019)
6. R.McIvor, M. McHugh, C. *Cadden International Journal of Public Sector Management (IJPSM).* **15 (3)**, 170-187, (2002) https://doi.org/10.1108/09513550210423352
7. M.E. Beliaeva *Global scientific potential (GNP).* **9 (18)**, 212-217, (2012)
8. Federal Law dated 09.02.2009 No. 8-FZ "*On providing access to information on the activities of state bodies and local self-government bodies.*" Art. 4. Retrieved from:http://ivo.garant.ru/#/document/194874/paragraph/24:0.
9. *Results of the survey "Poll 1. Portal of Rosreestr"* Retrieved from: https://rosreestr.gov.ru/site/feedback/polls/vote_result.php?VOTE_ID=2.
10. *Results of the survey "Poll 2. Obtaining information on the activities of Rosreestr".* Retrieved from: https://rosreestr.gov.ru/site/feedback/polls/vote_result.php?VOTE_ID=3.
11. Decree of the Government of the Russian Federation of December 22, 2012 N 1376 "*On approval of the Rules for organizing the activities of multifunctional centers for the provision of state and municipal services*" (with amendments and additions). Retrieved from: http://base.garant.ru/70290064/#ixzz6obkkuBVi
12. V. Weerakkody, Z. Irani, K. Kapoor, et al. *Inf Syst Front.* **19,** 285–300, (2017) ttps://doi.org/10.1007/s10796-016-9679-1
13. *Draft Concept for improving the provision of public services and the development of the MFC system.* Retrieved from: http://smev.belregion.ru/wp-content/files/Concep_MFC.pdf.
14. N.P. Pazdnikova, T.A. Pechenegina *International Journal of Economic Research.* **14(16)**, 389–398, (2017)
15. Z. Mingaleva, N. Pazdnikova, E. Mitrofanova. *Asian Social Science.* **10(24)**, 257–262, (2014)

Formation indicators of strategic actions for oil and gas corporations within the framework of sustainable development concept

Elena Lobova [1,*], *Svetlana* Zhukovskaya [2], *Alyona* Taskaeva [1], *and Lyudmila* Semina [1]

[1] Perm National Research Polytechnic University, Faculty of Humanities, Department of Economics and Industrial Production Management, 29 Komsomolsky Prospekt, Perm, 614990, Russia
[2] Perm State National Research University, Faculty of Economics, Department of Accounting, Audit and Economic Analysis 614990, Bukireva street, Perm, Russia

Abstract. Current trends in the global practice of forming sustainable development goals indicate that striving for innovative development is associated not only with investment activities, but also with tightening of environmental sanctions across the entire vertical of production and sales processes. Leading companies in the Russian oil and gas sector are prominent representatives of the industry and they shape their strategies in accordance with the UN Sustainable Development Goals. A study of the state of 6 companies of the oil and gas complex, which occupy stable positions in the markets of petroleum products and gas, is conducted in this paper. For the analysis, a methodology was chosen that covers the main directions in achieving the sustainable development goals: investment and innovation in technology, economic, social and environmental development. The materials of annual and consolidated reports on social, environmental and innovative development, the program of socio-economic and environmental development for 2015-2019 served as the information base for the formation of a sample of statistical indicators. The use of the benchmarking method made it possible to rank enterprises, find identity criteria, and evaluate possible strategic actions of oil and gas enterprises in achieving the integrated parameter of the Sustainable Development Index. On the basis of the obtained calculations, it is possible to form action strategies in the direction of "best practices" and adapt the worked out development programs relative to the general trend of achieving sustainable development of the oil and gas complex.

1 Introduction

The expert opinion of the majority of scientists involved in assessment and forecasting of innovative opportunities for implementing general schemes for industrial enterprises development [1], the problems of energy security in the framework of smart specialization strategies of developing countries [2], and the strategy of regions based on clustering the contribution of socio-economic results of industries [3] confirms the need to adapt the list of quantitative parameters of strategic development in the direction of sustainable growth. The system paradigm of the economic growth theory puts forward the priority of the problems of effective using natural energy resources on the innovative development path. Within the concept of sustainable development alternative energy sources, such as wind energy [4], solar energy [5-6], heat energy of land [7] and other types of energy resources [8] are increasingly promoted. In turn, the key indicators of sustainable growth of energy complex enterprises will allow us to assess the contribution to the achieving the main development directions of the sustainable development concept of the country.

Despite the fact that technical and technological innovations remain a priority in the methodology of the Energy Strategy until 2050 in order to achieve the main performance indicators of the oil and gas sector [9], its sustainable development institutions, both at the interregional and international levels, should fully reflect the latest trends in formation of the goal-setting system [10]. When setting goals for achieving sustainable development at the global level for assessing indicators in countries, special attention is paid to indicators formed on the integrated approach basis [11]. Such indicators are the corresponding indices allowing one to assess the dynamics of innovative (technical and technological), social, economic and environmental aspects. With all the variety of evaluation methods, there are still no clearly defined regulatory approaches for formation of indices for economic sectors.

The main hypothesis of the study is that the aspects of the sustainable development concept should be reflected in the strategic actions of the leading companies in the oil and gas economy sector.

To prove the proposed hypothesis, the following tasks are solved in the article: 1) cross-analysis of information from open sources was carried out and the main priorities in the strategy formation for the

* Corresponding author: elena.bykova555@yandex.ru

enterprise sustainable development were identified; 2) indicators for evaluating the indices of enterprise sustainable development were formed and calculated; 3) based on the benchmarking method, comparison of evaluation indicators was made and conclusions were drawn about possible directions for building strategic actions to improve the state of indicators.

2 Methodology

1.1 Justification of the method to select indicators for forming sustainable development indices

The need to develop indicators allowing one to parameterize the "sustainability" of advancement of civilization, countries, regions or individual market actors was noted in the "Agenda for the XXI Century", adopted at the Conference on Environment and Development in Rio de Janeiro in 1992 [12]. The relevance of this task was confirmed 20 years later at the UN Conference "Rio+20". One of the important decisions of the Conference is the need to develop sustainable development goals for the period after 2015, covering priority areas, and appropriate indicators to assess the process of achieving these goals [13]. The index system is widely represented by the leading international organizations of the United Nations, the World Bank, the OECD, the World Wildlife Union (WWF): the System of Indicators of Sustainable Development, the System of Integrated Environmental and Economic Accounting (Integrated Environmental and Economic Accounting) [14]; Human Development Index [15]; the System of environmental indicators that reflect modern technological and environmental challenges, the Ecological Footprint and the Living Planet Index) [16].

The fundamental difference between the indicators of sustainable development in comparison with their counterparts of the classical development model is consideration of environmental and social costs aimed at compensating for the cost of depleted natural resources and restoring the natural environment.

Summarizing the existing world experience in working out indicators of sustainable development, we can distinguish two approaches:

1) building an integral, aggregated indicator, on the basis of which it is possible to judge the degree of sustainability of socio-economic development. Aggregation is usually carried out on the basis of three groups of indicators: ecological-economic, ecological-socio-economic, or environmental;

2) building a system of indicators, each of them reflecting individual aspects of sustainable development. Most often, within the framework of the general system, the following subsystems of indicators are distinguished: economic, environmental, social, and institutional.

The choice of an approach to the development of indicators for sustainable development of oil and gas industry enterprises is advisable to form both on the basis of the comprehensive system of indicators including all the necessary assessment indicators for the 4 main areas of development, and taking into account complex aggregation in these areas, complementing the three-pronged approach of J. Elkington [17] with parameters for assessing technical and technological development.

1.2 Approach to formation of sustainable development indicators of oil and gas enterprises

The most prominent tools and methods for assessing sustainable development of an enterprise are:

- global 100 methodology [18];
- methodology for benchmarking sustainable development based on the GDS-Index [19];
- methodology of French Association for Standards, which proposes to take into account environmental indicators in the system of sustainable development based on ISO 14001: 2015 [20];
- in recent years, the approach proposed by the Russian Union of Industrialists and Entrepreneurs (RSPP) has been developed in Russian practice [21] and implemented in the framework of competitive cases of enterprises wishing to compete for the title of Leader of Sustainable Development on the electronic platform of the Moscow Stock Exchange. This format allows you to evaluate which of the UN goals a particular enterprise implements [22].

Based on the above methodological approaches, we will determine the list of indicators that meet the requirements for forming indicators as follows:

- Assessment of technical and technological stability (k1. 1. - shelf life, k1. 2. - return on assets, k1. 3. - return on fixed assets on sales profit, k1. 4. - return on fixed assets on net profit).
- Assessment of personnel sustainability (k2. 1. - production of profit from sales, production of net profit, k2. 2. - profitability of the payroll fund by profit from sales and k2.3. - profitability of the wage fund by net profit, k2. 4. - labor productivity).
- Assessment of sustainable commercial development (k3. 1. - revenue to cost ratio, k3. 2. - operating margin, k3.3. - business activity ratio, k3. 4. - return on sales by profit from sales, k3. 5. - return on sales by net profit, k3. 6. - turnover ratio of accounts receivable, k3. 7. - turnover ratio of accounts payable).
- Assessment of greening (k4. 1. - degree of environmental friendliness of products, k4. 2. - degree of environmental friendliness of the enterprise, k4. 3. - degree of total eco-efficiency of the enterprise (for all factors that determine eco-efficiency (water, air resources, solid waste disposal, etc., depending on the industry specifics)).

1.3 Methodology for the formation of indicators of strategic actions to achieve a high level of sustainable development

To choose an approach to the methodology formation, we define the term "sustainable development" as a single process, as a result of which there is a continuous improvement in the main financial indicators, an increase in the efficiency of investments in machinery and technology, full compliance with the requirements of environmental safety and development of social programs. The process of improving or worsening the rate of developing indicators can be implemented by calculating the indices based on the geometric mean [23-24]. The methodology for developing strategic actions will also use the principles of benchmarking [25-26] and multi-criteria analysis [27-28].

The sequence of calculations will be as follows:

1) Calculation of the selected relative indicators for the four groups of sustainable development indices.

2) Determination the partial indices of the selected indicators for the period from 2015 to 2019 by the formula:

$$I_{ij} = \frac{k_{ij}^1}{k_{ij}^0} \quad (1)$$

where k_{ij}^1 is the value of the selected relative indicator for the *i*-th group of indices obtained for the reporting period; k_{ij}^2 is the value of the selected relative indicator for the *i*-th group of indices obtained for the base period; *i* is the rank of the index group from 1 to 4; *j* is the rank of the relative indicator in the *i* - th group of indices.

3) Calculation of the average values of the partial indices based on the method of searching for the geometric mean I_{ij} for the period from 2015 to 2019 using the formula:

$$IS_{ij} = \sqrt[m]{\prod I_{ij}} \quad (2)$$

where *m* is the number of periods.

3 Results and Discussion

Cross-analysis of the data from reports on sustainable development, partial reports on financial development, implementation of environmental and social goals of the 6 largest oil and gas sector enterprises allowed us to implement paragraphs 1-3 of the proposed methodology.

The sample includes companies [29-34]. The results of the calculation of the partial indices are presented in Table 1.

Despite the fact that PJSC LUKOIL has a high rating in the aggregate index of Sustainable Development and is a leader among other companies, the results of calculations of partial and general differences indicate that this company has a leading position only in such partial indices as $I_{1.2}$, $I_{2.5}$, $I_{3.3}$, $I_{3.7}$, $I_{4.1}$, and index $I_{4.2}$ is the same for all companies.

Table 1. Values of partial indices of sustainable development

No.	indicator	weight (w)	PJSC LUKOIL	PJSC NOVATEK	PJSC ROSNEFT	PJSC TATNEFT	PJSC BASHNEFT	PJSC SURGUTNEFT
1	$I_{1.1}$	0,053	1,164	0,981	0,999	1,021	1,040	0,963
2	$I_{1.2}$	0,053	0,747	0,991	0,715	1,053	0,715	1,091
3	$I_{1.3}$	0,053	1,742	1,003	1,870	1,209	1,445	1,186
4	$I_{1.4}$	0,053	1,874	1,100	1,434	1,153	1,492	0,904
5	$I_{2.1}$	0,053	1,209	1,017	1,503	1,197	1,005	1,236
6	$I_{2.2}$	0,053	1,231	1,683	1,068	1,141	1,003	0,943
7	$I_{2.3}$	0,053	1,208	0,964	1,287	1,148	1,006	1,150
8	$I_{2.4}$	0,053	1,230	1,597	0,915	1,095	1,004	0,877
9	$I_{2.5}$	0,053	1,093	1,052	1,087	1,133	1,098	1,138
10	$I_{3.1}$	0,053	1,111	0,987	0,995	1,020	0,957	1,036
11	$I_{3.2}$	0,053	1,106	1,562	0,971	1,005	0,916	0,928
12	$I_{3.3}$	0,053	1,036	0,944	1,081	1,087	0,962	1,139
13	$I_{3.4}$	0,053	1,106	0,967	1,382	1,056	0,916	1,087
14	$I_{3.5}$	0,053	1,126	1,601	0,982	1,007	0,914	0,829
15	$I_{3.6}$	0,053	1,094	1,061	1,027	0,845	0,690	1,095
16	$I_{3.7}$	0,053	0,970	1,004	0,983	0,873	1,034	1,087
17	$I_{4.1}$	0,053	0,888	1,278	0,882	1,200	1,152	0,822
18	$I_{4.2}$	0,053	1,000	1,000	1,000	1,000	1,000	1,000
19	$I_{4.3}$	0,053	1,463	1,061	1,120	1,978	1,401	1,265
	SDI		**1,179**	**1,150**	**1,121**	**1,117**	**1,04**	**1,041**

4 Conclusion

Implementation of the continuous growth strategy requires companies to have dynamics in the planned and implemented indicators. The data obtained indicate the positive dynamics of all the partial indices of sustainable development of the selected major representatives of the oil and gas sector. The obtained values of the Sustainable Development Indices indicate the close positions of the companies. In the process of quantifying all the indicators or bringing them to a scale from 0 to 1, one can notice heterogeneity in the leadership positions according to particular criteria. This estimate is shown in Figure 1.

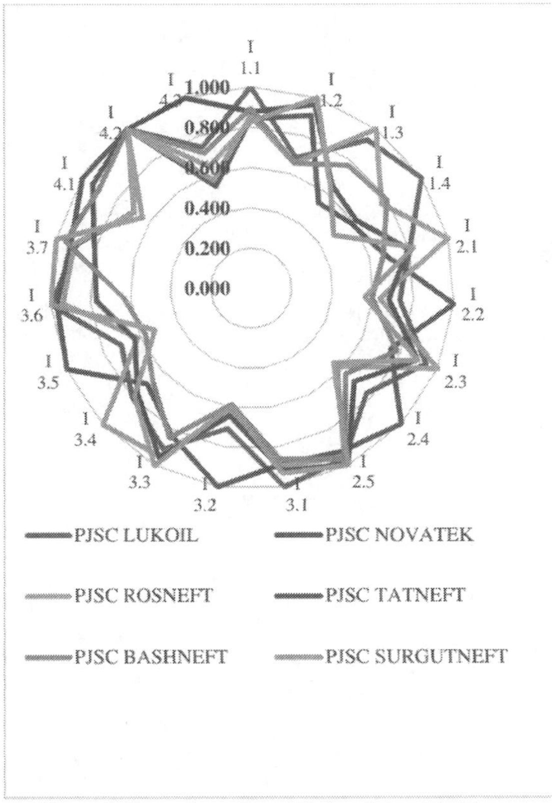

Figure 1. Comparison of quantified partial sustainable development indices.

Nevertheless, the results obtained allow us to form an approach to the gradual formation of strategic actions to further improve the indicators to the level of the best value for the industry segment. Namely, goal setting at the level of maximum parameters will allow us to form indicators of strategic actions to bring each of the companies, and the entire oil and gas sector-as a whole, to a new level of sustainable development.

References

1. I. Elokhova, L. Nazarova, Z. Mingaleva, *Insights and potential sources of new entrepreneurial growth*. Filodiritto Publisher, Bologna, 99-114. (2017)
2. D. Foray, X. Goenaga, JRC *Scientific and Policy Reports* (2013)
3. Y. Dubrovskaya, E. Kozonogova, A. Molodchik, Upravlenets-the manager, **10**, 64-73 (2019)
4. V. Klimenko, E. Fedotova, Thermal Engineering, **67 (6)**, 331-342 (2020)
5. B.V. Ermolenko, G.V. Ermolenko, Y.A. Fetisova, L.N. Proskuryakova. Energy, **137**, 1001-1012 (2017)
6. Zh. Mingaleva, N. Shpak. Thermal science, **19(2)**, 457-466. (2015)
7. K. Dehina, A.M. Mokhtari, B. Souyri, *Geothermics*, **87**, 101836 (2020)
8. A.Angeloudis, S.C. Kramer, N. Hawkins, M.D. Piggott, Renewable Energy, **155**, 876-888 (2020)
9. N. Vukovic, V. Pobedinsky, S. Mityagin, A. Drozhzhin, Z. Mingaleva. Sustainability (Switzerland) **11(17)**, 4629 (2019)
10. V. Kononov, O. Povarov, Proceedings, *World Geothermal Congress* (2005)
11. J. Weber, F. Gotzens, D. Witthaut. Energy Procedia, **153**, 22-28 (2018)
12. Agenda 21 Retrieved from: https://undocs.org/ru/A/CONF.151/26/Rev.1%28Vol.I%29
13. UN Conference "Rio + 20" Retrieved from: http://www.iblfrussia.org/a-conf.216-l-1_russian.pdf.pdf
14. *Integrated Environmental and Economic Accounting* Retrieved from: http://www.fao.org/3/bc966e/bc966e.pdf
15. *Human Development Index* Retrieved from: https://report.hdr.undp.org/part-2.html
16. J.A. Johnson, U. Baldos, T. Hertel, J. Liu, C. Nootenboom, S. Polasky, T. Roxburgh Global Futures: modelling the global economic impacts of environmental change to support policy-making. Technical Report, (2020) Retrieved from: https://www.wwf.org.uk/globalfutures
17. M. Kaptein, J. Wempe, Journal of Corporate Citizenship **1(2)**, 91-106 (2001)
18. Global 100 Retrieved from: https://www.corporateknights.com/
19. GDS-Index Retrieved from: https://www.gds.earth/wp-content/uploads/GDS-Index-2020-Methodology-1.pdf
20. ISO 14001: 2015 Retrieved from: https://www.boutique.afnor.org/norme/nf-en-iso-14001/systemes-de-management-environnemental-exigences-et-lignes-directrices-pour-son-utilisation/article/821532/fa176494
21. Russian Union of Industrialists and Entrepreneurs Retrieved from: https://rspp.ru/activity/social/?sphrase_id=164257
22. Moscow Exchange Competition Retrieved from: https://fs.moex.com/f/13161/etalon-2019-ar-rus.pdf
23. M. Grzebyk, M. Stec, Sustainable Development, **23**, 110–123 (2015)
24. T. Alferova, E. Shilova, E. Tretiakova, European Research Studies Journal, **18(3)**, 115-128. (2015)
25. S.C.L. Koh, A. Gunasekaran, S.M. Saad, Benchmarking: An International Journal, **12(4)**, 383–400 (2005)
26. M. Ungan, Benchmarking: An International Journal, **11(5)**, 504–520 (2004)

https://doi.org/10.1051/shsconf/202111600004

27. J.P. Brans, B. Mareschal, P. Vincke, *A New Family of Outranking Methods in Multicriteria Analysis* (1984)
28. J. Oberstone, *Management Science - Concepts, Insights and Applications* (1990)
29. official site PJSC LUKOIL Retrieved from: https://lukoil.ru/Responsibility/Sustainability
30. official site PJSC NOVATEK Retrieved from: https://www.novatek.ru/ru/development/
31. official site PJSC ROSNEFT Retrieved from: https://www.rosneft.ru/Development/
32. official site PJSC TATNEFT Retrieved from: https://www.tatneft.ru/ustoychivoe-razvitie/?lang=ru
33. official site PJSC BASHNEFT Retrieved from: http://www.bashneft.ru/disclosure/
34. official site PJSC SURGUTNEFT Retrieved from: https://www.surgutneftegas.ru/responsibility/

Modeling the process of innovative transformations at the industrial enterprise

Vladimir Patrushev[1*], and *Victor* Popov[2]

[1]Sorbent JSC, 6, Galperina Street, Perm, 614042, Russia
[2]Department of Management and Marketing, Perm National Research Polytechnic University, 29, Komsomolskiy Prospect, Perm, 614990, Russia

Abstract. The process of innovative transformations at an industrial enterprise is studied. The main features of the introduction of innovations into the production process are considered. A general model for carrying out innovative transformations at an industrial enterprise with a large share of high-tech production is proposed. The main elements of the proposed three-dimensional model include the basic processes of change, project and innovation management. Additionally, the models include such processes as: increasing innovation potential, improving internal communications through the introduction of digital technologies, increasing the efficiency of the production process by improving the production system. The applicability of the proposed model under real production conditions is investigated on the basis of Sorbent JSC, and the impact of the implementation of the innovation process on the overall economic efficiency of the enterprise is assessed.

1 Introduction

The continuous struggle to increase the economic efficiency of production is one of the foundations that can ensure high competitiveness, profitability and, as a result, the stability of the company. As international practice shows, in a modern economy characterized by increased competition, globalization and instability of sales markets, an effective strategy for increasing production efficiency is the introduction of innovations [1]. Hereinafter, the term innovation will mean any product or process introduced or implemented at an enterprise and leading to a significant improvement in the quality of an already produced product (or service), an increase in the efficiency of production processes, or the release of a completely new product with consumer properties different from previous ones [2]. Below are some examples showing the success of different companies through innovation:

1. Asian consumer electronics manufacturers such as Samsung Electronics Co., Ltd. are good examples of innovation to improve product quality. Until a few decades ago, Samsung Electronics was a domestic manufacturer, practically unknown outside of South Korea. The company's product line consisted of several dozen names, but their quality was low. However, the implementation of systemic reforms within the company in the late 90s, aimed at reducing the range of products while improving its quality, brought Samsung Electronics into the ranks of the world leaders in the production of electronic household products.
2. An innovative approach to improving the efficiency of production processes is clearly visible by the example of North American car manufacturers (General Motors, Ford Motor Company, Tesla, Inc. and others). Here, thanks to the robotization of production lines, it was possible to organize the production process in 24/7 mode with a small proportion of defective products and a reduction in human resources involved in production. As a result, a 30% increase in productivity and up to 50% reduction in production costs for automotive companies was achieved, which significantly increased the competitiveness of these companies in the global market.
3. And finally, an example of innovation for launching a completely new product is Apple Inc., which released in the early 2000s a number of unparalleled products and services that did not exist in the consumer electronics and music industries. As well, they created a niche of wearable electronics. As a result of the successful implementation of the innovation policy, the capitalization of Apple Inc. exceeded $ 1.044 trillion, which made it the most valuable public company in human history.

* Corresponding author: patrushev94@yandex.ru

The development of innovative production is an expensive and laborious process associated with high risks. At the same time, if successfully implemented it is innovation that can bring companies huge profits that are not available with the traditional material method of production or concentration of finance. That is why the economies of most of the so-called developed countries are essentially innovative, that is, economies based on constant technological improvement with the predominance of high-tech products in production and export. The added value of such products, as well as the technologies for their production, is very high, and the profit is created mainly due to the knowledge and skills of technologists and scientists. A good example is the economies of Japan, the United States and Germany, where spending on research and development of new technologies is about 2.7 - 3.5% of GDP [3]. At the same time, the share of these countries in the total export trade of high-tech products in the world market is 39% for the United States, 30% for Japan and 15% for Germany [3].

At the same time, the share of innovative production in Russia is still not large enough and lags far behind in terms of GDP expenditures on science-intensive developments and research and the share of exports of high-tech products from the countries listed above - leaders in introducing innovations into industrial production. Thus, according to Ref. [4], Russia's share of high-tech products in the world market is no more than 0.3%. At the same time, spending on research and development and innovation in Russia is about 1% of GDP [5]. That is why the modernization of the Russian economy, aimed at increasing the innovative component of industrial production, is one of the priority tasks facing us at the present time.

The purpose of this work is to model the process of introducing innovative transformations at a Russian industrial enterprise with high scientific and technical potential and its own industrial production. Also, the work evaluates the effectiveness of the proposed model in a real production process by the example of the company Sorbent JSC, located in the city of Perm. The period from December 2017 to December 2018 is considered. Based on the results of comparing real production indicators at the enterprise of Sorbent JSC with the expectations obtained in the modeling process, it is concluded that the proposed model is highly effective for the given type of enterprises.

2 Model description

Let us consider a general model for carrying out innovative transformations in an industrial enterprise, highlighting the main elements of the model, as well as the processes that must be taken into account when using this model in real production conditions. An important quality of a correctly constructed model is its generality, achieved by highlighting and taking into account the most important parameters and properties corresponding to real systems and, at the same time, ignoring all insignificant details making the research process more complex, but at the same time not significantly affecting the final result of modeling. In this case, the model must satisfy the following requirements: the correspondence of the model to the original (real) system in terms of the bundle of the most important qualities and characteristics, such as: universality, that is, the applicability of the model to the analysis and description of systems of the same type, which makes it possible to expand the application area of the model; accuracy, which characterizes the degree of coincidence of the data obtained during the simulation with the desired results.

Let us represent the general model of innovative transformations at a manufacturing enterprise as a combination of the following elements: analysis of the needs of the consumer market; research and development; prototype design and testing; improvement of production; marketing and sales [6]. In this case, the model implies a parallel arrangement of the listed elements in which several processes can occur simultaneously [7]. So, for example, the process of conducting scientific research occurs simultaneously with testing and improving production, the analysis of market needs does not stop even at the stage of selling the finished product. In addition, the model implies the existence of feedbacks connecting various elements (for example, market analysis and product implementation), which are at the same time the elements of accounting the constantly changing needs of the consumer (Figure1).

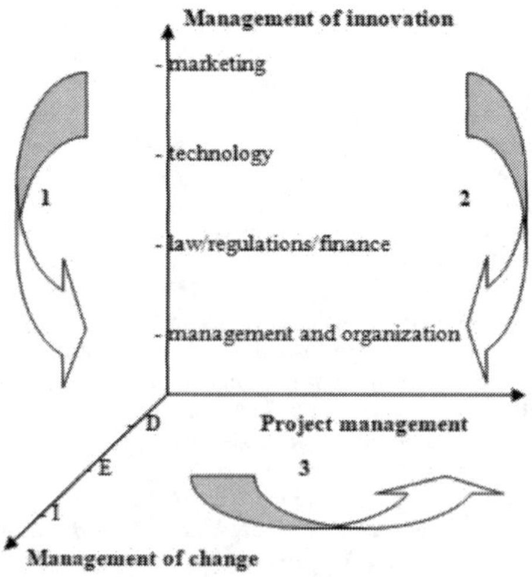

Figure 1. Model of the process of innovative transformations: D - diagnostics; E - engineering; I - implementation. Here, 1 – improvement of communications; 2 - improvement of the production system; 3 - development of the innovative potential of personnel.

 https://doi.org/10.1051/shsconf/202111600005

The model implies the creation of integrated product teams at the enterprise engaged in continuous analysis of the market for consumer goods and services, analysis of the internal characteristics of the enterprise, its production and scientific and technical potential, the availability of reserves and risks from innovative transformations, project management [8]. The model also implies the presence of processes for the recruiting and training of scientific personnel, designers, innovators, etc., planning a long-term personnel policy designed to ensure the continuity of the knowledge transfer process, selection of equipment and equipping laboratories and research centers. The model includes the production processes of the final product, as well as the search for optimal production parameters that ensure minimum production losses and, as a result, the optimal ratio between the cost of resources and time, on the one hand, and the quality and volume of production, on the other hand [9]. The model contains communication technologies, both within the company and in interaction with the external environment. The effectiveness of the implementation of the proposed model directly depends on the speed and volume of information transmitted between individual elements. Any disruption to the flow of information as a result of loss, distortion or delay of data entails additional losses in production due to underproduction or overproduction of the product, errors in strategic planning and/or scientific and technical research [10]. Here we emphasize the importance of the implementation and use of digital technologies (the so-called digitalization) as one of the components in increasing the efficiency of communication. At the same time, digitalization here does not mean a simple transfer of analog information to digital, but the creation of a fundamentally new, more efficient model for receiving, transmitting and storing information. As one of the components of the enterprise digitalization process, we note the development and implementation of artificial intelligence and machine learning technologies. These technologies, which have become widespread in recent years, can replace people in many processes related to the collection and analysis of information and promise in the coming years to revolutionize the tools available to companies for analyzing sales markets, optimization, forecasting and planning development strategies.

3 Results and discussion

An important part of this study is the use of the proposed model in real production conditions, as well as the assessment of the impact of the introduction of the innovation process on the overall economic efficiency of the enterprise. The company Sorbent JSC, located in the city of Perm, was chosen as a testing ground for validation the model of innovative transformations. Founded in 1939 as a plant for the production of activated carbon, the company Sorbent JSC is currently the largest developer and manufacturer of personal respiratory protection in the Russian Federation. The company also produces various types of activated carbon (more than 40 items), coagulants and systems for drinking water purification. The peculiarities of the enterprise include the presence of both its own scientific and technical and production lines. At the time of this study, the staff of the enterprise was 1200 people.

In accordance with the model described above, the following actions were carried out at the company Sorbent JSC:

1. Integrated product teams were created to analyze the current state of the enterprise, study the features of the organization of production and develop a strategy for the implementation of innovations. A preliminary analysis of the production process at the enterprise showed the presence of a number of labor-intensive and energy-intensive processes that affect the total production time and lead to noticeable losses. These processes, as a rule, referred to operations and activities based on the use of manual labor. It was also found that the presence of losses does not affect the quality of the final product, but leads to a decrease in the economic efficiency of the enterprise. As a result, the volume of manufactured products is less than the capacities available at the Sorbent JSC enterprise allow. To eliminate the identified production losses at the enterprise, it was decided to introduce an innovative production model based on the concept of lean manufacturing (LM). Here, we note that the choice of this concept is in good agreement with the goals and objectives of the Russian national project "Labor productivity and employment support", approved by the Presidium of the Council under the President of the Russian Federation for Strategic Development and Priority Projects (Minutes No. 12 dated September 24, 2018).
2. Further, diagnostics of all actions and processes that are potential sources of problems were made, the search for solutions aimed at correcting problems and increasing efficiency within the framework of the LM concept. Thus, partial or complete automation of workplaces was carried out, which entailed a significant reduction in the share of manual labor (while maintaining the total number of workplaces). Then the technology of visualization of the production process was introduced and the information centers necessary for this were created. This measure made it possible to create effective channels for control and management of production and management processes, and also ensured quick communication between employees of integrated product teams.

3. Finally, as a measure to increase production efficiency, a continuous flow of single-piece movement was organized in the finished product assembly areas, which made it possible to reduce the total time of the production process. Also, the storage conditions for finished products and components for their assembly were optimized.

Evaluation of the effectiveness of the innovative processes implemented at the company Sorbent JSC was carried out by comparing the main production indicators in December 2017 (i.e. before the start of the transformations) and in December 2018, when the main innovative elements were introduced into the production process and came into effect. The following indicators were considered: process time (PT), productivity, i.e. the amount of finished products produced by one employee per unit of working time and corresponding to the quality standards adopted at the enterprise and the share of work in progress (WIP). The following results were obtained:

1. As of December 2017, the manufacturing process took an average of 12 days. However, in April 2018, after the start of the innovative transformation program, this figure dropped to 5 days. Subsequently, the reduction in the process time continued and reached the value of 3 days as of December 2018. Thus, the improvement in the PT indicator during the study was 75%.
2. During the same time, there was an increase in productivity from 34 units per person per shift to 60. In other words, the change over the year was about 76.5%.
3. Also, positive dynamics was observed for the WIP indicator. Here, from December 2017 to December 2018, the reduction in WIP units was 2.75 - 3 times.

It is important to note that the above changes in production figures at Sorbent JSC were achieved as a result of the implementation of only some elements of the model of innovative transformation of enterprises proposed in this article. So, due to the time constraints imposed on this study, we did not investigate the processes of increasing the innovative potential of the enterprise associated with the recruiting and training of personnel, equipment for research and development departments, deep modernization of industrial equipment, and others. It is expected that the implementation of all measures provided by the model in the long term will lead to an even greater increase in the economic efficiency of the enterprise.

4 Conclusion

Thus, this work simulates the process of introducing innovative transformations at a Russian industrial enterprise with a high scientific and technical potential and its own industrial production. A system of measures is proposed to improve the efficiency of the production process by: managing changes, projects and innovations, increasing innovative potential, improving internal communications through the introduction of digital technologies, increasing the efficiency of the production process by improving the production system. Testing the model at the production facilities of Sorbent JSC shows the high efficiency of the proposed model. It was found that the implementation of innovative transformations in the production process at enterprises led to a 75% improvement in the time of manufacturing a product, as well as productivity value per employee per shift. At the same time, the share of work in progress decreased three times.

References

1. J. Schumpeter, *The theory of economic development.* (M.: Eksmo, 2007)
2. R. Boutellier, O. Gassmann, M. von Zedtwitz, *Managing Global Innovation: Uncovering the Secrets of Future Competitiveness* (3rd ed., Springer, 2008)
3. *Research and development management in Russian companies. National repor*t (Moscow: HSE University, 2011)
4. *Science. Innovation. Information society. Brief statistical collection* (M.: HSE University, 2009)
5. *Science of Russia in Figures,* 2010 (M.: TsISN, 2010)
6. GOST 57313-2016 *Innovation management. Innovation Management Guide.* (Moscow: Standartinform)
7. British Standard: 03.100.01 *Design management systems. Guide to managing innovation ICS* (2008)
8. *A guide to the project management body of knowledge* (PMBOK ® Guide) (5th ed., Newtown Square, PA: Project Management Institute, 2013)
9. V. Popov, G. Ostapenko, Management Mechanism for Implementation of Quick Response Manufacturing Concept the Enterprise, *Proc. 4th Int. Conf., Prague, University of Economics in Prague* (2016)
10. V.S. Patrushev, V.L. Popov, *Bull. PNRPU. Soc.-econ. sci.* **2**, 208 (2020)

Development of information support systems for production processes of contractor enterprises through an external electronic archive

Pavel Chashin[1], *Nadezhda* Soldatenko[2], *Pavel* Kuznetsov[2] *

[1] Iskra PJSC RMA, 28 Academician Vedeneeva Str., Perm, 614038, Russia
[2] Perm National Research Polytechnic University, 29, Komsomolsky Prospekt, Perm, 614990, Russia

Abstract. The main areas of activity of research and manufacturing associations include the creation of knowledge-based military and civilian products. Having unique competencies in the field of creating knowledge-based products for various fields of industry, research and production enterprises together with cooperative enterprises, including contracting partners, for many years have occupied a leading position in the markets of fuel and energy complex and other industries. Over the past five years, many industrial enterprises have gone from a critical decline in production and bankruptcy to a stable increase in work volumes, while the projects in the field of development of information support systems for industrial production processes have become highly demanded in the process of crisis management.

1 Introduction

The main areas of activity of research and manufacturing associations include the creation of knowledge-based military and civilian products. The PJSC Iskra Research and Manufacturing Association (hereinafter referred to as Iskra PJSC RMA) is not an exception of that kind. Having unique competencies in the field of creating products for the gas industry, Iskra PJSC RMA together with cooperative enterprises for many years has occupied a leading position in the market of the fuel and energy sector (hereinafter referred to as FES).

Projects implemented in the field of information support systems for production processes of enterprises are quite relevant, since after a critical decrease in the volume of work since 2016, the company is facing stable trends in the rapid growth of its market share and, accordingly, the volume of work performed by the enterprise for the fuel and energy sector.

The result of the above trend is an increase in the technical and administrative complexity of projects implemented by enterprises.

One of the main strategic goals of many enterprises, including Iskra PJSC RMA, is to maintain and increase its presence in the markets of high-tech products for the fuel and energy sector and other industries. To achieve this goal, using the existing unique competencies and high scientific and technical potential of employees, it is necessary not only to develop new types of products and win new markets, but also to increase the volume of products already delivered to existing traditional markets.

To maintain their position, successfully increase the volume of delivered products and develop in conditions of increased competition, the enterprises need to quickly respond to market changes and meet the needs of customers as much as possible.

This clearly requires the introduction of project management at enterprises, improvement of production efficiency, introduction of modern IT technology, professional skills of workers, and effective cooperation.

Today, about 80% of units and systems of products in the fuel and energy sector and other industries are manufactured at contracting enterprises, which in turn increases the risks of the enterprise failing to fulfill orders within the period stipulated by the contracts, and causes significant financial and image damage.

One important factor affecting the performance of the company's obligations by the contracting partner is timely information support of the production cycle with up-to-date documentation.

The goal of most industrial enterprises is to improve the information support system of the production process of the contracting enterprise by creating an external electronic archive (hereinafter EEA) and including up to 70% of contracting enterprises in a single information space.

2 Problem statement

Significant growth of production volumes in the fuel and energy sector and other industries, work with

* Corresponding author: pk.pnrpu@gmail.com

 https://doi.org/10.1051/shsconf/202111600006

cooperation taking into account the existing restrictions imposed in the Russian Federation, such as the Federal Law No. 223-FZ of 18.07.2011 "On procurement of goods, works, services by certain types of legal entities" and the Regulation on the purchase of goods, works, services of the Roscosmos state space corporation require the enterprise to build a new, efficient, unique process of information exchange with the contracting enterprise in terms of providing up-to-date documentation. The fact that contracting enterprises do not follow a single principle of documentation handling and use irrelevant information in the process of manufacturing product components and systems causes problems that directly affect the deadlines of product delivery to the customer.

3 Description of the approach to solve the problem

3.1 EFQM-based approach

The EFQM improvement model is the most acceptable to determine an objective realistic assessment of the current state of the approach, an area for priority changes, and to ensure effective processes in the organization.

Table 1 shows the description of the EFQM-based approach for improvement, enhancement and development, and Figure 1 presents the relationship of the approach to other approaches through criteria and criteria components within the EFQM model:

Table 1. Description of EFQM-based approach.

Criterion 5: Processes, products and services	Component of criterion 5a: Processes are designed and managed to create the added value for stakeholders
Component of the criterion: Perfect organizations manage a set of interconnected processes, including processes outside the organization, determine process owners and their roles and responsibility in developing, operating and improving processes	
APPROACH	
Name: Improvement of the information support system of the production process of the contracting enterprise through the creation of EEA. The purpose of the approach is to improve the existing process of production information support of the contracting enterprise for unconditional fulfillment of obligations on the terms and quality of the delivered products to the Customer. Implementation: creating EEA and involving 70% of contracting enterprises in the new process. The result of this approach is that the contracting company has up-to-date structured information for the organization of the production process. The approach is aimed at implementing one of the main strategic goals of the enterprise: – maintaining and increasing the presence of the enterprise in the markets of high-tech products for the fuel and energy sector and other industries in the interests of ensuring the energy security of the Russian Federation. Stakeholders outreach: Interest of owners in profit-making, leadership – in development of the enterprise and maintening the company image, consumers – obtaining finished products of the appropriate quality within the specified time frame. Relation to criteria: The diagram (Figure 1) presents the relationship of the approach to other approaches through criteria and criteria components within the EFQM model. Through criterion 5 : Processes, products and services the processes are designed and managed to create the added value for all stakeholders by managing a set of interconnected processes, including processes that exceed the organization limit (component of criterion 5a), the leaders ensure organization flexibility and effectively manage changes (component of criterion 1e), develop a strategy that is based on the understanding the needs, expectations of stakeholders and the external environment (component of criterion 2a), as well as the understanding of internal activities and capabilities of the organization (component of criterion 2b), ensure coordinated personnel activities, involve and empower, while recognizing that innovations may be applied to products, processes, marketing, organizational structure and business models (component of criterion 3c). Then, the management of relationships with partners and suppliers on a mutually beneficial basis and working together to achieve a mutual benefit and increase the added value to relevant stakeholders, providing each other with expertise and resources (component of criterion 4a) will improve perceptions and performance indicators for society (components of criterion 8a, 8b), personnel (components of criterion 7b), consumer (components of criterion 6b) and as a result, key business results (components of criterion 9a, 9b).	
DEPLOYMENT	
Responsibility for the approach lies with the heads of units involved in the formation of a set of necessary information. The approach is implemented through the involvement of units, while each unit generates and monitors information arising from its activities. Regulatory documents governing the approach are the corporate standard 061-2014 "QMS. Design, process and standardization documents. Accounting, storage, handling and modification procedures" and others. The achievement is that the contracting company has the necessary and sufficient information to organize the production process.	
ASSESSMENT AND REVISION	
The approach appeared in 1959 due to the need to organize interaction with the contracting company in the production cycle. The approach is regularly improved, the last stage in improving the approach was the electronic archive of technical documentation and a system for synchronizing databases of electronic archives of different enterprises at the enterprise. Similar approaches have been implemented at all leading foreign and domestic enterprises producing high-tech products. The approach is ranked as one of the most important for the effective implementation of projects in the enterprise, while in the process of implementation the approach may be supplemented.	

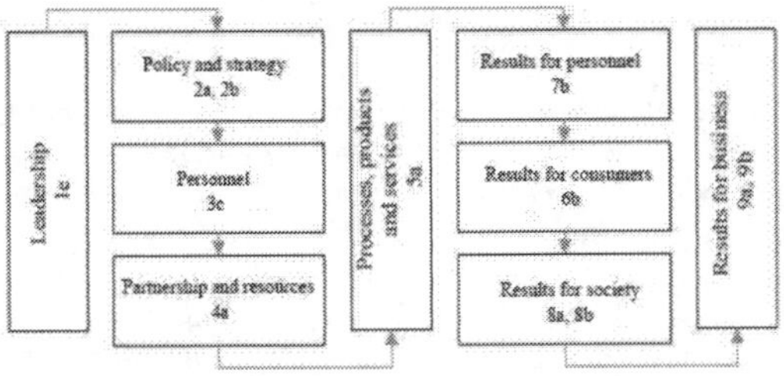

Fig. 1. Relationship of the approach to other approaches through criteria and criteria components within the EFQM model

Figure 1 shows that the considered approach "Improvement of the information support system of the production process of the contracting enterprise through the creation of EEA" is closely related to other approaches through the criteria and components of the criteria under the EFQM model. Thus, this approach is important for the company's performance and its improvement may significantly affect the final results of its activities.

3.2 RADAR-based assessment

Assessment by RADAR technology prior to the implementation of the project on information support of the production process of the contracting enterprise is presented in Table 2.

Table 2. RADAR Technology Assessment

Elements		Approach			Deployment			Assessment and Revision				
Features		Sound	Integrated	Total	Implemented	System	Total	Measurement	Study	Improvement	Total	Total score
				0			0				0	0
0%				5			5				5	5
				10			10				10	10
				15			15				15	15
				20			20				20	20
25%				25			25				25	25
				30			30				30	30
				35			35				35	35
				40			40				40	40
				45			45				45	45
50%				50			50				50	50
				55			55				55	55
				60			60				60	60
				65			65				65	65
				70			70				70	70
75%				75			75				75	75
				80			80				80	80
				85			85				85	85
				90			90				90	90
100 %				95			95				95	95
				100			100				100	100

3.3 Areas for improvement

Thus, according to the EFQM model, the assessment of the approach "Improvement of the information support system of the production process of the contracting enterprise through the creation of EEA" before the project is implemented makes 50%.

Strengths:

• approach exists, is standardized and regulated;

• approach supports the strategy to increase the efficiency of the strategic objective "Maintaining and increasing the presence of the enterprise in the markets of high-tech products for the fuel and energy sector and other industries in the interests of ensuring energy security of the Russian Federation".

Areas for improvement:

• insufficient indicators to measure the implementation of the approach;

• existing information support system does not ensure the timely transfer of up-to-date documentation to the contracting company;

• approach has several implementation options and does not allow for a flexible response to external changes.

The RADAR assessment made it possible to identify the areas for improvement. The implementation of the project to increase the efficiency of the information support for the production process of contracting enterprises in the manufacture of component parts of products will make the approach more flexible and adapted to external factors, as well as ensure the speed in providing up-to-date information to the contracting company.

4 Project

4.1 Project objective and tasks

Project objective: To increase the efficiency of the information support for the production process of contracting enterprises in the manufacture of component parts of products.

The main tasks of this project are as follows:

1) To analyze the current processes of providing contracting enterprises with up-to-date documentation;

2) To identify drawbacks of current processes and develop a new process;

3) To deploy EEA;

4) To perform initial loading of product components into EEA;

5) To organize and carry out pilot operation of the new process with 3-4 enterprises and contracting companies;

6) To put the process into industrial operation with the involvement of up to 70% of contracting enterprises.

Table 3 shows the assessment of the project objective for compliance with SMART criteria:

Table 3. SMART assessment of project objective

S	Specific	Increase the efficiency of the information support of the production process of the contracting enterprise in the manufacture of components of products by creating an external electronic archive
M	Measurable	Reduction of information transfer period to the contracting company by 75%
A	Achievable	By involving 70% of contracting enterprises in the new process
R	Relevant	The project is aimed at creating an "ecosystem" that unites contracting enterprises with common rules, integration standards and information resources
T	Timed	Project implementation period – 10 months

4.2 Project activities

As part of the task "To analyze the current processes of providing contracting enterprises with up-to-date documentation", only those contracting enterprises that produce units and systems for Iskra PJSC RMA within the fuel and energy sector were selected (63 enterprises) from the total number of contracting enterprises. Further, the schemes of existing processes "as is" of existing processes of document transfer to contracting enterprises were built: in hard copy – 33 enterprises, in soft copy using the FTP server – 29 enterprises and in soft copy using the synchronization of databases of electronic archives of enterprises – 1 enterprise.

As part of the task "To identify the drawbacks of current processes and develop a new process", the main drawbacks of existing processes were identified, the project team analyzed the alternatives and possible options for the new process. The new process "as must be the case" was proposed and created: EEA using a system for synchronizing databases of electronic archives and providing enterprises with an access to certain product components for contracting enterprises.

As part of the task "To ensure EEA deployment", the analysis of the necessary software for EEA organization was carried out. Procurement procedures were carried out, contracts for the supply of software were concluded, software was delivered and installed on the selected hardware platform.

As part of the task "To perform initial loading of product compositions in EEA", the EEA software and hardware platform was configured, product components with documentation sent earlier to contracting enterprises were selected and exported, import of product components into EEA was carried out, and a database synchronization system was set up.

As part of the task "To organize and conduct pilot operation of the new process with 3-4 enterprises and contracting enterprises ", 4 contracting enterprises were selected, instructions for work with EEA were developed, preliminary training was conducted, and access to the necessary product components was provided. At this stage, proposals for necessary improvements and automation were collected, the need was discussed and the corresponding measures were taken.

As part of the task "To put the process into industrial operation with the involvement of up to 70% of contracting enterprises", it is necessary to evaluate the results of pilot operation of the new process, decide on the transfer of the process into industrial operation, as well as to adjust existing regulatory documents.

4.3 Project risks

The likelihood of risks affecting the project was assessed and the necessary preventive measures were developed.

The results are shown in Table 4:

Table 4. Project risks

Event	Likelihood	Severity of consequences	Risk	Activities
Non-compliance of information systems capabilities with the stated requirements	0.2	0.1	0.02	Fit and gap analysis of system capability
Unwillingness of the contracting enterprise to work in the new process	0.5	0.1	0.05	Inclusion in P019 Regulation and contractual documents of work requirements in this process
Contractors do not have qualified personnel to work in the new process	0.5	0.1	0.05	Detailed instructions on how to work in the process, training and consultations
High workload of project team members	0.4	0.2	0.08	Connection of administrative resource, additional motivation
Lack of financial capacity to purchase the software package	0.8	0.5	0.4	Outreach to financial decision makers
TOTAL:		1	0.6	

4.4 Project performance assessment

The economic impact of the project was assessed. The payback factor was chosen to reduce the number of additional agreements to be executed due to the reason "Due to the change in the scope of work (adjustment of design documentation was made)" after procurement procedures and concluding contracts with contracting enterprises for the manufacture of units and systems of gas compressor units.

In 2020, additional agreements were concluded for the above reason with enterprises and contracting enterprises in the amount of 6,681 thousand rubles, which was provided for in the main contract. Thus, this amount may be considered the lost income for Iskra PJSC RMA.

As part of the project it is planned to involve 70% contracting enterprises in the new process, the lost volume of product sales could amount to 4,676,7 thousand rubles. As a result, the profit loss could amount to 2,745,7 thousand rubles.

The average payback period of projects to increase the efficiency of information support of production process of contracting enterprises is 1 year.

Based on the obtained efficiency data, it may be concluded that such projects are viable.

The RADAR assessment after the project implementation is presented in Table 5:

Table 5. RADAR Technology Assessment

Elements	Approach			Deployment			Assessment and Revision				
Features	Sound	Integrated	Total	Implemented	System	Total	Measurement	Study	Improvement	Total	Total score
0%			0			0				0	0
			5			5				5	5
			10			10				10	10
25%			15			15				15	15
			20			20				20	20
			25			25				25	25
			30			30				30	30
			35			35				35	35
50%			40			40				40	40
			45			45				45	45
			50			50				50	50
			55			55				55	55
			60			60				60	60
75%			65			65				65	65
			70			70				70	70
			75			75				75	75
			80			80				80	80
			85			85				85	85
100 %			90			90				90	90
			95			95				95	95
			100			100				100	100

Thus, according to the EFQM model, the assessment of the approach "Improvement of the information support system of the production process of the contracting enterprise through the creation of EEA" after the project implementation will be 80%.

5 Conclusion

The study provided the analysis of the current state of the EFQM-based approach and identified the areas for improvement, developed a project solution for improving the approach and carried out the analysis of the cost-effectiveness of its implementation, as well as the analysis of risks and their impact on the new approach.

Expected project outcomes:

1) Improvement of the approach according to the EFQM model calculated according to RADAR technology – 80%.

2) Reduction of the time to transfer up-to-date information to contracting companies by 70%.

3) Exclusion of hard-copy paperwork from the information support of the production process of contracting enterprises.

4) Improvement of the quality of manufactured units and systems at contracting enterprises due to the "ecosystem" of Iskra PJSC RMA, which unites 70% of contracting enterprises with general rules, integration standards and information resources.

5) Reduction of the cost of manufactured products within the fuel and energy sector.

The introduction of EEA and the involvement of contracting enterprises in the process will increase the number of projects implemented on time and within the budget, which, in turn, will lead to an increase in the competitiveness of the enterprise and an improvement in its image in conditions of fierce competition.

The implementation of the proposed project solution "Improvement of the information support system of the production process of the contracting enterprise" using the example of Iskra PJSC RMA is extremely relevant.

This solution may be applied to a given enterprise and to other industrial enterprises.

References

1. N.B. Akatov, M.M. Kustova *Development of competencies of managers of innovative projects within the framework of project-based learning programs: study manual.* (Perm: PRPU Publishing House, 2018)
2. V.M. Askinazi, V.F. Maximova *Investment analysis: study guide for undergraduate students.* (Moscow: Ewright, 2017)
3. V.V. Baranov *Processes of decision-making motivated by interests* (Moscow: FIZMATLIT, 2005)
4. E.A. Markaryan, G.P. Gerasimenko *Investment analysis: theory and practice: study manual.* (Moscow: KNORUS, 2017).
5. A.G. Gryaznova, M.A. Fedotova (Ed.) *Business Assessment* (Moscow: Finance and Statistics, 2008)
6. Yu.I. Popov, O.V. Yakovenko *Project Management* (Moscow: INFRA-M, 2011)
7. *Guide to Project Management Body of Knowledge (PMBOK Guide).* Fifth edition. (Project Management Institute, Inc, 2013)

8. *Guide to Project Management Body of Knowledge* (PMBOK Guide). Fourth edition. (Project Management Institute, Inc, 2008).
9. I.B. Rodionov *Project financing;* (Aleteya – Moscow, 2015).
10. Gordon Webster *Project Planning and Management for Managers* (Moscow: Business and service, 2006).

Technological innovations and the effect of 'dualism'

Oleg Sukharev*

Institute of Economics of the Russian Academy of Sciences, 117218 Nakhimovsky prospect, 32, Moscow, Russia

Abstract. The purpose of the research is to consider the effect of technological dualism for the Russian economy with an empirical demonstration that this effect has a different content. The research methodology consists of the neo-Schumpeterian theory of technological change, structural analysis, and empirical estimates that allow one to identify the characteristics of technological dynamics. The result of this research was a demonstration of the deviation from the classical effect of technological dualism of the Russian economy, which is caused by the existing structural and institutional characteristics of the functioning of activities various types. The main conclusion is that the classical recipes for stimulating technological innovation in the presence of the pseudo-effect of technological dualism will not become working tools with high action. We need systemic and institutional transformations that affect the incentives for technological renewal and the deployment of markets that accept products that embody the results of technological improvement and development. Thus, replacing the strategy of local technological innovations with a strategy of comprehensive incentives and large-scale technological innovations that improve the social parameters of life is an urgent task for the Russian economy.

1 Introduction

Technological innovations occupy the main place in the structure of innovations, largely determining further product, process and organizational innovations. From the point of view of Schumpeter's theory of development [1, 3, 5, 7-9, 11], technological innovations follow the logic of the effect of 'creative destruction' and 'combinatorial augmentation' [2,9]. Moreover, in 'creative destruction' some technological innovations displace other, when you 'combinatorial augmentation' of the integration of technology, but not usually in the form of coexistence of NBIC – technologies, as illustrative examples for combinatorics, and in the form of a new technological quality. Thus, the use of different technologies gives a completely different technological chain in the form of technological innovation. This is the fundamental difference between the scientific works of a number of authors, in particular, representatives of the Russian neo-Schumpeterian school [2] and the foreign neo-Schumpeterian school [9], which show the content and influence on the development of the combinatorial effect. However, in addition to the combinatorial effect, Schumpeter combinations compete with each other within each allocated group, as well as between groups, since, for example, technological innovations determine not only the product or processes, but also organizational changes and management. Thus, the overall structure of technologies represents a dynamically and flexibly changing technological framework of the economy. Moreover, along with the strengthening of technological influence within the framework of the combinatorial effect, there is also a competition of technologies, when some displace others. Labour-saving technologies (in the form of robots, artificial intelligence, computers, automata) displace labour that requires placement within the framework of labour-intensive technologies. Capital-saving technologies-provide an increase in the return on the capital used. In principle, they can take part of the labour released by labour-saving technologies. At the same time, the problem of labour placement in the field of labour-intensive technologies remains. The dynamic balance between the capital-intensive technologies of the manufacturing sectors and the labour-intensive technologies of other types of activity, which makes it possible to use the released labour force, is the classic content of the effect of technological dualism. It boils down to the co-existence of two classes of capital-intensive and labour-intensive technologies. The former release labour, the latter accept it. Thus, technological unemployment is levelled. With this adsorption of labour in the field of labour-intensive activities, there are objective restrictions for the introduction of labour-saving technologies in these sectors that accept the released personnel. Types of activities with capital-intensive technologies, such as recruitment, introduce labour-saving technologies that simultaneously increase the return on capital. This structural dynamics is determined by the initial technological parameters of the economy, but it also forms the vector of further structural and technological changes. Capital-intensive technologies have a labour-saving effect, cause technological unemployment and act in the direction of not increasing the level of income. The peculiar anti-social effect of such technologies is associated with a positive effect and the motive of the owner of the means

*Corresponding author: o_sukharev@list.ru

of production and this technology. Institutionally, this contradiction is not removed in any way, with the exception of technological dualism. The issues of equalization of technological effects both in the sectorial structure and in the interstellar structure (dependence on technology transfer) can't be considered fully resolved in economic science. This area of research has immanent uncertainty and practical multivariance, which makes it difficult to obtain accurate model solutions. Some aspect usually is not taken into account. Of course, the macroeconomic policy can also strongly influence the process of technological renewal and the technological structure. The relatively high cost of capital, even if it is highly worn out, will clearly hinder the introduction of capital-intensive and labour-saving technologies. The level of science and education provides an opportunity to learn new technologies and put them into production. Dualism can take on a very peculiar appearance when the technologies being developed are very different from those implemented and used. Independent structural dynamics, outside the technological context, expressed in the movement of labour from processing to other sectors of the economy (raw materials, services), accompanied by the compressed development of processing activities, can also hinder technological renewal in the direction of the introduction of capital-intensive, labour-saving technologies. Then this circumstance is the reason for the gap between the development and implementation of technologies, when technology can be possessed and have trained personnel, but it is not implemented. The technological gap of implementation is adjacent to the gap in capital-intensive and labour-intensive technologies, which institutionalizes the technological effect of dualism, giving it special characteristics that distinguish the technological dynamics of different countries (the technological structure, as a rule, is initially different).

Let's consider the technological dynamics and features of technological dualism on the example of the Russian economy. The purpose of the study is to identify the specifics of technological dualism for Russia, which forms the features of the dynamics of technological innovations. The research methodology is formed by structural and empirical analysis, application to the evaluation of the development and introduction of the most advanced technologies, which can be used to characterize technological innovations.

2. Research methodology. Technological dynamics and dualism

Technological innovations cover improved or completely new technology, and there can be two approaches to novelty. First, the technology may be new for a given organization, but not for the market as a whole, and it is not necessarily that it was developed by this organization. Second, the novelty can be provided for this particular organization, in the absence of a significant number of other organizations that use the same technology. Depending on the accepted form of accounting, as a result, there will be a discrepancy in the estimates of the magnitude of technological innovations. The latter are determined by the state of knowledge [11], however, not always R&D, high technologies, involving a significant amount of knowledge, determine economic growth [4, 6]. In addition, the widespread introduction of technological innovations, in particular digital technologies and automation, may not give the effect of productivity growth at all [8], revealing a high dependence on the created initial conditions for application [10] and the state of production technologies.

It should be noted that R&D does not always end with a positive result, not to mention that it turns into a highly improved or new technology. Of course, the array of R&D, after some time, determines the technological possibilities of development, but does not guarantee them. In this regard, it is not entirely convincing and reasonable to evaluate and distinguish the levels of technological activities by the share of R&D (such estimates are given in particular by Rosstat), since there is no direct relationship between the number of technologies and their level, quality, and volume of R&D conducted by costs, of course. The topic of the effectiveness of R&D and its implementation, as well as the development and implementation of technologies, includes an independent set of issues that depend on many circumstances, including the initial parameters of the operation of production. In our opinion, the accounting of technological novelty in its pure form will be most accurate if it is carried out in specific technological areas, types of technologies, adhering to a particular classification of technologies and innovations.

The technological dynamics of the economy can be studied from the perspective of two basic approaches. On the one hand, using an aggregate method that gives a general description of changes in technologies, including by type of technology. On the other hand, it is a structural method that reveals the current technological structure, as well as its impact on the development and other dynamic parameters of the economy, allowing us to assess the contribution of various technologies to the country's economic growth rate. Technological innovations according to the aggregate method can be evaluated by the number of firms and individual agents that implement technological innovations. The structural method of analysis allows to consider the structure of innovations (fundamental, improving, product, process, etc.), the contribution of each type to the growth rate, and the mutual influence of different types of innovations on each other. The whole range of new technologies included in the 'knowledge economy' can be estimated by the share of this sector and its contribution to the rate of economic growth. This will require identifying the "knowledge economy" sector itself along with other sectors in terms of an aggregate indicator, for example, value added, measuring its dynamics and contribution to the overall dynamics. At the same time, a separate task can be formulated as identifying the structure of the technologies of the 'knowledge economy' and its impact on the development of both this sector and the economy as a whole, including its dynamics. Of course, the aggregate approach assumes an assessment of technological

 https://doi.org/10.1051/shsconf/202111600007

dynamics by the number of different levels of technology development in each type of activity. Building such a structure can give a kind of portfolio of technological development, namely, to show how technologies are distributed within the framework of their unified classification and selected types for various types of economic activity. Such a portfolio, if it is obtained, will allow the distribution of technologies by type of work and technological saturation or concentration, since individual activities will take over most of the total composition of technologies. In particular, such activities, covering, for example, most of the technological innovations, will include processing, industry. Information processing, data retrieval, and financial transaction management technologies will be most common in services, including the financial sector.

Such a portfolio or "technological map" can serve as a reference point in assessing the quality of the technological novelty that arises and is used in the economy. Of course, the aspect of mutual determination of different technologies in different activities will also be very valuable, but its resolution will be facilitated by such a technology map or portfolio. The technological map provides a static view of the current technological structure by type of technology, and the portfolio allows you to set a number of optimization tasks, in terms of finding out the connections and technological dynamics. Portfolio elements are characterized by a certain impact, efficiency, amount of investment, risk, i.e., a set of parameters suitable for further quantitative research, empirical analysis, and model constructions in the framework of the tasks of the "portfolio choice", which in this case allows you to extend the possibility of the 'technology of choice' and development of the economic structure in the strategic dimension.

The technology portfolio or map (the difference between them as between the dynamics and the statics of the structure) also helps to find out the content of the effect of 'technological dualism', specific features for each country. To investigate this effect, including its identification, the following steps are required:

- identify areas of labour with capital-intensive and labour-intensive technologies (types of activities), create a technological map of labour, get a 'technology portfolio', identify firms and agents of innovators and conservatives in each type of activity and in general for the economy;

- to assess the composition of labour in the types of activities and the movement of labour from areas with capital-intensive, labour-saving technologies to the field of labour-intensive technologies;

- determine the amount of technological unemployment, that is, the unemployed part of the labour force due to the introduction of capital-intensive technologies;

- set the dynamics of capital and labour-intensive technologies input, highlighting the gap between the number of developed and implemented technologies;

- establish the number and dynamics of the introduction of advanced technologies, innovators and conservatives, highlighting the activities that they cover the most, and where such development or input is practically absent, with an explanation of the reasons for the current structure.

The final result listed empirical and model estimates (in the case of the creation of econometric models) is not only the identification of the effect of 'technological dualism', but the definition of what form it occurs – whether saturating the displacement of labour-intensive technology capital-intensive technologies (or back), or developed a dynamic equilibrium, and they co-exist without displacement (classic content dualism). In the latter case, the dynamics of technological renewal also stabilizes around a certain value and it becomes problematic to accelerate the change of technologies. The proposed method of analyzing technological dynamics allows not only to determine the depth and form of 'technological dualism', but also to identify pseudo-dualism, in which personnel from capital-intensive technologies are released not because of the introduction of new technologies of this class, but because of the loss of competitive positions of existing capital-intensive technologies. They are partially located in the field of labour-intensive technologies, thereby blocking technological renewal in these areas, as they delay the introduction of labour-saving (capital-intensive) technologies. Such input may occur, in particular, with the expansion of the use of digital technologies, but this class of technologies has the effect of increased involvement of the labour serving these technologies. This circumstance becomes a kind of barrier to saving labour, since with the introduction of new technologies, the volume of the object – information processed by them increases, as well as the tasks for using it and the requirements for its final quality multiply. These additional functions expand the applicability of labour.

Technological dualism becomes an institutional barrier for technological renewal, more precisely, for the dynamics of this process, since in some extent this renewal can occur, without a significant change in the structure of technologies at a selected and very short time interval. And the presence of the pseudo-effect of technological dualism fixes a completely different reason in the process of technological renewal – the introduction of technological innovations. The formation of an institutional barrier by the technological structure itself, as well as the pseudo-effect of the movement of resources between activities, makes it difficult to diffuse technologies. The observed technological dynamics is saturated, or is significant, but in the area of low values of overall manufacturability. For countries that are lagging behind in technological development, there is a vicious circle of technological lag.

A standard policy of promoting technological innovation may not break this vicious circle, as it is based on technological dualism or pseudo-effect, which make it difficult to diffuse technologies, and also includes backwardness or dependence in the development of science and education. The presence of a large gap between the created technologies and the implemented ones indicates the difficulty of technological diffusion, no matter where this gap is fixed - in the presence of a low or high number of advanced technologies being developed and implemented. Effective technological diffusion clearly implies an insignificant amount of such a gap. This question requires additional research that goes beyond the goal of this paper.

We will show some aspects of the presented methodological approach on the example of the Russian economy, revealing the presence of the pseudo-effect of technological dualism and the presence of significant structural and technological problems that complicate technological diffusion, that is, the penetration of even created technologies into production.

3 Discussion of the results. Technological changes and dualism in the Russian Economy

The effect of technological pseudo-dualism in the Russian economy can be easily shown by the example of the movement of personnel from the processing to the transaction and raw materials sectors (Figure 1, above). The technological efficiency of the sectors not only does not increase significantly, but even decreases it. Although the number of new technologies is growing (Table 1). Of course, the existing approaches in the field of statistical accounting don't allow to distinguish between capital-intensive (labour-saving) and labour-intensive technologies.

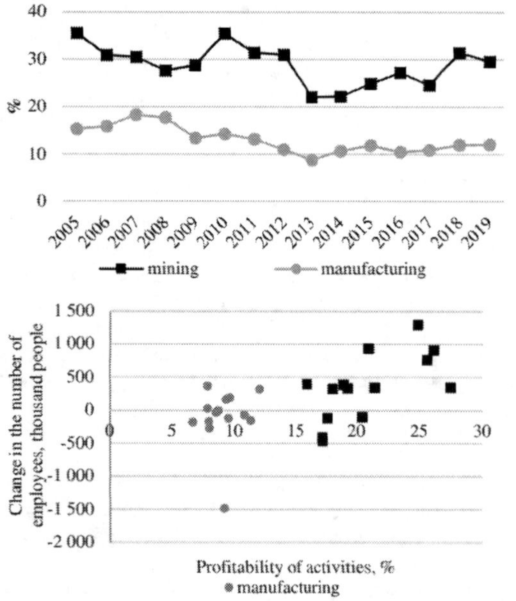

Fig. 1. Profitability of activities (mining and manufacturing), 2005-2019 (above); change in the number of people employed in the manufacturing and transactional and raw materials sector, 2006-2019 (below).

However, the transfer of labour from processing to other types of activity could be associated with the introduction of labour-saving technologies (capital-intensive), but in this case, in theory, the overall technological efficiency should increase, and in particular, this sector. Such a result was not observed in the Russian economy. The high difference in the profitability of mining and processing activities (Figure 1, left), apparently, was the main reason for the movement of labour resources towards the transactional and raw materials sectors. Non-capital-intensive technologies were the reason for the release of labour and its intersectoral movement.

The number of new technologies developed, which are considered advanced, is reflected in the dynamics of Table 1. There is an increase in production of 9 times, in processing - more than 2 times, in the production of machinery and equipment - more than 1.5 times. However, these overall positive figures of technological progress are offset by the presence of the described effect of technological pseudo-dualism, which leads to a slowdown in the creation of fundamentally new technologies (see Table 2). It should be noted that the overall level of adaptability (measured by the ratio of innovative to non-innovative products, works, and services) decreased during the years of increasing the number of these new technologies. This indicates the limited applicability of new technologies. In addition, we can talk about possible accounting problems of assigning emerging technologies to a new class and to fundamentally new advanced technologies.

Table 1. Number of advanced new manufacturing technologies developed for Russia by type of economic activity in Russia, 2010-2019.

type of activity / year	2010	2011	2012	2013	2014	2015	2016	2017	2018	2019
mining	5	7	12	12	23	17	25	23	21	49
manufacturin g	215	320	320	374	382	416	491	409	468	497
production of machinery and equipment (without production of weapons and ammunition)	31	34	47	62	31	33	40	37	44	51

Source: [12].

Table 2. The number of fundamentally new developed advanced production technologies by type of activity in Russia, 2010-2019.

type of activity / year	2010	2011	2012	2013	2014	2015	2016	2017	2018	2019
mining	0	3	2	3	2	1	0	0	1	2
manufacturing	1 6	18	16	24	32	26	32	33	34	35
production of machinery and equipment (without production of weapons and ammunition)	3	8	3	2	13	1	0	0	0	1

Source: [12]

https://doi.org/10.1051/shsconf/202111600007

Table 2 shows the changes in fundamentally new production technologies that have been developed in these types of activities in Russia over a ten-year period. In the field of civil production of machinery and equipment, the dynamics of technology development is not stable, and since 2015 there has been almost no positive movement in this area. Although manufacturing industries as a whole over a ten-year period increase this number of technologies 2 times, the basic industry-mechanical engineering is characterized, as we can see, by a high lag. Moreover, since 2016, manufacturing industries also do not show a significant increase in the number of fundamentally new technologies created. In the field of mining, we can talk about a stable situation, measured on average by the emergence of 1-2 fundamentally new technologies. Of course, it is difficult to say how many such emerging technologies should characterize the type of activity that can be considered the most acceptable for the technological development of the industry. The comparative dynamics of nanotechnologies will be a peculiar characteristic of the development of a new technological frontier, indicating the level of application of technologies (Table 3).

Table 3. Number of nanotechnologies developed and used, 2010-2019.

The number of nanotechnology	2010	2011	2012	2013	2014	2015	2016	2017	2018	2019
Developed	222	258	327	411	443	505	494	446	402	555
Used	354	526	748	907	937	1152	1166	1144	1258	1228

Source: [12]

According to table 3, the number of developed and used nanotechnologies increased from 2010 to 2015. Since 2016 it has been declining for three consecutive years, increasing only in 2019, while the number of nanotechnologies used decreases only in 2017, then in 2019.For this type of technology, the use not only exceeds the development several times, but this gap is steadily increasing over the considered time interval.

Figure 2 shows the ratio of used and developed advanced manufacturing technologies. We can see the superiority of use over development and at the same time a significant excess of new technologies purchased over the number of developed ones. The presence of such a large line gap in Figure 2 suggests that more domestic technologies are still used, which are outdated. The share of new technologies purchased relative to the technologies used is not large. Moreover, this ratio remains stable from 2011 to 2019. Given the crisis of 2020, we can assume, taking into account the inertia of technological innovations, that this ratio will remain approximately the same in the coming years.

If technologies are not constantly improved, new ones are not created, then automatically this process, due to its very deep nature, demonstrates the property of inhibition. In an economy that is not capable of a technological race, there is a technological lag at the system level, even if there are individual technological successes.

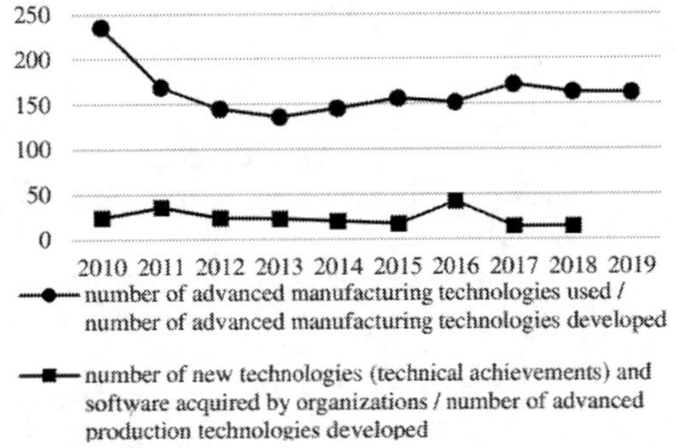

Figure 2. The ratio of used, developed and acquired advanced (new) technologies in Russia, 2010-2019.

4 Conclusion

Concluding the study, we note the following important positions.

First, technological dualism, which ensures the coexistence of two classes of technologies (capital and labour-intensive) and even a kind of support from one for the other, can be formed not because of technological renewal, but because of structural and inter-institutional problems of functioning and various types of activities that set the parameters for technological renewal and block technological development.

Second, the example of the Russian economy shows a pseudo-technological dualism. The released personnel from capital-intensive activities (processing) find their implementation in labour-intensive activities, but the reason for moving the resource is not the introduction of new technologies that save labour, but a high gap in the use of old and the emergence of new technologies, that is, a low level of technological renewal.

The revealed circumstance of the non-working classical effect of technological dualism, which itself arises in an already unfavourable situation of technological renewal, and the pseudo effect indicates the presence of serious problems, confirms the need to link the task of developing high-tech industries and R&D to the structural transformation of the economy. Moreover, the purpose of this transformation should be considered the deployment of the production of new types of products and a new stock and technological base for them. The findings suggest that standard measures to encourage innovation will always be palliatives in the current situation. They can lead to some revival, but they are not able to fundamentally change the process of technological innovation and its quality. The strategy of local technological innovation, implemented in Russia, and leading to some success, however, in the strategic perspective should be replaced by a strategy of diffusion of technological innovation, which has systemic consequences and improves the functioning of many activities at the same time. Here it makes sense to expand Schumpeter's vision of innovation to the area of social and managerial innovation, covering the implementation of government policy.

References

1. R. Nelson, S. Winter *Evolutionary theory of economic changes* (Finstatinform, 2000)
2. O.S. Sukharev *The theory of economic restructuring* (Lenand, 2016)
3. J. A. Schumpeter *Theory of economic development. Capitalism, socialism and democracy* (Eksmo, 2007)
4. S. Chen, G.H. Jefferson, J. Zhang China Economic Review, **22**, 133-150 (2011)
5. C. Ertur, W. Koch Interactions Journal of Economic Growth, **16**, 215-255 (2011)
6. P.M. Gil, O. Afonso, P. Structural Change and Economic Dynamics, **51**, 89-105(2019)
7. H. Hanusch, A. Pyka History of Economic Ideas, **15,** 23-41 (2007)
8. D. Horvat, H. Kroll, A. Jäger Procedia Manufacturing, **39**, 886-893 (2019)
9. K. Jati Applied Economics Letters, **8,** 81-84 (2001)
10. C.G. Machado, M. Winroth, D. Carlsson,P. Almström, M. Hallin Procedia CIRP, **81**, 1113-1118 (2019)
11. K.I. Westeren *Foundations of the knowledge economy,* 57-74 (2012)
12. *Rosstat.* Retrieved from: https://www.gks.ru/folder/11189.

Lean management as the innovative technology of an enterprise

Sergey Komarov[1*], *Mariya* Kudina[2], *Gulnaz* Suzdaleva[2], *and Dmitriy* Shishkin[2]

[1] Perm State National Research University, 15 Bukireva Str., Perm, 614068 Russian Federation
[2] Perm National Research Polytechnic University, 29 Komsomolski Str., Perm, 614990 Russian Federation

Abstract. The paper is devoted to the problem of management of innovative activity of an enterprise on the basis of Lean technology. The study shows that there are no generally accepted theoretical models for understanding this relationship and its unambiguous assessment. The paper discusses a pilot study of the relationship of Lean management factors and innovative activity for large Russian enterprises taking into account their life cycle and type of innovation. The results of the study demonstrate the correlation between lean production and innovative activity of the enterprise. The conclusions on the impact of the type of enterprise (business), the stage of its life cycle and the type of innovation on this relationship are innovative and important. The study is aimed at testing the methodology of future research, the purpose of which is to build a mathematical model of such a relationship.

1 Introduction

The main trend of modern management is to increase the innovative activity of the enterprise. The development of globalization, digitalization of production and customer orientation require innovative activity that increases the competitiveness and sustainability of the enterprise. Innovation is the introduction of a new product or service to the market, the introduction of a new production process, the development of a new business model, and the creation of new markets [1]. There are generally seven types of innovation [2]: technological innovation, process innovation, product innovation, innovation products, works, services, marketing innovation, organizational innovation, environmental innovation. Innovative development is carried out within the framework of the innovation process through the implementation of certain stages and business processes [3], and involves an extensive range of scientific, technological, organizational, financial, marketing, commercial activities and is only possible with the appropriate level of innovative activity of the enterprise.

Currently, there are many approaches to increase the innovation activity. One approach to increase the innovation activity of the company is the concept of "lean production". Lean management implies the identification and elimination of all types of losses and processes that do not generate value for the consumer [4]. The studies show that lean production methods lead to a reduction in the time of orders, rapid creation of new standards of work; optimization of orders leads to increased production "margin" [5, 6]. Lean production tends to be aimed at innovation in the improvement of production, but more recently, there is concern regarding the fact that it may contribute to the innovative development of the enterprise as a whole and increase its innovative activity.

Nevertheless, innovation experts are quite ambiguous in assessing the relationship between the methods of "lean production" and the innovative development of the company. On the one hand, lean production methods may be part of product and technological innovations, or through organizational and cultural changes at the enterprise lead to the formation of an innovative culture in the company. On the other hand, according to their purpose, lean production methods are aimed at optimizing the existing technologies. Therefore, according to a number of researchers, they "conserve" the current state of affairs and "slow down" the innovative development of the enterprise. This means that at the moment the problem is the relationship between the innovative development of companies and the methods of lean production: how does lean production affect the innovative development of the company?

Our hypothesis is that the use of lean production tools contributes to the innovative activity of the enterprise. At the same time, it is necessary to take into account the size of the enterprise, the stage of its life cycle, the type of innovations (horizontal and vertical), since different tools of lean production are required at different stages of the enterprise life cycle for different purposes.

The purpose of this paper is to analyze approaches to the interaction of lean production and innovative activity, as well as to test the tool of empirical study of the relationship between lean production factors and indicators of innovative activity. The purpose of this pilot study is to test the methodology based on the life

* Corresponding author: gulnaz.suzdaleva@yandex.ru

cycle of the enterprise and the type of innovation to verify our hypothesis during a further large-scale research project.

The objectives of this project are as follows: first, the analysis of the relationship between lean management and the innovative development of the enterprise, second, the analysis of the relationship between the factors of lean production and the factors of innovative activity, third, the consideration of the features of the use of lean production tools in order to increase the innovative activity of the company taking into account the life cycle and type of innovation. The solution of these problems will not only confirm the initial hypothesis of the study, but will also make it possible to formulate a mathematical model of the relationship between the factors of lean production and indicators of innovative activity, which will become the basis for managing innovative activity through the use of lean production tools.

2 Methods and Materials

2.1. Theoretical description

A fairly large number of studies are devoted to the relationship between innovative development and lean production. All of them may be divided into two groups.

The first group includes studies in which the relationship and mutual influence are evaluated quite positively. As a rule, these works emphasize the study of the influence of lean management on individual sides of the innovation process.

First, the use of lean management tools leads to increased customer orientation, the introduction of new services, and the formation of innovative leadership. These factors of lean production become the factors of increased innovative activity of the company. Thus, the study [7] shows that in companies, which business is based on the provision of traditional services, the use of lean production tools contributes to the development of customer-oriented business, the introduction of innovative ways of customer service. The work of Iranian researchers [8] raises the issues of how the innovative behavior of the company's employees is formed. The results show that there is a positive relationship between employee participation in lean thinking and the innovative behavior of employees, and that the introduction of lean production methods generally contributes to the formation of an innovative climate in the enterprise.

Second, lean production methods are seen directly as tools for managing innovation, more precisely, tools for the fastest commercialization of innovation. In the work [9], the authors of the study conclude that with the use of lean production tools, the original innovations in the new business may be introduced more efficiently. This is confirmed by the study [10], in which lean management methods are considered as tools for generating additional added value for product innovations.

Third, there are some works that consider the specific use of lean management in supply chain operations and improving methods aimed not at further waste reduction, but at achieving environmental effect. This concerns operating companies. Thus, the study [11] shows how sustainable lean production programs have contributed to improved performance.

Fourth, these are studies of the impact of lean production methods on the efficiency of innovative processes in production organizations. The research is very important here [12]. Researchers A.K. Moldner, J.A. Garza-Reyes, V. Kumar state that "little is known about such an influence"; therefore, in their study, three hypotheses were put forward about the negative, neutral and positive effect of lean production on the effectiveness of innovative processes. These large-scale surveys made it possible to formulate a refutation of the misconception of scientists and leaders that lean management and innovation are difficult to coexist. The conclusion is that the practice of lean production has a generally positive effect – from moderate to strong – on the introduction of additional and radical technological innovations in production organizations. Innovations, in turn, improve the operating performance of companies. Besides, the authors of the study proposed an integrative structural model, which is considered as a system of interconnected internal and external processes that together increase the innovation of the company. This model is open to additions, and different types of external and internal processes may be easily integrated into it. The general conclusion of this study is important – the integrative approach implies that lean innovation management stimulates the overall innovation of companies.

Adjacent to this are Lean-Startup efficacy studies [13; 14; 15; 16]. These studies are less optimistic because they note that Lean-Startup methodologies emerged precisely to reduce uncertainty of project-based innovation. The key point in applying these tools is to reduce the uncertainty of the innovation process, increase the soundness of decisions when introducing innovations and reduce the share of "dying" startups. However, in general, Lean-Startup management also proceeds from the direct positive impact of lean production on the innovative development of the company.

Finally, fifth, there are works considering the synergistic effect of lean production and innovative development. This concerns the difference in the goals of these management methods: innovation is aimed at the development of the enterprise, while lean production – at its sustainability. Therefore, a number of authors prefer to say that "lean processes may positively and negatively affect the company's sustainable performance depending on the state of its innovation capacity" [17]. A synergistic approach is also considered in another study based on the analysis of small and medium-sized enterprises [18]. Here, lean and sustainable innovations are considered as input criteria, and the economic, operational, environmental and social aspects of the enterprise are considered as output criteria for assessing its innovativeness. The main conclusion is that the combination of lean production and innovations helps to achieve sustainability of the supply chain.

Nevertheless, it cannot be argued that there is unambiguous evidence of the positive impact of lean production on the innovative development of companies. There are quite a few skeptics who consider this relationship not so unambiguously. Therefore, in the literature there is a second group of studies where this assessment is quite negative. Here, the main argument is that lean production is aimed more at stabilization and sustainability than at development.

At the same time, it is usually indicated that in modern management there is no clear idea of indicators of innovation and indicators related to lean production. This makes it impossible to clearly divide between these indicators and establish a correlation between them. Thus, according to [19], there are 4 groups of indicators, one of which is innovation, among the indicators of lean production. Other sources [for example, 12], on the contrary, consider the indicators of lean production factors as the indicators of innovative development. Besides, in all such studies, the original theoretical models themselves are different, and therefore the understanding of the relationship between lean production and innovative activity is different. This results in research findings being limited or even incompatible with other studies.

Moreover, the studies are often associated with different types of innovation, and this is an important success factor. For example, in the study [20], continuous improvement programs are associated with incremental innovations and process innovations. Lean production is more associated with gradual innovations than with radical ones [21]. Other studies consider product innovations. There is also little research on the relationship between the continuous improvement programs of Lean, Six Sigma and Lean Six Sigma, and the effectiveness of innovation.

Another aspect of the negative assessment of the impact of lean production on innovations is indicated in the work already reviewed [17]. The authors believe that different tools of lean production may act differently depending on the level of innovative development of the company. Therefore, it is quite possible, using different tools of lean production at different stages of the innovation process, to contribute to the growth of innovative activity by streamlining at initial stages and freeing up time and resources. From the point of view of the authors of this study, the most important issue is the determination of the balance and the point of view of using lean production tools in the company's innovation process.

Finally, there are works that deny the positive impact of lean production on innovative development associated with modern digitalization of production. Digitalization and lean production often contradict each other. Therefore, it is necessary to consider open process innovations. This does not mean the rejection of lean production methods at all, but, firstly, it indicates their subordination to digital technologies in the production process of the enterprise, and, secondly, the introduction of approaches to lean production in digitalization and transformation processes to accelerate them.

Based on the analysis of these works, it may be generally concluded that the negative assessment of lean production regarding innovation does not concern so much the lean production itself as a management technology, but its correct use in working with open and constant innovations. These questions are considered in the study of J. Craig, G. Pairaudeau, J. Pettersson [22], in the study [23]. The positive relationship between improvement programs and incremental innovations in products and processes shows that such improvements may lead to increased innovation opportunities.

However, the following should be noted. First, there are not enough empirical studies that clearly record the positive impact of lean production on innovative development. Second, there is no clear distinction between indicators of innovative development and factors of lean production. Hence the difference in the methodology of such studies is that their results are clearly verified on a certain empirical basis, but cannot be transferred and tested on other empirical material. Third, only some studies are devoted to large enterprises. And, fourth, there is no research that takes into account the life-cycle stages of the organization when establishing the relationship between lean technologies and innovation activities.

This is what determines the author's approach to the study of the influence of lean production tools on the innovative activity of the company.

2.2. Research methodology

Our hypothesis is that lean production positively affects the innovative activity of the enterprise. This hypothesis is based on the following studies [24, 18, 20] and our own study [25, 26].

The purpose of our study is to analyze the impact of lean management on the innovative development of companies, taking into account the type of innovation, on the one hand, and the life cycle of the company, on the other. Therefore, with the general methodology of the study based on the works [27, 20] other meaningful changes have been made to it. First, the list and content of judgments based on the preliminary classification of lean production factors and external and internal factors of innovative activity have been changed. Second, the difference is the testing of the research methodology at large industrial enterprises of the Russian Federation. Unlike small and medium-sized enterprises, for which innovative activity is not always the primary factor, for these enterprises innovative activity is the main factor of sustainability. Besides, only innovative enterprises participated in the study. This is fundamentally important for the development of Russian business, for which large enterprises are the driver of the innovative development of the national economy in general. Finally, the third distinguishing feature of our study is the recognition of the difference between horizontal and vertical innovations.

At the same time, we took into account the following limitations of this study. The innovative activity of the enterprise is an integral indicator determined by various factors. Meanwhile, we adhere, firstly, to taking into account only a limited list of factors of innovative activity (the presence of completed vertical innovations, the presence of completed horizontal innovations, the share of innovation costs, the speed of implementation of new ideas, the coefficient of introduction of new products, the competitiveness index of goods, the presence of the R&D department, the level of qualification and competence of personnel involved in the development of innovations, the resource potential of innovative activity. In this study, we do not define an integral indicator of innovative activity. This is explained by the subsequent task of our research project – provided that a quantitative model is obtained of the influence of a certain list of factors of lean production on indicators of innovative activity – the creation of a system for managing the increase of innovative activity by managing these factors. At the same time, this is a primary validation test of the conceptual model itself. It is important to understand the fundamental correctness of the model aimed at confirming our hypothesis.

The main purpose of the study is to determine the existence of a relationship between the level of innovative activity and the degree of use of lean production tools.

As a **pilot study tool**, an expert survey form was developed containing 20 statements based on the Likert scale (Likert scale – a starting point for the development of the tool was research [23, 18, 20, 26] – however, the shape of the tool was significantly refined): six statements characterizing the overall level of development and sustainability of the organization, nine reflecting innovative activities, and five statements on lean production tools. Besides, four questions were included in the questionnaire to identify the type of the enterprise: number of employees, industry, scope of activity, position of the respondent. Previously, the tool was validated ensuring the sustainability of judgments (the starting point for this was 80 negative and positive judgments). The indicators of innovative activity are taken from existing methods [28, 29, 30, 31]. As part of the pilot study, the experts were offered judgments that belong to three groups of factors influencing the company's sustainable competitive development, namely:

• fundamental factors, including conditions for sustainable management of the production system;

• factors determining the availability of lean production principles and tools in the company's production system;

• factors determining the presence of innovative processes and prerequisites for innovative development of the company.

Research method – an expert survey, which is optimal for collecting the necessary data (respondents must be professionals in the management and innovation of a particular enterprise).

Research object – innovative large industrial enterprises in various industries of Perm and Perm Territory.

Research subjects – factors of innovative activity, lean technologies.

Sampling description and calculation.

The representatives of Perm and Perm Territory enterprises took part in the survey: heads of enterprises, development directors, and employees engaged in innovation and production. The pilot sampling included 21 enterprises (optimal sampling for the study).

The sampling is available: the study involved organizations that were ready to disclose information (including commercial), as well as experts who had time to answer the questions in the questionnaire (in several stages). The sampling is quota-based: innovation and lean technologies at enterprises should have a decisive role; innovation is carried out at enterprises (any types of innovation, not only product innovations). The market in which experts work does not belong to the oligopoly or to the monopoly. The sampling included production enterprises from different industries.

Place and time of contact with respondents. The survey of experts was conducted in their offices, as well as through communication by e-mail and on social networks.

3 Results and Discussions

The results of the study show the development of innovative activity of the enterprise due to the implementation of lean production tools. Let us formulate key conclusions of the analysis.

1) The relationship between the factors of innovative activity of the company and lean production was revealed. It is established that it is positive, which is confirmed by the indicators of the variation coefficient $\gamma \leq 33\%$ specified in Table 1. Calculations were performed using the Integrated Analysis Methodology, which includes the collection of expert estimates and their statistical processing [32].

Table 1. Results of a comprehensive factor analysis of the relationship between sustainable management of the production system, innovation activity and lean production

Group of factor	Variation coefficient (γ), %	Upper quartile indicator and factors
Fundamental factors that include conditions for sustainable management of the production system	10.96	- Company's position at the "Growth" and/or "Stability" stage - Involvement of employees in the process of developing new products, ideas, processes
Factors determining the availability of lean production principles and tools in the company's production system	29.9	- Application of the set of lean production tools (flow mapping, visual control, 5S, etc.) - Leadership interest in the results of lean production in the company's production system
Factors		- Development and

determining the presence of innovative processes and prerequisites for innovative development of the company	24.8	implementation of vertical and horizontal innovations - Development of R&D Institute - Highly qualified staff involved in innovation development

The calculation was made according to the following formulas:

Mean-square deviation of estimates:

$$\sigma = \frac{\sqrt{\sum_{i=1}^{n} (X_i - \overline{X})^2}}{n} \quad (1)$$

where, n – number of experts; X_i – value of *i* expert.

Coefficient of variation:

$$\gamma = \frac{\sigma}{\overline{X}} \times 100, \% \quad (2)$$

2) The factor analysis suggests that fundamental factors (γ=10.96%) demonstrate the greatest convergence and significance in the relationship of the three groups of factors. The most important factors for ensuring the relationship between "Sustainable Management – Innovation Activity – Lean Production" are the following:

• in the group "**sustainable management of the production system**": position of the company at the "Growth" and/or "Stability" stage and the involvement of employees in the process of developing new products, ideas, processes;

• in the group "**factors determining the availability of lean production principles and tools in the company's production system**": use of a set of lean production tools; leadership interest in the results of lean production in the company's production system;

• in the group "**factors determining the presence of innovative processes and prerequisites for the innovative development of the company**": development and implementation of vertical and horizontal innovations; development of R&D; highly qualified staff involved in innovation development.

3) Fundamental factors affecting the level of innovation activity are the stage of the company's life cycle and the company's mainly horizontal (process) or vertical (revolutionary) innovations.

4) A possible confirmation of our hypotheses is the opinion of experts who note the potential of opportunities due to the application of lean management principles and tools, which stands in the position of rational distribution of all the resources available to the company. At the same time, the type of innovation is an important factor. The role of lean production tools is maximized when it comes to process and organizational, horizontal innovations. Besides, their role is minimal (but the use of tools is necessary) for the introduction of vertical, breakthrough innovations.

5) Life cycle and relationship of lean production and innovation activity. According to most experts, the streamlining and systematization of all processes, as a result of the use of lean production tools, allows freeing up temporary, information and material resources aimed at building the innovative potential of the company. This is also confirmed by the relationship between the major group factors and the second and third group factors, and there is a convergence in understanding the need to involve all staff in innovation processes and to introduce lean production philosophy and tools; to ensure leadership transparency; to welcome new ideas and projects (vertical and horizontal innovations).

6) Types of innovation activity and the relationship between innovation activity and lean production. The closest link between the indicators of innovative activity and lean production is observed at the "Stability" stage. Regarding the stages of the life cycle "Decline" and "Launch", the answers of experts are quite close in their values, i.e. they poorly demonstrate the relationship of the studied factors. At the same time, at the "Decline" stage, companies use vertical innovations to a minimum degree, compared to other stages. At all other stages, experts noted the same importance for both types of innovations (Figure 1).

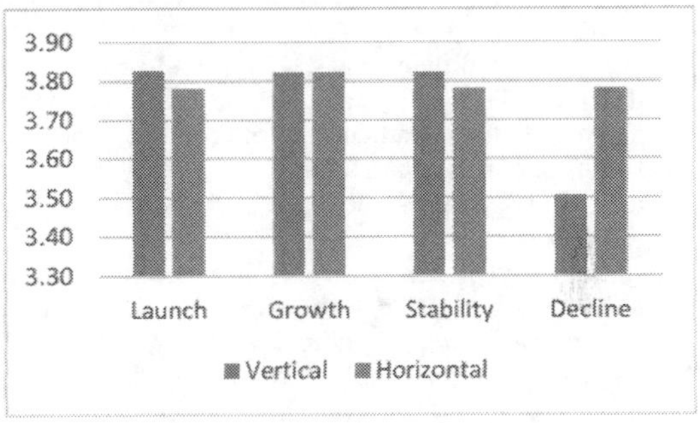

Fig. 1. Relationship between the life cycle phase and company innovation activities

7) The stage of development of the organization and innovative activity. It is important to take into account the development stage of the organization and the stages of the innovation process, since the goals and objectives of using lean production tools may differ depending on the level of development and level of innovation of the organization. Thus, at the "Launch" stage, the role of lean production tools may be minimal, while at the "Growth" and "Stability" stages it is necessary to organize processes to release resources.

4 Conclusions

The results of the study are largely determined by the research methodology and experts involved in assessments. The problem with such a study is that innovation experts may generally agree with the parameters for estimating innovativeness in the study, but as respondents they may disagree with the specific estimates (results) of the study. As the managers of a company, they may see a negative correlation between the actual commercialization of innovations and the obtained estimates. This does not question the correctness of the model, but raises the question of its applied significance.

Within the framework of the study it was important to understand the fundamental correctness of the model aimed at confirming our hypothesis. The results obtained in such a limited study (in case of positive validity of the conceptual model) make it possible to raise the question of the system of management of lean production indicators as tools for innovative development of the company. However, full confirmation of the hypothesis should be carried out in a larger sampling of industrial enterprises, and this goes beyond the scope of the present study. Based on this, it is possible to build a mathematical model of this relationship (DEA). Therefore, as part of further research it is necessary to expand the sampling, determine its size on the basis of statistical calculations taking into account such restrictions as industrial enterprises, size – large (300 or more employees), the presence of innovative activity, the introduction of vertical and horizontal innovations, the life cycle of the company. Sampling should be performed in the context of the life cycle stages of the organization, since the goals and motives for using lean production tools to increase the innovative activity of the enterprise at different stages of the life cycle may be different. Thus, the construction of a mathematical model of the relationship between lean production factors and indicators of innovative activity is the task of a future large study, the testing of which is covered in this paper.

In case of empirical confirmation of the initial conceptual model (mathematical dependence of lean production factors and indicators of innovative activity), the question may be raised about the system of management of lean production indicators as a tool of innovative development of the company. The management of this model may influence the choice of innovative development strategy. This is the next task of our research project.

References

1. Decree of the Government of the Russian Federation No. 2227-p "Strategy for the innovative development of Russia until 2020" (2011), retrieved from: http://static.government.ru/media/files/4qRZEpm161xctpb156a3ibUMjILtn9oA.pdf.
2. *Oslo Manual: Guidelines for Collecting and Interpreting Innovation Data* (2010), retrieved from:https://mgimo.ru/upload/docs_6/ruk.oslo.pdf.
3. V.L. Makarov, A.E. Varshavsky, *Innovation management in Russia: Issues of strategic management and scientific and technological safety*. (Moscow: Science, 2006).
4. J.P. Womack, D.T. Jones, *Lean Thinking: Banish Waste And Create Wealth in Your Corporation* (2003).
5. E. Dolzhenkova, L. Iurieva, *Proceedings of the international conference on trends of technologies and innovations in economic and social studies*, **38**, 132-138 (2017).
6. T. Goldsby. R. Martichenko, *Lean production and 6 Sigma in logistics. Logistics Optimization Guide*. (Moscow: Grevtsov Pablischer, 2016).
7. F.M. Morales-Contreras, P. Chana-Valero, M.F. Suarez-Barraza, A.S. Diaz, E.G. Garcia, *International journal of environmental research and public health*, **17 (15)**, (2020).
8. R.B.A. Hamid, M.D. bin Ismail, I.R.B. Ismail, *South East Asian Journal of Management*, **14 (1)**, 23-43 (2020)
9. S. Lobo, P. Samaranayake. *Benchmarking-an international journal*, **27 (5)**, 1633-1656 (2019).
10. A.C.O. Santos, C.E.S. da Silva, R.A. da Silva Braga, J.É. Corrêa, F.A. de Almeida, *Systems Engineering*, **23**, 281-293 (2020).
11. P. Ball, P. Lunt, *International journal of production economics*, **219**, 405-415 (2020).
12. A.K. Moldner, J.A. Garza-Reyes, V. Kumar, *Journal of business research*, **106**, 233-249 (2020).
13. D.S. Silva, A. Ghezzi, R.B. de Aguiar, M.N. Cortimiglia, C.S. ten Caten, *International Journal of Entrepreneurial Behavior & Research*, **26 (4)**, 595-628 (2020).
14. K. Still, *Technology innovation management review*, **7 (5)**, 32-43 (2017).
15. H. Edison, N.M. Smorsgard, X.F. Wang, P. Abrahamsson, *Journal of systems and software*, **135**, 69-87 (2018).
16. Z.B. Yordanova, *International journal of innovation and technology management*, **15 (2)**, 1850012, (2018).
17. M. Kumar, V.S. Rodrigues, *International journal of production economics*, **219**, 469-479 (2020).
18. D. De, S. Chowdhury, P.K. Dey, S.K. Ghosh, *International journal of production economics*, **219**, 416-430 (2020).
19. S.A. Darestani, N.H. Shamami, *Opsearch*, **56**, 717-738 (2019).
20. F.L. Lizarelli, J.C. de Toledo, D.H. Alliprandini, *Total Quality Management & Business Excellence*, **08**, 1-24, (2019).
21. A.B. Abdallah, S.E. Dahiyat, Y. Matsui, *Management Research Review*, **42 (2)**, 239-262 (2019).
22. C. Johnstone, G. Pairaudeau, J.A. Pettersson, *Drug Discovery Today*, **16 (1-2)**, 50-57 (2011).
23. M.C. Di Guardo, E. Marku, W.M. Bonivento, M. Castriotta, F. Ferroni, C. Galbiati, G. Gorini, M. Loi, *R&D Management*. (2021). DOI: 10.1111/radm.12453
24. M.G. Lins, L.P. Zotes, R. Caiado, *Total quality management & business excellence* (2019).
25. G.R. Suzdaleva. *System-integration model of management as the basis for increasing*

innovative activity of organizations. PhD dissertation (Yekaterinburg, 2014).

26. V. Potocan, *Innovative Issues and Approaches in Social Sciences,* **6** (**1**), 118-134 (2013).
27. P.K. Dey, C. Malesios, D. De, S. Chowdhury, F.B. Abdelaziz, *Bus Strat Env.,* **28**, 582-598 (2019).
28. Order of Rosstat No. 305 (as amended on 29.08.2013) "*On the approval of statistical tools for the organization of federal statistical observation of employment of the population and the activity carried out in the sphere of education, science and innovations*", retrieved from: http://www.consultant.ru/document/cons_doc_LAW_105874/.
29. V.A. Korolev, D.V. Kaplenkov, *Bulletin of Stavropol State University,* **70**, 137-142 (2010).
30. S.D. Anthony, M.W. Johnson, J.V. Sinfield, E.J. Altman, *The Innovator's Guide to Growth: Putting Disruptive Innovation to Work* (2008).
31. A.A. Nikolskaya, *Innovative activity of higher educational institutions: assessment and use in determining competitiveness.* PhD dissertation, (Ivanovo, 2012).
32. V. M. Mikhailova, O.A. Kuznetsova, A.V. Petrova, *International Scientific Journal,* **3** (**81**), 106-110 (2019).

Application of risk management principles in macroeconomic regulation

Uliana Aftakhova[1*], *Eugeny* Zhulanov[2] and *Aleksandr* Aftakhov[1]

[1]Perm National Research Polytechnic University, 614990 Komsomolsky Ave. 29, Perm, Russia
[2]Perm National Research Polytechnic University, 614990 Komsomolsky Ave. 29, Perm, Russia

Abstract. The article confirms the urgency of coordination strategic and tactical management of the country social-economic system on the basis of the principles and methods of risk management. It proposes an economic-mathematical dynamic model, describes an advanced complex of forward and backward linkages between national economy and its managerial system. This approach allows predicting and implementing development vector of a system more accuracy on the way to achieving its strategic goals. The authors' approach was tested on the base of the social and economic system of the Perm Region.

1. Introduction

The analysis of today's social and economic development programmes of the Russian Federation and its subjects revealed that only a list of social and economic indicators to be reached by the end of their implementation is accounted for. However, clearly defined tools for reaching specified strategic, let alone tactical, objectives are described rather sparsely, if at all. As a rule, only a list of budget allocation options is given, with performance evaluation mechanisms totally ignored [1, 2]. This stimulates the need for drafting a national policy that would facilitate making managerial decisions based on these mechanisms, as well as realistic planning and reaching tactical and strategic objectives in the social and economic system of the country [3].

The approach proposed by the authors implies a dramatic change in national policy in favour of principles and methods of up-to-date management. In particular, strategic and tactical planning and risk management. Thus, national policy assumes the aspect of preventive risk management, that is, a set of rules on making national policy decisions in risk management in the society by using specialized economic tools.

2. Literature review

The phenomenon of risks and instruments of their minimization does not lose its historical relevance. There are known attempts of prophets and fortune tellers of Ancient Rome and Ancient China to develop risk management techniques on the basis of ritual procedures. In particular, on the basis of the ancient Chinese "Book of changes" strategic state decisions were made for many centuries. It is this work today to be considered the first comprehensive risk management system. It is well known that behind the scenes of the Delphic Oracle there was a large group of priests working, collecting and analyzing various world information [2].

Despite such a long history of study, so far modern science has not developed an unambiguous theory of risk management. Since there is no common understanding of the "risk" category, the article authors define it as an activity related to the decision-making process to overcome uncertainty in the country's social-economic system.

Risk management is a procedure that is methodologically close to exact sciences [3]. Any social-economic system is described by its objective characteristics (parameters), which are inherent to it at any given moment of time. Therefore, the level of uncertainty characteristic of each parameter of this system will differ depending on the change in the external and internal environment. Thus, the strategic objective of risk management involves the fulfillment of a designated mission of the system, the achievement of its goals and objectives, the implementation of its strategic plan, and actions that assess the change in the state of the system in relation to the fulfillment of its mission. Ongoing risk management involves actions that actually ensure the implementation of mechanisms for day-to-day advancement of the system towards its goal. Thus, risk management system unites all actions that contribute to reducing the probability and magnitude of the system deviation from achieving its own strategic goal [3].

The paradigm of preventive risk management is based on the theory of integrative management, according to which the subject of management carries out navigation through risky landscapes. The movement trajectory of the object under control is a navigation strategy (ensuring stable development of the subjects of the Russian Federation and stable common economic growth, as a consequence) and navigation tactics (tasks

* Corresponding author: ulianaaft@yandex.ru

solved to achieve a strategic goal, based on the methodical forecasting tools, allowing to preventively minimize risks). The results of the study made it possible to identify the functional features of the preventive risk management system, which are presented in the table (see Table 1).

Table 1. Contents of functions of the preventive risk management system in the social-economic system

Function	Contents
Axiological	definition of the working population value and their subsequent generations on the basis of economic development potential of industrial production
Limiting	restriction of industry negative impact on the life quality of the population by improving working conditions and environmental activities of industrial enterprises
Stimulating	arrangement of favourable conditions for self-realization and maximizing the well-being of the entire population
Stabilizing	ensuring sustainable progressive industrial development by minimizing negative socio-economic and environmental impacts and intensifying innovation and technological modernization as a result of human potential
Compensating	compensation for the negative impact of industrial development on the life quality of the population
Evolutional	ensuring coevolutionary development of industry and civil society or mutual development of industrialization and civilization
Harmonizing	harmonization of systemically interconnected solutions of social, ecological, innovative, technological and branch tasks of regional and national economy development, minimizing damage to each of these spheres

These functional features defined the content of the management fad of risk management "Social-SOS". The proposed concept is to implement simultaneously three main approaches to the management of the social-economic system [4]:

- *social-strategic approach*, consisting of priority social and environmental goals identification while designing, creating and developing the production;

- *social-responsible approach*, assuming the determination of a mechanism to redistribute powers and responsibilities of social-economic system management subjects according to the criterion of the significance of change dynamics regarding the decisions taken by the subjects to achieve development social goals;

- *social-cost approach*, which provides designing of institutional mechanisms for the system development, adapted to real situation in world markets, arising from economic objectives of modernization in the long and short term, and based on the assessment of existing experience of negative social-economic, social-ecological and ecological-economic consequences of innovation-technological reformations in terms of money equivalent.

3. Methods

To implement effective policy, the authors developed risk assessment and management tools. They propose the formulation of a multi-criterion mathematical problem and the search for its optimal solution. Let there be a number of time intervals t_1^i, for each of which social risks are characteristic r_1^n. Every risk is described by a set of indicators j_1^m, quantitatively characterizing the interconnection harmony of the social and economic subsystems of the country. Thus, the state of the social-economic system at each moment of time can be recorded in the following form:

$$\{x_{ir}^{j}, t = 1 \dots i, r = 1 \dots n, j = 1 \dots m\}, \quad (1)$$

where x_{ir}^{j} is a set of indicators that characterize the state of the social and economic system during the period t.

To identify the risks affecting the social-economic system of the country, the authors' team conducted a survey in 30 subjects of the Russian Federation, 20 people were interviewed.

The general expert group covers 600 people and consists of two subgroups, surveyed in an equal proportion:

1) experts whose income level exceeds the average per capita per the subject of the Russian Federation;

2) experts whose income level does not exceed the average per capita per the subject of the Russian Federation.

The results of the survey made it possible to identify the following risks whose manifestation was noticed in all subjects of the Russian Federation studied (see Table 2).

Table 2. System of risks in the social and economic system of the country

No.	**Feature**
r.1.	Reduction in consumer demand
r.2.	Increase in professional morbidity
r.3.	Deterioration of conditions to found a family
r.4.	Deterioration of conditions to spend leisure time
r.5.	Increased environment pollution level

The level of the gravity of risk manifestations in the system is gauged based on the analysis of statistical data available to the general public. Threshold values of the authors' scale are defined based on the Cheddock correlation scale and are shown in table 3.

Table 3. Classification of severity level of risks manifestation

Correlation ratio	0-0.30	0.31-0.50	0.51-0.70	0.71-1.00
Dependence between social and economic processes	weak	moderate	noticeable	high
Severity level of risks manifestation	minor	moderate	significant	critical

In order to define strategic and tactical plans a quantitative interpretation of the results of social and economic system's functioning is called for that would comprise a set of the chief indicators characterizing its condition.

These indicators are identified by means of the econometric apparatus of a set of recursive equations (2) which consistently characterize the following:

1) the relation between the population size and standards of living in a Russian Federation subject;

2) the relation between the production volume of goods in the economic subsystem and the population size with standards of living in a Russian Federation subject;

3) the relation between the production volume of social services and the production volume of goods, population size and standards of living.

$$\begin{cases} P = a_{11} + a_{12} * LQ_i \\ V_{ind} = a_{22} * P + a_{23} * LQ_i + a_{24} * x_1 \\ S_s = a_{31} * V_{ind} + a_{32} * P + a_{33} * LQ_i + a_{34} * x_1 + \\ + a_{35} * x_2 \end{cases} \quad (2)$$

where P is the population in the region under study, million people;

LQ_i is the life quality of the population in the region under study;

V_{ind} is the volume of industrial production in the region under study;

S_s is the scope of services provided in the region under study;

a_{ij} is the required parameters of the regression equation in the system;

x_1 is state expenditure on the purchase of industrial goods in the region, million rubles.

The modelling is predefined by the principle of consistent mutual influence of the specified factors and systems. The developed model allows estimating the influence of the population's standards of living, which are affected by manifested risks, on the resulting indicators in the social and economic system. Consequently, a damage to the economy resulting from affected standards of living factors may be estimated with the help of this model. The damage is estimated as the difference between dependent variables in the recursive system before and after a standard of living factor is affected [5].

This approach allowed developing macroregulators (table 4), a set of risk management tools aimed at targeting factors of the population's standards of living and proportions of the national social and economic ,system. This made it possible to implement a new type of national policy based on preventive strategic and tactical management principles.

Table 4. System of macroregulators for country's social and economic system management [4]

Macroregulator name	Formula	Decryption
Value added tax graduated rate	$VAT = N_{a\ bas} * \frac{GRP_i}{GRP_{max}}$	$N_{a\ bas}$ – basic value added tax rate according to the Tax Code; GRP_i – gross regional product in the investigated region; GRP_{max} – maximum GRP
Industrial enterprises' state-regulated allocations for promoting staff health	$Z_D = P * D * \frac{RC}{ND} * \frac{TR}{GRP}$	Z_D – industrial enterprises' state-regulated allocations for promoting staff health, P – population; D – percentage of occupational diseases cases; RC – costs for of occupational diseases treatment in the investigated region; ND – number of cases of occupational diseases in the investigated region; TR – industrial output; GRP – gross regional product
Financial investment into school and pre-school establishment buildings	$GIS = \sum_{i=1}^{n} \left(\frac{P_r}{AS} - SN_i\right) * SIC$	GIS – public investment into school and pre-school establishment buildings; n – number of municipal districts; P_r – number of pupils in a district; AS – average capacity of one school; SN_i – number of schools in a district; SIC – investments into construction of one establishment building
Financial investments into culture and sport establishment buildings	$GIC = \left[\frac{P*AV*AC}{1000} - (RV + CN)\right] * CIC$	GIC – investments into culture and sport establishment buildings; AV – number of visits per 1000 citizens of the most populated region; AC – average capacity of 1 building; RV – number of visits per 1000 citizens of the investigated region; CN – number of culture and sport establishment buildings; CIC – investments into construction of one establishment building
Rate for the emission of 1 ton of pollutants in the atmosphere	$N_{ER} = N_{ERb} * \frac{\Delta Q_P - \Delta Q_{P\_min}}{\Delta Q_{P\_max} - \Delta Q_{P\_min}}$	N_{ER} – rate for the emission of 1 ton of pollutants in the atmosphere; N_{ERb} – basic rate for the emission of 1 ton of pollutants in the atmosphere; ΔQ_P – the number of uncaptured pollutants; ΔQ_{P_max} – pollutant emission in the most polluted region; ΔQ_{P_min} – pollutant emission in the most unpolluted region.

4 Results

The authors' proposed method was tested on the example of Perm Region. The econometric model obtained for this subject of the Russian Federation has the form:

$$\begin{cases} P = 5.627 * LQ_i \\ V_{ind} = -69.8984 * P + 499093.8 * LQ_i + \\ +5.262104 * x_1 \\ S_s = -0.54416 * P + 0.1102 * V_{ind} - \\ -235.86 * LQ_i - 0.0442 * x_1 + 1.35563 * x_2 \end{cases} \quad (3)$$

Analysis of the extent of risk manifestations in the region's social and economic system revealed that risks manifest themselves on a small scale in this subject of the Russian Federation. However, macroregulators presented in table 5 have been calculated in order to reduce them still more.

Table 5. Values for macroregulators aimed at levelling risks in the Perm Region in 2018, as defined by the authors

Macroregulator name	Value
Value added tax graduated rate, %	11.4
Industrial enterprises' state-regulated allocations for promoting staff health, in thousands of roubles (by the example of Lukoil Perm, LLC)	195.1
Financial investment into school and pre-school establishment buildings macroregulator, in millions of roubles	589.75
Financial investments into culture and sport establishment buildings regulator, in millions of roubles	-
Rate for the emission of 1 ton of pollutants in the atmosphere (by the example of ammonium), roubles per ton	15.27

After this the influence of the macroregulators on the population's standards of living was estimated, with key indicators characterizing the results of the social and economic system's functioning in the Perm Region reviewed. The values of potential opportunity costs for the Perm Region as a comparison between indicators before and after implementing the macroregulators can be seen in table 6.

Table 6. Results of the model application in Perm Region

Indicator	Variable value		Deviation	
	Before application	Forecasted value	absolute	relative
Life quality of the population	0.54	0.56	0.02	+3.70
Population	2611	3293	682	+26.12
Volume of industrial production	1644154	1728161	84007	+5.11
Scope of services provided	160204	173456	13252	+8.27
Severity level of risks manifestation	0.014	0.013	-0.01	-7.14

Following the analysis, the use of the tools not only reduces risks, but also reduces opportunity costs by transforming them into an additional source of developing the social and economic system of the Perm Region.

References

1. *Guide to Integrated Risk Management.* Treasury Board of Canada Secretariat, (2011).
2. V.N. Viatkin, V.A. Gamza F.V. Maevskiĭ *Risk-menedzhment: preventivnoe upravlenie.* (AP «Nauka i obrazovanie», 2013).
3. A.S. Williams, M.L. Smith, P.C. Young *Risk Management and Insurance* (McGraw-Hill, New York, 1995).
4. U.V. Aftakhova, E.E. Zhulanov *PNRPU sociology and economics bulletin,* **4** (2019)
5. U.V. Aftakhova, E.E. Zhulanov *PNRPU sociology and economics bulletin,* **5** (2018)

Innovation-oriented resource management model proposed on the basis of the methodology of economic assessment of resource capabilities of oil and gas industry contractors

Inna Vinokur[1*], *Svetlana* Ponomareva[1]

[1]Perm National Research Polytechnic University, 29, Komsomolsky ave., Russia, 614990, Perm

Abstract. The paper presents an innovation-oriented model of resource management, which was built on the basis of the methodology of economic assessment of resource capabilities of contracting companies of the oil and gas industry. Assessment and choice of resource capabilities of contracting companies is the most important scientific problem of large companies and enterprises of the oil and gas industry when purchasing material resources. Currently, the increase in the supply of the same type of resources exacerbates the problem of choice. Thus, a large number of algorithms, methods and other procedures were developed to select organizations that supply resources. However, there are not enough comprehensive tools to assess the feasibility of selecting a supplier taking into account factors that allow evaluating suppliers of a certain industry. The purpose of the study is to create an innovative resource management model based on the methodology of economic justification for choosing resource suppliers for oil and gas enterprises. The object of the study is enterprises and large companies of the oil and gas industry. The subject of the study is effective resource management based on the methodology of economic assessment of the resource capabilities of contracting companies.

1 Introduction

Scientific research provides a suggested list of basic criteria according to which a preliminary list is formed at the second stage of the above selection procedure: company size; cost competitiveness; domestic/foreign supplier; financial condition; availability of local (in this country) support; location; producer/distributor; market share; supplier relationship history; quality reputation; recommendations from other clients; the only (exclusive) supplier.

2 Materials and methods

In order to find new contracting companies for the supply of resources, the following well-known methods are mainly used: ranking method (the main criteria for evaluating the supplier are selected, then the employees of the procurement service and (or) other departments, possibly engaged experts, establish their importance by expertise. The ranking value for each criterion is calculated by producing the specific weight of the criterion for its expert point assessment (for example, for a 10-point system) for this supplier. Then the obtained ranking values are summed up according to all criteria and the final R_i ranking for a specific i supplier is obtained); cost estimation method or cost-factor method (mission method) (all possible costs and revenues are calculated for each supplier (logistics risks are taken into account). Then, from the set of options (missions), the most profitable (according to the criterion of total profit) is selected. In fact, this is a kind of method of ranking criteria by value; the method of dominant characteristics (focusing on one selected parameter (criteria). This parameter may be the lowest price, the best quality, the delivery schedule, inspiring the most confidence, etc. The advantage of this method is its simplicity, and the disadvantage is the ignorance of the remaining selection criteria); the method of preference categories (evaluation of the contracting company, including the choice of the evaluation method, depends on information obtained from many divisions of the company. Engineering services give their assessment of the supplier's ability to produce high-tech products and may competently judge its quality, the dispatching service reports on the timeliness of delivery of purchased material resources, production departments of the company – on the simplicity and convenience of direct use of material resources in the production process. The use of such a method implies the availability of information from many sources – a vast and diverse information base that considers each factor on an equal basis along with others, while for a company, perhaps some factor is key, for example, the ease of using products in the production process); the fuzzy set theory (the process of selecting a supplier is carried out on the basis of a wide range of evaluation criteria). The study develops a two-level hierarchical structure of a fuzzy model. The model uses fuzzy inference for intermediate variables to then pass

* Corresponding author: Innavinokur12345@mail.ru

clear values of these variables to systems at the subsequent hierarchy level).

The importance of selecting a supplier is explained not only by the activities of a large number of suppliers of homogeneous products, but also by the fact that the supplier must be, first of all, a reliable partner who participates in the implementation of its trading strategy. Supplier selection consists of the following main stages: 1) definition of customer requirements; 2) search for suppliers; 3) identification of potential suppliers; 4) preliminary assessment; 5) basic evaluation using the chosen methodology; 6) supplier selection; 7) monitoring and evaluation of works performed. One of the most important stages is "conducting a basic assessment using the chosen methodology". When making a decision at this stage, there are some difficulties, since suppliers have a different set of properties, which sometimes have a dual meaning: some indicators have good ranking, and others – bad.

3 Results and Discussion

In all the theoretical articles analyzed, the following are identified as the main evaluation criteria: price; price policy flexibility; terms of payment; product quality; availability of free production facilities; level; business activity of the enterprise; speed of supply of resources; supplier participation in innovative R&D; duration of the supplier's presence on the market; market reputation of the enterprise; company size; domestic/foreign supplier.

Scientists around the world are actively discussing the problems of supplying quality resources from various contracting companies, among the scientific works the contribution of the following authors should be highlighted: Andiani, F.M., Abid, F., Hendri, Girsang, A.S. Business intelligence for generating comprehensive report in electronic completion and handover [1]; Bigelow, L.S., Barney, J.B. What can Strategy Learn from the Business Model Approach [2]; Biskupska, D., Chandima Ratnayake, R.M. On the Need for Effective Lean Daily Management in Engineering Design Projects: Development of a Framework [3]; Cartwright, S., Davies, I., Archer-Brown, C. Managing relationships on social media in business-to-business organizations [4]; Colucci, M., Vecchi, A. Close the loop: Evidence on the implementation of the circular economy from the Italian fashion industry [5]; Dell'Era, C., Trabucchi, D., Magistretti, S. Exploiting incumbents' potentialities: From linear value chains to multisided platforms [6]; Frandsen, T., Boa, S.Ø., Raja, J.Z. Pricing structures for solutions: an exploratory study within the oil and gas industry [7]; Haaskjold, H., Andersen, B., Lædre, O., Aarseth, W. Factors affecting transaction costs and collaboration in projects [8]; Mejlænder-Larsen, Ø. A three-step process for reporting progress in detail engineering using BIM, based on experiences from oil and gas projects [10]; Nurcahyo, R., Dody, P.N., Yadrifil, Z.Z.Z., Habiburrahman, M., Wibowo, N. CRM dimensional analysis and ANP method for oil and gas industry sub-contractor in indonesia in the context of reducing customer complaints [11]; Ponomareva, S.V., Kutuzova, V.S. & Pavlovich, A.A. Calculation of Uncerta inties in Operating Strapdown Inertial Navigation Systems on Mobile Objects [12]; Ponomareva, S.V., Zheleznova I.V. Intrafirm Planning and mathematical modeling of Owner' s Equity in Industrial Enterprises [13]; Sund, K.J., Bogers, M.L.A.M., Sahramaa, M. Managing business model exploration in incumbent firms: A case study of innovation labs in European banks [14]; Wood, S. Developments in the HRM–Performance Research stream: The mediation studies [15]; Zulhafiz, W.M., Rahman, N.B.A. Unfair risk allocation in oil and gas upstream service contracts in malaysia: The necessity for oilfield anti-indemnity act [16].

In order to identify the most significant criteria, it is necessary to conduct an expert assessment under the conditions of the rating method. Let us use a 10-point scale (Table 1).

Table 1. Expert assessments of the weight of criteria

Criteria	Experts										Total
	1	2	3	4	5	6	7	8	9	10	
Price	8	8	5	9	7	9	9	6	7	9	77
Price policy flexibility	9	9	7	8	7	7	7	8	7	8	77
Payment terms	7	7	9	8	7	7	5	7	8	9	74
Product quality	9	9	7	9	9	7	7	8	8	6	79
Availability of free production capacities	9	9	6	8	8	6	6	5	6	6	69
Reliability level	9	9	10	9	7	6	6	7	7	8	78
Business activity of the enterprise	7	7	9	9	6	9	7	5	7	9	75
Delivery speed	8	8	6	8	9	5	6	6	5	7	68
Supplier participation in innovative R&D	6	6	2	6	8	3	3	2	3	2	41
Supplier market time	8	8	9	9	6	5	5	6	6	5	67
Market reputation of the enterprise	9	10	9	10	7	6	6	7	8	9	81
Company size	8	8	6	9	7	6	6	5	5	6	66
Domestic/foreign supplier	9	9	10	9	7	6	6	4	7	9	76

Let us calculate the average importance of the criterion: price – 7.7; price policy flexibility – 7.7; payment terms – 7.4; product quality – 7.9; availability of free production capacities – 6.9; reliability level –

7.8; business activity of the enterprise – 7.5; delivery speed – 6.8; supplier participation in innovative R&D – 4.1; supplier market time – 6.7; market reputation of the enterprise – 8.1; company size – 6.6; domestic/foreign supplier – 7.6.

The data in Table 1 show that the most important criterion for selecting a supplier is the "market reputation of the enterprise" – 8.1, and less important – "supplier/contracting company participation in innovative R&D" – 4.1. As a result of the analysis of supplier evaluation methods, the best supplier evaluation method was identified – fuzzy set theory. This method is complex for prompt processing of information and requires re-equipping the software product and computer base.

The method used at the moment by most oil and gas enterprises is a tender, and according to experts, it is not the best way to evaluate contracting companies. Based on the above arguments, it is possible to conclude that it is necessary to develop and form a method for evaluating resource contracting companies. The combination of fuzzy set theory and the ranking method may be very useful, because only ordinal estimation (ranking) is required, and at the same time the experts feel more comfortable with it, while the fuzzy set theory is suitable to face uncertainty in decision-making in this area.

Let *y* be a finite set of valid actions labeled as $\{y_1, y_2, \ldots, y_m\}$. *A* may be considered as a set of objects, solutions, alternatives, events or candidates.

Let us suppose that it is possible to analyze the effects of actions using a set of criteria. Then *C* is a set of criteria denoted as $C=\{C_1, C_2, \ldots, C_n\}$. The following stages were developed in an paper to rank the alternative using a new technique:

Stage 1 – projection of separation of alternatives. At this stage, alternatives (y_i) and criteria (C_j) should be ranked based on verbal language variables. For example, A_1 is much better (verbal language variable) than A_2, if, according to experts, A_1 is in much better situation than A_2. Thus, alternatives and criteria may be projected using verbal language variables: "Much better" (MB), "Better" (B) and "Equal" (E). If necessary, verbal language variables may be increased in projection.

Stage 2 – creation of the accessory function. Now that verbal language variables are used in the projection, it is necessary to define their membership functions with experts in the relevant research area.

Stage 3 – primary ranking (r'_j and r_j (y_i)). In each projection, we assign the value 1 to the first (best). The remaining estimates are calculated by adding the ratio values (fuzzy number) to the last. For example, if based on criterion 1 ((C_1), y_1 is MB than y_3 (MB' = (0.2; 0.7; 1)), $r'_1(y_1) = 1$ and $r'_2(y_2) = (1.2; 1.7; 2)$. This process must be repeated until r'_j.

Stage 4 – estimated distance ($d'_j(0, y_j)$). This process must be repeated according to all criteria.

Stage 5 – absolute ranking (R_j (y_i)). R_j (y_i) – average ranking for the *i* alternative and *j* criteria, such as $R_1(y_1) \leq R_2(y_2)$, if $d'_1(0, y_1) \leq d'_2(0, y_2)$.

Stage 6 – aggregation (R (y_i)). At this stage, the ranking method is used according to the absolute ranking of alternatives for each criterion: $R(y_i) = R_i$.

The final ranking is carried out according to $R(y_i)$.

This technique is easy to use in comparison with the fuzzy set theory. At the same time, it may study more data, and it is also more objective than the ranking method. The model may be used to evaluate the following indicators: objects, solutions, alternatives, events or candidates.

After the development of the new method, the ranking method, the fuzzy set theory and the new supplier assessment method proposed in the paper are comparable (Table 2).

Table 2. Comparative characteristics of contracting company selection methods

Characteristics of methods	1	2	3
Usability	3	2	3
Amount of information to be analyzed	1	3	3
Financial costs	3	2	2
Applicability to any enterprise	3	3	3
Time input	3	2	2
Objectivity	1	3	3
Total (Qij)	14	15	16

Table 2 shows that the generalizing measure of the characteristics of the methods indicates the best effectiveness of the method and amounts to 16 points.

Table 3. Summary of advantages and disadvantages of two contractor's resource evaluation methods

Tender	New methodology
Advantages	
– transparency of company information; – work with verified contracting companies; – payment delay.	– transparency of company information; – work with new suppliers identified in the market analysis; – self-selection of a supplier by meaningful criteria
Disadvantages	
– steep demand for entry requirements; – price dumping; – lack of supply market analysis of; – corruption; – absence of personal contacts; – complex and lengthy process between application and confirmation.	– relative complexity of calculations

In this case, the following subjects are also taken into account: leadership of the enterprise, employees of the procurement department, employees of the planning and economic department and employees of the legal department. The company of the Oil and Gas Industry identified the problem of choosing a contracting company. The current methodology (tender) does not quite suit the managers of the procurement department. After a survey of experts, a summary table of advantages and disadvantages of the two methods of evaluating a resource supplier was compiled: a tender and a new methodology for evaluating contracting companies (Table 3).

Let us consider the developed methodology for the company of the Oil and Gas Industry. At the first stage let us determine the list of resource suppliers for the company: Weatherford, Tatneft, Top-Co, ZERS, Art-Osnastka. At the second stage, let us evaluate the suppliers according to the previously selected criteria (Table 1) in accordance with the average importance of the criterion and determine the ranking of the criteria for evaluating the contracting companies for the supply of resources. The data was generated after expert evaluation of suppliers by the company's procurement department managers. The next stage is to match suppliers against the criteria and assign a primary ranking to each of them (Table 4).

Table 4. Primary ranking*

	y_1			y_2			y_3			y_4			y_5			y_6			y_7			y_8			y_9			y_{10}			y_{11}			y_{12}			y_{13}		
Weatherford	2	2.5	2.9	1	1	1	1.6	1.8	2	2.4	3.1	3.6	2	2.5	2.9	3.2	4	4.7	1.4	1.6	1.7	2.4	3.1	3.6	2.8	3.6	4.1	1	1	1	1	1	1	2	2.4	2.7	1.6	1.8	2
Tatneft	1	1	1	1.6	1.8	2	2	2.5	2.9	1	1	1	2.4	3.1	3.6	2.6	3.2	3.7	1	1	1	1.6	1.9	2.2	1	1	1	1.6	1.8	2	2.4	3.1	3.6	1.4	1.6	1.7	2.4	3	3.4
Top-Co Inc	2	2.4	2.7	2.2	2.6	3	1	1	1	1.6	1.8	2	1	1	1	1	1	1	2	2.4	2.7	1	1.1	1.2	1.8	2.2	2.4	2	2.4	2.7	1.6	1.8	2	1	1	1	1	1	1
ZERS	1.6	1.8	2	2.2	2.7	3.2	2	2.4	2.7	2	2.4	2.7	1.6	1.8	2	2	2.4	2.7	1.4	1.6	1.7	1	1	1	2.2	2.8	3.1	2	2.5	2.9	2	2.4	2.7	2.4	3	3.4	3	3.8	4.4
Art-Osnastka	2.6	3.3	3.9	2.2	2.8	3.4	2	2.6	3.1	2	2.5	2.9	1.6	1.9	2.2	1.4	1.6	1.7	2	2.5	2.9	2	2.5	2.9	1.4	1.6	1.7	2.4	3.1	3.6	2	2.5	2.9	2.8	3.6	4.1	2	2.4	2.7

*The primary ranking is assigned by experts (specialists of the company's procurement department).

Table 4 shows that the experts determined that the Weatherford is "much better" than other companies by the criteria of price policy flexibility, the duration of the supplier's presence in the market, the market reputation of the enterprise; Tatneft – price, product quality, business activity of the enterprise, supplier's participation in innovative R&D; Top-Co – payment terms, availability of free production capacity, reliability level, company size, domestic/foreign supplier; ZERS – delivery speed; Art-Osnastks – does not have a criterion to which a primary ranking would be assigned.

After determining the primary ranking, let us present the distance estimate for each criterion (Table 5).

Table 5. Distance estimation for each criterion

	y_1			y_2			y_3			y_4			y_5			y_6			y_7			y_8			y_9			y_{10}			y_{11}			y_{12}			y_{13}		
Weatherford	2.1	2.9	3.5	1.6	2.1	2.5	1.9	2.4	2.8	1.7	2.1	2.3	2.4	3.3	4	2.5	3.2	3.8	2.3	3.2	3.8	1.9	2.4	2.7	2.1	2.7	3	1.2	1.4	1.6	2.3	3.2	3.8	2.1	2.8	3.2	2.8	3.9	4.6
Tatneft	1.6	2.2	2.6	1.9	2.5	3	2.1	2.8	3.2	1	1	1	2.6	3.6	4.4	2.2	2.8	3.3	2.1	2.9	3.4	1.5	1.8	2	1.2	1.4	1.5	1.5	1.8	2.1	3	4.2	5.1	1.8	2.4	2.7	3.2	4.5	5.3
Top-Co Inc.	2.1	2.9	3.4	2.2	2.9	3.5	1.6	2	2.3	1.3	1.4	1.5	1.9	2.6	3.1	1.4	1.7	1.9	2.6	3.6	4.3	1.2	1.4	1.5	1.6	2	2.2	1.7	2.1	2.4	2.6	3.6	4.3	1.6	2.1	2.4	2.5	3.5	4.1
ZERS	1.9	2.6	3.1	2.2	3	3.6	2.1	2.7	3.1	1.5	1.7	1.9	2.2	3	3.6	1.9	2.4	2.8	2.3	3.2	3.8	1.2	1.3	1.4	1.8	2.3	2.5	1.7	2.2	2.5	2.8	3.9	4.6	2.3	3.1	3.6	3.5	4.9	5.8
Art-Osnastka	2.4	3.3	4	2.2	3	3.7	2.1	2.8	3.3	1.5	1.8	2	2.2	3	3.7	1.6	2	2.3	2.6	3.6	4.4	1.7	2.1	2.3	1.4	1.7	1.8	1.9	2.5	2.9	2.8	3.9	4.7	2.5	3.4	3.9	3	4.2	5

Now that the distance is determined, let us calculate the absolute values of the ranking with aggregation (Table 6).

Table 6. Calculation of absolute ranking values with aggregation

Contracting company	R_1 (A_i)	R_2 (A_i)	R_3 (A_i)	R_4 (A_i)	R_5 (A_i)	R_6 (A_i)	R_7 (A_i)	R_8 (A_i)	R_9 (A_i)	R_{10} (A_i)	R_{11} (A_i)	R_{12} (A_i)	R_{13} (A_i)
Weatherford	42	21	27	16	51	49	47	25	32	5.5	47	34	59
Tatneft	21	30	34	1	56	38	38	9.5	3	12	62	25	64
Top-Co Inc.	38	42	16	5.5	30	9.5	56	3	13	21	56	16	53
ZERS	30	42	34	9.5	42	27	47	3	23	21	59	44	65
Art-Osnastka	51	44	38	9.5	44	16	56	16	7	27	59	51	62

After calculating the distance and absolute value of the ranking with aggregation, it is necessary to represent the weighted normalized matrix (Table 7).

Table 7. Aggregation summary

	$R_1(A_i)$	$R_2(A_i)$	$R_3(A_i)$	$R_4(A_i)$	$R_5(A_i)$	$R_6(A_i)$	$R_7(A_i)$	$R_8(A_i)$	$R_9(A_i)$	$R_{10}(A_i)$	$R_{11}(A_i)$	$R_{12}(A_i)$	$R_{13}(A_i)$
Weatherford	0.4967	0.2546	0.3921	0.7398	0.5016	0.6986	0.4267	0.7949	0.7585	0.1302	0.3698	0.4193	0.4343
Tatneft	0.2484	0.3637	0.4938	0.0462	0.5508	0.5417	0.3450	0.3020	0.0711	0.2840	0.4879	0.3083	0.4711
Top-Co Inc.	0.4494	0.5091	0.2324	0.2543	0.2951	0.1354	0.5084	0.0954	0.3081	0.4970	0.4406	0.1973	0.3901
ZERS	0.3548	0.5091	0.4938	0.4393	0.4131	0.3849	0.4267	0.0954	0.5452	0.4970	0.4643	0.5427	0.4785
Art-Osnastka	0.6031	0.5334	0.5519	0.4393	0.4328	0.2281	0.5084	0.5087	0.1659	0.6390	0.4643	0.6290	0.4564

To estimate the range of alternatives, positive ideal and negative solutions are calculated for each criterion (Table 8).

Table 8. Positive and negative solution

	$R_1(A_i)$	$R_2(A_i)$	$R_3(A_i)$	$R_4(A_i)$	$R_5(A_i)$	$R_6(A_i)$	$R_7(A_i)$	$R_8(A_i)$	$R_9(A_i)$	$R_{10}(A_i)$	$R_{11}(A_i)$	$R_{12}(A_i)$	$R_{13}(A_i)$
A^+	0.6031	0.5334	0.5519	0.7398	0.5508	0.6986	0.5084	0.7949	0.7585	0.6390	0.4879	0.6290	0.4785
A^-	0.2484	0.2546	0.2324	0.0462	0.2951	0.1354	0.3450	0.0954	0.0711	0.1302	0.3698	0.1973	0.3901

After all of the above activities, all data must be compared to each other for each supplier to determine their ranking. Let us present the summary data of the methodology taking into account the ranking (Table 9).

Table 9. Summary of the methodology taking into account ranking

Name	d_i^+	d_i^-	R_i	R (A_i)	Ранг
Weatherford	0.6652	1.3956	0.6772	0.6772	1
Tatneft	1.2785	0.6402	0.3337	0.3337	4
Top-Co Inc.	1.2857	0.6091	0.3214	0.3214	5
ZERS	0.9181	0.9365	0.5050	0.5050	3
Art-Osnastka	0.8716	1.0737	0.5520	0.5520	2

The calculations confirm that according to the selected criteria the Weatherford is the best supplier. Let us present the innovative resource management model to the scientific community (Figure 1).

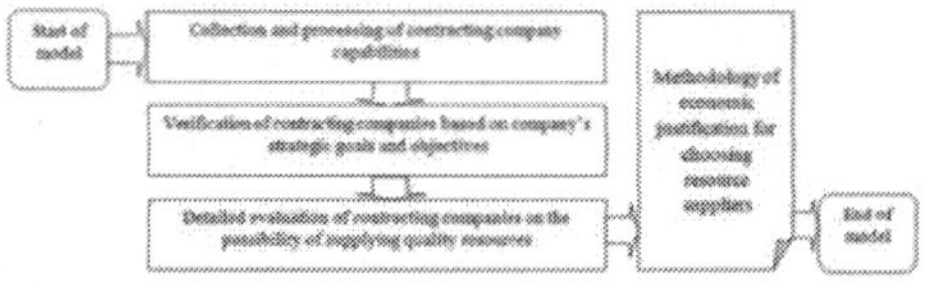

Figure 1. Innovative resource management model based on the methodology of economic justification for choosing resource suppliers for oil and gas enterprises

4 Conclusion

The authors of the paper proposed an innovation-oriented model of resource management compiled on the basis of the methodology of economic justification for choosing resource suppliers for oil and gas enterprises.

References

1. F.M. Andiani, F. Abid, ,Hendri, A.S. Girsang, *Advances in Science, Technology and Engineering Systems,* **5 (6),** 45-51(2020).
2. L.S. Bigelow, J.B. Barney, *Journal of Management Studies,* **58 (2),** 528-539.
3. D. Biskupska, R.M. Chandima Ratnayake, *IEEE International Conference on Industrial Engineering and Engineering Management,* **8978660**, 789-794 (2019)
4. S. Cartwright, I. Davies, C. Archer-Brown, *Journal of Business Research,* **125**, 120-134 (2021).
5. M. Colucci, A. Vecchi, *Business Strategy and the Environment,* **30 (2)**, 856-873 (2021).
6. C. Dell'Era, D. Trabucchi, S. Magistretti, *Creativity and Innovation Management,* **30 (1)** 31-46 (2021).
7. T. Frandsen, S.Ø. Boa, J.Z. Raja, *International Journal of Production Research,* **57 (22),** 6977-6992 (2019).
8. H. Haaskjold, B. Andersen, O. Lædre, W. Aarseth *International Journal of Managing Projects in Business,* **13 (1),** 197-230 (2019)
9. Ø. A. Mejlænder-Larsen *Engineering, Construction and Architectural Management,* **26 (4),** pp. 648-667 (2019).
10. R. Nurcahyo, P.N. Dody, Z.Z.Z. Yadrifil, M. Habiburrahman, N. Wibowo, *Proceedings of the International Conference on Industrial Engineering and Operations Management,* (March), pp. 605-613. (2020)
11. S.V. Ponomareva, V.S. Kutuzova, & A.A. Pavlovich. *J. Mach. Manuf. Reliab.* **49**, 723-730 (2020). https://doi.org/10.3103/S1052618820080099.
12. S.V. Ponomareva, I.V. Zheleznova *Journal of Physics: Conference Series.* P.032106. (2018)
13. K.J. Sund, M.L.A.M. Bogers, M. Sahramaa *Journal of Business Research,* **128,** 11-19 (2021)
14. S. Wood, *German Journal of Human Resource Management,* **35 (1),** 83-113 (2021)
15. W.M. Zulhafiz, N.B.A. Rahman, *International Journal of Business and Society,* **21 (S1),** 177-191 (2020).

SHS Web of Conferences **116**, 00011 (2021) https://doi.org/10.1051/shsconf/202111600011
ICSR 2021

Features of consumer value management of an innovative product

Oksana Gaifutdinova*, *Igor* Poroshin, *Elena* Sergeeva, and *Ekaterina* Artemova

National Research Polytechnic University, 29a, Komsomolsky prospect, Perm, 614990, Russia

Abstract. Today, much is being done in Russia to ensure that the country acquires the status of an innovation-oriented power that can occupy and maintain the world leading positions in the main, strategically important areas of the state's development. Successful development of any socio-economic entity operating in modern conditions is impossible without widespread usage of innovative technologies. New technologies are important in open innovation processes for enterprises, because due to them, they can meet customer expectations and achieve higher corporate financial performance. By creating new organizational connections to take advantage of emerging technological opportunities, entrepreneurs are restructuring the economy, driving change and overall economic growth. In addition to the focus on innovation, an important direction in the development concept of a socio-economic entity is solving environmental issues. More and more companies are thinking about creating environmentally friendly products with minimal impact on the environment. Such products have a higher attractiveness for the consumer and compete more successfully with foreign analogues. Creating an effective program of innovative growth in compliance with environmental requirements and standards is one of the most important factors in improving competitiveness and sustainable development of Russian enterprises.

1 Introduction

Cost management is essential to the success of any enterprise, because one of the most important goals of any company is to maximize its profits. In addition, an important goal of the enterprise is its own survival, and to this extent it seeks to minimize costs, therefore, the goal of good management is efficient usage of resources to maximize profits and maintain the enterprise competitive position. Successful combination of these two goals is achieved through innovation.

In general, innovation is a key factor that increasingly affects economic growth in all countries: these factors can be defined as particularly important circumstances of socio-economic development, as well as technological progress in countries and regions, in enterprises. The process of innovative development is becoming increasingly complex. Many new concepts and types of innovation have emerged, ranging from technological innovation to marketing innovation, and from closed to open innovation [1].

Eco-innovations that take into account modern business models, promising areas of business activity, improving products, as well as environmental and economic processes are becoming particularly popular.

Cost management includes all actions taken by management to reduce costs, without reducing efficiency or affecting customer satisfaction. Competent cost management requires qualitative assessment of the cost drivers in order to be able to successfully control and manage costs [2].

Cost management issues are particularly relevant for innovative products, as they are associated with a large number of additional costs and commercial risks [3]. This process relies on complex networks of partners and external stakeholders interacting through open innovation systems with inbound and outbound practices to address issues related to eco-innovation [4-7].

Implementation of production processes for producing innovative products requires a lot of economic resources. Volumes, types and conditions of their consumption are determined by the specifics of technological processes, implementation of which ensures production of goods with certain consumer characteristics. An important characteristic is the product price, which is determined by the volume of two main components-cost and profit. The value of the production cost depends on the physical and price characteristics of the resources used, determined by the activity of production processes. If the amount of profit is formed mainly under the influence of market conditions for the sale of manufactured products and, therefore, is less sensitive to managerial influences on the part of the enterprise, then the value of the production cost is the object of direct influence, which is sensitive to economic decisions taken at enterprises in relation to the choice of suppliers, the amount of necessary resources and production. Thus, the volume of resources used as well as the economic and organizational conditions for their consumption are of paramount importance for ensuring effective operation of the enterprise [8].

* Corresponding author: ogrigoryan@mail.ru

Another key area of developing modern economy and business is active creation of cluster structures of various types. Special emphasis is placed on clusters due to the fact that clusters can play the role of intermediaries for open innovations. There is a two-way relationship: open innovations of companies can determine creativity and complexity of a cluster, while at the same time, openness and creativity of a cluster determine the degree of open innovation of its members. This function is most fully performed by technoparks, which are a special kind of clusters and connect participants in various industries and sectors of the economy [9-12].

Clusters create a favorable environment for open innovations; therefore, cooperation within clusters helps to find sources of innovative development outside the enterprise, for example, by finding and combining ideas that complement existing R & D projects, and creating cooperative relationships with other enterprises and organizations [13]. At the same time, they regulate relations on transferring intellectual property in a certain way [14].

2 Methods

A cluster approach is used as a theoretical basis of the study. Clusters are recognized as an important factor of sustainable development because they can influence strategic goals of businesses and scientific units linked in clusters by common research or industrial goals, create social and relational capital based on trust, and stimulate innovation, including eco-innovation and technology.

The research methods are the calculation and analytical method, complex and factor analysis. The method of comparative analysis of efficiency of two operational variants of a gas turbine unit (GTU) for gas superchargers of main gas pipelines with capacity from 4 to 34 MW was also used.

As the study basis, data on the total costs of JSC "UEC-Aviadvigatel" - the design bureau for developing aviation gas turbine engines, as well as industrial gas turbine installations and power plants based on aviation technologies were used [15]. This company is a major manufacturer, as well as a main supplier of parts and components for operational maintenance of a gas turbine unit (GTU) for gas superchargers of main gas pipelines with capacity of 4 to 34 MW.

For the analysis, data were taken on the total costs for the GTU operation over the entire life cycle (30 years). The cost data is taken in current prices. The data is presented in table 1.

Table 1. Costs for the entire GTU life cycle, thousand rubles.

Costs for GTU Total		Total
Capital (for reconstruction)	Operational	
367 407,69	3 705 958,76	4 073 366,45

Table 1 shows that the GTU costs consist of capital and operational costs. Their total cost over 30 years is 4,073,366. 45 thousand rubles, of which 367,407. 69 thousand rubles go to capital expenditures and 3,705,958. 76 thousand rubles for operational costs.

The structure of GTU costs is shown in Figure 1.

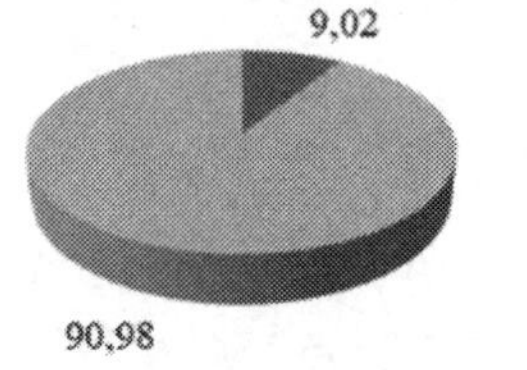

Figure 1. The structure of GTU costs, %

Figure 1 shows that operating costs account for the largest share of total costs (90.98 %), while capital costs account for the smallest share (9.02 %). Thus, the main emphasis in reducing the operating costs of the analyzed equipment should be made on those elements (nodes, parts) of the equipment that require the most attention and the most maintenance costs during the equipment operation. The following calculations are aimed at testing this hypothesis.

3 Results and Discussion

In the process of analyzing the studied enterprise capabilities to improve the efficiency of the resources used, as one of the ways to reduce the GTU capital and operating costs, the option of replacing the serial combustion chamber with a low-emission combustion chamber (LECC) was analyzed [16].

The low-emission combustion chamber (LECC) is an innovative product created in the process of scientific and technological cooperation of JSC "UEC-Aviadvigatel" specialists with foreign partners. Thus, JSC "UEC-Aviadvigatel" signed a contract with RGI (USA) for developing LECC for the GTE-25PA. Such cooperation is in the nature of cluster interaction, when experience and resources are shared on mutually beneficial terms [17]. As the basic design, a remote CC with a multi-module front device, previously tested in Aviadvigatel, was selected.

Creation of a low-emission combustion chamber (LECC) is one of the most important tasks facing Perm specialists, since LECC allows achieving a number of important environmental indicators. The use of LECC in modern gas turbine plants reduces the combustion temperature without using water, reduces emission of harmful substances into the atmosphere, which can significantly improve the characteristics of using gas turbine plants, including in the environmental sphere.

Thus, using LECC will ensure that Russian gas turbine equipment complies with the established Russian standards for emissions of nitrogen oxides (NOx) into the atmosphere: no more than 150 mg/nm3 for gas turbine gas transmittal units and no more than 50 mg/nm3 for power plants.

Calculation of the total costs for the GTU operation after the LECC introduction showed that during the entire 30-year operational period, these costs will amount to 3,745,454. 17 thousand rubles. The costs distribution by type (capital and operational) after the LECC introduction during the entire life cycle of GTU is presented in Table 2.

Table 2. Costs for the entire life cycle of GTU with LECC, thousand rubles.

GTU costs with LECC		Total
Capital (for reconstruction)	Operational requirements	
239 391,86	3 506 062,31	3 745 454,17

Table 2 shows that when using LECC, the GTU total cost costs will decrease and amount to 3,745,454. 17 thousand rubles. This reduction will occur both at the expense of part of the capital costs (about 35%) and at the expense of operating costs (about 6%). After the events, capital expenditures will amount to 239,391. 86 thousand rubles, and in turn, operating costs - 3,506,062. 31 thousand rubles. The absolute and relative changes in the GTU costs after introducing the LECC are presented, respectively, in Tables 3 and 4.

Table 3. Absolute change in the GTU costs after LECC introduction, thousand rubles.

Costs	GTU	GTU with LECC	Absolute deviation
Capital costs	367 407,69	239 391,86	-128 015,83
Operation costs	3 705 958,76	3 506 062,31	-199 896,45

Table 4. Relative change in GTU costs after LECC introduction

Costs	GTU, thousand rubles.	GTU with LECC, thousand rubles.	Relative deviation, thousand rubles
Capital costs	367 407,69	239 391,86	-34,84
Operational costs	3 705 958,76	3 506 062,31	-5,39

According to Tables 3 and 4, it can be concluded that capital costs will decrease by 128,015. 83 thousand rubles, or by 34.84%, and operational costs will be reduced by 199,896. 45 thousand rubles or by 5.39%.

Thus, after introducing LECC at GTU, both capital and operational costs are reduced, which can be considered effective from the point of view of the economic approach.

As for the ratio of capital and operational costs, comparative analysis of the new structure of total costs for the GTU operation showed that after the introduction of LECC, the largest share in the total costs is still occupied by operational costs, accounting for 93.61% of the total volume, and, capital costs is only 6.39%. The updated cost structure for GTU with LECC is shown in Figure 2.

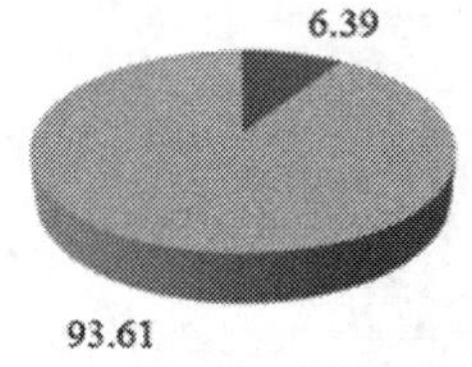

Figure 2. Cost structure for GTU with LECC, %

The second stage of the analysis of the innovative product effectiveness– LECC-was the analysis of the consumer value of LECC [18].

According to the approach adopted in this study, the consumer value of the implemented LECC technology, defined as the difference between its utility, value in consumption and the costs of its development and acquisition, is the higher the greater success is achieved in managing the cost price items for its creation.

However, the calculations showed that the actual cost price of LECC does not allow implementing this event at the moment. The difference between the current and planned cost of LECC is shown in Table 5.

Table 5. Difference between the real and planned cost price of LECC, thousand rubles.

Planned cost price of LECC	Real cost price of LECC	Excess of cost price
20 087,76	73 798,99	53 711,24

At the same time, the planned cost price is understood as the cost price of the LECC, which will allow it to successfully compete with the currently used CC. According to the expert assessment of the specialists of JSC "UEC-Aviadvigatel", the cost increase of using LECC should not exceed 30% of the cost price of the serial CC; otherwise implementation does not have any pronounced economic effect.

At the same time, as can be seen from Table 5, in order to optimize the cost of GTU by implementing the new LECC technology, it is first necessary to reduce its current cost by 53,711,24 thousand rubles.

This problem is caused by strong overestimation of a number of cost items when calculating the cost of manufacturing LECC. Figure 3 shows the main cost items of the current cost price of LECC.

Figure 3 shows that the main materials and purchased components occupy more than half of the cost price structure (54.44%), followed by general production costs (17.12%) and basic wages (13.48%). The lowest share in the total cost structure is taken by additional wages (2.88%) and insurance premiums (4.94%).

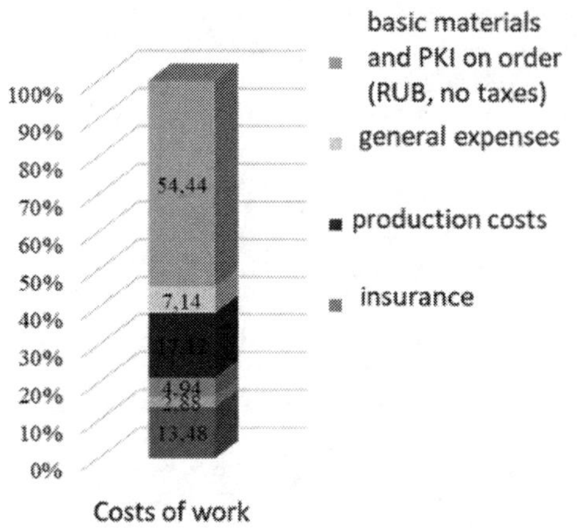

Figure 3. Composition and structure of costs in the cost of LECC, %

Measures to reduce the cost of LECC should be based primarily on such an element of costs as material costs, since their share in the total cost is the largest and amounts to 54.44%.

Therefore, the main task of the design department of JSC "UEC-Aviadvigatel" is further technological developments for the production of LECC from the point of view of managing individual cost items. At the same time, cost management focuses on reducing or eliminating waste, reducing the cost of materials used, effective pricing, and strategic efficiency improvements.

4 Conclusion

Recently, companies have been actively adopting new business models based on changes that allow them to compete in environmental and / or sustainable market segments. As a result, there is a dynamic growth in the importance and popularity of clusters both as a business model used for organizing economic activities and as an instrument of state economic policy.

This study was devoted to analyzing the possibility of widespread introduction of an innovative product in the oil and gas industry - a low-emission combustion chamber (LECC). The introduction of LECC into the GTU structure plays an important role in the development of enterprises in the sector, and can also have a strong impact on the competitiveness of the company that produces it - JSC UEC-Aviadvigatel.

Firstly, the use of LECC reduces the cost of the GTU life cycle. This is of great importance in the competition for the consumer by increasing the consumer value of the product.

Secondly, this development is innovative and allows us to significantly improve the performance, including in the environmental sphere, which is a significant factor for industrial enterprises today.

All the facilities of PJSC Gazprom are planned to be equipped with a low-emission combustion chamber (LECC) produced by UEC-Aviadvigatel [19], which will significantly reduce the harmful load on the atmosphere. Within ten years, it is planned to supply all the facilities of PJSC Gazprom with new equipment.

Thirdly, a significant number of enterprises are considering the possibility of introducing eco-innovations to simultaneously create economic and environmental value, which is achieved through the use of LECC.

Management technology for the development and creation of LECC is important from the point of view of interaction and exchange of experience within the cluster with other enterprises (foreign and domestic). It also plays an important role in the environmental impact by reducing the negative impact on the environment. It improves the technical and economic usage parameters through application of innovative technologies. However, it has a high cost of creation. Therefore, the issue of reducing the cost price of LECC is important for further development and possible research within the project.

References

1. R. Baneliene, B. Melnikas, R.Strazdas, E Tolocka. *Terra Economicus* **16(4)** (2018)
2. J. Nkeobuna, N. Ugoani *International Journal of Business, Economics and Management* **6(4)**, 232-247 (2019)
3. N. Akatov, Zh. Mingaleva, I. Klackova, G. Galieva, N. Shaidurova *Management Systems in Production Engineering* **27(4)**, 250-254 (2019).
4. L. E. Valdez-Juárez, M.Castillo-Vergara *Journal of Open Innovation: Technology, Market, and Complexity, MDPI, Open Access Journal* **7(1)**, 1-19 (2020)
5. A.V. Timiryasova, L.N. Safiullin, T.V. Kramin, R.A. Grigoriev *Mediterranean Journal of Social Sciences* **6(1S3)** 96-100 (2015)
6. P.-Y. Nie, C. Wang, L.-K. Lin *Area* **52(4)** 731-740 (2020)
7. M. M. Ospanov, A. T. Uskelenova, K.Kh. Shadiyev *Industrial Engineering & Management Systems* **19(1)**, 59-69 (2020)
8. M.A. Sellitto, C.G. Camfield, S. Buzuku *Sustainable Production and Consumption* **23**, 94-104 (2020)
9. Zh. Mingaleva, N. Shaidurova, V. Prajová *Management Systems in Production Engineering* **26(4),** 241-245 (2018).
10. T. Rayna, L. Striukova *International Journal of Entrepreneurship and Innovation Management* **19(5-6)**, 285-288 (2015)
11. X. Chen, E. Wang, C. Miao, L. Ji, S.Pan *Sustainability* **12(7)**, 2848 (2020)
12. C. Mo, C. He, L. Yang *Economics Letters* **188**, 109003 (2020)
13. A.M. Kowalski, M.Mackiewicz *J. Open Innov. Technol. Mark. Complex* **7(1),** 21 (2021)
14. Zh. Mingaleva, I. Mirskikh *Procedia Social and Behavioral Sciences* **81** 329-333 (2013).

15. Portal of industrial cooperation of the Perm region Retrieved from: https://prompermkrai.ru/catalog/mashinostroenie/informatsiya-o-predpriyatii-ne-zapolnena-14-02-2019-08-47-36
16. A. G. Ishkov, V. G. Nikitin, R. S. Kashapov, A. O. Prokopets, A. N. *Poychelyuzny Gasovaya Promyshlennost* **750(1)** 92-97 (2017).
17. T. Barnes, S. Raynor, J. Bacchus *Journal of Strategy and Management* **5 (1)**, 81-102 (2012)
18. S. Chakrabarty *Journal of Strategy and Management* **13(4)**, 535-550 (2020)
19. *Kommersant* Retrieved from: www.kommersant.ru/doc/4112646

Problems of intellectual property accounting in industrial enterprises in Russia

Mikhail Gorodilov*, *Marina* Naam, *Yulia* Murashova, and *Tatiana* Ketova

Perm State University, 15, Bukirev Street, Perm, 614000, Russia

Abstract. This paper analyzes important accounting and financial problems associated with identification, determination of the moment of recognition, proper assessment of intellectual properties that are part of intangible assets, determination of their useful life, as well as the search for relevant methods of their depreciation. According to the generalization of theoretical and their own practical experience, the authors propose the approaches to solve these problems, designed to increase the level of reporting reliability for stakeholders, as well as to ensure proper accounting and control of these objects in order to provide their safety and rational use at industrial enterprises.

1 Introduction

The organization of high-quality accounting of intellectual properties (IP) of an industrial enterprise provides the opportunity to protect the legal rights of an enterprise to the results of intellectual activity (RIA). It also predetermines a high level of competitiveness of an enterprise, as well as the prerequisites for its development. The results of intellectual activity received in the course of the performance of staff labor functions are subject to the risks of unauthorized access, use, theft, disclosure and destruction. The leveling of these risks is achieved by standardized work in the identification of positive RIA and the protection of the rights of a company by patenting (in cases stipulated by law) or in the mode of production secret (know-how).

Non-recognition of the received RIA as an intellectual property in cases when accompany has the right to get it in accordance with the law, leads to the increase in the above mentioned risks. A company reasonably has the right to accept IPs for accounting even in cases where the main costs of their creation have already been recognized in earlier periods. Accounting and financial support of the IP management system is designed to form objective information in accordance with the interests of various user groups. However, the presence of unresolved theoretical and methodological issues makes it difficult to get objective information. Some of these issues, as well as the solutions proposed by the authors, will be studied in this paper.

* Corresponding author: mag@econ.psu.ru

2 Methods

The research was based on the study of the methods for manifestation of intangible assets and intellectual property assets in the financial reports of companies in comparison with the approaches set out in International Financial Reporting Standards, national standards of Russia, as well as special rules of law for the protection of intellectual property. The authors used the publications of leading researchers in the field of accounting and evaluation of intangible assets, regulatory reference materials, special literature, internal regulatory documents of enterprises, data of financial, managerial and statistical accounting and reporting, analytical and informational materials published in the Russian and foreign press and presented in the Internet as the information basis for the research. They also used their personal materials obtained as a result of direct study of the state of the problems of accounting of IPs at specific industrial enterprises in Russia.

3 Results and Discussion

The issues of the identification of intangible assets, their evaluation and proper reflection in the financial reports of companies, including industrial enterprises, are controversial in the scientific and special literature.

Solving the issue of protecting the positive results of intellectual activity (RIA) is one of the primary tasks of an industrial enterprise in terms of managing them. Positive RIA are subject to recognition as an IP and protection through the introduction of a trade secret regime (know-how) or through patenting.

The moment when the decision is made may not coincide with the moment when the cost capitalization starts for various reasons. This can be either an employee error or a later discovery of the RIA's usefulness to the enterprise. In any case, this is not a reason for refusing to recognize a positive RIA as an IP object and reflect it as part of the integral assets, according to the accounting rules. The primary task is to protect the IP.

A number of scientific studies indicate the relevance of these issues for industrial enterprises in different countries. In the presented scheme of know-how management in manufacturing industries, N. D. A. Majid et al. emphasize the importance of identifying information that constitutes a know-how, determining the duration of the trade secret regime and the method of its protection [1]. These decisions aimed at IP protection affect the moment when a positive RIA is recognized as an IP object and, consequently, affect the recognition of an IP in accounting: its initial assessment, useful life, and disclosure of information in the financial statements.

The question of choosing an IP management strategy is no less relevant. The enterprise can protect the IP or share it with other enterprises. The scientific literature examines three main IP management strategies:

- a "defensive" strategy aimed at preventing the spread of knowledge and creating barriers to competition,
- a "collaborative" strategy aimed at collaborating with other organizations,
- an "improvised" strategy in which a firm protects its IP without a clear goal [2].

A number of researchers conclude that the most promising IP management strategy is a collaborative strategy: cross-licensing, patent pooling, etc. [2, 3]. The chosen IP management strategy may also have an impact on the recognition of IP in accounting. So, if an enterprise plans and sees real opportunities for the transfer of IP rights to IP use under a licensing agreement, then the royalty method can be chosen to estimate the initial cost of an IP recognized as part of the integral assets. On the other hand, if the company seeks to

protect the received RIA from other organizations as much as possible, or has no reason to believe that it will be in demand, then the use of another method will be required to estimate the initial value of the intangible assets, and the royalty method will not be applicable.
Thus, we observe that the recognition and initial valuation of intangible assets is ultimately influenced by factors that do not depend on the financial services of the enterprise, and to which, on the contrary, financial services need to adapt.

Another aspect of this issue is highlighted in the work of Z.-M. Zadorozhnyi, V. Yasyshena, who point to the problem of the unified approaches lack to the definition of the concept of "intangible asset" at accounting, financial, research and regulatory levels. According to their opinion, the difficulty of accounting of intangible assets is in the fact that by their nature they are diverse and include different objects. Due to the fact that today various intangible assets exist and continue to be created, as well as due to differences in economic, legal, accounting and valuation approaches to understanding of their essence, there is a significant gap between the accounting methodology of intangible assets and the modern needs of a market economy [4].

This is also true for Russian industrial enterprises, where work with intangible assets is carried out, as a rule, by several departments at once. Thus, the R&D department is engaged in the identification of positive RIAs and their formalization in the form of the object of intellectual property. In their work, these divisions are guided by the requirements of the legislation in the field of intellectual property protection and do not operate with the concept of "intangible asset". At the same time, the concept of "intangible asset" is used in a different context in another unit responsible for accounting of intangible assets - in the financial accounting department of an industrial enterprise. In the department responsible for tax accounting, a concept of "depreciable property" is used. However, pursuing different goals in their activities, these departments must coordinate their actions, since late recognition of RIA entails the need to recognize intangible assets and determine its value, which, in turn, may have tax consequences in order to calculate a reasonable tax burden.

It is necessary to note that in the practice of national enterprises, proper attention is not paid to the identification of IPs, which leads to their still low share in the structure of assets. In order to confirm this conclusion, the authors analyzed the financial reports of the 50 largest Russian companies in dynamics for 2015-2019, presented on the websites of Russian Statistic Service (for 2015-2018) and the Federal Tax Service (for 2019). The analysis results are presented in Table 1.

The results of the analysis of indicators in dynamics presented in Table 1 clearly indicate a stable trend of the increasing share of intangible assets in the total value of assets of Russian companies over a 5-year period.

Table 1. Results of the analysis of financial reports of the 50 largest companies in Russia

Indicators	**2015**	**2016**	**2017**	**2018**	**2019**
Total assets, billion rubles	51,680	55,785	59,318	64,907	67,338
Including in the field of industrial production	*38,054*	*39,886*	*42,798*	*47,136*	*48,573*
The total value of intangible assets, billion rubles	110	127	152	224	249
Including in the field of industrial production	*51*	*62*	*75*	*99*	*103*
Share of intangible assets in total assets,%	0,21	0,23	0,26	0,34	0,37
Including in the field of industrial production	*0,13*	*0,16*	*0,18*	*0,21*	*0,21*

At the same time, even in TOP-50 companies that apply International Financial Reporting Standards due to state requirements (as the largest, socially significant organizations the shares of which are quoted on stock exchanges) and, therefore, having an idea of the existing approaches to the evaluation and accounting of intangible assets, do not aspire to make significant efforts to increase capitalization precisely at the expense of intangible assets. The share of intangible assets in the total amount of assets, although increased in 2015-2019 from 0.21 to 0.37%, but still it remains negligible. (For the largest organizations in the sphere of industrial production, the situation is even more critical: the share of intangible assets by the end of 2019 does not exceed 0.21%).

For example we compare: in the financial reports of the Siemens group at the end of the reporting period ended on September 30, 2019, the intangible asset value (excluding goodwill) was € 9,800 million, which was 6.5% of the total assets, equivalent to € 150.248 million.

At the same time, W. W. Choi et al., provide evidence that the financial market positively evaluates intangible assets disclosed in the financial statements. The results of the study confirm the requirement that intangible assets are reflected in the balance sheets of companies [5]. Disclosure of information about intangible assets in the financial statements has a positive effect on attracting additional financing from investors by enterprises [6]. The study of the role of industry factors and intangible assets in the activities of companies in Colombia also shows that intangible assets play an important role among the factors that determine the effectiveness of specific companies. Evidence supports the sustainability of intangible assets as drivers of competitive advantage in an evolving environment [7].

C. Eckstein's research shows that the problems associated with the measurement and recognition of intangible assets are international [8]. The applicable accounting standards have a major impact on the recognition and measurement of intangible assets. T. Dinh et al., analyzing the accounting of intangible assets in Australia according to national standards and IFRS, conclude that the introduction of stricter accounting rules for Intangible assets in accordance with the IFRSs reduces the quality of the set of information for investors [9]. However, despite the fact that the question of the degree of application of International Financial Reporting Standards (IAS, IFRS) in terms of the identification of intangible assets is debatable, in our opinion, the direct application of IFRS in Russian industrial enterprises will currently bring positive results. This applies to the identification of intangible assets for the purposes of their capitalization in the company's statement of financial position, and their valuation, which is confirmed, among other things, by the empirical data provided in table 1. It should be noted that in the current Russian legislation there is a rule that allows the direct application of IFRS in the event that any issue is not described in RAS.

The specifics of the Russian industrial enterprises activities determine the composition and structure of intangible assets. A significant proportion of the intangible assets are positive RIAS recognized by IP, which are protected by a patent or the introduction of a trade secret (know-how) regime. For them, timely identification and reliable assessment of their value, at which they will be reflected in accounting and financial statements, is of great importance.

The issues of the valuation of intangible assets and, in particular, IP, have not lost their relevance for more than 40 years. Approaches to IP valuation are traditionally considered in the economic literature in the context of the net asset index impact. The first significant disputes over the evaluation of intangible assets were triggered by a series of major acquisitions in the 1980s. The most notable deals, such as KKR/RJR Nabisco, Philip Moris/Kraft и Nestlé/Rowntree, were performed at prices that were five times higher than the book value of net assets calculated from the balance sheet. However, the results of such acquisitions did not always have positive consequences and in some cases a company was overvalued.

Nestlé Company paid £2.5 billion for the acquisition of Rowntree, which was significantly higher than the net asset value of the company and 2.5 times the capitalization of Rowntree before the takeover. As D. Hyde noted, Nestlé saw the promise of synergies in R&D, products, administration and sales, resulting in economies of scale. This is the example of the highly intelligent integration of two companies into a competitive business [10]. However, not all acquisitions were successful. For example, the "deal of the century" to acquire RJR Nabisco at a price significantly higher than the book value of net assets did not bring the expected profit for the New York company Kohlberg Kravis Roberts & Co. (KKR).

The concerns expressed in the financial environment regarding the reliable valuation of the value of the intangible assets, provided the need for serious research. On the one hand, the difficulty of determining the value of intangible assets is related to the uncertainty of their future benefits and useful life [11]. On the other hand, this is due to the conscious desire of enterprises to manipulate the value of intangible assets and the cash flows generated by them. A number of studies confirm that the impact of IP transactions on the income tax burden is one of the factors influencing the decision on the place of registration of intellectual property rights in MNCs [12, 13].

It is necessary to note that in the modern experience of Russian industrial enterprises, the method of accounting of IPs at historical cost, which consists of actual and documented expenses for the creation and (or) acquisition of assets, prevails. Among other aspects, this approach leads to an incorrect assessment of the value of the net assets of an industrial enterprise, which may turn out to be both unreasonably overestimated and underestimated.

The current IFRS provides accounting of the objects of intellectual property at fair value, which can be determined using one of the following approaches stipulated by International Valuation Standards:

- market-based;
- cost-based;
- income-based.

At the same time, intangible asset accounting at their fair value poses a serious problem for administrators and accounting professionals around the world [14].

Generalization of the performed research shows the following picture of the applicability of approaches to the assessment of the value of intellectual property (Figure 1).

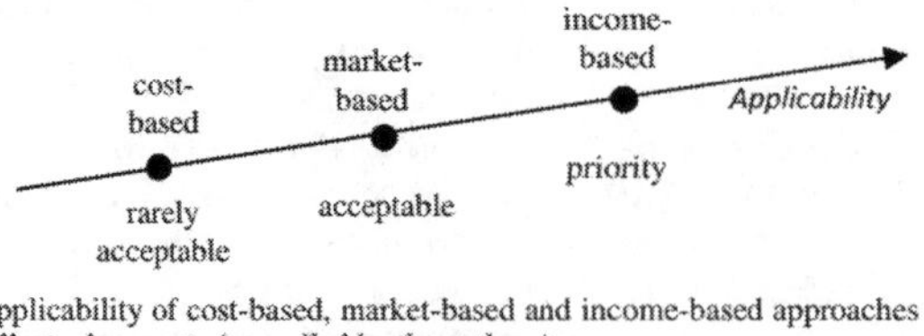

Figure 1. Applicability of cost-based, market-based and income-based approaches to the determination of the fair value of intellectual property (compiled by the authors)

The problem of the determination of the proper value of the intellectual property for the purposes of accounting and reflection in financial reports is not the only one. Among other problems that the authors identified based on a practical study of the state of intangible asset accounting at a large industrial enterprise as a part of a research project (contract No. 29052018, June 27, 2018, concluded with Perm State National Research University), there were the following ones:

- **The problem of the determination of the moment of the beginning of capitalization of intangible assets** in financial report**.** Thus, according to the paragraph 52 of IFRS 38, in order to assess whether an intangible asset created independently by an organization meets the recognition criteria, the process of the creation of an asset should be divided into two stages: research stage and development stage. At the same time, if, within the framework of an internal project aimed at the creation of intangible assets, a company can not separate the research stage from the development stage, then a company takes into account the costs of such a project as if they were spent only at the research stage (p. 53, IFRS 38). Some RIA can be obtained already at the research stage. However, such a result can not be recognized as an intangible asset for accounting purposes, since at this stage not all criteria for the classification of the received RIA as intangible assets are fulfilled. Thus, for RIA received by a company during the research phase, the possibility of future economic benefits can not be estimated. In this regard, according to the paragraph 54 of IFRS 38, an intangible asset that is the result of research (or the implementation of the research stage within an internal project) can not be recognized. Research costs (or the research phase of an internal project) are recognized as an expense when they occur. RIA received by a company at the stage of development is recognized as intangible assets if it is possible to simultaneously demonstrate the conditions of paragraph 57 of IFRS 38;
- **the problem of the differentiation of capitalized costs that form the value of the object of intellectual property itself,** reflected in the future as an intangible asset at fair value is non-capitalized R&D costs occurred in the current period on the one hand and other costs, but on the other hand the return of it will be carried out in future periods. Thus, the costs of positive RIA can be capitalized, while the costs incurred by a company at the research stage can not be capitalized, regardless of whether a positive result is obtained. For example, the costs associated with activities aimed at the production of new knowledge should not form the initial cost of intangible assets, even if new knowledge was produced at the research stage and a report on it was made. A derivative is **the problem of intellectual property accounting identified by a company as a result of inventory;**
- **the problem of the determination of the relevant useful life and the depreciation method.** Thus, the useful life of an IP as an independent inventory item according to the rules of financial accounting in Russia is determined when it is accepted for accounting as part of an intangible asset by a company independently or by expert means. Since the composition of intangible assets is diverse, it is advisable for a company to determine this period in a differentiated manner at the first stage. At the same time, taking into account the dynamism of market development, high rates of scientific and technological progress, the authors consider it possible to establish shorter useful life for intangible assets. Then, the current accounting procedure offers several methods to calculate the depreciation of intangible assets. During the choice of a method to calculate the depreciation for intellectual property, it is recommended to pre-group them by types (groups), for example, to allocate exclusive rights to utility models, exclusive rights to an invention, etc., which can be included in separate groups for the purpose of their own choice, relevant depreciation method etc.;

- the problem of differences in approaches to the formation of the initial value of intellectual property for the purposes of financial and tax accounting. This problem is a consequence of the problem of the determination when the capitalization of costs begins. In the conditions when an industrial enterprise conducts research and development as a part of contracts with external customers, the costs associated with these works are charged to the cost of works specified by this contract. However, during the work for third-party customers, an enterprise can obtain a positive result of intellectual activity (RIA), which is not required under the terms of the contract, but is valuable to a company itself. In the experience of Russian industrial enterprises, such positive RIAs are usually called "side" ones. As a rule, they represent know-how (production secrets) and are protected by the introduction of a trade secret mode. However, often at the moment when a positive result of the so-called side RIA becomes obvious to an industrial enterprise itself, the costs of research and development within the framework of a contract with an external customer are already attributed to the cost of sales and can not be adjusted. At this point, the difference in the goals of the activities of different departments of an industrial enterprise is most clearly manifested. Thus, the R&D department seeks to recognize a positive RIA as intellectual property as soon as possible after the process of identification in order to protect the rights to it. With rare exceptions, this department is not interested in the evaluation of know-how. At the same time, they do not seek to recognize the result of intellectual activity as positive until sufficient evidence of its applicability is given. Thus, this department in some cases does not promptly report the need to capitalize costs.

The financial accounting department needs to determine the initial cost of intellectual property for recognition as a part of intangible assets. However, it is no longer possible to determine the initial cost as the sum of the costs of the creation of an asset, since the moment when the capitalization of costs began was not determined in a timely manner. The omission of the start of capitalization of costs and the recognition of a positive RIA as intellectual property later leads to the need to determine the fair value of assets (sometimes even in 1-2 years after the result is obtained). As it is mentioned above, the income-based approach is very rarely applicable and it is considered the least reliable in global practice.

The application of market-based approach to the determination of the fair value of know-how, or other objects of intellectual property, is usually limited due to the lack of an active market and information about similar transactions. In the end, the most applicable method is discounted cash flows within the income-based approach to the determination of fair value. The application of this method causes a rather high cost of an intangible asset in financial accounting. The tax accounting department, on the contrary seeks to evaluate intellectual properties at the lowest cost, since, according to the rules of the Tax Code of the Russian Federation, such identified IPs are subject to recognition at market value, which is included in non-operating income in the period of asset recognition. Thus, the cost at which intellectual property will be recognized as depreciable property for tax accounting purposes directly affects the tax burden of an industrial enterprise. The tax accounting department faces a situation where, on the one hand, the market value of IPs is not determined due to the absence of an active market. On the other hand, it faces the situation when the only alternative approach is to determine the fair value using the income approach according to IFRS rules, the application of which leads to a significant increase in the tax burden enterprises.

The described situation is a "conflict of interest" within an industrial enterprise, which consists in the fact that one does not see a positive side RIA at the moment when it was necessary to start cost capitalization, and, accordingly, costs do not begin to be capitalized on time. As a result others face the need to determine the fair value of an intangible asset and still others must resolve the issue of the increase in the tax burden due to the unplanned recognition of the object of intellectual property in non-operating income at fair value.

However, just as a tax accounting department may search for any way to reduce the fair value of intellectual property to decrease the tax burden, so a financial accounting department may search for a way to increase the fair value by any means in order to form a picture of a more favorable financial position of a company in financial reports. The solution to this problem is to ensure the coordinated work of all departments of an industrial enterprise (departments of research and development, financial accounting and tax accounting).

In order to study and solve the above mentioned problems, the authors propose to turn to IFRS, requirements of national accounting standards and accumulated experience of Russian and international companies.

Conclusion

Thus, the following trends in the study of the accounting issues of the objects of intellectual property can be determined:

- The direct application of IFRS in Russia provides wider opportunities for the timely recognition and reliable assessment of intellectual property assets as a part of intangible assets;
- The recognition of RIA results in the need to recognize intangible assets and determine its value, which in turn can significantly affect the tax burden of a company;
- Having different goals, the specialists responsible for the preparation of reports may tend to overestimate the value of intangible assets, while the specialists responsible for calculation and payment of taxes tend to underestimate the value in order to achieve the desired parameters of taxation, thus regulating the tax burden. It is important to understand that the assessment of identified intangible assets at fair value is a mandatory process to provide reliable indicators of financial statements and, in particular, the indicator of net assets;
- The most applicable methods for the evaluation the object of intellectual property are the methods based on the calculation of the present value, implemented in the framework of the income-based approach;
- Among the problems of intangible asset accounting, the problems of the determination of the proper value of intellectual property should be highlighted as well as the determination of the moment of the beginning of capitalization of intangible assets, differentiation of capitalized costs that form the value of intellectual property itself and non-capitalized costs, determination of the relevant useful life and depreciation method, determination of the initial value of IPs for the purposes of financial and tax accounting, etc.

References

1. N. D. A. Majid, A. M. Shariff, R. Rusli, K. I. Azman, Procedia Engineering, **148**, 1089 (2016)

2. M. Grimaldi, M. Grecob, L. Cricelli, Journal of Business Research, **123**, 156 (2021)

3. T. Grzegorczyk, The Journal of High Technology Management Research, **31** **(1)**, (2020)

4. Z.-M. Zadorozhnyi, V. Yasyshena, Marketing and Management of

Innovations, **1**, 132 (2019)

5. W. W. Choi, S. S. Kwon, G.J. Lobo, Journal of Business Research, **49 (1)**, 35 (2000)
6. M. Labidi, J. F. Gajewski, Research in International Business and Finance, **48**, 426 (2019)
7. V. Andonova, G. Ruíz-Pava, Journal of Business Research, **69 (10)**, 4377 (2016)
8. C. Eckstein, Accounting Forum, **28 (2)**, 139 (2004)
9. T. Dinh, H. Kang, R.D. Morris, W. Schultze, Journal of international financial management & accounting, **29**, 247 (2018)
10. D. Hyde, J. Ellert, J. P. Killing, European Management Journal, **9**, 1 (1991)
11. O.M. Vakulchyk, V.V. Fesenko, A.V. Kaluhina, Accounting and Finance, **3 (85)**, 6 (2019)
12. R. Griffith, H. Miller, M. O'Connell, Journal of Public Economics, **112**, 22 (2014)
13. T. Neubig, S. Wunsch-Vincent, International Journal of Innovation Studies, **2**, 120 (2018)
14. M. Agurto Roque, G. Chavez-Cruz, S. Chuquirima-Espinoza, Revista conrado, **14**, 23 (2018)

https://doi.org/10.1051/shsconf/202111600013

Development of a digital model for public administration quality of real estate in the Russian Federation based on international experience

Olga Gvozdeva[1*], *Elena*Kolbneva[2],*Yulia* Sinitsa[1], *Marina* Smirnova[1], *Ilya* Chuksin[1]

[1] State University of Land Use Planning, 15, Kazakova St., Moscow 105064, Russia
[2]Voronezh State Agrarian University named after Emperor Peter the Great, 1, Michurina St., Voronezh 394087, Russia

Abstract. The authors studied the issues of recent changes in the system of state management of real estate.According to the authors it is necessary to use digital technologies of big data, neurotechnologies, artificial intelligence, new production technologies, and distributed registry systems for the formation of a digital economy.This will allow defining digitalization as the main source of long-term economic growth. The authors analyzed scientific research on the competition for global digitalization trends and determined the following: mobile Internet, e-commerce, artificial intelligence, however Blockchain technologies and visual identity are considered to be more promising and breakthrough trends. In order to assess the growth rate of the digital economy in the Russian Federation, the authors conduct the analysis in the studied area on the basis of quantitative and qualitative indicators of foreign leading countries. The authors come to the conclusion that the main result of digitalization is the increase in revenue, the increase in the speed of provision of services and their quality and the decrease in resource intensity.

1 Introduction

In the course of recent changes, the system of public administration of real estate is considered as the main factor indicator of development in digital economy taking into account its digitalization.The process of digitalization of the economic sectors of the Russian Federation is aimed at to achieve economic, social and technological effects that allow the country to provide a decent level of development and a stable competitive environment in the global market. During the formation of digital economy, the process of changing the traditional forms of established economic relations into digital form starts to prevail, which creates the undoubted potential of digital economy to increase the viability and security of the country.

It is necessary to take into account the quantitative and qualitative indicators of foreign countries in the studied industry and assess the state of the country in the international arena for an independent assessment of the growth rate of digital economy in the Russian Federation.For example, in the United States, the volume of digital economy is quite high, significant gross domestic product (GDP) is explained by intensive investment in digitalization, including attracting funds from the state and consolidated budget.Today, according to statistics from the relevant services, the share of digital economy in the United States is 10.9%, in China - 10%, in the EU - 8.3%.

The high specific indicator of the United States is explained by the introduction of large-scale information and digital solutions in digitalization system, as well as the development of its own projects "Alibaba" and "Huawei" [1].

The opinions of Russian experts in the field of the growth rate of digital economy in the Russian Federation agree that the country is lagging behind the leading countries in integrated development. In particular, according to the Digital McKinsey expert group, digitalization is a source of long-term economic growth, driven by the tools presented in Figure 1:

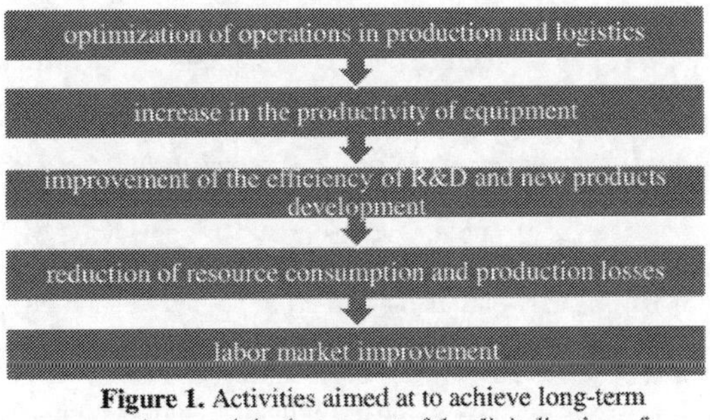

Figure 1. Activities aimed at to achieve long-term economic growth in the context of the digitalization of economy

For the formation of digital economy, it is planned to use digital technologies of big data, neurotechnologies,

*Corresponding author: chuksin-99@mail.ru

artificial intelligence, new production technologies and blockchain systems [2].

On the basis of Rostelecom Company, scientific research was carried out to find global trends in digitalization. The trends were the following: mobile Internet, e-commerce, artificial intelligence, while more promising and breakthrough trends were Blockchain technologies and visual identification of a person, capable to identify and to establish the identity of a consumer of a service. In the global ranking of the development of digitalization trends, the Russian Federation ranks the 11th place [3].

In the studies [3], the main problems of the introduction of digitalization are presented, including (Figure 2):

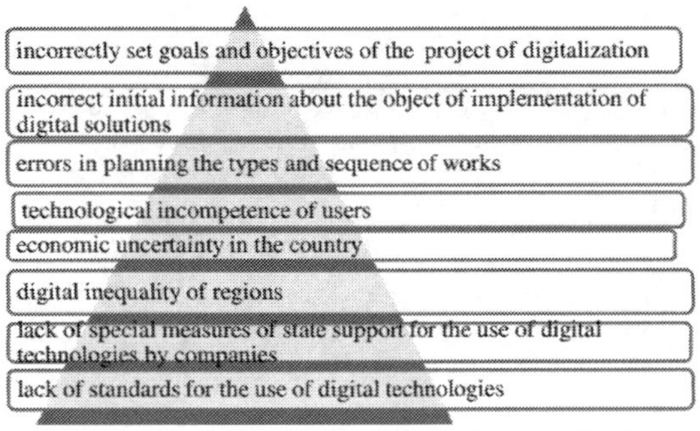

Figure 2. Main problems of introducing digitalization

The purpose of digitalization for the next 15-20 years is to take a leading position in the global economy through digital industry transformation and a competitive national digital industry.

2 Materials and methods

The importance of the study of the category of quality of public administration, both in general and in particular is explained in many scientific works that define quality as a basic key characteristic of public administration that has an impact on the development of the state.

In 2008, under the direction of V.V. Markin and A.V. Ostashkov, the studies on the the toolkit of the quality strategy of public and municipal administration within the framework of the modern concept of public administration were described in one of the first domestic publications on the identification of the concept of public administration quality.In this paper, the quality of public administration is understood as a different range of characteristics of the state and its departmental institutions, based on the unity of the three branches of government - legislative, executive, judicial, and transparency and observance of the rule of law.Ultimately, the quality of public administration is manifested in its direct impact on the socio-economic development of Russia and its regions.

A narrow understanding of public administration is determined in terms of the effectiveness of administration, the effectiveness of its organization and the rationality of institutions and functions of the executive branch.

E.V. Talapina, points out [4] that the quality of public administration can not be objectively assessed within the state apparatus itself. It is necessary to develop a quality assessment system that takes into account the satisfaction of participants in public administration - citizens, civil society institutions and business structures. The category of public administration quality can include three approaches (Figure 3):

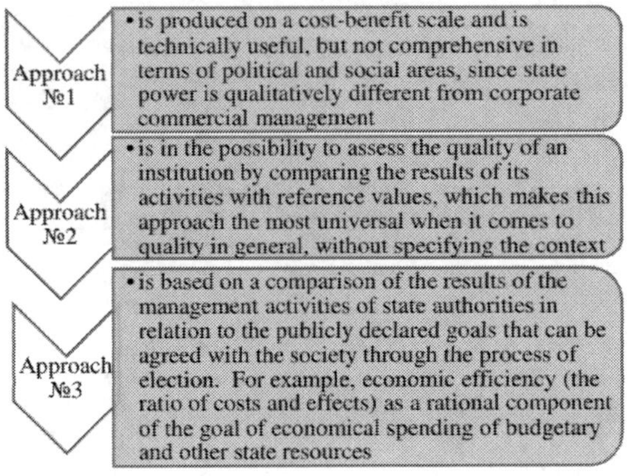

Figure 3. Approaches to the category of quality of public administration

In the world scientific literature, the concept of public administration quality is closely related to two prevailing concepts of public administration over the past decades: New Public Management and Good Governance.

The main goal of the New Public Management model is to increase the efficiency and effectiveness of the public administration system.

World Bank experts N. Manning and N. Parison [5] determined several goals of transformation, common for all states, related to (Figure 4):

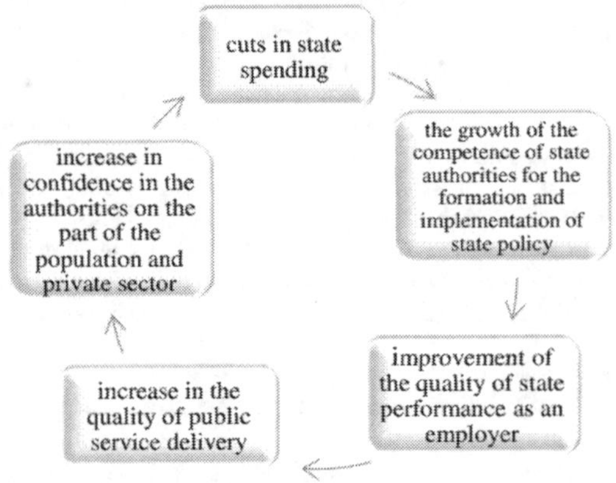

Figure 4. Ways to the goals of state transformation based on developing sectors of digital economy

New Public Management and managerial principles of state governance are not general due to the fact that they move into the background "social mission" of the state, as a result of which, citizens are considered not as a source of political power, but as consumers of state and municipal services.

According to D.G. Krasilnikov [6] the concept of Governance, which is an independent management model, is defined as the content of administrative reforms that began in Western countries and spread to the rest of the world. Itcontains the parameters of the ideal management model, which can be used in the analysis to determine the degree of deviation from the sample, i.e. in terms of management quality.

The most common international assessment of the quality of national governance is the Governance Research Indicator Country Snapshot (GRICS), calculated every two years since 1996 for almost 200 countries. The indicator is developed on the basis of more than several hundred variables, indices and research results that serve as indicators of good governance of government bodies and include:

1) the mechanisms for the selection and management of government activities;

2) the ability of the state to clearly and concretely formulate and define the policy of the state;

3) respectful attitude of citizens and the state to the institutions that manage social and economic interactions in society.

Six GRICS indexes include (Figure 5):

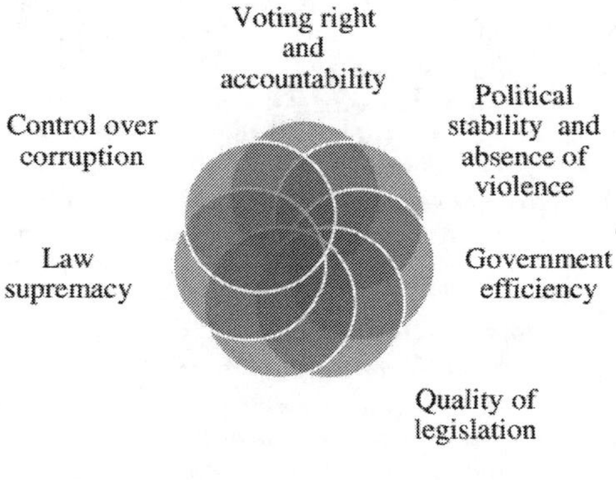

Figure 5. GRICS indicators

Among other indicators of the quality of public administration, the majority of indicators are based on a methodology to assess it with a narrower approach. For example such indices are presented in Figure 6:

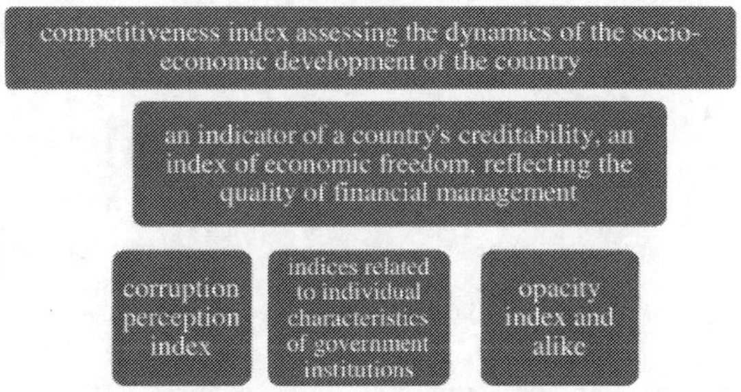

Figure 6. Indicators of the quality of public administration

The Growth Competitiveness Index (GCI), which is supported by the World Economic Forum and includes the assessment of the quality of public institutions, is quite popular.The index includes nine components, among which the state of public institutions stands apart, expertly assessed by such parameters as the provision of property rights, the state of ethics and the level of corruption, undue influence (favoritism, etc.) and government efficiency. All these parameters relate to the basic conditions for the provision of economic growth.

The Corruption Perception Index (CPI) is a composite index developed by the international non-governmental organization Transparency International, which has been calculating CPI annually since 1995 using a methodology based on the results of its surveys. The main purpose of this index is to present the current state of the level of corruption in the country. The Corruption Perception Index is based on data from 17 different surveys and studies conducted by 13 independent institutions among entrepreneurs and local analysts, including surveys of the population (both its citizens and foreigners).

At the same time, Russian researchers [7] note that in Russia international methodsto assess the effectiveness of public administration are not used fully and systematically during the determination of the quality of public administration.

The dynamics of GRICS can be demonstrated according to the data of the World Bank [8] and in Figure 7.

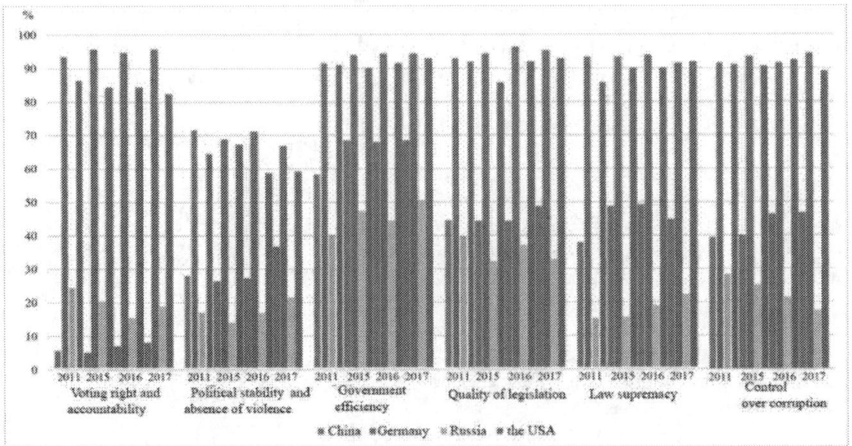

Figure 7. World Bank Percentile Ranking of Governance Indicators for the period 2011 - 2017 [8]

After the analysis of the ranking of leaders in terms of indicators, we can conclude that Germany has the highest quantitative values by all indicators. The indicators of the Russian Federation are significantly lower in comparison with other countries throughout the period.For example, in 2017, Germany exceeded Russia by 5.1 times in terms of voting rights and tax reporting and 2.9 times in terms of quality of government regulation.

Thus, we can say that in Russia there are serious problems in the formation of public administration in all the studied indicators.

The indicators of the quality of public real estate management were classified into 2 groups and 11 subgroups (Figure 8):

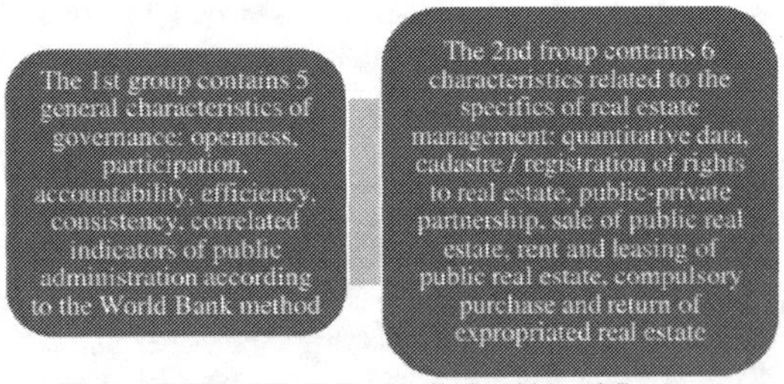

Figure 8. World Bank Percentile Ranking of Governance Indicators for the period 2011 - 2017 [8]

The figures 9 and 10 show the assessments of the presented characteristics of the quality of state property management in the context of indicators of subgroups. They show that in Russia there was a rather low level of quality of state property management.

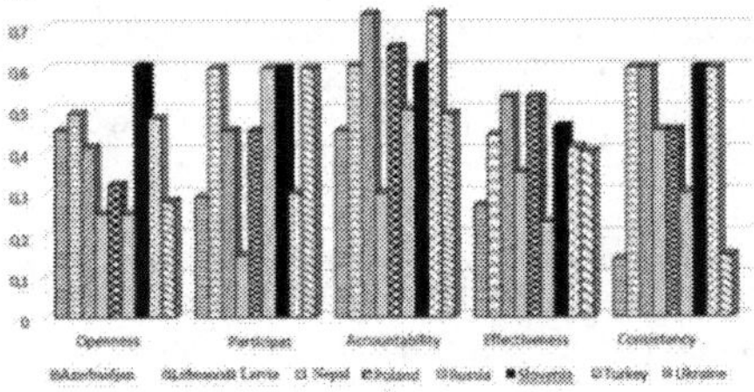

Figure 9. Assessment of the characteristics of the 1st group of quality management of public real estate [9]

According to the studies of indicators of the quality of public administration in the field of real estate (Figures 9-10), the definition of such indicators has significant drawbacks associated with the lack of quality information and methodological flaws that relate to the overall suitability of real estate.

The problems of the quality of public administration in general in the Russian Federation have a large number of issues and are not properly studied, namely, the key categories and theoretical provisions of public administration of real estate are not fully revealed.

Nowadays in accordance with the list of state programs of the Russian Federation, approved by the order of the Government of November 11, 2010 No. 1950-r, 44 state programs are provided. These state programs are represented by five program blocks. At the same time, the effectiveness of the implementation of government programs is 78-95% [9].The main goals of the program are in the context of the increase in the efficiency and quality of public administration in terms of federal property and material reserves [9].

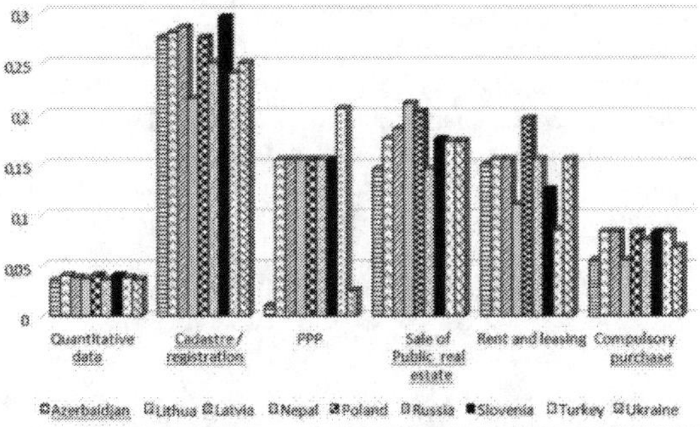

Figure 10. Assessment of the characteristics of the 2nd group of quality management of public real estate [9]

In the Russian scientific literature, the developments on qualimetric and vector analysis of quality are also actively carried out, including in the field of real estate management, which often requires expert methods and multicriteria analysis [9].

3 Results and Discussion

According to the research results of the Boston Consulting Group, the expected digitalization results of public administration are presented in Table 1 [9].

Table 1 - Forecasts of the results of digitalization in public administration [9]

Indicators and types of services	2018	2019	2020
Number of government services for business in electronic form, mln. in year	2,028	2,253	2,504
Number of free hours due to digitalization, mln. hours per year	17	22	34
Cost savings of the state and business due to digitalization of public services, trillion. rub. in year	2,85	3,87	5,71
government cost saving	0,21	0,22	0,23
business cost savings	2,63	3,65	5,48

The digitalization of cadastral registration in Western Europe, the USA and Canada has become the ground of the information base for various types of registers (real estate, forestry, water, environmental and urban planning).In most economically developed foreign countries, the cadastral registration is the unity of the fate of a land plot and real estate objects that are firmly connected with land. This relationship allows consumers of digital services of land information systems to economically and efficiently carry out the processes of transferring ownership of real estate. It also allows the subjects of state management of real estate to manage processes in the studied sector of the country more effectively [10].

According to expert opinions, the implementation of the program "Digital Economy in the Russian Federation" in state management of real estate can be achieved with the following measures (Figure 11):

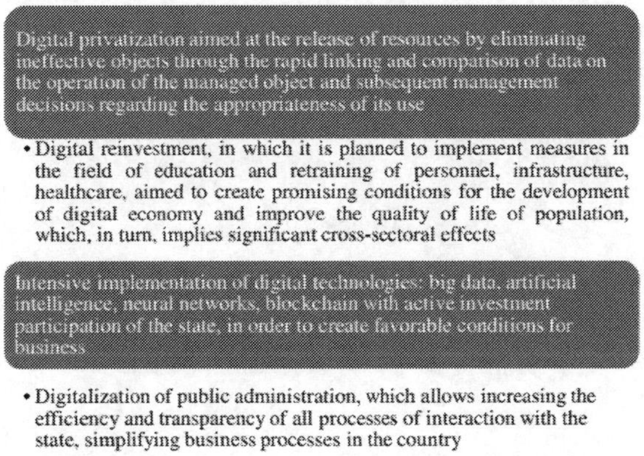

Figure 11. Activities carried out within the framework of the program "Digital Economy in the Russian Federation"

In order to implement the goals of digitalization of public administration, the technologies of the electronic document management system are used.

In order to improve the efficiency of the use of information on real estate, the Federal Service for State Registration, Cadastre and Cartography has introduced a digital project for the analysis and processing information, as well as comparison with the data contained in the register of federal property. This project has the mechanisms for analytical processing of information, which allow reducing the number of errors in data and carrying out systematic work to update information [10]. It is planned to develop a big data system in order to identify the reserves to increase the efficiency of federal property use [10].

For the implementation of the program "Digital Economy of the Russian Federation", the center of strategic developments has already proposed the idea of "State as a platform", in which the digital transformation of public administration should be based on a targeted basis of quality improvement, efficiency, consumer satisfaction, where the key factor is the well-being of citizens and economic effect.

The improved approaches to accounting and registration imply the introduction of Blockchain technologies designed to optimize the process of the registration of real estate and become the basis for the creation of open registries (Figure 12). The advantage of this technology is the ability to exclude intermediaries during accounting actions on the example of state registration of real estate.

Blockchain-technologies
- distributed database organization system. Its work is based on the distribution of data between blocks, which can be located on computers of thousands of users in different parts of the world.
- any project created on the basis of blockchain technology can be compared to a database that allows storing the same information on many different servers
- each transaction in such a scheme is decentralized, which excludes the possibility to change or falsify data

Figure 12. Systems ofBlockchain-technologies

We propose to create a server room in the central office ofthe Federal Service for State Registration, Cadastre and Cartography.

For this, it is necessary to involve two specialists for the stable operation of ASIC-miners in the server room and one employee representing the call center for technical support of territorial departments.

The salary of an employee responsible for the mining machine is 40,000 rubles per month and the specialist representing the call center earns 20,000 rubles per month, therefore, 40,000*2 + 20,000 = 100,000 rubles per month. Total costs for wages will be 100,000* 12 = 1,200,000 rubles in year.

For the work of specialists, we need the buy ASIC-miners. The cost of one ASIC 2TH / sTerraMinerIV is 187,000 rubles. We need to buy 2 ASIC-miners in the server room; therefore it is necessary to get 187,000*2 = 374,000 rubles. It is also necessary to buy 2 power supplies; the cost of one is 10,000 rubles. Their total costs will be 20,000 rubles.

Electricity costs depend on the server power consumption, kW/h; power consumption of the ASIC-miner, kW/h; cost of electricity, rubles/kWh; amount of time.

One device consumes 0.96 kW/h, therefore, 0.96*24 = 23.04 kW/day is consumed by one ASIC-miner. The project needs 2 installations, 23.04*2 = 46.08 kW/day.

ASIC-miners work throughout the year, seven days a week, the cost of electricity is 3.8 rubles Kw / h. As a result, the total cost is 46.08 * 3.8 * 365 = 63,913 rubles per year.

To organize a server room, we need: one personal computer which costs 50,000 rubles; a server box - 3,000 rubles; a switch - 3,000 rubles and power cables 7 * 600=4,200 rubles; unlimited Internet at a cost of 1,500 rubles per month, as a result, 1,500*12 = 18,000 rubles per year is the Internet costs.

Another cost is depreciation charges.

$$A_{PC} = (C_{pc} * H_d), (1)$$

whereC_{pc}– cost of 1 PC, rub. In our case it is 50 000 rublesand the cost ofpower supplies 20 000 rubles;

H_d– depreciation rate of fixed assets per year, %.

The cost of depreciation charges amounted to 14,000 rubles per year.

Basic costs are calculated using the following formula:

$$З_{BAS} = З_{TOT} + З_{SOC} + З_{MAT} + З_{D} + З_{OTHER}, (2)$$

$З_{BAS}$ = 1 200 000 + 374 000 + 120 000 + 63 913 + 50 000 +10 200 + 18 000 + 14 000 = 1 750 713 рублей.

The total cost of the development of a server room will be about 2 million rubles. It is necessary to note that these costs are one-off.

With the use of this technology, it is planned to conduct about one thousand transactions per month, 12,000 transactions per year. The deal will be approximately up to 2 million rubles. Basically, up to 3% are taken by intermediaries for transactions. On average, we will be able to save about 700 million rubles a year on the services of intermediaries.

That is, the use is really reasonable and economically profitable.

The stratified representation gives the understanding of the concept of digitalization of public administration in the system of coordinates of its quality, which allows reflecting its numerous features from the point of the satisfaction of needs [11].

The digitalization stratum model allows creating an information base of public administration utility indicators and form digital models of controlled situations, taking into account the subject-object composition, goals and quality indicators.

Table2. Stratum model of digitalization of public administration in the context of its quality

Stratum	Stratum area	Stratum content
Stratum 6: Idea	quality supply of public administration in the context of digitalization	organization of interaction between government entities using distributed software in order to ensure its quality
Stratum 5: Research	digitalization of public administration of R&D	creation of economic and mathematical models of real estate, models of formation, implementation of the quality strategy of public administration of real estate for the development of IT projects
Stratum 4: System representation	ecosystem of digital platforms of government	informatization of the digital ecosystem of public administration, formation of interaction between digital platforms based on the regulatory code of the digital space, quality requirements of public administration
Stratum 3: Conctruction	database of real estate of the Russian Federation	creation of an information space based on the Unified State Register of Real Estate to coordinate the interaction of government entities for structuring and storing data on real estate
Stratum2: Technologies	cloud, fog and dew computing, Blockchain technologies, big data processing, neurotechnologies	processing and mining the entire set of big data on the functioning of real estate, extracting information from data on real estate
Stratum 1: Embodiment	information communication system	information interaction of government entities in the Internet space

4 Conclusion

The main result of digitalization is the increase in revenue, the increase in the transparency of activities, the increase in the speed of provision of services, the decrease in resource intensity and the improvement in the quality of products and services. According to experts, mobile technologies, industrial automation, virtualization technologies, digital design and modeling have a significant impact on the results of digitalization.
Digitalization gives the opportunity to improve the standard of living of the population in accordance with employment opportunities, the increase in purchasing power, the increase in social participation of citizens in the management of processes that improve the convenience of living, the development of digital services, public services and national security [12].
The above mentioned review, showing a low rating of the quality of public administration in Russia, suggests that the management of real estate in the Russian Federation reflects problems in the field of quality supply.

As it is noted in the Good Governance Guide 2014, "good governance" leads to better decisions, helps local governments meet their legislative responsibilities, and, in fact, provides an ethical framework for governance. This allows assessing the work of the government as an effective management process.

The use of digital technologies of big data, neurotechnologies, artificial intelligence and new production technologies will improve the management system based on the digitalization of economy sectors.

References

1. V.G. Svetashov*Monitoring the quality of real estate management processes*(Moscow, 2017)
2. M.G. Rodionov Bulletin of the Siberian Institute of Business and Information Technologie, **1**, 25, 62-66 (2018)
3. T.I. Leonova, Yu.A. KalazhokovaFundamental research, **6,** 1, 143-147 (2015)
4. E.V. Talapina*Modernization of public administration in the information society* (Moscow, 2015)
5. N. Manning, N. Parison.*Public administration reform: international experience* p. 35–41 (Moscow, 2003)
6. D.G. Krasilnikov, O.V. Sivintseva, E.A. Troitskaya*ARS ADMINISTRANDI,* **2**, 45–62 (2014)
7. V.V. Markin, A.V. Ostashkov*Monitoring of state and municipal services in the region as a strategic tool for improving the quality of regional management: experience, problems, recommendations* (Moscow, 2008)
8. Report of World bank: Worldwide Governance Indicators, Retrieved from: http://info.worldbank.org/governance/wgi/#reports.
9. M. Gross, R. Zróbek*Land Use Policy*, **49**, 352-364(2015)
10. A.A.Varlamov, S.A.Galchenko, O.V. Gvozdeva, I.V.Chuksin International agricultural journal, **63**, 5 (377), 69-72, (2020)
11. S.N.Danilova Management of the balanced development of multifunctional centers for the provision of state and municipal services: abstract dis. candidate of economic sciences: (St. Petersburg, 2019)
12. O. V.Gvozdeva, M. A. Smirnova, I. V.Chuksin, M. V. Shakirova Moscow Economic Journal, 12, p.48(2020)

Organization behavior control of spatial cluster-network integration of the oil and gas complex

Leonid Kozhemyakin, *Maria* Osipova*, and *Vladislav* Nikitin

Perm National Research Polytechnic University, 29, Komsomolsky prospekt, Perm, 614990, Russia

Abstract. The article presents, substantiates and tests the author's methodological tools based on a set of economic and mathematical models reflecting the processes of formation, development, state and organization behavior control of spatial cluster-network integration of the oil and gas complex. The object of the research is the interregional and intersectoral interactions of the oil and gas complex, considered by the example of the regions of the Volga Federal District of the Russian Federation. The authors understand spatial cluster-network integration as a production pool of mutually beneficial cooperative ties of autonomous business structures based on related interdependent types of economic activity and forming a common space of socio-economic interests regardless of industry and territorial affiliation. The organization behavior control of spatial cluster-network integration requires taking into account the interests of all interested stakeholders, which can be represented using the structural-functional model of the four-link spiral concept. This organization behavior control approach, based on the cluster-network nature of spatial integration, will strengthen territorial and sectoral relationships; a harmonized synthesis of disparate resource potentials will contribute to the rationalization of their use; form the competitive advantages of business structures, the region and the country as a whole.

1 Introduction

A slowdown in economic growth, an increase in sanctions pressure on the domestic system, and comprehensive digitalization require the search for new approaches to creating, strengthening and assessing the effectiveness of industrial development organization behavior control. In this case, an important role is played by the organization behavior control of spatial cluster-network integration, in both intra-inter-sectoral and inter-regional interaction.

The purpose of the article is to test the author's tools based on a set of economic and mathematical models that reflect the processes of formation, development, state and organization behavior control of spatial cluster-network integration of the oil and gas complex. The object of the research is the interregional and intersectoral interactions of the oil and gas complex, considered by the example of Volga Federal District of the Russian Federation (hereinafter – VFD). The subject is instrumental and methodological support for improving the organization behavior control process of spatial cluster-network integration of subjects of the oil and gas industry.

The analysis of the theories of spatial development of socio-economic systems of different levels allowed the authors to formulate the following definition: spatial cluster-network integration is a production pool of mutually beneficial cooperation ties of autonomous business structures based on related interdependent types of economic activity and forming a common space of socio-economic interests, regardless of branch and territorial affiliation [4].

At the same time, the organization behavior control of spatial cluster-network integration requires taking into account the interests of all interested stakeholders, which can be represented using the structural-functional model of the concept of a four-link spiral. This organization behavior control approach, based on the cluster-network nature of spatial integration, will strengthen territorial and sectoral relationships; a harmonized synthesis of disparate resource potentials will contribute to the rationalization of their use; form the competitive advantages of business structures, the region and the country as a whole. Consequently, the increase in cluster-network integrations is a fresh direction in the development of the regional economy.

2 Materials and Methods

Organization behavior control of these integration processes is impossible without appropriate instrumental and methodological support for assessing and predicting the development of cluster-network interaction. The author's developments are based on the works of domestic and foreign authors [3, 9, 11].

Graphically, the methodology and main indicators in this work are presented in Figure 1.

The author's methodological approach is based on the concept of "cluster load index", which shows the region's predisposition to the creation or development of existing cluster-network integrations, also indicates the presence

* Corresponding author: osipova.mu@mail.ru

of industry dominance in the region, thereby predetermining the subject as a region with one or another (other) types of economic activities. In the long term, the cluster load index makes it possible to assess the direction of the region to select strategic benchmarks.

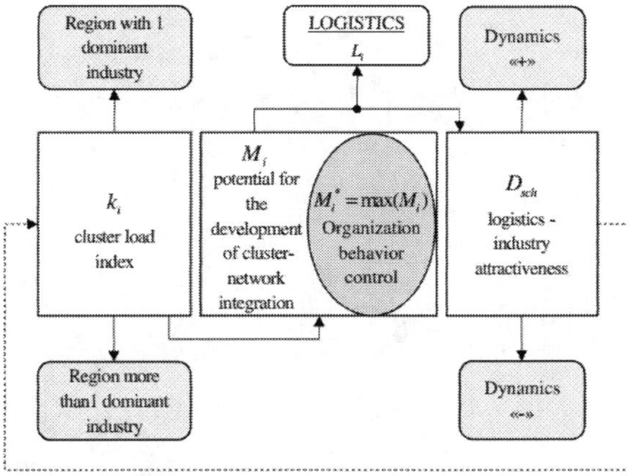

Figure 1. Methodology for managing spatial cluster-network integrations

The proposed toolkit allows for a broader analysis of the integration development of business structures, takes into account the sectoral specifics of regional development. It also takes into account the differences in statistical methods of analysis according to All-Russian classifier of economic activities for the analyzed period 2005-2019. This leads to a more informative, accurate calculation and forecasting of systems development [8]. This methodological approach includes several steps.

Step 1. A quantitative assessment of the parameters of the cluster load index is made according to Eq. 1.

$$k_i = \sqrt{\frac{\sum_{j=1}^{n}\left(\frac{w_{ij}}{W_j} - \frac{\overline{w_{ij}}}{\overline{W_j}}\right)^2}{E}}, \quad (1)$$

where k_i is cluster load index;

w_{ij} is specific indicator of j-th sub-industry of shipped products for any type of economic activity in i-th region;

$\overline{w_{ij}}$ is average value of j-th sub-industry of shipped products for any type of economic activity in i-th region for selected year;

W_j is specific indicator of j-th sub-industry for Russia as a whole;

$\overline{W_j}$ is the average value of the subsectors of shipped products for any type of economic activity for Russia as a whole for the selected year;

$i \in I,\ I = \overline{1,m},\ m$ is number of regions;

$j \in J,\ J = \overline{1,n},\ n$ is number of all subsectors for the selected period; E is multiplier calculated by Eq. 2:

$$E = \frac{n_g^*}{\max(n)}, \quad (2)$$

where n_g^* is the number of current subsectors for the selected year.

Multiplier $E \in (0;1]$ is calculated to correctly determine the index due to differences in the editions of the methodology for including industries according to OKVED for the selected period 2005-2019. The more informative and more accurate calculated cluster load index is achieved by decreasing the amplitude of the monotonic function under the root in Eq. 1. That is, taking as a dimensional space all indicators calculated for a different time period with a different number of sub-sectors according to the corresponding All-Russian classifier of economic activities, a single dimensional space is formed n (R^n).

In paper [5], dependence k_i, cross regional product and profit of enterprises are considered. For training a neural network and subsequent modeling of dependence and forecast, the cluster load index k_i is normalized.

The disadvantage of formula (1) is that the resulting values are unlimited on any domain of definition. To solve this problem, rationing is proposed.

The normalization can be performed in different ways, but in the current work it is proposed to use a function similar to the Gaussian function, which takes values from 0 to 1, which has the following form (Eq. 3):

$$f(x) = \frac{1}{\sqrt{2\pi}} e^{-\frac{(\alpha_k x - \beta_k)^2}{2}}, \quad (3)$$

where α_k, β_k - empirical parameters. Then, in our case, the normalized cluster load index $k_{i\,norm}$ will take the following form (Eq. 4):

$$k_{i\,norm} = \frac{1}{\sqrt{2\pi}} e^{-\frac{(\alpha_k k_i - \beta_k)^2}{2}} = \frac{1}{\sqrt{2\pi}} e^{-\frac{\left(\alpha_k \cdot \sqrt{\sum_{j=1}^{n}\left(\frac{w_{ij}}{W_j} - \frac{\overline{w_{ij}}}{\overline{W_j}}\right)} - \beta_k\right)^2}{2E}}, \quad (4)$$

where $k_{i\,norm} \in \left(0; \frac{1}{\sqrt{2\pi}}\right]$.

Step 2. Formation of spatial cluster-network interactions within the framework of the creation of an integral cluster presupposes interregional interaction of two or more regions capable of developing a single strategy aimed at increasing the competitive potential of the parties, while optimal sectoral and inter-sectoral localization should be attractive to all interested stakeholders: the state, business structures and the population [6].

The degree of development of the infrastructure basis for interactions plays an important role, especially the presence of the development of the transport sector, which contributes to the optimization of logistics flows. The formalization of the logistic component can be represented in the form of Eq. 5.

$$L_i = \omega_1 L_i^{Exb} + \omega_2 L_i^{Ast}, \quad (5)$$

where L_i is traffic density in i-th region;

L_i^{Exb} is density of public railways (kilometers of tracks per 1000 km^2 of the territory);

L_i^{Ast} is density of public roads with hard surface (kilometers of tracks per 1000 km^2 of the territory);

ω_1, ω_2 are weight coefficients, where $\omega_1 + \omega_2 = 1$, $\omega_1 = \frac{L_i^{Esb}}{L_i^{Esb} + L_i^{Ast}}$, $\omega_2 = \frac{L_i^{Ast}}{L_i^{Esb} + L_i^{Ast}}$.

In the same way, we normalize for L_i; then the normalized indicator $L_{i\,norm}$ is calculated as follows (Eq. 6):

$$L_{i\,norm} = \frac{1}{\sqrt{2\pi}} e^{-\frac{(\alpha_k L_i - \beta_k)^2}{2}} = \frac{1}{\sqrt{2\pi}} e^{-\frac{\left(\alpha_k \cdot \left(\omega_1 L_i^{Esb} + \omega_2 L_i^{Ast}\right) - \beta_k\right)^2}{2}} = \frac{1}{\sqrt{2\pi}} e^{-\frac{\left(\alpha_k \cdot \left(\frac{L_i^{Esb\,2} + L_i^{Ast\,2}}{L_i^{Esb} + L_i^{Ast}}\right) - \beta_k\right)^2}{2}}$$

$$L_{i\,norm} \in [0;1] \tag{6}$$

The author's methodology is based on the calculation of the integral indicator – "potential of development of cluster-network integration" (hereinafter CNI), which reflects the relationship between the cluster load index of a region and the development of its traffic flows. A quantitative estimate of the parameters of the CNI potential is calculated by Eq. 7.

$$M_i = k_{i\,norm} \cdot L_{i\,norm} = k_{i\,norm} \cdot \frac{1}{\sqrt{2\pi}} e^{-\frac{\left(\alpha_k \cdot \left(\frac{L_i^{Esb\,2} + L_i^{Ast\,2}}{L_i^{Esb} + L_i^{Ast}}\right) - \beta_k\right)^2}{2}} \tag{7}$$

where M_i is development potential of cluster-network integration.

The transformed author's model can be formalized in the form of Eq. 8.

$$M_i = \frac{1}{\sqrt{2\pi}} e^{-\frac{\left(\alpha_k \cdot \sqrt{\sum_{j=1}^{n} \left(\frac{w_{ij}}{W_j} - \frac{\overline{w_{ij}}}{\overline{W_j}}\right)} - \beta_k\right)^2}{\frac{2n_g^*}{\max(n)}}} \times \frac{1}{\sqrt{2\pi}} e^{-\frac{\left(\alpha_k \cdot \left(\frac{L_i^{Esb\,2} + L_i^{Ast\,2}}{L_i^{Esb} + L_i^{Ast}}\right) - \beta_k\right)^2}{2}} \tag{8}$$

Thus, the value of the potential for the development of cluster-network integration, assessed using the author's methodology, can allow all interested stakeholders to make balanced organization behavior control decisions aimed at enhancing or restraining their joint efforts in the field of these interactions. Thus, there is an opportunity for business structures and regions to increase their attractiveness through the optimal use of resource potential.

According to the authors, the target function of the model, taking into account the most important challenges of the current stage of development, among which, during the development of the model, the following were highlighted:

– increasing the efficiency and effectiveness of the development parameters of all socio-economic systems;

– implementation of optimal accumulation, redistribution and use of resource potential in conditions of harmonization of interests of all stakeholders.

These organization behavior control requirements are implemented through the formalization of the target setting - the search for the maximum potential for the development of cluster-network integration, under a certain system of restrictions (Eq. 9):

$$\begin{cases} M_i^* = \max(M_i) = \max\left[k_{i\,norm} \cdot L_{i\,norm}\right] = \\ = \max\left[\frac{1}{\sqrt{2\pi}} e^{-\frac{\left(\alpha_k \cdot \sqrt{\sum_{j=1}^{n} \left(\frac{w_{ij}}{W_j} - \frac{\overline{w_{ij}}}{\overline{W_j}}\right)} - \beta_k\right)^2}{\frac{2n_g^*}{\max(n)}}} \times \frac{1}{\sqrt{2\pi}} e^{-\frac{\left(\alpha_k \cdot \left(\frac{L_i^{Esb\,2} + L_i^{Ast\,2}}{L_i^{Esb} + L_i^{Ast}}\right) - \beta_k\right)^2}{2}}\right] = \\ = \max\frac{1}{\sqrt{2\pi}}\left[e^{-\frac{\left(\alpha_k \cdot \sqrt{\sum_{j=1}^{n} \left(\frac{w_{ij}}{W_j} - \frac{\overline{w_{ij}}}{\overline{W_j}}\right)} - \beta_k\right)^2}{\frac{2n_g^*}{\max(n)}}} \times e^{-\frac{\left(\alpha_k \cdot \left(\frac{L_i^{Esb\,2} + L_i^{Ast\,2}}{L_i^{Esb} + L_i^{Ast}}\right) - \beta_k\right)^2}{2}}\right], \\ w_{ij} \in \left[\tilde{w}_{ij} - \delta_j; \tilde{w}_{ij} + \delta_j\right] \subset [0;1], \\ \sum_{j=1}^{n} \delta_j = 0, \\ i \in [1;m], \\ i \in [1;n]. \end{cases} \tag{9}$$

where M_i^* is the maximum potential for the development of cluster-network integration;

$\overline{w_{ij}}$ is the specific indicator of j-th sub-industry in the volume of shipped products by the type of economic activity "Manufacturing" of i-th region for the calculated year;

δ_j is the point neighborhood $\overline{w_{ij}}$.

Due to the fact that the organization behavior control impact has a certain time lag and the subjects are not immediately able to reorient themselves, it is especially difficult to change the sectoral specifics of regional systems, therefore the authors made the assumption that δ_j -neighborhood varies in the range of +/–0.05 (+/–5% of current values $\overline{w_{ij}}$) with the condition $\sum_{j=1}^{n} \delta_j = 0$.

Step 3. Analysis of the long-term perspective of cluster-network integration is one of the significant criteria for enhancing the interest of stakeholders in cluster-network integration.

According to the authors, one of the key indicators based on retrospective data that can become the primary guideline for making managerial decisions is logistic-sectoral attractiveness (hereinafter – LSA), expressed by a dynamic change in the integral indicator – "the development potential of cluster-network integration", this indicator is calculated according to Eq. 10:

$$D_{sch} = \frac{M_i^*\big|_t - M_i^*\big|_{t-1}}{M_i^*\big|_t}, \qquad (10)$$

where D_{sch} is the indicator of logistics and industry attractiveness;

$M_i^*\big|_t$ is the current value of the potential of the functional orientation of the region;

$M_i^*\big|_{t-1}$ is the previous value of the potential of the functional orientation of the region.

Values M_i^* for each region in the work is estimated in increments of 1 year.

Thus, the proposed author's methodology makes it possible to more comprehensively assess the long-term prospects of cluster-network integration, serves as one of the criteria for activating the interest of stakeholders.

3 Results and Discussion

The author's approach and methodology were tested in the regions of Volga Federal District (VFD), initial indicators are presented in the official publications of Federal state statistics service (Rosstat) [6,7]. As shown by the analysis, Volga Federal District as a region is focused on the manufacturing industry, therefore, the cluster load index is calculated for the manufacturing industry (the type of economic activity "manufacturing").

In the work, the calculation of the index is presented using the example of Perm Krai and Volga Federal District in dynamics for the period under review of 2005-2019. If the cluster load index grows, therefore, the entity is positioned as a region with one dominant sectoral strategy - the manufacturing industry. In our case, there is a harmonious change in the index indicators relative to Volga Federal District, which indicates the presence of another industry that makes a significant contribution to the formation of the socio-economic component of the region.

In Perm Krai, as well as in Volga Federal District, one of the leading directions is the oil sector. Figure 2 shows the prevalence of the oil and gas complex both in Volga Federal District and in Perm Krai.

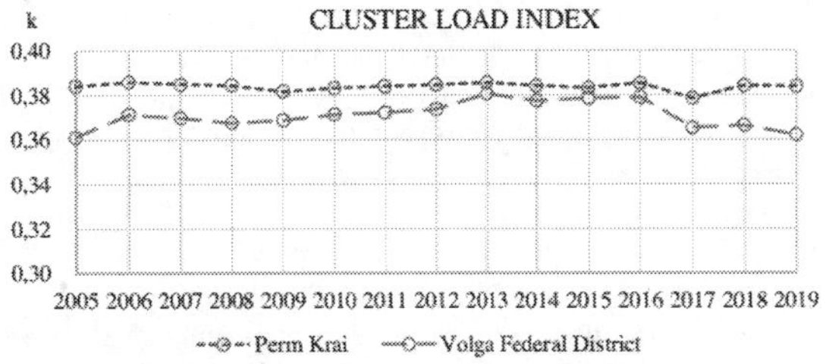

Figure 2. Relationship between the cluster load index of Perm Krai and Volga Federal District

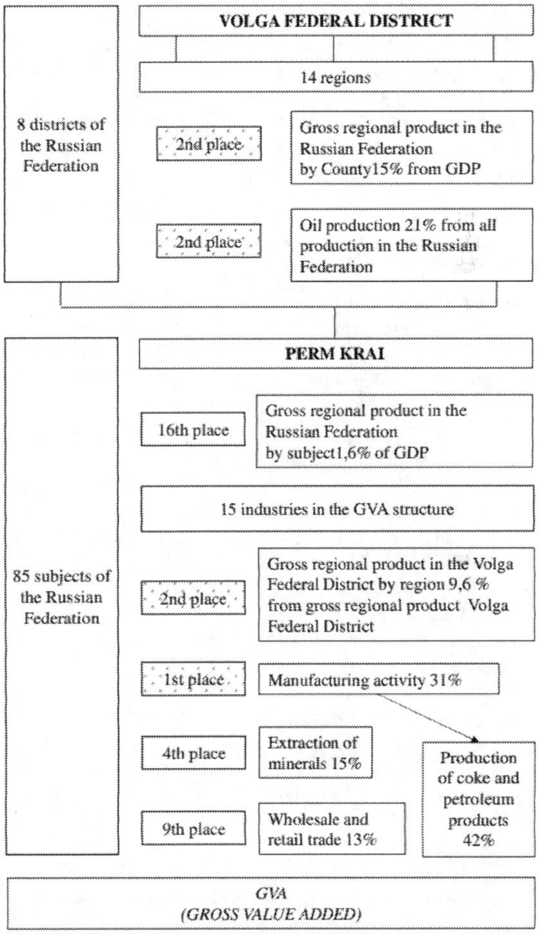

Figure 3. Sectoral indicators of Perm Krai and Volga Federal District

The cluster load index of the Perm Territory for the same period from 2005 to 2019 has three most obvious periods of change: 1) from 2005 to 2013 the cluster load index of the region decreased from 4.5 to 3.1, although it can be seen that it had a period of slight growth at that time; 2) from 2013 to 2016 the cluster load index of the region slightly increased to 3.4 in 2014, and then adjusted downward to 3.2 in 2016. that is, during this period, the cluster load index of the region practically did not change significantly and had a value in the interval from 3.05 to 3.4; 3) from 2016 to 2019. The cluster load index of the region increased from 3.1 to 4.4 and had a period of rapid growth from 2016 to 2017, and also had a slight decline in 2018, possibly caused by the economic crisis of 2018.

The diagrams also show that Perm Krai increases the average value of VFD cluster load index, i.e. in Perm Krai there are already all the opportunities for the development of existing clusters and the creation of new ones within the manufacturing industry, associated mainly with oil refining, since the oil sector plays one of the main roles in the overall development of both the entire district and Perm Krai (see Figure 3).

For example, in case of modeling changes in the dynamics of growth of logistic-sectoral attractiveness, we have negative dynamics, it is necessary to exert a stronger influence on the redistribution of sub-sectors, thereby expanding the range of changes (for example, +/–0.05 (+/–5% of current values $\overline{w_{ij}}$)).

4 Conclusion

The analysis of the development of the regions of the Volga Federal District, the processes of cluster-network integration made it possible to reveal the importance of methodological and organizational tools in the processes of their development; the most important place is also given to the choice of a strategy corresponding to the current socio-economic situation of regions and industries. The inefficiency of spatial cluster-network integration is caused by the following reasons:

– advanced directions of cluster policy are financed primarily from the state budget (mainly from the federal budget), the advantages of the project-oriented approach are not used fully enough;

– poor coordination of stakeholder actions requires stimulating the development of public-private partnerships and related infrastructure, including programs for the development of human resources of relevant ministries and departments;

– the absence of a unified methodology for assessing the effectiveness of spatial cluster-network integration, adopted at the state level, complicates the implementation of industrial and regional policies, and the adoption of appropriate strategic decisions.

The presented author's methodology can be applied to all regional socio-economic systems, used by enterprises of any field of activity and all stakeholders.

In this paper, it is hypothesized that if instead of k_i we use a neural network for training M_i (this approach is presented in [5]), then control is reduced to the fact that, due to the monotonicity of the function (Eq. 5). Somehow affecting the logistics infrastructure, expressed by the density traffic flows L_i we influence on k_i, which in turn predetermines the level of GRP and the profit of enterprises-potential participants in cluster-network integration.

References

1. V. I. Borisevich, P. S. Geizler, B. C. Fateev. *The economy of the region* (Mn.: BSEU, 2002)
2. L. S. Markov. *Clusters: formalization of relationships in non-formalized production structures* (Novosibirsk: IEOPP SB RAS, 2006)
3. D. A. Mikhalev. *Modeling of processes of formation and development of regional industrial clusters*: dis. Cand. Ekon. Sciences. (Ivanovo, 2015)
4. V. P. Pervadchuk, M. Yu. Osipova, L. V. Kozhemyakin Economics and Entrepreneurship. **10** (111), 978-983, (2019)
5. L. V. Kozhemyakin Applied Mathematics and Control Sciences **4,** 137-152 (2020) DOI: 10.15593/2499-9873/2020.4.09
6. Federal State Statistics Service for the Perm Region. *Perm Krai in numbers: A short statistical collection/Territorial body of the,* Moscow, (2018).
7. FSSS. *Regions of Russia: Socio-economic indicators* stat. sb. Rosstat. Moscow, (2018)
8. K. M. Carley Simulation Modelling Practice and Theory, **10**, 253-269, (2003)
9. B. Lengyel, I. Iwasaki, M. Szanyi Journal of Economics, **2 (51),** 149-167, (2010) doi: doi.org/10.15057/18777
10. C. Moler, Numerical Computing with MATLAB, Electronic edition: The MathWorks, Inc., Natick, MA, SIAM, Philadelphia, (2004)
11. Ö. G. Sölvell, Lindqvist & C. Ketels *The Cluster Initiative Greenbook* (Stockholm: Ivory Tower, 2003)
12. R. Teigland & G. Lindqvist European Planning Studies, **15 (6)**, 767-786, (2007)

Management of generation Z staff in the digital economy

Yaroslav Kraidenkov*, and *Lyudmila* Sviridova

Nizhny Novgorod Branch of the Higher School of Economics, 25/12, Bolshaya Pecherskaya str., Nizhny Novgorod, 603155, Russia

Abstract. This article is a continuation, which examines the difference between generations and their common qualities for forming a successful team, but with a focus on generation Z. Materials on the comparison of generations X and Y are published in the journal Vestnik PNIPUV in the article Team building in the digital economy [1]. This article presents an analysis of the current state of generations X and Y with the relevant research points and recommendations for team building of personnel of generations X and Y. In 2019, at the time of the publication of the first part of the article, our team started a project to conduct their research in the part of Generation Z. Participants in this study were young people born between 2003 and 2005. The study ended in January 2021. A total of 1,14 people were interviewed during the period: high school students, university students, and first-and second-year students of universities in the cities of Moscow, Samara, Nizhny Novgorod, Novokuibyshevsk, Syzran.

1 Introduction

According to experts from the Boston Consulting Group (BCG), the Fourth Industrial Revolution is now taking place. Its features include the following:

a) formation and development of fully automated industries interacting with each other and with consumers using artificial intelligence and cloud services.

b) exponential growth of the number and scale of innovations, generated solutions of non-relevant industries.

c) exchange of ideas and experience within the framework of the onset of the digital economy and changing the perception of the company-client links.

This revolution was first mentioned by German industrialists at the Hanover Fair in 2011. Then it was said that the fourth industrial revolution is now beginning, where Germany must play a leading role. This statement is not justified, since from the creation of the European Union, the German economy has been acting as its "locomotive". To accelerate the internetization of production, the industry development strategies "Platform Industry 4.0" and the state program "Industry 4.0" have been developed and launched. This is also the predicted beginning of this trend for our economy.

2 Materials and methods

Any development begins with a foundation. Let us take a look at the main stages of the Industry 4.0 (Fig.1).

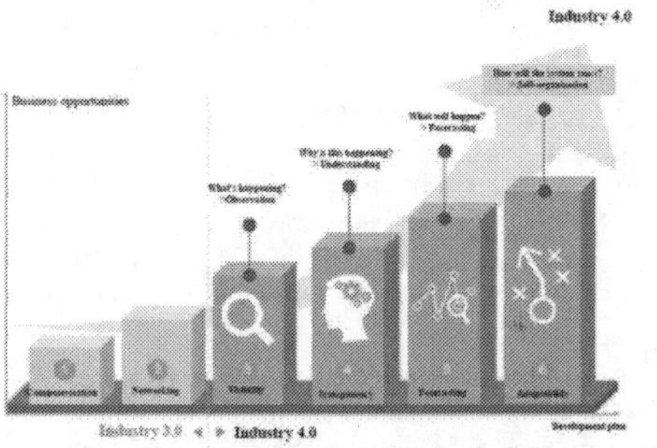

Fig. 1. Main stages of the Industry 4.0

The development and subsequent formation of the Industry 4.0 requires the results of the third industrial revolution (Fig. 1., first two stages).

Stage 1. Computerization.

This term refers to the provision of all major components of production with means for real-time digital control. The use of modern equipment initially allows switching to digital control, in the event of the use of outdated or analogue equipment, its modernization or replacement is necessary.

Stage 2. Networking. This implies the combination of isolated technologies into a single shared environment. Network interaction allows combining automatic design and production procedures with process control tools, organizing remote maintenance, etc.

Stage 3. Visibility. This concerns the creation of a digital alter ego of an enterprise for correct forecasting of situations in the future and rapid response in the present. The main difficulty in this aspect is not so much technical, but the reliability of data collection both in case of multiple data sources and in the absence of the ability to collect without human participation.

Stage 4. Transparency.

* Corresponding author: kraydenkov@gmail.com

Stage 5. Forecasting.

Stage 6. Adaptability.

Stages 4, 5, 6 may be separated for description only, since their implementation is interconnected. Transparency means linking digital display to analytical systems. It (link) allows competently extracting knowledge, patterns from the data array. This opens up the possibility of adapting analytics to production technologies (forecasting). Forecasting ability opens up the possibility of automating functions related to adapting a business to changing external conditions (adaptability).

When entering the path of the fourth industrial revolution, it is necessary to change the mentality of each individual: from performer to managers and the whole company. The cultural and social atmosphere that makes it possible to realize the advantages of the Industry 4.0 is based on two major milestones:

1. Ready for change. Openness to innovations, continuous professional growth (this concept must be distinguished from the career growth), commitment to change.
2. Free social interaction at all levels. It is ensured by a democratic style of leadership, the opportunity to openly express views and active participation in the evolutionary process.

3 Results and discussion

Analyzing the current situation with the generation Z in Western companies and comparing it with the situation with the Generation Z in Russia, we should address the peculiarities of the mental model of behavior of employees of the Generation Z in Russia.

The modern Generation Z requires a special approach in organizing their professional activities in Russian companies, which is based on the focus on emotional orientation in the management of employees of the Generation Z.

The study carried out by successful executives and companies revealed that emotionally intelligent or competent leaders achieve far greater results than those who rely more on intellectual intelligence or management knowledge. According to the results, "emotionally reasonable companies" understandable to people are increasingly superior to companies without such understanding. Today, the representatives of the Generation Z have already entered the market. The time has come to understand the peculiarities of the mental model of behavior of the Generation Z for the successful management of its development and effective management in Russian companies.

In 2019, at the time of the release of the first part of the paper, the authors started a project to conduct their research in part of the Generation Z. The participants in this study were young people born between 2003 and 2005. The study ended in January 2021. In total, 1 thousand 14 people were interviewed during the period: high school students, students of secondary educational institutions and first and second year students of higher educational institutions from Moscow, Samara, Nizhny Novgorod, Novokuybyshevsk, Syzran.

The previous paper presents the analysis of the current state of the Generations X and Y with the corresponding research areas and recommendations for team building of the Generations X and Y [1, 2, 3].

Some publications reflect the disadvantages of the theory of generations, below are two main ones and the description of how they were taken them into account in this study.

1. *The theory of generations was created by American scientists following American experience, but the United States is large in order to equally describe generations from respectable areas of California, and, for example, residents of Iowa, where farmers historically lived.* In our study, we expanded the geography of the sampling from Syzran to Moscow, while trying to maintain the same type of groups of young respondents (for example, the interviews were conducted in ordinary general academic schools: not lyceums and gymnasiums).

2. *In addition to the historical circumstances forming the personality, it is influenced by a huge number of other factors that may completely offset the first.*

Table 1. Study of the current state of generations.

Research area	Generation		
	X	**Y**	**Z**
What does it mean to be happy?			
Enjoy the respect of people	13%	7%	0%
Have a good and interesting job	33%	20%	4%
Have a good family	27%	33%	31%
Feel free	7%	13%	31%
Living well, in abundance	20%	27%	35%
What is the means of achieving the goals?			
Initiative, entrepreneurship, ability to express oneself	17%	22%	46%
Respect for traditions	3%	3%	4%
Personal efforts	6%	6%	27%
Mutual assistance, trust in people	11%	11%	0%
Family	19%	19%	0%
Professionalism	22%	17%	8%
Material security	8%	8%	4%
Responsibility	14%	14%	12%
Conditions for success?			
Good education	22%	22%	15%
Material security	19%	17%	4%
Useful contacts	11%	14%	19%
Luck (occurrence, concurrence of circumstances)	8%	8%	7%
Freedom to fulfil one's needs and abilities	14%	19%	43%
Legality as a state-mandated procedure	3%	6%	0%
Stability in the country	17%	11%	4%
Respect for moral principles	6%	3%	8%

Note:
Generation X born in 1962-1981
Generation Y born in 1986-2002
Generation Z born in 2003-2005-present

In order to reduce the statistical error from the influence of other factors, we increased the sampling to 1 thousand 14 people, which made it possible to reduce the level of influence of such factors as family, environment, financial level on the results of the study.

The authors decided to conduct their research for several reasons: firstly, to try to identify differences in generations that may affect the work efficiency of a team compiled from different generations.

Secondly, to determine the values of generations and find the causes of conflicts between their representatives.

Thirdly, to conduct studies of our citizens from our cities with our mentality in order to eliminate the influence of the differences in Western and Russian culture.

The format of earlier studies by our colleagues (described in the previous part) was taken as the basis for the present study to comply with the correctness of data comparison in the future.

Let us give a brief description of generations paying more attention to the Generation Z as the representatives of this generation are already beginning to enter the age of labor relations.

The characteristics of the generations X and Y is given in the previous paper [1]. Below are the main abstracts for these generations and their comparison with the Generation Z.

During the study, the roles of participants were determined according to the classification of R.M. Belbin.

Let us briefly describe the roles for this classification.

Type: Implementer.

Character traits: Conservative, executive.

Positive qualities: Leadership potential, practicality, energy, self-discipline.

Acceptable disadvantages: Lack of flexibility, resistance to unverified ideas.

Type: Coordinator.

Character traits: Calm, confident, manageable.

Positive qualities: Good listening skills, ability to consider and evaluate the merits of all proposals. Strong motivation to achieve the goals.

Acceptable disadvantages: Common in terms of intelligence and creativity.

Type: Motivator.

Character traits: Very nervous, responsive, dynamic.

Positive qualities: Internal drive and willingness to fight inertia, inaction and self-deception.

Acceptable disadvantages: Tendency to discontent, irritability and impatience.

Type: Idea generator.

Character traits: Individualist, serious, unorthodox.

Positive qualities: Giftedness, ingenuity, intelligence and knowledge.

Acceptable disadvantages: "Has head in the clouds", underestimates the practical details or the need for a protocol.

Type: Provider.

Character traits: Extrovert, enthusiastic, inquiring, sociable.

Positive qualities: Contacts people well and develops something new, shows resilience in a difficult environment.

Acceptable disadvantages: Loses interest in work when its initial attractiveness passes.

The study revealed the following correlation of roles (total study participants – 1,014).

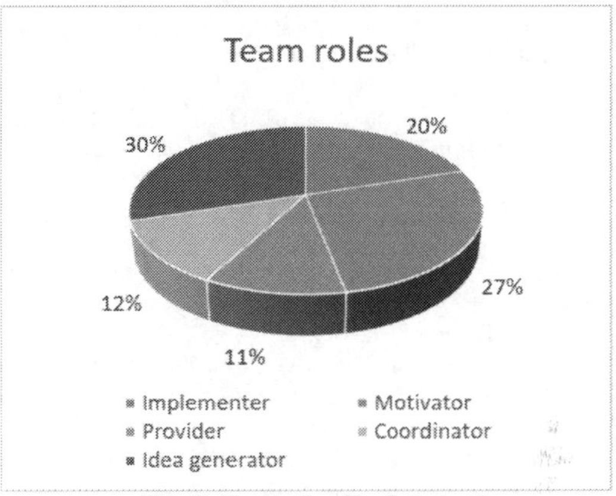

Fig. 2. Generation Z role behavior

The main findings of the study are as follows:

Similar to past generations, the new generation continues to correlate family values and wealth with the criteria of happiness, but the significance of feeling free has significantly grown. At the same time, recognition by people does not matter at all for the representatives of this generation.

As a means of achieving goals, the modern generation does not principally consider such tools as professionalism, family, mutual assistance and trust. The main focus of the generation is on initiative, entrepreneurship, personal efforts and responsibility. The main condition for success for the young generation is freedom to realize its needs and abilities. At the same time, the importance of education, material security and stability of the outside world has significantly decreased, compared with both the Generation X and the Generation Y.

Against the background of the above, the results on determining roles in the team are quite self-contradictory. The respondents least want to be motivators and generators of ideas, although it is these roles that are inherent in the need for self-realization and freedom. Denying the importance of professionalism, education and respect among people, the representatives of this generation more want to be implementers and coordinators in teams.

This contradiction makes it possible to put forward a hypothesis – either this generation has not yet decided on its roles due to its low age, or it, in principle, does not want to be determined and does not see any meaning in this. As a result, some company executives conclude that it is easiest to not deal with the employees of the Generation Z. In fact, this new generation includes thinking employees, ready to work with excessive inclusion in activities, capable of solving super-complex tasks and ready to implement ambitious projects in a rapidly changing external harm. Only a competent approach is needed in managing the employees of the Generation Z.

Recommended approaches to the human resources management of the Generation Z

The companies, especially in the field of information technology, are interested in the efficiency and professionalism of the activities of the employees of the Generation Z.

The personnel market in IT until 2020-21 has maintained stable indicators of demand for personnel. The development of IT technologies, their applicability in previously atypical areas for them and the automation of processes hold IT specialists as one of the most popular in the labor market. The number of vacancies for remote work and development outsourcing has significantly increased.

The employees of the Generation Z are characterized by the rejection of higher education and the ability to obtain the necessary knowledge from sources that they have been taught to use since childhood. Besides, they are skeptical of the curricula of higher educational institutions and choose exactly what they really need. The combination of initiative and entrepreneurship as a means of achieving the goal (expressiveness – 47%) gives them entrepreneurial abilities. Moreover, for the sake of realizing their goal, they are ready to work from a position of excess activity, performing more in the process of activity than is prescribed by the job description.

Every new employee of the Generation Z who comes to the company must understand that for him personally the work of the company is a chance to succeed. In this, the employee of the Generation Z relies solely on his own efforts, these efforts shall only be addressed in the right direction of innovative technologies and improving the efficiency of the company. To this end, HR service of the company needs to implement a number of measures taking into account the peculiarities of the mental behaviors of this generation:

– to assign to a new employee Z a mentor – a professional with significant production experience in the company, the so-called "carrier of corporate culture ";

– a professional mentor together with HR employees conduct an interview with a novice about the prospects for development and advancement within his personal career and possible prospects overcoming his misunderstanding of mutual assistance and lack of trust in people.

Practice shows (Rosatom – Greenatom, LLC VF Tanker of the Volga Shipping Company) the feasibility of creating a site that forms a new employee's understanding of the career path, for example, the analysis of the Career Tree with key competencies necessary for career growth for each type of career:

– each new employee of the Generation Z is included in the career planning system;

– the company should cultivate a talent management program in which to recognize, evaluate, develop, rotate are one of the main areas of activity in the effective management of employees of the Generation Z.

– HR service of the company assigns a development manager to each new employee, who assesses the level of his adaptation to the new operating conditions, performs monthly and then quarterly monitoring at the initial stage;

– it is desirable to create a retention system with new employees of the Generation Z taking into account the main indicator of the mental model: "Freedom to realize your needs and abilities" while achieving successful adaptation.

It is impossible to cheat and manipulate them. Quite often, the attempts to focus their activities from the standpoint of generally accepted laws and regulations (KPI system and grades) acceptable to categories X and Y are unsuccessful, result in failures for the HR service and the categorical exit of employees of the Generation Z if they were not explained their preferences from using this system openly and intelligently.

To retain a new employee of the Generation Z it is necessary to give him the opportunity to fulfill his great desire to become a realizer or integrator in projects with mandatory personal growth through increasing his personal competencies, self-development and self-actualization in project management, team building, effective communications, time management, stress management. All this should be included in the personal development plan of the new employee of the Generation Z. Thus, it is possible to eliminate the loss of interest in the work of employees of the Generation Z when its initial attractiveness passes. One important feature of personnel management in large Russian companies may lead the Generation Z employees to spontaneous exit. According to the results of the international study of personnel engagement (2011) by Gfk Tusnmark, uncomfortable working conditions in large corporations are a feature of Russian business in companies where the number of employees is 1000 or more. So, pressure from the company to work for many hours in a row is 33%, and stress at work is 47%, while in Western companies these figures are 23% and 37%, respectively.

The employees of the Generation Z distinguish 2 performance management poles that sufficiently provide control from the professional mentor and at the same time form a sense of freedom and trust in the employee so necessary due to his mental model of behavior. Control is possible with the help of trackers, which act as a regulator in the teamwork, as well as between employees, and between an employee and a mentor. At the same time, it is possible to monitor working time, physiological indicators, control the desktop, content, as well as control the history of the browser. Such sensitive and not obvious control is easily accepted by the employees of the Generation Z. However, it is obvious that at any time the employee's activities may be monitored, and this deprives the employee of the opportunity to violate the requirements imposed on him.

Today, due to the circumstances associated with the pandemic, many employees work in remotely using instant messengers and cloud technologies. The personal responsibility of staff members is of great importance in these circumstances. Despite the absence of a tight schedule, a system of trust in the employee is formed, when the result is monitored and not the process of activity itself. A state of continuous presence at work is formed, always in communication, which, nevertheless, allows perceiving this situation as a feeling of independence and the opportunity to live our lives, which is extremely important for Z-category employees: "We believe that you know how and what to do and trust you".

4 Conclusion

Continuous communication, open discussion with the employees of the Generation Z in organizing all complex issues are the main tools of HR managers and mentors. At the same time, the employees of the Generation Z do not need a lot of communication. This is especially important when setting a task. The instructions should be short with a specific follow-up algorithm and specific indicators of the desired result for all items of the proposed action algorithm. All this should be communicated to the employee of the Generation Z verbally and in writing. Communication should be short, more through social networks and video services.

The employees of the Generation Z need a qualified and competent professional – a leader who must constantly demonstrate his "advancement" in modern means of communication so that not to lose confidence and respect as a leader. But this is not enough for the employees of the Generation Z. They are oriented towards the principle of "equality" considering it possible to behave "on equal terms" with their direct supervisor if they realize their own high level of professionalism and competence. Often, they do not recognize the existence of traditional levels of management and mandatory subordination to the immediate manager. All this must be explained to the employees of the Generation Z quite patiently and logically.

Even if there is no "hot news" on the agenda it is necessary to hold scheduled meetings, individual interviews on specific results of activities and possible achievements of high results in activities for the employees of the Generation Z. It is necessary to make sure that the employees of the Generation Z feel continuous support and understanding at all levels of management provided absolute transparency and openness of information. Even with the slightest concealment of information, the employees of the Generation Z may permanently lose confidence in company management [4, 5].

The assessment of professionally important and personal qualities of the employees of the Generation Z showed that high motivation to achieve successful results in the activity is formed in the process of implementing ambitious projects, which are based on complexity on the verge of real opportunities and appropriate remuneration (Freedom to realize their needs and abilities – 43%, Initiative, entrepreneurship, ability to express oneself – 46%). Often the employees of the Generation Z accept the "Principle of long reins". At the same time, under no circumstances should the immediate leader loose these "reins".

Besides, it should be understood that the existing system of benefits and compensation for most employees, including the Generations X and Y, may be absolutely indifferent to the Generation Z.

The assessment of professionally important and personal qualities of working employees of the Generation Z showed that high motivation to achieve successful results in the activity is formed in the process of implementing ambitious projects, which are based on complexity on the verge of real opportunities and appropriate remuneration (freedom to realize their needs and abilities – 43%, initiative, entrepreneurship, ability to express oneself – 46%).

The means of achieving these goals include initiative, entrepreneurship, the ability to express themselves, and orientation to their own efforts. Here it is important to consider the competent formation of their internal motivation to achieve successful results in their activities through the formation of high self-esteem and a high level of goals [6, 7].

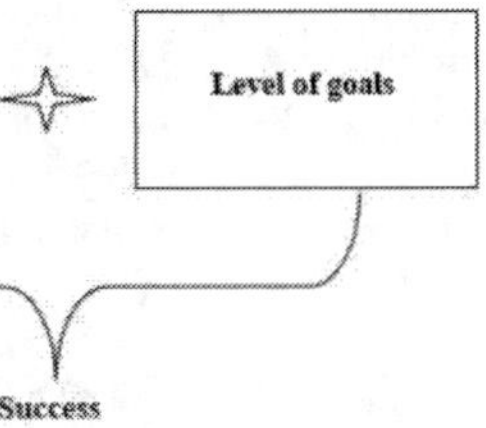

Fig. 3. Internal motivation for successful performance

The combination of a high level of self-esteem when the employee of the Generation Z sees himself as capable of achieving significant success, taking a leading position and displaying perseverance and tenacity in achieving difficult tasks, active, able, entrepreneurial, and high-level goals, when the employee of the Generation Z has a large goal in life formed for several years and constantly achieves it or the main goal for life, a number of intermediate goals that facilitate the achievement of the main, then he becomes a good specialist, active, entrepreneurial, may be a leader, harmonious and productive in work, and all these combined "condemns" the employees of the Generation Z to success.

$$\text{Self} - \text{esteem} = \frac{\text{Success}}{\text{Aspirations}}$$

Self-esteem is a positive part of the self-concept, i.e. the system of views and ideas of an employee of the Generation Z about himself. How to increase self-esteem? Reducing aspirations is an unconstructive path. The development of self-justifying trends (external causal attribution) is absolutely inappropriate and not characteristic of the employees of the Generation Z. Psychological barriers lead the individual to deconstructive behavior (aggressiveness, social apathy, "red eye syndrome").

The orientation of the employees of the Generation Z to the success associated with the development of the personality, i.e. the way to realize himself, self-actualization, makes them effective and competent, which in turn increases the efficiency and creates the key and leading competencies of the company, thus making it successful in conditions of fierce competition.

References

1. Y.A. Kraidenkov, L.V. Sviridova, *Bulletin of PRPU. Socio-economic sciences.* **4,** 314-324, (2019)
2. N.V. Korzh, E.V. Shanina *Social Sciences. Sociology,* **1(37)** (2016)
3. G.B. Kosharnaya, N.V. Korzh, *News of higher educational institutions. Volga region. Social sciences,* **1 (33),** 2015
4. A. Mersino, *Emotional intelligence for project managers. Practical Guide* (Moscow: Mann, Ivanov and Ferber, 2018)
5. A. Edmondson, Teaming: How Organizations Learn, Innovate, and Compete in the Knowledge Economy (Moscow: Sberbank Library, 2016)
6. *Building Leaders for the Next Decade: How to Support the Workplace Goals of Gen X, Gen Y and Gen Z* Retrieved from: https://www.insead.edu/centres/emerging-markets-institute/research
7. *Brave New Workplace: A Look at How Generations X, Y and Z are Reshaping the Nature of Work* Retrieved from: https://www.insead.edu/centres/emerging-markets-institute/research

Social identification factors of managers and entrepreneurs

Evgeniia Lazukova*, and *Guzel* Seletkova

Perm National Research Polytechnic University, 29, Komsomolsky ave., Perm, 614990, Russia

Abstract. The social identification of managers is divided into deep and shallow. The authors give the comparative analysis of the social identification of managers and entrepreneurs through its types, as well as through the determination of these types by various factors. Among entrepreneurs, the shares of those who have and those who do not have a conscious, deep identification with their social and professional group are approximately the same. The same situation is among managers. In addition, the authors studied the influence of gender, age, material situation, level of education, secondary employment and other factors on the determined types. Gender does not affect these types of social identification. The real factors also include the additional work, age, the use of competencies, satisfaction with the results of activities and membership in professional associations.

1 Introduction

The scientific literature examines the impact of institutional [1] and cultural [2] differences on the development of entrepreneurship. Akerlof and Kranton [3] introduce the concept of identity into the function of economic effectiveness. Identity plays an important role in the explanation of the motivation to choose entrepreneurial activity with the subsequent influence on the economic future of a person [4].

The strategic business decisions are determined by a personality of an entrepreneur, his reaction to changes, therefore, the identity of entrepreneurs is important for the survival of business [5, 6, 7, 8, 9, 10, 11]. An entrepreneur and an enterprise arise as a function of constant experience, while an enterprise creates an entrepreneur, as an entrepreneur creates an enterprise [12]. Thus, it is necessary to note the importance of the personality of an entrepreneur and social identity and identification in particular.

There are scientific studies that reveal the connection between the social identification of entrepreneurs by various motives and results of entrepreneurial activity [13, 14, 15, 16, 17, 18], with the way of thinking [19] and the meaningfulness of entrepreneurial experience [20, 21]. The start of your own business involves changing roles and building the identity of the role of a founder of a company [22, 23]. Scientists also note the importance of the study of identity not only as fixed characteristics, but also as a process and its mechanism [24].

Some researchers note that entrepreneurs, when starting and running a business, try on different roles that become a part of their identity. During the study of the process of the choice of a role and formation of an identity, three mechanisms are distinguished: "the perception of an entrepreneur as a person who "takes off his hats", the discovery of a new meaning (new role identity) within the enterprise and imprinting of role identity" [25].

2 Materials and Methods

The process of identification with a professional group can be deep or shallow. Our empirical sociological research among leaders and entrepreneurs in Perm conducted in 2019-2020 showed what it depends on.

The social identification of these two groups was divided into formal (shallow, associated with mentioning a profession/position only to a direct question about it) and into a conscious one (deep, where a profession/position is also mentioned in the answer to the question: "Who am I?"). In accordance with this, 4 types of respondents were singled out according to social identification: "formal leader", "perceived leader", "formal entrepreneur" and "perceived entrepreneur".

We made an attempt to find out the influence of various factors on the identification of managers and entrepreneurs comparing these two socio-professional groups.

3 Results and Discussion

The analysis of the research results did not reveal the influence of th gender of respondents on the type of social identification. However, the influence of other socio-demographic and socio-economic factors was revealed (Table 1).

According to the research data, as the professional experience increases, the share of the type of "perceived manager" grows (from 27 to 40%) and the share of the type of "formal entrepreneur" decreases (from 20 to 13%). The proportions of the remaining types do not significantly depend on professional experience.

* corresponding author: e.las@mail.ru

Table 1. Influence of various factors on the types of social identification (in%).

Factors	Conscious entrepreneur	Formal entrepreneur	Conscious manager	Formal manager
Gender				
Male	32	19	30	19
Female	37	16	33	14
Age				
Up to 39 years old	28	13	38	21
40 and older	43	19	28	10
Professional experience				
Up to 5 years	36	20	27	17
Over 5 years	31	13	40	16
Education				
Secondary	40	20	20	20
Vocational secondary education	50	22	16	12
Higher education and incomplete higher education	30	17	36	17
Marital status				
Married (including unregistered marriage)	34	19	34	13
Not married	33	14	30	23
Secondary employment				
...is present	28	18	34	18
...is not present	42	17	28	13
Professional associations				
...members	20	20	20	40
...not members	35	17	33	15
Satisfaction with the results of their activities				
...satisfied	36	18	33	13
...not satisfied	29	18	21	32
In case of other sources of income ...				
...agree to work	24	9	38	29
...not agree to work	38	21	30	11
Job...				
... interesting and important, regardless of payment	39	17	36	8
... - is important, but there are things much more interesting than work	31	14	32	23
... is the main source of living	33	21	29	17
Competences...				
...are being realized	33	17	33	17
... are not being realized	0	50	33	17
Attitude towards a profession				
A profession is the best or one of the best	35	12	34	19
A profession is neither better nor worse than others	31	22	31	15
A profession is worse than others	67	0	33	0
Attitude to market conditions				
Negative	24	6	29	41
Neutral	24	23	35	18
Positive	42	15	32	11
Attitude towards a role of state				
State assistance to one degree or another is allowed	41	21	19	19
State assistance to one degree or another is not allowed	31	16	38	15
Succeed in the professional field, ...				
... considering laws, goals of business and employees	31	17	34	18
... using any means	44	22	25	9
The decision of your children to go into business ...				
...would be approved	36	16	30	18
... would not be approved	12	15	58	15
... is found difficult to answer	52	19	19	10

The level of education has some influence on the studied types. Among those with a low level of education (among those who graduated from school or college/technical school), the most common type is the "conscious entrepreneur" (40 and 50%, respectively). At the same time, among those who have got or are getting higher education, the most popular type is the "conscious mamnager" (there are more than a third of them - 36%).

The absence of marriage increases the chances of getting into the "formal leader" type. In relation to other types of social identification, the influence of marital status was not revealed.

Secondary employment has some influence on the studied types. The presence of additional work significantly increases the chances of getting into the type of " conscious entrepreneur" (from 28 to 42%) and slightly reduces the chances of getting into the type of " conscious manager" (from 34 to 28%).

Table 1 also presents the influence of membership in professional associations on the types of social identification. Membership in professional associations reduces the share of « perceived» types (from about 35 to 20%), and also more than doubles or increases the share of « formal types » within the statistical error (of the « formal manager » type from 15 to 40% and of the « formal entrepreneur » from 17 to 20%).

The satisfaction with the results of their activities affects the characteristics of the identification of managers and entrepreneurs. It has a beneficial effect on social identification, because it increases the possibility to transform into "conscious" types when the "formal" types are reduced or unchanged.

Among those willing to work in the presence of other sources of income, the most common type is a "conscious manager" (38%); among those who do not agree to work in this case, the type of "conscious entrepreneur" is most widely represented (38%); this, apparently, refers to the high importance of the material side of labor for conscious type, which is less typical for those who are identified as a manager.

During the study of attitudes towards work, influencing the types of social identification, some patterns are determined. The types that have a stable identification with the socio-professional group ("conscious entrepreneur" and " conscious manager") demonstrate not so strong, but rather stable dependence on the factor. These typological groups among respondents who are interested in work and emphasize the importance of work are the most representative (a concious entrepreneur - 39% and a concious manager - 36%) in comparison with similar groups of respondents who have a different attitude to work (there are things that are much more interesting than work - 31% and 32%, respectively; work is a source of living - 33% and 29%, respectively).

Whereas the respondents who belong to the typological groups « formal entrepreneur » and « formal manager », on the contrary, are more inclined to demonstrate their attitude to work as a source of income (21% and 17%, respectively) in comparison with similar groups of respondents with a different attitude to work (there are things that are much more interesting than work - 14% and 23%, respectively; work is interesting and important - 17% and 8%, respectively).

These patterns can be interpreted as the evidence of a more pronounced terminal labor motivation (in comparison with instrumental motivation) in typological groups with a deeper social identification ("conscious entrepreneur" and "conscious manager"). Typological groups with shallow social identification ("formal entrepreneur" and "formal manager"), on the contrary, have a greater tendency towards instrumental labor motivation (in comparison with terminal motivation).

The types of identification are associated with the implementation of existing competencies. In the group of those who implement their existing competences (knowledge, skills, and abilities) in their profession, "concious entrepreneurs/ managers" prevail (one third or each type); while among those who do not their competences, the type "formal entrepreneur" dominates (it presents a half). Consequently, a deeper identification with an entrepreneur and a manager is associated with the application of their knowledge, skills, and possessions.

The contradictory results are revealed during the study of the influence of such a factor as attitude to the profession on the types of social identification. Among the respondents who have a positive or neutral attitude to their profession, there are practically no fluctuations in the representativeness of typological groups. The exception is the typological group of "formal entrepreneurs", which is more representative in the case of a neutral assessment of their profession than a positive one (22% and 12%, respectively).

However, there is a significantly larger share of the type of "conscious entrepreneur" among respondents who have a negative assessment of their profession (67%) in comparison with those who have a neutral (31%) and positive (35%) assessment. It is assumed that such a shift in the assesments should be regarded not as evidence of a stable trend, but as an isolated case due to the small number of respondents who have a negative attitude to their profession, which makes these patterns statistically insignificant. In this case, individual representatives of the typological group show a pronounced instrumental motivation, since, under the condition of clear negative attitude towards their profession, the respondents demonstrate a stable identification with their social group.

According to the results of our research, the types of the considered identification are determined by the relation to market conditions. Those representatives with a negative attitude to the market are mainly refer to the type of "formal manager" (41%); as a rule, the carriers of a neutral attitude are the representatives of the "concious manager" type (35%).

It is necessary to note that among those who positively relate to market conditions in Russia, the representatives of the "concious entrepreneur" type prevail (42%), which is explicable both from the point of view of objective dependence, the possibility and efficiency of running a business on the level of development of market relations and above mentioned dependence, which shows that the type of conscious entrepreneur is the most representative among the respondents with the highest level of well-being.

Among the respondents who believe that the state assistance to its citizens is permissible, the type of "conscious entrepreneur" prevails twice over other types (41%). We think that this feature of Russian entrepreneurship in many respects contradicts the specifics of its activities related to the readiness to take initiative. Among the supporters of the responsibility of citizens themselves (not state) for the life and well-being of an individual, the type of "concious mamager" has a greater value (38%).

We can assume that the presence of this regularity in the type « concious manager » can be explained by the strong connection of this type with such characteristics as diligence (Pearson's correlation coefficient 0.74); the ability to find a common language with other people (coefficient 0.72); as well as the average relationship with assessments of organization (coefficient 0.68); self-confidence (coefficient 0.66); the ability to lead people (coefficient 0.61). Thus, the representatives of this typological group probably place greater responsibility on a person than on the state.

There are different types of respondents among those who have different ways to achieve success in their professional sphere. The modal group among those who use any means to achieve their goals in the professional sphere is the type of "conscious entrepreneur" (44%). Among those who are not ready to " stop at nothing," 2 types prevail, represented by about a third: "a conscious manager" and "a conscious entrepreneur" (34 and 31%, respectively). Moreover, among those who claim that they achieve their goals by any available means, the proportion of types of « entrepreneurs » is greater, and on the contrary the types of «managers» are less. Apparently, an unconditional orientation towards success is more typical of businessmen, including against a particular individual. Among those who responded positively to the question of the hypothetical decision of their children to become entrepreneurs (as well as among those who found it difficult to answer this question), the most common type was the "conscious entrepreneur" (36 and 52%, respectively). In the first group, it seems that these are entrepreneurs who are satisfied with their professional activities, and in the second group are those who find advantages and disadvantages in it.

Most of the representatives of the "conscious manager" type would not approve the attitude of their children towards this question. Apparently, the rejection of entrepreneurship led them to management activities. It is worth noting that shallow identification of the type of "formal entrepreneurs" significantly reduces their share among those who would approve the decision of children to do business (from 36 to 16%) and among those who have not formed an opinion on this issue (from 52 to 19%).

At the same time, Oliver notes that the formation of entrepreneurial identity must begin at an early age, since the entrepreneurial characteristics of parents and peers have a significant impact on the formation of identity [4].

4 Conclusions

In the activity of a manager, his identification with a professional group plays a special role. According to the study, the following conclusions can be drawn.

The additional work and secondary employment changes the proportion of types with deep identification: for entrepreneurs it increases, for managers it decreases. The increase in age slightly affects the depth of identification as it increases the proportion of types of entrepreneurs, while reducing the proportion of types of managers. Membership in professional associations increases shallow identification with one or another social and professional group and the use of competencies and satisfaction with the results of activities increases deep identification.

References

1. S. Estrin, T. Mickiewicz. Small Business Economics. **37,** 4, (2011). doi: 10.1007/s11187-011-9373-0
2. R. Shinnar, O. Giacomin, F. Janssen Entrepreneurship Theory and Practice. **36(3)**, 465-494. (2012). doi: 10.1111/j.1540-6520.2012.00509.x.
3. G.A. Akerlof, R.E. Kranton. Quarterly Journal of Economics. **105(3)**, 715–53. (2000)
4. O. Falck, S. Heblich, E. Lüdemann CESifo Working Paper Series. **2661**. (2009)
5. E. Fauchart, M. Gruber. Academy of Management Journal, **54(5)**, 35-57. (2011)
6. L. N. Chaplin, D. R. John. Journal of Consumer Research, **32(1)**, 119-129. (2005)
7. M. Levy, P. Powell. *Strategies for growth in SMEs: The role of information and information systems.* (Oxford, UK: Information Systems Series (ISS), Elsevier, 2005)
8. R. K. Singh, S. K. Garg, S. G. Deshmukh Benchmarking: An International Journal, **15(5)**, 525-547. (2008)
9. D. Vargas, R. G.-T. Rangel Technological Forecasting and Social Change, **74(1)**, 90-99. (2007)
10. E. E. Powell, T. Baker Academy of Management Journal, **57(5)**, 1406-1433. (2014)
11. C. Slade. *Creating a brand identity: A guide for designers.* (London, UK: Laurence King Publishing, 2016)
12. M. H. Morris, D. F. Kuratko, M. Schindehutte,, A. J. Spivack. Entrepreneurship: Theory and Practice, **36(1)**, 11-40. (2012)
13. G.A. Alsos, T.H. Clausen, U. Hytti, S. Solvoll. Entrepreneurship & Regional Development, **28(3)**, 234-258. (2016)
14. M.E. de la Cruz, A.J.Verdú Jover, J.M. Gómez Gras. European Research on Management and Business Economics, **24(2)**, 90-96 (2018
15. M. Gruber, E. Fauchart. Academy of Management Journal, **54(5)** (2011)
16. J. Musona, K. Puumalainen, H. Sjögrén, A. Vuorio. MDPI, Open Access Journal, **13(2)**, 1-37 (2021)
17. Ph. Sieger, M. Gruber, E. Fauchart, T. Zellweger. Journal of Business Venturing, **31(5)**, 542-572 (2016)
18. M. S. Cardon, J. Wincent, J. Singh, M. Drnovsek. Academy of Management Review, **34(3)**, 511-532 (2009)
19. R. Korte. Advances in Engineering Education, **7(1)** (2018)
20. S. Korhonen, T. Leppäaho. Journal of International Entrepreneurship, **17(3)**, 355-388 (2019)
21. P. Gregori, P. Holzmann, M.A. Wdowiak. Journal of Business Research, **122(C)**, 488-501 (2021)
22. H. Hoang, J. Gimeno. Journal of Business Venturing, **25(1)**, 41-53 (2010)
23. M. Cardon, D. Gregoire, C.E. Stevens, P. C. Patel. Journal of Business Venturing, **28(3)**, 373-396 (2013)
24. C. M. Leitch, R. T. Harrison Entrepreneurship & Regional Development, Taylor & Francis Journals, **28(3-4)**, 177-190 (2016)
25. B. D. Mathias, D. W. Williams Journal of Business Venturing, **33(3)**, 261-277 (2018)

Dynamics of competitive forms of social partnership and their management in modern Russian society

Valery Levchenko*

Perm National Research Polytechnic University, Komsomolsky Avenue,29,Perm,614990, Russia

Abstract. The author proposes to consider the process of optimization of social partnership management with regard to its competitive forms, relying on fairly representative and fairly numerous results of specific sociological studies carried out over more than three decades. The structure of competitive forms of social partnership and its changes over several decades in modern Russian society are analyzed.The main determinants defining the development of competitive forms of social partnership are presented. The main consequences (effects) of various competitive forms of social partnership for the productivity of activities, interpersonal relations and personal development are revealed.In particular, it is found that the strongest positive effect on practically all aspects of human life is provided by cooperative forms of social partnership and passive forms of competitiveness and individualism have the greatest negative effect. It is proposed to use the revealed patterns of competitive forms of social partnership in order to optimize management both in the system of social partnership and social interaction in general.

1 Introduction

Social partnership with regard to socio-economic analysis can be presented as a kind of manifestation of subject-subject relations, which can take various forms [1; 2].Accounting and optimal use of various forms of social partnership play a fairly important role in the management system in modern society.One of the most universal options for the analysis of the forms of social partnership is their consideration from the standpoint of competition [3; 4; 5].

The author propose to consider competitiveness as a category that has particular types of its manifestation, as a special form of «co-activity» of people, carried out through the implementation of mechanisms for comparison, validation and regulation by the subject of activity in the process of exchange with the environment, serving as the basis for the changes in his activity and the quality of interconnection with other subjects and which is a consequence of two ambivalent properties typical of a social subject: the property, expressed in the desire to distinguish oneself and the property, manifested in the tendency to interact [3; 6; 7]. At the same time, it is proposed to single out four main competitive forms of social partnership as particular manifestations of competitiveness on such grounds as the direction and intensity of interaction: competitiveness, altruism, cooperation, individualism. The competitive form of social partnership is an expression of one of two tendencies inherent in a subject - the demonstration of the private interest of a subject (when a subject pursues his private, individual goals, usually against the interests of other subjects) in the process of comparison, evaluation and regulation of his activities.The opposite competitive form of social partnership can be considered as altruism and it is also a one-sided manifestation, but already of a different tendency inherent in a subject of social interaction (a tendency towards integration, subordination to a group, often against oneself and one's interests). Theoretically, the optimal competitive form of social partnership is cooperation, which is a mechanism for mutual stimulation of the activity of subjects in the process of comparison, evaluation and regulation of their activities, when the maximum possible (but not necessarily the same) simultaneous realization of two opposite tendencies inherent in a subject in relations of social partnership occurs: self-realization of a subject and the realization of common interests.

In other words, subjects achieve their goals only when other partners also achieve them, when maximization of mutual benefits takes place and not their equality.Another competitive form of social partnership is individualism, when a partner achieves his goals without any connection with the achievements of others [3; 8]. There is also an intermediate competitive form of social partnership, when the subjects of competitive relations are usually inferior in something to each other, but at the same time they ensure the partial realization of their goals and interests.

2 Results and discussion

The research on the basis of the Department and Laboratory of Sociology of Perm National Research Polytechnic University which is conducted regularly from 1979 to 2017 covered more than 13,000 people. The competitive forms of social partnership in various spheres of life (socio-political, socio-cultural, leisure,

* Corresponding author: levv66@mail.ru

etc.) were studied, but the main attention was paid to their peculiarities in professional sphere.

The analysis of changes in the competitive forms of social partnership over the past three decades has revealed the following patterns. Firstly, almost all types of competitive forms of social partnership have been constantly determined for more than three decades. If we summarize all the data for the specified period of time, then the distribution of competitive forms of social partnership can be represented as follows (Figure 1).

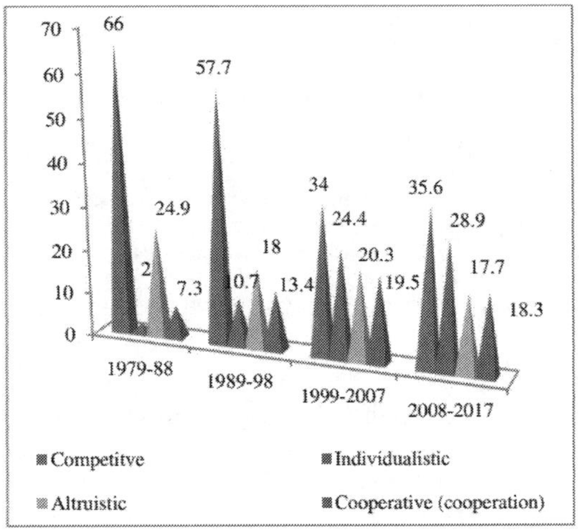

Figure 1. Dynamics of competitive forms of social partnership in modern Russian society from 1979 to 2017,%

Secondly, if we analyze specific changes in the competitive forms of social partnership for a given period, then it is necessary to note the predominance of competitive forms throughout the entire time period with a gradual decrease in their level (first of all, passive forms of competition), a certain decrease in passive (reactive) forms of altruism, a gradual increase in active cooperation and a confidently significant increase in the number of respondents with individualistic positions.

We also should pay attention to one more tendency to change the competitive forms of social partnership - a significant consistent increase in active forms of competition and, at the same time, a methodical reduction in passive forms of social partnership (Figure 2).

It is possible to note that the results of our research generally coincide with the limited data on the distribution of competitive forms of social partnership obtained by G.E. Belitskaya [9] and J. Reikovski [10, p. 29–32], and reflect the main trends of their changes in modern society. In particular, J. Reikowski notes that "in modern society: Polish, Russian and other former countries of "socialist direction", there is a tendency towards the complete destruction of the collectivist orientation and the progress of individualization" [10, p. 29-32]. G.E. Belitskaya revealed that in modern Russian conditions the predominant type of personal consciousness of social problems is competitiveness («I am the subject» - «society is an object»), which is an «active, empirical, optimal type» with a predominance of individual values [9, p. 60–62]), clarifying and specifying the outlined changes. Thus, the turning point (that is, when the most significant changes in the development of competitive forms of social partnership took place) can be considered as the time from 1991 to 1998. Since 1999, a certain stabilization in the state of competitive forms of social partnership begins.

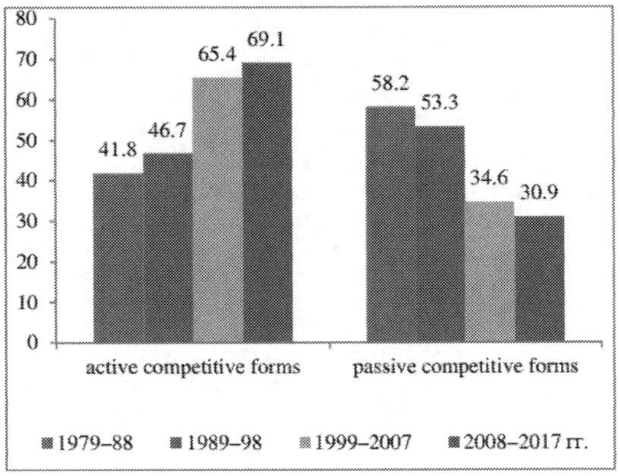

Figure 2.Distribution of active and passive competitive forms of social partnership from 1979 to 2017,%

It is precisely these periods in the development of Russian society that are considered to be as follows: transitional (crisis) - the beginning and mid-1990s, and the beginning of stabilization (the beginning of the XXI century). Thus, competitive forms of social partnership may be one of the most accurate and unified indicators of social development of society at different levels.

Thus, the proposed approach to the consideration of this phenomenon can be used to optimize the process of management (regulation) of social interaction at all levels (from interpersonal to interstate) and in all areas of joint activity (taking into account the characteristics of each of the competitive forms of social partnership allows more accurate and a differentiated approach to the problem of the improvement of the organization of joint activities).The analysis of the development of competitive forms of social partnership in various social groups also allowed concluding that competitive forms of social partnership can be considered as a significant and sufficiently sensitive to the influence of various factors (from factors of the macroenvironment to intrapersonal) of differentiating criterion of intragroup structuring.

The results of these studies indicate that competitive forms of social partnership are not just one of the indicators of the formation of joint relations, but are one of the most accurate and sensitive indicators of this social (and socio-psychological) phenomenon. It was found that competitive forms of social partnership are one of the leading determinants of social (group) structuring. Moreover, competitive forms of social partnership are at the same time one of the main sources of the development of joint relations in society (due to the internal contradictions inherent in them), and one of the most accurate and reliable indicators of the level of development of joint relations.

Thus, at the initial stage of the development of joint relations (the stage of scattered, unformed social relations, the stage of diffuse group), such a competitive form of social partnership as competitiveness is most often spread. Such a competitive form of social partnership as altruism usually prevails at the intermediate, marginal level of the formation of joint relations (a stage of association). At the highest stage of development of social relations (in a highly cohesive social group, collective, team), such a competitive form of social partnership as cooperation often prevails. First of all, a higher level of development of social relations is determined by a higher integrity of functioning in subjects of such a competitive form of social partnership as cooperation. Since this very form of social partnership (cooperation) has more complete (holistic) involvement of the essential properties of the subject of competition - simultaneous maximum involvement in social partnership relations of not one, but both tendencies typical of the subject of cooperation.

Thus, the internal contradiction of competitive forms of social partnership, which has constant asymmetry, acts as one of the main sources of self-development and joint relationships. Cooperation, a the most progressive competitive form of social partnership largely determines the development of the most progressive form of social relations - communitarianism, « real collectivity », a real team and the highest stage of development of social partnership.

Such objective and subjective factors as the state of the wide social environment at the level of society as a whole, the form of ownership of the organization, social and professional status (content of work, profession), level of education, involvement in power relations (participation in management, political activity, etc.), value orientations of subjects, personality traits etc. have the most significant effect on the formation of competitive forms of social partnership [3].

The performed analysis of the effect of various factors made it possible to clarify the system of the main determinants of each competitive forms of social partnership to a certain extent. In particular, competitiveness is most intensively formed and developed among employees of state-owned organizations (in comparison with organizations of other forms of ownership), among women (in comparison with men), among young people (in comparison with older age groups), among employees with secondary education (in comparison with workers with higher education), among respondents with a predominance of such life values as "career" and "personal business qualities", among respondents with radical forms of political activity and such personality traits as high "irritability", high "sociability", a high level of "self possession", lack of "shyness", with a high level of "need for achievement".

The development of competitiveness is negatively affected by joint-stock ownership, professional membership in a group of specialists or managers, predominance of orientation towards work or profession, traditional legitimate forms of political activity, and such personality traits as "masculinity".

The formation of individualism is positively influenced by the state form of ownership of an enterprise, an uncertain position of participation in management, belonging to a group of women workers with a secondary education, paternalism, indifference to politics and low level of "spontaneous aggressiveness". Membership in a professional group of specialists or leaders, radical forms of political activity, a high level of "need for achievement" negatively affect the development of individualism.

Joint-stock ownership of an enterprise, participation in management, belonging to a group of older men with secondary general or higher education, employees or managers, predominance of orientations towards "work, profession" or "career", legitimate forms of political activity, pronounced "masculinity", a low level of "irritability" and "irritability" have a positive effect for altruism. The state form of ownership of an enterprise, young age (up to 30 years), belonging to a professional group of workers, political passivity and some personality traits (high "reactive aggressiveness", "irritability", "sociability") negatively affect the formation of altruism.

The development of cooperation is positively influenced by joint-stock ownership, participation in management, belonging to a social group of men, specialists or managers with higher education, orientation towards "work, profession", "culture, education", "personal business qualities", relying on "their own forces (not state assistance)", legitimate forms of political activity, a high level of personality traits such as "self-possession", "masculinity", "need to achieve", a low level of "irritability".

The following factors as state ownership, membership in a professional group of employees, radical forms of political activity, a high level of "reactive aggressiveness", "sociability" have a negative impact on the formation of cooperation.

More important patterns were identified during thestudy of the influence of competitive forms of social partnership on performance, interpersonal relationships and certain aspects of personality development. In addition, the scheme used for the analysis of competitive forms of social partnership made it possible not only to bring some clarity on one of the most confusing problems of competitivenessi.e. the issue of the consequences of the effect of competitive forms, but also to clarify significantly the differentiating effect of each of the twelve competitive forms of social partnership in various spheres of life such as the productivity of activity, joint group existence and personal development.

It is found that such a form of competition as active cooperation is determined by the strongest optimizing and unambiguous positive influence on all aspects of the life of a subject such as the effectiveness of his activities, interpersonal relations and the state of a personality. Other competitive forms of social partnership are not only inferior to cooperation in terms of the strength of their impact, but also do not have such unambiguous consequences in various spheres of life. Thus, competitive forms of social partnership have a rather strong, although second largest optimizing effect on objective performance indicators after cooperation.

At the same time, the subjective assessment of these consequences, as well as the impact on the socio-psychological climate of the group and personal development, is lower than similar indicators in subjects with a mindset for cooperation and altruism. Altruistic forms of social partnership, being inferior to cooperation and competitive forms of social partnership in terms of their impact on objective performance indicators, nevertheless constantly rank second place in terms of their beneficial effect on the socio-psychological climate, interpersonal relations and personality development.

That is, the subjects with an orientation towards altruistic forms of social partnership are characterized by a more optimistic subjective position in relation to their activities and social environment. The influence of such a competitive form of social partnership as individualism on the productivity of activities and the socio-psychological state of a group is relatively weak and is inferior to the strength of the influence of other forms of social partnership. We can also pay attention to some increase in the negative impact of the individualistic form of social partnership on the indicators of personal adaptation.

There is also a stronger positive effect on the performance, interpersonal relations and personality development of active competitive forms of social partnership compared to the influence of similar passive competitive forms of social partnership. Thus, it can be noted that the most controversial competitive form of social partnership in terms of its consequences is passive altruism and predominantly negative effects are typical of such passive competitive forms of social partnership as competitiveness and individualism.

At the same time, the passive competitive form of social partnership has a more negative effect on the socio-psychological state of a group and the passive individualistic form of social partnership - on the well-being of an individual [8]. The discovered differentiating effect of various competitive forms of social partnership can be explained by their inherent essential characteristics: the completeness of the implementation of essential properties of a subject of competitive forms of social partnership and the level of their intensity. In turn, the revealed patterns of functioning of competitive forms of social partnership can (and should) be used to optimize the management of social and interpersonal interaction.

3 Conclusion

It turns out that competitive forms of social partnership are one of the essential differentiating features of intragroup structuring and social development, which, on the one hand, is sensitive to the influence of macro- and microconditions. On the other hand, competitive forms of social partnership have a strong and fairly unambiguous effect on the main spheres of human activity (performance, social relations and personality development). They can reasonably be considered as one of the indicative and universal factors of the social and socio-psychological development of a particular subject of interaction and one of the main sources of activity of subjects of joint activity.

References

1. N.A. Kolesnikova, Nesterov Yu.A. *Socio-economic and humanitarian journal of Krasnoyarsk State Agrarian University,* **1(7),** (2018)
2. D.G. Pruitt *Negotiation behavior*(New York: Academic Press,2013)
3. V.V. Levchenko, Bulletin of the Peoples' Friendship University of Russia, Psychology and Pedagogy, **2**, 32-36, (2010)
4. R. Axelrod, *The evolution of cooperation, New York, Perseus Books Group,* (2006)
5. A.G. Shmelev, *Productive Competition. Experience of constructing a unifying concept, Institute «Open Society»,* **55**, (Moscow: Magister, 1997)
6. B.D. Bonta, *Psychological Bulletin,* **121(2)**, 299-320, (1997)
7. M. Deutsch, *The Handbook of Conflict Resolution: Theory and practice,* 3rd Edition (San Francisco: Jossey-Bass, 2014)
8. V.V. Levchenko, O.V. Yuryeva, E.N. Rastorgueva, J.S. Neverova, L.Y. Talipova, *Espacios,* **40(25)**, (2019)
9. G.E. Belitskaya, *Psychological journal,* **15(4),** 54–63, (1994)
10. J. Reikovski, *Journal of Psychology,* **14(5)**, 23–34, (1993)

Determining the potential of organizational structures models of a high-tech cluster

Elena Lobova*, *Vladimir* Postnikov, *Stanislav* Rybin, *Irina* Reshetnikova

Perm National Research Polytechnic University, 29, Komsomolsky Prospekt, Perm, 614990, Russia

Abstract. As a rule, during the choice of the most viable organizational structure of a cluster, traditional methods of the assessment of economic efficiency are used, taking into account the influence of financial, production and technological risks. The high level of the fluctuations of social, political, economic and technological impacts of the intercountry level has a strong impact on the result of their production activities. The peculiarities of this influence include the emergence of new institutions of interaction between the subjects of a cluster, the growth of uncertainty within the system and a change in the level of specificity of the assets of its participants. The task to find new tools for the assessment of the potential of economic security of a cluster is becoming urgent. It is necessary to find such tools which would take into account the ability to adapt to technological and geopolitical challenges in the future. Within the framework of this study, the authors developed an algorithm for the assessment of the potential of alternative models of the organizational structure of a high-tech cluster. The important methodological aspect of the process of forecasting the level of adaptability of alternative cluster structures to external changes is the condition for the availability of institutional support to promising areas of technical development. The developed algorithm presupposes the further development of the methods of system analysis and the formation of Institutional Atlases for the parameters of the assessment of the opportunistic behavior risks of subjects or synergistic effects. The use of this algorithm allowed drawing some important intermediate conclusions about the unequal composition and the influence of dominant institutional factors in modeling different organizational structures of a high-tech cluster.

1 Introduction

Cluster innovation structures are a considerably young subject for research in economics. The implementation of the sustainable development paradigm has a comprehensive impact on the formation of innovative clusters. It presupposes a flexible and deep transformation of technological development directions taking into account the global tendencies of the enhancement of environmental standards, post-crisis trends towards economic stabilization, the complication of the principles of constructing managerial tasks and special integral cultural values.

The relevance of the research topic is in the fact that in order to form breakthrough innovative technologies that determine the vector of scientific and technological progress, it is necessary to build such models of interrelations that would have the maximum potential for the implementation of the set goals, taking into account the factors formed by the general paradigm of the development of the country. The purpose of the study is to develop an algorithm for the assessment of the potential for the implementation of the paradigm of sustainable development in the process of modeling alternative organizational schemes of high-tech clusters.

The object of the research is the system of interconnections, management institutions and interests of subjects in the organizational model of a high-tech cluster. The subject of the research is the process of the assessment of the level of cluster adaptability to the conditions of the implementation of the paradigm of sustainable development.

2 Research methods

The main problem during the choice of an adequate system of indicators for the assessment of the performance of a cluster is the need to form such a metric system, which will reflect the main strategic guidelines for the development of the country and the region in scientific, technical and technological breakthroughs providing environmental and economic security. Unfortunately, the appearing disagreements associated with the risks of the changes in the geopolitical goals of strategic development [1, 2], crisis phenomena and structural changes in economy [3], the emergence of new determinants of economic growth [4] and technical and technological requirements for entities implementing the concept of Industry 4.0. [5, 6] are mainly conceptual in nature and not always clearly parameterized. The process of formalization of the parameters is complicated by a number of specific properties of the object of research itself: firstly, each high-tech and innovative cluster is a unique institution in terms of its functional and competence composition.

* Corresponding author: elena.bykova555@yandex.ru

The result of the activity may affect some separate target task of a scientific and technical nature [7] or a complex one related to the strategy of smart specialization [8]. Secondly, the strategic goal-setting of clusters should be focused on comprehensive indicators of regional development [9, 10]. Thirdly, the proposed methods, which are most adequate for the task of modeling cluster structures [11, 12, 13, 14] are focused on the target indicators of industries and do not take into account the potential of the "best practices" of the subjects of a cluster. One of the concepts that most fully reflect the relationship between the indicators of the development of the regional level of management and the parameters of the formation of the competitive advantages of an enterprise is the concept of EFQM [15, 16, 17].

3 Research results

The task of modeling the structure of high-tech cluster institution with a high potential for the implementation of strategic goals is associated with the assessment of the possibility to adapt such a system to the emerging factors of the paradigm of sustainable development and "smart" economic growth.

In order to implement the algorithm, it is necessary to simulate alternative options for the production and supply structures of a cluster. Further, the assessment is carried out according to the following algorithm:

At the first stage of the algorithm:

1. An array of assessment indicators for each designed structure is formed;
2. The process of bringing the assessment indicators to a single parameterization system is carried out, the array of reduced indicators $\{q_{ij}\}$ is formed, where i is the number of the model, j is the number of the indicator;
3. A preliminary integral estimate for each cluster model without taking into account the influencing factors S_{ij} is formed;

At the second stage of the algorithm:

4. The principle of the assessment of the compliance of indicators with economic institutions of safety is formed determining their value in the process of production and marketing activities. To do this, we introduce In_{ij} indicator, which includes the number and value of the institutions used in the value chain:

$$In_{ij} \quad i=\overline{1,m}\ ; j=\overline{1,k}, \tag{1}$$

where k- is the number of institutions in accordance with the composition of the institutional atlas (in a cluster industrial structure, it was 53 institutions [18,19]);

5. The transformation of the array of indicators by the value of the composition and value of institutions will be made according to the following formula:

$$q_{ij}= > S_{ij}NI_{ij} \times In_{ij}$$

$$NI_{ij} = \begin{cases}1\\0\end{cases}, \tag{2}$$

where NI_{ij} - the presence of an institution in the provision of the parameter q_{ij} (this indicator is an expert assessment of 0 or 1)

The third stage of the algorithm includes two tasks:

5. The formation of an approach to assess the level of adaptability of indicators to the strength of the influence of external factors of the institutional environment for each parameter relative to the maximum possible potential level of influence for each model:

$$KO_{ij} = \frac{S_{ij}}{S_{ij\,max}} \text{ по } j = \overline{1{,}16} \tag{3}$$

6. The integral assessment of the potential of each organizational model, taking into account the influence of factors of the external and internal environment of a cluster:

$$QI_i = q_{ij} \times KO_{ij} \tag{4}$$

7. The assessment of the level of the model potential based on the comparison of integral estimates with and without the influence of external and internal institutional factors according to the following formula:

$$\Delta_i = Q_{mi} - QI_i \tag{5}$$

Table 1. Results of the relative point assessment of the characteristics of institutions

Assessment characteristics of institutions Q_{ij}	**M 1**	**M 2**	**M 3**
Set block of indicators			
1. Growth in research and development and other projects	85	100	37
2. Growth in the volume of investment costs of organizations	38	100	32
3. Growth in output per employee of an organization	30	68	100
4. Growth in the volume of innovative products shipped by organizations	13	100	75
5. Growth in the total revenue of organizations	71	79	100
6. Growth in the number of small innovative enterprises	75	60	100
7. Growth in the number of patented organizations - participants in the results of intellectual activity	100	68	96
Match block of indicators			
1. Leadership	61	71	51
2. Staff	61	71	51
3. Strategy	51	61	61
4. Partnership and resources	61	61	61
5. Processes, production, services	61	71	41
6. Results for staff	51	71	51
7. Results for consumers	51	51	41
8. Results for society	51	51	41
9. Business results	61	71	51
Total Q_{ij}	**920**	**1154**	**989**

The developed algorithm was tested on the problem of modeling three systems:

Model 1 – the implementation of a value-added system in the format of an innovation cluster,

Model 2 - a system of implementation in the format of a technological cluster

Model 3 - in the format of a strategically formed production cluster, taking into account innovative formations of business processes.

As a result of the quantification of the initial data formed on the basis of the expert report of the Consulting Company using the EFQM model, a table of basic values was created (Table 1)

The expert assessments of the importance of an institution in the process of the implementation of models in the formation and support of Q_{ij} indicators were carried out on a 5-score scale and are presented in Table 2.

Table 2. Institutional Score for the Basic Institutional Atlas

№	Institution	M1	M2	M3
Management institutions				
1	Institute for Industrial Policy	5	5	5
2	Institute of Foreign Economic Activity	0	5	1
3	Institute of Priority Areas	4	5	4
4	Self-government institute	5	4	4
5	Institute for Development Programs	4	5	5
6	Security Institute	5	5	5
7	Institute of guarantees	4	5	5
8	Institute of income	5	2	2
9	Institute of Management	0	5	5
10	Responsibility Institute	5	5	5
11	Institute of production activities	5	5	5
12	Institute for Strategic Planning	4	5	5
13	Activity Control Institute	1	5	4
14	Institute for Examination of Scientific and Technical Projects	1	5	5
15	Institute for Intercompany Planning	5	5	5
16	Institute for the organization of production activities	5	5	5
17	Institute for the organization of information activities	5	5	5
18	Institute of material incentives for activities	4	2	2
19	Institute for non-material incentives for activities	4	5	5
20	Institute of Technology	5	5	5
21	Business Consulting Institute	1	5	4
22	Institute for Research Activities	2	4	5
23	Institute for design development	5	5	5
24	Trial Production Institute	5	3	1
25	Customer Service Institute	4	3	1
26	Legal Defense Institute	5	5	5
27	Public welfare institute	4	3	5
28	Institute of Club Benefits	2	5	3
Institutions of Resource Use				
29	Institute for the Application of Information Resources	3	4	5
30	Institute for the Application of Intangible Assets	2	5	5
31	Institute for the Application of Material Resources	5	3	2
32	Institute for the Application of Labor Resources	5	5	5
33	Institute for the Application of Financial Resources	5	4	4
34	Institute for Licensing and Certification	5	4	4
Institute of interaction				
35	Institute of contractual relations	3	5	4
36	Institute of state orders	5	5	5
37	Counterparty Search Institute	5	2	2
38	Institute for Communication Activities	4	4	4
39	Institute for the Integration of Science and Business	3	4	5
40	Institute of Informal Relations	4	5	3
41	Benefit transfer institute	3	5	4
42	Institute for interaction organization choice	5	5	3
43	Institute of inter-firm cooperation	5	5	3
44	Institute of Transport and Logistics Systems	5	5	3
45	Transfer Price Institute	1	5	3
Institutions of external influence				
46	Market Institute	5	2	2
47	Institute of Currency and Export Control	1	5	1
48	Institute of Taxation	3	3	3
49	Property Institute	5	3	3
50	Institute of Education	3	5	5
51	Technology Transfer Institute	3	5	5
52	Institute for Diffusion of Innovation	3	5	5
53	Institute for the formation of added value	1	5	3
	Total	**196**	**234**	**207**

The maximum number of points for all institutions is 265 points, therefore, if the model receives:

1) 0-88 points then the influence of institutions is insignificant

2) 88-176 points, then the level of influence of institutions is average

3) 176-265 points then the level of influence of institutions is high

Then, according to the developed algorithm, the assessment was carried out and the composition of institutions that directly affected the implementation of indicators was assessed.

4. Results and discussion

The obtained values of assessment are shown in Table 3 and indicate a positive deviation of the integral indicator in all models. This means a more significant impact of the institutional environment of a cluster on the activities of the enterprises included in it in comparison with expert assessments of risks incorporated in the socio-economic assessment of each project.

Thus, the values of the deviations make it possible to talk about the high level of the institutional potential of the second model, but the presence of a large number of institutions forms a high uncertainty in the effectiveness of links between business processes. Model No. 1 can be characterized as a low level of uncertainty and a high degree of elaboration of new innovative mechanisms of interactions.

Table 3. Assessment of the model potential of production and supply system of high-tech cluster

Models of production and supply system	QIi	Qi	Δ
Model 1	934,17	920	14,19
Model 2	1 364,13	1 154	210,40
Model 3	1 081,20	989	92,09

Model No. 3 demonstrates a high level of potential, despite the fact that it is based on the business processes of an industrial cluster. The introduction of new technologies of remote access in business processes leads to a slightly higher level of risks (compared to Model No. 1).

In conclusion, it is necessary to note the importance of the focus on the processes of mutual influence of the factors of the global paradigm of sustainable development and the parameters for the assessment of the potential of high-tech clusters. The introduction of innovative forms of business processes for remote access to data, the commercialization of projects through the structures of technologic and industrial clusters will help national enterprises to take a leading position at the global market of high-tech products.

Acknowledgments

The research was performed within the framework of the basic part of the state task of the Ministry of Science and Higher Education of the Russian Federation to Perm National Research Polytechnic University (topic No. FSNM-2020-0026).

References

1. B. Szetela, G. Mentel *Economic Annals* **161 (9-10)**, 31-34 (2016)
2. G. Bachnera, J. Mayera, K.W. Steiningera, A. Anger-Kraavid, A. Smithc, T.S. Barkerc, *Ecological Economics,* **172**, 1-14 (2020)
3. Zh. Mingaleva, A. Gataullina, *Middle East Journal of Scientific Research* 12 (11), 1535-1540 (2012).
4. R. Anand R, K.C. Ceng, S. Rehman S, L. Zhang, *Potential Growth in Emerging Asia, IMF Working Paper* **14/2**, (International Monetary Fund, Washington, DC. 2014)
5. D. Mourtzis, S. Fotia, N. Boli, P. Pittaro *International Journal of Advanced Manufacturing Technology* **97 (1-4)**, 91-103 (2018)
6. L. Feng, X. Zhang, K. Zhou Eurasip *Journal on Wireless Communications and Networking* (2018)
7. A. Lesnik, Zh. Mingalyova Economy of Region **3**, 190-197 (2013)
8. D. Foray, X. Goenaga *The goals of Smart Specialization: S3 policy brief series (JRC Scientific and Policy Reports).* Brussels: European Commission (2013)
9. O. Andreeva, V. Postnikov, *Journal of Advanced Research in Law and Economics* **8(8)**, 2324-2333 (2017)
10. M. Delgado, M.E. Porter, S. Stern *Research Policy* **43(10),** 1785–1799 (2014)
11. J. Bezdek, R. Ehrlich, F. Wull, *Computers & Geosciences,* **10 (2–3)**, 191–203 (1984)
12. Y. Dubrovskaya, E. Kozonogova, A. Molodchik *Upravlenets-the manager* **10**, 64-73 (2019)
13. E. Kozonogova, D. Kurushin, Y. Dubrovskaya *Scientific Visualization,* **11(5)**, 126-141 (2019)
14. Zh. Mingaleva, O. Gayfutdinova, E. Podgornova, *Engineering and Technology*, **58**, 1041-1051 (2009)
15. A. Escrig, J. Bou, I. Beltrán *Universia Business Review* **50**, 72-105 (2016)
16. A. Escrig, M. Segarra, B. García, I. Beltrán *International Journal of Production Economics* **200**, 1-14 (2018)
17. E. V. Popov, R.A. Semyachkov *Journal of Management Policies and Practices* **2(3)**, 57-70 (2014)
18. E. Bykova *Innovation Management, Entrepreneurship and Corporate Sustainability, Proccedings of the 4th International Conference*, Prague (2016)

Management of operational and calendar planning of experimental and serial machine-building production based on the use of petri nets

Dilyara Malikova[1*], *Evgeny* Zhulanov[2], and *Evgeny* Slashchev[1]

Izhevsk State Technical University named after M.T. Kalashnikov, 7, Studencheskaya street, Izhevsk, 426069, Russian Federation
Perm National Research Polytechnic University, 29, Komsomolsky prospect, Perm, 614990, Russian Federation

Abstract. The article deals with the management methods of serial and experimental production based on the application of Petri net modelling methods in the operational and calendar planning of the manufacturing of machine-building products. The proposed management model, in contrast to existing analogues, allows making a set of management rules, analysing the relationships formed in the process of pilot production and its structure, as well as organizing optimal production according to the criterion to minimize the costs.

1 Introduction

The development of high-tech production and competitive products at enterprises and organizations of the military-industrial complex (hereinafter referred to as MIC) is of strategic importance for the Russian economy. On the one hand, this increases the country's defence capability and the level of development of industry technologies. On the other hand, it allows enterprises and organizations of the military-industrial complex to increase Russia's export potential by increasing its share in the international arms market.

However, the high achievements are based on serial and experimental production, which, as a result of repeated experimental production of machine-building and instrument-making products, allows working out the manufacturing technology. All this requires large material costs, which increase the cost of production, the size of which depends on many factors, including the lack of network infrastructure, which determines the degree of involvement of structural divisions in the management of the production process. In addition, accumulating the best practices of the domestic and foreign practice of using low-waste and non-waste technologies in the manufacture of competitive products and improving the processes of organizing serial and experimental production of defence industry enterprises the integration and management support is essential, taking into account the principles of lean production. This support is implemented in the form of a mechanism for managing serial and experimental production (hereinafter referred to as the SEP), organizational and economic set of tools that has its information basis, purpose, tasks, direct and feedback links with the objects of influence at the defence industry enterprises. [3**Ошибка! Источник ссылки не найден.**] However, the organization of the coordinated application of this tool is based on human management decisions, which, combined with the insufficient digitalization of the analysed and managed production processes, does not allow obtaining the optimal organizational variation of the SEP with minimal costs. As a result, there is a need to apply special methods of operational and calendar planning of these processes that can provide the fastest and most cost-effective management decisions in the SEP. As a basis for the organization of such management, it is proposed to use the methods of Petri net modelling. Its application in the operative calendar planning based on the proposed prerequisites should cover the entire complex of processes and characteristics that reflect its essence, including the design of life cycle processes, the composition of the product, the sequence of manufacturing parts, changing the structure of the product, supply and technological operations.

2 Prerequisites used for modelling operational and calendar planning using Petri Nets

The main features of Petri nets [1] are the ability to reflect the parallelism, asynchrony, hierarchy of the simulated objects and lifecycle management processes.

A Petri net is formally represented as a set of the following form [3, c.4; 5, c. 41**Ошибка! Источник ссылки не найден.**]:

$$N = (P, T, F, H, \mu_0),$$

where P is a finite non-empty set of positions (states or places);

T is a finite non-empty set of transitions (events);

$F : P \times T \rightarrow \{0, 1, 2...\}$ - input function (input stream);

* Corresponding author: Dil-ma@mail.ru

https://doi.org/10.1051/shsconf/202111600019

$H{:}T\times P\rightarrow \{0, 1, 2...\}$ - output function (outgoing stream);

μ_0 – initial marking.

The graphical structure of a Petri net is represented by a directed bipartite graph [4, c. 255], where the set of vertices is divided into two non-crossing sets (see Figure 1) - edges: input/output

The Petri Net (Figure 1) is characterized by the following sets:

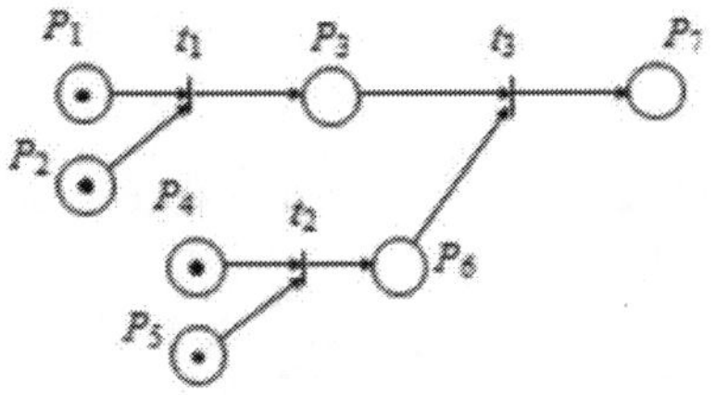

Figure 1. the Petri Nets

$P = \{P_i / i = 1,..., 7\}$ – positions; $T = \{t_j / j = 1,..., 3\}$ - transitions from one planning stage to another one;

$F = \{P_1t_1, P_2t_1, P_4t_4, P_5t_4, P_6t_3\}$ - arcs of outputs from plan positions;

$\mu_0 = (1, 1, 0, 1, 1, 0, 0)$ – initial plan marking.

The transition from one marking to another is carried out by triggering transitions (or initiating transitions, i.e. starting work on a stage). The formal condition for triggering the transition compares the number of labels in P_i with the number of arcs between P_i and t_j. For a Petri net (see Figure 3), t_1 is considered an excited transition, since there are two labels at positions P_1 and P_2, and two arcs take off the positions.

$$\mu(P_0) - F(P_0t_j)\geq 0, \forall_{P_i} \in P, \quad (1)$$

Starting the transition as a whole replaces the initial marking with a new one according to the following rule:

$$\mu'(P_c) = \mu(P_c) - F(P_i t_j) + H(t_i P_i), \forall_{P_i} \in P, \quad (2)$$

where P_c is the number of labels in one opposition, i.e., as a result of triggering the transition t_j, $F(P_i t_j)$ labels are excluded from all its output positions and $H(t_i P_i)$ labels are added to each input position.

The movement of the labels from one position to another shows the sequence of the process. These actions can be widely used, for example, when modelling the movement of orders in various management structures of organizations and production processes of operational and calendar planning.

When modelling the processes of operational and calendar planning, the production of a part using Petri Nets is mainly used in two ways.

The first method is characterized by the fact that the vertices-circles P_i denote real-valued sets (parts, materials, tooling, equipment, etc.), and the transitions t_i– energy connection-actions on them (see Figure 1).

The second method is characterized by the fact that the vertices-circles P_i denote the state of real-valued sets and energy sets, and the transitions t_i–the boundaries of the beginning and end of the process (production stage) (see Figure 2.).

In the presented Petri net, the vertices P_3 and P_5 - interpret the movements of the product in stages:

P_3– completing the product with a subproducts P_1, transferring it to the production area and making the development batch P_4;

P_5– package P_2 of the development batch P_4.

P_7– shipment of the development batch P_4 with a set of permits P_6

Vertices P_1 and P_7 show the operational and calendar planning of the product production process

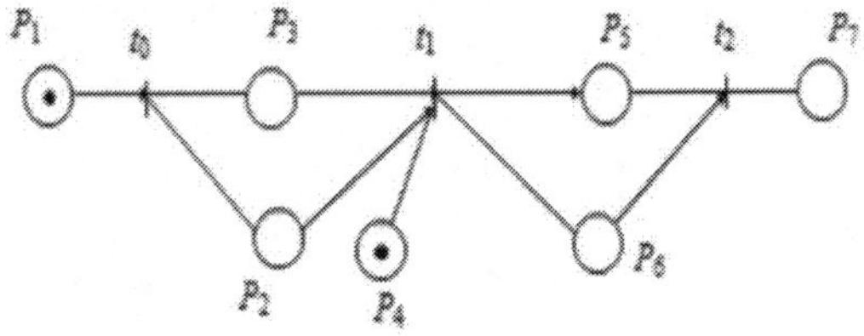

Figure 2.Petri Net when assembling two parts P_1 and P_4 using an industrial robot P_2 and a press P_6.

The mathematical model of the process (the movement of labels) can be represented by a matrix:

	P_1	P_2	P_3	P_4	P_5	P_6	P_7
μ_0	1	0	0	1	0	0	0
μ_1	0	1	1	1	0	0	0
μ_2	0	0	0	0	1	1	0
μ_3	0	0	0	0	0	0	1

A Petri net can also be represented as a time chain if all real-valued sets are discarded from it (Fig. 3):

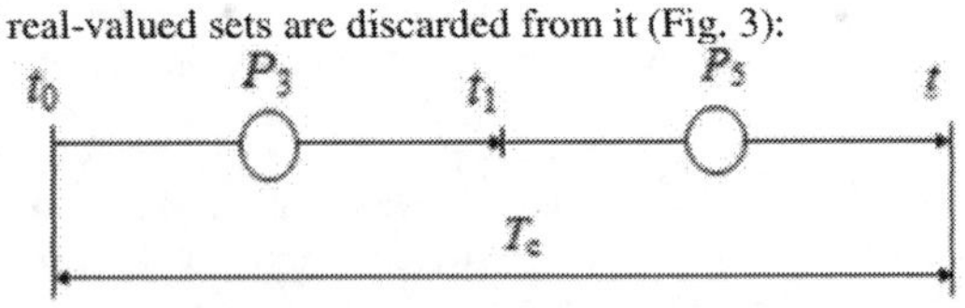

Figure 3. The temporal chain

From the temporal chain, the cycle time of the process T_C and the efficiency can be determined taking into account the influence of random deviations.

2 The system of calculation of statements in operational and calendar planning

Planning progress modelling is based on the following assumptions.

1. The intended service of the machine, drawings, specifications and technical conditions are mastered by the technologist.

2. The manufacture of the machine and its parts is provided.

3. The subject of planning was compared with others to find a similar technology. No analogue can be found.

4. Typical loading of the enterprise with the use of equipment and without equipment is known.

One of the main tasks that must be solved in operational and calendar planning is the task of determining the sequence of production and the optimal division of the product according to the stages of preparation of assembly units of parts.

The main requirements for the plan installation sequence are the terms of time limits and access restrictions. The terms of the time limit are determined by the allocation of the boundary terms of the contract and the links with the manufacture of other products.

When planning, the condition for restricting the access of one part (product) of the planned batch to another ends with the relation part X_1 restricts the access of part X_2 to the place of manufacturing: $X_1 \langle\rangle X_2$, where $\langle\rangle$ is the sign of the restriction relation.

The restriction in the movement of production in all directions of some parts (products) by others means that there is a movement restriction ratio between the items: $X_1 \langle\rangle X_2$.

The main requirements for the planning procedure based on the conditions of time limits and access are expressed in the following provisions:

1. If one part (product) restricts another, then it is assigned to the base and installed earlier than the one for which it is the base. Otherwise, this provision can be formulated as follows: if one of the parts (product) is the base, then it is installed in the plan of the first:

$$\frac{a_1 \vee a_2,\ (a_2 = O_1 \rightarrow b_i \,\&\, O_j)}{a_2 < a_1.}, \tag{3}$$

2. If one part (product) restricts access to the production of another, then the limited part (product) is installed in the plan before the one that restricts it:

$$(a_1 \langle\rangle a_2) \rightarrow (a_2 < a_1). \tag{4}$$

3. If there is a set of parts (products), they are identified as one part (product) and they are installed and planned simultaneously:

$$(a_1 + a_2 + a_3 = K) \rightarrow (a_1 \,\&\, a_2 \,\&\, a_3 = C), \tag{5}$$

where K is a set; C is the connection of the prefabricated product to the plan.

4. If the part (product) is installed in the plan with the basic part (product), then initially it is focused on the complexity of manufacturing relative to the basic one, and then it is included in the operational and calendar planning.

$$I \rightarrow B \,\&\, P, \tag{6}$$

where I is the installation of the part in the plan relative to the base one.

All of these provisions are implemented only if the terms and access restrictions are defined. [2]

3. Basic rules for modelling a production plan using information and computing systems

Based on the evaluation of statements in the design of operational and calendar planning, a new grammar of algebraic expressions is revealed, which defines the rules for constructing formulas from planning objects using operation signs and separators. The main actions that interpret the planning processes with propositional variables, signs of addition, subtraction, and transformation techniques are considered.

In the planning algebra, the regularities and rules for constructing algebraic expressions from the order execution progress and its transformation are revealed, while the order execution sequence must be preserved and the number of actions determined.

The main regularities of the algebra of operational-calendar planning of machine production are as follows:

- commutativity (permutability):

$$3a_1 + 2a_2 + a_3 = 2a_2 + 3a_1 + a_3, \tag{7}$$

where a is the propositional variables (0 or 1) denoting the name of the parts in the order with the indication of the product number in the index, and the coefficients denote the priority of the order of manufacture;

- distributivity (distributiveness):

$$2(2(2a_1 + a_2) + a_3) + a_4 = 2(4a_1 + 3a_2 + a_3) + a_4 = \\ = 6a_1 + 5a_2 + 3a_3 + a_4, \tag{8}$$

where in the formulas, when opening the brackets, the coefficients are not multiplied, but added, which is a difference from the traditional law of distributivity;

- anti-associativity:

$$3a_1 + (2a_2 + a_3) \neq (3a_1 + 2a_2) + a_3. \tag{9}$$

The rules for constructing formulas using operation signs and separators also have their characteristics, for example: addition and subtraction depend on the ratio of the order of concentration and differentiation of actions during planning.

The addition or production of an order with a strict sequence is carried out by the production of products:

a) different:

$$a_1 < (a_1 + a_2) < (a_1 + a_2 + a_3) \rightarrow 3a_3 + 2a_2 + a_3; \tag{10}$$

b) identical:

$$(3a_1 + 2a_1 + a_1) \rightarrow 3{,}3a_1, \tag{11}$$

where the coefficient 3.3 shows both the number of actions and products.

If the same orders in the plan have a coefficient before the brackets indicating the order of production, then only the integer part of the number changes, for example:

$$5(3a_1 + 2a_1 + a_1) = 8{,}3a_1. \quad (12)$$

If it is necessary to include positions in the production plan, consisting of a strict sequence of new products, the differentiation is determined by the implication:

$$a_1 < 2(a_1 + a_2) + (a_1 + a_2 + a_3) \to 2(2a_1 + a_2) + a_3. \quad (13)$$

The subtraction of the product at the concentration or the exclusion of the product from the plan in a strict sequence is modelled by the formula:

$$(3a_1 + 2a_2 + a_3) - a_3 = a_3 \oplus (4a_1 + 3a_2) \\ - 3a_2 = a_3 \oplus 3a_2 \oplus 5a_1. \quad (14)$$

Subtraction, which simulates the process of excluding a product from the plan when differentiating in any sequence, is characterized by the following transformation:

$$(2(2a_1 + a_2) + a_3) = (4a_1 + 3a_2 + a_3) \\ - a_3 = a_3 \oplus (5a_1 + 4a_2) - 4a_2 = \\ = a_3 \oplus 4a_2 \oplus 6a_1. \quad (15)$$

It should be noted that when subtracting, the products do not disappear, but occupy the initial position in the disassembled form with coefficients indicating the number of actions.

The product or synchronous parallel planning of identical parts is expressed by implication:

$$(a_1 \& a_1 \& a_1 \to a^3) \quad (16)$$

The introduction of an absolute value denoting the removal of all separators and priorities of the planning order is determined by the expression:

$$|(2(2a_1 + a_2) + a_3)| = |(4a_1 + 3a_2) + a_3| = a_1 + a_2 + a_3. \quad (17)$$

By entering an absolute value, the number of items in the plan is determined by the number of actions on the item in the plan.

All these principles are interdependent and mutually dependent. They reflect objective and stable relationships that are manifested in operational and calendar planning.

This is followed by the calculation of technical and economic indicators of the operational and calendar planning process.

On the basis of complex modelling, the following technical and economic indicators are calculated:

1. From the modelling of the evaluation of statements, the type of planning is determined as sequential.
2. When drawing up the scheme, the following parameters were easily identified: the sequence of operations, the composition of assembly units and sets of parts, as well as the time of operations.
3. When simulating the process with the help of a diagram, the following parameters are determined:

a) The load factor of jobs is determined according to the formula:

$$p_l = \sum_{i=1}^{m} T_i / T_{ci}, \quad (18)$$

where $\sum T_i$ – is the total time of formation (t) of items,

T_{ci} – production time (t) of items.

For each workplace of the production line, the coefficient value will be:

pl= 60/65 = 0.9, p2 = 80/100 = 0.8,

p3 = 40/110 = 0.38.

b) Maximum number of sets of items in the workplace (work in progress) N=3 (see line tl);

c) The average volume of work in progress (p) for the entire line is calculated using the formula:

$$n = 1/T \sum_{i=1}^{m} t_i * n_i, \quad (19)$$

where: T is the simulation period,

ti– the time during which there were i items at the workplace. For the initial approximation:

n1 = 1 /65*(30* 1 +25*2+5*2)=90/65= 1.5 items

For the initial approximation:

n2=1 / 100*(60* 1+30*2)

d) Downtime in the workplace due to irrational - differentiation of technological operations of operational and calendar planning occurs when the fourth product is being formed. Their value can be determined by the formula:

P=(timax-t(i+l))*rn, (20)

where timax is the maximum release time or clock cycle, in particular for this example, timax=20min,

t(i+l) – time to operate on the workstations, except for the first operation,

m – the number of manufactured products.

For this line, the average amount of downtime due to irrational differentiation of operational and calendar planning is equal to:

Rsr=(timax+t(i+1) * m / Tc

or

Rsr=(20-10)* 4/110= 40/110=0.4 min.

Note that the amount of downtime due to irrational differentiation of operational and calendar planning should tend to zero.

e) The average length of the queue of sets of parts waiting to be assembled and connected

$$L = 1/T_i \sum_{i=0}^{L} t_i * l_i, \quad (21)$$

where: t_i – the time during which there were L_i items in the queue.

For this line, queues are formed at the first and third workplace

Ll = l /65*(25*1+5*2)-30/65=-0,25L3=l/100*(30*l)=0.3 items

f) The waiting time for the sets of items in the queue will be:

$$W = 1/m\sum_{i=1}^{m} w_i, \qquad (22)$$

where: w – is the waiting time for a set of product items.

For this line, it is equal to:

Wl = l/4*(5+10+20)=35/4= 8,9 min.

W2=l/4*(30)=7.5 min.

Production line capacity:

N = m /Tc = 4/110 = 0.036 pcs/min.

All the main technical and economic indicators are summarized in Table 1.

Table 1. Production line indicators

No.	Indicators	Calculation formulas	Quantitative value
1	Coefficient downloads	n0=nl+n2+n3	0.6
2	Release cycle products	tB=timax	20 min
3	Efficiency	Ph=m*60 /Tc	2.2 pcs / h
4	Average volume of - work in progress	n=n1+n3/2	1.35 sets
5	Length of the queue of sets of parts	Lsr=Ll+ L3/2	0,28 sets
6	Queue waiting time	Wsr=Wl+ W3 / 2	8.2 min.
7	Average value of downtime	Rsr=(tirnax+ t(i+l)m/Tc	0,4 min.
8	Product output volume units per cycle	m	4 pcs.
9	The complexity of manufacturing a single product	$t = \sum_{i=1}^{m} ti$	0.91 hours

It should be noted that to build the considered model, it is not necessary to know the physical nature of the process.

4 Conclusion

The article considers the author's approach to managing the operational and calendar planning of the SEP of a machine-building or instrument-making product based on the use of Petri Nets. The purpose of this control is to obtain a motion matrix that allows optimizing the implementation of the plan in dynamics based on the analysis of system parameters interrelated processes of operational and calendar planning of the SEP.

This author's approach develops the possibility of automating the process
of planning at the stage of serial and operational production, allows minimizing the number of human errors in the process of operational and calendar planning and reducing the amount of work in progress, which is covered by errors in calculations. The result of the application of the author's approach to management is also a reduction in the consumption of material resources in the process of manufacturing experimental samples of machine-building and instrument-making products.

References

1. D.M. Malikova, E.S. Slashchev , S.A. Shilyaev, R.F. Shaikhov, V.G. Osetrov *Journal of Advanced Research in Dynamical and Control Systems*, 12 (2 Special Issue), 391–395, (2020)
2. D.M. Malikova, E.S. Slashchev, E.P Istomin., M.R. Vagizov, O.E. Kolbina *IOP Conference Series: Earth and Environmental Science*, **507(1)** 012018 (2020)
3. D. V. Isaev *Financial analytics: problems and solutions*. **20 (254)** 2-12 (2015).
4. P. V. Skorodumov *Economic and social changes: facts, trends, forecast* **4(34)** 253-259 (2014)
5. A. A. Khanova, A. A. Muntyanova, K. I. Averyanova Science Bulletin of the NSTU **71(2)**, 39–58, (2018)

Impact of information-infrastructure mechanism of self-development management at industrial enterprises on employee engagement: empirical evidence

Denis Markov [1,*], *Nina* Markova [1], and *Galina* Chernovalova [1]

[1] Perm National Research Polytechnic University, 29, Komsomolsky prospekt, Perm, 614990, Russia

Abstract. The work deals with the problems of a self-development organization considered as a self-organization system without hierarchy and employee position. Such kind of management is realized by self-management teams having full authority, responsibility and focus on the result. An individual and group motivation mechanism uses the employee potential to achieve one's ambition targets. The employee and one's potential are engaged with a maximum power in all fields of activity. Enterprises management explores employee engagement to estimate and manage this process. Each manager has enough cases when employee goes far beyond job responsibilities to achieve the result. This employee is engaged taking an initiative and informal responsibility. Engagement is a multidimensional construction, which demonstrates employee wishes to invest self-resources (knowledge, skills, competences, emotions and power) in enterprises activity. Engagement is a positive moment implying active involvement, loyalty and direct work. A lot of engagement investigations are aimed at management, result and employee's participation in organization management. The study of employee's initiative is a part of employee's engagement according to infrastructure and information factors at industrial enterprises in Perm region described in the article.

1 Introduction

Enterprises have to provide extended reproduction of internal resources mostly at the moment of unpredictable changes in all life activity spheres. It means instrument determination for initiating employee's initiative behavior based on behavioral motivation, initiative and high engagement. This human-oriented organization is a self-development structure. Self-development and self-organization concepts create a successful activity base and long-life leadership organizations of future, such as 3M, Oracle, Zappos, Haier, Amazon and others.

We describe self-organization as an enterprise process internal transformation without external impact on effective management of changes. And self-development is as a vision transformation, which defines the strategic management result with the self-organization process [1-2].

The enterprise process of internal transformation provided by engagement of the employee involves high emotional relation to enterprises [3-6]. Engagement employees make more effort to do work effectively and achieve the strategic target.

The first terminology «Engagement» was used in academic papers by W. Kahn in 1990 [4]. In 1997 C. Maslach and M. Leiter [5] continued Kahn research. In 2004 W. Schaufeli [7], in 2008 A. Saks [8], in 2008 W. Macey and B. Schneider [9] described «Engagement» terminology.

There are some differences in engagement descriptions, but all authors mark key characteristic of engagement – positive mental state which characterizes vitality, loyalty and job concentration. Profit from employee's engagement is described by loyalty, commitment, talent, effort maximization, share of enterprise values to achieve enterprise targets. Engagement employees invest not in their job, but in enterprise success [10-11].

Engagement is a measurable category, which uses different measurable models: «Job Demands – Resources» [12], Aon Hewitt [13], «ECOPSI consulting» [14]. Engagement combines rational thinking, emotions, intention and behavior for creating optimal productivity. There are three aspects of engagement: «Tell», «Stay» and «Try».

We created an information-infrastructure mechanism (IIM) of self-development enterprise management (Fig. 1) which consists of realization conditions, a logical and temporal structure and information-resource support [15].

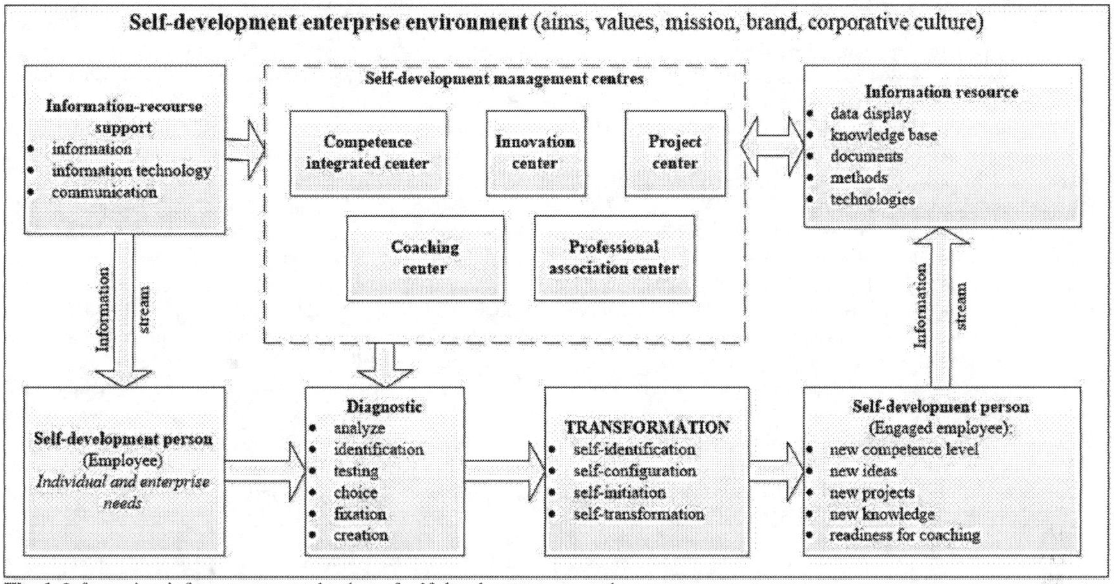

Fig. 1. Information-infrastructure mechanism of self-development enterprise management.

Enterprise culture environment consists of corporative culture, brand, values, mission, aims. Culture environment determines the mechanism of realization of condition synchronized with personnel needs. The IIM structure consists of an integrated competence center, innovation center, project center, professional association center and coaching center.

The self-development center personnel have special professional competence. They have management information on the enterprise task at the concept, strategic, tactic and operation levels. These personnel have management authority and perform special functions using information.

Self-development centers provide accumulation of personnel knowledge and experience, trying and choice of the best practice, maximum personnel engagement, personnel motivation, employee's needs and joint achievement of enterprise aims in the shortest possible time.

Entering and outgoing information on enterprises is processed by self-development management centers: analysis, identification, testing, choice, fixation, creation. It means information management occurs by the degree of importance and new methods of prompt knowledge obtaining from key personnel, knowledge structuring and creation of the navigation system to increase competitiveness, flexibility and adaptivity of the enterprise.

These mean information stream organization for conditions creating accurate and fast management decisions, quality and fast specialist training, starting a self-development mechanism.

IIM rises personnel engagement, starts self-development and self-organization processes: self-identification, self-configuration, self-initiation, self-transformation initiate the self-development enterprise mechanism [16].

Employee's engagement is business success and competitiveness key factors. Conditions that increase competitiveness can be created using special management practice [17].

2 Theoretical background and investigation hypotheses

IIM rises employee's engagement. The mechanism of the elements' impact on employee engagement is different. There are key elements that strongly influence employee's engagement. Enterprises increase employee's engagement, improve self-development environment, achieve goals faster at the lowest cost if they work with mechanism's components purposefully.

Taking into consideration the results of the previous studies of employee engagement, the authors put forward the following hypotheses (H).

H1: Self-development management centers, self-development environment, information, information technologies and communications positive impact on employee's engagement.

H2: Self-development management centers and self-development environment which consist basically of IIM components influence employee's engagement strongly. Others components influence slightly because they are structural components.

H3: Impact of IIM components on employee's engagement depends on the employee's category gained at enterprises. If the employee's category is higher in the enterprise management structure, then IIM components influence highly. If the employee's category is lower in the enterprise management structure, then IIM components influence less.

3 Methodology and data

The «2S System» methodology developed by Perm National Research Polytechnic University [2] is the base of the current empirical investigation. Self-development can be realized in different directions that are practically unreal to predict. Thus, IIM activity's result can be determined by IIM component's impact on employee's engagement directly.

IIM impact assessment on employee's engagement is realized by a latent construction (R) which consists of the following variables (Fig. 2):

- frequency of request for the enterprise knowledge base unilaterally (R1);
- frequency of doing activities in self-development management centers unilaterally (R2);
- frequency of taking an initiative (R3).

The questionnaire survey of Perm region employee's enterprises was realized in 2020 for hypotheses validity. There were 294 respondents: 62.5% - workers, 29.3% - specialists (engineers and office), 8.2% - managers. In the questionnaire survey there were questions which were grouped about IIM components: self-development management centers, self-development environment, information, information technologies and communications (Table 1). Likert's scale is used for evaluation answers: 1 – minimum score; 5 – maximum score.

The Partial Least Squares (PLS) method is used for evaluation of hypothesis statistics validity. The PLS method is used for creating complex variables models [18]; it determines the impact between dependent and independent variables [19-20]. We joined independent variables in latent constructions: I – information, IT – information technologies, E – self-development environment, Ct – self-development management centers, Com – communication (Fig. 2). An investigation scheme (Fig. 2) is used for creating models: General – all employees, Work – workers only, Spec – specialists only, Manage – managers only.

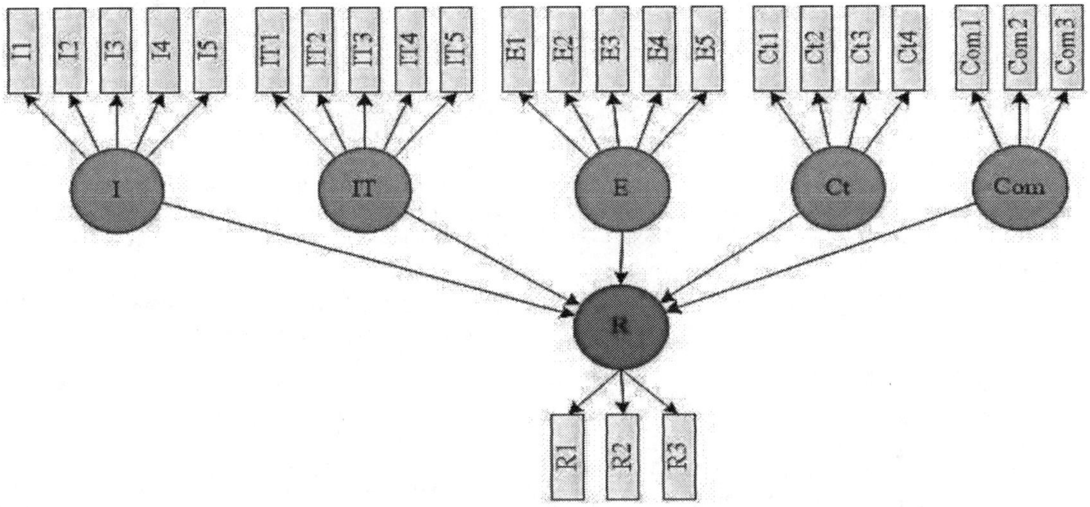

Fig. 2. Investigation scheme of employee's engagement.

Latent constructions are made preliminarily by the Least Squares (LS) method. The latent construction validity is evaluated using Principal Components Analysis (PCA). We created a dependent latent construction «Employee engagement» – R and five independent latent constructions (IIM components): I – information, IT – information technologies, E – self-development environment, Ct – self-development management centers, Com – communication. Latent constructions validity is evaluated by Cronbach's Alpha [21-22] and latent constructions of Outer Loadings elements. Cronbach's Alphas and Average Variance Extracted (AVE) of created latent constructions are values that are sufficient for validity (Table 1).

Creating the model and calculation are realized in the Smart PLS program [23].

4 Results

We have summarized the results (Table 1) and obtained generalized models' characteristics (Table 2).

Table 1. Latent constructions characteristics.

Characteristics	Name	Model							
		General		Manage		Spec		Work	
		Outer Loadings	Mode	Outer Loadings	Mode	Outer Loadings	Mode	Outer Loadings	Mode
Employee's engagement	**R**								
R-square		0.642		0.724		0.643		0.671	
Cronbach's Alpha		0.845		0.843		0.835		0.855	
Average Variance Extracted (AVE)		0.763		0.762		0.751		0.775	
Frequency of sending request to enterprise knowledge base unilaterally	**R1**	0.863	3	0.818	4	0.831	3	0.887	3
Frequency of doing activities in self-development management centers unilaterally	**R2**	0.899	3	0.912	3	0.909	1	0.900	3
Frequency of taking a self-initiative	**R3**	0.857	5	0.886	4	0.858	5	0.853	4
Enterprise self-development management centers		**0.400**		**0.635**		**0.584**		**0.331**	
Cronbach's Alpha		0.781		0.832		0.728		0.795	
Average Variance Extracted (AVE)		0.610		0.665		0.572		0.623	
Enterprise self-development management centers' awareness	**Ct1**	0.824	3	0.775	1	0.855	1	0.823	3
Enterprise self-development information place's awareness	**Ct2**	0.580	1	0.851	1	0.305	2	0.645	3
Professional associations involvement	**Ct3**	0.849	4	0.738	4	0.856	3	0.861	4
Enterprise self-development directions' awareness	**Ct4**	0.838	3	0.890	4	0.855	3	0.811	3
Enterprise self-development environment		**0.289**		**0.025**		**0.095**		**0.385**	
Cronbach's Alpha		0.761		0.802		0.747		0.769	
Average Variance Extracted (AVE)		0.515		0.549		0.498		0.525	
Enterprise values awareness	**E1**	0.773	4	0.769	5	0.786	5	0.763	4
Enterprise values sharing	**E2**	0.750	4	0.747	5	0.695	5	0.772	4
Enterprise values delivering	**E3**	0.796	5	0.760	4	0.759	5	0.826	5
Harmonization of employee and enterprise's needs	**E4**	0.643	1	0.775	1	0.639	1	0.629	3
Enterprise self-development program availability	**E5**	0.606	1	0.647	1	0.636	1	0.605	1
Information		**-0.104**		**-0.545**		**-0.149**		**-0.049**	
Cronbach's Alpha		0.836		0.913		0.848		0.823	
Average Variance Extracted (AVE)		0.591		0.733		0.542		0.583	
Speed of receiving required information	**I1**	0.861	4	0.933	4	0.901	4	0.829	4
Required information volume	**I2**	0.876	4	0.904	4	0.899	4	0.866	4
Required information understandability	**I3**	0.733	4	0.766	4	0.767	4	0.683	4
Required information place's awareness	**I4**	0.750	4	0.820	5	0.567	5	0.800	4
Required information employee's awareness	**I5**	0.586	5	0.848	5	0.419	5	0.609	5
Information technology		**0.180**		**0.706**		**0.139**		**0.165**	
Cronbach's Alpha		0.832		0.844		0.766		0.856	
Average Variance Extracted (AVE)		0.593		0.592		0.525		0.628	
Information technology completeness for working	**IT1**	0.796	4	0.801	4	0.778	5	0.805	4
Information technology rapidity for working	**IT2**	0.795	4	0.709	4	0.794	4	0.819	4
Information technologies and resources access	**IT3**	0.745	4	0.851	5	0.638	5	0.779	4
Information technology relevance	**IT4**	0.799	4	0.713	4	0.813	4	0.815	4
Information technology direction	**IT5**	0.712	3	0.762	3	0.565	4	0.741	3
Communication		0.184		0.283		0.266		0.138	
Cronbach's Alpha		0.882		0.865		0.915		0.861	
Average Variance Extracted (AVE)		0.809		0.788		0.855		0.783	
Communication procedure formalization (delivering enterprise values and corporate culture to employees)	**Com1**	0.891	4	0.851	4	0.917	4	0.880	4
Communication procedure formalization (receiving an initiative from employees)	**Com2**	0.932	4	0.952	5	0.959	4	0.913	4
Communication procedure formalization (feedback)	**Com3**	0.874	4	0.855	3	0.898	3	0.861	4

In Table 2 the green color shows a middle or highly positive impact, red color – middle negative impact and white color – low impact (negative or positive).

Table 2 presents impact values of IIM components on employee's engagement:

Enterprise self-development management centers highly influence the employee's engagement in all models among all IIM components.

Enterprise self-development management centers highly influence employee's engagement in terms of managers; average influence is on specialists and the lowest – on workers. It is caused by more creative job and better self-realization at the «Manage» level than in case of «Spec» and «Work». Specialists work more creatively than workers do. Workers work functionally, they solve specific tasks only without creating, and self-development management centers influence their needs less.

Table 2. Models characteristics.

Latent construction	General	Manage	Spec	Work
R-square	0.642	0.724	0.643	0.671
Outer Loadings:				
Enterprise self-development management centers	0.400	0.635	0.584	0.331
Enterprise self-development environment	0.289	0.025	0.095	0.385
Information	-0.104	-0.545	-0.149	-0.049
Information technology	0.180	0.706	0.139	0.165
Communication	0.184	0.283	0.266	0.138

Enterprise self-development environment influences employee engagement positively, but less than self-development management centers do in general including in each model. There is a reverse dependence: environment impact on engagement workers is more than that of specialists and managers; it is greater for engagement specialists than it is for workers. It is caused by a more self-sufficient level of specialists and managers than that of workers. They determine their tasks frequently independently than workers do, who works by directions. Managers and specialists understand and share corporative goals, values and culture; they have higher education and cultural level. They form enterprise's self-development environment themselves, but workers depend on one.

The initiative communication procedure influence employee's engagement a little positively in all models. It is more important for managers and specialists because their job is more creative and they have communications that are more creative. Communication environment allows more opportunity for initiative activities. For workers this IIM component is formal. They are informed about it quite well, but apply rarely.

Information affects negatively employee's engagement in all models. It is caused by very excessive information for all employees. They prefer to solve professional tasks and to avoid additional information. Employees think that work with information increases labor expenditures. Particularly there is a tendency in the «Manage» model. Managers do office job with information and additional data increase their job volume. Additional information is additional job which requires new activities which can be not paid.

Information technologies influence employee's engagement greater than that of managers. They work with information and strongly depend on information technologies. IT determines the quality of manager's activity. Specialists use IT for supporting their activity. IT should be sufficient for activity and no more. Workers do not use IT in their activity.

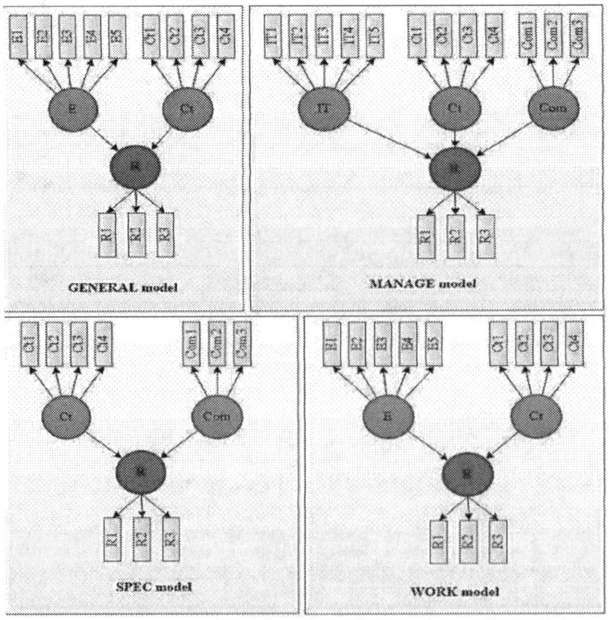

Fig. 3. Models of IIM components' influence on employee's engagement.

Table 3. Models characteristics.

	General	Manage	Spec	Work
R-square	0.606	0.654	0.625	0.644
Outer Loadings:				
Enterprise self-development management centers	0.443	0.431	0.615	0.382
Enterprise self-development environment	0.413			0.498
Information technology		0.346		
Communication		0.280	0.272	

4 Conclusions

We have used investigation results (Table 2) and created impact models of IIM components for employee's engagement in each employee's category (Fig. 3). We determined their characteristics (Table 3).
Model characteristics (Table 3) prove the H1 hypothesis. Enterprise self-development management centers and enterprise self-development environment influence positively on employee's engagement.

The H2 hypothesis is proved. Enterprise self-development management centers and enterprise self-development environment are basic components, but environment impact of enterprise self-development is too little for managers and specialists. Other components influence slightly.

The H3 hypothesis has been proved partially. Almost all of IIM components influence stronger employee's engagement if the employee's category is higher in the enterprise management structure with the exception of enterprise self-development environment. Enterprise self-development environment's impact rises with the reduction in the employee category level.

We can recommend improvement for employee's engagement:

- for all employee: to create self-development management centers as information centers which use the intellectual product of enterprise employees, starts and supports enterprise development processes;
- for workers: to form enterprise self-development environment, to deliver enterprise goals, values and culture, to coordinate enterprise goals and needs with worker's needs, to create self-development worker programs;
- for specialists: to formalize enterprise values and corporative culture delivering procedures and receiving an initiative;
- for managers: to implement actual, high-performance information technologies with managers, requiring to provide necessary access, to formalize enterprise values and delivering corporative culture procedures and receiving an initiative.

References

1. J. Schumpeter, *Theory of economic development: a study of entrepreneurial profit, capital, credit, interest, and the business cycle* (Progress, Moscow, 1982)
2. A.V. Molodchik, *Theory and practice of a self-developing organization* (Inst. the Economy, Yekaterinburg, 2001)
3. F. Laloux, *Discovering the Organizations of the Future* (Mann, Ivanov, and Ferber, Moscow, 2016)
4. W.A. Kahn, *The Academy of Management Journal*, **33 (4)**, 692-724 (1990)
5. C. Maslach, M. Leiter, *The truth about burnout* (Jossey-Bass Publ., 1997)
6. D.A. Markov, N.A. Markova, V.L. Popov, *Lean and Quick-Response Manufacturing* (PNRPU, Perm, 2018)
7. W.B. Schaufeli, A.B. Bakker, *Journal of Organizational Behavior*, **25**, 293–315 (2004)
8. A.M. Saks, *Journal of Managerial Psychology*, **21 (7)**, 600–619 (2006)
9. W.H. Macey, B. Schneider, Industrial and Organizational Psychology, **1(1)**, 3-30 (2008)
10. G. Gemma Robertson-Smith, C. Markwick, *Employee Engagement. A review of current thinking* (Inst. For Employment Studies, Brighton, 2009)
11. N.A. Markova, D.A. Markov, *Bulletin of the PNRPU. Social and Economic Sciences Series*, **4**, 329-341 (2016)
12. A.B. Bakker, M.P. Leiter, *Work engagement: a handbook of essential theory and research* (Psychology Press, New York, 2010)
13. *Official website of AXES Management*, Retrieved from: https://axes.ru/
14. *The official website of the consulting company in the field of personnel management ECOPSI Consulting*, Retrieved from: http://www.ecopsy.ru/
15. G.A. Chernovalova, *Fundamental research*, **3**, 217-223 (2015)
16. N.B. Akatov, *Managing the transition to self-developing innovative organizations: theory and practice* (PNRPU, Perm, 2012)
17. S.V. Komarov, *From self-organization to self-development: changing the management paradigm* (Inst. of Economics of the Ural Branch of the Russian Academy of Sciences, Yekaterinburg, 2013)
18. C.M. Ringle, S. Wende, J.-M. Becker, *SmartPLS 3. Boenningstedt* Retrieved from: http://www.smartpls.com
19. K.K. Wong, *Marketing Bulletin*, **24**, 1-32 (2013)

20. K.K. Wong, *The Magazine of the Marketing Research and Intelligence Association,* **6**, 20-23 (2010)
21. LJ. Cronbach, *Psychometrika*, **16**, 297-334 (1951)
22. N. Shmitt, *Psychological Assessment*, **8 (4),** 350-353 (1996)
23. *Official version of the SmartPLS software product, free access 30-day Trial version,* Retrieved from: https.:smartpls.com

Educational strategies of Russian managers as a factor of personal success and innovative development of society

Natalia Vyatkina*

Perm National Research Polytechnic University, 29, Komsomolsky prospekt, Perm, 614990 Russia

Abstract. The current state of Russian society and economy is characterized by the contradictory processes that impede innovative development. One of the most important subjects of innovation and modernization is the social and professional group of managers. A significant factor in the success of managerial decisions leading economy and society to the development and growth based on innovative technologies is the level of education of managers. The paper reflects the relationship between the level of educational capital of middle managers and professional and personal successes that promote the development of society.

1 Introduction

The most important direction of the strategic development of Russian society is the innovative development of economy. These processes are based on the high level of human resources management, the modernization of management institute and the growth of the efficiency of management decisions.

The role of qualified and highly educated managers is especially growing in the post-industrial economy, where intelligent technologies, information and knowledge form the production of goods and services.

At the same time, there are a number of serious obstacles to innovative economy in Russia. Firstly, labor productivity in our country is significantly lower than that of the leading countries in economic development: "According to the Organization for Economic Cooperation and Development (OECD), preliminary estimated labor productivity in Russia amounted to $ 26.5 per hour in 2017. In terms of labor productivity, Russia was inferior to all OECD countries, except for Mexico, where it was $ 21.6. The average for the OECD is $ 54.8, while the labor productivity of leaders Ireland and Luxembourg was about $ 99 per hour" [1]. In addition to other reasons, the insufficient level of education of modern Russian managers plays a significant role in this aspect.

Secondly, the complex, changeable and unpredictable modern social reality determines the symmetrical characteristics of the business environment and economic processes. This is reflected in the VUCA model [3] - an acronym that conveys the meanings of a challenging and unstable situation in global economy (VUCA: volatility, uncertainty, complexity, ambiguity), which becomes even more relevant during the Covid-19 pandemic.

This situation determines the demand for innovative management approaches, solution of non-standard management tasks, new management technologies, ability of managers to consider general situation, predict and prevent threats, apply a creative approach and form effective teams. This social and business environment places particular demands on leaders operating in the conditions of ever-aging knowledge, data and concepts. Within the framework of the concept of lifelong education modern successful managers, competing and operating in an uncertain market, must constantly learn and improve not only the level of professional knowledge, but also develop their soft skills - social skills and personal competencies.

Managers as a social and professional group and the representatives of the "new middle class", engaged in strategy issues at enterprises, from the very beginning of their profession, demonstrated characteristics setting them apart from other social groups. The direction to a high level of their education and the education of their children, selectivity in social environment and location of residence, interest in public life, preservation and improvement of their own resource characteristics (social ties, educational level and physical condition) play an important role in this issue [2]. American researchers of the group of leaders W. Warner and D. Rossides also emphasized the pronounced direction of managers to a high level, placing this group at the upper level of the middle class in their multi-class social systems [4].

Thirdly, the challenges of society and the market presuppose an appropriate level of educational programs for managers at different levels: an

* Corresponding author: zxcx1@yandex.ru

interdisciplinary integrated approach, focus on breakthrough innovative technologies reflecting the latest scientific knowledge, compliance of the content of education with the needs of social actors (employers, business and government). At the same time, educational structures in Russia do not fully meet these requirements.

2 Results and Discussion

What are the reasons why the education of managers is the most important factor in the transition to an innovative economy?

The innovative transition presupposes the intellectualization of economy: the increase in knowledge intensive technologies for the production of goods and services in order to increase the efficiency of the economic sector of society. New industries, new products and new technologies allow increasing productivity, automating labor, reducing production costs, increasing profitability and investment attractiveness of business and the country in general.

The innovative economy provokes positive changes in the life of society: new categories of goods and services more fully satisfy the existing and emerging needs of persons. New jobs for qualified specialists are emerging, the quality of life of people is improving, the food, economic, and defense security of the country is increasing and the investment rating is increasing.

It is obvious that such transformations of society and its economic subsystem are impossible without a high level of education of managers in all spheres of production: public administration, industry, services, financial and banking sectors, digital ecosystem of society. The creation and maintenance of a high level and the dissemination of new knowledge in society is the basis of the post-industrial economy and innovation.

In this regard, the authors designate the most important problem of Russian society: professional education for managers of a varying quality [5].

Thus, the group of professional managers in Russia is complex in terms of educational status. Some managers do not have enough competence, which creates problems in the identification of their social and professional status.

Firstly this is explained by the fact that the manager group includes the representatives of different generations. Depending on age, the group includes the representatives who began their career in Soviet times and the representatives of generations of the 90s, 2000s, 2010s with different levels of education, professionalism, set of values and life experience [6].

The system of higher professional education majoring in "Management" is heterogeneous, and, despite the massive graduation of specialists in this area, it often does not meet the market and time requirements for the quality of education.

Business education (MBA) is enough expensive. High-quality training in this area is available only in the largest cities of the country, a small part of the professional community, while "using classrooms to educate working managers is a good idea, but pretending that you are creating managers from those who never managed anything is a real fraud ... " [7].

It is necessary to note that the management paradigms which are popular in Russia are based on Western management concepts, where the manager's career and professional achievements are closely related to the concept of personal life success. Therefore, the issues of the level of education of Russian managers remain very topical.

The educational capital refers to the objective resource characteristics of a person (along with financial situation, social status of parental family, residence in an urban or rural area).

The general trends in the relationship between the level of education and the direction towards success, reflected in a number of modern Russian studies, are as follows:

- A significant impact of personal resource availability on the direction to the achievement of life success, an active life position, new higher levels in personal and professional life [8];

- A direct dependence of the level of education of a person and intergenerational educational mobility (the desire of a person to increase his educational capital in comparison with his parents) [9];

- The higher the level of education of the individual, the stronger the achievement values and strategies of behavior are manifested.

The study of the values of middle managers, conducted in 2017-2020 allowed us to reveal that the representatives of this socio-professional group stand out in terms of the level of educational capital from the average Russian person.

In the professional group of middle-level managers, not only persons with higher education predominate, but also a high level of education of the closest people (spouses, friends, as well as a higher level of education of parents) was noted.

Middle-level managers noted good level of education as one of the main factors for the achievement of success in life in Russia (57%), along with such factors as the presence of connections (56%), a relevant in-demand profession (33%), career (28%) and material wealth (32%).

It is revealing that Russians in general choose material wealth as the main factor in the achievement of success, and managers, playing the role of "cognitariat" or "knowledge workers" [10] (E. Toffler's term as opposed to "proletariat"), note that it is a high level of education that contributes to success in life in modern Russian conditions. This can be considered as an indirect sign of the transition of Russia to a post-industrial economy, where the cognitive components plays a leading role in the development of society and economic processes and the educational level, knowledge and professional qualifications of a person become the main social competences.

To a large extent, the level of education and cognitive activity predetermine the ability of a person to successfully adapt to a changing social environment [11], which affects the achievement of success. We also note the fact that education associated with professional success, which is the most important sphere of life and achievements, is considered as a factor of success [12].

With regard to the studied socio-professional group of middle managers, we note that in terms of educational capital, its representatives are in a more advantageous position in comparison with Russians in general.

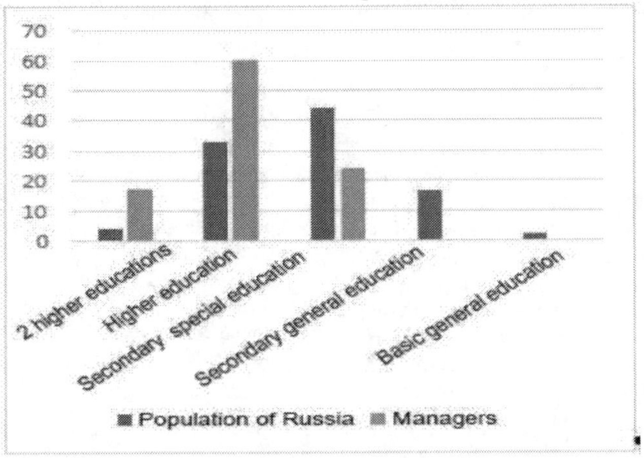

Figure 1. Comparison of the educational level of middle managers and the population of Russia

In addition to the prevalence of people with higher education, researchers also note a high educational capital of the closest people of middle managers (spouses and friends, as well as a higher level of education of parents in comparison with other professional groups) [13]. At the same time, while some authors note that the educational level of Russian managers in terms of professional management knowledge is insufficient [13] (in comparison with Western colleagues-managers), other studies, on the contrary, show the increase in the educational level of managers in dynamics [14].

Managers choose the level of education as the dominant path leading a person to success, demonstrating high adaptability, a desire to increase knowledge and improve their professional competencies.

Here is a quote from the interview of a manager: "My path to success is education and self-education. Even if you have a decent higher education, you should not stop. Therefore, it is important to continue training through seminars, conferences, symposia and exhibitions" (male, 29 years old, territorial manager of an investment company).

At the same time, in modern Russian society, the mechanisms of social mobility and achievement of success are not balanced and contradictory, which is confirmed by the fact that managers call social ties a similar factor for the achievement of social success, along with education.

The level of education is an important parameter for the assessment of the level of general and professional competencies of any social and professional group. When it comes to a group of managers, this social characteristic becomes especially important. The employees of companies and organizations applying for a managerial position must meet certain criteria in terms of education, knowledge and professional experience.

A high level of education is necessary for a leader and is justified by the economic feasibility and market requirements for the quality of management.

The study showed that about 25% of middle managers do not have higher education. In comparison with the group of professionals (doctors, teachers, engineers) they have a lower level of education. This is explained by the fact that the profession of "manager" is quite young for Russia and with the development of the market economy. These positions were occupied mainly by persons who, first of all, had pronounced personal leadership qualities (ability to lead, initiative, proactivity).

With the development of society and economic agents, strict requirements begin to be imposed on leaders, including the level of education. This leads to the fact that the level of education of managers is constantly increasing. In 1994, 52% of Russian leaders had a higher education; in 2018 this level of education was already demonstrated by 83% of managers. More than a quarter of middle managers (27%) received 2 higher educations trying to meet the requirements of employers and increase their value in the labor market.

The significantly higher level of education of managers in comparison with the average for the considered industries is also confirmed by the following data obtained in the study: among managers, in principle, there are no individuals with basic and general secondary education, the majority of managers (80%) have higher education, 40% more people than the industry average have two higher educations.

The study of leaders also demonstrates their values, aimed at the high importance of education in their lives. The research shows that 70% of middle managers respond that they have a need to improve the level of education, professional skills and qualifications. Accordingly, 65% of managers are going to raise the level of education in the near future. 71% of respondents note that education is the most important intangible value and the main means to realize their life tasks.

Managers consider the improvement of their educational level as a specific practical value noting that it leads to higher wages.

The purpose of the improvement of education and its implementation are organically embedded in the overall strategy of professional and social growth and development, characteristic of middle-level managers. Middle managers are characterized by an active life and professional position, focused on achievement, career, improvement of their standard of living and professional positions.

On the other hand, such a strategy is stimulated by the labor market, when enterprises and organizations operating in conditions of constant crisis and high competition impose a high level of demands on managers, indirectly linking the level of professionalism of a manager and his level of education.

3 Conclusion

Thus, the study showed that the educational level of Russian middle-level managers significantly exceeds the educational level of employees in the related industries and Russia in general. At the same time, the educational level of managers is growing every year and the values of education are dominant in the system of their values. This reflects the process of intellectualization and professionalization of the social and professional group of managers and indicates a gradual transition to a post-industrial society.

Let us determine the main areas of education for Russian managers:

- The concept of lifelong education, which allows permanent formation and development of new competencies;

- Shorter and more concentrated MBA programs (as opposed to the classic 2-year programs) with the focus on a specific field of activity: finance, marketing, HR, etc.

- Segmentation of managers in training by levels (top managers, middle managers, line managers) and by area of specialization;

- Improvement of education within the framework of their own training centers, corporate universities, created directly in a particular organization or an enterprise;

- On-the-job education (online or with the help of modular programs).

If the above-mentioned tendencies in the educational capital of Russian managers continue, then the conditions for the development of the production of goods and services necessary for the innovative development of the Russian economy will become possible: the development of proactive analytics and forecasting losses from unforeseen changes in the economic conditions, the expansion of analytical functions with the introduction of CRM systems at all stages of the functioning of organizations and enterprises, the use of modern management technologies using automated solutions (CRM, ERP, project management systems, ECM, WMS, CAD, CAMS and other specialized systems, specialized software), continuous training of management at all levels in order to adapt the level of qualifications of managers to changing uncertain conditions of society and markets.

References

1. *Russia is 3.8 times behind Ireland in terms of labor productivity,* Retrieved from: https://www.rbc.ru/economics/05/02/2019/5c5872889a794725eb8d815e
2. J. Sullivan, *VUCA: the New Normal for Talent Management and Workforce Planning,* Retrieved from: https: //www.ere.net/vuca-the-new-normal-for-talent-management-and-workforce-planning/
3. E. Lederer, J. Marshak *The new middle class.* (N.Y. 1937)
4. W.L. Warner (Ed.). *Yankee City.* (New Haven & London: Yale Univ. Press, 1963)
5. D.W. Rossides *The American class system. An introduction to social stratification.* (Boston: Mifflin, 1976)
6. L.Yu. Shemyatikhina *Education and Science* **1** 63–71 (2008)
7. O. V. Kryshtanovskaya *Sotsis.* **8** 17–49 2002
8. G. Mintzberg *Managers are required, not MBA graduates. A hard look at soft management policies and managerial training.* (Moscow: Olymp-Business, 2008)
9. O. I. Shkaratan, V. A. Bondarenko, Yu. M. Krelberg, N.V. Sergeev *Social stratification and its reproduction in modern Russia* (Moscow: SU-HSE, 2003).
10. V.S. Magun, M.G. Rudnev *Educational policy.* **9-10 (47-48)**, 65–72 (2010)
11. E. Toffler *Metamorphoses of power* (Moscow Publishing house ACT, 2003)
12. A.S. Gottlib *SotsIs.* **7** (2001).
13. N.V. Nesterova, D.A. Bezenkov *Bulletin of SUSU. Series: Social Sciences and Humanities,* **7 (47)** 142-146 (2005)

14. S.A. Inyasevsky *Journal of Sociology and Social Anthropology*. **XI (3),** 77–99, (2008).
15. Yu.G. Byuraeva *Bulletin of TSU*. **340,** 48–52, (2010).

Increasing company innovation in Russia: developing employee engagement through leadership and knowledge culture

Ekaterina Baidina*, *Natalya* Molodchik, *Natalya* Nagibina

Perm National Research Polytechnic University, Komsomolski pr., 29, Perm, Russian Federation

Abstract
The article examines the impact of knowledge culture and transformational leadership on employee engagement and company innovation. The partial least square is applied to the data based on the questionnaire survey of the top and middle managers of 110 different Russian enterprises. The study reveals a significant positive impact of employee engagement on company innovation, as well as an empirical evidence of significant positive influence of knowledge culture and transformational leadership on employee engagement.

1 Introduction

In the post-pandemic period, for their strategic stability, an increasing number of companies are studying the factors of the formation and development of innovative activities of personnel. Innovations in the form of breakthrough ideas allow obtaining new products, improving services, increasing the efficiency of business processes and thereby ensuring the company's competitiveness [1]. The combination of organizational ideas, people involved and effective technology in any business delivers meaningful and sustainable results. An increase in engagement by 5% in a year gives a growth in the company's revenue by 3% [2].

In modern management, engagement is called a business asset, which is comprehensively studied and enhanced. In this study, the authors suggest that transformational leadership and a culture of knowledge increase employee engagement in production activities, which in turn has a positive impact on the growth of innovations. The influence of these factors will be tested empirically.

The purpose of the presented article is *"to identify the influence of the culture of knowledge and transformational leadership on the formation of personnel engagement in order to increase the number of innovations in the organization."*

2 Methodology of the study

The authors use structural equation modelling in particular partial least square technique in order to estimate the impact of leadership and knowledge culture employee engagement and its further influence on company's innovation. Partial least square allows modeling unobservable latent variables of the study, which are constructs o interest that cannot be measured directly. In order to measure such constructs, the researcher should build a theoretical model and describe it with the help of several observable variables [3]. It means that latent variables are measured indirectly and measurement errors should be taken into account by describing results. For this study the authors construct 4 latent variables, namely, innovation, engagement, transformational leadership and knowledge culture. They were identified through the questionnaire addressed to top management of companies. The latent variable is represented by the system of equations (1):

$$\begin{cases} x_1 = \alpha_1 + \beta_1 X + \varepsilon.x_1 \\ x_2 = \alpha_2 + \beta_2 X + \varepsilon.x_2 \\ x_3 = \alpha_3 + \beta_3 X + \varepsilon.x_3 \\ x_4 = \alpha_4 + \beta_4 X + \varepsilon.x_4 \end{cases} \quad (1)$$

where α_i is a constant; β_i is the regression coefficient; and $\varepsilon.x_i$ is the error term.

The smartPLS software is applied to estimate the system of several equations simultaneously. As a result, the researcher obtains regression coefficients, variation and covariance of independent variables. The validity and reliability of the model is evaluated with the help of Cronbach's alpha (CA) and Average Value Extracted (AVE). In particular, the coefficient of Cronbach's alpha should be greater than 0.8. In this case one can state a high internal consistency of latent variable meaning that questions (metrics) describe one phenomenon. For Average Variance Extracted the estimates should exceed 0.6 in order to state a high discriminant validity of latent constructs.

The authors use previous theoretical and empirical studies for developing metrics which are included in the questionnaire survey. The top and middle managers of Russian enterprises participated in the study. The final database consists of 110 observations which is equal the number of enterprises. Descriptive statistics show that more than 75% of respondents had been working at the enterprise for more than 3 years. Considering industry

* Corresponding author: katerina-a-d@mail.ru

characteristics, most of the enterprises are from manufacturing, namely 64%, the service sector is presented by 22%, the IT industry - 9% and trade representatives are 5%.

According to the criteria of the number of employees, 32% of enterprises can be classified as small and medium-sized, 65% are large and 3% are super-large enterprises. It should be noted that almost all enterprises that participated in the survey are actively engaged in innovative activities in all types of innovations. At the same time, product and technological innovations stand out, the volume of which increased for most of the surveyed enterprises.

During the survey, the presence of knowledge culture, transformational leadership, employee engagement, company innovation was assessed on a 5-point Likert scale. This scale identifies "1" as the absence of this practice at the enterprise, "2" means rare use of this practice, "3" - the practice is used from time to time, "4" corresponds with regular usage of the practice, and "5" - the practice is used on an ongoing basis. The data collection was organized in the period from August 2019 to February 2020. The survey was shared with the help of the Google Forms toolkit. This instrument allows sending a link to the online form to the respondent and then make data analysis.

3 Theoretical and empirical justification of latent constructs

3.1 Innovation

The experience of successful companies proves that it is innovative performance that ensures sustainable development in the future.

A lot of studies [4] are devoted to the definition, composition, structure, measurements, mechanisms and risks of innovation and innovation potential management. From the standpoint of strategic and HR management, the study of the sources and drivers of innovation development made it possible to determine the vector of the formation of the innovation environment. The involved employee, being an innovator [5] / inventor / entrepreneur [6], initiates innovation activities in the organization, thus increasing the number of innovations.

The model was based on technological, product and managerial innovations [1]; marketing innovations were not taken into account, since in the calculations the factor load for this element turned out to be below the norm. The selected elements (types of innovations) have high discriminant validity (AVE = 0.720) and are internally consistent (CR = 0.806).

In any business, meaningful and sustainable results will be provided by a combination of organizational ideas, people involved and effective technologies [2].

3.2 Employee engagement

The logic of the formation of metrics for assessing engagement is based on the concept, forms of manifestation, conditions and factors affecting engagement.

Employee engagement theory has revolutionized management practice. More than 30 years have passed since 1990 when W. Kahn [7] published his seminal article on personal involvement in work.
In his research, he concludes that based on their own experience in the work environment, employees make decisions about the degree of their involvement.

Since then, interest in engagement research has grown every year. To improve the competitiveness of a company in a rapidly changing world [8], researchers are constantly studying the phenomenon of involvement. To date, a staggering array of theories, concepts, definitions and measurements have been developed [9]. Kahn W. [7] considers involvement from the perspective of personal role participation: cognitive, emotional, psychological expression of the personality, the true self. Shuck B. [10] examines engagement from the perspective of attitudes towards work tasks and assignments. In contrast to burnout, engagement refers to an activated, positive state of mind directed toward work assignments.

Bailey C., Madden A., Alfes K., Fletcher L. [11] pooled the results of 214 studies, presenting the meanings, antecedents, and outcomes of engagement for the organization and well-being of the employee. The authors identified six different conceptual approaches to engagement, as well as a theoretical rationale for engagement within the "job requirements - resources" framework with the dominant concept of "work engagement" by the Utrecht Group. Five groups of factors preceding involvement have been identified: 1. psychological states; 2. working design; 3. leadership; 4. organizational and team factors; 5. organizational interventions.

Therefore, the formation of observable variables for assessing engagement was based on the following models:

- Gallup. The famous Q^{12} Gallup Access questionnaire contains 12 questions [8]—elements describing the fundamental psychological requirements for unlocking human potential.

- Culture Amp [12]. Since 2011 and by the time of the study in 2018 (1200+ companies from more than 120 industries), in the "Standard" format, the following main ones were identified: training and development, leadership, focus on service and quality.

- Kincentric. According to Kincentric data, worldwide engagement is stable (65%) and has increased by 1% in recent years (data from 2019 [2]). The questionnaire is based on a concept containing a bunch of 3 indicators: speaks, stays and aspires, and includes 6 questions, 2 for each indicator. Speaks

(speaks positively about the organization, conversing with colleagues, potential employees and clients), stays (feels a sense of belonging and wants to be a part of the company), aspires (is motivated to make additional efforts to achieve success in his work and for the company as a whole).

In the model, the latent construction "Engagement" is defined by three variables. They are shown in Table 1. These elements have high discriminant validity (AVE = 0.639), are internally consistent (CR = 0.716) and reflect the high explanatory power of each of the elements of the latent construct.

The authors' research interest is based on Kahn's finding that engagement is not static; it fluctuates with the employee's experience, which is influenced by organizational factors. Consequently, the results of engagement primarily depend on the actions of the organization, management and culture; the practical meaning of studying organizational factors of engagement is determined by leadership and the culture of knowledge.

3.3 Transformational leadership

As it was indicated earlier, involved employees are a condition for the formation of a competitive advantage of enterprises [13]. Also, leadership is a necessity for the successful operation of enterprises in a modern competitive environment [14].

Rabiul M.K. and Yean T.F. confirmed that transformational leadership style significantly affects employee engagement [15]. The transformational leader influences the involvement of subordinates through inspirational communication with them [16].

Researchers Bouwmans M. [17], Wang B. [18], Lee Y. [19] and others confirm the positive relationship of the transformational leadership style with the high independence of employees, joint decision-making and feedback-oriented behavior. Therefore, to build engagement, transformational leaders empower their subordinates to achieve the long term company's goals and take into account the views of others.

Employee engagement refers to a transformational leadership style and manifests itself in leadership actions exemplifying entrepreneurship, responsibility and innovation.

The latent construction of "leadership" is defined by four elements in our empirical research questionnaire that have been converted into statements for analysis. The results of empirical testing are presented in Table 1. They show that the selected elements have high discriminant validity (AVE = 0.688), are internally consistent (CR = 0.849) and reflect the high explanatory power of each of the questions of the latent construction.

Leadership behavior, which manifests itself in the form of mentoring, providing more independence to employees, orientation towards actions, both in the interests of employees and the organization as a whole, allows forming the involvement of personnel.

3.4 Knowledge culture

The culture of the organization is one of the important management tools in competitive organizations, capable of creating a corporate climate with trusting relationship, openness and participation [20]. The culture of knowledge strengthens the common goals of employees and the organization, supports fruitful work with the team, allows to determine the areas of staff development, taking into account the current competitive situation, participates in the formation of conditions for the growth of human capital.

Table 1. Factor loads of metrics included in latent constructions

Metrics	Factor Loads
Innovation (CR = **0.806**, AVE = **0.720**)	
Management innovation	0.816
Product innovation	0.859
Technological innovation	0.870
Engagement (CR= **0.716**, AVE = **0.639**)	
Employees, on their own initiative, make suggestions for improving the company's products or services	0.775
Despite personal interests, employees voluntarily strive to achieve the desired results with additional effort	0.843
Most employees require minimal management oversight to achieve the desired results	0.778
Transformational Leadership (CR= **0.849**, AVE = **0.688**)	
Leaders know how to inspire their subordinates to solve difficult work issues	0.797
Leaders prefer to consider the opinions of the people around them	0.833
Leaders empower employees to achieve the strategic goals of the organization	0.827
Leaders demonstrate innovative and entrepreneurial patterns, and take responsibility for their decisions	0.861
Knowledge culture (CR= **0.892**, AVE = **0.699**)	
The organization approves and supports the desire of employees to learn	0.821
The organization approves and supports teamwork	0.875
The organization encourages employees to take initiative and generate ideas for product and manufacturing process improvement	0.820
The organization endorses the independence of employees in defining and achieving their goals	0.867
There is a high degree of trust between people in the company	0.793

In the questionnaire, the authors chose 5 metrics for the analysis of the latent structure "Knowledge culture", while the choice of metrics was based on the experience of previous studies and relied on the peculiarities of the culture of knowledge of organizations that manage their competitiveness. These 5 metrics are in Table 1. The authors considered these constructions to be sufficient for studying the knowledge culture, relying on the principle of optimality when creating a questionnaire.

The latent construct "Knowledge culture" was tested in terms of validity and internal consistency, and their values were obtained: AVE = 0.699, CR = 0.892, respectively. Testing of the latent construction was carried out on the example of employees of various Russian enterprises. The results are in Table 1.

Knowledge culture directly affects employee engagement, as evidenced by many studies.

A jointly supportive organizational culture and excellent leader performance have significant effects on professional engagement and job satisfaction according to Meng J. and Berger B. K. [21].

Studies by Jena L.K., Pradhan S., Panigrahy N.P. [22] confirm that engagement establishes "a clear connection between the employee and the organization's activities" and provides a better understanding of the organization's goals, as argued by Hughes J.C. and Rog E. [23].

Farrell M.A. found that a learning orientation, which is part of organizational values, has a positive effect on organizational engagement [24].

Table 1 contains 4 latent constructs "Engagement", "Knowledge culture", "Transformational Leadership" and "Innovation" with their decoding metrics and an indication of the factor load.

4 Results of the study

Correlation analysis and one-factor models showed that knowledge culture and transformational leadership have a positive and significant impact on employee engagement, while engagement, in turn, has a significant impact on company innovation.

Further, the authors searched for the optimal configuration of studied variable interactions with a joint effect on innovation. Different models were tested and the final decision was taken considering the following issues. Firstly, theoretical justification and empirical evidence of previous studies regarding the relationship of the studied variables were taken into account. Secondly, the authors have chosen the final model based on the results of one-factor model and correlation analysis.

The final model was significant and indicated high quality with the Standardized Root Mean Residual (SRMR) equaled 0.08. Table 2 shows the estimated coefficients for the model.

One could see that the influence of knowledge culture (0.506) is twice as strong as the influence of transformational leadership (0.234) on the employee engagement of Russian enterprises. Together, they explain around 48% of the variation in the engagement variable. In turn, the variable "employee engagement" explains 23.6% of the variation in the dependent variable "innovation".

Table 2. Results of model estimation

	Original Sample (O)
Engagement -> Innovation	0.486
Knowledge culture -> Engagement	0.506
Leadership -> Engagement	0.234

Thus, we obtained empirical evidence about the positive and significant impact of knowledge culture and transformational leadership on employee engagement in order to increase company's innovation.

5 Conclusions

In this study, a hypothesis was put forward about the influence of employee engagement on the growth of the number of innovations in the organization due to the formation of a culture of knowledge and leadership.

Four latent constructs were tested with fifteen element questions including: innovation; involvement; culture of knowledge; transformational leadership. For adaptation to a Russian-speaking audience, the questions-elements were selected from the works of English-speaking authors. The high discriminant validity and internal consistency of all investigated constructs confirmed the results of the empirical analysis. This confirms the contribution of the authors of this article to the empirical knowledge of the elements of "Innovation", "Engagement", "Knowledge culture" and "Transformational Leadership" for the academic community in order to conduct further research in this area.

The work shows that the culture of knowledge has the greatest influence on the involvement of personnel. Leadership has a direct and positive, but smaller impact on employee engagement. There is no doubt that a small sample makes the resulting empirical knowledge limited. Accordingly, this factor should be taken into account when formulating conclusions and practical recommendations.

The results of the study of the formation of involvement due to the positive influence of organizational mechanisms reflect the data on 110 Russian enterprises.

Both businesses and researchers can use the findings for their own purposes. Top management of the companies could take into account these empirical results by implementing human resource management practices with the aim to increase company innovation. Scholars could apply proposed research framework to the analysis of other samples and provide comparative analysis.

References

1. Oslo Manual. *Recommendations for collecting and analyzing innovation data.* Third edition. Translation from English. Second revised edition. M. 31-107. (2010)
2. *Kincentric. Culture&Engagement. Wired for Engagement. Improving Engagement Through Better Hiring.* Retrieved from: https://www.kincentric.com/-/media/kincentric/2019/november/culture/wired-for-engagement.pdf
3. O.V. Mitina. *Experimental Psychology,* **1**, 142 (2008)
4. Perm National Research Polytechnic University Publishing. *Schumpeterian Readings.* The Ninth International Scientific Research Conference Proceedings. 241 (2020)
5. World Bank. Retrieved from: https://data.worldbank.org/indicator/NY.GDP.MKTP.KD.
6. I.A. Schumpeter. *Economic development theory. Capitalism, socialism, democracy.* 864 (2007)
7. W. Kahn. *The Academy of Management Journal.* **33 (4),** 692-724. (1990).
8. *The Power of Gallup's Q12 Employee Engagement Survey* Retrieved from: https://www.gallup.com/access/323333/q12-employee-engagement-survey.aspx
9. W.H. Macey, B. Schneider. *The Meaning of Employee Engagement. Industrial and Organizational Psychology.* **1(1),** 3-30 (2008).
10. B. Shuck. *Human Resource Development Review,* **10**. 304–328 (2011).
11. C. Bailey, A. Madden, K Alfes, L. Fletcher. *International Journal of Management Reviews.* **19**, 31-53 (2017)
12. *Insights from Culture Amp's 2018 employee benchmarks* Retrieved from: https://www.cultureamp.com/blog/2018-employee-benchmarks/
13. S.L. Albrecht, A.B. Bakker, J.A. Gruman, W.H. Macy, A.M. Sachs *J. Organ. Eff. People are performing.* **2 (1)**, 7-35 (2015).
14. O. R. Mahdi, M. K. Almsafir.. *Procedia - Social and Behavioral Sciences.* **129**, 289-296 (2014).
15. M. K. Rabiul, T. F. Yean. *International Journal of Hospitality Management.* **92**, 102712 (2021).
16. *Transformational Leadership: A Quick Start Guide: Growing, Supporting, and Empowering People to Amplify Change in Organizations.* Retrieved from: https://static1.squarespace.com/static/571faf00c2ea510eafddb70b/t/5c1f087121c67c0d67c426a8/1545537650781/2018-Forum-Transformational-Leadership.pdf
17. M. Bouwmans, P. Runhaar, R. Wesselink, M. Mulder. *Teaching and Teacher Education.* **65**, 71-80 (2017).
18. B. Wang, J. Qian, R. Ou, C. Huang, B. Xu, Y. Xia. *Social Behavior and Personality: An international journal.* **44 (7)**, 1201-1208 (2016).
19. Y. Lee, J. Kim. *Public Relations Review.* **47 (1)**, 101998 (2021).
20. M. Molodchik, C. Jardon. *The Learning Organization.* **22 (6)**, 306- 316 (2015).
21. J. Meng, B. K. Berger. *Public Relations Review.* **45 (1)**, 64-75 (2019).
22. L. K. Jena, S. Pradhan, N. P. Panigrahy. A*sia Pacific Management Review.* **23 (3)**, 227-234 (2018).
23. J.C. Hughes, E. Rog. *International Journal of Contemporary Hospitality Management.* **20 (7),** 743-757 (2008).
24. M. A. Farrell. *Marketing Bulletin.* **10**, 38-51 (1999).

Innovative ideas of the intelligentsia in the field of management and zoning of the Urals territory

Gulsina Selyaninova*

PNRPU, 29, Komsomolsky prospekt, Perm, 614990, Russia

Abstract. The article is devoted to the characteristics of changes in the management and administrative-territorial structure of the Urals under the regime of A.V. Kolchak (November 1918 - July 1919). These changes took place with the active participation and on the basis of the ideas of the intelligentsia involved in power in this period of Russian history. The administration and zoning of the Urals territories under A.V. Kolchak's regime bore a succession character in relation to the previous regime of a "democratic counter-revolution" that existed during the reign of the Provisional Regional Government of the Urals (August - November 1918). However, the size of the subordinate territories under the Supreme Ruler was more ambitious, which also determined the need to solve more significant administrative tasks and changes in the area of zoning. The regions became large territorial units. They were divided into provinces, counties and volosts. S.S. Postnikov was the head of the Ural region, and the Perm province was ruled by Chistoserdov N.P. Administrators were appointed to rule the counties, who managed to hold a congress of the counties governors of the Perm province in February 1918.

1 Introduction

The coup in November 18, 1918, which ended the activities of the Directory, put an end to the lingering illusions of the Russian intelligentsia about the possibility of democratic development. The coup by A.V. Kolchak led to the establishment of a new political system in the territories of Siberia and the Urals, which were not subject to Soviet power. Being a supreme ruler, A.V. Kolchak possessed the highest military and civilian power.

Almost within several months after A.V. Kolchak had come to power, there was a restructuring of the management system and the administrative-territorial structure in the subordinate territories [1]. Representatives of the intelligentsia who entered the structures of state power in the Urals and Siberia took an active part in it.

2 Materials and methods

This study is based on an analysis of archival documents from the State Archive of the Perm region and the State Archive of the Sverdlovsk Region. Textual analysis of archival sources is complemented by methods of intellectual history and structural-functional analysis.

3 Results

Large territorial units were called "krais". The post of the Chief of the Territory was established, who was subordinate to the Minister of Internal Affairs. In the Urals, the Samara-Ufa and Ural Krais were created, but the previous administrative-territorial division into provinces, counties, volosts was also preserved. The territories of the Perm province made up the Ural Krai, but Perm was captured by the Kolchakites only on December 24, 1918. Soon after that, by the end of January 1919, the military and civil administration of the region was formed.

In February 1919, the Zemstvo department of the Ministry of Internal Affairs began to collect information on the territorial division of volosts and counties. This was due to the fact that "the population of some districts arbitrarily, without permission, distributed the boundaries of the former administrative divisions" due to the "inconsistency of existing borders with new conditions, with the emergence of new economic and cultural centers, to which the population naturally gravitates" [9, p. 5].

Based on the petition of the Chief Head of the Ural Krai, the Cherdyn district zemstvo and the Troitsko-Pechora, Savinoborsk, Ust-Shchugorsk volosts of the Ust-Sysolsk district of the Vologda province, the head of the Ministry of Internal Affairs decided to annex the above listed volosts to the Cherdyn county (district) of the Perm province, since the moment of liberation from the Bolshevik the province were outside the

* Corresponding author: gsina@mail.ru

https://doi.org/10.1051/shsconf/202111600023

administrative connection with the center of the Ust-Sysolsk district, occupied by the Bolsheviks [5, p. 8].

During the offensive of the Red Army in summer 1919 and the abandonment of Perm in the department of general affairs of the Ministry of Internal Affairs, the question of reorganizing the Perm province and dividing it into two units - Perm and Yekaterinburg - was discussed [6, p. 1].

During the functioning of the "white" regime by A.V. Kolchak, the management of the Urals territories was carried out by the Chief of the Ural Territory, who was stationed in Yekaterinburg. This position was introduced by order of the Supreme Ruler on December 10, 1918 [8, p. 18].

The Council of Ministers adopted a special resolution on the issue of the reorganization of administrative bodies in the Urals - "On the abolition of the chief commissioner post for the Urals region of and the administrative bodies established by the regional government of the Urals". The Chief of the Ural Krai was instructed to determine the gradualness and order of this reorganization. The resolution, signed by the Chairman of the Council of Ministers P. Vologodskiy on December 20, 1918, was approved by Admiral A.V. Kolchak on December 28 [3, p.33].

The forty-eight-year-old nobleman, mining engineer with the rank of collegiate councilor Sergei Semenovich Postnikov, who had previously worked in the Urals in the Theological Mining Society from 1914 to 1918, became the Chief of the Ural Krai. On August 9, 1918, he was appointed Minister of Trade and Industry of the Siberian Government and at the same time was a commissioner of the Siberian government under the Council of VOPU [15, p. 120]. After being appointed as a Chief of the Ural Krai, S.S. Postnikov also retained the position of assistant minister of trade and industry of the Russian government [15, p.131].

S.S. Postnikov was appointed to the post by the decree of the Supreme Ruler. Colonel Domontovich, the commander of the 1st Ulan regiment, became his assistant and deputy [8, p. 20].

Order No. 7 of the Chief of the Ural Krai of January 3, 1919 stated that he should be contacted on issues beyond the competence of subordinate bodies of local and provincial administration, and "not subject, remembering that they are responsible for supporting the state and destroying its enemies, and that every minute of delay through diverting them from their direct work can be disastrous for Russia" [8, p. 18].

The Chief of the Ural Krai was the successor of the departmental chief directorates of the Provisional Regional Government of the Urals, which, having been created in August 1918, functioned until December 1918. The chief executives (ministers) of the Provisional Regional Government of the Urals were replaced by commissars and other officials.

The first orders of S.S. Postnikov touched upon the issues of reorganization of power structures in the Urals. The right to determine the procedure for this reorganization was granted to him by a decree of the Council of Ministers of December 20, 1918 [7, p. 33]. In a telegram dated January 22, 1919, S.S. Postnikov expressed the demand for general issues to be resolved, having previously agreed on them, and also notified that the approval of certain persons in different positions is made by the Minister of Internal Affairs through the Chief of the Ural Krai [7, p. 22].

On December 31, 1918, the main department of agriculture and state property of the VOPU was disbanded. The Perm Department of Agriculture and State Property became its successor [8, p. 19]. Instead of the chief executive, a labor commissar for the Ural region (krai) was appointed (both terms are found in the documents). It was N.M. Obukhov [8, p. 17].

On January 1, 1919, the main department of public education was disbanded [8, p. 17]. V.M. Anastasyev, who retained the position of director of the Yekaterinburg School of Art and Industry, was appointed a trustee of the Orenburg educational district [11, p. 27].

The question connected with the seat of the Committee (Congress) of the Constituent Assembly members was decided by S.S. Postnikov in favor of Yekaterinburg [8, p. 14]. Former Deputy Chief Executive Officer of Public Education VOPU A.M. Rozov, the teacher of the Yekaterinburg real school, and the assistant professor of the Ural Mining Institute, the teacher of the Yekaterinburg Teachers' Institute M.N. Khitrin were appointed to the posts of district inspectors [8, p. 13].

By order of the All-Russian Government, the Main Directorate of Mining Affairs of the Urals was disbanded, and in its place the Ural Mining Directorate was restored, headed by the Chief Head of the Ural Mining Plants, Mining Engineer Alexander Andreyevich Ivanov. But the Ural Industrial Committee continued to function in its former composition under the chairmanship of the mining engineer A.E. Gutta [8, p. 10].

At the disposal of S.S. Postnikov, there were funds for the reorganization that were issued against receipt to newly appointed persons. So, V.M. Anastasyev, still in the capacity of the chief manager of the VOPU, was given an advance payment "for the maintenance of personnel under the authorized representative of the Ministry of Public Education and for economic expenses in the amount of four thousand rubles" already on January 3, 1919 [11, p. 26]. Funds were also allocated to address issues that can hardly be called priority ones. On January 20, 1919, 7 thousand rubles were allocated for the purchase of collections of antiquities from the museum of the art and industrial school to the trustee of the Orenburg educational district V.M. Anastasyev [11, p. 27].

The head of the Ural Krai was directly subordinate to the governor of the Perm province. The governor of the province during the reorganization of the management system under A.V. Kolchak began to be called the provincial commissar, in his subordination were the governors of the counties (former county commissars) [10, p. 2]. Nikolai Pavlovich Chistoserdov was appointed to the post of governor of the province. In August 1918, he was the head of the administrative department of the Yekaterinburg garrison, and after the Kolchak coup - "assistant to the commissioner for the maintenance of state order and public peace" in Yekaterinburg [16, p. 3; 18, p. 5]. Mikhail Vasilyevich Kukaretin, who for a long time was an indispensable member of the Perm provincial presence, became the assistant-deputy governor of the province [2]. By the way, it should be noted that before January 4, 1919, that is, before N.P. Chistoserdov, he served as governor of the province. By the end of January 1919, the office of the governor of the Perm province and the staff of the provincial government were formed. Provincial veterinary inspector A.I. Yutsevich; provincial medical inspector P.P. Shipilin; provincial engineer E.I. Artemov, provincial architect E.E. Pehler were appointed to their positions [15, p. 108]. On March 5, 1919, P.I. Danilov was appointed chief of the city police [15, p. 126].

Managing the Perm province, N.P. Chistoserdov dealt mainly with various socio-economic issues, including, for example, land management issues, the tasks of supplying towns with fuel, building materials, etc. [14, p. 263].

Since Perm remained the provincial center, and the Chief Head of the Ural Krai was in Yekaterinburg, there was some inconvenience in management. At the end of January 1919 S.S. Postnikov agreed in Omsk on the transfer of the governor of the province to Yekaterinburg. However, the intervention of the commander of the 1st Central Siberian Corps, Lieutenant General Pepelyaev, canceled the relocation of the provincial presence [17, p. 3, 6]. Pepelyaev's order aroused the indignation of S.S. Postnikov, who stated that the chief heads of the region, being the highest representatives of the supreme power at the local level, are exclusively subordinate to the Supreme Ruler. And the transfer of the province administration to Yekaterinburg was agreed with him, as well as with the Chairman of the Council of Ministers and the Minister of Internal Affairs [17, p. 5, 5rev.]. But the protest hung out in the air.

All branches of civil administration within the Perm province were subordinated to the chief of the Ural Krai [8, p. 18].

Managers were appointed at the head of the counties. In mid-February 1919, a congress of county managers of the Perm province was held. S.S. Postnikov recommended that the issue of volost zemstvos should be comprehensively considered at it [12, p. 23]. Before the congress, the county managers in their telegrams outlined a range of problems that needed to be discussed. Among them were the development of regulations on county governors, the staff and credits of the chancellery, relations with other public institutions and control over them, the differentiation of the functions of military and civil authorities in the frontline zone, etc. [12, p. 19, 26].

The congress was chaired by the governor of the province, N. P. Chistoserdov. In addition to him, his deputy M. V. Kukaretin, permanent members of the provincial presence N. E. Andriolli, N. P. Chernyaev took part in the work of the congress; the governors of the counties arrived from their places: Perm – A. N. Ilyin, Kungur – A. I. Yushkov, Solikamsk – A. S. Komarov, Krasnoufim – A. N. Nefedov, Shadrin – E. N. Sumarokov, Irbit – M. A. Atmakin, Verkhotursky – E. K. Bekman, Cherdynsky – M. Ya.Kazantsev, Yekaterinburg – V. A. Startsev and Kamyshlovsky – A. N. Chechetkin [13, p. 1].

At the congress, the Regulations on the Chief head of the Ural Krai were announced, according to which "the latter is granted the right to make orders before the publication of the grounds for such in the legislative order" [13, p. 8]. All decisions of the city and zemstvo district self-governments must be submitted to the Governor of the province for consideration. The participants of the congress were also familiarized with the still valid Regulations of the All-Russian Provisional Government (Directory) on county commissars who received the status of county managers under A.V. Kolchak. They were assigned the rights and duties of the leaders of the nobility, the chairmen of the administrative committees responsible for the safety of citizens, the chairmen of the evaluation commissions, as well as the chairmen of the zemstvo administrative presence in charge of land use questions [13, p. 1rev.].

A heated discussion was caused by the question of dividing the districts of the Perm province into categories in accordance with the salaries of the content received. The Congress found it most fair to divide the counties (uyezds) into the following categories: the 1st category – Perm and Yekaterinburg with the counties, the 2nd category – Kungur, Krasnoufimsk, Solikamsk, Cherdyn, Verkhoturye, Irbit, Kamyshlov and Shadrinsk with the counties [13, p. 3]. Later, in March 1919, taking into account this decision, the Council of Ministers adopted the "Schedule of dividing the localities of the provinces of Orenburg, Perm and Ufa and the Turgay region into categories for the production of maintenance for employees of government agencies" [4, p. 9].

The consideration of the issue of the relationship of county managers with the police and military authorities was relevant for the current moment. It was decided that the military commandants are only agents of the commissioner for protection, and the investigative commissions should belong to the civilian department, but a special law is needed about them, the release of which is expected soon [13, p. 3]. The congress discussed the question of the states of the county governors' offices; in the organization of county congresses, it was stated that it was necessary to adhere to the decisions of the Provisional Government of 1917, since there were no orders to abolish them [13, p. 4rev.].

Some of the county managers spoke in favor of restoring the normal activities of the volost administrations and granting them a broader right to monitor the activities and record-keeping of the volost zemstvos. This question was raised due to the fact that

https://doi.org/10.1051/shsconf/202111600023

the zemstvos were turning from self-government bodies into part of the administrative apparatus with police functions, and, therefore, it was necessary to extend the supervision of county managers to the entire range of activities of the volost self-governments. N.P. Chistoserdov explained that the county managers already had the right to audit all the bodies of the zemstvo's self-government, since it belonged to the provincial commissar, and "the latter can entrust the right to audit the county commissar" [13, p. 5, 7, 8rev.].

The county managers sought to be able to impose administrative penalties on the rural administration bodies – the village elders. This requirement was met [13, p. 6].

The congress approved the staffing of the office of county managers, which included a secretary, two assistants, a registrar, two scribes, one typist, a messenger, a watchman, an accountant, a veterinarian and a veterinary paramedic, and medical personnel [13, p. 11, 11rev.]. The expenses for the maintenance of the states had to be "requested through the governor of the province by submitting special estimates" [13, p. 13].

4 Discussion

Although Russian historians have turned to the study of public administration and zoning in the Urals and Siberia during the Civil War, here we can note mainly works related to the analysis of changes in the state structures of Siberia, such as the study of A.N. Nikitin [1]. The question of the ideas and activities of the intellectuals under the Kolchak regime in transforming the system of administration and zoning in the Urals remained outside the scope of existing studies.

5 Conclusion

Thus, when forming the bodies of state power and local self-government under the regime of A.V. Kolchak, in the sphere of zoning, the authorities relied primarily on the innovative ideas of the intelligentsia, who actively formed the system of management and administrative-territorial structure of the Urals and sought to create stable conditions for peaceful life in its territories.

The management and zoning of the territories of the Urals under A.V. Kolchak was of a continuous nature in relation to the previous regime of "democratic counter – revolution" that existed during the rule of the Provisional Regional Government of the Urals (August-November 1918). However, the size of the subordinate territories under the Supreme Ruler was more extensive, which determined the need to solve more significant administrative tasks and changes in the field of zoning. The regions became large territorial units. They were divided into provinces, counties, and volosts, and county managers were appointed at the head of the counties.

References

1. A N Nikitin □.□. Statehood of "white" Russia: formation, evolution, and collapse. (Moscow: National Institute of Business, 2004)
2. Liberation of Russia (Perm). 1919. 1 Jan
3. State Archive of the Perm region, form R-746, inv. 1, doc. 1
4. State Archive of the Perm region, form R-746, inv.1, doc. 3
5. State Archive of the Perm region, form R-746, inv. 1, doc. 8
6. State Archive of the Perm region, form R-746, inv. 1, doc. 10
7. State Archive of the Perm region, form R-746, inv.2, doc. 1
8. State Archive of the Perm region, form R-746, inv. 2, doc. 2
9. State Archive of the Perm region, form R-746, inv. 2, doc. 3
10. State Archive of the Perm region, form R-746, inv.2, doc. 4
11. State Archive of the Perm region, form R-746, inv.2, doc. 11
12. State Archive of the Perm region, form R-746, inv.2, doc. 13
13. State Archive of the Perm region, form R-746, inv.2, doc. 14
14. State Archive of the Perm region, form R-746, inv.2, doc. 28
15. State Archive of the Sverdlovsk region, form P-1956, inv.1, doc. 1
16. State Archive of the Sverdlovsk region, form P-1956, inv.1, doc. 2
17. State Archive of the Sverdlovsk region, form P-1956, inv.1, doc. 44
18. State Archive of the Sverdlovsk region, form P-1957, inv. 1. doc. 15

Effectiveness of participation in public social partnership: new views and judgments

Alena Arbuzova[1], *Natalya* Pazdnikova[2,*]

[1]PNIPU, 29, Komsomolsky ave., Perm, 614990, Russia
[2]PNIPU, 29, Komsomolsky ave., Perm, 614990, Russia

Abstract. The paper is devoted to the problems of public social partnership, the importance of the participation of the government, society and business in it is highlighted. The aim of the paper is to demonstrate the need to study the relations of social partnership and the importance of evaluating the effectiveness of the participation of government, society and business in them. Relations between the partnership actors are developing unpredictably, but the directions of state policy demonstrate the vector that is set in Russian practice. The authors analyze the existing theories in the field of social partnership and highlight the theory of stakeholders. The categories "public social partnership" and "public management of labor resources" are formulated. A system of indicators in the actor context is proposed, which assesses the level of participation of each of the partnership actors in the author's ideology of evaluating the effectiveness of participation. The authors' conclusions support the fact that the effectiveness of actors' participation cannot be assessed without a clear understanding of the social policy mechanisms used, as presented in the Russian regulatory framework.

1 Introduction

There is much talk these days about involving ordinary citizens more definitively and directly in the policy process. Dialogue, deliberation and citizen engagement are considered as the remarkable features of the current forms of public social partnership as attempts of creating collaborative basis for public involvement processes compete with more traditional top-down approaches. Public deliberation, a defining concept of deliberative democracy theory, is experiencing a renaissance among both scholars and policymakers. Televised town halls are now commonplace during election campaigns. Citizen dialogues have been used to elicit informed opinion and to probe for shared public values in conjunction with major policy reform initiatives.

Communications between different public organizations and citizens take an important place in the process of reducing the degree of information asymmetry and ensuring successful co-production and co-delivery of services. A well-functioning modern society requires that state employees treat sympathetically and respectfully citizens and vice versa. However, effective communication is necessary to maintain a healthy relationship and build mutual trust that facilitates co-production.

Citizen participation has been sourced through participative forms such as online technologies, social media and neighborhood meetings. Gordon & Mugar (2020) acknowledge that the use of digital platforms to engage citizens in matters that affect is on the rise.

The interaction of society and government has always been considered as a vector of modern state policy, but such interaction has not always had certain forms.

2 Materials and methods

There are a plurality of evaluation approaches and methods determining the effectiveness of participation in public social partnership, but according to Abelson J. (2006) they can be divided into three main groups:

1) user-based which assumes different participants have different goals and that evaluation must take them into account;

2) theory-based which is driven by theories and models of public participation and applies normative evaluation universally to any public participation effort;

3) goal-free evaluation which is not constituted by any stated goals and is conducted in the absence of any theory. Most evaluation studies to date fall under the user-based category. Another distinguishing feature of public participation evaluation is the emphasis on either process or outcome evaluation.

3 Results and Discussion

The problem of assessing public social partnership participation is connected, firstly, with the fact that the task of defining the end-point of a participation exercise for purposes of measuring effectiveness is often unclear.

* Corresponding author: pazdnikovan@mail.ru

Possibility of measuring the social impacts of the process, which takes substantial period of time, is difficult to disentangle from other events that are influential to the policy process, may be limited.

Some specialists offer to measure the effectiveness of participation in public social partnership using approaches based on perceived benefits/costs, which are crucial for uninterrupted involvement in collaborative attempts. In cases when involvement is considered as beneficial, it can be regarded with more participation, attendance, committee membership, and member recruitment (El Ansari and Phillips, 2004). The benefits of such involvement are multiple; due to this they can be distributed among all stakeholders, including users, professionals and organisations, hence personal, community and organisational benefits are important. On the personal level, benefits include recognition, enjoyment, skill enhancement improved control, activity, and satisfaction.

At a level of society participation unites the community initiatives (Naylor et al., 2002), and enlarges the volume of community resources that can support the credibility of attempts.

Benefits are defined as performance and satisfaction. In case if we measure it in non-economic terms, we are able to use such notions as "improved decision-making" and "competitiveness". Participation benefits are not easy to define, measure and assign monetary value to, but they could be used to urge politicians of a necessity to change.

The costs of participation are the negative consequences of participating. Absolute costs are determined as the upfront price of a product or service; while opportunity costs are benefits forgone by tying up resources in directly one particular use and withdrawing them available for alternative uses. A good service with an absolute cost of zero (i.e. free) may still have opportunity costs, e.g. time spent volunteering which could otherwise have been in paid employment.

Measuring the costs/benefits of participation has been met with skepticism and suspicion, or conversely interest and enthusiasm.

As far as there are not any properly tested measurement tools, another area of attention that need to be studied arises.

Most of the outcome-oriented empirical research has focused on assessing the impacts of public participation processes on a range of citizen participant attributes. These studies have reported: increased levels of interest and knowledge of public issues; improved capacity for future public involvement; increased propensity for social bond formation; and improved trust of fellow citizens. There has been less research about the direct impacts of public participation on the policy process and political decision-making. What has been produced offers mixed and ambiguous results (e.g., public deliberation can produce outcomes that influence policy but the conditions under which this occurs are not easily identifiable). However, more recent literature finds a strong association between the widely accepted decision outcomes and 'processes in which agencies are responsive, participants are motivated, the quality of deliberation is high, and the participants have at least a moderate degree of control over the process' (Beierle and Cayford, 2002).

A conceptual map of public participation evaluation can be depicted as it is shown in figure 1. The scheme presents different groups in the process of interaction as well as the stages of the process itself.

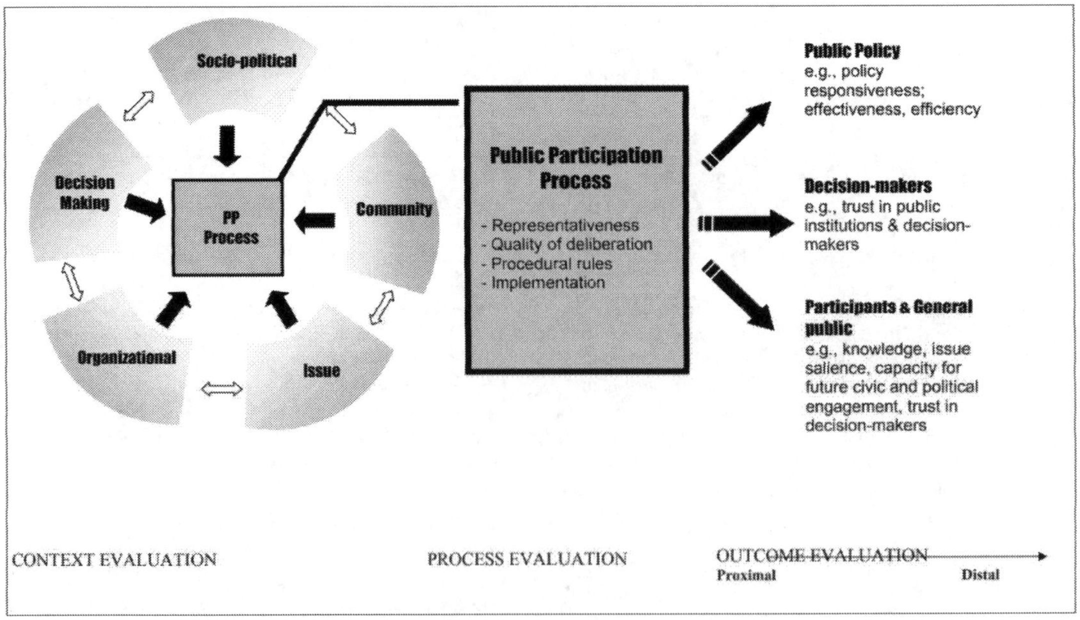

Figure 1: A design of Public Participation Evaluation

With the introduction of the term "social partnership" in the field of labor relations in the Labor Code of the Russian Federation, the forms of interaction were defined. According to Russian legislation, social partnership is a trilateral relationship between the government, employees (or employees ' representatives) and employers, where trade unions that represent the interests of employees play a special role.

However, when it comes to the public management of the territory's labor resources, the social partnership may acquire a slightly different understanding and form. The fact is that the socio-economic development of regions and their municipalities, as well as the solution of socially significant problems – all this is determined by the resource potential, including labor. In this regard, the labor potential of the territory gives its "fruits" in the case of partnerships between the "power" actor, the "business" actor and the "society" actor, since each of the parties receives its own benefits from such interaction.

Before formulating the concept of "public management of labor resources", it is necessary to identify the category of "public social partnership", which can be understood as a relationship between equal partners aimed at meeting common interests and solving socially significant problems. In general, identification should be carried out within the framework of the "subject-subject" paradigm, and for this purpose, in order to determine the demarcation of the actors of the "power-business-society" partnership, the goals and benefits of each should be established.

Power. The ability of the authorities to assess the state of the labor potential of the territory and to conduct various forms of dialogues with the actors "business" and "society" in order to identify the main problems and suggest ways to solve them. At the same time, the practical interest of the authorities is to increase the level of labor potential through the implementation of project and non – project forms of interaction.

Society. The main advantage for the society from the position of public social partnership means public participation in public administration, and the diverse interests of the population contribute to the choice of directions for the development of the territory as a whole and participation in socially significant events and projects. In this connection, it is possible to distinguish the goals inherent in society in social partnership - the acquisition of social goods of the appropriate quantity and quality; participation in the public control of social processes; awareness of the situation and free movement to places where conditions are more favorable, etc. However, the key goal of the company is to maximize the satisfaction of its own needs and participate in the search and further planning of measures to meet them. At the end of the XX century, Freeman R.E. formulated the "stakeholder theory" or stakeholder concept, based on the understanding that it is necessary to take into account the interests of stakeholders, that is, people and organizations.

Business. The activities of firms, large companies, and individual entrepreneurs are focused on the creation of goods, the provision of services, and the implementation of economic activities, which ultimately forms the country's GDP. That is why business acts as an employer for the population, and its interests are always aimed at obtaining financial benefits (income). Based on this, the goals of business from participation in social partnership are to extract profit and opportunities for its growth from cooperation. Returning to Freeman's theory, we note that stakeholders have a significant influence on business decisions and, in his opinion, private companies should build relationships with all stakeholders.

Thus, in pursuit of their own goals and interests, various actors automatically enter into agency conflicts, the solution of which is possible in the social partnership dialog box, but evaluation problems still exist.

Assessing the effectiveness of actors' participation in social partnership requires not so much the necessary methodology as an adequate system of evaluation indicators.

We see the following system of indicators in the actor context by the example of the Russian Federation (Table 1).

Table 1. Indicators of government, society and business participation in social partnership

Indicators	2019	2018	2017
Power (authorities)			
Mortality of the working-age population, per 100 thousand people.	470,0	482,2	484,5
The number and composition of the labor force, thousand people.	75398	76190	76285
Production of the gross domestic product, billion rubles.	166056	186754	195925
Higher education coverage, %	36,8	33,5	32,9
Professional education coverage, thousand people.	3006	2931	2852
Average monthly income per capita, thousand rubles.	35249	33178	31897
Unemployment rate	5,4/4,6	4,8	5,2

Society			
Minimum wage, (rubles per month)	11280	9489	7500
Number of trade union members, million people	19,9	20,7	25,1
The value of the subsistence minimum per capita, rubles per month	11606	10 328	9673
Number of initiative budgeting projects, pcs.	130	102	89
Total number of territorial public self-government (TPSG) projects received	140	116	103
Business			
Net financial result (profit minus loss), RUB million	9036848	12400336	15758426
Share of profitable organizations, %	64,1	66,9	73,5
Number of PPP projects, pcs.	179	403	654
The volume of investments in public-private partnership (PPP) projects, billion rubles.	12,67	20,54	23,96

Obviously, the list of indicators depends on the territory under study, its features and factors of the external and internal environment. Clearly, the ideology of evaluating the effectiveness of participation in social partnership of such actors as: "power", "society", "business" can be presented as follows (Figure 2).

The presented matrix of the effectiveness of participation in the social partnership of actors shows the evaluation vectors, in which the initial stage is the analysis and assessment of the problems of civil society. Starting this assessment, we should not forget that at each new stage of development, problems are caused by the situation and the peculiarities of the socio-economy and politics.

In turn, the identified problems determine the setting of goals and objectives for the actors in the social partnership, the achievement of which is evaluated by a system of balanced indicators (performance), which determines the effectiveness of the participation of each actor, in the end.

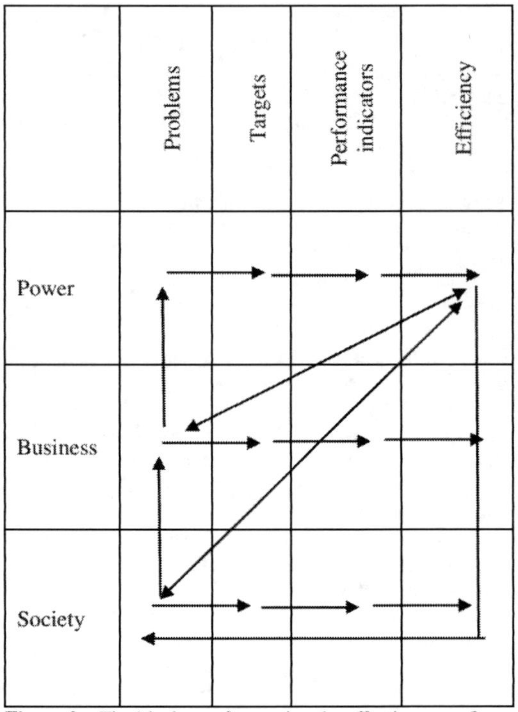

Figure 2 – The ideology of assessing the effectiveness of participation in social partnership

In turn, achieving the maximum effect of participation in social partnership for each actor depends on the vector and directions of social policy and the mechanisms that are defined in it. In our opinion, the mechanisms of social policy can be differentiated in different ways (institutional, socio-economic, organizational, etc.), but all mechanisms should be determined by the regulatory framework of the Russian Federation.

Thus, the socio-economic mechanisms include:

- budget programs of social design of the population and NGOs;
- provision of grants and subsidies for solving social problems;
-provision of tax benefits to society and business
– projects of territorial public self-
government – projects of initiative budgeting of citizens;
-mechanisms and projects of public-private (municipal-private) partnership;
– funds of local communities;
- volunteerism and volunteering programs;
- mechanisms and programs for strengthening the family, etc.

Institutional and organizational mechanisms include:

- institutions for regulating charitable activities;
– public chambers;
– activities of youth parliaments;
– mechanisms for citizens ' appeals;
– various institutions of business associations (unions of industrialists);

– public hearings in the budget process;
– public control and public expertise;
– institutions for human rights, for the rights of the child;
– anti-corruption mechanisms, etc.

4 Conclusion

Generally, there is a lack of economic assessment of participatory processes and social interventions. This is probably due to the complexity of participatory processes, resistance to the use of economic methods by practitioners or decision-makers, lack of appropriate data, in addition to the cost of undertaking proper economic evaluations.

Of course, the practice of social partnership requires constant regulation by the authorities, but the interest of the participants themselves is obvious, since the solution of socially significant problems for them is possible only together. Thus, the development of public social partnership is possible, firstly, with the implementation of new optimal mechanisms of social policy; secondly, with the presence of social communications between all actors of the partnership; thirdly, the introduction and development of brand-oriented business; fourthly, stimulating the interest of participants through various types of benefits provided to them.

References

1. J. Abelson, *Assessing the Impacts of Public Participation: Concepts, Evidence, and Policy Implications* (Canadian Policy Research Networks Inc., Ottawa, 2006)
2. El Ansari Walid, Ceri J Phillips, Health Promotion Practice, **5(1)** (2004)
3. T. Beierle, J. Cayford, *Democracy in practice: Public participation in environmental decisions.* (Resources for the Future, Washington DC, 2002)
4. T. Kryshtaleva, The World of economy and management. **17,** 3 (2017)
5. V. Galaganov, *Organization of the social welfare authorities* (Academia, Moscow, 2018)
6. M. Guslova *Organization and content of social work with the population* (Academia: Moscow, 2018)
7. O. Osipova, *Actualization of the social partnership of the university and the enterprise* (LAP Lambert Academic Publishing, Moscow, 2018)
8. A. Ollerenshaw, Local Government Studies, **43**, (2017)

Passionarity management: adapting the model to modern conditions

Yulia Karpovich[1], and *Natalya* Pazdnikova[1*]

[1]PNIPU, 29, Komsomolsky ave., Perm, 614990, Russia

Abstract. Public administration is now quite often identified with state and municipal administration, while the sign of identity between these concepts cannot be put, since public administration is the result of the development of state and municipal administration, its "higher stage of evolution". The article describes the situations when traditional ways of public management do not work that requires the development of qualitatively new methods of management or adoption of already developed once to the current conditions. The article is aimed to show the need to study the peculiarities of passionarity management as a perspective trend of public management. The authors analyze the existing theories in the field of passionarity management. The authors 'conclusions support the fact that the effectiveness of public management could be determined by the level of management passionarity.

1 Introduction

The changes taking place in the culture could be revealed through the process of creation new institutions, as well as new values in the modern society. While the economy undertakes the globalization trends year by year, the cultural capital is also growing and accumulated, creating different social forms.

Each era has its own effective management models. However, with the change of management paradigms, the approaches to choosing the most effective management tools also change. With the increasing complexity of public relations, those managers whose leadership and business qualities surpassed those of average managers, the most widespread and widespread, are beginning to play an increasingly important role [1].

In modern conditions, the need for a new quality of territorial management subjects that strive to form a high competitive tone and economic stability of the region in the long term is becoming increasingly clear. The existing scientific and practical approaches, which describe the formation of a new kind of relations in the public administration field, do not meet the requirements of large-scale and continuous social and economic innovations, as regards the expanded reproduction of new corporate knowledge.

The successfully implemented local initiatives have been as a rule indigenously trigged and directed to mobilization of local resources, diversification of different types of economic activity, introduction of new products on the market, improving skills and creation of new organizational forms for these activities.

Nowadays the intelligentsia as a social class, due to its weakness, cannot influence either social development or the economic ones. It has not still sought participation in contests for elected government positions, and it happens not due to the lack of money required for the participation, but because of uncertainty about achieving success. The same reason prevents this class from taking part in the struggle for power, although it is scientific class who could scientifically justify economic and political decisions, describe the best alternative variants and possible final result. The problems of the effectiveness of public administration in the regions actualize the issues of developing new approaches and tools in the management of territories. However, the search for and development of new tools and approaches is complicated by a number of factors, which should primarily include:

- terminological and conceptual uncertainty of the basic concepts of the control system[2];
- limited research on the nature and manifestation of passionarity in the field of public management.

2 Literature review

The problem of passionarity under the conditions of decreasing of creative activity of the personnel, already at the early stages of ontogenesis, has increasingly become the subject of scientific research recently. From a psychological standpoint, this problem is studied by V.Y. Bogdanov, who made a comparative analysis between the personal typology of L.M. Gumilyov and the psychological typology of character developed by K. Leongard. M. Kovalenko studied interconnections of passionarity, temperament and motivation [3]. K.G. Frumkin, tries to explain passionarity through the theory of sexuality by S. Freud (passionarity as manifestation of sublimated unconscious drives) etc.

Thus, Y. Beglov, the author of the information theory of thinking, proposes to determine passionarity from the standpoint of person's conscious understanding of their actions. "Passionarity is a predisposition to choose

* Corresponding author: karpushki@mail.ru

behaviors with poorly predicted effects instead of the well-predicted ones but with unsatisfactory after-effects. Such predisposition is the property of the current state of thinking which can be caused by innate features of the brain, evoked by behavior stereotypes and personal experience".

A broad definition of this term is given by modern dictionaries of foreign words, namely: passionary (from the Latin passio – suffering and even passivity, but also affect, passion; the one who possesses a higher activity, passion, whereas a passionarian (passion-filled) is an active person who has considerable inner energy. Derivatives in European languages differ in connotation. The British added a new meaning to passionarity: for them passion is an outbreak of anger; for the Poles it is rage, fury; passionant in Romanian is a person capable of carrying along and exciting others; for the Dutch, Danes, Germans and Swedes passion is just an excitement; in Ukrainian, the Latin passio is interpreted as passion (пристрасть). In everyday wide use, they call it any strong desire and even a weak drive. Thus, the etymology of this term, which is so polysemantic and diverse, is being specified. When appeared, this concept triggered various reactions in the scientific world: from admiration to hard criticism. Even in modern conditions the interpretation of the term has different variants in accordance with the authors' scientific preferences. For example, D.G. Davydov in his studies equates the concept of passionarity with the notion of vigorousness, proposed by M.P. Karpenko (from the English vigorous – active, cheerful, full of energy).

The great number of individual's actions is dictated by the instinct of self-preservation. Passionarity, on the other hand, has the opposite vector, as far as it perceives people to sacrifice themselves in the name of a great idea. When the aim of a person at the level of instinctive impulses to safe life is less than passionarity impulses, such a person is called a passionary one. As far as currently we have no instruments and methods of measuring the magnitude of passionarity, we can only talk about a tendency to increase or decrease a degree of passionarity.

Passionaries have a great potential of influencing the life of society. As it was proven by the historical facts society in different times could be in different states. If the society is in an equilibrium condition, there are no circumstances that are able to made a person to change his or her behavior, and no social imperatives are observed to strive voluntary death for the sake of their own aims. People with the abilities to be engaged in science and arts lead a quiet life build houses, cities. It's the time when harmonious individuals live in a stable society, feats and creative take-offs are valued, but there are only a few people capable to implement them in real life.

Using this assumption we can conclude that passionaries are ready to support society in a state of norm. It supposes that there is no passionary tension in the society. If any deviation happens, which leads to a situation when generally accepted imperative is viewed as something odious, society understood that it is in need of someone who will provide it an opportunity to survive and live in peace to the best of one's duties, and this idea is supported by the majority of the population.

When society is going through a time of serious reforms, involving the processes of creating a completely new social system, its passionarity of society is manifested in high level of public activity, which means the readiness of population for heroic deeds. In such a state, a part of society represents as much more than the whole. This is the elite of society, which could be called "the source of passionarity".

3 Materials and methods

Public management is characterized by a more extensive subject structure in comparison with state and municipal forms of government, since it involves the active participation of not only local authorities but also the population in the process of creating conditions for the balanced development of the territory. In this context, special attention should be paid to the special characteristics of proactive local representatives' community, who are able both to take an active part in the public administration and to involve other people in this process with the aim of raising the life quality of the population at the place in which they live [4, p.199].

The process of effective management is impossible without the development and implementation of new ways and methods of management, since the tools that previously gave tangible success lose their effectiveness over time and vice versa. Creativity in the public administration system is becoming more and more popular. The development of effective management tools involves the active use of the human factor, psychological and emotional aspects of motivation for making management decisions. After all, public administration is currently characterized by an active transformation of the sphere that adapts to changes in the external environment of the region, into a sphere that independently forms modern trends in regional development, which leads to a change in the nature of interaction between public administration bodies and the regional environment.

Such a characteristic feature of modern management can be reflected by a concept, introduced by L. Gumilyov, as passionarity. According to his classical definition, passionarity is "the ability and desire to change the environment, or, translating into the language of physics, to break the inertia of the aggregate state of the environment" [5, p. 266]. The "passionate impulse of behaviour" is directed sometimes against the instinct of both individual and collective self-existence. It gives us the right to say that it has the vector that is opposite to basic instincts. Super-passionarity, according to L. Gumilev, is the ability to reject personal needs for the sake of interests of one's nation, to make sacrifices for achieving highest national goals, collectivism, multiplied by heroism and working capacity.

Passionarity (in a broad sense), according to scientists, is an integrative characteristic of the personality whose qualities are directed towards self-improvement and improvement of society and are based

 https://doi.org/10.1051/shsconf/202111600025

on the personality activity and have social significance: purposefulness, ability to overcome obstacles and ability to cope with overstrain; complementarity, leadership, adaptability, developed intuition, constructive aggressiveness and mature emotionality. However, it is quite natural that over time the meaning of the concept of passionarity has transformed and now its meaning differs from the one that was implied by L. M. Gumilyov.

It also happens that passionarity refers to the ability of a person (or group) to super-effort, super-tension, while the measure of passionarity is considered the relative number of such individuals in the society under consideration. In modern works, the manifestations of passionate behavior in relation to individuals, but not to society, are called vigor.

Passionarity is characterized by an exaggeration of the priority of the target settings, which reduces the obstacles that stand in the way of achieving them. This often explains the "revolutionary effectiveness" of management results, since this approach ignores traditional constraints, thereby expanding the boundaries of the search for new opportunities. The result of using this approach in management is management decisions of a fundamentally new quality level in comparison with the classical methods of public administration, which most fully take into account all existing limitations. Passionarity here manifests itself in the fact that the manager does not always manage to stay within the framework of rational decisions [6, p.18]. And public administration, as is well known, is not always based on the principles of economic efficiency, in some cases, social aspects prevail over cost (economic) ones.

4 Results

The researchers emphasize that the adaptive requirements of modern public administration call for leaders who take responsibility and do not expect that they will receive a revelation or instructions from higher management [7, p.192].

The transfer of the concept of passionarity from the plane of an individual to the sphere of public administration does not consist in a simple transfer of properties to a broader object, but in the special nature of their manifestation in such complex systems as public ones. This is due to the fact that in social systems, the self-organization of the individual is supplemented by social organization, since people acting in society are guided by their personal motives and values [8, p.67]. At the same time, the basis of leadership should be the process of interpersonal passionary influence that unfolds between the leader and the rest of the corporate community.

Thus, the formula of the ideal system of public administration in the logic of L. Gumilyov's ethnogenesis is as follows: "If a certain optimal ratio is established between the most important types of members of an ethnos - harmonious individuals (in whom passionarity and instinct are balanced) and passionaries (in whom passionarity is stronger than instinct) - the system is almost irresistible" [9, p.30].

L. Gumilev separated the society into two different groups of people:

1) Passionaries who could be described as charismatic and powerful people having their own clear vision about the future perspectives of the society development as well as the strong will to achieve goals. They are able to involve other individuals into action and strongly motivated at achieving certain aims.

2) Subpassionaries who themselves are not able to completely adapt the social system. That is why they need additional support, but can be easily mobilized into the revolutionary actions under direction of passioners.

Revolutionary changes led by passioners are possible to occur if a situation of total instability has developed or in case if the existing institutional set-up is been reforming, and the process of reforming is accompanied with a high degree of dissatisfaction in a society.

The quantity of subpassionaries raises when general ethnic activity, which also could be called “common identity basis” reduces. The share of this group can be compared with entropy-eroding cornerstones of an existing system, which gives passionaries chances to improve the system more easily.

Passionaries, playing role of so called “energy bombs”, attract and direct activity in other people with greater energy and, eventually, in "normal" people with lower energy as well. According to results of survey passionarians who are active creators of history make up a small part of society – 3-5% [10, p. 59].

The decline in the overall level of passionarity of the system, associated with a decrease in the number of passionaries, leads to the fact that in the main areas of public administration (politics, the army, the civil service, ideological institutions, and so on), the so-called "harmonious individuals" begin to dominate, whose passion impulse is equal in magnitude to the impulse of the instinct of self-preservation, or sub-passionaries, whose passion impulse is less than the impulse of the instinct of self-preservation. These types of personality begin to occupy vacant managerial positions due to the lack of passionaries with the help of family ties and the system of clannishness, which is very common in Russia. This tendency of these "new people" in management creates new patterns of behavior, which leads to the simplification of the public administration system and the inhibition of its development, taking into account the needs of society.

If we try to classify different types of passionarity according to its orientation, we can distinguish two key types as regards the criterion of constructiveness and destructiveness of passion activity: adequate passionarity and «shadow» passionarity.

Adequate passionarity is manifested in the high activity of the individual to achieve goals that contribute to his self-actualization, spiritual and personal development [11]. Even if the passionary deliberately sacrifices one area of his life in the name of embodying the idea of another area of his life (for example, refusing to start a family because of complete immersion in a profession or going to a monastery in the name of spiritual goals, etc.) - as long as such actions do not have a clear destructive impact on the personality of the passionary this will be considered a manifestation of adequate passionarity.

Shadow passionarity is shown in the implementation of ideas connected with destructive and self-destructive actions. Fortunately, shadowy passionaries are less common in society: usually, a person who is prone to antisocial and immoral behavior does not have a clearly expressed "idea of evil" and not inclined to show a strong will of gathering people around the idea.

It raises the question about defining and measuring the qualities by which we can choose passionaries or subpassionaries. As far as we can describe passionarity as a complex of socially and economic valuable features allowing person to increase useful for the society and the enterprise abilities, it could be reasonable to mention among such features the following ones:

- Strong-willed qualities. The theory of passionarity characterizes a passionate person with a "long will". And although the goal often remains illusory, a passionate person under the influence of an irresistible desire for activity, easily overcomes any obstacles that are encountered on the way to achieving his plan, finds ways and means to achieve the goal.
- Adaptability. A passionate person is capable of conquering new areas, developing them. Therefore, he or she has a high level of adaptation to new conditions, as well as the ability to change conditions for himself, improving the new environment. Its ability to adapt is assessed as overadaptation.
- Sacrifice. Passionate leaders lead a bright but short life; they are often diagnosed with an inclination to take risks, low level of value in their own lives, combined with an experience of passion and impulsivity.
- Aggressiveness. The passionary theory in some sources is called the theory of aggression, because associated with the conquest of territories, with the conduct of wars and revolutions.

5 Conclusion

Leaders have always were key figures in changing a society. Revolutionary changes need charismatic leaders – so-called passioners. Only such type of people is able and has a strong desire to lead crowds to new achievements and to direct people and institutions to changes. Leaders, not managers, implement changes in politics and business. The success if their ideas depends on their ability to create new teams as well as on the managerial process to leverage and add substance to their visions and energy. Passionarity manifestation of the leaders as the ability of a personality to change the environment and themselves, the need to overcome difficulties. Passionarity is a dynamic feature of the psyche that occurs as a result of passionarity push, and forms a certain number of people, within the population, who tend to act.

Today in the world and, in particular, in Russia, the so-called "creative class" is being formed, consisting of people involved in solving important technical, innovative, scientific and other tasks based on independent thinking, creative abilities and a high level of education. This is a special rational type of people for whom knowledge, individual characteristics and personal merits (career) are important.

Development of activity as internal energy is possible on condition that there is imagination, creativity and courage, self-confidence, self-reliance, purposefulness and ability to prove own opinions.

In terms of management passionarity means a set of personality qualities that provide an active life position of the person. Passionarity is a characterological dominant, an irresistible inner desire (drive) (conscious or, more often, unconscious) of activities aimed at achieving a certain aim (often illusory).

Management based on a high level of performance discipline in the creation of goods and services is the foundation of a traditional organization. In contrast to this form of management, the model of passionate leadership, based on highly developed leadership capital and characterized by a relevant "leadership network", is now beginning to prevail in more promising self-learning organizations.

To overcome this tendency of the gap between the priorities of public power and the needs of society, passionarity is needed, which allows managers to rise above their personal interests. Passionarity will give an impetus to new tools and methods of management, that is, the preservation and development of the nation.

In the context of public administration, it is passionarity that provides the "beneficiary" (the state) with the implementation of their tasks by involving human capital in the management process. With regard to the applicability of the passionary management style in Russia, socio-cultural factors and especially the morphology of passionarity need careful empirical testing in relation to specific socio-structural and socio-territorial conditions.

The most actual problems for public management today are creativity, team building, facilitation and moderation, transformation of beliefs, emotional competence, leadership, etc. Passionarity becomes the driving force in achieving the public management goals, a positive change in the aggregate state of the internal environment. This becomes the key to success.

The need for passionaries gives rise to the question - are passionaries born or become? In what way is passionarity formed?

The assumption that passionarity is inherited has not been scientifically proven, but the fact that it is contagious is known and proven. Thus, ordinary citizens who are in proximity to any epicenter begin to behave as passionary ones. If a person goes away for a certain distance, they behave again as usual. This phenomenon is called passionary induction and is actively used by experts. Obviously, the first two ways have nothing to do with upbringing. Passionary induction will be understood as changes in the mood and behavior of people in the presence of more passionary individuals. Consequently, such a property of passionary personality as passionary induction can be recognized as the key one.

References

1. L. Puzyrevsky, *Bulletin of the St. Petersburg University of the Ministry of Internal Affairs of Russia,* **4 (60),** 2013.
2. O. Glushakova, *Public management of the processes of ensuring the quality of life in socio-economic systems* (Pub. House of the Siberian Academy of Finance and Banking, Novosibirsk, 2016)
3. N. Havrysh, K. Krytiy, *Development trends in pedagogical and psychological sciences: the experience of countries of Eastern Europe and prospects of Ukraine* (Baltija Publishing, Riga, 2018)
4. Yu. Karpovich, *Proceedings of the XIX All-Russian Scientific and Practical Conference with International Participation* (2020)
5. L. Gumilev, *Ethnogenesis and the Earth's biosphere* (Azbuka-klastika, St. Petersburg, 2002)
6. V. Nikishkin. *Practical marketing,* **1**, 95 (2005)
7. R. Kheifets, D. Lauri, *Work of the leader. Leadership.* (Alpina Business Books, Moscow, 2006).
8. Yu. Podkorytov, G. Vinokurov, *Strategic stability,* **2 (43)** (2008)
9. A. Vladimirov, *Beachhead,* **4-5** (2015)
10. Y. Zymyna, *Pedahohycheskoe obrazovanye,* **1** (2007)
11. N. Vyunova, M. Larskikh, I. Strebkova, *Vestnik VSTU,* **3-2** (2014).

Assessing the decoupling of economic growth from environmental impacts in industrial regions of the Russian Federation: an impact decoupling approach

Anna Pepelyaeva*, and *Aleksandra* Krutova

Perm National Research Polytechnic University, 29, Komsomolsky prospect, Perm, 614000, Russia

Abstract. The article analyzes the dynamics of the economic activity level in Russian industrial regions and its impact on the main environmental elements (water, air and soil) by calculating the decoupling coefficient. In the study a hierarchical cluster analysis was carried out, resulted in the formation of seven regional clusters, five of which had a significant share of the industrial sector in the gross value-added structure. Decoupling coefficients were calculated for 64 constituent entities of the Russian Federation belonging to different types of industrial clusters. The ecological and economic balance assessment of industrial regions growth showed the decoupling effect presence for most of them. The movement towards environmentally sustainable growth is more evident in the regions of the industrially diversified cluster. The analysis of three decoupling coefficients showed that in the vast majority of industrial regions the growth rates of pollutant emissions into the atmosphere and the growth rates of polluted wastewater discharges into surface water change at a lower rate than per capita GRP, which indicates a "green trend" in the region's economic activity. However, the decoupling coefficient, showing the dynamics ratio of production and consumption waste and GRP per capita, has negative value in 37.5% of regions. The most problematic one in terms of this indicator was the Trade and manufacturing cluster, including 21 constituent entities of the Russian Federation. According to the authors, it may be related to the trade sector of these regions.

1 Introduction

Dramatic climate change, declining biodiversity, and land degradation are just a few of the challenges humanity facing in an effort to improve its well-being. Extensive economic growth due to increased consumption and pressure on the environment has already led to irreversible consequences. The situation calls for fundamental changes in production and consumption patterns on the part of humanity. The post-industrial stage of economic development implies not only its structural and sectoral restructuring, a change in the technological basis of production, or an increase in the number of high-tech industries, but also its greening. Economic growth at the national or regional level, including the growth of industrial production, should be considered together with national or regional environmental safety, which is the main condition for the long-term socio-economic development of the territory.

In 2011, the United Nations Environment Program (UNEP) called for a decoupling of economic growth and environmental impact (UNEP, 2011). Decoupling occurs when the growth rate of economic activity is higher than that of resource use or environmental impact (relative decoupling), or there is an increase in economic activity while decreasing resource use or reducing the environmental impact associated with this economic activity (absolute decoupling) [1].

The term "decoupling" in environmental economics was proposed by the OECD in 2000 in "Indicators to Measure Decoupling of Environmental Pressure from Economic Growth Organization for Economic Co-operation and Development" [2] and means "the process of sustainable, long-term, predictable and controlled trends decoupling in economic growth, consumption of natural resources and environmental pollution at all stages of the product life cycle, without redistributing environmental risks in time and space" [3, p. 4].

The decoupling effect can be achieved through a conscious transition to a circular economy, which involves the transformation of linear material flows into cyclical ones through a combination of intelligent infrastructure and product design, standardization, reuse, recycling and recovery [4].

The emergence of the decoupling effect remains one of the most important growth trends in the region's industrial sector. The issues of using decoupling for the study of environmental and economic problems in Russia at the macroeconomic level are considered in the studies of Bobylev S.N., Zakharov V.M., Mikhailova E.G., Prokopyev M.G. and others [5-7].

The assessment of ecological and economic progress at the regional level through the calculation of the decoupling index or decoupling factor was carried out in the studies of such Russian scientists as Akulov A.O., Anikina I.D., Anikin A.A., Bashirova A.A., Gorbunova

* Corresponding author: alexkrutova@yandex.ru

https://doi.org/10.1051/shsconf/202111600026

V.A. S., Zabelina I.A., Kalashnikova I.V., Kirillov S.N., Kozhevnikov S.A., Korobitsin B.A., Lebedeva M.A., Pakina A.A., Samarina V.P., Trushkova E.A., Shkiperova G.T., Filippova K.V., Fesenko R.S., Fomina V.F., Fomin A.V., Chumakov V.N., Yashalova N.N. and others [8-19].

As a part of the methodological approach for assessing decoupling of growth trends in the global, national or regional economy and the corresponding ecosystem, resource decoupling and impact decoupling are assessed. Resource decoupling reflects the efficiency of using natural resources at all stages of the production process, impact decoupling shows the use of waste-free technologies and the efficiency of treatment facilities [20]. The presence of resource decoupling and impact decoupling indicates a movement towards circular economy.

In this study, the authors assessed the presence of impact decoupling in regional industrial-type clusters in the Russian Federation.

2 Materials and methods

Research hypothesis: the sectoral structure of the Russian regions influences decoupling indicators.

The aim of the study is to identify and analyze the effect of decoupling in the ecological and economic activities of Russian industrial regions.

The methodology proposed by the authors for assessing the impact decoupling in the constituent entities of the Russian Federation of industrial type includes several stages.

At the first stage, the Russian regions were grouped according to the structure of gross value added (GVA) based on the method of hierarchical cluster analysis.

Clustering was carried out on the sectoral structure of GVA basis according to OKVED 2 established in the Russian regions in 2017. Seventeen types of economic activity were identified: sections P, O, and Q were combined, T and U were excluded from the analysis, since they do not consider the enterprises economic activities in the region.

As a result, 7 clusters of the Russian regions were identified: "Industrial-diversified", "Trade-manufacturing", "Logistic", "Agrarian-touristic", "Manufacturing", "Mining-social" and "Mining" [21, p. 95-108].

For further analysis, 5 industrial-type clusters were taken (regions included in the logistics and agrarian-touristic clusters were excluded from further research).

At the second stage, indicators are proposed to assess the decoupling effect through the Decoupling Index or the Decoupling Factor.

In this study, three decoupling factors will be calculated according to general formula 1.

$$D = 1 - \frac{EP_t/DF_t}{EP_0/DF_0} \qquad (1)$$

where DF0, DFt (Driving Force) is an indicator of economic growth in the base and current periods; EP0, EPt (Environment Pressure) are indicators of the negative impact on the environment in the base and current periods.

As an indicator of economic growth (DF), the study will use GRP per capita (rubles/person). As an indicator of the negative impact on the regional ecosystem for the calculation of D1, the EP1 indicator is proposed as the emissions of pollutants into the atmosphere per capita (tons/person); for calculating D2 the EP2 indicator is used as the discharge of polluted wastewater into surface water per capita (m3/person); for calculating D3 the indicator EP3 is used as production and consumption waste per capita (tons/person).

The choice of indicators is conditioned by the following requirements: objectivity; importance of the indicator; quantifiability; relevance, i.e. the indicators structure and composition should correspond to the structure of the "Environment Pressure" category and be sufficiently indicative for regional assessment; indicators reliability and availability, the possibility of comparison in dynamics taking into account regional specifics.

In addition, the choice of the proposed indicators of negative impact on the regional ecosystem is due to the fact that the anthropogenic pressure on the environment is carried out in all its subsystems: air, land and water. The use of per capita indicators in the study allows for more objective interregional comparison.

At the third stage, the data collection and processing for the regions of the selected clusters, calculation and assessment of decoupling coefficients (D1, D2, D3) were carried out.

Since the indicator of "production and consumption waste" in the regional context has been published by the Federal State Statistics Service once every two years since 2011, and the data on GRP for 2019 have not yet been published by Rosstat, the study period was limited by the time period of 2011-2017. In addition, the Republic of Crimea and the city of Sevastopol were excluded from further analysis, as they were not part of the Russian Federation at the beginning of the analyzed period.

3 Results and discussion

Table 1 shows the calculated decoupling coefficients for five clusters of the Russian industrial regions.

The positive value of the decoupling coefficients (D_1, D_2, D_3) indicates the presence of the decoupling effect, the multidirectional dynamics of the regional economy growth and pressure on the territory ecosystem, the excess of economic growth rates over the environmental pressure growth rates, and negative value exceeds the environmental burden growth rates over economic growth. That could be interpreted as obtaining a lower economic result on the amount of resources spent, accompanied by deterioration of the environmental situation.

Table 1. Impact decoupling coefficients (*Di*) in Russian industrial-type regions *

Regional grouping by clusters	D_1	D_2	D_3
Industrial-diversified cluster (10 regions)			
Belgorod region	0,31	0,37	0,30
Kursk region	0,33	0,84	0,33
Republic of Karelia	0,24	0,22	0,24
Murmansk region	0,45	0,43	0,42
Republic of Tatarstan	0,33	0,61	-0,56
Udmurt Republic	0,19	0,59	0,15
Samara region	0,42	0,37	0,46
Perm territory	0,25	0,44	0,25
Krasnoyarsk region	0,90	0,58	0,25
The Republic of Khakassia	0,85	0,63	-1,83
Trade and manufacturing cluster (21 regions)			
Bryansk region	0,39	0,58	-0,14
Oryol region	0,23	0,41	-0,38
Stavropol region	0,39	0,46	-0,80
Pskov region	0,11	0,45	-5,73
Voronezh region	0,41	0,52	-0,43
Tambov region	0,34	-1,82	-2,40
Kaliningrad region	0,48	0,30	-0,64
Rostov region	0,34	0,48	0,47
Republic of Adygea	0,19	0,53	0,11
Penza region	0,39	0,50	0,23
Altai region	0,76	0,06	-2,02
Novosibirsk region	0,16	0,48	-42,51
Ivanovo region	0,39	0,52	0,43
Kostroma region	0,22	0,46	0,47
Moscow region	0,38	0,51	-5,30
Smolensk region	0,36	0,46	-0,79
Tver region	0,27	0,45	-0,93
Moscow city	0,32	0,41	-0,22
Saint Petersburg	0,31	0,54	-0,18
Kurgan region	0,38	0,46	-0,02
Saratov region	0,32	0,65	-0,01
Mining and social cluster (9 regions)			
Komi Republic	0,49	-0,54	0,27
Kemerovo region	-0,16	0,53	0,06
Tomsk region	0,34	0,42	0,53
Irkutsk region	0,85	0,52	0,23
Chukotka Autonomous Okrug	0,39	0,63	0,16
Republic of Sakha (Yakutia)	-2,60	0,71	-0,31
Magadan region	0,45	0,74	-1,85
Astrakhan region	0,66	0,72	0,69
Orenburg region	0,42	0,46	0,21
Manufacturing cluster (21 regions)			
Vladimir region	0,36	0,47	0,56
Nizhny Novgorod region	0,42	0,51	0,43
Sverdlovsk region	0,45	0,54	0,42
Yaroslavl region	0,47	0,54	-0,03
Republic of Bashkortostan	0,24	0,46	0,66
Leningrad region	0,32	0,27	-0,58
Ryazan region	0,47	0,47	0,27
Kirov region	0,38	0,69	0,46
Chuvash Republic	0,18	-1,27	0,12
Ulyanovsk region	0,34	0,41	0,29
Arkhangelsk region	0,55	0,48	-1,05
Volgograd region	0,31	0,61	0,06
Mari El Republic	0,35	0,54	0,35
The Republic of Mordovia	0,25	0,65	0,28
Kaluga region	0,26	0,54	0,28
Tula region	0,52	0,57	-0,62
Lipetsk region	0,43	0,47	0,56
Vologda region	0,40	0,36	0,28
Novgorod region	0,29	0,81	0,42
Chelyabinsk region	0,53	0,53	0,28
Omsk region	-1,77	0,47	0,44
Mining cluster (3 regions)			
Sakhalin Region	0,15	0,44	0,33
Tyumen region (including Khanty-Mansi Autonomous Okrug and Yamalo-Nenets Autonomous Okrug)	0,55	0,35	1,00
Nenets Autonomous Okrug	0,59	0,02	-39,64

*Calculated according to [22, 23, p.43, p. 492]

According to Table 1, a number of conclusions can be drawn.

For all the analyzed clusters, the coefficient D_1, reflecting the result of comparing the growth rates of pollutant emissions into the atmosphere from stationary sources per capita with the growth rates of GRP per capita, has a predominantly positive value. This may indicate the regions movement towards the introduction of circular economy principles in matters related to air pollution.

It should be noted that the total emissions of air pollutants in 2017 compared to 2010 changed insignificantly, the decrease was less than 1%. Also, insignificant changes took place in the sectoral structure of air emissions, where the emissions from manufacturing enterprises dominate (Figure 1).

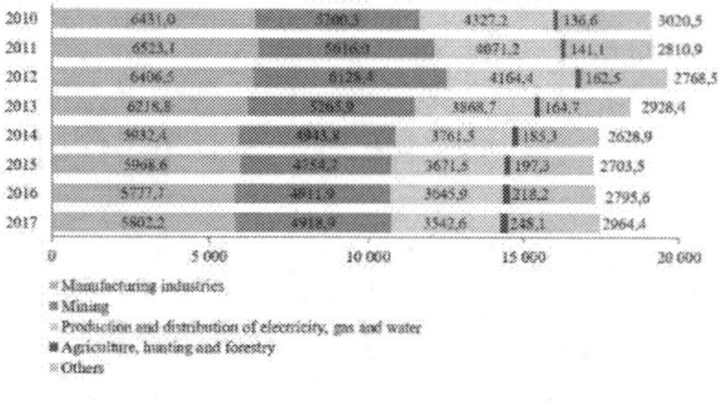

Figure 1. Distribution of emissions (thousand tons) from stationary sources by type of economic activity, 2010-2017 in the Russian Federation [24]

According to the report "On the state and protection of the environment of the Russian Federation in 2017" [24], a number of measures were implemented in 2017 to reduce emissions of pollutants into the atmosphere. The total amount of financing used for their implementation amounted to 59.8 bln rubles. Among the main production activities noted in the report, it is worth mentioning the improvement of technological processes, an increase in the efficiency of existing treatment facilities, the commissioning of new treatment plants, sources of pollution, etc.

Thus, we can state the effectiveness of the implemented actions from the circular economy perspective, the excess of the growth rate of economic activity over the growth rate of pressure on the environment.

Coefficient D_2 shows the growth rate ratio of polluted wastewater discharges into surface water per capita with the growth rate of GRP per capita. As in the case with the D_1 coefficient, positive values prevail in all analyzed clusters.

It is due to the fact that for the period of 2010-2017 the volume of wastewater discharge into surface natural water bodies of the Russian Federation decreased by 13.4% and amounted to 42,575.74 mln m^3 in 2017 [24]. This trend may be associated with an increase in fixed capital investments during this period aimed at putting into operation wastewater treatment facilities and water recycling systems by 43% [24].

It is worth paying attention to the volume distribution of water protection capacity in 2017 between federal districts (Figure 2). The largest volume is in the Siberian Federal District. For regions of the Siberian Federal District, the D_2 coefficients have positive values, while in the regions with relatively smaller capacities for water resources protection, there are also negative values of the D_2 coefficient.

A decrease in the volume of wastewater discharge, along with the obtained positive values of the decoupling coefficient D_2, also indicates the effectiveness of the implemented measures, a decrease in pressure on the environment per unit of economic effect.

The coefficient D_3 values, reflecting the ratio of the production and consumption waste rates per capita with the growth rates of GRP, differ significantly from the previous indicators values. It should be noted that the regional coefficients values depend on their belonging to a particular cluster (Figure 3).

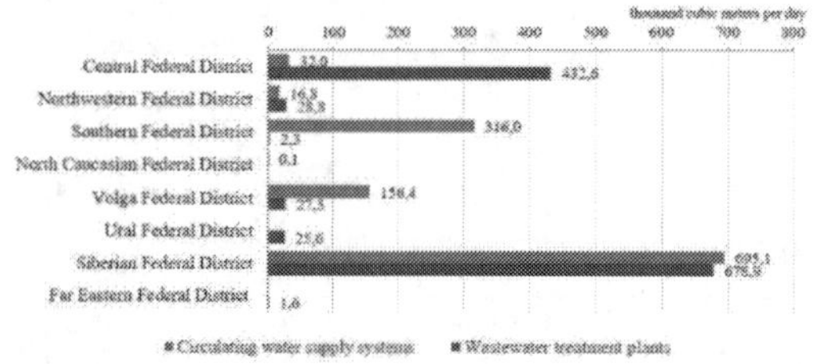

Figure 2. Distribution of capacities commissioning volume for the protection of water resources in 2017 by federal district in the Russian Federation [24]

Figure 3. Distribution of industrial regions of the Russian Federation by decoupling coefficient D_3

For an industrially diversified cluster including 10 regions, with prevailing share of manufacturing industries, a high share of the mining industry, the average level of the trade and social sphere, positive values of decoupling coefficients were obtained, except the Republic of Tatarstan and the Republic of Khakassia. Also, this cluster is characterized by a relatively small range of the decoupling coefficient values variation, the difference between the maximum and minimum values within this cluster is 2.29. On the one hand, it shows the homogeneity of the regions included into the cluster from the standpoint of the environmental impact dynamics per unit of economic result, and the improvement of the cluster regions in the field of circular economy on the other hand.

Similar conclusions can be drawn regarding the Manufacturing cluster and Mining-social cluster. Manufacturing cluster including 21 constituent entities of the Russian Federation is characterized by a predominant share of the manufacturing industry, an average level of trade and social sphere, the absence of mining industry. The mining-social cluster includes 9 regions, in the GVA structure of which the mining industry share prevails, it has an average share of the social sphere and an insignificant share of other types of activity. Positive values of the coefficient D_3 and a relatively small range of variation of these indicators also prevail here.

The "Trade and manufacturing cluster" including 21 regions is characterized by a high share of trade and manufacturing industry, and the absence of the mining industry. The cluster is the most interesting one for further research. The predominance of the decoupling coefficient negative values indicates that the growth rates of production and consumption waste significantly exceed the growth rates of GRP.

Since in three clusters of regions - "Industrial-diversified", "Manufacturing" and "Trade and manufacturing", manufacturing industries occupy a significant share (the average share amounted to 23%, 30.6% and 18.2% respectively) in the sectoral structure of economic activity, and a negative coefficient D_3 is mainly observed in the regions of the Trade and manufacturing cluster, it can be assumed that with improvement of technological basis in the manufacturing sector of the Russian regions, the indicator of waste may grow due to the trade sector.

The best available technology approach in the implementation of environmental and industrial policies is the measure to prevent environmental pollution and good feasibility management of sector-specific environmental impacts. It is actively introduced at industrial enterprises with the most negative impact on the environment. Since January 2015, the requirements for environmental regulation have been established according to Federal law No. 219-FL [25] based on technology standards that should not exceed the technological indicators of the best available technologies. However, considering the application areas of available technologies, it is revealed that trade is not included in the list of BAT application areas, which may indicate the lack of requirements for processes, technologies, services and criteria for achieving environmental objectives.

4 Conclusion

The practical significance of this study is the impact decoupling analyses of the Russian regions, taking into account the sectoral structure, which can be used to develop a strategy for regional sustainable growth.

As a result of the study, for 64 regions of the Russian Federation included in industrial clusters of different types, three decoupling indicators were calculated for the period of 2011-2017: the ratio of changes in the volume of emissions into the atmosphere, the discharge of polluted wastewater into surface water, the volume of waste generation per GRP unit.

According to the first two indicators (D_1, D_2), positive values are observed in most regions of industrial clusters, which indicates the movement of these regions towards circular economy and ensuring regional environmental safety.

However, according to the indicator assessing the dynamics of production and consumption waste per capita and per capita GRP, in 37.5% of regions there is a negative value of decoupling factor, and the most problematic for this indicator was the "Trade and manufacturing cluster", in which 88% of regions received negative values. According to the authors, it may relate to the trade sector of these regions.

At the next stage of the study, the resource decoupling of the selected regions will be assessed, and more detailed analysis of the most problematic and successful regions of each cluster will be carried out in order to identify the factors of the largest (negative or positive) impact of economic practices and instruments of regional industrial policy on regional ecosystems to improve them.

References

1. E. Sanye-Mengual, M. Secchi, S. Corrado, A. Beylot, S. Sala, *Journal of Cleaner Production*, **236** (2019)
2. Indicators to Measure Decoupling of Environmental Pressure from Economic Growth Organization for Economic Co-operation and Development, Retrieved from: http://www.oecd.org/officialdocuments/publicdisplaydocumentpdf/?doclanguage=en&cote=sg/sd(2002)1/final
3. V.P. Samarina, *National interests: priorities and security*, **2** (2015)
4. Ellen MacArthur Foundation official website Retrieved from: https://www.ellenmacarthurfoundation.org/
5. S.N. Bobylev, V.M. Zakharov, *On the way to sustainable development of Russia: bulletin*, **60** (2012)
6. E.G. Mikhailova, *Economic and Social Changes: Facts, Trends, Forecast*, **11** (2018)
7. M.G. Prokopyev, *Market economy problems*, **2** (2020)
8. A.O. Akulov, *Economic and social changes: facts, trends, forecast*, **4** (2013)
9. I.D. Anikina, A.A. Anikin, *Regionalnaya ekonomika. Yug Rossii*, **7** (2019)
10. A.A. Bashirova, *Management of economic systems: electronic scientific journal*, **10** (2016).
11. I.A. Zabelina, *Economic and Social Changes: Facts, Trends, Forecasts*, **12** (2019)
12. M.A. Lebedeva, *Actual problems of modern economy*, **6** (2018)
13. S.N. Kirillov, A.A. Pakina, N.I. Tulskaya, *Science Journal of VolSU. Global Economic System*, **19** (2017)
14. S.A. Kozhevnikov, M.A. Lebedeva, *Problems of Territory's Development*, **4** (2019)
15. B.A. Korobitsyn, *Biosphere compatibility: people, region, technologies*, **21** (2018)
16. V.N. Chumakov, R.S. Fesenko, V.S. Gorbunova, *Bulletin of Education and Science Development of the Russian Academy of Natural Sciences*, **22** (2018)
17. G.T. Shkiperova, *Economic analysis: theory and practice*, **43** (2014)
18. V.F. Fomina, A.V. Fomin, *North and the market: the formation of the economic order*, **1** (2018)
19. N.N. Yashalova, *Regional economy: theory and practice*, **39** (2014)
20. Y. Wu et al., *Journal of Cleaner Production*, **190** (2018)
21. V. Karpushkina et al. *Problems of economic security: challenges and experience in the digital*

economy (Publishing Center of SUSU, Chelyabinsk, 2019)
22. Rosstat Environmental protection in Russia 2018: Statistical collection Retrieved from: https://gks.ru/bgd/regl/b18_54/Main.htm
23. *Regions of Russia. Socio-economic indicators. 2020*: Statistical collection (Moscow, 2020)
24. Rosstat *State report "On the state and protection of the environment of the Russian Federation in 2017"* Retrieved from: https://gosdoklad-ecology.ru/2017
25. Federal Law N 219-FZ "*On Amending the Federal Law" On Environmental Protection "and Certain Legislative Acts of the Russian Federation"*. Retrieved from: http://www.consultant.ru/document/cons_doc_LAW_165823/

https://doi.org/10.1051/shsconf/202111600027

How to manage public opinion: theoretical and applied tasks of the sociology of public opinion

Viktor Burko*

Perm National Research Polytechnic University, 29, KomsomolskiProspekt, Perm, 614990, Russia

Abstract. The article deals with theoretical and applied problems that exist today in the field of interaction between government bodies and public opinion. At the end of the last century, the existence of the sociology of public opinion was questioned by many Russian sociologists, who considered public opinion to be a kind of state of public consciousness that did not have specific scientific categories. Such a theoretical paradigm made it difficult to find ways to effectively interact with public opinion. At the same time, many problems of the sociology of public opinion about which the many authors spoke at one time still exist today. In particular, the problem of determining the role and place of the subject of public opinion remains important, especially in the field of interaction between power, management structures and the population. Today, in the era of digital communication, the importance of an innovative approach to building a representative sample is growing, when surveys are increasingly conducted online and the results of such a survey cannot always serve as the basis for making effective management decisions. And, finally, there remains an important problem of improving the professionalism of public opinion researchers (pollsters) working in the fields of both political and industrial marketing. The author makes an attempt to draw the attention of management specialists to the need to take these problems into account in cooperation with specialists representing the field of sociology of public opinion.

1 Introduction to the problematic

In the ordinary consciousness of the inhabitants of Russia, the concept of *public opinion* has long been associated with sociology and sociological polls. This is both good and bad.

Good, because more than 30 years ago the words *sociologist* and *sociological* sounded exotic in the speech of Russians. And now these concepts have become familiar to the majority of the population.

But this is also bad, because the study of public opinion is not the study of all the processes taking place in society, which sociology has been doing for over 200 years and having accumulated a rich arsenal of knowledge about society. Another negative side of the confusion between the concepts of *public opinion research* and *sociology* is that the process of studying public opinion is seen by many only as conducting a population survey on topics of interest only to a sociologist or the customer of such a survey. Although, in fact, professional preparation and conduct of a public opinion poll requires knowledge of both theoretical and applied aspects of the sociology of public opinion. Without this knowledge, it is impossible to professionally compose a survey questionnaire to study **public opinion**, build an accurate (in terms of quality and number of respondents) sample for conducting a survey, prepare interviewers (questionnaires), conduct the survey itself and, finally, correctly process the results. At the same time, we do not belittle the dignity of specialists who specialize only in conducting mass polls (by the way, abroad they are more often called not sociologists, but *pollsters*. Highly professional pollsters are specialists who provide great assistance to politicians, businessmen and scientists. However, in our opinion, if a good sociologist can become a good pollster, then even a highly professional pollster cannot always be a good sociologist.

The existence of the sociology of public opinion as a branch of sociology is still a subject of discussion among sociologists, many of whom consider public opinion to be only one of the forms of public consciousness. From our point of view, the sociology of public opinion has every reason to be considered a special section of sociology along with the sociology of labor, education, etc., since it has its own field, object and subject, about which one of the begetters of the study of public opinion in Russia, B.A. Grushin [1]. We can also refer to the opinion of another well-known Russian sociologist Yu.A. Levada, who, like B.A. Grushin, was among the organizers of the All-Union (now - Russian Public Opinion Research Center -VCIOM). Yu.A. Levada, in one of his public lectures on the website Polit.ru, called the study of public opinion "lesser sociology" [2].

The main thing that distinguishes a professional sociologist from a pollster is an understanding of the nature of the phenomenon of public opinion. This is how, for example, A.A. Oslon, who successfully combined both professions in one person—pollster and sociologist: "... If you think about it, the concept

* Corresponding author: victor-burko@yandex.ru

of"public opinion"turns out to be not so simple. On the one hand, it implies heterogeneity, mosaicism, since it combines the opinions of many people ("how many men, so many minds"). On the other hand, public opinion can exert pressure as if it were the will of some special and very influential instance (would B.N. Yeltsin have gone into early retirement without the influence of public opinion?)" [3].

In one of his articles, B.A. Grushin stated about two fundamentally different approaches to the study of public opinion—**traditional and sociological** [4]. The first approach involves the study of public opinion at the level of the operational consciousness of the population (probes aimed at studying the changing problems of politics, economics, culture, etc.). The second approach aims to fix some stable relations in the mass consciousness that influence the formation of public opinion on a particular problem.

The history of Soviet and Russian sociology developed in such a way that Russian sociologists during the existence of the USSR paid the greatest attention in their research and scientific and practical developments to the second, in Grushin's terminology, sociological approach. The traditional approach to the study of public opinion was widely developed in Russia only in the late 1980s after the formation of the Russian Public Opinion Research Center (VCIOM) and the formation of numerous services for the study of public opinion in various regions of Russia [5,6]. Although the first large-scale developments in the sociology of public opinion began in the USSR back in the 60s, they were mainly of a fragmentary-sectoral or search nature. It is hardly legitimate to speak about the massive use of the results of public opinion polls in the system of political administration.

The appearance in the media of the results of all-Union and then all-Russian polls, the participation of sociologists in election campaigns of various scales, and the development of the sociology of marketing have exacerbated the problem of the place and role of the sociology of public opinion in political processes. This is especially true of those sociologists who began to actively engage in public opinion polls.

An effective solution to this problem is hampered by the presence of the following negative factors. First of all, this is the absence of fundamental scientific and methodological works on the sociology of public opinion. Over the past 30 years, only two works by domestic researchers of public opinion have appeared in Russia, in which the subject of this branch of sociology is thoroughly and deeply analyzed: the works of M.K. Gorshkov and D.P. Gavra [6, 7]. For the sake of fairness, it should be noted that in the same period in the Russian sociological scientific journalism, several periodicals appeared, in which scientific and applied problems of the study of public opinion are considered [8, 9, 10]. Another serious reason that impedes the effective use of the results of public opinion research in the practice of social management is the sociological illiteracy of the "users" of public opinion. And although the situation here has begun to change in recent years, it seems that cardinal changes are not coming that early.

And, finally, one cannot but reckon with such an important factor as the current state of Russian society, which is characterized by the instability of social, economic and political processes. Discussing the epistemological aspects of public opinion on the website of the Public Opinion Foundation club in the early 2000s, B.A. Grushin wrote: "I must say that in connection with my birthday, it's scary to say, on my 75th birthday, the Levada Center on its Internet site addressed me with a message, which, among other things, included the following words: "Boris Andreevich once wrote in his hearts that there is no public opinion in this country, but it was he who did almost more than anyone else to ensure that this public opinion exists." Thank you, of course, for such a high assessment of my work, but I would like to object. Boris Andreevich wrote this not in his hearts, but, as they say, of sound mind and firm memory. At the same time, I formulated two maxims: in the affirmative form "Not every poll of the population is a poll of public opinion itself" and in the interrogative form "What do we really study when we think or pretend that we are studying public opinion?" In this regard, I would like to turn to the questions of Viktor AleksandrovichBurko and partly Nikolai Katsuk regarding the place in social knowledge, in sociology, of the so-called population polls. **The fact is that population surveys are an independent branch of social knowledge that does not coincide with sociology. It arises at the junction of sociology proper, social psychology, demography, mathematics and a number of other sciences** (highlighted by myself, V. Burko). It is instrumental in nature and provides a solution to all sorts of purely pragmatic tasks in the field of politics, economics, culture, etc. **On the other hand, the study of public opinion as one of the modes of mass, I emphasize, not public, but mass consciousness is carried out not only in the form of polls and is a section of sociology or social philosophy, along with other sections: the sociology of the family, politics, etc.**" (highlighted by V. B.) [11].

All of the above obliges us to turn to the main scientific categories of the sociology of public opinion: its nature, object, subject, structure and functions.

2 Methods of Research

The work uses general scientific methods of dialectics, analysis, synthesis, analogies, as well as elements of the theory of comparative analysis. Also, the method of analysis of documents (texts) presented in monographs of the pioneers of public opinion research in Russia was used, and content analysis of speeches in the media by pollsters, representatives of domestic public opinion research services.

3 Results and Discussion

The subject of the sociology of public opinion, definition. Discussions about the subject of the sociology of public opinion are central to the debate between sociologists and pollsters. The recently published work of G.B. Yudin contains an assertion that "sociological science does not engage in public opinion research and, moreover, regularly sharply criticizes such research" [12]. Statements of this kind among domestic sociologists were cited even at the time of the beginning of the practice of polls in the USSR, i.e. from the middle of the XX century and, as we can see, they take place in today's discourse. Particularly active discussions about the subject of sociology of public opinion were conducted among Soviet sociologists in the 60s and 70s, when such prominent sociologists as B.A. Grushin, A.K. Uledov, Yu. Vooglaid, V.S. Korobeynikov, V.A. Mansurov, M.K. Gorshkov and others. A peculiar result of those discussions on the subject of this branch of sociology was summed up by D.P. Gavra in his monograph [7]. The author examines the subject of sociology of public opinion from three sides: epistemological, sociological and ontological. If we generalize these three approaches to the nature of public opinion, then we can formulate the definition of the subject of public opinion as follows: **public opinion is an interested-value evaluative attitude (activity) of a social subject to information-accessible, debatable, information-accessible objects that affect his interest, functioning in the spiritual or spiritual-practical form. Public opinion is a specific social institution that ensures the inclusion of this activity in the processes of social regulation, playing a special role in this system, having its own functions and the nature of interaction with other institutions, capable of functioning in all spheres of the society** [7].

The object of public opinion is the discussion of approaches. Among the categories considered in this section, the category of the object of public opinion appeals to sociologists the least discussion. Even in his monograph "Opinion about the World and the World of Opinions", which is basic for our presentation, B.A. Grushin very deeply and thoroughly examined the concept of an object of public opinion and formulated the main approaches that allow the researcher to single out among the many facts, events, phenomena and processes that surround us, those that can be called objects of public opinion, as opposed to simply objects of attention of a social researcher. Grushin, summarizing the approaches of his colleagues in the field, formulated the following criteria for attributing an object of public opinion: public interest, discussion and competence. These criteria did not cause much discussion among Russian researchers of public opinion. We only note that later in his work, which was discussed above, D. Gavra proposed to clarify the formulation of the third of the named criteria, proposing to name its information accessibility, which is more consistent with the characteristics of the object of public opinion as opposed to the subject.

Who can be the subject of public opinion? This category has caused, perhaps, the greatest controversy and discussion among sociologists specializing in the study of public opinion. The reason for this is the existence of public opinion not only in the field of sociology as a science, but also, mainly, in the field of politics as a political institution. Researchers of public opinion constantly face the following questions: what social community can be called the subject of public opinion? at what stage does such a social community become an actor of political action? How does the formation and "disintegration" of the subject of public opinion take place? And many other questions as well concerning the role and place of public opinion in the life of any society. The question about the subject also has an important methodological significance, the answer to this question in applied sociological research is especially important, since in this case the subject of public opinion simultaneously acts as an object of empirical research, a source of necessary information, and in practice represents a general population to which the researcher's conclusions apply.

The main disputes and debates in the discussion of the problem of defining the subject of public opinion occur around two positions: can any social group adhering to a certain point of view on the object of public opinion be considered a subject of public opinion, or can only the "popular majority" (Gorshkov) be the subject of public opinion? Selkova E.P. in her recent work made an attempt to summarize the numerous discussions about the subject of public opinion, considering this problem both in historical and substantive terms [13]. In it, the author, among other things, cites the point of view of D.P. Gavra, noting his synthetic approach to defining the subject of public opinion, calling it a subordinate-pluralistic approach. The essence of this approach is that, on the one hand, it proceeds from a pluralistic paradigm that recognizes the potential for a plurality of subjects of public opinion. On the other hand, it requires the observance of a certain subordination in the structure of the subject and proposes the people and the popular majority as the leading and main subject.

Thus, highlighting the subject of public opinion in a specific sociological study, we must consider it in two aspects: sociological and political. In the first case, we conduct a sociological analysis of all groups of respondents who adhere to one point of view or another on the object of public opinion (Grushin calls such an analysis differentiating and integrating). In the second case, it is necessary to additionally conduct a politological assessment of the degree of influence of all subjects of public opinion on the socio-political situation affected by the object of public opinion.

The study of the functions of public opinion is the most important aspect of the sociology of public opinion. Public opinion, being one of the key factors in the life of society, performs many functions, the significance of which is not always taken into account by researchers. Pollsters and politicians, when conducting opinion polls, most often investigate cognitive and evaluative functions. It is important for them to know, firstly, what the population, the public think about this or that fact, phenomenon or event, and, secondly, how the population estimates the significance of this fact, phenomenon or event for themselves. The sociological approach to assessing the functions of public opinion consists in analyzing the entire spectrum of functions inherent in public opinion. In the Russian sociology of public opinion, there are several approaches to the classification of the functions of public opinion. The most complete systematization of all possible functional possibilities of public opinion was given in his monograph by D.P. Gavra [7]. The author, first of all, proposed to classify the groups of functions into epistemological and sociological, institutional and non-institutional. Further, depending on the goals and objectives of the study, the author proposes to distinguish a large number of functions: from the simplest in content (for example, evaluative) to complex ones, such as, for example, the functions of socioregulation.

A very important research stage in the process of analyzing the functions of public opinion is the stage of compiling tools for obtaining sociological information aimed at assessing the manifestation of a particular function. Grushin called this stage the logical connection between function and expression. This means that it is necessary to formulate evaluative indicators not only from the point of view of formal logic, but also taking into account what function of public opinion the judgment is aimed at. For example, the judgment: "Do you trust the governor of the region"allows you to implement the evaluation function and the nomenclature function. And the judgment: "What needs to be done to increase the degree of public confidence in the authorities?"allows the analytical and constructive functions to manifest.

4 Results and Conclusions

When, in the distant 80s of the last century, the pioneer of the study of public opinion in Russia B.A. Grushin published the article on two approaches to the study of public opinion, at about the same time a famous French sociologist Pierre Bourdieu gave a lecture, the text of which he later published with a somewhat provocative title: "Public opinion does not exist" [14]. Bourdieu considered the main arguments for such a statement to be politicization in public opinion polls and the low professional, in a sociological sense, level of pollsters. The above-mentioned monograph by G. Yudin also examines the problems of studying public opinion in a modern interpretation [12]. All these works are united by the fact that the authors, each in his own way, actualize the theoretical and applied problems that still arise before the researchers of public opinion.

1. If we talk about theoretical problems that are today in the center of attention of sociologists studying public opinion, then, as mentioned above, the basic problem here is **the lack of research interest** among Russian social scientists in the nature of the phenomenon of "public opinion". And although publications related to public opinion constantly appear in the domestic media, almost all of them concern only the presentation of the results of one or another public opinion poll. It is difficult to recall works of a theoretical nature, which would offer a new view of nature, the object or subject of public opinion. The only exceptions are the few speeches and articles of well-known Russian sociologists, journalists and political scientists, in which an attempt is made to analyze the features of the manifestation of the phenomenon under consideration. Some attempt to draw attention to this problem is, perhaps, the above-mentioned work of G. Yudin, which is more political than sociological in nature. At the same time, we must pay tribute to the author, who devoted a significant part of his work to a very important applied problem of the sociology of public opinion—the representativeness of the results obtained in the course of public opinion polls. In this regard, we note that a great deal of work on familiarizing Russian researchers of public opinion with foreign authors is carried out by VCIOM and Public Opinion Foundation, with the assistance of which many translated monographs have been published [see, for example, 15, 16, 17]

2. The most important problem of the sociology of public opinion, which is both theoretical and applied in nature, is the problem of determining the **subject of public opinion**. In the previous section, we have already touched upon this problem, noting its dual nature—sociological and politological. G.B. Yudin devotes a significant place in his work to this problem. The author in great detail in the historical and substantive aspects examines two main approaches to the analysis of the influence of public opinion on the life of society: from the point of view of democratic and bourgeois theories. Figuratively speaking, the difference between these two theories lies in the placement of a punctuation mark at the end of the famous expression "Vox Populi - Vox Dei". The supporters of the democratic theory put an exclamation mark here, and the supporters of the bourgeois theory put an interrogative one. In other words, the first theory is the desire to transfer power into the hands of the people, and the second is to make politics rational and built on respectful discussion. This contradiction is of a dialectical nature and does not have an unambiguous solution—in different historical periods, for different states, this or that structure of public power is more acceptable. And here the role of social scientists (sociologists and politologists) is very important, as they can help politicians outline more effective tactics and strategies in making managerial decisions.

3. Another very important problem in the practice of studying public opinion is the problem of determining the **method of obtaining sociological information**. At the beginning of the XXI century B.A. Grushin put this problem at the center of attention of Russian sociologists [18]. And although since that time the public consciousness of Russians has undergone serious changes towards greater stability and competence in understanding the society in which they live, new "pitfalls" have arisen making it difficult to obtain in the course of public opinion polls reflecting the true point of view of the population on this or that socially significant problem. These "pitfalls" include: the socio-political conditions in which the public opinion survey takes place, the professionalism of the developer of the public opinion survey program, the presence (absence) of the necessary method and survey instruments, the skill of the interviewer (questionnaire).

4. And, finally, the problem of paramount importance in Russia today is the **cooperation of researchers of public opinion with social and political institutions of the country**. Active steps towards each other will give a synergistic effect of both scientific and applied nature [19, 20]. And the main condition for the success of such cooperation, in our opinion, is the interest of management structures at all levels in receiving and analyzing the results of public opinion polls.

Today, no one needs to be convinced that public opinion exists and can be measured. All the more responsibility sociologists and pollsters assume when they begin to study public opinion. The truth and adequacy of the results of their public opinion research depends only on them, on their understanding whether they correctly identified the object of public opinion, what functions of public opinion are supposed to be revealed in the course of the research, how the general and sample populations that are selected for the survey will relate, whether the method is public opinion research is the most effective for obtaining adequate results and, finally, to what extent the results of the survey will help to solve the problems of managing the social and economic development of the country.

References

1. B.A. Grushin*Opinion about the world and the world of opinions*. (Moscow: Praksis, 2011).
2. Yu.A.Levada*"What sociology can and cannot do"* Retrieved from: https://polit.ru/article/2005/01/04/levada/
3. A.A. Oslon*Journey into the world of public opinion*Retrieved from: https://bd.fom.ru/report/map/gur050108
4. B.A. Grushin*SotsIs*,**4**, 53–58 (1988)
5. J. P.Willerton,L. Sigelman,*Journal of Communist Studies,* **7(2),** 217–234.(1991)doi:10.1080/13523279108415085
6. M. K.Gorshkov *Public opinion: History and modernity*. (Moscow: Politizdat, 1988)
7. D.P. Gavra*Public opinion as a sociological category and social institution*(SPb.: ISEP, 1995).
8. Wikipedia *Bulletin of public opinion*Retrieved from: https://ru.wikipedia.org/?curid=1249743&oldid=105744043
9. *Monitoring of public opinion*Retrieved from: https://book.wciom.ru/zhurnal_monitoring/
10. *Social reality*Retrieved from: https://corp.fom.ru/socreal
11. *Personality: Grushin Boris Andreevich*Retrieved from: https://bd.fom.ru/report/cat/sci_sci/sotsiologiya/gur050106
12. G.B. Yudin*Public Opinion, or the Power of Figures*(SPb.: Publishing House of the European University at St. Petersburg, - 174 p).
13. E.P. Selkova*Theory and practice of social development,***6**, 84-86 (2012)
14. P. Bourdieu *Sociology of politics*. (Moscow, 1993)
15. G. Gallup, S. Ray. *The pulse of democracy. How public opinion polls work.*(Moscow: VCIOM, 2017)
16. W.Lippman,*Public opinion*. (Moscow: Institute of the Public Opinion Foundation, 2004).
17. J. Zaller,*The origin and nature of public opinion*. (Institute of the Public Opinion Foundation, 2004)
18. B.A.Grushin,*Nezavisimayagazeta*. (2000).
19. V.A. Burko*Proceedings of the XVII (dist.) All-Russian scientific conference in memory of prof. Z.I. Feinburg,* Perm National ResearchPolytechnic University, pp. 50-55 (2019)
20. A. Efimova, D. Strebkov*The International Spectator, 1–19,*(2020) doi:10.1080/03932729.2019.1700040

Monetary and informational form of capital as the basis of investment support for the inter-phase transition to an innovative economy

Oksana Butorina*, and *Yulia* Starodumova

Perm National Research Polytechnic University, 29 Komsomolsky prospect, Perm, 614990, Russian Federation,

Abstarct. Within the framework of the identified predominant processes that form the uniqueness of inter-phase transitions, the article highlights the accumulating processes of system transformation. At the same time, within the framework of the transition from the neo-industrial to the digital economy, informatization as a conjunctural process, based on the mass introduction of digital technologies, contributes to the formation of the monetary and information form of capital. Based on the analysis of the definitions of monetary and informational forms of capital, two approaches can be identified. The first of them is narrow, from his point of view, the MIFC is investment resources in real assets through digital and IT technologies. The second approach is broad, it allows us to interpret MIFC as an independent form of capital associated with the self-increasing cost of investments in the development of information and labor resources to create new high-tech products that can increase surplus value at each intra-phase transition of the post-industrial megacycle.

1 Introduction

The current development of the national and global economy can be defined as a transformational period from industrial to post-industrial development. Its uniqueness is in the complex interweaving of processes characteristic of the final phase of the industrial megacycle and the nascent post-industrial megacycle. At the same time, the final cycle is called the industrial megacycle by us because its dynamics reflect the essential nature of the industrial era, its heyday fell on the period of domination of the II-IV technological order with a total time duration of more than 100 years (1835-1940), while starting from the 60s of the XX century, during the transition to the V technological order, the conditions for a new stage of the industrial economy – neo-industrial, which is considered by us as the final stage of the industrial era, was formed.

We define the postindustrial megacycle similarly as a cycle associated with the formation and development of a post-industrial society and its economy. As the authors point out they are characterized by the following processes:

- the production of services is of priority importance, which is associated with its ability to form a larger surplus product than the industrial "core";

- the importance of information as an economic resource increases significantly, it becomes the main transformative production resource;

- the formation of profit largely depends not on the efficiency of the production sector, but on the effectiveness of intellectual and managerial activities;

- increasing the importance of the human factor as a resource that can accumulate information and create new knowledge;

- the predominance of venture capital businesses that can quickly adapt to changing business conditions and consumer preferences, functioning "ahead of the curve";

- introduction of new forms of investment support for economic development.

The processes outlined above are the cross-cutting nature, and they manifest themselves with varying degrees of intensity in each phase of the post-industrial cycle.

2 Results

Previously published works of the authors proved that the post-industrial megacycle can include the phases of the classical four-phase cycle, each of which is characterized by a corresponding transition from one type of economy to another. It should be particularly noted that in each phase, processes are formed and developed that give uniqueness to the interphase transition within the post-industrial megacycle. Thus, transition from neo-industrial to digital economy (Depression phase) characterized by science-intensive technotronic industrialization of the economy, based on the creation of new forms of business, markets, and consumers, supported by the technological capabilities of the digital economy; comprehensive renewal and transformation of productive forces based on digital technologies in the areas of electronic document management, electronic accounting systems, electronic

* Corresponding author: ok.butorina@yandex.ru

data warehouses, the use of GRM, the creation of corporate social networks, the expansion of the use of ICT in production as part of the automation process, robotization, computerization, in management, communication and entertainment, as well as in the field of electronic payment systems in the framework of e-commerce [14, 18]

Recovery phase (transition from digital to innovation and information economy) характеризуется informatization and innovatization of the economy and society as a whole, based on the venturization of business, the formation of the monetary and information form of capital, large-scale commercialization and socialization of information, the continuous mass generation of innovations, significant changes in the structure of the economy, based on the growth of the share of neo-industrial, innovative, information RCTEA [10, 11]

Lifting phase (transition from innovation and information economy to neo-economy) характеризуется informatization, generation of knowledge and innovations with a strong feedback between innovations and directions of their use, assuming the priority of information as an economic resource, individualization of knowledge, generating the importance of the information sector in the innovative development of national and global economies, globalization of business, investment and labor flows [7, 9, 12]

Based on the presented table, the following conclusions can be drawn:

The third is that modern development can be characterized by a transition from the neo-industrial to the digital economy, the completion of which can be determined by increasing the efficiency of production by optimizing costs and increasing profitability, expanding the range of benefits created in the field of virtual and digital economy, the beginning of structural changes in the economy – increasing the share of the neo-industrial sector in the created national wealth, transformation of the dynamics and income structure of all participants in the production process.

If we take into account the presence of recurrent dependencies between processes that also have a cyclical nature, within the framework of the transition from the neo-industrial to the digital economy, we can assume that informatization as a conjunctural process, based on the mass introduction of digital technologies, contributes to the formation of the monetary and information form of capital.

Of particular note is the lack of unity of interpretations of the concept of "monetary and informational form of capital", which required a more detailed differential analysis of the components of this definition. In the most general form, capital is considered to be money that can generate income (in other words, it is money that performs the function of accumulation). Depending on the source of accumulation (the possibility of increasing), it is customary to allocate different forms of capital. Despite all the variety of forms in the framework of this work, the priority of the research was given to monetary and informational forms.

Next, we will present the most common definitions of the monetary form of capital in the modern economic literature:

- the common denominator to which the cost of capital in the form of any asset is made. In monetary terms, the value of both physical and human capital is formed. Monetary capital, or capital in monetary form, is an investment resource. By itself, monetary capital is not an economic resource, it does not directly participate in production, and it is used to buy factors of production [19];

- one of the components of economic capital, based on the realization of the objectified material state in the form of financial resources [13];

- this is money intended for the acquisition of physical capital, in other words, invested in the business, a working source of income in the form of means of production: machinery, equipment, buildings, structures, land, stocks of raw materials, semi-finished products and finished products used for the production of goods and services [16].

To concretize the author's definition of the monetary and informational form of capital, we consider it appropriate to compare the most well-known interpretations of information capital:

- this is the accumulated investment in the IT sector, technological, scientific and technical, socio-economic and spiritual information used to generate income (technology is the basis of this capital) [19].

- a new form of intellectual capital associated with the formation of information and communication technologies. It acts as a new self-increasing value, characterizes the efficiency of the use of production factors, among which the most important is the intellectual resource. New knowledge is being created, and a new system of economic relations is being formed within the framework of the digital economy [3];

- a special form of expression of intellectual capital, characterizes the process of accumulation and dissemination of information and experience, the development of socio-economic foundations for the dissemination and transfer of knowledge. Information capital becomes the core of the "knowledge management" system, which leads to the creation of corporate universities and centers of scientific and technical information, and the transformation of organizations into self-learning systems [4];

- this is the available capital of organizations that implement information services [21];

- this is a part of non-financial assets associated with the formation of certain information resources [2, p. 46];

- this is a set of renewable information resources that simultaneously act as a capitalized stock and flow, which have a consumer and commercial value and bring income to their owner when involved in economic turnover [6, p. 13];

- these are databases, information systems that provide knowledge routing and rapid implementation of new developments [5, p. 10];

- these are information systems, technologies,

infrastructure, and an IT resources management system that are necessary for the implementation of the strategy [1, p. 10];

- these are databases, information systems, communications, and technologies. The basis of information capital is business applications, infrastructure and services [17, p. 221];

- intangible capital, which includes: knowledge capital, human capital, organizational capital [6, p. 136-153];

- this is capital, which includes technological, scientific and technical, social and economic, spiritual information used in economic turnover to generate income [6]

Based on the presented table, when determining the essence of the monetary and information form of capital (MIFC), two approaches can be distinguished. The first of them is narrow, from his point of view, the MIFC is investment resources in real assets through digital and IT technologies. According to I. N. Sokolov, these include crowdfunding, crowdlending, crowdinvesting and P2P lending, and other alternative financing platforms [15].

According to financial analysts, these platforms are essentially an alternative financial tool for attracting capital to startups and small businesses from a wide range of micro-investors.

Crowdfunding is a type of investment that is very similar to charity. For the money invested in the project, you get some kind of gift or privilege. For example, you may be sent a thank-you letter, sent a product under development before the official start of sales, or given a lifetime discount on the purchase of the product. The bonus depends on the imagination and capabilities of the person who raises money for his product [20].

Crowdlending is already investment-lending. Through the financial platform, you "credit" the entrepreneur. With this type of investment, the profit can reach up to 30% per annum.

There are two types of crowdlending:

p2p-lending ("peer to peer") - an individual lends to an individual

p2b-lending ("peer to business") - an individual lends to a company (legal entity)

Crowdlending is very convenient for business, money is provided without bail and with an interest rate lower than the bank rate.

One of the most popular platforms for crowdlending abroad is Lendingclub.com. In Russia, the Town of Money (townmoney.ru) and Stream (investor. potok. digital) are known [20].

Crowdinvesting is lending to small businesses. If a company that needs investment has shares, then it makes each investor its own shareholder. If the company does not have shares, then the investor is offered other options for registering a share in the company. This financial instrument combines the advantages of both crowdlending and venture capital investment, since by distributing the "loan" necessary for the company among several donors, the chance of collecting the final amount increases.

A well-known global platform for crowdinvesting is Crowdcube.com. The Russian equivalent is Starttrack.ru [20]

The second approach is broad, it allows us to interpret MIFC as an independent form of capital associated with the self-increasing cost of investments in the development of information and labor resources to create new high-tech products that can increase surplus value at each intra-phase transition of the post-industrial megacycle.

3 Discussion

The presented author's definition of the MIFC allowed us to present the process of creating self-increasing value as follows:

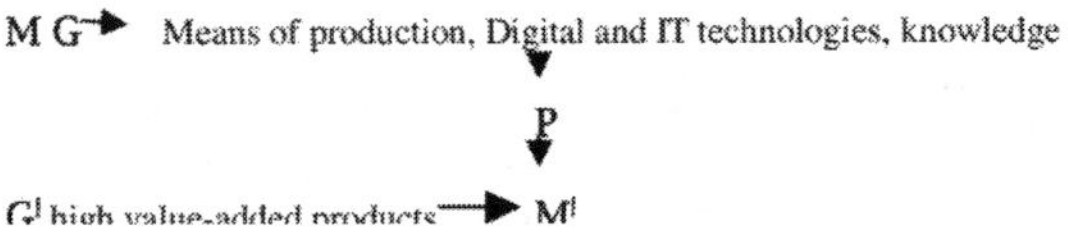

The connection at the first stage of the cycle the creative personality with information and communication technologies and other information resources leads to the possibility of creating new products that have the properties of innovation. The production of such products at the second stage of the cycle (the production stage of the information capital cycle) is associated with the creation of a new product with special information properties that have a high added value.

The essence of the functioning of the MIFC allowed us to identify its additional properties:

- the existence of an accumulation effect that occurs after a certain time period;

- relatively lower mobility compared to other forms of capital;

- the need for permanent improvement of its quality content;

- a high level of personification associated with a person as a basic carrier and translator of information;

- accumulation of technical, economic, organizational, technological and managerial content in it;

- its manifestation in the process of production of final goods is the costs of production and labor costs;

The implementation of the essence of the MIFC and its distinctive features allowed us to specify the conditions for the formation and development of the MIFC.

- increasing competition between global producers in the field of information and communication technologies;

- the need for reorganization;

- development of competition between manufacturers of innovative products;

- increasing the role of information and communication technologies in economic development based on the creation of appropriate infrastructure;

- improvement of the regulatory system in the field of digital economy [4.].

It should be particularly noted that the large-scale use of MIFC can help accelerate the transition from the industrial era to the information and innovation phase of the post-industrial megacycle based on:

- increments of knowledge, when scientific knowledge and unique experience and special skills of a person become the main factor for ensuring economic growth and sustainable economic development;

- ensuring the stability of the reproduction process at all levels of the economic system;

- formation of new global investment flows;

- introduction of new financial institutions based on digital and IT technologies.

4 Conclusions

Thus, within the framework of the prevailing processes that form the uniqueness of inter-phase transitions, we can distinguish the accumulating processes of system transformation, such as: general informatization, structural transformation, innovatization, capitalization of labor resources, service production, science-intensive transformation of productive forces, the formation and mass use of new forms of investment support for economic development. At the same time, within the framework of the transition from the neo-industrial to the digital economy, informatization as a conjunctural process, based on the mass introduction of digital technologies, contributes to the formation of the monetary and information form of capital. Based on the analysis of the definitions of monetary and information forms of capital, two approaches can be distinguished in determining the essence of the monetary and information form of capital (MIFC). The first of them is narrow, from his point of view, the MIFC is investment resources in real assets through digital and IT technologies. These may include crowdfunding, crowdlending, crowdinvesting, and P2P lending, as well as other alternative financing platforms. The second approach is broad, it allows us to interpret MIFC as an independent form of capital associated with the self-increasing cost of investments in the development of information and labor resources to create new high-tech products that can increase surplus value at each intra-phase transition of the post-industrial megacycle. The essence of the functioning of the MIFC allowed us to identify its additional properties:

1) the presence of the accumulation effect that occurs after a certain time period;

2) relatively lower mobility compared to other forms of capital;

3) the need for permanent improvement of its qualitative content;

4) a high level of personification associated with a person as a basic carrier and translator of information;

5) accumulation of technical-economic, organizational-technological and managerial content in it;

6) its manifestation in the process of production of final goods is the costs of production and labor costs.

References

1. G. Astratova, E. Burnasheva., V. Shupletsova *Information capital and its evaluation. Marketing Research Company* Retrieved from: http://www.kpilib.ru/article.php?page=325
2. V. Bauchenkov *Economic and Humanitarian Sciences*, **8** (2014)
3. Yu. Bashin *Bulletin of the Moscow State University of Instrument Engineering and Informatics. Series: Socio-economic sciences*, **18** (2009).
4. A. Bushuev, L. Lezhenina *Science studies*, **4** (2012)
5. A. Zinovev *Development of innovative activity of the enterprise on the basis of management of intangible assets:* PhD dissertation (Moscow, 2007)
6. S. Dyatlov, T. *Scientific Works of DonNTU. Series: economic* **103-1** 137-143, (2018), Zyablyuk, N. Titova *Bulletin of the Moscow University*, **6,** 2 (2016)
7. Yu. Korechkov, L. Lezhenina *Bulletin of Eurasian Science*, **3** (2018)
8. *Economic theory on the threshold of the XXI century: Neoeconomics.* (Moscow: Yurist. 2001).
9. R. Nizhegorodtsev *The Economy of Innovation* (Moscow: Ruscience, 2016)
10. P. Lemeshchenko *Neoeconomics: subject certainty and theoretical contours* Retrieved from: https://www.bsu.by/Cache/pdf/351783.pdf
11. Ya. Popova News and Society. Economy. Retrieved from: http://fb.ru/article/33482/innovatsionnaya-ekonomika
12. V. Radaev, *Economic Sociology*, **4** (2002)
13. T. Romanova Digital Economy in Russia Retrieved from: http://fb.ru/article/334484/tsifrovaya-ekonomika-v-rossii.
14. I. Sokolov, *Pravo. Management. Marketing*, **1** (2018)
15. *Reference book* Retrieved from: //https://spravochnick.ru/ekonomika/kapital/.
16. R. S. Kaplan, D. P. Norton M. Pavlova. *Strategic maps: transformation of intangible assets into tangible ones* (Moscow: Olymp-Business, 2007)
17. I. A. Sushkova The transition of Russia to a neo-industrial model of economic development in modern conditions Retrieved from:: http://vuzirossii.ru/publ/neoindustrialnoj_modeli/34-1-0-5341

18. *Financial analysis* Retrieved from: https://1-fin.ru/?id=281&t=1372
19. O. Chulanova Materials of Afanasiev readings. **1 (18),** (201)7.
20. N. Shitukhina, *Scientific Bulletin of the SIM,* **2**, (2018)

Modern city: the system of local government and support of small and medium-sized businesses

Evgeniy Buchwald*

Institute of Economics of the Russian Academy of Sciences, 32, Nakhimovsky ave., Moscow, Russia

Abstract. The article analyzes issues related to the impact of various reforms in the sphere of Russian self-government on the development of Russian cities, especially in the context of new territorial administration systems. As the initial premises of such reforms, we consider, first of all, the constitutional innovations of 2020, which in one way or another affected the economic and legal foundations of Russian local self-government, in particular, through the approval of a single institution of "public power" in the country. The concept of "public authority" is intended to embody the unity and interaction of state and municipal administration in solving the problems of socio-economic development of territories. Further, in the context of the development of cities and their management systems, the tasks of preparing new "Foundations..." of state policy in the field of local self-government are discussed. Emphasis is placed on the need to strengthen the role of municipal management structures in supporting small business forms. The importance of specifying the place and role of various types of municipalities in the practice of strategic planning is shown.

1 Introduction

The development of modern urban entities in the Russian Federation – from the largest megacities to medium-sized and small urban settlements – is directly determined by the effectiveness of the functioning of local self-government systems operating in them. In turn, this effectiveness largely depends on how effectively local policies for the development and support of small and medium-sized enterprises (SMEs), which generate jobs and the bulk of local budget revenues, are implemented. Today, the role of municipal management in solving the problems of developing and supporting the SME sector of the Russian economy is small. The positive change in this situation is associated with the improvement of the institutional and legal foundations of Russian local self-government and the strengthening of its economic, primarily financial and budgetary component.

The economic and legal foundations of Russian local self-government have not remained unchanged in recent years. However, it is safe to say that 2020 brought a number of important changes in this regard, which give reason to count on new positive developments in this direction. Let us note two important points here.

First of all, we should pay attention to the initiative put forward by Russian President Vladimir Putin in January 2020 at a meeting of the Council for the development of local self-government to prepare a draft of the new "Fundamentals of state policy in the field of local self-government development for the period up to 2030". These "Fundamentals..." should replace a similar 1999 document, which remains in force for the time being. The second of the most important events of 2020 related to the prospects of Russian local self-government is the introduction of amendments and additions to the Constitution of the Russian Federation, approved by the Law on amendments to the Constitution of the Russian Federation of March 14, 2020 [1].

These and other changes and additions to the institutional and legal framework of Russian local self-government should be recognized as very relevant. Indeed, since the adoption of Federal law No. 131 on the general principles of organizing local self-government in the Russian Federation and the implementation of the municipal reform initiated by it, a large number of problems have accumulated, the unresolved nature of which hinders the effective functioning of Russian local self-government. To some extent, the "defects" of Federal law No. 131 and the reform itself were eliminated by amendments and additions to this legislative act. Some progress in this direction was achieved due to the above-mentioned amendments to the Constitution of the Russian Federation. However, a number of systemic issues concerning self-government as the most important institution of Russian society and the state still remain open [2, 3].

This is largely due to the very limited contribution that local municipal governments in cities can make to national efforts to develop and support the SME sector in the Russian economy. This situation inevitably determines the trend of declining effectiveness of state support for SMEs, which could not be overcome due to some "cosmetic" adjustments in the system of institutions of this support. As a result, the SME sector of the Russian economy in terms of the number of SMEs and the number of employees has been "stomping around" for 10 years,

* Corresponding author: buchvald@mail.ru

https://doi.org/10.1051/shsconf/202111600029

and its share in the country's GDP also does not show sustainable growth [4, 5].

2 Materials and methods

The analysis of this situation and its causes is complicated by the low reliability and constant delay in publishing statistical data on the SME sector of the Russian economy, especially at the regional and local levels. This also applies to information on the amount of state support to SMEs, where there is no final data on the amount of funds allocated for this support, in particular, in the context of the Federal budget, as well as regional and local budgets.

In 2019, after a long period of "silence", Rosstat published data on the share of SMEs in Russia's GDP (21.9% for 2017). According to previously submitted data from the accounts chamber of the Russian Federation, the contribution of SMEs to the Russian economy in 2014 was 19.0%; in 2015 - 19.9% and in 2016 - 21.6%. However, already in 2018, Rosstat recorded a decrease in the share of SMEs in Russia's GDP to 20.2%. This is, in fact, the level that appeared in expert forecasts of the early 2000s.note that despite some differences in national criteria for SMEs, its share in the GDP of developed countries is 50-60%. So, in the UK, this figure is 51%; in Germany-53%; in Finland-60%; in the Netherlands-63%. The passport of the current national project of Russia "Small and medium-sized enterprises and support for individual entrepreneurship" States that the share of SMEs in the country's GDP will reach 32.5% by 2024. However, in this situation, especially taking into account the negative results of 2020, this target indicator seems almost unattainable [6].

3 Results and discussion

One of the main reasons for the stagnation of the SME sector in the Russian economy should be considered the low effectiveness of measures to support it at all levels of state and municipal administration. If at the sub-Federal (regional and municipal) level, the main reason for this situation can naturally be considered the scarcity of funds for such support, as well as its huge differentiation by region, then at the Federal level, institutional instability in the SME support system had a significant negative effect. This refers to the constant succession of replacements in the Executive authorities or in other institutions that were charged with a leading role in the implementation of the national policy of development and support of small forms of management in the national economy. It should be noted, that such a leapfrog with the parent institution in the framework of the state (federal) policy on SMEs was negatively projected on the regional level of management. For example, after the creation of the State Committee for the Support and Development of Small Business in 1995, similar structures were created in many regions of the Federation, with which the State Committee for the Support and Development of Small Business sought to establish "vertical" interactions. After the abolition of this Federal Executive body in 1998, similar regional structures were significantly reduced, and the practice of clear "vertical" interaction on this issue practically disappeared for a long time. The municipal level of government (with the exception of a few of the largest cities – capitals of the constituent entities of the Russian Federation) has never been fully included in this vertical at all.

Thus, among the main problems of the state policy on SMEs should be considered its "vertical imbalance" in terms of resources used (an ever-growing share of Federal resources in the total amount of state support for SMEs) and institutional instability. The way out of this situation should be considered the restoration of an institutionally and economically balanced vertical in the system of state support for SMEs. The basis for this can be the principle of unity of public power and interaction of all its components, which is now enshrined in the Constitution of the Russian Federation.

It is characteristic that following the approval of the Institute of "public power", the Government of the Russian Federation initiated a reform of the leading Federal development institutions, largely focused on strengthening and interaction of government bodies within the framework of a single system of "public power". Development institutions are one of the instruments of socio-economic policy of the state, focused on solving those tasks that most often cannot be fully provided by market mechanisms of management and require state intervention. In many respects, this concerns the need to implement measures of state and municipal policy for the development and support of SMEs and, of course, its orientation to the requirements of an innovative economy [7].

Development institutions as an instrument of socio-economic policy of the state have a number of features and advantages that are significant in terms of their potential contribution to the development of the SME sector of the Russian economy. First of all, as an instrument of strategic management, development institutions can function in concert as a creature of the Federal, regional or municipal (city) level. This specificity can be fully projected on the role of specialized development institutions in solving the problems of SME development and support. All development institutions involve close cooperation and co-financing both on the part of the Federation and its subjects, and often on the part of the municipal management level. At the same time, this co-financing from each level of the budget system can be carried out both in direct form (investments in production and infrastructure facilities; subsidies and grants to residents), and in indirect form (tax and customs benefits to residents). Finally, all effective development institutions, including those dealing with the development and support of SMEs, rely on the active use of public-private and municipal-private partnership mechanisms.

However, the realization of these advantages is currently hindered by the fact that the concept of "development institution" is not defined uniformly by a legislative act; the legal basis of existing development institutions is regulated by various laws and even separate bylaws.

It is no coincidence that recently the activities of development institutions (at least in relation to federal-level development institutions) have attracted serious criticism both from official structures (for example, the accounts chamber of the Russian Federation) and from individual groups of experts. First of all, it has been consistently pointed out that budget injections into development institutions do not have the proper effect in terms of increasing the volume of economic activity, investment, jobs, and income at all levels of the country's budget system. Nevertheless, practice has shown that there are very few or no alternative tools for solving the problems of boosting the national economy and its SME segment, other than development institutions. In this regard, the government of the Russian Federation has abandoned the previously imposed moratorium on the creation of new SEZs (special economic zones); the list of TASEDs (territory of advanced socio-economic development) is being expanded, including in single-industry towns of Russia. For example, 2020 was a record year for the number of new SEZs: 8 such new institutions were created. Currently, there are 36 SEZs operating in the country, which attracted 830 residents with a declared investment volume of more than 1 trillion rubles. However, it seems that positive developments in the activities of development institutions, if at all, are taking place very slowly.

In this regard, in November 2020 the government of the Russian Federation has announced one of the most notable reforms regarding development institutions as organizational structures that play the role of a "guide" from the strategic priorities of the state in the socio-economic sphere to their practical implementation on the ground. Until recently, there were 40 federal-level development institutions operating in the Russian Federation in the form of various funds, banks, and corporations. The number of regional and municipal development institutions is measured in hundreds. All of them were created to solve various tasks, including those directly related to supporting small and medium-sized businesses, although in reality all effectively functioning development institutions should somehow give an impetus to the spread of small forms of management.

In the course of this reform, a large institutional block was formed on the basis of the State Corporation VEB.RF, which will manage the Corporation for the development of small and medium-sized businesses; the Russian export center; the Russian Agency for export credit and investment insurance; RUSNANO; the SKOLKOVO Foundation; the Fund for assistance to the development of small forms of enterprises in the scientific and technical sphere; the Fund for infrastructure and educational programs; the industrial development Fund. Some of the functions of the eight liquidated development institutions have been reallocated between VEB.RF and federal executive authorities. In particular, such institutions as JSC "Special economic zones"; the Fund for the development of single-industry towns; the Russian Fund for the development of information technologies; Rosinfokominvest; the Agency for technological development; the Agency for the development of human capital in the Far East and the Arctic; The Fund for the development of the Far East and the Arctic; the Agency for attracting investment and supporting exports of the Far East were liquidated (Fig. 1)..

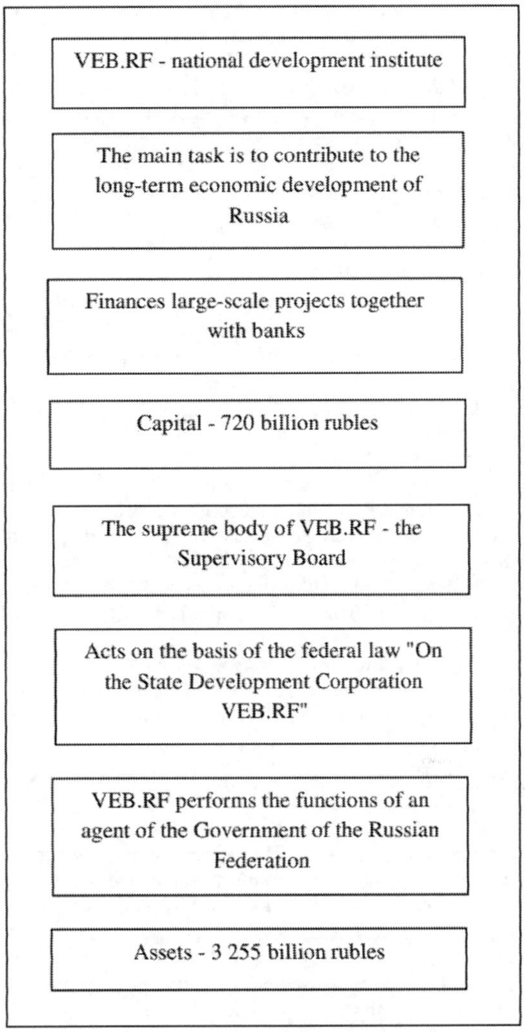

Fig. 1 Characteristics of VEB.RF

In particular, such institutions as JSC "Special economic zones"; the Fund for the development of single-industry towns; the Russian Fund for the development of information technologies; Rosinfokominvest; the Agency for technological development; the Agency for the development of human capital in the Far East and the Arctic; The Fund for the development of the Far East and the Arctic; the Agency for attracting investment and supporting exports of the Far East were liquidated.

The main focus of this stage of reforms is to increase the socio-economic effectiveness of development institutions, which can lead to an increase in their contribution to the development of the Russian economy as a whole, and to a more consistent solution for the development and support of SMEs. One of the conditions for moving forward on this path is to restructure the functioning of these institutions on the principles of unity and interaction of all levels of public power in the country. In other words, one of the main tasks in the activities of Russian development institutions is to really unite or integrate the efforts of Federal, regional and municipal government bodies on the basis of public authority in solving the most important economic tasks, including the development of small and medium-sized businesses. At the same time, we are talking about combining their financial, as well as organizational, information and human resources; about forming a unified legal and methodological framework for all issues of socio-economic policy that are solved with the participation of development institutions.

In the field of development and support of SMEs, the main task of development institutions as the Federal, regional and municipal level is that, on the one hand, to help SMEs' "start-ups" to attract investment for the development of its business, and on the other hand encourage every startup was understandable and effective growth strategy, having a good marketing and other support. The role of development institutions is also to avoid getting SMEs into certain critical situations that can put entrepreneurship on the brink of survival. It is obvious that such effective monitoring of SMEs can be implemented only at the municipal level. At the same time, it is important to form a practice in which the interests of SMEs are realized not only through specialized institutions of its support, but also through all other institutions of both sectoral and spatial development located in the respective territories. This will significantly increase the effectiveness of the state-municipal policy of SME support [8].

At the same time, it should be borne in mind that strengthening the contribution of development institutions to solving key issues of SME development and support depends not only on the quantitative and qualitative performance indicators of these institutions, but also on how clearly this range of tasks is positioned in the institutional and instrumental framework of strategic planning practice. However, now the institutional and instrumental apparatus of strategic planning in the Russian Federation does not have a clear structuring and goal orientation. Thus, the same goals of state policy in relation to SMEs are implemented through a system of state programs, federal and regional projects, Federal and regional development institutions, etc. of course, this situation creates additional difficulties for positioning development institutions in the system of instruments of development policy and support for SMEs. In this sense, it is extremely important that participation in the development and support of SMEs becomes not just a "burden" for the entire range of development institutions, but also relies on a clear definition of the tasks that each of these institutions should focus on, according to its general functional specialization and the specifics of the region of its spatial localization [10].

4 Conclusion

To solve this problem, it is important to record the following conclusions. First, it is desirable to develop and adopt a single Federal law "On development institutions of the Russian Federation", including specifying the role of these institutions as one of the instruments of SME development policy and support. Secondly, the role of "development institutions" as one of the instruments of SME support policy can be effectively implemented only on the basis of federalization of this policy and balanced distribution of the economic resources available to this policy at all levels of public authority [9, Pp. 97-108]. It is necessary to expand the economic and other opportunities of the municipal level of management to support SMEs, including by implementing targeted municipal programs, activating the activities of municipal "development institutions" aimed at developing and supporting SMEs, primarily in the largest cities of Russia [11].

Acknowledgements

The article was prepared with the support of the Russian Foundation for basic research in the framework of scientific project No. 20-010-00120-A "Small and medium-sized entrepreneurship as a tool for regulating the spatial structure of the Russian economy".

References

1. E.M. Buchwald *Federalism.* **1**, 16-20, (2021)
2. O.N.Valentik *Innovations. The science. Education.* **28,** 114-121, (2021)
3. E.V.Kiskin *Federalism.* **4 (100)**, 107-126, (2020)
4. S. V. Nekrasova *Russian economic bulletin* **3 (3)**, 116-120, (2020)
5. E. M. Buchvald *Bulletin of Vladimir state University named after A.G. and N.G. Stoletov. Series: Economic Sciences,* **1 (23)**,114-128 (2020)
6. A.M. Morozova *Scientist's Rostrum,* **10,** 370-373, (2020)
7. E.M. Kobozeva, T.V. Vorobeva *Innovative economy: prospects for development and improvement,* **35 (47)**, 40-47 (2020)
8. A.V. Polyanin, Yu.P. Soboleva, V.V. Tarnovsky *Bulletin of Omsk University. Series: Economy.* **18(1),** 32-44, (2020)
9. Yu. V. Vertakova *Economic Sciences.* **13(4)** (2020)
10. Z. Mingaleva, N. Pazdnikova, E. Mitrofanova *Asian Social Science*, **10(24)**, 257–262 (2014)
11. O.V. Butorina, N.P. Pazdnikova, Y.V. Karpovich *Espacios* **39(27)** 2018

"Western Gates of the Urals": geocultural branding of a small city

Sergey Dianov*, *Yuliya* Dianova, *Lilia* Basarieva and *Olga* Poluyanova

PNRPU, 29, Komsomolsky prospekt, Perm, 614990, Russia

Abstract. The article presents an analysis of the image resources of Vereshchagino -a small town in the Perm region, which should be used in the process of implementing the strategy of its geocultural branding. In popular scientific works and mass media, the city is called the "Western Gate of the Urals", meaning that the Vereshchagino railway station is a key point, bypassing which, freight and passenger trains leave the Ural region. In 2023, Vereshchagino station will celebrate 125 years since the moment of its foundation. At the beginning of the XXI century, the dynamic development of the urban environment in Vereschagino was interrupted. The municipal authorities missed the opportunity to pursue a policy of territorial marketing branding, to actively promote the brand images of the town in the Ural geocultural space. Today, Vereshchagino has accumulated typical problems for a provincial town – demographic, communal, and socio-cultural ones. Urban communities express dissatisfaction with the visual and aesthetic appearance of the urban environment. According to the author's point of view, the development and implementation of the Vereshchagino geocultural branding strategy is a progressive solution that allows creating creative industries in the city and ensuring sustainable socio-economic development. As a methodological basis, we used the conceptual ideas of the creative action of the British scientist Ch. Landry, as well as the postulates of the theory of geocultural branding of the city by the Russian researcher D. N. Zamyatin.

1 Introduction

The history of Vereshchagino began at the turn of the XIX-XX centuries. During this period of time, the construction of a railway transport network was launched on the territory of urban and rural settlements of the Ural region. In 1895, on the site of the future town of Vereshchagino, the construction of one of the sections of the Perm-Kotlas railway began, including construction of a terminal, a railway station and a locomotive depot. By 1898, the main works on the construction of objects were completed, the station was named "Ocherskaya", and in 1903 it was renamed as the "Voznesenskaya" station [1]. The opening of the movement of passenger and freight trains on the Glazov-Kama section took place in November 25, 1898. In 1907, the settlement formed at the station was named "Alexandrovsky" [2]. In the spring of 1916, the Ministry of Railways granted the request of railway workers and station employees to rename Voznesenskaya station and the village of Alexandrovsky to Vereshchagino. In the articles published earlier, the author's point of view was stated, according to which this initiative was recognized as one of the first collective creative actions in the history of Vereshchagino, indicating that people understand the visible and invisible meanings of creativity and social asceticism [3].
During the Soviet period, there was a steady socio-economic development of Vereshchagino. In 1924, the settlement became the administrative center of Vereshchaginsky district, then in 1928 an urban-type settlement, and finally, on June 19, 1942, Vereshchagino received the status of a town of district subordination. The population had grown from 6 thousand in 1931 to 22 thousand people in 1959 [4]. In the second half of the XX century, a complex of city-forming enterprises was established there: a knitting factory, a bakery plant, a dairy plant, a reinforced concrete plant, a steam locomotive repair plant, etc. In the 1950s-1980s, the urban space had been actively developing, and residential multi-apartment construction was carried out at a rapid pace. It is important to note that in the 1990s Vereshchagino showed positive indicators of economic and socio-cultural development. In contrast to other Perm towns, the demographic situation here was stable, by 2001 the number of citizens reached 25.2 thousand people. People from rural areas moved to the city, specialists came with families from Perm in search for better opportunities to maintain material well-being. In 1993, the book called "Vereshchagino - Western gate of the Urals" by local historian V.G. Melchakov was published [4]. In 1995, a monument to the Russian battle artist V.V. Vereshchagin was erected on the station square. Traditions of weaving and technologies of working with textiles were implemented in the factory production of knitted clothing, the production of which was carried out at a high level of artistic skill. We express the opinion that at the turn of the 1990s-2000s, Vereshchagino had unique conditions for implementing

* Corresponding author: sadianov@gmail.com

programs for brand identification of local image resources. Creative urban communities should be engaged in promoting Vereshchagino brand images in the Ural region. The geobrand "Western Gates of the Urals" could serve as an organic basis for the creative scenario of the urban environment development. However, the recent history of the town has shown that the municipal authorities and local communities were not ready to implement innovative programs. The geocultural potential of Vereshchagino was not in demand and was not used in the current plans of socio-economic development. In turn, this led to serious negative consequences. Let's describe some of them.

First, the city has ceased to be an attractive and comfortable place to live. By 2007, its population had decreased to 22.6 thousand people [5]. As of January 1, 2020, 21.2 thousand people already lived in Vereshchagino [6]. The average life expectancy in 2019 was 65 years (for men - 59 years, for women - 73 years) [7]. The outflow of the population, primarily young people and highly qualified specialists, can be explained by the traditional reasons for a provincial city: difficulties in finding employment (especially in the state and municipal service), the unavailability of innovative educational and medical services, the underdevelopment of the cultural and business sphere, the slow pace of commissioning new housing, the lack of creative industries, etc. Second, the working part of the population of the Vereshchaginsky urban district is only 37%, the overall unemployment rate is estimated at 7% [8]. There is a fairly high level of "otkhodnichestvo" (temporary, most often seasonal work outside the place of permanent residence). According to the authors of this study, this figure is equal to 25%. Citizens who have permanent residence in Vereshchagino prefer to work outside of it. As the researcher of this problem Yu. M. Plyusnin notes, at present there are at least 15-20 million people of such workers in Russia. According to him, a third of all families in the province live on latrines, that is, a type of economic activity that is almost not taken into account by official statistics [9]. Third, in the urban environment of Vereshchagino, the problems of municipal economy were exposed: a high percentage of worn-out housing and communal networks, unsatisfactory condition of road infrastructure, unsettled garden and park areas, desolation of suburban areas, etc. In winter, in the city you can observe snow-covered benches and garbage bins, icing of pedestrian paths, untimely clearing of snow from house territories, etc.

These data allow us to conclude that today the implementation of the strategy of geocultural branding of the city is vital for the city of Vereshchagino and its population.

2 Materials and methods

The theoretical and methodological basis for this work is fundamental and applied research in the field of cultural studies and geography, history and ethnography, economics and sociology, design and technical aesthetics, museum studies, as well as the works of local historians and specialists in the field of urban studies.

The solution of the problems considered in the article is based on the concept of geocultural branding of the city in the form in which it is presented in the scientific works by Doctor of Cultural Studies Dmitry Nikolaevich Zamyatin [10], as well as on the theory of the creative city by the British scientist Charles Landry, in particular, on his vision on creative action [11]; [12].

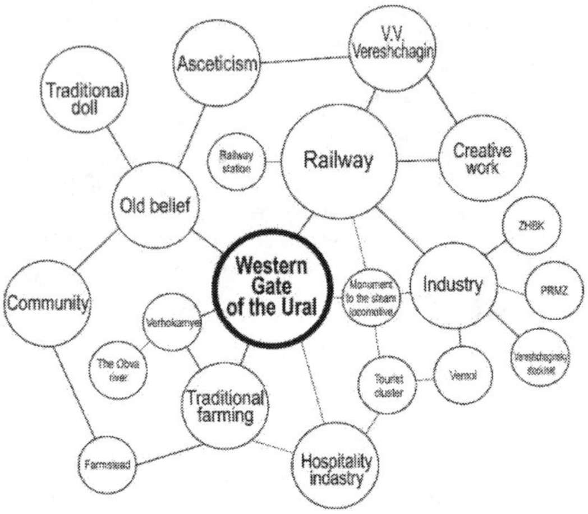

Figure 1. Geocultural scheme of Vereshchagino

So, geocultural branding of territories is a project-network activity aimed at the applied use of the geoculture of the territory (regional identity, historical and cultural heritage, archetypal geographical images, local myths or cultural landscapes) in order to form and promote a socially significant and effective (attractive) territorial image [13]. A city, a municipal district, or a rural settlement can be represented as a purposeful, detailed, structured image. And this image of the territory can be built, formed as a geocultural brand. The territory's geocultural brand is its strategic asset.

Based on this definition, it becomes obvious that the visual and semantic framework of the urban environment is formed on the basis of a certain geoculture of the city. In this concept, we include a phenomenal core (urban legends, legends, unofficial toponymy, folk-spoken ("live") speech, places of memory), around which cultural realities are built, which have not lost their symbolic and semantic meaning in the maelstrom of urban events. Both individually and in combination with each other, cultural realities can act as cultural codes of the city or as the basis for the strategy of urban cultural policy. In those of them, where the subjects of the mutual influence of space and culture, nature and human activity find expression, the attributes of a geocultural resource are clearly traced. The distinctive feature of the latter is that it is organically woven into the space of the city, its dynamics and changes are the focus of attention of local communities and individual urban residents. In the rhythms of everyday life in urban life, geocultural resources are endowed in the minds of citizens with earth-like features.

Ch. Landry said that "Each city can find its own form of creativity, which is unique. The very nature of the city will contribute to the development of creativity, which is due both to the existing niches in it and to the hybrid ideas that arise on this basis. The cultural strategies that we offered to the cities were aimed at solving complex problems. The success or failure of the strategies depended less on the material assets of the city, and more on the ideas and approaches to solving problems" [12].

It seems that the tactic of gradual deployment of the city's geocultural potential can be an effective solution. At the first stage, urban communities identify a list of local geocultural resources, identifying those that can become image resources in the course of project activities. As a rule, a large city is characterized by the presence of three or four geocultural resources, for a small city – one or two. At the second stage, creative actions with a geocultural resource are tested in an urban environment. It should be noted that at this time, creative solutions, techniques, and technologies are being worked out, the combined result of which is to determine the parameters of the future promising model of a creative city. The third stage is the development of the city's geocultural branding strategy and the beginning of its implementation. The fourth stage is the achievement of intermediate results in the process of becoming a model of a creative city. The fifth stage is the presentation of the city's creativity on a Russian and global scale.

3 Results

First of all, we will present the author's geocultural scheme, which shows the image resources of Vereshchagino town (Figure 1).

In this geocultural scheme, it is proposed to recognize the creative idea "Vereshchagino – the Western Gate of the Urals", which is well known to citizens, as a geocultural brand of Vereshchagino. It was already noted above that the prospects for launching this brand were outlined in the late 1990s. From the local periodicals of recent years, the discourse about the "Western Gate of the Urals" has entered the content of social media platforms [14]. Let's list the image resources that make up the conceptual basis of a geocultural brand:

- Railway and its infrastructure (railway terminal, platform, station, station square, railway bridge), which have been developed in a historical place and serve as a place to receive guests of the city, including tourists and sightseers.

– JSC "Vereshchaginsky PRMZ "Remputmash" is a city-forming object, founded in 1961 on the basis of a locomotive depot that has existed since 1899. About 700 people work at this enterprise. It should be noted that the company specializes in the repair of track equipment (freight cars, wheel sets), as well as in the manufacture of self-propelled snowplows. We note also that, when at the end of 2018, the Perm plant named after Shpagina was closed, part of its production capacity was transferred to the Vereshchaginsky PRMZ. Today, the museum-socio-cultural space "Shpagin's Factory" is located on the squares of the former Perm plant.

- Railway symbols: the monument to the locomotive FD21-3092 (built in 1941), the sculpture "Railway Worker's Doll" on the pier territory (2019), etc.

- Monument to the battle artist V. V. Vereshchagin on the station square (sculptor A. A. Uralsky, 1995).

– House of Culture of railway workers. It was built in 1928 and was originally called the "Club of the October Revolution". It hosts festive events, community meetings, competitions and festival programs.

The crest of Vereshchagino also reflects the image resources of the territory, such as the railway, the richness of the regional nature, golden brushes for creativity).

In addition, the geocultural brand "Western Gates of the Urals" could be supported by appealing to such cultural institutions of the city district as:

- library and museum organizations;
- educational organizations and local history centers;
- religious organizations;
- city equestrian club.

At the same time, a serious problem hindering the launch of geocultural branding should be recognized as the lack of a developed sphere of the hospitality industry in Vereshchagino. The first hotel in the city was the Priuralie Hotel, opened in the fall of 1968. However, its number of rooms (12 rooms), as well as the number of rooms of several mini-hotels, currently does not allow to accommodate large groups of tourists and sightseers. Accordingly, without the creation of a city network of the hospitality industry, the task of developing a tourist cluster seems almost impossible.

We will highlight several priority tasks that need to be achieved in the process of implementing the Vereshchagino geocultural branding strategy:

- establishment of creative sites in the town for the cultural activities of urban residents and their creative groups (for example, the open-air museum "Gates of the Urals", where it is possible to place different variations of gates, doors, keyholes, etc.);

– construction in Vereshchagino of the outdoor environment and parks of themed sculptures with small architectural forms (e.g., puppet images of urban fair, steam locomotive from different materials – wood, glass, plaster, foam, etc.);

– introduction of creative visual appearance of Vereshchagino in accordance with art geocultural concept of the brand (e.g., styling benches, advertising structures, exterior public buildings and houses);

- development of a positive image of Vereshchagino in the media and information space of the country as a city "open" for people of creative professions;

- creation and development of the tourist cluster "Obyorskoe porechye - the center of traditional crafts".

It is important to try to involve Vereshchagino as a social partner of JSC "Russian Railways", its subsidiary – JSC "Perm Suburban Company", in the process of implementing geocultural branding. Today, an electric train of the latest type "Lastochka" runs from the Perm II station to the Vereshchagino station. With the successful launch of the tourism cluster and the development of the hospitality industry in the city, the West Ural Express project can become a real creative action. Tourist groups will be able to get to Vereshchagino in just 1.5 hours. JSC "Russian Railways" can also assist in the organization of exhibition areas for the placement of cars and dining cars of different historical eras.

4 Discussion

The authors of this article held a round table called "Creative City: cultural environment and creative potential of modern Perm" on the 13th of February, 2020, in the framework of the VI Open Regional Championship "Young Professionals" (WorldSkills Russia). This event was held at Perm Construction College. Some of the issues discussed were devoted to the category "Creative initiatives of the population of small towns in the Perm region". So, L. P. Bazarieva, relying on her practical experience of visiting cultural events in Vereshchagino, Ocher and Bolshaya Sosnova in 2019, brought up for discussion the problem of the lack of creative industries in small towns. Among those present at the round table was a delegation from Vereshchagino. The chief bibliographer of the Vereshchagin Central district library named after V. G. Melchakov, I.A. Sazonova, said that the local communities of the city district are waiting for changes. In the city, the bus stops on the bus routes look the most expressionless. Creative people try to produce creative products within the framework of different competitions and festivals. One of the most popular events among the citizens is the contest "Madame-Maslenitsa", where children's initiative groups under the guidance of masters create unique images of the doll "Maslenitsa". In turn, the representative of the Women's Council of the Vereshchaginsky city district, N.A. Sushcheva, expressed the opinion that Vereshchagino has image resources for putting into practice the original model of the creative city.

As a result of the round table, its participants adopted a number of practical recommendations, including:

- initiation of a series of events (round tables, panel discussions, Internet forums) on the topic "Geocultural branding of Perm and the territories of the Perm Region» by university and professional creative communities in 2021-2022;
- competitions for creative projects to transform the urban environment (street spaces, pedestrian sidewalks, squares and alleys) hold by municipal authorities, regularly develop a grant system to support creative projects of talented young people (schoolchildren, students, young professionals);
- support of the author's creative initiatives on geocultural branding of small towns of the Perm region, which are based on the historical and cultural phenomena of the Perm land.

On March 31, 2020, S.A. Dianov, Yu.V. Dianova and L.P. Bazarieva planned to hold a round table in Vereshchagino on the topic "Creative action for the socio-cultural environment of Vereshchagino: scenarios of creativity". The partner was the Vereshchagin Central District Library named after V. G. Melchakov. However, this event did not take place due to the introduction of restrictions in the Perm Region in connection with the coronavirus pandemic. The round table is scheduled to be held in June 2021.

5 Conclusion

The conceptual idea of the geocultural brand for the city of Vereshchagino "Western Gates of the Urals" is based on the phenomenal core of the geoculture of its territory. Historical railway infrastructure, dynasties of railway workers, symbols of the labor feat of railway workers during the Great Patriotic War (1941-1945), the name of the outstanding artist V. V. Vereshchagin, who visited the pier village thanks to the highway in 1904, books of the local historian V. G. Melchakov and modern poets living in Vereshchagino are unique resources that should be used when launching the town's geocultural branding. The process of educational activities is especially important, conveying the importance of geocultural branding of the city to the young residents of Vereshchagino as the most active and mobile category of citizens.

References

1. S.G. Volegova *About the date of the city of Vereshchagino. History of renames and statuses (2017).* Retrieved from: https://nashural.ru/mesta/permskij-kraj/vereshagino/
2. G.A. Litovchenko*The road through the ages: about the Perm railway workers: from event to event* (Pushka, Perm, 2003).
3. S. Dianov, Yu. Dianova, *Man and Culture,* **4** (2020).
4. V.G. Melchakov *Vereshchagino – the western gate of the Urals* (Regional Creative Center, Perm, 1993).
5. Prmstat *Estimation of the permanent population of the Perm Region in the context of municipalities as of January 1*, 2007 (Permstat, Perm, 2007).
6. Permstat *Estimation of the population of the Perm Region as of January 1*, 2020 Retrieved from: https://permstat.gks.ru/folder/33429.
7. *Average life expectancy of the population of the Vereshchaginsky city district* Retrieved from: http://veradmgo.ru/about/#mo-element-region-demografiya (2020).
8. *Population structure of the Vereshchaginsky city district.* Retrieved from: http://storage.inovaco.ru/media/project_mo_181/41/ea/21/81/c7/3a/struktura-naseleniya-na-01012020.pdf.
9. Yu. Plyusnin, A. Pozanenko, N. Zhidkevich, *The world of Russia. Sociology. Ethnology,* **1**, 24 (2015).
10. D. Zamyatin, *Geocultural branding of cities and territories: from theory to practice. The book is for those who want to design and create other spaces* (Aleteia, St. Petersburg, 2020).
11. C. Landry, *The Creative City: A Toolkit for Urban Innovators* (Earthscan, London, 2000).

12. C. Landry, *The creative city* (Classics-XXI century, Moscow, 2011).
13. D. Zamyatin, *Modern problems of service and tourism,* **2,** 9, 2015.
14. A. Pustovalov, L. Nekrasova, *Eurasian Humanitarian Journal,* **1,** 2019.

Innovative aspects of feedback delivery in management process by applying emotional intelligence

Elena Rastorgueva*

RINO PSNIU Center for Psychological Education, Perm National Research Polytechnic University, 29, Komsomolsky Prospekt Str., Perm, 614990, Russia

Abstract. This article focuses on examining new specific practical recommendations for delivering feedback in the process of staff management in organizations by applying the resources of emotional intelligence. Managers' I-statement mistakes, imperceptible at first sight, are analyzed and explained from the point of view of their psychological correctness; ways of correcting them are presented. In detail, with real-life examples of the feedback in the organizations, it is explained how a leader can take into account the psychological difficulties of feedback perception and increase its effectiveness. Thus, the following mistakes of the feedback are analyzed: the feeling the one giving feedback is not mentioned at all; his feeling is generalized; it is determined in an unspecified way or expressed to a person in general; the person's behavior, which evokes a feeling, is not described at all or is described in an unspecified way; it is not specified what exactly has caused a feeling in a communicative situation; a conclusion on the person's actions and the person in general as a generalization of his actions; exaggerating; identifying a trait of character of someone's actions as a generalization of his actions; emphasis; describing internal processes of another person; expressing another person's attitude to something or someone; a subjective evaluation; interpreting a person's actions; an instruction (lecturing, sermonizing); a tirade; a mockery; an accusation; comparing a person with someone; a condemnation; a threat, an insult, a humiliating statement; condescending phrases; a wish; a request. Correct I-statements are a universal mechanism of self-regulation, self-support.

1 Introduction

Emotional management or emotional-intelligence-based management is one the recognized modern approaches in management [1–10]. However, this approach cannot be seen as deeply developed in its practical aspect in management and psychology.

On the one hand, there is enough experimental evidence accumulated in modern science regarding the positive effects that emotional intelligence as well as effective feedback has on management success in organizations, on the development of students', prospective consulting psychologists' and psychotherapists' professional skills [11–23]. It is commonly believed that a leader sets the dominant tone for an entire organization in dealing with their own and other people's emotions, which improves the interpersonal relationships in a team, the quality of work environment and employees' social skills [1–10, 17–23]. Nowadays, it is acknowledged that the ability to be conscious of one's feelings, to reflect on them well and to express them openly and naturally in one's behavior is an internal condition of a person's individuality self-actualization (along with its other internal conditions) [26, 33–35].

On the other hand, emotional intelligence in management and team work can be used effectively and reasonably only if a manager themself has a high emotional intelligence. It is known that it requires a manager to specially develop and master necessary skills [1–10].

2 Materials and methods

Not every manager realizes the importance of working with one's and employee's inner feelings. A manager may have a mindset and illusion of purely rational staff management, whereas the irrational factors are not only present, but then also remain really uncontrolled. In this case, managers especially lack their own psychological proficiency for effective team management. A manager's habit of ignoring and setting aside their own feelings (as well as subordinates' emotions and feelings), "maintaining their composure" is far from being the ideal of successful self-regulation. The accumulation of one's feelings may cause unreasonable actions (emotional explosions) at an unpredictable moment, when it is no longer a person oneself who chooses to express their feelings, but the feelings explode uncontrollably determining the person's actions. Feelings and emotions which have not been realized at the right time are known to affect the person's conscious life, their professional decisions, the manifestations of the unconscious through wrong actions and neurotic symptoms, chronic diseases, all of which significantly

* Corresponding author: elnigor@mail.ru

influence the process and results of work and communication [32]. As revealed by the practice of corporate psychologists working at organizations, the complexity of communication and interpersonal relationships between staff members may significantly impede work interaction and reduce the efficiency of employees' work [10]. Feedback is often present in the communication between colleagues and between manager and subordinates, it is necessary, but in fact it often complicates work interaction, even though it is supposed to have many positive functions [31].

On the one hand, feedback in the communication process has traditionally been a subject of psychological studies for solving the problem of optimizing communication in various areas of human life and activity [11–16]; on the other hand, there is a need to study more deeply the use of feedback in the process of staff management.

Seeing that the term feedback (**FB**) has multiple meanings, it is important to delineate the approach to studying it [25, 24, 29, 30]. We shall use a complex, narrow-sense definition of FB [29, 30, 24]. In other words, when we speak of FB from a manager to subordinates, FB means information from the manager on a subordinate, on how the manager sees their words and actions, their personality in general and interpersonal communication with them.

In actual everyday communication, the functioning of FB is often connected to many circumstances which limit its effectiveness [24]. It is known that under the conditions of a social perception training, these limitations are overcome and the FB effectiveness increases [24]. First of all, the intentional delivery of FB intensifies, its adequacy and informativeness increase; the perception of FB in a training also intensifies, probably due to minimizing value judgments, increasing the use of self-descriptive judgments, increasing feedback specificity [29]. E.V. Lushpaeva has determined that socio-psychological trainings create specific conditions for developing self-reflection owing to an extensive FB system [27].

In the Feedback intervention theory, A.N. Kluger and A.S. DeNisi propose five theses, three of which, it would appear, are extremely important in management: a person's attention is limited; under regular conditions, attention is directed according to the level of a target in the hierarchy; interactions accompanied by FB shift the attention locus, thereby affecting behavior [11, 12]. The authors have made a significant generalization that in case of FB, individual differences have not been an important explanatory principle of its effectiveness [12]. Thus, feedback directs person's attention and what is relevant is the determination not of its individual principles, but of the general ones.

Feedback occurs extensively in psychotherapy practice and helps clients to know and explore the "blind spots" in their psyches [33–35]. Here, it is important to remind the following principles of FB effectiveness in psychotherapy: the readiness of a client to perceive feedback, the absence of value judgments on the part of a psychotherapist, the concreteness of FB and its directedness at a particular person [33–35].

It appears that this accumulated knowledge on FB may and must be realized and used in everyday work communication, and not only within the framework of psychological work with people, but also to establish a more effective FB delivery by managers in the process of staff management.

Now we shall examine and compare two ways of delivering FB best known in psychology nowadays: You-statements and I-statements.

Ways of delivering feedback

1. You-statement. It is the most common way of giving FB in everyday life.

It is structured according to the formula: *You + are like this.*

A person who gives FB notices various psychological qualities of another person, says 'you are' and names some quality (a characteristic, personality trait, state, feeling, attitude, motive, purpose, value, etc.).

Examples: 'You are very conscientious'; 'You are disorganized'; 'You are too careless'; 'You are too persistent'; 'I think you are sensitive'; 'I may be wrong, but you seem upset'; 'I see you are not being sincere now'; 'Your work is the most precious thing for you'; 'You have an amazing talent for upsetting us all'.

A person's psychological defense is triggered when they get negative FB, when someone mentions qualities which are unpleasant personally to them or which are condemned and seen as negative in society (for example, 'irresponsible', 'impudent', 'silly', etc.). From the point of view of S. Freud's theory of the unconscious, a person tends to automatically (unconsciously, without thinking) dismiss, that is, to defend themself from, any disagreeable information about themself and negative emotions associated with it [32]. In this case, FB is not very effective. When it is important to make a person hear, reflect on themself and their behavior, it often proves impossible to achieve using You-statements, as a person does not perceive such information attentively and seriously, does not reflect on it deeply, does not want to. Using You-statements has a great risk of ruining a relationship (they are considered conflict-generating), because in this case communication is associated with strong negative emotions. The game of 'you are like this and you are like that' may continue in further communication or, conversely, may end abruptly by breaking off relationships.

2. I-statement. A less common way of delivering FB in everyday life.

It is structured according to the formula: *I (or to me) + feel (or perceive) such (specific) feeling (or sensation, state) + about such (specific) external action of another person in the communication process.*

A person who delivers FB, by means of reflection, defines a feeling (or sensation, state) which they experience at the present moment of communicating with another person, and says 'I (or to me)', names the feeling, then specifies by which external action of another person the feeling (or the sensation) has been triggered.

Examples: 'I admire you when you finish all the necessary work so quickly'; 'I feel tense when I see so many errors in your work documents'; 'I am happy to see how many clients you have managed to attract today'; 'I get upset and angry when your tasks have not been completed in a week'; 'I feel uncomfortable when people come close to me and stare into my eyes'.

I-statements are a more effective and more psychologically correct way of delivering FB as compared with You-statements, because a person receiving FB does not dismiss information about themself automatically, can consider it more attentively within themself, reflect on their behavior and actions. I-statements are far from being a 'magic pill to solve all troubles' (they cannot replace requests, wishes, explanations, demands, punishments, control, support, motivation and other essential interactions of a manager with employees). They are not aimed at making another person do something, but they may give a chance to change a person's actions and behavior in general by their own intention, as well as to improve the communication process. Furthermore, I-statements allow a person who gives feedback to deal with their own emotions and feelings by expressing them directly.

Acquiring the skill of making correct I-statements requires not only training, but also deep understanding and awareness of numerous mistakes made when attempting to use I-statements, even when a person is familiar with its formula. These mistakes may be involuntary, imperceptible to a person. Thus, in real everyday life, You-statements and I-statements are most often mixed together in our speech, of which we are also poorly aware. Unfortunately, the mixing and mistakes often completely outweigh the efficiency of FB.

It is impossible to completely avoid You-statements in the communication process (the more so since they perform a positive function when identifying positive qualities—improve self-esteem of a person who gets FB, support them, truly make them happy, inspire, give strength and energy). However, if and when we want someone to hear and consider our FB, reflect on themself, use it for self-knowledge and improvement of their work and behavior, it is important to learn, even after having made unconscious You-statements, and notice the mistake to convert it into I-statements.

The problem in mastering I-statements and their effective use is a common wrong view that simply using the pronoun "I" and expressing one's feelings can be considered an I-statement (for example, 'I am mad at you'). On the contrary, there are numerous social stereotypes identified in literature which limit or completely exclude any verbal expression of one's feelings and emotions in work relationships and make it impossible for a person to use I-statements at work [10].

We shall now examine particular FB mistakes in using You- and I-statements (and ways of correcting them) often seen in the professional life of managers, and explain them from the point of view of psychological correctness. These mistakes have been identified during our work as a psychologist for training employees at several commercial and state organizations. The psychological work has been performed with employees of different departments (psychological trainings and psychotherapeutic dialogs). The following research methods have been applied: the method of observing everyday communication between managers and subordinates (the object of observation is feedback from a manager to a subordinate), the method of analyzing and summarizing FB mistakes. The examples below illustrate only potentially problematic situations. FB mistake correction in every case is conducted by bringing a statement to the full and precise I-statement formula (given above). Thus, general recommendations for delivering FB are given, regardless of the individual qualities of communication partners.

3 Results and discussion

Feedback mistakes

1. The feeling is not mentioned at all.

Examples of the mistake: 'You have ruined our platforms'; 'You have not even discussed this issue with me personally'; 'You are not looking at me when talking'.

These FB examples only describe the facts of what is happening, the specific external actions of another person. However, a person who gets FB may easily ignore these facts or interpret them in their own way, unless it is said how exactly the facts are seen, how they affect another person and the work communication process. It is necessary to specify and add a feeling which is experienced at the present moment by a manager who delivers FB.

Correction: 'I am mad at you for presenting your material in diagrams on the platform'; 'Too bad you have not even discussed this issue with me personally'; 'I feel uncomfortable that you are not looking at me when talking'.

Feelings and emotions are a person's subjective evaluation of things that happen. Much more often, managers tend to accumulate and hold back most of their feelings by their will, and when the restraint mechanisms no longer work, they tend to uncontrollably and inadequately (more intensely than a trigger in the situation) release the emotions in the form of You-statements on their subordinates, which causes even greater negative social consequences. It is important to remember that there are no right and wrong feelings. It is also necessary to understand that even short-tempered people will be less offended by a contextual personal feeling (about oneself) than by a negative generalized judgment about them. Moreover, it is easier to express a less intense feeling, consciously choosing the way to express it, than to express a very intense feeling, if it has been accumulating for a long time. It is more reasonable to direct one's will not at holding back (hiding) feelings, but rather at choosing a more appropriate creative way of expressing one's feelings, taking into account the communicative situation, the qualities of a communication partner, the relationship with them, the importance of maintaining the relationship, one's own qualities (goals, values, resources, etc.).

2. The feeling is not specified:

- only the negative or positive pole of a feeling is expressed; a person who gets FB may interpret a statement in their own way and misunderstand the feelings of a person who gives FB. **An example of the mistake:** 'It is unpleasant when you are not looking at me'. Correction: 'It is unpleasant, uncomfortable when you are not looking at me';

- a feeling is expressed metaphorically; when expressing feelings in metaphors, everyone implies something personal which may be absolutely unclear to another person. **An example of the mistake:** 'I am down in the mouth because of your words'. Correction: 'I am down in the mouth, upset because of your words';

- only the degree of a feeling is expressed. An example of the mistake: 'It is unbearable when you make your report for such a long time'. **Correction:** 'I begin to worry and get angry when you make your report for such a long time'.

In all the examined cases, it is necessary to specify and identify one's feeling arising at the present moment.

3. Generalization and expressing feelings about a person in general.

An example of the mistake: 'You annoy me'. Correction: 'It annoys me that you laugh when we are discussing this important issue', 'It surprises me that you still have not submitted the report'.

It is necessary to specify a person's external action which evokes the feeling. It is very useful to identify feelings in each particular situation and express them for a particular reason. Too often, subordinates interpret the manager's feelings and words about them literally and wrong because of vague expressions used by the manager, and they react according to their literal interpretation. Eventually, it becomes more and more difficult to fix the relationships and to communicate further. Furthermore, feelings about a person may become not only more intense, but also more contradictory, if they have been accumulated after different situations, and then it becomes objectively more and more difficult for a manager to express them.

4. The person's behavior, which evokes a feeling, is not described at all.

An example of the mistake: 'I get tense when I see you'. Behavior, actions or their elements, external expressions (or, conversely, the absence of desired behavior, actions and expressions of another person) are not stated, nor are characteristics of the environment created by them. It remains unclear what has caused the feeling. **Correction:** 'I get upset if you do not greet me when we meet (get angry that you do not bring the final version of your work for so long)'. Here, a subordinate's external action, which causes the feeling, is specified.

5. Another person's behavior is inadequately described as a generalization of their actions.

Examples of the mistake: 'I feel offended if you are not with me'. 'It's a shame it happened like this', 'Your eccentricities make me laugh'.

It is necessary to specify a person's external action which evokes the feeling. **Correction:** 'I feel offended if you do not share your thoughts during a meeting', 'It's a shame it happened like this, that you have not come to talk to me personally', 'The way you have just rapidly and abruptly waved your hands when talking makes me laugh'.

6. It is not specified what exactly has caused a feeling in a communicative situation, when it does not concern the behavior of a FB recipient directly, but concerns the communicative situation.

An example of the mistake: 'I am sorry it is so'. A specification is needed. **Correction:** 'I am sorry that your computer broke right at the time of the important negotiations'.

7. Conclusion on the person's actions and the person in general as a generalization of their actions Examples of the mistake: 'I am shocked that you are unable to negotiate', 'It is awful that sorting out things with you is impossible'. Correction: 'I am shocked that you often lecture your colleagues', 'It is awful, painful that you end the conversation as soon as we start talking tensely'. It is again important to specify a person's external action which causes the feeling.

8. Exaggerating another person's actions as a generalization of their actions (or words). In this case, there is a dogmatic tone, a negative generalization in describing another person's actions. An exaggeration usually contains such words as **'everywhere', 'never', 'nowhere', 'in no way', 'nothing', 'at all'**, etc. Even if a person has done something well and right once in their life (usually, they remember it perfectly), a 'never' is already a distorted description of the person and reality. Because of the fact that a person who delivers FB is not objective, one does not wish to listen to them any longer. One inaccurate word causes a protest, a defense reaction, FB is no longer perceived and all I-statements become ineffective.

An example of the mistake: 'It disgusts me that you are **always** shouting' Correction: 'It disgusts me that you are shouting now (speaking so loudly, expressing your thoughts so loudly)'.

9. Identifying a person's quality (psychological quality, character trait) as a generalization of their actions. As said before, a negative generalization about a person provokes a psychological defense and makes them imperceptive to FB. Every person tends to unconsciously dismiss any negative information about themself, automatically, without thinking (see above). Moreover, it is not known for sure whether what is expressed in FB as a psychological trait is more due to the qualities of a person who gives FB or a person who gets it.

An example of the mistake: 'I am mad at you for being so naive'. **Correction:** 'I get mad when you just take the suppliers' word for it'.

10. Emphasis—a word or phrase which emphasizes the seriousness of a negative generalization about a person, it may increase the negative feelings of a person who gets FB and complicate its acceptance.

An example of the mistake: 'It amazes me **how** slow you are'. Correction: 'It amazes me that you make mistakes in such simple calculations'.

11. Describing internal (psychological) process of another person. The internal psychological processes of another person (sensations, perception, memory, thinking, imagination, desires, motives, needs, goals, meanings, values, etc.) are invisible to an eye, so one

 https://doi.org/10.1051/shsconf/202111600031

cannot know and judge them from outside with full confidence; they cannot be proven because they occur within another person. If one speaks of the internal processes incorrectly, a person may not take it seriously, and in case of a description mistake they completely dismiss the entire FB, hence it becomes ineffective.

Examples of the mistake: 'I am upset that you do not understand me (that you haven't tried hard, do not work hard, do not understand, do not see, do not hear, do not remember your responsibilities, do not recall, do not want, do not appreciate)'. Correction: 'I am upset that you haven't even asked how I am'.

12. Expressing another person's attitude to something or someone—another person's attitude is part of their internal processes, so it is always only an assumption. We cannot see, truly know another person's attitude, we should not assert our assumption unless they tell us about it. Seeing that this mistake is very common and important, it is examined separately from all the other internal processes (see section 11).

An example of the mistake: 'I am angry that you are so suspicious of me'. Correction: 'I am angry that you check my computer'.

13. Expressing a subjective evaluation, an opinion on the behavior of a person who gets FB. Many arguments and explanations can be found for any opinion, and different people have different opinions (I think that..., in my view..., one may..., one may not...), other people may disagree, they are not obliged to agree. Expressions of one's opinion often turn into a sermon or discussion, and FB is not perceived and slips away, if a person does not agree with the expressed opinion. Of course, every person and especially a manager must be able to express their opinion on all sorts of things. But when delivering FB, it is a mistake to express one's evaluation or opinion on a person's behavior, because it decreases the effectiveness of this FB. It is better to deliver FB separately from expressing an opinion.

An example of the mistake: 'You talk rudely to your colleagues', 'I am shocked that one can offend a young woman like that'. Correction: 'It surprises and upsets me when I hear you saying to your colleague: "you are not welcome here"...', 'I am shocked that you insult a young woman'.

14. Interpreting person's actions—explaining the words or actions of another person to whom FB is given, when the meaning of their words and actions is not obvious and can be understood differently, including understanding the true deep reasons of person's actions.

Too often, interpretations trigger psychological defense, are perceived negatively and distract a person from hearing something about themself which they could analyze on their own if they wanted to. In psychotherapy, there are many requirements for using the interpretation technique by a psychotherapist, so that it may be perceived, be useful and effective for a client. In everyday life, interpreting another person (becoming their "psychologist", analyzing them) without their permission, without agreeing to have a private conversation, is a trespassing of another person's boundaries. If colleagues have agreed on this (it is preferable to agree anew in each new communicative situation), the interpretation must be separate from other FB.

An example of the mistake: 'I am upset that problems at home may have made you so cold-hearted'. Correction: 'I am upset that you refuse to speak with your colleague and discuss the situation'.

15. Instruction, lecturing, sermonizing—direct or implicit, polite or rude instructions, lecturing and sermonizing decrease the FB efficiency. If there is no agreement with an adult person that they will be taught, it is unnatural and inappropriate, another person is not obliged to listen to anyone. The feeling of protest stops a mature person from perceiving such FB.

An example of the mistake: 'It's important that you learn to listen to people'. Correction: 'I am offended that you often interrupt me'.

16. A tirade—an explosive expression of emotions about what one may or may not do, directed at another adult person. It annoys and distracts most people from the FB content. One must launch into a tirade also separately from giving FB, if a person who gives FB needs it (thinks that they may release their emotions) or it happens in an uncontrolled way.

An example of the mistake: 'It's outrageous, I'm sick of them playing computer games at work'. Correction: 'I'm tired of waiting for you to return to work'.

17. Mockery — something exaggeratedly good is said about a person, while a negative sense is implied. Mockery irritates most people, thus distracting them from the FB content.

An example of the mistake: 'I can see you are a real "champion"!'. Correction: 'It surprises me how many things you do at the same time and I worry that it may affect the quality of work'.

18. Accusation, blaming—it is always difficult to evaluate something objectively: whether certain traits and actions leading to negative consequences (such as alcoholism) are a person's fault or trouble, or both, or a fault and trouble shared by somebody else. Any suggestions or direct words about guilt (even if the guilt is obvious for a person themself and for everybody) trigger psychological defense and, consequently, impede the perception of FB, complicate its acceptance. One must decide what is more important at the moment: to accuse or to deliver FB so that a person may hear.

Examples of the mistake: 'I am sad that we couldn't meet because of your fault', 'Don't you understand that I'm waiting for your initiative?'. Correction: 'It's sad that you refused to meet me yesterday', 'It annoys me that you don't give any proposals on how to increase our profits'.

19. Comparing a person with someone or something—when a person in general (or their certain quality, feeling or action) is compared, the value of their own uniqueness (which is an important psychological fact) is questioned, a person who gets FB is reluctant to perceive this information and the FB effectiveness decreases.

 https://doi.org/10.1051/shsconf/202111600031

An example of the mistake: 'It offends me that you are talking down to me, like my father'. Correction: 'It offends me that you are talking so harshly to me'.

20. Condemnation—an expression evoking shame, a feeling which is very difficult to handle and which denies something good and valuable in a person at the moment. As a rule, it causes a strong psychological defense and insensibility to FB.

An example of the mistake: 'It is insufferable, it's shameful to live like you'. Correction: 'It is insufferable, your lies disgust me and make me feel desperate'.

21.Threat, insult, humiliating statement—a normal person is definitely not obliged to listen to FB, take it seriously and consider it after being threatened, insulted or humiliated. It is important to decide in every communicative situation whether one wants to humiliate a person or to make sure they hear what one has to say about them.

An example of the mistake: 'Too bad you are brainless'. Correction: 'Too bad you haven't considered the facts'.

22. Condescending phrases cause a natural wish of an adult person to protect themself and their point of view and not to take FB seriously.

An example of the mistake: 'I am shocked, you dare to contradict me...'. Correction: 'I am shocked that you refuse to act the way I tell you to'.

23. Wish and request are only indirectly considered FB, because they are information about a person who gives FB. Wishes are often insufficient, and feelings more clearly express our view of a person, it is necessary to state them. It is also better to express a wish concerning another person specifically enough and separately from FB, otherwise attention of a person who gets FB is directed at doing or not doing what is wished or requested by another person, evaluating it, instead of being directed at self-reflection.

An example of the mistake: 'I get angry, I would like you to understand me', 'I am so uncomfortable, I ask you to stop talking near me'. Correction: 'I am angry that you call every hour asking confirming questions', 'I feel uncomfortable, uneasy to overhear you talking about too personal experiences'.

4 Conclusion

1. Compared with You-statements, I-statements are more correct and sustainable way of delivering FB in subject-subject interactions in communication, creating conditions for the maximum possible effectiveness of FB in the management process.

2. There are numerous FB mistakes in everyday communication, most of which remain unnoticed and unconscious despite knowing the differences between You- and I-statements. These ways of delivering FB differ not only in their pronouns, it is necessary to take into account all the elements of the I-statement formula: I (or to me) + feel (or perceive) such (specific) feeling (or sensation, state) + about such (specific) external action of another person in the communication process.

3. Deep understanding of particular FB mistakes by managers gives opportunity to correct them under the conditions of real management communication, apart from socio-psychological trainings and psychotherapeutic space.

4. Effective FB improves the quality of the self-knowledge process of a person who gets FB, expands opportunities for correcting their behavior and work. And for a person who gives FB, correct I-statements are a universal mechanism of self-regulation, self-support, as well as developing the qualities of a mature person.

6. Using correct I-statements is one of the ways of developing emotional intelligence at an organization for its proper functioning and increasing the effectiveness of a team.

References

1. R. Bar-On, *The Bar-On Model of Emotional-Social Intelligence* (ESI) (2006)
2. H. Gardner, *Frames of mind: The theory of multiple intelligences* (2011)
3. D. Goleman, *Emotional intelligence* (1995)
4. D. Goleman, *Emotional intelligence: Why it can matter more than IQ* (1995)
5. D. Goleman, *Working with emotional intelligence* (2000)
6. J.D. Mayer, P. Salovey, & D. Caruso, *Psychological Inquiry*, **15**, 3 (2004)
7. J.D. Mayer, P. Salovey, & D. Caruso, *American Psychologist,* **63**, 6 (2008)
8. J. D. Mayer, R. D. Roberts, & S. G. Barsade, *Annual Review of Psychology*, **59**, 507–536 (2008)
9. K. Nordström, J. Ridderstrale, *Funky Business Forever: Capitalism for Fun* (2008)
10. S. Shabanov, A. Aleshina, *Emotional Intelligence. Russian practice* (2014)
11. *A.N. Kluger, A.S. DeNisi, Academy of Management Executive*, **14,** 1, 129–139 (2000)
12. *A.N. Kluger, A.S. DeNisi, Psychological Bulletin,* **119**, 2, 254–284 (1996)
13. *K*.M. Dyason, D.C. Shanley, Analise O'Donovan & S. LowChoy, *Psychotherapy Research,* **30**, 3 (2019)
14. N.E. Winstone, E.G. Hepper & R.A. Nash, *Educational Psychology* (2019)
15. B. Harks, K. Rakoczy, J. Hattie, M. Besser & E. Klieme, *Educational Psychology: An International Journal of Experimental Educational Psychology,* **34**, 3 (2014)
16. F. Weck, Y. M. Kaufmann & V. Höfling, *Psychotherapy Research,* **27**, 4 (2016)
17. Ph. Gragg, *Legal Reference Services Quarterly,* **27**, 2-3 (2008)
18. Sh. A. Scott-Halsell, W. Saiprasert & J. Yang, *Journal of Teaching in Travel & Tourism,* **13**, 4 (2013)
19. L. Andonian, *Occupational Therapy in Mental Health,* **33**, 4 (2017)

20. H. Odaci, F. I. Değerli & N. Bolat, *British J. of Guidance & Counselling*, **45**, 5 (2017)
21. L. Andonian, *Occupational Therapy in Mental Health,* **33,** 4 (2017)
22. E. J. Austin, D. H. Saklofske, & S. M. Mastoras, *J. of Psych.* **62**, 1, 42-50 (2010)
23. K.A. Barchard, *Educational & Psychological Measurement,* **63**, 840-859 (2015)
24. A.B. Belov, *Theoretical and Experimental Psychology,* **5**, 2, 81-90 (2012)
25. V.I. Gordienko, *Communication on the way to encyclopedic knowledge*, 25–28 (2006)
26. E.N. Gorodilova (E.N. Rastorgueva), *Self-actualization and its connection with integral personality* (2002)
27. E.V. Lushpaeva, *Development of reflection in communication by means of social and psychological training* (1989)
28. L.A. Petrovskaya, *Competence in Society: Socio-psychological training* (1989)
29. L.A. Petrovskaya, *Interpersonal Perception in the Group*, 152-168 (1981)
30. O.V. Solovyova, *Psychology of Communication: Humanity Training*, 44–45 (2007)
31. E.N. Stremousova, *The effectiveness of feedback in the interpersonal relationships of employees of the organization* (2010)
32. S. Freud, *The Psychopathology of Everyday Life* (2015)
33. E. Shostrom, L. Brammer, *Therapeutic Psychology. Basics of Counseling and Psychotherapy* (2002)
34. I. Yalom, *The Gift of Psychotherapy* (2018)
35. I. Yalom, *Existential Psychotherapy* (1999)

Theoretical aspects of problem solution in production systems

Vitaly Sergeevich Borovik[1,*], *Dmitry Anatolevich* Skorobogatchenko[2], *Vitaly Vitalievich* Borovik[3], *Oksana Vladimirovna* Procenko[3] and *Anastasiya Vitalievna* Borovik[3]

[1] Volgograd Science and Technology Center, Komsomolskay 12, Volgograd, Russia
[2] Volgograd State Technical University, Akademitheskay, 1, Volgograd, Russia
[3] Department of Capital Construction of the State Public Institution of Volgograd Region «Directorate of Highways», Zukova 62, Volgograd, Russia.

Abstract. The article attempts to present conceptually the definitions of the terms a "problem situation" and a "problem". The problem, goal, strategy, means and conditions are analyzed in time, in interconnection and interdependence. A scheme for the solution of a problem based on the digital strategy model is presented. The one-to-one correspondence of the elements of the sets within the implementation of the corresponding measures connected with the solution of the problem is analyzed as an illustration of the way out of the problem situation. The model is analyzed which shows the time sections of the Minkowski space for the control vector, taking into account the reaction of the control vector to the influence of external and internal factors in time within the production system. The conclusion is made about the evolutionary reaction of the system to the emergence of a problem situation, which is formed for its self-preservation, prevention of system integrity violation and the possibility of automating the process of problem solution on the basis of a digital model.

1 Introduction

Joseph Schumpeter in the introduction to his work [1] writes: "This is because, whatever the field, the problems and methods that are in use at any given time embody the achievements and carry the scars of work that has been done in the past under entirely different conditions. The significance and validity of both problems and methods cannot be fully grasped without knowledge of previous problems and methods to which they are the (tentative) response". An important prerequisite for the successful solution of problems is their adequate definition in time.

Taking into account all the variety of information reflecting the study, research and resolution of various problems, as a rule, there are no clear indications of what, in fact, the authors see the problem, for example [2-8]. Hence, it can be assumed that at present time, intuition plays a leading role in understanding the essence and content of what can or should be seen as a problem. However, intuitive understanding does not contribute to emphasis of the main thing, i.e. a controlled result.

It is very difficult to concentrate resources accurately enough, to balance their consumption in the most critical areas connected with the problem solution. Unclear wordings disorient the workers. Unclear understanding prevents the identification of the main direction of work, balance in the use of resources, justification and critical analysis of problem solution. This situation affects the way in which the activities are coordinated, leading to a significant waste of resources. As J. Schumpeter writes " ... the highest claim that can be made for the history of any science or of science in general is that it teaches us much about the ways of the human mind. To be sure, the material it presents bears only upon a particular kind of intellectual activity. But within this field its evidence is almost ideally complete. It displays logic in the concrete, logic in action, logic wedded to vision and to purpose" [1]. The absence of a clearly defined problem does not contribute to an unambiguous formulation of the goal and strategy for its achievement. According to Aristotle: "As the most important and dominant science, which all other sciences, like slaves, do not dare to contradict, should be called wisdom of the science of the purpose and the good (as for their sake there is something else)" [9].

The concepts that have developed to date on this topical issue do not fully meet the requirements of consistency. The aim of the work is to study the problem, goal,

*Corresponding author: borovikv@mail.ru

strategy, means and conditions in time, interconnection and interdependence.

2 Research methodology

Aristotle suggested the division of problems into three groups : practical, theoretical and those which serve to achieve any truth [10]. In practice, situations constantly arise, the resolution of which with the help of available means, knowledge and experience is complicated [11]. In such cases, it is customary to say that there is a problem. The problem, like the task, comes from a problem situation [12, 13]. The appearance of a problem situation is connected with the emergence of information (Fig. 1) [14,15].

The analysis of a problem situation creates the prerequisites for its transformation either into a task or into a problem. For our further research, we will consider such a division of the problem situation sufficient for all the variety of interpretations of the problem concept [16, 17].

In our opinion, in production systems, the problem situation manifests itself in information about the discrepancy between the actual result and the desired one, or real conditions and desired means. Inconsistency in such systems arises under the influence of information, which, as a rule, has the status of a document.

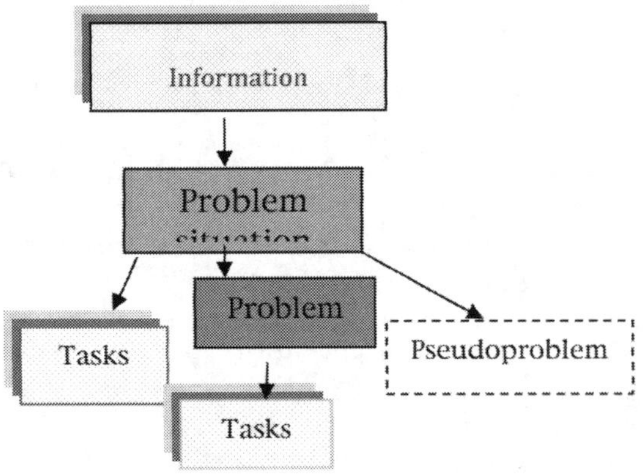

Fig.1. Structural diagram of the process of problem situation transformation

The types of problems, depending on the field of activity, can be scientific, industrial, social, political, etc. [17]. In terms of the scale of influence, production problems can be local, regional, sectoral, state, global, etc. Each type and scale of influence of a problem situation requires its own individual approach to its solution.

For our study, the problems associated with the production activities of enterprises, for example, road construction industry, are of interest. As a rule, an inaccurate definition of the goal, areas, scale of influence, errors in the process analysis of the conditions, means and time in which a problem situation arose, leads to a pseudo-problem [19]. The methodology and techniques for problem solution isolated from problem situations have been developed in sufficient detail, for example in the paper [20].

The process of recognition, isolation and solution of production problems has not been sufficiently developed. The available recommendations present the descriptive part of the problem in sufficient detail, without defining it, for example in the paper [21]. In our opinion, a problem situation manifests itself in the course of comparing information about the results of production process at the current moment with information about the results that had been planned earlier. The discrepancy between the values of the compared parameters and the absence of a solution at a given moment of time causes a problem situation if the parameters of this discrepancy go beyond the limits established earlier.

For example, in accordance with the contract between an owner and a road construction company, a road was built with a five-year warranty. Three years later, it was documented that in terms of its transport and operational state the road does not meet regulatory requirements. The warranty was not honored. It was established that the transport and operational state of the road does not correspond to the planned transport and operational state and, in terms of its parameters, goes beyond the previously established limits. A problem situation arose.

The occurrence of the problem is associated with information about the discrepancy between the results of production activities at the current moment and the results that are planned for a long term period, also if this discrepancy goes beyond the established limits. In this situation, the time becomes a sign of the distinciton between a problem situation and a problem.

For example, the management of a road construction company may have two main ways out from the problem situation mentioned above.

1. To solve the problem, i.e. develop a repair project for the restoration of the transport and operational state of the highway and carry out a set of works in accordance with the project. It takes time t and resources to get out of a problem situation $\sum_{i=1}^{n} X_i$.

2. To refuse to honor the warranty obligations, which will create the preconditions for a problem, because the customer can submit the relevant documents to the arbitration court and he is likely to win the case with subsequent sanctions. In addition, for the road construction company in the future, it may also be difficult to conclude contracts, etc. It will take time t_1 and resources to solve the problem $\sum_{i=1}^{n} X_i^{'}$. At the same time, as a rule, $t< t_1$ and $\sum_{i=1}^{n} X_i < \sum_{i=1}^{n} X_i^{'}$.

The concept what we suggest is consistent with the position of the author [22], who, emphasizing the method for problem identificaiton, notes that it is based on the "assessment of the state of a controlled system founded on the results of scenarios of self-development and controlled development." In the authors' opinion, the "scenario results" can be seen as information about possible predicted events, i.e. planned for a future period. Thus, a common feature of a problem and a problem situation is information about the inconsistency of the results. The difference between a problem situation and a problem lies in the fact that the occurrence of a problem situation is connected with the appearance of information about the discrepancy between the planned results and the results available in the current period of time, and in case of a problem it is about the results of the current period and the forecasted one. It should be emphasized that the emergence of a problem situation is strictly connected with the goal, strategy, conditions, means and time. A change in goal and strategy can lead either to a task or to a problem if the problem situation is ignored.

Research results

A strategy for problem solution is understood as a plan of actions coordinated by enlarged indicators in space and time, which in a certain way ensures the achievement of the main goal on the basis of adequately used conditions and means.

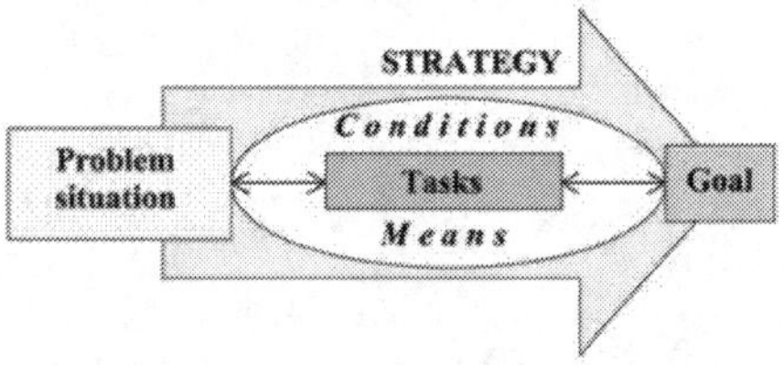

Fig 2. Diagram for a problem solution based on a strategy

The need to solve the problem is associated with the emergence of a problem situation. Objectively, the emergence of a problem situation is connected with the process of production development and progress.

In order to get out of the problem situation, it is necessary that each finite set of available factors A (conditions, means) corresponds to a finite set of necessary factors B, which ensure the solution of the problem (Fig. 2). Then we can write

$$\sim , \qquad (1)$$

If the sets n do not intersect each other, i.e.

$$A_n \cap A_{n^l} = \varnothing \text{ then } n \neq n^{/}, \qquad (2)$$

and the sets n also do not intersect each other, i.e.

$$B_n \cap B_{n^,} = \varnothing \text{ at } n \neq n^{/}. \qquad (3)$$

For all n An~Bn . Then each

$$a_0 \in \bigcup_{n=1}^{m} A_n , \qquad (4)$$

matches a single element

$$b_0 \in \bigcup_{n=1}^{m} B_n. \qquad (5)$$

The correlation of each element of the set A according to some rule to one and only one element of the set B and when correlating each element of the set B to one and only one element from the set A, then

$$\bigcup_{n=1}^{m} A_n \sim \bigcup_{n=1}^{m} B_n. \qquad (6)$$

One-to-one correspondence (6) in the implementation of appropriate measures connected with the solution of the problem can be considered as evidence of the way out of the problem situation. Information about the problem situation gives an impetus for collecting, systematizing and analyzing information that allows us solving a problem or defining the problem. For further analysis, it is important to note that the problem situation after the division into tasks and problems may arise again when solving a task or a problem. The emergence of problem situations, problem solution and forecasting problems is a constantly evolving process.

As a rule, such situations are aggravated by the innovative development of production, in conditions of a shortage, primarily of time, as well as information about new technologies, materials, and processes. The desire to find a particular time in the development of the processes occurring in the production system in terms of advanced technology, leads to a number of assumptions that deserve serious attention [24,25,26].

J. Schumpeter pointed out the possibility of modeling production processes in time, mentioning the introduction of a time parameter into the production function: "If we want to base our theory of production on the theory of the Jevons-Böhm-Bawerk-Taussig "roundabout" process, we can introduce the notion of time directly into the production function by writing $x = \psi(v_1, v_2 \ldots v_n; t) \ldots$» [1].

Let us turn to the production function (we consider it as a digital model), for example, in the following form [28]:

$$Y = C_0 \prod_{i=1}^{n} X_i^{\alpha_i} \qquad (7)$$

where Y is the calculated index (for example, the amount of work, etc. in physical-material or value terms); $_i$, $i = \overline{1,n}$ - factors (for example, fixed assets, materials, labor, energy and qualifications of personnel) that determine the parameters Y (in physical-material or value terms); $_i$, $i = \overline{i,n}$ characterize the contribution of $_i$ into Y; C_0 is a coefficient reflecting the influence of factors which were not taken into account by the model.

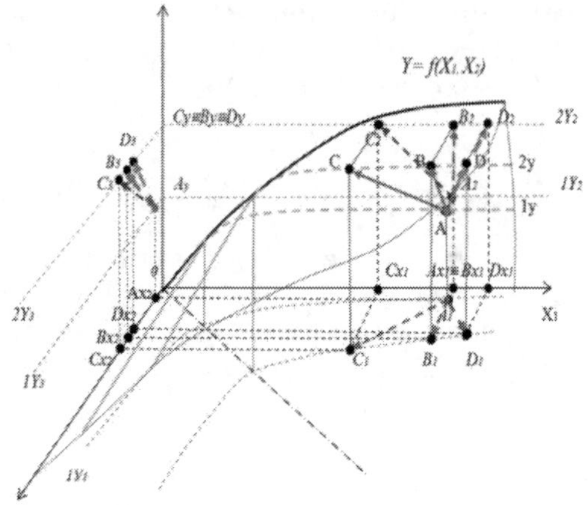

Fig. 3. Three-dimensional model based on production function, displaying a surface illustrating control options when solving the problem of output from $1Y$ to $2Y$.

For example, a digital model of the type $Y = 1.132 \cdot X_1^{0.763} \cdot X_2^{0.438} \cdot X_3^{0.211} \cdot X_4^{0.176} \cdot X_5^{0.076}$ was obtained and it characterizes the hypersurface of the produciton process under study. Using the obtained digital model, it is not difficult to determine the mathematical model of the control strategy. For graphic interpretation of a digital model in three-dimensional space, it is advisable to choose the most influenced factors, for example, X_1 and X_2. (Fig. 3), [28]. Curves $1Y$ and $2Y$ connect points with the same numerical values of Y. Their projections $1Y_1$, $2Y_1$, $1Y_2$, $2Y_2$, $1Y_3$, $2Y_3$ are isoquants.

In Figure 1, point A, has a numerical characteristic, for example, the amount of work $1Y$. The $2Y$ the scope of work can be achieved by introducing, for example, a new technology, an infinite number of combinations of X_1 and X_2. For example, let'is analyze points B, C and D, illustrating three options for the achievement of the amount of work, characterized as $2Y$. Depending on the purpose, conditions and means, one or another variant of the combination of the use of resources is selected. From the mathematical point of view, the optimal strategy for reaching $2Y$ should be recognized as the vector **AB**, since $\mathbf{AB} \perp 2Y$ and it is the shortest distance between the two curves Y_1 and Y_2, as a perpendicular to the tangent at point B.

It should be noted that the use of a digital model makes it possible to unambiguously determine the optimal control strategy.

Let us exemplify the emergence of a problem situation by the development of production function in time. Let us suppose that the beginning of some production process is fixed at point A. Time is associated with the consumption of resources and their movement in space, connected with the production process. The "past" of the process coincides with its beginning, and time "flows" along with the process of resource transfomraiton in space [28].

Hence the conclusion is that time conditionally flows in the same direction as the changes of the observed object. Motion, as a generalized concept, is usually characterized by vector quantities and it can be assumed that the "proper" time of the process will be collinear and co-directional with the "displacement" vector. Let us exemplify this graphically using the geometric interpretation of space and time (Fig. 4) [18, 28 - 31].

In real production conditions, control in space and time begins from point A (from the achieved result) in the direction of point B (the planned result, for example, production of a good). The actual geometric development of the process occurs over time A' from to B'. It is worth mentioning that this process corresponds to a certain time or the planned time of commodity production. Depending on the processes associated with the control developing in time point B of the vector **AB** it describes a rather complex trajectory (Fig. 4).

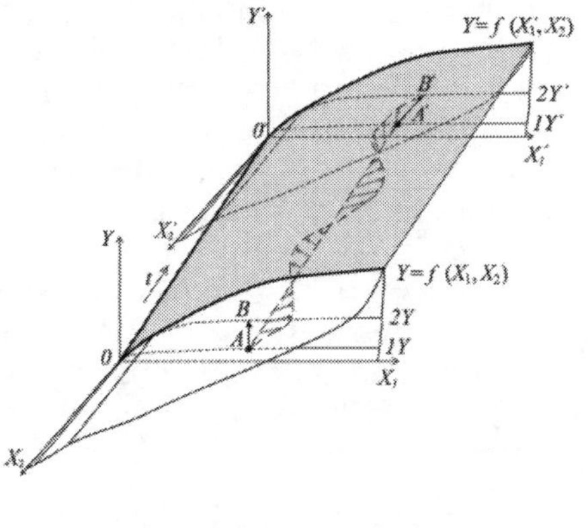

Fig. 4. A model exmplifying the time sections of the Minkowski space for the control vector **AB** (transformations **AB** into **A'B'**), taking into account the reaction of the control vector to the influence of external and internal factors in time in the production system

It is explained by the fact that the system of mobile equilibrium tends to change in such a way as to minimize the effect of external influences [32]. The influence of a set of factors developing in time, i.e. changing conditions, changing means, delays in the reaction of the control system to changes in the production process, including overregulation [33], and other reasons, the trajectory of the movement of the control vector in time in a general form can be presented for example, in the form of a wavy surface (Fig. 4).

As it can be seen from Figure 4, the position of the vector **A`B`** is not the shortest distance between $1Y$ and $2Y$, and it indicates a problematic situation.

4 Discussion of the results

In a problem situation, it is necessary to clearly identify either the problem or the task. To identify this means, first of all, to define accurately the task or problem. Most clearly, this can be done using the the strategy as the main goal of the enterprise or an organization. The emergence of a problem situation is associated with a goal.

If, in order to achieve the goal set previously at a specific point in time, the discrepancy of conditions or means is found out, then this indicates the occurrence of a problem situation. Innovative development involves setting goals that determine a more intense way of functioning of the organizational and economic system. As a rule, innovations are based on a more efficient use of available funds (resources, including time) [34] and production conditions. Therefore, we can state with confidence that innovative development creates the preconditions for the emergence of problem situations and objectively contributes to the production.

It is extremely important for practice to determine the signs and properties of a problem situation. It is necessary to identify the basic requirements that play an important role in understanding and definintion of problems, as well as effective methods for their solutions. A detailed study of these issues allows us separating at an early stage the real problem from the pseudo problem.

A problem situation is transformed into a task if a way out of this situation is found with the help of information about the available means and conditions for problem solution (Fig. 1, 3). In contrast to the task, the problem is perceived as a contradictory situation. It is characterized by the need to perform certain actions, for which at a particular moment in time there is no necessary information about possible options for the transformation of funds or conditions for its solution. The emergence of production problems as a rule is associated with the lack of information about the possibility of getting out of the problem situation by means of the solution of the problem. To obtain such information, as a rule, it is necessary to perform a set of special actions, for example, research and development, etc. [35].

For example, a problematic situation can be connected with the supply of stone materials to facilities under construction, which are significantly far from railway communications. You can get out of this problematic situation by solving the problem, i.e. using road transport.

If we stop using road transport as a very expensive method of delivering materials, then the problematic situation will be transformed into a problem. One of the ways to solve the problem could be, for example, the use of local soils as the base of the pavement. However, there is no information on the appropriate effective technology that meets the regulatory requirements at a specific point in time for specific conditions in the enterprise. To solve this problem, it is necessary to conduct research, tests, production experiments, develop technology, regulatory documents, aggre documents, approve papers, purchase the necessary equipment, etc. A combined option is possible, to conduct a complex of research and development work simultaneously with the solution of the problem, i.e. using vehicles to solve the problem.

Taking into account the desire of any system of mobile balance to minimize the effect of external influence, evolutionary protection in the form of self-preservation has been formed in the production system. However, the influence of unresolved internal and external problems can lead the system to an unsatisfactory state, to a violation of the integrity of the system.

The integrity of the system is a fundamental non-reducibility of the properties of the system to the sum of the properties of its constituent elements and the non-deduction from the last properties of the whole; dependence of each element, its properties and relations in the system on its place, functions, etc. within the whole [36].

In dynamics, this means that the impact on one element of the system (or some of them) necessarily causes a reaction, a change in other elements (dynamic integrity). For example, if the plans of an enterprise are integrated into a system, then a change in demand for products will entail adjusting the tasks for supply, for labor, perhaps for capital investments, cost and other indicators [37].

However, the violation of the integrity of the system can also be expressed by the transition of the system to a new qualitative state. In the system, under the influence of resources coming from outside, there is a slow quantitative accumulation of insignificant changes. This occurs up to a certain limit, beyond which a radical change in its state is observed, which can be carried out almost instantly, in a jump.

The system temporarily finds itself in an unstable state, "loses its memory," and the nature of its subsequent development is determined only by those random factors that act on the system at that moment. In order to get out of it, the system has the following opportunities: degradation, destruction, involution or self-organization, complication, evolution. Quantitative changes turn into structural or qualitative ones, and the whole process of system development can be presented as a series of slow and abrupt changes replacing each other. Each of these changes can be presented as a transition of the system to a new qualitative state [38]. This fact reveals a contradiction. On the one hand, there is a desire for self-preservation, and on the other, there is the need for improvement.

5 Conclusion

An attempt is made to conceptualize the content of the definitions of problem situation and problem. As a starting point, a significant difference between the concepts of task, problem situation and problem is accepted. It is shown that the problem, goal, strategy, means and conditions should be analyzed in time, in interconnection and interdependence. The possibility of problem solution based on a digital strategy is shown. A one-to-one correspondence of the elements of the sets was established during the implementation of appropriate measures connected with the solution of the problem, as evidence of the way out of the problem situation. A model was developed that exemplified the time sections of the Minkowski space for the control vector, taking into account its response to the influence of external and internal factors in time in the production system. It is found that the evolutionarily system response to the occurrence of the problem situation is created for preservation, prevention of disruption of the integrity of the system. The manifestation of the contradiction in the trend of the system for self-preservation on the one hand, and the need for its improvement on the other is shown. The digitization of the process of getting out of a problem situation creates the prerequisites for its automation.

References

1. J. A. Schumpeter *History of Economic Analysis.* (London: Taylor & Francis e-Library, 2006)
2. *Proceedings of the international scientific and technical conference.* (Bryansk State Engineering and Technological Academy. Bryansk. 1998).
3. Problems of technology improvement of material resources saving. *Interuniversity collection of scientific papers of the Kazakh Polytechnic Institute named after V.I. Lenin.* (Alma-Ata, 1987).
4. *Problems of improvement of planning and forecasting of capital construction.* Interuniversity thematic collection of works of the Leningrad Civil Engineering Institute. (L .: 1981).
5. E.E. Rumyantseva *New Economic Encyclopedia.* (Moscow: Infra-M. 2006).
6. *New Economic Dictionary.* (Moscow: Institute of New Economy. 2006)
7. V. Lukyanova *Problems of technical regulation reform. Standard and quality* **12** (2006)
8. *Problems of the construction complex of Russia. Materials of the 10th interhnational scientific and technical conference at the 10th international specialsized Exhibition "Construction and utilities - 2006"* (Ufa, 2006).
9. Aristotle. *Works in four volumes. Volume. 1.* (Publishing house of socio-economic literature "Mysl", 1976)
10. Aristotle. *Works in four volumes. Volume 2.* (Publishing house of socio-economic literature "Mysl". 1976)
11. I.G. Gerasimov *Research structure.* (Moscow, "Mysl", 1985).
12. *A Brief Psychological Dictionary.* (Moscow, Publishing house of political literature, 1985).
13. *Philosophical Encyclopedic Dictionary.* (Moscow, Soviet encyclopedia, 1983)
14. D.A. Molodtsov *Stability of the principles of optimality.* (Moscow: Science. 1987).
15. *Logical dictionary-reference.* (Moscow, Science. 1975)
16. *Types and levels of problem formulation.* Retrieved from: http://www.trizminsk.org/e/248009.htm.
17. V S Borovik *Transportation Research Procedia.* **20,** 74–79, (2017)
18. P.A. Vasev, *Proceedings of the 12th International Conference on Computer Graphics and Machine Vision GrafiKon.* P. 431–437 (Novgorod, 2002).
19. I.G. Gerasimov *Research structure.* (Moscow: Mysl., 1985).
20. A.G. Teslinov *Development of control systems: methodology and conceptual structures.* (Moscow: "Globus". 1998)
21. F. Heiligen *Formulation of the problem of posing problems in: Cybernetics and systems '88,* Available at: http://pespmc1.vub.ac.be/Papers/Problem-Formulation.html.
22. Z. K. Avdeeva *Methods for the formation of strategies for solution of semi-structured problems based on cognitive models.* (Abstract of the thesis for the degree of technical sciences) *Institute of Control Sciences of the Russian Academy of Sciences.* (Moscow, 2006).
23. R. R. Stoll 1967 *Sets, Logic and Axiomatic Theories.* (W.H. Freeman and Company. San-Francisco and London, 1967).
24. V.P. Kazaryan *Concept of time in the structure of scientific knowledge.* (Moscow: Publishing house of Moscow State University, 1980).
25. Smolin Lee, *The trouble with physics: the rise of string theory, the fall of a science, and what comes next.* (Houghton Mifflin, Boston, 20016).
26. M.T. Zinaliev *Physics of Time* (LAP LAMBERT Academic Publishing, 2015)
27. G.B. Kleiner *Production functions. Theory, methods, application.* (Moscow: Finance and Statistics, 1986)
28. V.S. Borovik, V.V. Borovik, *Eastern-European Journal Enterprise Technologies* **3/5 (81)**, 4-10, 2016. Doi:10.15587/1729-4061.2016.69383
29. H. Minkowski, *Geometrie der Zahlen.* (Leipzig-Berlin: R. G. Teubner, 1910)
30. A. Einstein *The essence of the theory of relativity.* (Foreign Literature Publishing House, 1955).
31. A.B. Fashchevsky *Graphical representation of four-dimensional space.* Retrieved from: http://khd2.narod.ru/authors/fshvsky/spc4d.htm.
32. A.A. Bogdanov *Tecktology: A General Organizational Science.* (Moscow, 1989)
33. V.S. Borovik, A.V. Borovik. *IOP Conference Series: Materials of Science and Technology,* **832**, 012044 (2020) doi:10.1088/1757-899X/832/1/012044.

34. V.S, Borovik, V.V.Borovik, *Eastern- European Journal of Enterprise Technologies.* **6,** 10(84), 41-48(2016) DOI: 10.15587/1729-4061.2016.86535.
35. A. Kondakov *From the materials of the "Intellectual Club".* (St. Petersburg, 1997).
36. *System integrity.* Retrieved from: https://economic_mathematics.academic.ru/4939/
37. L.I. Lopatnikov *Economic and mathematical dictionary: Dictionary of modern economic science.* 5th ed. (Moscow: Delo, 2003).
38. *Synergetics as a universal scientific paradigm.* Retrieved from: http://knowledge.allbest.ru/biology/2c0b65625b3ac68b5d53b88421306d27_0.html.

Theory of analysis of the innovative ecosystems development

Evgeny Popov*, *Victoria* Simonova, and *Igor* Chelak

[1] Ural Institute of Management - a branch of the RANEPA, 66, 8 Marta Str., Yekaterinburg, 620144, Russian Federation

Abstract. The ecosystem approach makes a significant contribution to understanding the current transformations of socio-economic life. New end-to-end technologies have a high potential to minimize the transaction sector in order to focus on meeting the real needs of people and firms and forming economic, political, cultural, and value institutions for sustainable development and justice for all. It seems that the technological breakthroughs of recent years, as well as the reverse pandemic economic dynamics, provide ample opportunities for the spread of innovative ecosystems as a new form of coordination of production activities. The unique characteristics of ecosystems as predominantly self-organizing communities allow us to identify a number of conceptual approaches to the analysis of their development: network, system, digital, stakeholder, and transactional. The objectives of the article do not include their detailed presentation. We show the applicability of approaches for ecosystem analysis. Based on the description of the options and possibilities for the applicability of these approaches, we came to the conclusion that the nature of innovative ecosystems is dual. First, the latter allow us to model (structure, systematize) interorganizational, inter-firm relations in a new way, to identify ecosystem features of communities, industrial and regional complexes. Second, ecosystems need to be recognized as a new form of transactional coordination of economic activity. They are alternatives to institutions such as the firm, the market, and hybrid forms. The increment of scientific knowledge consists in the implementation of the main goal of the study - the systematization of methods for analyzing the development of innovative ecosystems based on the formation of ideas for ecosystem analysis of innovation management.

1 Introduction

Digitalization, networking of business and social life have become the dominant trends in the development of both the service and commodity economy over the past decade. And in 2020, perhaps the most "destructive innovation" in history appeared-a total lockdown. The current epidemiological situation is turning the world economy into a reverse one. The waves of the pandemic directly affect the growth and slowdown of production, entrepreneurial and innovative activity and consumption, migration from the offline environment of activity to online and back. Unstable, turbulent conditions require economic agents to be flexible, adaptable, and quickly change the types of transactions. Such advantages are provided by ecosystems that allow you to build business models outside of rigid frameworks, hierarchical coercion, clear boundaries, in some cases - without contractual registration of relations.

The scientific ideas that were at the origin of the application of the ecosystem approach were more inclined to use analogies with natural self-organizing ecosystems in economic analysis than modern publications. Among the latter, there are studies aimed at finding methods for managing ecosystems as a response to the risks of their possible unpredictable development trajectories [1, 2]. It is worth recalling that one of the program articles of James Moore, who introduced the concept of ecosystems into economic terminology, is called "Predators and prey: The new ecology of competition" [3]. It is noteworthy that the main economic concepts: development, growth, are also introduced from the natural sciences. Interdisciplinarity allows us to take a fresh look at economic phenomena. Perhaps in the future, the economy will be considered exclusively from the point of view of the interaction of ecosystems as a concept that most fully reflects the relationship between economic entities and environmental factors [4].

A number of authors do not share the optimism about the possibility of shifting the focus of economic analysis mainly to the ecosystem approach. L. Ramenskaya points out that the use of the prefix "eco", in particular, in the analysis of innovative systems, is not always justified [5].

The ecosystem research paradigm expands the traditional understanding of economic relations as an interaction of independent actors. The essence of the concept of innovative ecosystems (IES) is the position that all actors within conditional territorial boundaries are somehow interconnected [3, 6], through strong and weak, implicit interactions (by analogy with the physical nature of objects in the Universe).

* Corresponding author: chelak@mail.ru

On the other hand, the authors of newly emerging economic models and concepts are forced to respond to the rapid platformization and centripetal ecosystem processes occurring in the business environment, and to include ecosystem concepts in the arsenal. Thus, D. Bryukhanov and co-authors, analyzing the popular EFQM model, point out that "in contrast to the traditional concept of a competitive environment, [in the EFQM model] the economic activity of an organization is considered as an ecosystem, which includes consumers and producers, partners and competitors, and other interested parties" [7].

An important aspect of the nature of the socio-economic ecosystem, in contrast to the company as an artificially created (artificial) system, is mainly independent generation, self-organization, self-development [8]. Inversive relationships [9] in the ecosystem and hybrid forms of coordination arise spontaneously, not hierarchically, and in this way there is a certain similarity with the market. But an ecosystem is not a spontaneous market or a particular case of a hybrid. It is more a cognitive analysis tool than a management tool. Given the enormous potential of ecosystems to reduce transaction costs in modern conditions [10], we will later discuss in more detail the hypothesis of the place of ecosystems in the set of forms of economic coordination.

2 Features of innovative ecosystems

From a number of ecosystem paradigms, we will distinguish two concepts of understanding the innovation ecosystems. The first is based on the recognition of any business activity (and the ecosystem in which the activity is carried out) as innovative: set up for renewal, innovation, constant changes in all areas of production activity [11]. This approach is closer to the Schumpeterian understanding of innovation as the basis of entrepreneurship [12]. The second concept focuses on a relatively narrow understanding of innovation, and considers the innovation ecosystem as a condition for the development and commercialization of innovations as new products and services, "new value propositions" [1]. This approach focuses on research and marketing activities, development institutions, and the infrastructure of innovative ecosystems. In general, these paradigms do not contradict each other and can be considered as a partnership. But despite the variety of approaches to understanding the phenomenon under consideration, there is a lack of systematization of approaches to the analysis of the innovative ecosystems development in existing studies. And this is quite understandable from the perspective of the formation of an ecosystem paradigm that has every chance to become a new theory of ecosystems, along with such well-established systems of scientific knowledge as the theory of the firm or the theory of markets. Hence, the purpose of this study is to contribute to the developing theory of ecosystems by systematizing methods for analyzing the innovative ecosystems development based on the formation of ideas for ecosystem analysis of innovation management.

3 Materials and methods

Applying the concept of innovative ecosystems, researchers focus on the conditions, interaction factors, focused on the creation of novelty. Logical, structural, and systematic methods are the starting methods for studying the development of ecosystem analysis of innovation process management. Complements the set of analysis of scientific publications. The research materials are scientific developments concerning the relationship between economic entities - stakeholders of ecosystems: firms, households, the state and other structural elements of the socio-economic environment. The paradigmatic basis of the research was system theory, institutional theory, concepts of network interaction and stakeholder relations, as well as a new type of economic interaction – digital (Econotronics).

4 Results

Table 1 shows the main methods of ecosystem analysis that allow us to collectively consider the most significant parameters of innovative ecosystems.

Table 1. Methods of ecosystem analysis

Methods of analysis and sources of the theory of analysis of the development of innovative ecosystems	**Fundamentals of the analysis of the development of innovative ecosystems**
Network potential theory [13] [14]	The presence and characteristics of links between the elements of the ecosystem can be assessed through the network potential of the elements (firms, organizations, communities)
System theory [15] [4]	The analysis of ecosystems is based on the discreteness/ continuity of space and time
Theory of stakeholder interaction [16]	Ecosystem analysis is carried out by identifying the pool of ecosystem stakeholders and assessing their mutual influence
Digital Potential theory [17]	The analysis is carried out by assessing the digitalization of inter-organizational interactions in the ecosystem
Transactional configuration theory [18] [19] [21]	The analysis can be carried out by the ratio of asset specificity and transaction costs, since, on the one hand, ecosystems occupy an intermediate position between hybrids and the market, on the other hand, they can be distinguished as a separate form of coordination

Source: author's development

5 Discussions

Network theory (the concept of network readiness) is primarily aimed at analyzing the potential for convergence, integration, and permanent communication of the entrepreneurial core of the innovation ecosystem (firm) with partners of the same order. Graphically, the network effect of inter-firm interactions manifests itself on the plane, in the structures of non-hierarchical configurations. The formation of partnerships in graphs strengthens the innovative potential of the network, and the inclusion of the interaction environment in the objects of analysis forms the vision of the network ecosystem. Research in one of the most" breakthrough " industries in the modern economy (the information technology sector [14]) has shown that networking can generate significant innovative effects. This approach has the potential to develop in the context of the information society evolution, based on the fact that information is an essential resource, especially from a behavioral and psychological point of view (in particular, to overcome negative institutional effects - reducing distrust, eliminating information asymmetry).

Of course, the analytical concepts presented here work in practice in conjunction with other paradigms. Thus, the network potential shows the dynamic, structural aspect of ecosystem interactions, one of the priority goals of which is to reduce transaction costs, and in this regard, the network approach brings us closer to the scheme of Oliver Williamson (Figure 1), where the network as a mechanism of economic coordination is a pure hybrid, combining both market and firms forms of contracting.

The systematic approach to the analysis of innovative ecosystems, developed in the research of G.B. Kleiner and his colleagues, is characterized by a broad historical scientific and methodological base, which allows us to consider the spatial evolution of socio-economic ecosystems as dynamic, open communities. It should be noted that the ecosystem theory itself was originally formed due to the theory of systems [20]. Laid down in the basis of the author's systematic approach Kleiner's universal concept of "space-time" introduced fundamental physical phenomena into economic research that allow us to study innovation ecosystems from the point of view of "intra - and intersystem, horizontal and vertical, interperiodic and interzonal coordination of innovation participants" [15]. The system approach suggests considering ecosystems from the perspective of a tetrahedron: objects-environment-processes-projects, deployed in specific communities - clusters, platforms, networks, incubators. Thus, the network concept also finds its application in the analysis methodology under consideration.

The stakeholder approach is characterized by increased attention to the conditionally external environment of economic activity, which accelerates and intensifies the generation of innovations (due to the presence of dense, strong or implicit, weak links with various elements of the environment). The approach allows to eliminate inefficient attitude to stakeholders as to the periphery of business processes. On the contrary, the novelty of the ecosystem paradigm is to emphasize the conceptual role of the environment of the central element of the ecosystem (for example, firms, platforms, communities). Consumers, competitors, suppliers, government authorities, socio-political institutions, mass media and social media, the scientific and educational sector, and infrastructure organizations play the most prominent role in the environment. The function of equal participants of the ecosystem is postulated for the stakeholders. Without them, it is difficult to explain the nature of the innovative effects of the company's activities. Within the framework of the concept of mutual influence of stakeholders on each other, the nature of self-organization, self-development of the innovation ecosystem can be explained, as well as its capacity can be calculated. The power of the IES is manifested in two ways. The element capacity depends on the typology of stakeholders, their number, and the prevailing transactions. Innovation power describes the ability of an ecosystem to produce and distribute productive innovations.

The emerging transformation of the post-industrial information economy into a digital one leads to the need to develop up-to-date research algorithms. The theory of digital potential can be fruitful in the study of ecosystems due to the fact that the rapid formation of the latter in recent years is due to significant technological progress and economic potential (especially in terms of cost reduction) of information and communication technologies, which is clearly shown by the examples of the largest players in the digital economy, whose ecosystems were formed from seemingly ordinary applications: local social networks (Facebook), search services (Google), online stores (Amazon, AliExpress). Digital business models in the presence of competent management, marketing, and not yet detectable the energy of ecosystem self-organization are able to combine sources of added value, consumers (no matter what-goods, services, content), other stakeholders, groups of influence in a virtual form, but real in economic effects configuration.

Digitalization is not limited to automation and a focus on big data. This is where the deployment chain of most end-to-end technologies develops. For example, Big data analysis requires conditions for automatic data collection, machine vision, artificial intelligence, distributed ledgers (blockchain) for data transfer and protection, cloud technologies for storage, advanced communication systems and Internet communications, digital media for collecting feedback and reflection. Of course, the availability of technological capabilities is not the determining and only condition for the platform growth of the innovation ecosystem. For consumers, the economic opportunities that modern platforms provide are valuable. The hype effect and the search for factors that influence the rapid growth of the popularity of individual ecosystems are also highly important (for example, where is Facebook and its main competitor in the 2000s, MySpace?).

The digital approach to ecosystem analysis makes it possible to adjust management strategies, evolve marketing techniques, and prepare for new technological

https://doi.org/10.1051/shsconf/202111600033

perspectives (it is difficult to imagine how a quantum computer will change the digital and real world).

The institutional design of coordination in economic activity is determined by the ratio of benefits and losses from the corresponding configuration of transactions. To determine the subject of the transactional approach to the innovation ecosystem, the concepts of transaction costs and asset specificity are introduced into the analysis tools.

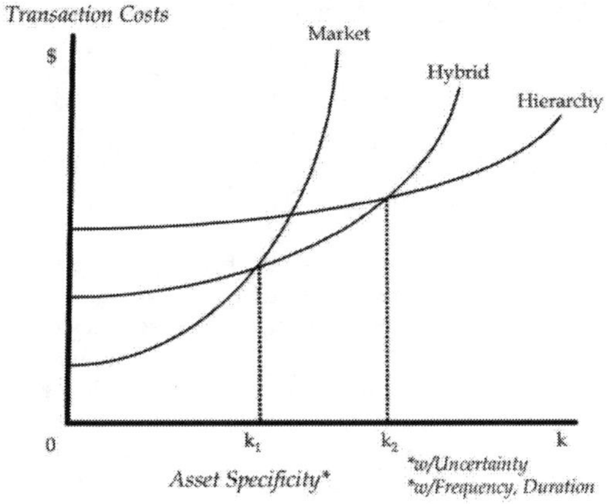

Figure 1. Transaction configuration forms
Source: [18]

Ecosystems, along with other forms of collective activity, decide on the effective use of resources and the rules for this kind of activity. Markets, firms — well-established institutions in the economy. In contrast, ecosystems have not yet received an unambiguous interpretation, including an institutional one. Where is the basis here? What type of contract between stakeholders prevails? Starting from the theory of Oliver Williamson, we propose to consider the forms of coordination broadly, including in the sphere of analysis both the core of innovative ecosystems (firms) and the elements of the environment in which they operate, the environment itself. Therefore, a cluster, alliance, partnership, or network (forms that do not claim to include the environmental component within their boundaries) can be classified as hybrid forms of coordination as synthetic configurations that combine the characteristics of a market and a firm, and ecosystems can be separated into a unique class. On the other hand, if the firms configurations of economic activity clearly gets the characteristics of the ecosystem due to the effect of minimizing transaction costs and reducing the risk of depreciation of specific assets, which is clearly manifested in platform digital ecosystems, is it methodologically correct to classify it as a firm or ecosystem? From a practical point of view, the application of the institutional approach allows us to solve the issues of minimizing the transaction costs of choosing and configuring institutions. But in the real conditions of economic life, modern forms and models of global corporations are difficult to assess from the point of view of the classical "firm – market" scheme. What, for example, is a platform like Alibaba? To what configuration is it appropriate to refer the SBER system, which includes completely different sectors of the economy and business models?

From the point of view of the number of subjects of management of forms of transactions, the ecosystem can be considered hybrid, because here, unlike the market, there can not be an unlimited number of subjects of management and, unlike the firm, there can not be a single dominant. From the point of view of the frequency of transactions, the ecosystem is located on the border of constantly occurring and rarely occurring transactions. As a result, it is possible to determine the place of the ecosystem in the scheme of forms of coordination of economic activity only after the fundamental establishment of the nature of the ecosystem itself, its essential features. If the ecosystem is understood as a certain community of stakeholders, in the center of which the analyzed firm, enterprise, company is a hybrid with a set of management subjects more than one. If an ecosystem is interpreted as an environment of interactions, institutions, and infrastructure components (for example, a regional innovation ecosystem), then Oliver Williamson's three-tier set of alternatives is not fully applicable, and the ecosystem has a higher-order configuration than a hybrid. With this understanding, we can state the need to supplement the set of alternatives for the transactional configuration with this large fourth element.

6 Conclusions

Scientific research and business practice, including social entrepreneurship, indicate a high potential for the development of innovative ecosystems in the near future (both regional ones, where the authorities are the conditional center of orchestration, and entrepreneurial ones).

The new leaders of the economy are ecosystems. Examples are the most expensive corporations-Microsoft, Apple, Amazon, Alphabet (Google), Alibaba, Facebook. And these examples inspire other companies to transform and become ecosystems. In Russia, the flagships of ecosystem evolution are SBER and Yandex. Enterprises of ROSTEC, ROSCOSMOS, and ROSATOM corporations are showing interest in ecosystems. The ecosystem approach in the near future has every chance to change the concept of the company as the main business configuration.

These approaches allow both individually and jointly (systemically) to provide researchers with a variety of tools for analyzing the development of innovative ecosystems.

The ecosystem approach has a characteristic feature, in contrast to the concepts where the company is at the center of the analysis and the achievement of outstanding performance primarily depends on this company, the main methods of analysis are concentrated on it. The emphasis in the ecosystem is shifting to the formation and development of interactions, connections of elements, transactions of stakeholders, and "background" environmental influence.

The methods of the theory of ecosystem analysis are applicable for:

• mapping, institutional and structural design of the ecosystem being created, aimed at developing innovative dynamics,

• assessment of the level and capacity of self-developing innovation ecosystems,

• development of antitrust policy tools for platform ecosystems,

• self-analysis of economic entities, assessment of their place in ecosystems.

The study was carried out with the financial support of the Russian Foundation for Basic Research in the framework of scientific project No. 20-010-00333.

References

1. 1 M. Talmar, B. Walrave, K.S.Podoynitsyna, J. Holmström, A.G.L. Romme, Long Range Planning, **53**,4 (2020)
2. 2 S.M. Lee, S. Trimi, Journal of Business Research, **123** (2021)
3. 3 J. F. Moore, Harvard Business Review, May/June (1993)
4. 4 G.B. Klejner, Jekonomicheskoe vozrozhdenie Rossii, **1 (59)** (2019)
5. 5 L.A. Ramenskaja, Upravlenec, **11, 4** (2020).
6. 6 M.V. Alikaeva, L.O. Aslanova, A.A. Shinahov Proceedings of VSUET, **82**, 3 (2020).
7. 7 D.Ju. Brjuhanov, C. Fan, G. Van, Schumpeterian Readings, The Ninth International Scientific Research Conference Proceedings, 36-47 (2020)
8. 8 A.I. Tatarkin, Jekonomika regiona, **4 (36)** (2013)
9. 9 D.A. Sevostyanov, Inverse analysis in management, Novosibirsk, IC NGAU "Zolotoy kolos" (2017)
10. 10 G. Kleiner, V. Karpinskaya. Transition of Firms from the Traditional to Ecosystem Form of Business: The Factor of Transaction Costs, CRFMELD 2019, LNNS 110 (2020)
11. 11 R.J. Rabelo et al., IFAC-PapersOnLine 48-3 (2015)
12. 12 J.A. Schumpeter, Theory of economic development. Capitalism, socialism and democracy, Moscow, Eksmo (in russ.) (2008)
13. 13 E.V. Popov, K.A. Semjachkov, V.L. Simonova, Journal of economic theory, **1** (2017)
14. 14 J.S. Sayles et al., Environ. Res. Lett, **14** 093003 (2019)
15. 15 G.B. Klejner, S.E. Shhepetova, G.A. Shherbakov, Economics of Contemporary Russia, **4** (79) (2017)
16. 16 E.V. Popov, V.L. Simonova, I.P. Chelak, Innovations, **6** (260) (2020)
17. 17 S.Y. Barykin et al., J. Open Innov. Technol., **6**, 124 (2020)
18. 18 O. E. Williamson, The economic institutions of capitalism: firms, markets, relational contracting, NY, Free press (1985)
19. 19 A. Shastitko, Voprosy Ekonomiki, **8** (2016)
20. 20 O. Granstrand, M. Holgersson, Technovation, 90-91 (2020)
21. 21 A.A. Panova, Economic policy, **13**, 4 (2018)

Theoretical aspects of innovation management

Dmitry Fedulov1, *, *Alexander* Pobedin2

[1] South Ural State University, 76 Lenin Ave, Chelyabinsk, 454080, Russia
[2] Urals Institute of Management - Branch of the Russian Presidential Academy of National Economy and Public Administration, 66, 8 March St., Ekaterinburg, 620066, Russia

Abstract. In recent decades, the need to theorize and practise innovation has become apparent to many domestic and foreign companies. The process of introducing innovation mechanisms differs from company to company. Businesses and innovation theory experts do not yet agree on the methodology for implementing innovation effectively. The increasingly competitive environment of the world encouraged many companies to move towards innovative development. Innovation became an important tool for sustainable development for many companies in an unstable and rapidly changing external environment. Innovation is used to create a new product or develop a process to increase a company's profits and strengthen its market position. The term "innovation" has a different understanding today, depending on the area of application: firms or sectors. The article examines the theoretical concepts of innovation.

1 Introduction

In the 21st century, the need for innovation has become apparent to many domestic and foreign companies. The processes of globalization and integration of the world economy facilitated an active transition to innovative development in many sectors of the economy. Innovation has become an important tool for progressive development for many companies in an unstable and rapidly changing business environment. Innovation is used to create a new product or develop a process to increase a company's profits and strengthen its market position. The process of introducing innovation mechanisms differs from company to company, but a common motive for introducing innovation into business practices is the increasing competition between companies in an increasingly unstable and volatile external environment. The term "innovation" has a different understanding today, depending on the area of application: firms or sectors. Many researchers believe that innovation should contribute to meeting the public interest and the basic needs of citizens. To ensure the achievement of the desired results, innovative companies need to assess the resources required and the tools needed. The article examines the theoretical concepts of innovation.

2 Results and discussion

At the beginning of the 21st century, the need to theorize and practise innovation has become apparent to many domestic and foreign companies. The process of introducing innovation mechanisms differed from company to company. Businesses and innovation theory experts do not yet agree on the methodology for implementing innovation effectively.

Conway and Steward linked innovation to the practice of successfully introducing a new product or idea into the marketplace to provide a solution to a firm's existing problems [1]. Innovations included new processes, ideas, services and products designed to improve the functioning of the organization's structural units, reduce costs and improve quality overall.

Sakalyte and Bartuseviciene linked innovation firstly to internal development and research based on the company's own knowledge and developments, and secondly to the use of the experience of innovations applied by other companies. Internal development and research enable the application of new methods to the production process, provides access to new markets and enables the efficient application of production processes to busy markets [7].

Many researchers have repeatedly emphasised the paramount importance of innovation [2,4,5,11].

Researchers of innovations define them as an incentive for the development of the economy of countries, society and scientific and technological progress in general.

Let's look at the existing rating of countries by the level of innovation, presented in the figure below [12]

* Corresponding author: d373@rambler.ru

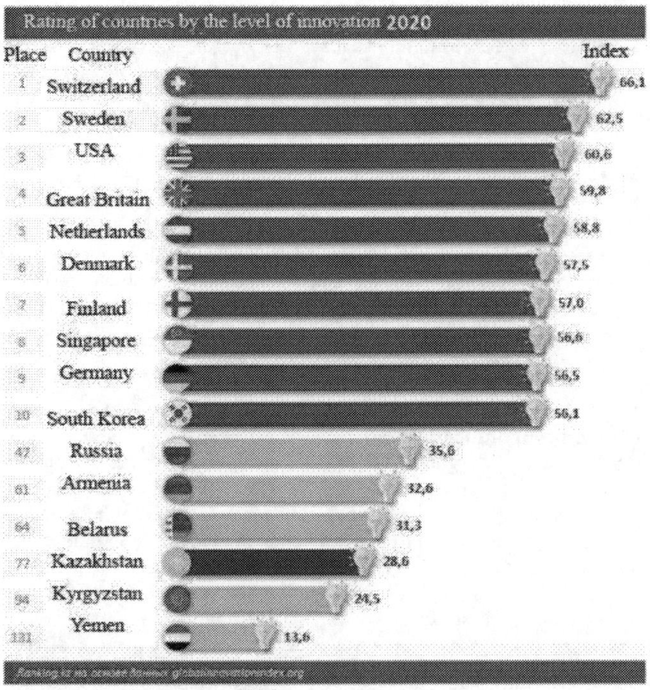

Fig 1. rating of countries by the level of innovation 2020

As we can see, the leading positions in the ranking in terms of innovation implementation are occupied by developed countries of the world. It is in them that special attention is paid to the development of innovations in various areas.

Researchers of modern innovations distinguish the following areas [13]:

- Biotechnology
- Architecture
- Computer technology
- Automotive industry
- Interaction with the surrounding world
- -Energy
- Medicine
- Scientific potential
- Robots
- Nanowear

Scientific American magazine and experts of the World Economic Forum highlighted the following innovative technologies) [14]:

- Bioplastics from plant waste
- Social robots
- Metalinzy
- A potential cure for Alzheimer's
- Smart fertilizers
- Joint telepresence
- Blockchain against food poisoning
- Safe nuclear reactors
- Storing data in DNA
- Network energy storage devices.

For this reason, innovation activities, combining aspects such as entrepreneurship, creation and learning, are regarded as production activities, providing a shift from the previous one to a new, higher level. This refers to a product, a system or a process. These transformations aim to meet the differentiated needs of society and to ensure a competitive environment with all market actors.

Schumpeter and Drucker understood innovation as an area of entrepreneurial activity focused either on taking advantage of a new technology (process) that was not previously used, or on fundamentally transforming the product using inventions [16].

Love and Roper defined innovation as a process that includes various activities, such as distinguishing a new product, creating, producing it, and executing a marketing strategy[9]. These researchers distinguish the following stages of new product creation: recognition (authentication) of the newly created product, design and packaging of the new product, consultancy services related to product development and promotion of the product in the market. As can be noted, these steps are quite time-consuming and require a fair amount of existing knowledge.

In summary, the process of introducing innovation drives radical change in an organization. Innovation is used to create a new product or develop a process to increase a company's profits and strengthen its market position.

The need for innovation research has now become apparent to many in the scientific community, business and policy makers, as innovation today provides a significant competitive advantage for companies and ensures the growth of economic wealth in society. Researchers around the world today use many models of innovation, based on different approaches.

* Corresponding author: d373@rambler.ru

Most research scholars use Schumpeter's classical theory of production factors as their base, arguing that market structure and size are the main innovation motives. Subsequently, the more in-depth research into different industries and the specifics of knowledge management found the theory of production factors discussed above to be irrelevant. It is generally accepted that environmental factors such as training and additional knowledge, as well as management techniques, are more important than the specifics of each individual industry. That is why innovation is considered to create new markets, reduce production costs, improve product quality and increase production output. Successful innovation in the production process ensures increased profitability[17].

An important aspect characteristic of the concept of innovative development is the speed of innovation. This can be considered from the following perspectives. First, speed of innovation corresponds to the initiation of the innovation process and the introduction of a newly created product on the market. Innovation is critical in the current volatile and unstable markets, increasing product differentiation and rapidly changing customer preferences. The above-mentioned reasons make it necessary to launch new products faster than competitors [11]. Secondly, the speed of innovation in production relates to the company's ability to adopt new technology as quickly as possible. The greatest benefit from innovation will accrue to the companies that are the first to adopt the innovative mode of production, considering the level of risk involved to be acceptable [11].

According to Hong, innovation activities allow companies to move to a fundamentally new level by significantly improving a product or process, while ensuring an increase in market share and profits [6]. The economic theory considers a situation in which a firm that undertakes a radical innovative transformation of its production process or technology can become a monopolist for some time, maximizing profits and increasing its market share. Arias-Aranda believes that innovation provides a sufficient level of stability for a company over a sufficiently long time and to win in an ever-increasing competitive struggle[1].

Research on contemporary types of innovation highlights the importance of innovation development, which relies on important components of the innovation system such as institutions, buyers and suppliers. This is why it is so important for an innovative company to have proven partners [8]. Other researchers believe that the benefits of a company's innovation activities come from bringing a new product to market quickly. Sometimes it is possible for companies that use an innovation strategy in their operations to create new markets at the expense of the significant advantages arising from it [11].

Considering innovation as a set of actions has become relevant in recent decades. It is important to have a continuous innovation process involving the analysis, creation and introduction of new products and services to the market. This involves approaching innovation as a system, as many actors are involved in the process. Considering this systematic approach, we note that the development of this approach depends on a number of factors: methods for promoting the innovative product or service, the specific technology used in the production process, the interaction between buyers and distributors of innovative products, the specific characteristics of the product, the market environment and other external environment factors.

There is currently no consensus among researchers of modern innovation categories regarding the distinction between process innovation and product innovation, as many academics believe such a division is arbitrary. The distinction between process innovation and product innovation may be relevant if there is a comparative analysis of service and industrial companies. The strongest link between process innovation and product innovation is evident when considering service businesses rather than industrial companies. This is due to the specific functioning of the service sector, which involves an immediate and direct sense of the quality of the service provided by customers [4].

The development of innovative products is paramount to economic growth. This is why it is necessary to study existing publications on the subject to understand the correlation between business profitability and innovation, and to recognize the need to develop a company's innovative directions to ensure stability and sustainability in the future.

The innovative way of development provides the information necessary for the emergence of new technologies. Information and innovative production technologies enable new companies to become market leaders not only at the country level, but also on a global scale, creating groundbreaking products that have never been seen before. Innovative production technologies enable new companies to take a significant share of the market, meeting the special needs of customers by producing rare goods and services.

A close relationship between business and innovation is key to the success of companies. The ability to take risks and innovate in the production process are key ingredients for a successful business. The successful implementation of innovative concepts in the operations of various companies shows that small firms, which are the most flexible and adaptable to implement anything new, have the most success. By innovating, small firms have a significant impact on markets, helping to transform them[15].

The concept of innovative production is based on a creative approach that involves the ability to think in new ways, to see the prospects for change in existing technologies and processes and the new opportunities that are available to companies when they embark on a journey of innovation. The innovative way of development changes the production and management methods and the organizational structure of the company [10]. We note that the innovative way of development ensures that the company achieves a meaningful competitive advantage in the marketplace.

Innovative companies apply tools to improve the production process, achieving a meaningful competitive advantage not only in national markets, but also internationally. Modern information technology is also significantly expanding the horizons of innovative companies, enabling the necessary communication between the company and its customers, as well as providing the necessary knowledge about customers and competitors.

The choice of innovative development gives a company a competitive edge through the introduction of new technology, making it possible to produce products with special properties that are not available on the market. The ability to manufacture customized products comes from an innovative approach based on the latest scientific advances. By ensuring the possibility of producing unique products, the company achieves maximum customer loyalty, as the product cannot be bought from other competing companies. The overall value of an innovative company increases relative to its competitors, and it can take a leading position in the international market. But the above-mentioned opportunities are only available to companies that consistently follow an innovative path of development and offer unique products and a high level of service [7].

Innovative development paths are not only important for manufacturing companies, but also individual countries. This is the reason why the concept of innovative development now exists internationally as a key factor in economic growth. There is a view that it is possible to consider the processes of production globalization and internationalization, as well as entry into international markets, as innovation [7].

We should note that any company that chooses an innovative path of development and introduces innovations into the production process should strive to maximize the capitalization of investments in innovation not only in the long term, but also in the medium and short term. A company needs to develop a certain organizational structure to foster innovation. This organizational structure will facilitate the acquisition of necessary information to support innovative development and the development of innovative products and services. These transformations should subsequently lead to a strengthening of the company's market position, margins and market share.

Ganotakis and Love suggest using the Innovation Value Chain (IVC) as a convenient mechanism for indicating the strengths and weaknesses of companies that have embarked on innovation. This tool is suitable for companies operating in the high-tech sector and developing innovative, unparalleled products [5].

By offering unparalleled, innovative products to the market, high-tech companies contribute to the competitiveness of individual industries and the economy as a whole.

When considering the implications of the innovation value chain for a company, we note an increase in overall productivity, a clearer view of bottlenecks, the ability to trace multiple connecting elements throughout the production process and the ability to target financing more effectively as a result of the innovation. As a result, the company has the opportunity to maximize the benefits of innovation once the product or service has been introduced to the market.

3 Conclusions

Thus, we have considered the essence of the innovative development path of a company, the impact of innovation on economic growth and the increase in the level of well-being of society as a whole. Innovation is a tool for improving a company's economic efficiency and strengthens its position not only in national but also in global markets. Innovation contributes to scientific and technological progress, the development of society as a whole and global competitiveness. As we have noted in this paper, innovation today is perceived as a rather voluminous concept, so we have looked at existing concepts of innovation. The article presents different theoretical approaches to the creation of innovation.

References

1. D. Arias-Aranda et al. *European Journal of Innovation Management.*, **4**(3) (2001)
2. M. Bogers, J. West *Creativity and Innovation Management.*, **21**(1) (2012)
3. S. Conway, F. Steward *Managing and shaping innovation.* (Oxford University Press, USA, 2009)
4. F. Damanpour, *Journal of Management Studies.*, **46**(4) (2009)
5. P. Ganotakis, J. H. Love, *Journal of Product Innovation Management*, **29**(5) (2012)
6. S. Hong, et al. *Journal of Economic Surveys.*, **26**(3) (2012)
7. G. A. Knight, S. T. Cavusgil *Journal of International Business Studies.*, **35**(2) (2004)
8. K. Laursen, A. Salter *Strategic Management Journal.*, **27** (2006)
9. J. H. Love, S. Roper *Industry and Innovation.*,**16**(3) (2009)
10. G. T. Lumpkin, D. Dess, *Academy of Management Review* **21**(1) (1996)
11. Prajogo D. I. *Knowledge and Process Management*, **13**(3) (2006)
12. *Global innovation Index. 2020* Retrieved from: https://www.globalinnovationindex.org/gii-2020-report#
13. *Agency for Innovation and Development of Economic and Social Projects* Retrieved from: http://www.innoros.ru/
14. *Ten innovative technologies that will change our lives* Retrieved from: https://trends.rbc.ru/trends/innovation/5df26cc19a79477250ac8483
15. H. K. Steensma, et al. *The Academy of Management Journal.*, **43**(5) (2000)
16. O. Stripeikis, J. Ramanauskas, *Management theory and studies for rural business and infrastructure development*, **1**(25) (2011)

https://doi.org/10.1051/shsconf/202111600034

17. E. Webster *Economics of Innovation and New Technology*, **13**(8) (2004)

To the question of the prospects for the formation of an innovation ecosystem in Kazakhstan

Zulfia Magrupova [1], *Kashamida* Tazhibekova[1], *Aigerim* Shametova[1,*], *Kulzira* Abdraimova[1], and *Kyrmyzy* Tursyngaliyeva[1]

[1] Karaganda technical University, 56, Nazarbayev Avenue, Karaganda, 100003, Republic of Kazakhstan

Abstract. The article presents the results of research on innovation activity based on the analysis of the directions of innovative development of Kazakhstan and suggests a set of measures, the implementation of which will allow building a connected system of measures for the formation of an innovative ecosystem. The main focus is given to improving the efficiency of innovation activities based on an institutional approach, which is aimed at the formation of a stable system of relationships between participants in the innovation cycle, each of which is individually interested in improving competitiveness based on the use of innovations. The formation of an innovation ecosystem also includes the development of two innovation clusters as the basis for the formation of a knowledge-based economy. Based on the analysis provided, we concluded that the organization of innovative infrastructure did not provide the required pace of innovative development of Kazakhstan for a number of reasons. To date, the Republic has successfully implemented state programs that aim to create conditions for the transition to the fourth industrial revolution. The focus on innovative development remains important and mandatory for achieving the goals set out in the program documents.

1 Introduction

To date, the country is successfully implementing an industrialization Program, and the Third modernization of Kazakhstan has been launched. The development Strategy "Kazakhstan-2050" was adopted, which aims to create conditions for the transition to the fourth industrial revolution. However, we must clearly understand that without relying on science and innovation it is impossible to achieve it. Understanding this important factor contributed to the development of a state program for the development of the National innovation system in Kazakhstan (Programme, 2005-2015). Thus, the focus on innovative development remains important and mandatory for achieving the goals set out in the program documents, which consist in forming an open-type innovation system based on the use of domestic and foreign scientific potential and technologies. The formation of a national innovation system involves the formation of relevant institutions, the role of which in the development of innovation is crucial. In addition, already on the basis of an institutional approach, the task of forming a stable system of relations between all participants in the innovation process is being implemented, taking into account the interests of each.

2 Problem Statement

The national scientific and innovation system - as a set of interrelated legislative, structural and functional components, ensures the innovative development of the Republic of Kazakhstan. NIS is considered both as a set of economic entities and as a system of relations between them.

The national innovation system cannot be imported, the Republic of Kazakhstan creates and develops it independently. The main thing is to implement selection of priority scientific and technical directions and the mechanism of their implementation. The model of turning knowledge into innovation is changing. If it was a linear structure in the administrative-command model, then the market model uses a network structure, the elements of which are the academic sector, the applied research cluster, the public sector, the corporate sector, and the main research cluster. The result in the form of new knowledge, ideas, innovations, innovations can be obtained at any stage of the innovation cycle. Market model NIS includes relatively independent sectors of the economy, the relationship between which is determined by market relations, but with the regulatory role of the state.

3 Research Questions

The paper analyses the differences between one model of building a NIS (market economy) and another (administrative-command line), which consist of the following.

• In the context of the creation or transformation of new knowledge, a separate economic operator the

* Corresponding author: gsina@mail.ruashametova.74@mail.ru

subject is focused on the realization of their own interests and values.

• The main thing in the innovation process is the system of relations formed between its participants.

The ongoing processes of regionalization in the economy have created prerequisites for the formation of the regional structure of NIS as an independent subsystem of innovation management. Speaking about RIS, we note that their functional capabilities are individual, since their composition and structure depend on regional specialization, level of education, innovation and economic potentials. The European Commission has defined RIS as "networks of firms forming a local production cluster together with institutional infrastructure" (Russel M. G., 2014201.4). RIS is a multi-level system that includes:

the first level - system-wide, which can be defined as a set of public and private educational and scientific organizations that create and disseminate innovations in the region;

the second level - the branch level, which is a functional subsystem of the regional branch economy;

the third level - institutional, which is considered as a spatially organized subsystem of NIS. Let us denote the material and territorial factors of the allocation of RIS TO an independent NIS subsystem.

Mandatory designation condition RIS as a separate subsystem is the concentration of innovative business structures and institutions within certain territorial boundaries with the appropriate organization of structural links between them and the formation of a system of relationships throughout the research and production cycle. Thus, the spatial concentration of economic entities and the territorial division of labour by region determine the characteristics of RIS and the opportunities for their growth and development. From the point of view of macroeconomics, we can conclude that RIS is a manifestation of the prevailing conditions and factors of production in the region, which determine both internal and external factors for the development of innovation potential. RIS can be represented as a multi-level complex structure in which subsystems can be distinguished with hierarchical links between them (vertical links), and elements combined in blocks (horizontal links) (Rudich S. B., 2017). RIS management presupposes the existence of an object and a subject of management. The object of management is R&D, transfer and commercialization of technologies, the subject of management is economic entities engaged in innovative activities, state and local authorities that regulate the sphere of innovation throughout the scientific and production cycle. And already the interaction of subjects and objects of RIS occurs in the markets for creating innovations and inventions, capital investments and investments, scientific and technical personnel, and innovative infrastructure (Demidov V. A., 2008). The need for the formation of RIS is determined by the field of R&D and the availability of investment sources of development that require state regulation of innovation activities through the adoption of legislative and regulatory acts and the implementation of managerial functions in this area.

3.1 Formation of an effective regional innovation system

To increase the role of regions in the development of innovative business activity, state support for the innovation ecosystem is implemented, which is carried out on the basis of the principle of decentralization. This principle assumes that regions independently form regional innovation policies and finance small and medium-sized business projects. Such policy contributes to the development of inter-regional competition in obtaining budget financing, and on this basis - to the formation of an effective regional innovation system, which includes, on the one hand, an innovative infrastructure, and, on the other hand, a logically linked set of tools for stimulating innovation activity on the part of local authorities. At the same time, M&E budgets include expenditures on financing the most promising innovation projects and supporting the activities of innovation infrastructure. Central government agencies and development institutions finance the costs of providing methodological and informational support at the expense of Republican budget programs. (GAZINDUSTRY, 2019). The reorientation of the structure of scientific grants and programs for the needs of the state program for industrial and innovative development is directly related to the development of the Law "on the commercialization of the results of scientific and (or) scientific and technical activities", which contains mechanisms for financing the implementation of innovations in production. In 2015, it was approved, and in addition, the Rules for conducting expert reviews of commercialization projects and the rules for financing projects for commercialization of the results of scientific and (or) scientific and technical activities (hereinafter referred to as RSSTA) were adopted. In total, from 2016-2018, in the priority areas of the State program for industrial and innovative development (hereinafter referred to as SPIID) 3 grant financing competitions for commercialization were held. According to the results of competitions in accordance with the decisions of the national tax service, 158 projects in the following areas are financed with an annual budget of 5.4 billion tenge: Agribusiness – 27 projects, Biotechnology - 22 projects; Information technology - 20 projects; Healthcare - 13 projects; food Production - 12 projects; Chemical industry - 12; Construction materials - 10 projects; Energy - 10 projects; machinery and equipment Production - 9 projects; Metallurgy - 8 projects; Ecology - 6 projects; geological exploration - 3 projects; production of other finished products - 3 projects; Nuclear industry - 2 projects; Tourism - 1 project. At the moment, 9 projects have been completed. All projects are aimed at solving strategic problems of the state, while interdisciplinary projects solve the problems of several industries. The end result of each funded project is a new and / or improved product, product or service that is ready to enter the market. By the end of 2019, the following interim results of the implementation of RSSTA commercialization projects were achieved: (Investments, 2019):

- total income from the sale of innovative products increased from 800 thousand at the beginning of the year to 3.7 billion tenge;
- volume of tax deductions increased from 1 billion to 2.3 billion tenge;
- more than 450 scientists successfully implement businesses based on scientific inventions;
- export volume reached 113.8 million tenge;
- 1,240 jobs were created.

61 RSSTA commercialization projects have reached the production stage. At the end of the three-year sales cycle, by 2021, a socio-economic effect will be achieved in the form of achieving the following indicators: the volume of products sold is more than 15 billion KZT; the amount of tax deductions – more than 5.5 billion tenge; the number of newly created jobs – more than 1,260; new types of products - more than 253.

3.2 Creating an innovative ecosystem

An innovative ecosystem is formed, which includes the development of two innovation clusters as the basis for the formation of a knowledge-based economy: the Astana business campus of Nazarbayev University and the innovation cluster Park of innovative technologies. For the implementation of specific production projects, the innovative technologies Park attracts local and foreign high-tech companies, it houses research centres and laboratories for joint research projects and development work, as well as their further commercialization. In order to implement this step, the Law of the Republic of Kazakhstan "on the innovation cluster "Park of innovative technologies" dated June 10, 2014 was amended, providing for mechanisms to encourage the attraction of high-tech multinational companies. According to the Strategy for the development of the Innovative cluster "Park of innovative technologies", it is planned to create 5 technology development centres together with TNCs. At the moment, 4 centres have already been created.

In order to continue the development of the innovation ecosystem, within the framework of the Law of the Republic of Kazakhstan dated December 26, 2018 "on amendments and additions to certain legislative acts of the Republic of Kazakhstan on employment", legislative amendments to the Tax code providing for a preferential tax regime came into force in 2019. Start-ups, including foreign ones, are exempt from paying four main taxes (IPN, CPN, VAT, social tax) in order to reinvest these funds in the development of projects. More than 273 IT companies have already acquired the status of Astana Hub residents. The Technopark also has a special visa and labour regime for foreign employees – specialists from other countries can easily obtain a work visa to Kazakhstan for a maximum of 5 years. At the moment, the startup ecosystem created on the basis of Astana Hubincludes more than 273 IT companies registered as participants who have the opportunity to receive tax, labour and visa benefits.

The volume of attracted investments in domestic startups in the company in 2020 amounted to 38 billion tenge, of which 11.7 billion tenge was attracted by startups Astana Hub. To ensure interaction with large domestic enterprises, investors and international partners, the KPMG Digital Village corporate innovation development Center Village, QazAngels business angel club were launched, agreements were reached with Plug & Play, Amazon Web Services, 500 startups, etc. Investments were attracted from Germany, Singapore, Switzerland, South Korea and the United States. Overall, the Astana Hub has already launched 352 projects (Investments, 2019).

From July 2018 to April 2019, 183 startups completed an incubation program based on the Astana Hub. From February 2018 to May 2020, 138 startups completed the acceleration program. Among them they are convenient round-the-clock Tastamat mail terminals Tastamat for issuing parcels from online stores, the Clockster workplace access control system Clockster and kid security GPS bracelets that allow parents to track their child's location.

A number of innovative startups have already reached the international level. A project at the level of world discovery can also be called AbsCrypt - a program for encrypting large media files, which works much faster than its competitors. Among innovative startups Astana Hub that attracted the attention of investors- Billboard Video. This is a platform for intelligent notification of the population through media boards. Smart billboards display useful information for road users, such as navigation, weather conditions, traffic congestion, and any other relevant messages.

Among the projects is the startup INUI, which builds gaming stadiums in the digital space. By betting on the algorithm, the project solves the problem of fraud in eSports. More than $3 million was invested in the development of the eSports platform by investors from Switzerland, South Korea and the United States. The beta version of matchmaking is scheduled to launch in early 2020 simultaneously worldwide (BCG, 2019).

In 2017, ACF PIT launched an International acceleration program "Startup company Kazakhstan" for the search, acceleration and financing of startups in order to create a critical mass of high-tech SMEs in the Republic of Kazakhstan. As a result of the advertising campaign conducted in Kazakhstan, the CIS countries and other foreign countries, such as Norway and Poland, about 3,000 applications for participation in the program have been received since 2017. In order to attract venture co-investments in projects "Startup company Kazakhstan" in 2017, ACF established a joint Kazakh-American venture investment Fund GVA Alatau Fund. The managing partner is Tech Garden Ventures, LLC (since May 2019). In 2018-2019 (three program streams "Startup company Kazakhstan") GVA Alatau Fund supported 96 innovative start-up projects worth about 990 million tenge in the following areas: Industry 4.0, FINTECH, Smart city, New materials. Having implemented the first stage of creating a professional startup environment through supporting startups at the early stages, the Fund's focus is consistently shifting towards supporting more mature startups, including: portfolio companies, in priority areas of industry development in the Republic of Kazakhstan (TSTP "industry 4.0").

The IT quarter in Almaty (created at the expense of private investment) is 100% full: 4 laboratories with TNCs, more than 50 companies, 40 startup teams are located.

According to the corporate acceleration program, in 2017, 10 companies of the FEZ "PIT", in 2018 – 10 corporations of the Republic of Kazakhstan passed corporate acceleration in Silicon Valley in the USA, in 2019 – more than 10 companies of the Republic of Kazakhstan passed acceleration in South Korea. Since 2020, the main task of the corporate acceleration of ACF Tech Garden is to train and raise awareness of Industry 4.0 technologies using the best international and domestic practices (GAZINDUSTRY, 2019).

4 Purpose of the Study

The purpose of the study was to analyze and evaluate the directions of state support for the innovation ecosystem Kazakhstan based on the principle of decentralization

Prior to the adoption of the program of industrial and innovative development of Kazakhstan for 2020-2025, the innovative infrastructure of Kazakhstan included entities created with the participation of state budget funds (Republican, local budgets, funds of organizations with state participation), as well as service development institutions created by the state.

However, this infrastructure did not activate the interaction between its subjects and did not implement the function of providing a full range of services for participants in the innovation process. Organizations of innovative infrastructure did not provide the required pace of innovative development of Kazakhstan for a number of reasons:

- territorial dispersion of regions with innovative potential and insufficient coverage of the innovation infrastructure of the regions;
- need to create an infrastructure to support both sectoral and regional innovation development processes;
- relatively low number of innovative initiatives at the first stage of industrial and innovative development of Kazakhstan, which does not require a large number of innovative infrastructure facilities in one region;
- need to comply with the principle of expediency of expenditures and their compliance with the capabilities of the state budget;
- selection of the most optimal forms of innovation infrastructure organizations to avoid duplication of individual functions;
- actual absence and urgent need to create an information support system for the interaction of all elements and participants of the NIS.

In accordance with the adopted programs of industrial and innovative development of the Republic of Kazakhstan, it is planned to create innovation support centres in each individual region of Kazakhstan on the principle of "one window". They are equipped with the necessary infrastructure and resources to promote innovative projects and assist innovators, as well as have an appropriate mechanism for monitoring and evaluating the results of innovation policy programs.

5 Research Methods

Theoretical concepts and methodological developments of domestic and foreign scientists on the development of innovative ecosystems served as the theoretical basis for scientific research, and in addition, general scientific methods were used, including the system approach, network theory, expert assessment method, classification of species, comparison method, quantitative analysis methods, and some others.

Periodical, statistical, analytical and translated literature on the research area was used as information resources.

6 Findings

In the modern world, the level of development of the state depends on achievements in the field of innovation. Innovations should be characterized by novelty, applicability in any field of human activity, the ability to quickly implement in the market, and the ability to bring economic and other types of effect. In Kazakhstan, there are prerequisites for reorienting the state's priorities to support innovation. For successful and rapid transfer of new scientific knowledge to the economy, it is necessary to create a modern effective national innovation system and achieve a high national innovation capacity of society, following the example of economically developed countries. For the formation of an innovative economy of the country, the following is required: creation of innovative technologies; a set of systematic measures that ensure a rapid change of marketable products that are ahead of competitors. It includes the implementation of the innovative principle: "everything that works is already outdated"; priority solution to the problem of high quality of products; transition to" innovative" education, with the development of creative abilities, the formation of a national innovative mentality, innovative culture.

Public financing and regulation of innovation processes play an important role for the effective development of innovation activities. The innovation process should be funded from various sources. State regulation is a prerequisite for the successful development of innovation activities. It is necessary to carefully coordinate the activities of all participants in the innovation process, using the levers of the market economy: it is necessary to lower the credit rates of banks, attract financial resources from the private sector, involve more enterprises and organizations in the innovation process, thereby reducing the economic risk of innovation, expanding the potential opportunities for innovative development of the country's economy.

https://doi.org/10.1051/shsconf/202111600035

7 Conclusion

The loan Agreement between the Republic of Kazakhstan and the World Bank on the implementation of the joint project "Stimulating productive innovations" (hereinafter referred to as the Project) was signed and ratified. Pursuant to the order of the Deputy Prime Minister of the Republic of Kazakhstan dated October 31, 2019 No. 17-12 / 2533, the ICRIAP of the Republic of Kazakhstan was designated as the project administrator, and on the basis of the act of acceptance and transfer in 2020, it was transferred to the ICRIAP of the Republic of Kazakhstan. The change of administrative subordination will allow implementing a number of measures:

1 PR. company concerns the need for digitalization of industrial enterprises by the example of best practices;

2. promotion of the Smart Industry Management Platform (SIMP) among it companies and industrial enterprises;

3. informing industrial enterprises about Industry 4.0 technologies;

4. conducting educational events on Industry 4.0 technologies;

5. acceleration of mature solutions for industry;

6. promotion of IT companies participating in the Smart Industry Management Platform (SIMP). To this end, the center for technological development (competencies) in the industrial sphere (MMC) of the Republic of Kazakhstan is being created. ACF is actively negotiating with TNCs (McKinsey&Co, Siemens, Honeywell, Parallel, etc.).

This, in turn, will allow us to optimize the work on further attracting high-tech companies, registering a joint venture, implementing production projects and startups, as well as supporting more mature projects, including portfolio companies

References

1. V. A. Demidov, N. N. Lebedeva, O. S. Oleinik *Regional innovation system: potential and development trends (*Publishing house Volga state University, Russia, 2008).
2. *Investment as a key driver of Kazakhstan's economic growth: new enterprises, advanced technologies and job creation*, Review, Retrieved from: /https://primeminister.kz/ru/news/reviews/investicii-kak-klyuchevoy-drayver-rosta-kazahstanskoy-ekonomiki-novye-predpriyatiya-peredovye-tehnologii-i-sozdanie-rabochih-mest (2019)
3. M.G. Russel et al. *Business Angels and Investment Organizations as Networked Co-creators of the Finnish Innovation Ecosystem* Triple Helix IX International Conference "Silicon Valley: Global Model or Unique Anomaly? (2011)
4. GAZINDUSTRY. *New industrial policy of SPIID 3.0 / JSC "Kazakhstan centre for industry and export"*, Retrieved from: https://almaty.atameken.kz/uploads/content/files/%D0%9F%D1%80%D0%B5%D0%B7%D0%B5%D0%BD%D1%82%D0%B0%D1%86%D0%B8%D1%8F%20QAZINDUSTRY%20for%20businesses.pdf. (2019)
5. BCG. JSC "Samruk-Kazyna". *Magro-economic prospects of Kazakhstan and new investment horizons in digitalization.* Retrieved from: https://sk.kz/upload/iblock/c8c/c8c46093ecd03383b2294ab438a7c083.pdf
6. S. B. Rudich Ekonomicheskie nauki, **41**, (2017)
7. *Program on formation and development of the national innovation system of the Republic of Kazakhstan for 2005-2015* Retrieved from: http://adilet.zan.kz/rus/docs/P050000387_#1.

Survival of innovative technology startups: the case of the technet regional corporate accelerator

Olga Andreeva[1*], *Vladimir* Postnikov[2]

Perm National Research Polytechnic University, 29, Komsomolsky prospect, Perm, 614990, Russia

Abstract. The article discusses the question of how the participation of technology startups in the corporate accelerator affects their survival. We compare the activity of technological projects over 3 years according to the Startup Barometer and the results of the work of the corporate accelerator "Big Exploration" for 2018-2020. The survival rate of accelerator projects interacting with the UEC corporation is by 20% higher than that of projects developed independently. This research was partially supported by the Ministry of Education and Science of the Russian Federation to Perm National Research Polytechnic University (topic No. FSNM-2020-0026).

1 Introduction

Any startup has similar problems at the initial stage of development. Such problems are difficulties in finding sources of financing for the project, conducting market research, determining a group of consumers or a market place. One of the effective tools for the startups development, actively used in the last 15 years, is an accelerator. The accelerator is a program that ensures the accelerated development and implementation of innovative projects that are in the early stages of development through the implementation of intensive mentoring, investment support for projects. In recent years, a new type of corporate accelerators has been actively developing, which create opportunities for small startups to interact with large industrial customers.

Studies of world experience in works of Alänge S., Steiber A. (2018), Becker B., Gassmann O. (2006), Birley S. (2005), Pauwels C., Clarysse B. (2016), Kamenskikh M. (2018) presented the problems associated with the development of new technologies and products. It is effective to use gas pedals as an institution of networking of startups, industrial enterprises, universities, authorities and external institutions.

Corporate accelerators have gained active development in recent times in the world and in Russia. The corporate accelerator is a mechanism for achieving the company's goals, providing accelerated implementation, development of startups that have passed the test of the "viability" of projects (Kamensky M.A. (2018), Postnikov V.P., Gulyaeva M.M. (2019). The corporate accelerator is considered as a way to stimulate industrial innovations with the help of startups. Within the framework of the corporate accelerator, educational events are held; company resources and mentoring support are provided in various areas.

The empirical study presented in this paper aims to statistically assess the survival outcomes of startups. The development of technology startups in Russia can be considered an element of the national strategy since 2015, when the documents and formats defining the "Platform of the National Technology Initiative" were finally determined. The goal of this Platform was to create conditions for the development of a national innovation ecosystem, high-tech products. Officially, the Platform began work in 2018, after Dmitry Medvedev had signed a decree, so the analysis of startups has been also performed starting from this period.

The study examines the results of the startups development that passed through the Great Exploration accelerator and worked as part of the accelerator with large companies, such as the United Engine Corporation in the format of a corporate accelerator.

The purpose of the study is to determine the dynamics of the corporate accelerator startups survival rate, as well as to assess the degree of influence of large industrial partners on the formation and development of startups.

The empirical basis of the study was the projects of the accelerator "Great Exploration" for the period of 2018-2020. The main focus of the study is on the analysis of projects of the TechNet regional corporate accelerator, which includes projects in the field of advanced production technologies, digital design and part modeling, facilities, technologies, systems, industries, additive technologies, development of new materials and coatings, industrial robotics, sensors and photonics. The key industrial partners and industry experts in the direction of the TechNet accelerator "Great Exploration" are traditionally the enterprises of the aerospace cluster of the Perm Territory (Andreeva O., Postnikov V. (2017), in particular, the enterprises of the United Engine Corporation. The experts, who are representatives of the UEC, formed a group of projects,

* Corresponding author: oleandrperm@gmail.com

which were further developed with the help of the corporate accelerator and became the object of monitoring.

2 Literature Review

Factors contributing to the development and strengthening in the market of technological startups are a separate subject of discussion and research in different countries. World statistics show that no more than 1% of innovative developments reach the stage of real sales. The rest of the projects stop developing or close down, not having found enough consumers. The reasons may be both low adaptation of product characteristics to consumers demands and difficulties in product and team managing, insufficient knowledge and skills in working with the market, planning skills. Evaluating the factors that allow a technology team to develop its project is of interest to various groups of researchers who study the relationship between certain team actions and the result. Caliendo M., Goethner M., Weißenberger M. (2020) describes a series of studies on the reasons that determine entrepreneurial persistence. According to their findings, startup survival correlates quite strongly with both the individual characteristics (human capital and personality) of the founder and the business characteristics of the founder. Eliakis S., Kotsopoulos D., Karagiannakiand A., Pramatari K. (2020) also consider the personal aspects of technology startups founders and employees, focusing on the organizational performance of the entrepreneurial team.

A large-scale quantitative study by Goraczkowska J. (2020) formulates the hypothesis that a startup's participation in a technological incubator or university business incubator contributes to its higher probability of conducting R&D and implementing product and technological innovations. This is confirmed by the author through a study in which the sample includes 1,058 industrial enterprises located in two Polish regions.

Often researchers note that the stimulating factor for the emergence of innovation for new enterprises is the interaction with corporate accelerators. Crişan, E.L., Salanţă, I.I., Beleiu, I.N., Bordean, O.N., Bunduchi, R. (2019) monitored papers that identified mechanisms explaining how accelerators work in general and what role they play in supporting entrepreneurship and innovation. Key elements of innovation development included: validating ideas and products; providing product development and model learning; providing support to increase startup market access and growth; and supporting innovation.

Urbaniec, M., Żur, A. (2020) conducted a study confirming that corporate accelerators are a source of innovation. Participation in such accelerators can be used to develop entrepreneurial market logic. The authors use qualitative methods to explore the motivations behind interactions of corporations with startups and identify the corporate advantages and disadvantages of such a business model.

A local study on corporate startup support programs was conducted by T. Gutmann, D.Kanbach, S.Seltman, (2019). In this in-depth study, with an inductive research design, the subject was the SAP Industry 4.0 startup program. The authors used qualitative methods to study a recently established program of corporate accelerators at one of the world's largest enterprise software companies. The study resulted in a list of benefits for startups participating in corporate acceleration programs related to rapid time-to-market, accelerated sales, and knowledge development.

S. Kurpjuweit, S. M. Wagner (2020) note that collaboration between established companies and startups is increasingly seen as a component of the corporate strategy. N. Del Sartoa, D. A. Isabelleb, A.Di Mini (2020) examine the relationship between participation in an accelerator program and startup survival. The authors analyzed 38 startups from five Italian accelerators and compared them to a control group of 38 Italian startups developing independently. Using qualitative comparative analysis (using fuzzy sets), the authors show that participation in accelerator programs by itself does not affect company survival.

The effectiveness of corporate accelerators is also studied in detail by Russian researchers. A. Popov (2020), examining the structure of corporate demand for high-tech startup solutions, finds that if a startup's business model is focused on large business as its main customer segment, such startups should be ready to have their own resources not only for the deal-making process, but also for technical finalization of their own solution based on the corporation's recommendations. Yu.N. Alexandrin and Yu.V. Gorbova (2020) analyze the specifics of the development of different of accelerators (state, corporate, private and university) in Russia. They note that Russian accelerators are not only developing similar to global trends, but also have better business models compared to other countries. In addition, it was found that corporate accelerators are the most actively developing format at the moment.

The survival rate of startups in different industries is assessed mainly by practices (investors, experts and analysts of various funds). For example, Tsaplin E. and Kosova O. (2019) evaluate the usefulness of accelerators for IT companies through case analysis. They found that participation in an accelerator develops entrepreneurial qualities, helps to avoid some mistakes, but does not give guarantees of survival in the market.

Stayton J., Mangematin V. (2019) look at the mechanisms by which accelerators help new technology ventures minimize start-up time. They highlight this factor (rapid product launch) as key to the survival of a startup. The authors also describe acceleration mechanisms, among which are the provision of such types of resources as offices, the opportunity to develop R & D, a network of expertise and industry contacts, special regimes and funding sources, and the provision of intensive mentoring.

3 Design and Methods

A comparative data analysis was done to assess the dynamics of the technological startups development that participated in the Great Exploration accelerator. The first stage aggregated information on technological entrepreneurship in Russia, collected by the StartupBarometer project.

The second stage monitored the lifecycle of startups that participated in Great Exploration in 2018-20. A selection funnel was constructed, which all projects that applied undergo. Then 62 formalized interviews were then conducted with projects that applied to the TechNet track. Of these, 39 projects underwent general acceleration, and 23 participated in the corporate accelerator of PJSC UEC (United Engine Corporation). The interviews included four sets of questions about how the project has developed over the years. This data was analyzed in terms of assessing project survival and determining the impact on their current state of interaction with UEC as a strong industrial partner.

4 Results

The first stage of the study involved the analysis of startup data collected by the StartupBarometer project. The data collection was done by questionnaire survey of technology startup participants from 2018 to 2020. Over three years, there has been an increase both in the composition of participants and in the variety of technological areas represented by participants (Table 1).

The method of data collection remained unchanged. All these years it has been a survey, including questions to make a portrait of the startup founder, access the team, the level of business development, ways to sell, and sufficiency of financing.

Each year the questionnaire is supplemented with useful questions, but the main questions remain unchanged. The table shows that the data grouping system has changed somewhat over the course of the study. The 2019 survey showed that almost 80% of respondents are interested in working with corporations. Entrepreneurs meet representatives of large companies at industry events (27%) and get contacts through networking (25%). 15% of founders participate in competitions and corporate accelerators for this purpose.

In 2019 corporations were seen as a strategic partner by 41%, as an investor by 12%, as a client by 37%. According to the results of the survey in 2020, it was found that 80% of entrepreneurs (the same as the year before) are already working or are considering working with corporations. At the same time, 68% of startups say that corporations currently do not help to develop their businesses. When clarifying the degree of interest of corporations in working with startups, 69% of founders accessed it as average or below average. And 11% believe that corporations have absolutely no interest in such cooperation.

The age of startups as of the date of the survey was also accessed. In 2018, the average age was 3 years. In 2019, the age distribution was as follows:

– 1 year - 20%;
– 2 years - 14%;
– 3 years - 13%;
– more than 4 years - 53%.

In 2020, there was a slight increase in proportion in each age segment:

– 1 year - 28%;
– 2 years -18%;
– 3 years - 15%;
– more than 4 years - 39%.

A review of the projects age shows that about half of them have remained active for four years or more. Data from 2019 highlighted projects older than 5 years, their share was 15%.

The results of the first stage are interesting because they show a cross-section of the situation in the same time interval as the analysis of the Great Exploration projects. This does not mean that startups older than 3-4 years have high financial indicators; rather it shows that "physical survival" can be estimated with a coefficient of about 0.4.

The second phase of the study is a study of the development of projects that were selected by the Great Exploration experts for acceleration and passed the demo day.

To understand the scale of involvement of technology projects in acceleration over three years, we consider the project selection funnel presented in Table 2. The indicators in Table 2 show a 3.1-fold increase in the number of applications to the accelerator over 3 years, while the number of applications in the TechNet direction has increased 2.5 times over 3 years. There has also been a 46,4% growth in the number of projects selected and passed the acceleration program over 3 years, including a 44,4% growth in the TechNet direction. The outpacing growth in the number of incoming applications leads to increased competition among projects for entry into the accelerator.

While in 2018 37.3% of projects were selected for the acceleration program, in 2019 – already 25.0%, and in 2020 – 17.7%. In the TechNet direction, 50.0% of applications were selected for the acceleration program in 2018, 38.3% were in 2019, 29.2% were in 2020. Thus, the competition to get projects into the acceleration program has almost doubled. It is possible to replace the higher percentage of projects involved in acceleration in the TechNet direction compared to other areas, which indicates a higher quality and prospects of projects aimed at working with corporate partners.

In order to assess the survival rate of projects, such characteristics of startups as their existence in 2021, their getting additional funding and active cooperation with industrial partners were considered (see Table 3).

Table 1. Characteristics of StartupBarometer participants

Year	Number of respondents	Main directions	Structure (development stage)
2018	300	Marketplaces, FinTech, SaaS.	9% - idea 49% - prototype and MVP 42% - sales and growth
2019	500	Produsts for business, AI-technologies, BigData, EdTech, IoT, Marketplaces.	14% - idea 54% - prototype and MVP 32% - sales and growth
2020	620	SaaS-projects, AI, EdTech, Hardware, Software, BioTech, Auto&Transportation, MedTech, Marketplace, Robotics	5% - idea 51% - prototype and MVP44% - sales and growth

Authors on the basis of StartupBarometer

Table 2. The selection funnel for technological projects of the Great Exploration accelerator in 2018-2020

No.	Index	2018		2019		2020		2018-2020	
		Total	incl. TechNet	Total	incl. TechNet	Total	incl. TechNet	Total	incl. TechNet
1	Number of applications received, un	150	36	296	47	462	89	908	172
2	Number of projects admitted prior to selection, un	134	29	186	37	341	58	661	124
3	Percentage of projects admitted to selection, %	89.3%	80.6%	62.8%	72.3%	73.8%	65.2%	72.8%	72.1%
4	Number of projects selected for acceleration, un	56	18	74	18	82	26	212	62
5	Percentage of projects selected for acceleration, %	37.3%	50.0%	25.0%	38.3%	17.7%	29.2%	23.3%	36.0%
6	Number of projects on the demo day, un	42	10	40	9	64	18	146	37
7	Percentage of projects presented on demo day, %	28.0%	27.8%	13.5%	19.1%	13.9%	20.2%	16.1%	21.5%

Authors on the basis of the Great Exploration

Table 3. Analysis of the survival rate of Great Exploration accelerator startups for 2021

No.	Index	2018			2019			2020			2018-2020		
		Total Tech Net	With out UEC	UEC	Total Tech Net	With out UEC	UEC	Total Tech Net	With out UEC	UEC	Total Tech Net	With out UEC	UEC
1	Number of projects that have undergone acceleration, un	18	10	8	18	12	6	26	17	9	62	39	23
2	Relevant projects for 2021, un	8	3	5	15	9	6	20	12	8	43	24	19
3	Percent of survival startups , %	44.0 %	30.0 %	62.5 %	83.3 %	75.0 %	100.0%	76.9 %	70.6 %	88.9 %	69.4 %	61.5 %	82.6 %
4	Projects that received grant funding, un	6	1	5	13	8	5	7	5	2	26	14	12
5	Percentage of projects funded against accelerated projects, %	33.3 %	10.0 %	62.5 %	88.9 %	66.7 %	83.3 %	26.9 %	29.4 %	22.2 %	41.9 %	35.9 %	52.2 %
6	Projects working with industrial partners, un	6	1	5	8	5	3	16	10	6	30	16	14
7	Share of projects against projects that have undergone acceleration,%	33.3 %	10.0 %	62.5 %	44.4 %	41.7 %	50.0 %	61.5 %	58.8 %	66.7 %	48.4 %	41.0 %	60.9 %

Authors on the basis of the Great Exploration

The funnel of projects through the stages of the accelerator is shown in Fig. 1.

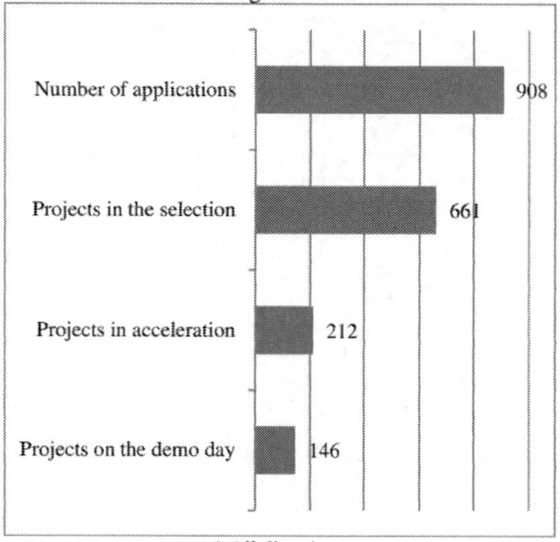

a) All directions

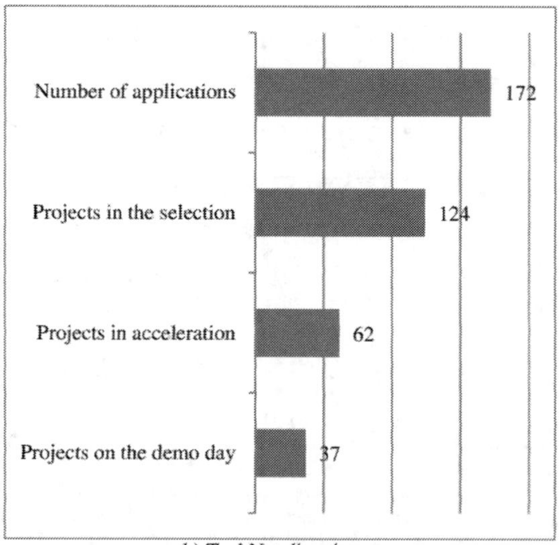

b) TechNet directions

Fig. 1. Funnel of projects of the accelerator of the Great Exploration for 2018-2020

According to Table 3, the survival rate of startups for 2021 out of those that passed the accelerator program in 2018-2020 is 69.4%. At the same time, the survival rate of startups that worked within the regional corporate accelerator of the United Engine Corporation (UEC) is higher than that of other startups (82.6% versus 61.5%). This suggests a positive impact of corporate acceleration programs on startup survival relative to standard acceleration programs. Over the 2018-2020 period, only 4 projects (17.4%) among projects working with UEC closed, and 19 projects continue to function, while among projects not interacting with the UEC, 15 projects (38.5%) among projects not interacting with UEC closed, while 24 projects continue to function.

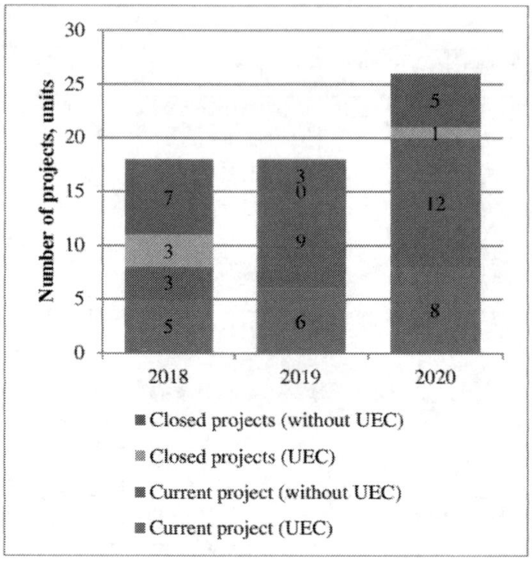

Fig. 2. Dynamics and structure of survival of projects in the TechNet direction of the Great Exploration accelerator for 2018-2020.

The analysis of projects that received grant funding also demonstrates the positive impact of corporate acceleration programs. The percentage of funding raised by startups that worked within the regional corporate accelerator programs is higher than that for the rest of the startups (52.2% versus 35.9%). Interaction with the United Engine Corporation within the acceleration program had a positive impact on startups in getting their products to market (30.4% versus 20.5%) and on their interaction with industrial partners (60.9% versus 41.0%).

Thus, the study showed that startups work with large industrial partners within the framework of corporate acceleration has a positive impact on development and survival in the future. First of all, this is associated with a higher quality of expertise on the part of partners at both the selection and acceleration stage. Working with an industrial partner allows the startup to gain quality industry expertise in technology and market, gain access to modern enterprise laboratories and the opportunity to conduct pilot tests, and subsequently to better form a value proposition to bring the product to market.

5 Conclusions

The issues of startup development, factors of their growth and survival, and conditions stimulating their success are of interest to both theoretical researchers and experts, and investors. Crişan, E.L. et al. (2019) reviewed 98 works over 15 years (2004 to 2019) on the impact of accelerators on startup development that identifies mechanisms to support innovation through acceleration programs. StartupBarometer theoretical research and examination of data on startup characteristics and growth dynamics has provided direction for analyzing data on technological projects

* Corresponding author: oleandrperm@gmail.com

participating in the Big Exploration accelerator. The research, conducted on projects of the Technet direction, provides an initial overview of the development dynamics of startups that have passed the regional corporate accelerator and interaction with large industrial partners within it.

A comparison of the results of technological startups that have undergone corporate acceleration in cooperation with UEC and without enterprises it allows us to draw the following conclusions:

– The corporation within the accelerator provides detailed technological expertise focused on its market or segment. Such expertise also guides the startup to the technological cycle that exists in this field.

– The project receives specialized industry feedback from the corporation, allowing it to focus specifically on this industry and production.

– Participation in a corporate accelerator does not accelerate the launch of the product on the market, due to a more detailed elaboration of requirements for the product, so that it can (if it is positively prepared) qualitatively integrate into the complex industry production chain.

Further research in this topic will focus on the impact of the development stage and startup portrait on survival and an analysis on the rest of the accelerator areas is envisioned.

References

1. S. Alänge, A. Steiber, *Triple Helix,* **8,** (2018). https://doi.org/10.1186/s40604-018-0053-9
2. B. Becker, O. Gassmann, *Journal of Technology* Transfer, **31** (2006). https://doi.org/10.1007/s10961-006-0008-6
3. S. Birley, *Journal of Business Venturing,* **1,** (2005). https://doi.org/10.1016/j.technovation.2015.09.003
4. C. Pauwels, B. Clarysse, *Technovation,* **50–51,** (2016), https://doi.org/10.1016/j.technovation.2015.09.003
5. M. Kamenskikh *Journal of Advanced Research in Law and Economics,* **9,** (2018). https://doi.org/10.14505//jarle.v9%202(32).15
6. M.A. Kamenskikh *Regional economics: theory and practice,* **9,** (2018). https://doi.org/10.24891/re.16.9.1725
7. M.M. Gulyaeva, V.P. Postnikov *Regional economy: theory and practice,* **4,** (2019)
8. O.Y.Andreeva, V.P. Postnikov, *Journal of Advanced Research in Law and Economics,* **8**, (2017). https://doi.org/10.14505/jarle.v8.8(30).02
9. M. Caliendo, M. Goethner, M. Weißenberger *Journal of Small Business Management,* **58,** (2020), http://dx.doi.org/10.1080/00472778.2019.1666532
10. S. Eliakis, D. Kotsopoulos, A. Karagiannakiand , K. Pramatari, *Administrative Sciences.* **10,** (2020). https://doi.org/10.3390/admsci10030039
11. J. Goraczkowska *Oeconomia Copernicana.* **11,** (2020). https://doi.org/10.24136/oc.2020.032
12. E. L. Crişan, I.I. Salanţă, I. N. Beleiu, O. N. Bordean, R. Bunduchi, *Journal of Technology Transfer,* **46,** (2019) https://doi.org/10.1007/s10961-019-09754-9
13. M. Urbaniec, A. Żur, (to be published) *International Entrepreneurship and Management Journal,* 2020. https://doi.org/10.1007/s11365-020-00646-1
14. T. Gutmann, D. Kanbach, S. Seltman, *Problems and Perspectives in Management,* **17,** (2019). http://dx.doi.org/10.21511/ppm.17(3).2019.18
15. S. Kurpjuweit, S. M. Wagner, *California Management Review,* **62,** (2020) https://doi.org/10.1177%2F0008125620914995
16. N. Del Sartoa, D. A. Isabelleb, A. Di Minina *Technovation,* **90–91,** (2020) https://doi.org/10.1016/j.technovation.2019.102102
17. A.N. Popov *Bulletin of REU named after G.V. Plekhanov.* **17,** (2020). http://dx.doi.org/10.21686/2413-2829-2020-4-36-47
18. E.V. Tsaplin, O.A. Kosova, *Information Society.* **4-5**, (2019).
19. J. Stayton, V. Mangematin, *J TechnolTransf* **44,** (2019). https://doi.org/10.1007/s10961-017-9646-0
20. Yu.N. Alexandrin, Yu.V. Gorbova *Economics: theory and practice.* **3,** (2020). http://econ.kubsu.ru/images/3_2020.pdf
21. H. Meyer, J. Sowah, *UBI Global World Rankings of Business Incubators and Accelerators* 2019-2020 http://dx.doi.org/10.13140/RG.2.2.16066.53441

Scientific organization of innovative labour

Andrey Kuznetsov[1*], *Nikolai* Zakharov[2], and *Marina* Perfiljeva[2]

[1]Udmurt Branch of the Institute of Economics, Ural Branch of the Russian Academy of Sciences. Izhevsk. Russia
[2]Herzen State Pedagogical University of Russia. Saint Petersburg. Russia

Abstract. The industrial revolutions that have passed so far have been an alternation of "machine revolutions" and "labor organization revolutions". The third industrial revolution, which created the modern digital communication world, became a prerequisite for the formation of a new organization of work. This new organization is, first of all, the organization of labor of workers involved in the development of a new product, or the organization of innovative labor of workers. The study of the types of innovative labor shows that at least three models need serious improvement, which the authors have assigned the following names: "design" model, "competitive" and "creative". The scientific organization of innovative labor involves identifying for each model: necessary and sufficient working conditions, including everyday and professional comfort; determination of conditions for the formation of motivational guidelines adequate to their activities among members of collectives engaged in innovative work; developing a clear incentive program based on legal and simple principles of rewarding and punishment.

1. Introduction

Klaus Martin Schwab, founder and executive Chairman of the Davos Forum, in his book "The Fourth Industrial Revolution" [1] revealed the impressive prospects of the coming Fourth Industrial Revolution. The perspectives presented by the outstanding modern German scientist have stimulated worldwide active scientific research interest in the problems of global technogenic changes and in the peculiarities of human activity, human labor in the coming new era.

Many of his views on the near future, on global technogenic transformations, for us, living at the turn of the XX/XXI centuries, do not cause any doubt, because before our eyes, within two to three decades, within one generation lifespan, there were amazing cardinal changes (from the queue to the library – to the Google search engine, from the dial-tone telephone to the smartphone, from the processing of data arrays using a calculator to Excel tables and SPSS – and this is not a complete list of the most amazing changes). And we are quite ready to expect such changes in the future.

Moreover, in order to clearly present the long-term prospects shown by the German economist Klaus Schwab, it is necessary to stand firmly on the scientific foundation providing the vision of these prospects. This means that it is necessary to have a clear picture of "road signs", and maybe obstacles awaiting us in the future. The best way to understand what to expect ahead is to assess the patterns of the path travelled. In this regard, let us consider the main patterns of the previous industrial revolutions.

2. Problem statement and methodology

According to the authors, the following industrial revolutions should be considered:

Zero Industrial Revolution – the emergence of the *manufactory*. The manufactory played an important role: based on labor division, manual labor was organized, using the simplest hand tools. And due to the labor organization of workers armed with such tools, productivity has significantly increased.

The manufactory, which provided a new organization of labor and increased productivity, formed the production need for a constant increase in energy supply. The first machines appeared precisely in manufacturing - first, powered by humans and animals, then by wind and water.

The First Industrial Revolution is the emergence of powerful machines *using steam, oil, coal, and then nuclear decay energy*.

The machines using the powerful energy contained in fossil resources have provided a sharp rise in productivity.

In this regard, it should be noted that despite the appearance of new tools – machines with high energy capacity, the organization of labor remained the same – manufacturing. Such kind of labor organization was not suitable for production, equipped with machines using energy incomparable with manufacturing production. Consequently, there was a need for a new labor organization required for machine production. Thus, there was a need for a new industrial revolution.

*Corresponding author: alkuznetsov 63@mail.ru

The Second Industrial Revolution is mainly and essentially the emergence of a *scientific labor organization* and the immediate introduction of scientific developments into production. The foremost examples of the Second Industrial Revolution are the moving assembly line introduced by Henry Ford in the United States in 1908 and the steel rolling mill by Karol Adamiecki in Russia in 1901. Behind the vivid examples is a real revolution, a mass transition to scientific forms of labor organization in all industrial countries, no matter what they are called – capitalist or socialist, democratic or totalitarian. The labor organization based on scientific achievements is the essence of the Second Industrial Revolution.

The twentieth century is a century of scientific and technological development. The very name reflects the direct connection of technology and science, the constant improvement of technical systems based on scientific research. This was the basis for the fundamentally new "machines" creation, and their further widespread production – electronic computers. And along with the electronic computers, the development of technical communication tools begins.

The Third Industrial Revolution is taking place at the end of the twentieth century. It was vividly described by the American economist Jeremy Rifkin in his book "The Third Industrial Revolution" [2]. The result of this revolution is the transition to natural energy sources. And *digital technologies* become the essence.

The Fourth Industrial Revolution, according to Klaus Schwab and his critics, is associated with the processes *of man displacing from the production sphere and replacing him with complex machines*.

This is the most general outline of the sequence of industrial revolutions.

Proceeding from this description, there is every reason to conclude that there is a pattern in the dynamics of industrial revolutions - "machine" revolutions alternate with "organizational" ones:

The "manufacturing" labor organization (or the *Zero* Industrial Revolution) causes the appearance of *production machines* (which is actually the *First* Industrial Revolution), and these, in turn, stimulate the emergence of the "*scientific labor organization*" (the main characteristic of the *Second* Industrial Revolution), which leads to the appearance of "*digital and communication machines*" (that is, to the *Third* Industrial Revolution).

There is every reason to believe that after the Third Industrial Revolution there should be another "organizational" revolution, by analogy with the "manufacturing" and "scientific" ones. And it is this industrial revolution that will be the condition for the revolution that Klaus Schwab called the Fourth.

Consequently, we will call this "organizational" revolution – revolution 3.5, i.e. "3.5 Industrial Revolution", "3.5 Industry" and consider its influence at present.

3. Problem solution

It is an important fact that at the beginning of the twentieth century, the share of "time for product development" was about 20% and 80% was the technical production process. And at the beginning of the XXI century, everything is the opposite – 80% [3] is the share of time spent on product development today, while 20% of the time is spent on production. This indicates that the production process is quite perfect. But the process of product development needs improvement and the creation of a new labor organization.

Previous "organizational revolutions" have improved physical labor. And today, this work is increasingly being transferred to robots and manipulators. But there is another human labor that needs to be organized – the development of a new product, which is called "innovative" nowadays. Its fundamental point is that either one person or a group of people develop a new product without any specific technical task, but aiming at finding "something" new that will attract consumers (or in some other way provide the customer with a competitive advantage).

To date, the arsenal of experience in organizing the work of developers or project teams is not rich. A lot is reported about the working conditions of Google and Mail employees [4]. There are work descriptions of scientific research teams [5]. There is an increasing number of publications about the labor organization of research and development teams aimed at nuclear bomb creating, both in our country and in the United States [6] – very interesting information about the scientific teams work organization.

This interesting and diverse information requires special analysis, therefore, it is necessary to continue the study of various types of developers labour.

Let us consider the variants for organizing the work of developers:

Variant 1. "Rationalization", or improvement of the production process.

Variant 2. A "project" is essentially a search for a solution to the stalemate that an enterprise or institution is experiencing or approaching to. The task of the project team is to find a new promising path.

Variant 3. "Race". A striking example is the implementation of nuclear programs in the United States and the USSR, as well as space projects in the 60s between the countries.

Variant 4. "Surprise" - creating a product that will surprise and interest many people. The activity of the "Apple" company serves as a notable example of this variant, although there are many examples in other areas – for example, artistic creativity: the release of a new film, book, musical composition, painting.

Let us consider these variants as models for organizing the work of developers (or using the modern terminology – the organization of "innovative" activities).

Variant 1, or the first model, which we will call "*rationalization*". The only model for organizing the work of developers (or innovative work), which is fully developed and fully applicable. This could be exemplified by the activity of Henry Ford to improve the car production line, and which he described in his book "My Life" [7]. The system of rationalization activity applied in the Soviet Union since the 30s [8]. And, of course, Japanese quality control circles [9].

In its broadest sense, the process of rationalization activity is as follows:

- First, it is a creation of an organizational system motivating employees to make proposals to rationalize and improve the existing production system.
  - this motivation system can be clearly *ideological* in nature, as in the USSR,
  - or be permeated by the *corporate* unity spirit, as in Japan,
  - or *business* spirit, as in the United States,
  - but in all cases, it provides the employee with a motivational mood, which can be expressed by the formula: "*I am interested in solving this problem*". It is necessary to pay attention to the fact that the rationalization activity "will not work" if such a motivating system is not developed beforehand, which will create a socio-cultural environment that sets the rationalization mood [10];
- secondly, under the influence of a motivating mood, the employee constantly searches for and discovers the *problem*;
- thirdly, based on his professional competence, he finds a *solution* to overcome the problem;
- fourthly, the employee *applies* to the competent authority, which evaluates it and, in case of positive assessment, usually ensures the implementation of this application;
- fifthly, the employee receives a *reward* for his "innovative suggestion";
- sixthly, the employee-innovator by his example encourages other employees to innovate.

This is the basic scheme – and therefore it is the same for the USSR, the USA, and Japan, although each country had its own specific cultural differences. To date, this model is developed and can be quite successfully applied. Its main focus is to improve the existing production process.

Variant 2, or the second model, we will call it "*project*". This model becomes relevant when a company needs to diversify or find new ways of development. This happens when the company finds itself at an impasse or is approaching it. A separate case is when a newly created company is looking for its own path.

Usually, several project groups are formed to find new directions in the hope that at least one group will open up an interesting perspective.

Project team activity process:

- working conditions are characterized by:
  - *time of execution*. The activities of the project team are time-limited. The validity period of such a group is determined by the relevance of the product being created. If the need for the product becomes irrelevant, the group is dissolved.
  - production needs, and therefore proceed from the capabilities of the enterprise. So the working conditions can be neither "luxurious" nor "ascetic", but they should:
    - first, motivate the group *to achieve the result*, and exclude other motivating factors:

Example (case): a specialist was invited to the project group, for whom the living conditions and provision within the project group were significantly higher than usual ones, so his main focus became not a professional activity, but the preservation of these living conditions. The opposite example, a specialist experiencing household discomfort will not be able to focus on work.

- secondly, professional tools should be adequate for the task to resolve the problem.

A hypothetical example, a surgeon can operate with a knife, but the result will be better if he uses a set of surgical instruments produced by a reputable company.

- thirdly, representatives of the project group should have unrestricted access to their workplace (without regime restrictions). In this case, the company should adapt to the mode of the project group (and not vice versa).

- motivational system:
  - the company should have a clear motivating factor that encourages employees to join the project group and actively participate in the creation of an innovative product. The motivating factor is the reward expected by the employee, and the one that everyone needs individually (for someone it may be participation in the profit from the project, for others the Star of Hero of Labour – everyone expects his own reward);
  - on the other hand, future members should be aware of the losses extent, if their group fails.
- the problem that needs to be solved should be clearly stated to the project group;
- the group is formed on the basis of solving the problem

Organizing the work of the project group is very expensive, and the result would be reached in the unlikely case (because not all groups can solve the problem they face). This is the main stopping factor in the creation of project groups when the company management faces problems of further development.

The third Variant, or the competitive model – in a certain sense, is a variation of the second model, it is almost the same project group, but put in a competitive environment. Nuclear projects, the space race between the United States and the USSR are the actions of project groups in competition with each other. The components of the labor organization are similar to the compounds of the second model, but a very important component is added – the "espionage" of the competitor achievements. The results of this "espionage" can serve as research material for the activities of the project group, influence the improvement of labor organization, and act as an intellectual and labor incentive. That is, the obtained "intelligence" material can open up new solutions for researchers (intellectual incentive), and information about the achievements of a competitor can encourage higher labor activity (labor incentive).

This model is even more expensive than the second one. And it also does not guarantee success.

Variant 4, or "*creative model*". This model can be represented by variations of both the activities of a group and an individual employee. This model includes all the components of labor organization inherent to each model described above.

Its main difference from other models is that it cannot be created for a certain period of time, it is effective only if it acts constantly.

The motivational core of this model is in principle similar to other models, and the main factor is the internal mood of the participants to solve the addressed problem.

Working conditions of the creative worker or group are essential within this model. To date, there are different opinions regarding the working conditions for such activities. Examples of labor organization for creative workers are diverse. Many examples demonstrate that the working conditions of creative workers are such that they do not experience even the slightest discomfort in anything. There are opposite examples demonstrating that ascetic, even "Spartan" working conditions are typical conditions for creative activity.

These are some examples lacking a scientific basis for the phenomenon that certain conditions can affect the effectiveness of creative work.

Therefore there is a need for assessing the impact of working conditions on creative work. This is the first research goal sought by the authors of this article.

Among the parameters of working conditions, there is, in our opinion, the most important – "tools for creative work".

The term "tools for creative work" is enclosed in quotation marks since relevant clarifications are required. The creative workers in their activities use a variety of tools, sophisticated equipment and devices which are called the working tools by the authors of the article.

We differentiate the tools into two types: tools necessary for activities according to a given "algorithm" and for activities without any established algorithm, we will call them working tools for "search". With their help, the creative workers perform search activities and do not perform work according to a given algorithm (although search activities do not exclude individual procedures performed according to the algorithm).

To study the work of a creative worker, it is necessary to investigate the influence of the "tools of a creative worker" factor on the effectiveness of his activity. For this purpose, we will single out the parameters of the study on our phenomenon: "tools for creative work".

So, the parameters of the study on the" tools for creative work" phenomenon include the following components:

1. "adequacy" of the tools to this creative activity:
1.1. creative worker proficiency in tools necessary to solve the problem under study;
1.2. these tools are among the best for solving similar tasks;
2. accessibility of tools – to what extent the tools are available to the creative worker, or what are the limitations in accessibility:
2.1. due to the specifics of the enterprise mode;
2.2. due to the peculiarities of the financial responsibility of the creative worker;
2.3. due to the inability of the company to acquire the necessary "tools" - the lack of the necessary funds;
2.4. due to the uniqueness of the tools and the existing queue for their receipt, or any other distribution system;
3. professional development of a creative worker focused on mastering modern "tools".

Along with the working conditions, an equally important phenomenon is the motivation of creative workers. There are opposite views in public opinion. On the one hand, the creative worker is motivated by the work itself and ignores other incentives. On the other hand, a creative worker will be efficient only when all his needs and even "whims" are instantly satisfied.

This is existing in public opinion idea on the motives and incentives for creative work. The main conclusion here is the following –the motives for creative work have not been studied yet. Therefore, the third research task of the authors is to investigate the motives for creative work, and on the basis of this – to identify effective managerial incentives for creative work.

This is the set of tasks for the study of innovative labor by modern socio-economic sciences.

4. Conclusion

Proceeding from the patterns of industrial revolutions, the authors believe that the next Fourth Industrial Revolution, justified by K. Schwab, should be preceded by an "organizational" industrial revolution, which can be conditionally called "3.5 Industrial Revolution", "3.5 Industry". This revolution should be based on the scientific organization of innovative work.

The study of the types of innovative labor shows that today only one type of innovative labor – rationalization activity –fully corresponds to the Scientific Organization of Labor. Or activities for the improvement, rationalization of technological processes.

The system of rationalization activity was formed in all industrial countries, acquiring its own cultural specifics. That is why "Japanese quality control circles", "Ford innovations", "Soviet innovators" seem so different. But all of them solved the same problem, and despite the external differences, they had similar organizational structures of the labor incentive system. This innovative work, aimed at the scientific

https://doi.org/10.1051/shsconf/202111600037

improvement of technologies, based on scientific achievements, has been constantly self-improved. That is why this model, without experiencing Future Shock, entered the XXI century and feels completely free, as in the twentieth century.

The other two models – project and competitive – need to be improved, under the modern realities.

But their improvement depends on the improvement of the oldest type of innovative work – creative. And its study involves a number of research goals related to the assessment of the impact on creative work:

First, the study of working conditions;

Secondly, the study of the "tools of the creative worker";

Thirdly, the study of current incentives and motivations.

References

1. K. Schwab, *The Fourth Industrial Revolution.* (New York: Crown Publishing Group, 2017).
2. *J. Rifkin, The Third Industrial Revolution. Palgrave MacMillan, 2011*
3. A. Chechulin, O. Roy, S. Shelonaev, E. Knyazeva, *Conference Proceedings: The impact of Industry 4.0 on job creation 2020.*(Trenčín: FSEV TnUAD), pp. 106-112
4. A. Chechulin, A. Kuznetsov, V. Grakhov, M. Perfiljeva, *Science and technology.* **19(2)** 125-129 (2020)
5. R. Rodes, *The Making of the Atomic Bomb.* (NY. Paperback, 1986)
6. H. Ford, *My Life And Work.* (Paperback, 2008)
7. *All-Russian Society of Inventors and Rationalizers* Retrieved from: http://www.ros-voir.ru/
8. W. Deming, *The Essential Deming: leadership principles from the father of quality.* (NY, McGrow-Hill, 2013)
9. N. Zakharov, *The organizational behavior of civil servants.* (Moscow, 2021)

Structural management models of technological innovations

Vladislav Gusev*

Institute of Control Sciences by V.A. Trapeznikov of the Russian Academy of Sciences, 65, Profsouznaya str., Moscow, 117997, Russia

Abstract. In order to define the parameters of structural innovations of the technological core of the economic system, a formalized criterion of the effectiveness of these innovations has been proposed, a model of the technological core has been developed, as well as mathematical methods of its analysis. The developed model uses the cross-sectoral balance sheet of the national accounts of the economy. The analysis of the model consists in calculating the preferred structure of the technological core and calculating plans for its phased development.

1 Introduction

The development of the economic system over the long term is always linked to qualitative changes in the composition and organization of its basic component, the technological nucleus, which includes all significant economic activities and changes. While the traditional economy has the characteristic of sustainability, the development process has been unsteady. Innovation is a prerequisite for development. For example, according to Joseph Schumpeter [1], a useful innovation or technology, while gaining popularity, initially unsettles the system, allowing for the greatest profit until it reaches the whole system, when there is a new equilibrium state at which profit becomes minimal. And that brings new innovation, and so on.

Innovation can range from the introduction of new technologies based on previously unknown phenomena and principles to structural changes in the technological core within existing, traditional technologies. The latter type of innovation is possible when the proportions of economic activities do not provide sufficient macroeconomic indicators, such as those of comparable but more successful countries. In this sense, one can speak of the imbalances of the technological core. In addition to comparative analysis, these imbalances can be determined by analyzing the mathematical model of the technological core of the economy and determining its optimal structure.

The introduction of structural innovation is countered by the stability of the established structure of the technological core. This sustainability is due to the routine nature of business, the absence or inefficiency of large-scale investment projects. Overcoming the persistence of existing imbalances and moving towards a more forward-looking structure requires certain structural innovations supported by project investment activities. As a result of their implementation, the system becomes a new stable state with preferred macro characteristics.

Defining the parameters of the structural innovations of the technological core of the economy requires formalizing the criterion of the effectiveness of these innovations, developing a model of the technological core, as well as mathematical

* Corresponding author: gusvbr@mail.ru

methods of its analysis. The model is consistent with the System of National Accounts and is based on a cross-industry balance sheet. The analysis of the model consists in calculating the preferred structure of the technological core and determining the parameters of the stability of its states.

2 Materials and Methods.

2.1 Structural Innovation Performance Indicators

The effectiveness of structural innovation can be measured by the relative value added of these innovations per unit of direct cost. This makes it possible to use as a measure of the efficiency of the technological core of the economy; its productivity - the share of value added per unit of direct cost. Using a closed-loop input-output model of the Leontiev type [2], the productivity of the technological core of the economic system shall be determined by analogy with the efficiency factor in the technique as $\pi = Y/Z$, where Z is the intermediate input, Y is the value added (GDP), V - gross output. Denote material intensity $a = Z/V$, gross output $V = Y + Z$. Then:

$$\pi = (V-Z)/Z = 1/a - 1.$$

Since different states of the economic model have different levels of productivity, the choice of the most productive model can be considered.

The productivity potential of the technological core of an economic system is defined for the economic model as:

$$\pi^c = \max_{V,Z} \pi .$$

The multipurpose production model [3] defines the output multiplier (the indicator of the productivity of the economic system) as a function of the structural proportions of output and prices of output and service industries. The maximization of this output-to-input ratio determines a balanced output-to-price structure that corresponds to a reproduction regime, where growth shares for all products and services are the same.

It is assumed that Z_{ij} is the direct input of industry j on output of type i products or services, V_j is the output of products and services of type j. On the basis of these data a_{ij} - the coefficients of unit input are calculated:

$$a_{ij} = Z_{ij} / V_j ,$$

which form a process matrix A.

The input-output model can be represented by the following ratio:

$$Z_i(t) = \sum_{j=1}^{n} Z_{ij} = \sum_{j=1}^{n} a_{ij} V_j(t),$$

or

$$Z = AV ,$$

from where the stationary reproduction process can be represented by a ratio:

$$V_i(t) = \gamma_i \sum_{j=1}^{n} a_{ij} V_j(t),$$

where γ_i is a multiplier of the output of the industry i, which denotes the excess of industry output over inputs. Denote γ - the minimum by industry multipliers:

$$\gamma = \min_i \gamma_i .$$

The formulation of the optimization problem for the output structure vector V with the minimum output multiplier maximum criterion γ is as follows:

$$\max_{V_i} \gamma$$

with technological restrictions on production:

$$V \geq \gamma AV .$$

This task allows defining the equilibrium system of releases.

In target setting, when final consumption, exports and accumulation are formally part of an activity, the stationary regime allows a change in the share of output in those areas. Otherwise, the values $\gamma_i - 1$ are the shares of output of industries used for final consumption, exports and savings. To maintain these shares equal to the reference values $\overline{\gamma_i} - 1$, where:

$$\overline{\gamma_i} = V_i(t_0) / \sum_{j=1}^{n} a_{ij} V_j(t_0) / \max_j \gamma_j$$

the optimization problem for the release structure will take the form of:

$$\max_{V_i} \gamma,$$

with condition

$$V_i(t) \geq \gamma \cdot \overline{\gamma_i} \sum_{j=1}^{n} a_{ij} V_j(t), \text{ or}$$

$$V \geq \gamma \overline{A} V ,$$

where

$$\overline{A} = \text{diag}(\overline{\gamma}) A .$$

Statement 1. If the matrix A is square, the solution to the problem of finding an equilibrium output system is its eigenvector.

The release growth condition is:

$$V(t) \geq V(t-1) .$$

The measure of productivity can be derived from this objective as:

$$\pi = \gamma - 1$$

and represents the share of value added in the value of intermediate consumption under the equilibrium technological development of the economic system. In equilibrium mode we have:

$$a = 1/\gamma .$$

Computation of the eigenvector of the matrix corresponding to the maximum eigenvalue λ, which ensures the stable equilibrium state of the technological core, can be done by the iterative method.

Statement 2.

Let A be a non-negative matrix whose maximum modulo eigenvalue is λ, |λ|<1. The eigenvector x of this matrix satisfies the equation:

$$A x = \lambda x.$$

The eigenvector x with a modular maximum eigenvalue is searched by an iterative procedure:

$$\lambda^k = \|Ax^k\| / \|x^k\|,$$

$$x^{k+1} = \lambda^k A x^k$$

where k is the iteration number. Procedure breakers are conditional by

$$\|x^{k+1} - x^k\| < \varepsilon,$$

where ε is the given computational accuracy (ε=0.001). The eigenvalue λ of A is estimated as:

$$\lambda = \lim_{k\to\infty} \lambda^k$$

and the eigenvector x

$$x = \lim_{k\to\infty} x^k.$$

This result shows that the inertial process of reproduction of the technological nucleus has a stable equilibrium with a fixed point equal to the eigenvector x.

2.2 Method of evaluation new technologies by productivity index

Since

$$V_i(t) = Z_i(t) + C_i(t) = \sum_{j=1}^{n} a_{ij} V_j(t) + C_i(t),$$

where $C_i(t)$ is the final consumption of the service , the unit cost factors are calculated by the formula

$$a_{ij} = Z_{ij} / V_j = Z_{ij} / (Z_j(t) + C_j(t)).$$

The new technology with number ν adds to the cross-sectoral balance sheet:

- a new line of its costs for other services and industries;
- a new column of this technology in other industries and final consumption.

Then the performance criteria of the new technology will be the fulfillment of inequalities:

$$\pi^{\nu}(t) \geq \pi(t),\ t = 1,2,..,T.$$

$$\pi^{\nu}* > \pi*$$

2.3 Conversion of Output Indices

When the output volumes change, the unit cost estimates a_{ij} change too. In order to capture the results of the change the volumes V_i in the previous cycle, the direct cost factors are converted:

$$\overline{a_{ij}} = a_{ij} \cdot V_i / V_j$$

The transformation operations of the process matrix in current prices to the matrix in relative prices and vice versa are considered.

We denote D – diagonal matrix with diagonal $V_1, V_2, .. V_n$:

$$D = \text{diag}(V);$$

C is a diagonal matrix with a diagonal $1/V_1, 1/V_2, ..1/V_n$;

then $\overline{A} = DAC$.

Statement 3. If all $V_i \neq 0, i = 1,..,n$, then a transformed matrix with coefficients $\overline{a_{ij}} = a_{ij} \cdot V_i / V_j$ has the same eigenvalue as matrix A, and the eigenvector is equal to the original one up to the multiplier D.

Let us call this process matrix transformation D a deformation transformation. To move to the prices on a relative scale (in proportion to the prices), the eigenvector of outputs V^0 on an absolute scale is used by deforming the process matrix:

$$A^1 = C^0 A^0 D^0, \text{ where } D^0 = \text{diag}(V^0),\ C^0 = (D^0)^{-1}.$$

In this case, the eigenvector of the process matrix becomes unit after deformation. The eigenvalue will be maintained and the eigenvector $v*$ of the matrix A^1 in relative scale can be converted to absolute scale by conversion:

$$V* = \text{diag}(V^1) v*.$$

In the calculation of values V_i, the values are interpreted as volume indices of the services to be performed, and the limitations on them in the assumption of non-discharge are in the form of

$$V_i(t) \geq 1, i = 1,...,n.$$

2.4 Indicative projection of structural innovation

It is not possible to instantly implement a change in the structure of the releases, making them equilibrium. In order to determine the most rational plan for the development of the industry, the following local objective may be used:

$$\max_{V_i} \gamma,$$

technological restriction on production:

$$V_i(t) \geq \gamma \sum_{j=1}^{n} a_{ij} V_j(t),$$

and the condition $\theta > 1$ of the output growth at a rate of one stroke:

$$V_i(t) \leq \theta \cdot V_i(t-1), i = 1,...,n.$$

By repeating the procedures for finding an optimal solution and converting the matrix of direct costs from tact to bar, we get an indicative multi-stage plan-forecast of the joint development of branches of the technological core of the economy. The indicative plan calculation procedure uses absolute and relative output values. If V^1 - current price vector, then in the first step the process matrix deformation is used to shift to relative output v^1:

$$A^2 = \text{diag}(V^1)^{-1} A\, \text{diag}(V^1),$$

the corresponding conversion is reversed for intermediate computing v:

$$V = \mathrm{diag}(V^1)v.$$

Then the problems of finding a vector of relative volumes of output are solved:

$$\max_{v^i} \gamma$$

with technological constraint:

$$v^k \geq \gamma^k A^k v^k,$$

and the condition of relative output growth $\theta, \theta > 1$ at one stroke rate:

$$I \leq v^k(t) \leq \theta \cdot I,\ k = 1,2..,$$

where I is the unit vector. Then the following statements can be used to calculate the indicative projection plan for joint industry development.

Statement 4. The sequence V^i for a finite number of bars goes to the proper vector of process matrix A, and the evaluation γ^i goes to the proper number of that matrix.

Note. When converting process matrix A with matrix deformation $D = \mathrm{diag}(v^*)$, the solution to the planning problem becomes trivial: $v = cI$, where $c \geq 1$. That is, when achieving technological equilibrium, the structure of the output does not change further.

Statement 5. If there is a solution to local problems at $i \geq 1$, the quantity of indicative releases in absolute units of the following type can be obtained:

$$V^i = \prod_{j=i}^{1} \mathrm{diag}(v^j) \cdot V^0.$$

Statement 6. If for all steps the coefficient is constant, the relative increase in output from a certain clock becomes equal. This property is similar to the central property of optimization models of economic dynamics [4].

If x - the resulting eigenvector, then for y≠x the value of the estimate $a = \|Ay\| / \|y\|$ can be reduced, this corresponds to an increase in the productivity estimate π.

In order to find an output vector that maximizes the productivity of the technological kernel, we will apply a gradient descent procedure. Let A be a positive matrix with non-negative components. Express material content $a = Z/V$ in terms of the technology nucleus model:

$$a = \frac{\|Ax\|}{\|x\|},$$

where the vector norm is calculated as the average:

$$\|x\| = \sum_{i=1}^{n} xi / n.$$

Productivity assessment can be presented as:

$$\pi = 1/a - 1.$$

Statement 7.

We denote I - unit vector, E - diagonal unit matrix, h - step value, Minimization procedure:

$$\min_{x_0 \leq x \leq x_1} a(x).$$

The gradient descent method is:

$$x^{k+1} = \max(x0, x^k + h\frac{(aE - A^T)I)}{\|x\|},$$

where k is the iteration number.

The choice of the step h determines the speed of convergence of the process: the more it is the greater the speed of convergence. However, infinite enlargement does not accelerate convergence.

The process of optimizing the output structure produces a sequence of incremental assessments of the productivity of the technological core, accompanied by a sequence of increments in the output of certain industries. At the same time, the increase in output from other industries does not lead to an increase in productivity assessment. The rate of convergence depends on the size of the step h: first, when the step is increased, the speed is increased.

2.5 Expenditure accounting for non-critical, but significant industries

We consider a list Ξ of activities of significance for the state, but with small values of equilibrium indices of output and not belonging to the whole «bottleneck» (non-critical, relevant branches: science, culture, social sphere, safety, ecology): $i \in \Xi$.

An additional constraint to the indicative planning task is:

$$V_i(t) \geq \xi_i V_i(t-1),\ i \in \Xi,\ 1 \leq \xi_i \leq k.$$

Fig. 1 shows the dependence of the output of the three branches (computing, communications, education) on the cost of research with the optimal output distribution.

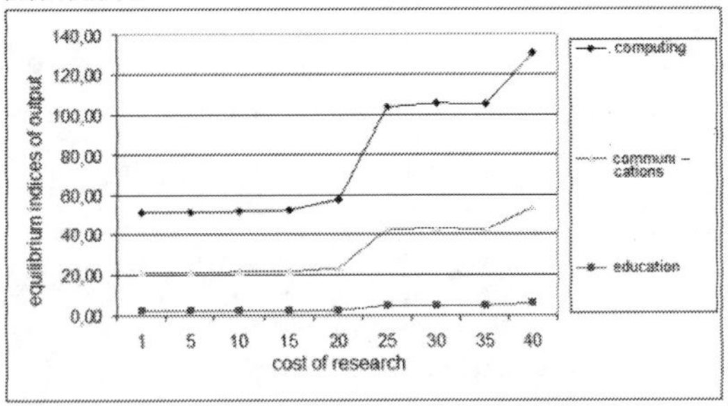

Fig. 1. Impact of research expenditure on technological equilibrium

This correlation shows that the multiplication of research costs across a wide range has little impact on the equilibrium of inputs and outputs in other industries. The swings at the 20 and 35 levels are explained by the reserve of the industry's imperfections and by the fact that only a 20- and 35-fold increase in the cost of science, respectively, could change the structure of other industries output and require a further increase in the gross output of the economy as a whole.

 https://doi.org/10.1051/shsconf/202111600038

3. Results. Calculation of the projection plan

The following is a graph of the evolution of the productivity factor of the technological core in the balancing of output at successive stages of indicative planning, where the lower bound of the ratios of output was 1 and the upper limit of the variation in the ratios of output was $\Delta\theta = 1.5$ to the stage.

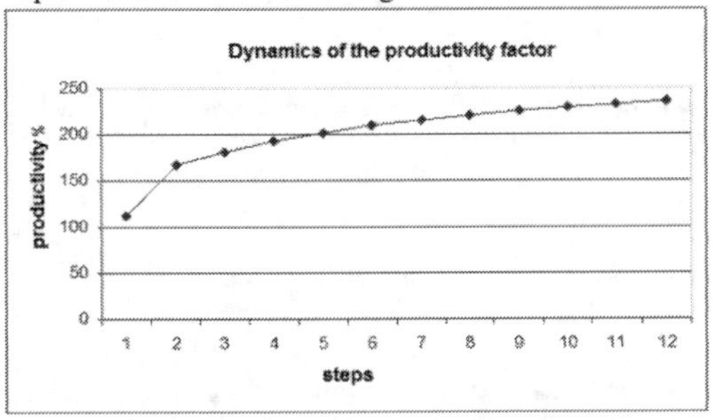

Fig. 2. Changes in the productivity factor of the technological core by optimizing the production proportions at successive stages of indicative planning.

The following is a graph of the indicative development of the proportions of output of some industries in solving the problem at successive stages of indicative planning for data on the Russian economy.

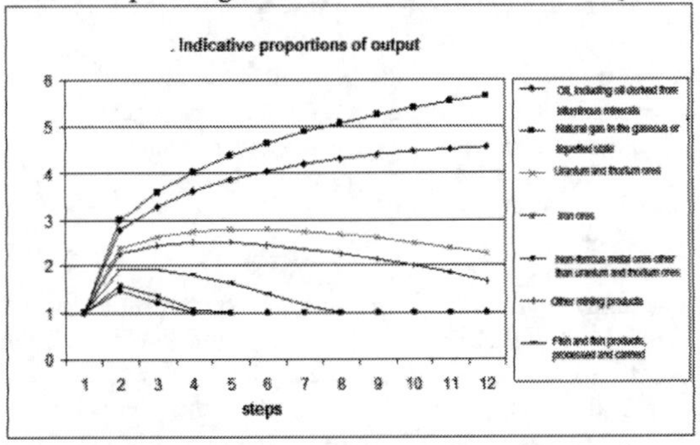

Fig. 3. Indicative development of proportions of output of some industries at successive stages of indicative planning for data on the Russian economy.

Figures 2 and 3 demonstrate the central characteristic of the [4] development model of the technological core of the Russian economy, namely, that output reaches a constant level with sufficient time of operation.

Conclusion

Despite its high potential for development, the modern economy of the Russian Federation is facing crisis phenomena. At the macro level, these include low GDP growth, the economy's critical dependence on oil and gas exports, exchange-rate volatility and undervaluation, a small share of manufacturing, dependence on external sanctions, and poor governance. These factors contribute to the under-realization of the potential of the technological core of the economy [5]. The results show the possibilities for improving the economy's efficiency, based on a gradual change in the structure of its technological core. Based on Rosstat's data, possible growth in economic productivity could be more than doubled. The practical realization of this possibility should be linked to the formulation of strategic plans for the development of the economy. And, in addition to the selection of priority areas for the development of the technological nucleus, it requires the application of adequate methods for forecasting multisectoral dynamics. This takes into account all major aspects of economic activity: stock formation, accumulation, final consumption of the State and households, export-import flows [6, 7]. Planning at a new level also involves appropriate institutional arrangements [8].

References

1. J. Schumpeter, *Quarterly Journal of Economics.* 23, 213-232 (1908)
2. V.B. Leontiev, *Economic essays. Theories, studies, facts and policies. (Essays in economics. Theories, theorizing, facts, policies), Translation into Russian* (M.: Politizdat, 1990)
3. V.B. Gusev et al, *Indicative planning and implementation of regional policies* (M.: Finance and statistics, 2007)
4. P. Samuelson, R. Dorfman, P. Solou *Linear Programming and Economic Analysis* (McGraw- Hill Book Company, 1958)
5. V.B. Gusev, Management Problems. 3, 18-25 (2007),
6. P. Samuelson. *Economics. V 1* (M.: IHP «ALGON» VNIISI, 1992).
7. A.A. Petrov, I.G. Rashalov, A.A. Shananin *Experience of mathematical modelling of economics* (M.: Energatopublishes, 1996)
8. V.I. Antipov *GOSPLAN. Yesterday, today, tomorrow* (M.: Concept, 2019)

Management of development of innovative ecosystems in reclamation

Irina Yurchenko[1*], *Mikhail* Bandurin[2], and *Inna* Bandurina[2]

[1]All-Russia Research Institute of Hydraulic Engineering and Reclamation by the name of AN Kostyakov, Bolshaya Akademicheskaya str., 44, Moscow, 127550 Russia
[2]Kuban State Agrarian University named after IT Trubilin, Kalinina str., 13, Krasnodar, 350044Russia

Abstract. In the context of the coming digital globalization of the economy, an important factor in increasing the level of development of ameliorated agriculture is the widespread introduction of innovative agricultural technologies and agricultural production management into the practice of the domestic agro-industrial complex. The methodological basis of research is based on information and analytical methods, methods of expert assessments, system analysis and synthesis. Based on the results of the studies and the performed assessment, the role of innovation systems in the structure of the agricultural production economy and the experience of their application in the ameliorative sector of the economy in comparison with developed foreign countries are characterized. The influence of innovations on the efficiency and productivity of labor is shown and proposals are presented to reduce and eliminate the technological backwardness of agricultural production on reclaimed lands, consistent with the trends in the development of technologies for the production of advanced industries of the domestic and world economy. The main stages of the formation of production management systems, which provide support for the assignment of control actions, are presented. Reasons are revealed that hinder the full use of innovative technologies in agricultural production, associated with the lack of free funds among small-scale agricultural producers and their proper understanding of the strategic importance of innovations in the production process. A prognostic analysis of the innovative development of agricultural production based on "intelligent agricultural technologies", is presented which unites objects via the Internet for receiving and exchanging information from built-in services.

1 Introduction

The need for a timely and effective solution to one of the most important and priority tasks of the agro-industrial complex of the Russian Federation—the achievement of the best world-class indicators by domestic agricultural production—led to the industry's interest in the development of an innovative system for precision crop production management, which is a powerful toolkit for raising the economic level of agricultural technologies [1]. In this regard, studies devoted to the precise management of reclaimed agroecosystems are very relevant and in demand.

In modern realities, examples of the successful use of precision control of technological processes on reclaimed lands in domestic agricultural production are clearly not enough. Traditionally, agricultural production has been the most conservative sector of the economy. Low profits, high levels of risks of failure to achieve the planned economic results due to the difficult to predict variability of the cost of consumed material and technical resources and commodity products, a multi-tier chain of suppliers, etc. are the factors that hinder the intensity of the inflow of private investment and require constant government support from both domestic and foreign agricultural producers [2]. In the latter case, the share of state support can exceed 30%.

At the same time, it becomes clear that the economy of the domestic agro-industrial complex is now confidently striving for the achievements of successfully developing Russian production sectors [3]. For the period from 2012 to 2018, the growth of Russian agricultural production amounted to 15%, which is three times higher than the growth rate of the country's gross domestic product, which reflects the market value of all domestic final goods and services, which amounted to no more than 5%. But this is significantly lower than the volume of supplies of agricultural products from the world's leading exporters, which is largely due to the imperfection of crop farming agricultural technologies [4].

The existing level of use of creative resource and energy-saving operations, including in terms of the effective use of irrigation water, the use of innovative reclamation solutions, leads to the lag of Russian indicators of efficiency and labor productivity from those of world leaders. The grain yield in Russia does not exceed 29.2 c/ha, which is much lower than the yield

* Corresponding author: chepura@mail.ru

of grain crops achieved in many countries with high-tech crop production systems (in the EU countries it amounts to 40 c/ha, 56 c/ha in China and 60 c/ha in the USA) [5].

The national innovation system of each country, including the structure, goals, functions, is formed and transformed depending on both the level of economic development of the state, the initial state of the scientific and technological complex, personnel potential and mentality of workers, and the established strategic priorities of the state innovation policy and the tasks to be solved [6].

Forcing the technological backwardness of agricultural production on ameliorated lands is a complex multicomponent problem. Its modern solution is linked to the transition of agricultural production to technological structures of a higher level 5.6 (according to the classification of the state innovation strategy) from the actually achieved level 4. The transition of agricultural production to new innovative technological paradigms will require the implementation of new technical, technological, organizational and other solutions that characterize the distinctive features of these paradigms: biotechnology, artificial intelligence, the use of the Internet and an integrated high-speed communication system, etc. [7].

Combining operational data with intelligent digital applications that process information in real time opens up new prospects for increasing the validity of decisions on the appointment of corrective actions, implementing the control of natural factors, multivariate calculations, analysis of an expanded set of indicators, forecasting, modeling, assessing the consequences of decisions made and other methods of high-precision formation of agricultural technological processes, ensuring the feasibility of developing innovative agricultural production management systems on reclaimed lands.

2 Materials and methods

The purpose of this work is to consider the evolution of the formation of innovative systems for managing agricultural production on reclaimed lands in the domestic agro-industrial complex in comparison with developed countries of the West, the United States of America, Canada and other states, as well as to assess their role in the structure of the economy of agricultural production. The implementation of this goal required the solution of the following tasks:

- carrying out an analytical assessment of the formation of digitalization of production processes and production in the sphere of land reclamation of the domestic agro-industrial complex;
- determination of the structure and degree of saturation of the world market with innovative farming technologies;
- consideration of conceptual approaches to the development trends of precision management systems for the reclamation regime of the agroecosystem in the era of global digitalization of the country's economy.

The novelty of the research lies in the development of a scientific substantiation of the need and feasibility of digitalization of the management system of the reclamation water complex based on technologies for solving the fundamental problems of agricultural production formed by the information computer digital platform of the melioration sphere. The practical significance is in determining the fundamental provisions for the development of a functional structure and organizational measures for the introduction of powerful tools for full-factor automation of procedures for managing the reclamation regime of agroecosystems.

The methodological basis of the performed research work was formed by the information and analytical method, expert assessments, system analysis and synthesis [8].

3 Results and discussion

The Western European, or Euro-Atlantic, model of the national innovation system, widespread in many Western European countries, with their well-known world-class scientific centers dating back many years, corresponds to the strategy of increasing innovation. The model contains all phases of the innovation life cycle, covering fundamental and applied science, high-quality education, the creation of innovations and their mass commercialization [9].

A certain recession in the European economy, which occurred after the devastating Second World War, made most countries to abandon expensive scientific research requiring mega-installations (the so-called "big science") even where such work had previously been carried out. European researchers focused mainly on relatively inexpensive, but priority research in biology, chemistry, pharmacology, and other areas that are effective in practical implementation and commercialization.

At that time, in foreign practice, much attention was paid to improving the systems of assistance in making managerial decisions, the basis of which was the automation of operations and procedures for assigning control actions [10]. The beginning of this process was laid by the formation of computer management information systems (MIS), which guaranteed the development of the requested reports, generated according to the preferences of the decision maker. Beginning in the sixties of the last century, the formation of model-oriented digital systems of assistance in the appointment of control actions has been noted.

In 1970-1980, commercial digital technologies were practiced [11] intended for:

- the formation of a proper system of data for the management of economic entities in the procedures for assigning control actions (Executive Information System, EIS);
- the provision of collective solutions to a single problem (group decision support system, GDDS);
- the solution of integral interdepartmental problems that required theoretical training from the manager and the practice of its implementation in various production sectors of the enterprise (organizational decision support system, ODDS).

The nineties of the 20th century were marked by the development of specialized DSS with the functionality of online analysis and data transformation in the Online mode for the implementation of fundamental key decisions (Online analytical processing, OLAP). Analysis of the observed trends in the field of improving management and assessing the priorities of the evolution of digital technologies in the field of agricultural reclamation revealed the advantages of DSS as the most appropriate system meeting the requirements of operational and strategic management of information and economic processes in the specified sector of the economy [12].

The integration of DSS-based software into the monitoring of the state of agroecosystems contributes to the formation and evolution of a priority approach to innovations in agricultural production—technologies of precision (highly accurate) farming. The implementation of these new studies and the emergence of fruitful results in the field of control theory in general and in the field of optimization of production process control, in particular, is associated with the powerful development of electronic and computer technologies and was initiated by the urgent need of oil, gas and food production, water supply, energy supply and other sectors of the economy in the evolution of management of a spatially distributed infrastructure of remote objects. The results obtained were initially used in crop production by successful municipalities to automate irrigation of lawns, golf courses, etc., and subsequently to create an automated process control system for irrigation systems in agricultural production.

The indicators of the Euro-Atlantic countries leading in terms of economic development demonstrate an enviable stability. Their place in the ratings changes only slightly and they, invariably, confirm their status as world leaders. Currently, in the countries of Western Europe, there are two quite distinct trends in the formation of innovative technologies: on the one hand, self-reliance, and the integration of the scientific and innovative potential of individual countries into a single European research space, on the other. Applied research is supported by both state grants and projects carried out at the expense of funds and in the interests of large corporations. The contribution of regional authorities, in which interests innovative projects are being developed, is also significant.

An analysis by the University of Nebraska-Lincoln established the ongoing interest of farmers in precision farming technologies. The work analyzed the results of a questionnaire survey of 126 participants of the Nebraska Extension exhibitions in 2015 [13].

The report notes:

- the highest level of application of digital technologies for collecting and analyzing soil samples (98%) and high-speed Internet (94%);
- active usage of productivity control and accounting systems, farmland mapping, GPS navigation technologies (more than 80%);
- the demand for systems for the differentiated application of fertilizers (68%);
- the growing popularity of satellite and aerial photographs, the use of which is approaching 30%.

It should be recognized that there is a fairly high degree of implementation of innovative farming technologies in the world market, primarily in the American and European markets. In Europe, the priority of introducing new technologies into agricultural production belongs to Germany. China and India have achieved good results in mechanization and automation of crop production. By 2050, it is planned to increase the specified target world market to $240,000 million, which indicates the impact of innovative digital technologies on the development of agribusiness [14].

The factors that caused such global changes in the organization of agricultural production include:

- a decrease in the growth rate of the world agricultural market to 3% per year, which negatively affects the provision of food to the growing population of the planet [15];
- shortage and complete depletion of resources after extensive development of agricultural production, primarily land and water [16];
- development and availability of systems for collecting, analyzing and processing data that provide control, study and assessment of the current state of agroecosystems and forecast the consequences of management decisions, which has a positive effect on saving finances, labor costs and time [17].

The currently observed next stage of innovative development of agricultural production is caused by the processes of digitalization of the economy, marking the end of the "analog" period of agricultural production and the formation of the era of digitalization and mass automation of business processes, due to the state scientific and technical policy.

Following the example of the technology achievements of the Industry 4.0 project, we can talk about the Agrocomplex 4.0 technology revolution, which promises to ensure the productivity of farmland that exceeds the maximum results from the introduction of mechanization, the use of herbicides and the use of genetically modified seeds. The basis of the functional structure of this complex is "intelligent (smart) agricultural technologies" integrated as part of AIoT projects (Agricultural Internet of Things), which unite objects via the Internet to receive and exchange information from built-in services [18].

The goal of the AIoT project is to automate the entire set of agro-technological operations for the cultivation of agrocenoses. The constant components of the project include AIoT platforms [19], which unite a group of technologies that are used as a basis for creating a specific and specialized system of digital interaction for the development of industry web applications [20]. The platform software complex ensures the operation of remote devices for receiving data and assigning control actions, regulates the receipt and storage of data, as well as their information security.

There are clearly not enough examples of effective use of the full-fledged capabilities of intelligent precision technologies in domestic agricultural production. Only sufficiently large enterprises can carry out a full-fledged re-equipment of production. Small-scale producers, which make up the bulk of economic entities in agricultural production, lack both free funds and strategic thinking due to their local integration into a single global production process of the agro-industrial complex [21].

According to the estimates of the Grand View Research group, in 2017 the value of the intelligent agricultural production market in the Russian Federation was $221,800K, which slightly exceeded 1.2% of the value of a similar market in the world. Among the domestic organizations involved in the implementation of specialized software on the IoT platform, Rightech and JSC "Component" stand out which effectively transform the indicators received from different sensors [22]. Rigtech and kSense IoT platforms automate the monitoring of road transport and agricultural machinery, storage and processing of agricultural products, monitoring the state of agrocenoses and managing livestock.

The developments of ExactFarming company are focused on managing the productivity of phytocenoses and the profitability of agricultural production using the ExactFarming online service/mobile application. The Smart4agro program complex from Alan IT implements a choice of control actions in the field of agricultural production, monitoring the state of farmland based on a cloud service with geoinformation and analytical support. IoT design in the Russian Federation is also carried out by Strizh Telematics and LEIS LLC.

The world leader in the development of digital technologies for transport and agricultural production is held by Iteris Inc (USA), which provides consulting support on weather forecasting, the use of irrigation water, the state of soil fertility and phytocenosis, implementing software and hardware tools for the precision management platform of agricultural production of its own original design and mobile application ClearAg.

Leading participants in the smart farming market abroad are also companies of the United States (John Deere, Trimble, Raven Industries, AGCO Corporation, Ag Leader, Autonomous Solutions), the Netherlands (CNH Industria), Germany (CLAAS) and Canada (Farmers Edge).

3 Conclusion

Thus, the transition to intelligent agricultural production is becoming inevitable, although it is proceeding more slowly than it should be. This makes it obvious the need for the timely development of the concept of a digital platform for a reclamation water management complex, providing services for solving the main production problems in the field of agricultural reclamation.

References

1. N. Yusupbekov, F. Adilov, F. Ergashev *Journal of Automation, Mobile Robotics & Intelligent Systems* **11(3)** 53-57 (2017)
2. E. V. Kuznetsov, A. E. Khadzhidi, Y. A. Poltorak, M. Kuznetsova *EurAsian J. of BioSciences* **13(1)** 315-321 (2019)
3. J. Carr *Publishing house Secret of the firm* (2005)
4. D. V. Kozlov *Water Resources,* **28(2)** 215-219 (2001)
5. M. A. Bandurin, I. F. Yurchenko, I. P. Bandurina *International Multi-Conference on Industrial Engineering and Modern Technologies, FarEastCon 2019*, 8933970 (2019)
6. I. V. Bogachev *Forum Cnews Transcript of the speech on November 14, 2013 Moscow* (2013)
7. V. I. Olgarenko, G. V. Olgarenko, I. V. Olgarenko *Melioration and water management* **6** 8-11 (2013)
8. T. Safronova, S. Vladimirov, I. Prikhodko, A. Sergeyev *E3S Web of Conferences* **210** 05014 (2020)
9. D. A. Markov *Izvestiya of the Russian State Pedagogical University. A. I. Herzen* **12(90)** (2009)
10. F. K. Abdrazakov, S. S. Orlova, T. A. Pankova, E. N. Mirkina, O. V. Mikheeva *Journal of Interdisciplinary Research* **8(1)** 154-161 (2018)
11. L. V. Kireicheva, N.P. Karpenko *Eurasian soil science* **48(5)** 524-532 (2015)
12. Report "Digital Russia: New Reality" (2017)
13. M. Porter Competition *M: Ed. House "Williams"* (2003)
14. S. Vladimirov, I. Prikhodko, T. Safronova, E. Chebanova *E3S Web of Conferences* **175** 12010 (2020)
15. V. I. Olgarenko, I. V. Olgarenko, V. I. Olgarenko *IOP Conference Series: Materials Science and Engineering,* **698(2)** (2019)
16. L. V. Kireicheva, O. A. Zakharova *Eurasian Soil Science,* **35(9)** 990–995 (2002)
17. B. P. Chesnokov, O. V. Naumova, V. A. Strelnikov, F. K. Abdrazakov, B. A. Tronin *International Journal of Applied Engineering Research,* **11** 2140-2144 (2016)
18. M. A. Bandurin, V. A. Volosukhin, A. V. Mikheev, Y. V. Volosukhin, I. P. Bandurina, *IOP Conference Series: Materials Science and Engineering,* **327(2)**, 022010 (2018)
19. T. Jaegers *Eurostat Statistics in focus* (2013)
20. Vater Hendrick, Bail Elena. *Haufe-Lexware mbH&Co* (2013)
21. P. Watson, N. Chileshe, D. Maslow A *Construction Industry Development 2nd Postgraduate Conference –Cape Town South Africa* 129-141 (2004)
22. M. A. Bandurin, V. A. Volosukhin, A. V. Mikheev, Y. V. Volosukhin, I. P. Bandurina *International conference on mechanical engineering, automation and control systems* 327 (2017)

Social orientations as a factor in the accelerating differentiation of the information society

Mikhail Ermakov*, *Natalia* Okonskaya, and *Vasiliy* Stegniy

Perm National Research Polytechnic University, 29, Komsomolskiy prospect, 614000, Perm, Russia

Abstract. In this article, a sociological analysis of the "social orientations" category is presented as a methodological principle of separation of external (social attitudes) and internal (value orientations) incentives of communication. External incentives, drawing a person into technological dependence on the NBIC-technologies used, cause a break in humanitarian relations, hinder the development of self-identity, and lead to a decline in social differentiation. In opposite, human capital, as the main element of the novelty of production relations in the information age, enters into a deep hidden contradiction with the power of management and the alienation of the individual abilities and needs. Subjective factors, expressed by the human value orientations, in the conditions of new opportunities opened up by informatization and digitalization of all social institutions, receive global advantages for maintaining harmony and progress in the society development, despite the objective instability of irreconcilable contradictions between large social groups. For empirical identification of qualitatively new subjective factors of social differentiation multiplication, the author's typology of personalities is used, which unites actors into typological groups distinguished by social orientations. Terminal type, distinguished by the dominance of terminal (ultimate semantic) values and motives in the personality structure; inert type, the core of the "information poor"; instrumental type, occupying an intermediate position. Primary and secondary research made it possible to predict new opportunities for using the institution of education, in addition to the objective potentials of society self-development in the case of multidirectional differentiation.

1 Introduction

An incorrectly selected methodology results in the research made in a theoretically undetectable area, and any hypothesis, remaining verifiable, becomes unprovable. Differentiation of subjects is a complex process, and due to its multidimensionality, this concept has received opposite interpretations. In those theories where the study of social ***experience*** prevails, differentiation is considered in its negative aspect, since the multidirectional activity that forms the basis of any experience corresponds to the loss of the collective effect, destroying coherence and effectiveness of any social institution. On the contrary, differentiation acts as a fundamental law that guarantees stability in those studies that focus on fundamental forecasts of changes in negative situations, in order to find ways to resolve deep social contradictions. Consistency and organization are the foundations of differentiation and adaptation of living beings. This biological fundamental law acquires a new systemic complexity in the information society. "There is a need to introduce synthetic concepts that combine the sides of the dichotomy: deterministic chaos (ordered disorder; necessary, natural randomness), self-organization (randomly arising order); fractality (discrete continuity, integral part) … It leads to the idea of the feedbacks existence in the system … in case of complex systems" [1, p. 31], to which the social human-sized system undoubtedly belongs.

This general specificity of the complex systems development, derived on the basis of general scientific disciplines (synergetics, in particular) in their unity with ontology, that is, a philosophical picture of the world, made it possible to abandon the one-sided technicist methodology that was popular for researchers of global problems in the middle of the 20th century and outdated by the 21st century. The authors of the article hypothetically conclude that a society without increasing differentiation (segregation of individuals into isolated groups that destroy the objective processes of the social labor division, social stratification, as well as the inward processes of the individual all-round abilities development) is a degrading society. Whatever research methodology we use, the fundamental approach to progress as a complication based on the open systems self-development will remain a development criterion, while the system closedness and its simplification is a model of collapsing integrity.

It is much more effective to calculate the risks of social entropy, which, according to Forrester, "is a social movement turned into a disorderly (heat) form" [2, p. 361], considering the leveling of the main internal personal potentials of participants in social processes. Without denying the growing danger of wasting human abilities in the event of their disunity, political or civil

* Corresponding author: mikhail_ermakov@mail.ru

confrontation caused by a "manifested identity" [2, p. 368-369], we insist that the main reason for the decline in social potential is the hidden growth of animal biological motives of human behavior in their technical reformatting. This is how the man's alienation from his abilities, discovered by the classics of Marxism, manifests itself today through the inability to control the results of his labor.

2 Researching the problem

- Differentiation is a universal characteristic, since any activity has its own definite and stable carrier. As there is no energy outside the carrier, there are also no social relations outside the person and social institutions.
- If a person does not guarantee the stability of his social space-time boundaries, then external environmental influences, often negatively directed towards the interests of an individual person, become a guarantee.
- The unification of people in various social institutions is a normal process of increasing social differentiation, since the person essence is the complex of all social relations.
- In all cases of participation in social institutions, a person as a representative of a collective subject can be a carrier of communicative, strategic, normative or dramatic action according to the typology of J. Habermas [3]. At the same time, instrumentally rational, value-rational, affective or traditional actions can be distinguished in human behavior, if we use the typology of social actions by M. Weber [4].
- The increase in differentiation, motivated from within, coincides with communicative activity, partly intersecting with the instrumentally rational activity, partly with the value-rational activity. For the sake of simplicity, let us dwell on the communicative type of social action, and then the problem of the meaning of activity, or social orientations, arises most acutely.
- Social orientations are an element of a bunch of concepts such as social attitudes (an externally determined element of consciousness of a collective subject, actor) and value orientations [5]. In contrast to meaningful value orientations, social attitudes are multiple, superficial, do not contain unity, and easily fit into any rational-technological environment, since they combine a method of achieving goals and a means of achieving goals. The destruction of moral values, or the internal way of self-development, leads to the washing out of the individual person role from all social institutions, leading to a fall in social differentiation, to the destruction of the complexity and social potential of society, replacing it with the biological potential of adaptation, survival, in adaptation to the NBIC environment requirements.

If in the days of early capitalism the only social group that gained in relation to the majority were the owners of the means of production, today, a practical benefit is technology, an artificial environment, which, due to the relitivization of social relations, blocks the growth potential of human capital as the main element of the information age novelty. Schumpeter [6] called this social problem the power of management, working in conjunction with big business and an alienated labor force. Modern society is technocratic and exists due to the spread of false consciousness. A social system with a predominance of a technical and economic basis instead of human capital guarantees the collective subject integrity even in critical situations, protecting against painful crises of spirituality by symbolic replacement of knowledge potential with the wealth of the information environment. However, crises of spirituality indicate the presence of spirituality! The gain in the rationalized use of the means of labor occurs due to the sacrifice of the value orientations in favor of communicative activity. The losers are small and medium-sized businesses, which, ideally, are employers for 80% of participants in social production. "The weight of GDP is still produced by modern capitalist employers, and judging human capital in terms of the employees' number would be an excessive expansion of the "human capital" concept. We offer a criterion for such participation of an employee in the profit production of small or medium-sized businesses, when the employer pays a share of the profit as a monetary reward to a particularly valuable employee, exceeding the amount of the cost of his labor. If an employee receives only wages, he cannot be considered human capital, since in order to provide himself with the means of life, he gave in hiring his capabilities and his needs up to the alienation of labor" [7, p. 102]. With the breakdown of humanitarian relations caused by technization, the abundance of material and cultural resources seemingly endless and durable may disappear at any moment, as the philosopher Jose Ortega y Gasset warned in his famous work "Rise of the Masses": "easily attainable benefits are based on certain and not easily attainable human qualities, the slightest shortage of which will immediately scatter a magnificent structure to ashes" [8, p. 319].

Today, not only Russia, but the whole world is in just such a situation predicted by the Spanish philosopher. The economy has suffered in many ways since the massive lockdown, but the main one of this ways is the wage earner. According to qualified forecasts, "in Russia at the end of 2020 the decline in GDP may exceed 5%, and the number of the poor may rise from 18 million to 20-21 million people. In these conditions, all the problems and limitations of development are aggravated» [9, p. 18].

At this moment of irreversible and dangerous loss of well-being, the subject participating in the collective subject ceases to critically perceive the substitution of concepts, moving further and further from control over social interests in his inner world. Just as the industrial society was perceived by sociologists of the Frankfurt School as the greatest danger of personal leveling, so the information society demonstrates this trend even more clearly, since the digital world has multiplied the alienation of artificially created information from user-subjects. The essence of these dangers is when the possible wealth, the absolute well-being of society is accompanied not only by environmental and military crises, but begins to balance on the brink of self-destruction at the intersection of all global crises - the point of loss of self-identification by a person as a result of the replacement of human capital with symbolic capital. This is a communication activity devoted to NBIC-technologies. Instead of the contractual practices of a subject with a subject (collective or individual), when the main instrument of agreement became language, which is important for reaching agreement with all participants to achieve a common goal, a technological deceptive, symbolic artificial world, a world without a subject, absorbing a person, pays him compensation in the form wages, and it does not matter whether it is large (for managers) or small. If in the process of cognition as internal differentiation "the subject has not mastered certain areas of complex system knowledge, using the terminology of Academician V.S. Stepin - human-sized (humanitarian, legal, moral, philosophical and sociological, etc.), then the result may be an irreversible inability to see systemic objects associated with the problems of man and society. So, moral problems, even voiced and spoken, may not "hurt a living"; Forgetting the history of one's native country also refers to this phenomenon of spiritual blindness caused by the lack of fundamental humanitarian knowledge. The atavistic features of rationality are rooted precisely in everyday experience, when opinions replace knowledge, while relativism, subjectivism are in demand in all areas of human practice, except for the narrowly professional technological sphere, where the leading role is played not by the subject himself, but by new technologies" [10, p. 21-22].

3 Results and discussion

With the threat of unemployment, the presence of a paid job makes a person adaptively, affectively and rationally subordinate to external regulators-differentials. As a result, differentiation falls by exactly as much as the value-rational sphere of activity in people's behavior has decreased. We can trace this data on the basis of sociological research.

The possibilities of a collective entity (state, political institutions, religious organizations, trade unions) to standardize and normalize behavior, on the one hand, and knowledge, on the other, are sharply limited.

Firstly, because of the depopulation, when the projected indicators of an increase in the birth rate, an increase in average life expectancy, and a decrease in mortality in the field of cardiovascular and oncological pathologies diverge from reality in a depressingly significant way. In 2017, a full-fledged depopulation began, when the number of deaths exceeded the number of births by 136 000 people [11, p. 6].

Secondly, the planned indicators of economic growth (as, for example, a serious upward spurt of the economy with an increase in GDP per capita by 50% provided for in the Decree of the President of the Russian Federation of May 7, 2018 "About national goals and strategic objectives of the Russian Federation development for the period up to 2024") are not implemented or are partially implemented [11, p. 5].

Thirdly, the subjective area of relations also needs to be taken into account due to the availability of the digital sphere of communication. Of particular note is the cognitive functions of social media and media. Even education is becoming increasingly dependent on the flow and directions of the digital environment created with the help of NBIC technologies. The social environment is in a dissociated, disordered state.

The abstractness of the digital environment, used as a source of information to replenish the cognitive functions of a person, does not contain unity by definition: the abstractness of concepts, words, natural human language is an irreplaceable partial!

Social orientations of the modern personality are multifaceted and contradictory. As we assume, one of the main factors leading to the inconsistency of the personality social orientations is that some of the instrumental and avoidant values and motives are blocked by the personality as socially disapproved, but at the same time their latent influence on the personality orientations remains [12, p. 116]. The most active subjects of economic, political and socio-cultural institutions, which we unite through the concept of "intellectual owners", can resist this.

"A person endowed with intellectual property is protected more than a person endowed with a thing (where one's own personal participation is accidental, outwardly). He is protected not only from material distress, but also from the interference of destructively oriented people. Deception, envy, and competition can only be overcome by a constructively oriented organizer of his own life. It is assumed that an exploited person is brought to the dangerous line of losing himself, may lose the meaning of life" [13, p. 64]. Meanwhile, "when a creative person - an intellectual owner - is included in any social structure - a family, a professional team, a party, an informal group - their mutual enrichment and transformation occurs, which can lead to the emergence of new intellectual owners, personalities" [13, p. 63]. And it doesn't matter whether large capitalist property becomes a bulwark for the development of an intellectual property owner, or the employer is a small businessman, or the state.

The significant potential for self-development, self-renewal of the Russian intellectual community is growing in the social institution of the self-employed, which contributes to the private property development in Russia on the basis of the so-called "Smallest entrepreneurship" [14, p. 822]. This definition, which is neither basic nor generally accepted, nevertheless emphasizes the differentiating power of this practice, since it makes it possible to engage in several types of economic activities with the help of one's own means of production at once, while often not even breaking with previous work as employees. Since the institution of the self-employed combines practical and the cognitive spheres of differentiating activity, we assume this phenomenon is an optimal soft way of correcting the latent influence of the technological blocking of the individual self-awareness.

The main advantage of the self-employed is the ability to work for oneself, freedom, the ability to independently make decisions about employment and time [14, c 830]. More than 330 thousand people registered as self-employed in Russia [14, p. 833].

The institute of education can promote the development of the self-employed and the development of intellectual property owners among the so-called the workforce in its industrial social form, merging in function with the new high technology, but in a modernized, individualizing version, bypassing test practice, leveling the deep differentiation of knowledge. "Professional knowledge acquired in universities and colleges becomes obsolete within 2–3 years of work. In this regard, investment in education should be a necessary element of the labor strategy" [15, p. 263]. This strategy is essentially based on the law of value, since profit growth is guaranteed where a larger share of capital is concentrated in living labor V in comparison with materialized labor C. Investing in education increases the share of V without attracting additional labor, since human knowledge capital through educational practices is steadily and significantly growing. The social gain in educational investments, which is also manifested by a differentiating effect in labor functions, is also large. Numerous risks associated with the processes of the market economy formation, which include social insecurity, unemployment, the risk of losing social and labor status, go away precisely because the employer chooses high professionals who are ready to regularly change and improve their qualifications. "Analysis of Russian and foreign educational statistics indicators shows that Russia is one of the most highly educated countries in the world. Thus, 96% of Russians under the age of 64 have tertiary education and about half of Russians have secondary and higher professional education. In addition, 94% of Russian workers have not lower than complete secondary education, which is significantly higher than the average for the economically developed OECD countries, where it is 75%. According to this indicator, Russia is more than twice ahead of Brazil and more than four times ahead of China" [15, p. 267]. These indicators, which do honor to a Russian who holds a high level for preserving the best traditions of the Soviet economy era, can explain the high resistance to social factors of instability.

We can talk about the presence of the social orientations influence on numerous areas of personality activity, based on the results of empirical research. At the same time, we note that the representations organized through the individual social orientations and the corresponding individual behavior models of the person as a subject of economic, political and sociocultural subsystems are not chaotic and multidirectional. On the contrary, there are regular connections between them. As a result, the constructed "picture of the world" of a person, which is based on his social orientations, determines the position of this person in the global social space. Thus, the social orientations of the individual act as the system-forming criterion of social stratification in the conditions of the emerging information society (a criterion of a higher order than the traditional ones: income, power, education and prestige), setting new boundaries of social differentiation.

Basing on the materials of the author's research from 2012 to 2016 ("Social representations of the personality about the information society (on the example of PNRPU)" (2012 and 2016, N=1246, Perm); "The status of youth in modern Russian society" (2015-2016, N=2238, Perm)) by the example of students it can be stated that typological groups, distinguished by social orientations, indicate the outlines of social differentiation of the emerging information society. The most harmonious type, corresponding to the characteristic changes in the institutional system, is the terminal type, distinguished by the dominance of terminal (final) values and motives, labor and personal, in the structure of the personality. Representatives of this type, to a greater extent than representatives of other typological groups, are characterized by an orientation toward education as a basic value, a desire for self-education and self-development throughout life, as well as the possibility of creative work.

In addition, representatives of the terminal type are more aware of the information society at the level of self-esteem as well as in the analysis of the characteristics of the society being formed that they assess. Representatives of this type more often emphasized the importance of education in the organization of entrepreneurship, higher assessing the share of enterprise costs for high technologies and the importance of regular advanced training of employees as the main way to achieve a competitive advantage. Representatives of the terminal type are much more likely point to the prospect of increasing the labor and specialization division level, seeing the future in highly specialized professions, which also corresponds to the existing dynamics of changes in the labor market, shows the level of awareness and adaptability of representatives of the type. A similar situation is observed when assessing changes in the social structure of society, where representatives of the terminal type more often note the prospect of the structural organization of society around the mass "middle class". At the same time, the view of the social and labor sphere among representatives of the terminal type is more personality-oriented than among representatives of other groups of typology, they believe that new technologies will only facilitate human labor, without a person they are useless. We assume that in this position it is possible to trace the appeal of the terminal type representatives to creative work, which is the basic element of the system of labor values for this type, and this work cannot be automated even with the development of technology. When assessing the prestige of labor activity, the terminal type, less often than other groups of typology, indicates the importance of the wages level and the prospects for career growth.

The inert type, on the contrary, can be assessed as the least adapted to changes in society, risking, in the conditions of growing social differentiation, being "out of work", forming the core of the "information poor". This type is characterized by an "avoidance" orientation, a tendency to conformism, high inertia and an almost complete lack of orientation towards self-development. The key goal for this type is to minimize risks.

Representatives of this type can be assessed as the least aware of the information society and the institutional changes that accompany its formation. Moreover, this assessment is valid both when analyzing the answers of representatives of this typological group about various aspects of the functioning of society and changes in it, and on the basis of their self-esteem.

Describing this typological group, we also note that its representatives are more inclined to regard advertising as the main expense item of an entrepreneur. This indicates an increased attention to this information channel and, therefore, dependence on external channels of influence. The inert type, more often than other groups of typology, considers the acquisition of patents and other intellectual property as a way to achieve a competitive advantage, and also the inert type is more inclined to predict the replacement of hard human labor with a "machine". When assessing the prestige of labor, representatives of the inert type more than other types note the importance of high wages, additional benefits and the managerial nature of activities. Significant is "the opportunity to work under the guidance of a professionally competent leader ...". These assessments present a projection of the characteristics of an inert type: conformity, lack of inclination to creative and innovative activity, focus on labor activity with minimizing costs. The assessments of the inert type representatives are quite reactionary; they ignore the "weight" of a qualified worker in the production of an information-intensive product and the growing demands for the level of training of an employee and his education in a transforming society. In addition, the inert type is rather pessimistic about the prospect of changing the social structure of society, anticipating an increased conflict.

When assessing the complementarity of typological accompanying the formation of the information society, we can note clear differences. According to this criterion, the terminal and inert types act as a kind of antagonists in relation to each other, where the terminal type is the most optimal, corresponding to the requirements of the new society for the individual and the worker. The instrumental type occupies an intermediate position. Despite the alienated nature of labor inherent in this type, the representatives of this group demonstrate a sufficient level of awareness, highlighting and correctly assessing the changes inherent in the transitional stage in various social institutions. In addition, the assessments of instrumental type representatives in most cases correspond to the median distribution, it can be interpreted as evidence of the presence of instrumental orientations in modern student youth, albeit in a latent form, which is a legacy of the industrial period. Representatives of the instrumental type consider various forms of activity, work in general and higher education, in particular, through the prism of instrumental motivation and needs.

Taking into account the studied patterns of the social orientations influence on the representations and behavior of the individual, the correspondence to institutional changes, it turns out to be possible to outline the foundations of social differentiation of the emerging information society. As a result, the use of the author's typology makes it possible to analyze adaptation for various social subjects or even entire countries. Thus, the results of approbation of the methodological approach to the secondary analysis of the European Social Survey data, a cross-cultural study of attitudes, values and behavior of the European countries population, look extremely interesting.

The methodology of the European Social Research, in conjunction with the author's typology, makes it possible to distinguish in the studied population groups belonging to the instrumental, terminal and avoidant / inert types and also to expand the typology, highlighting

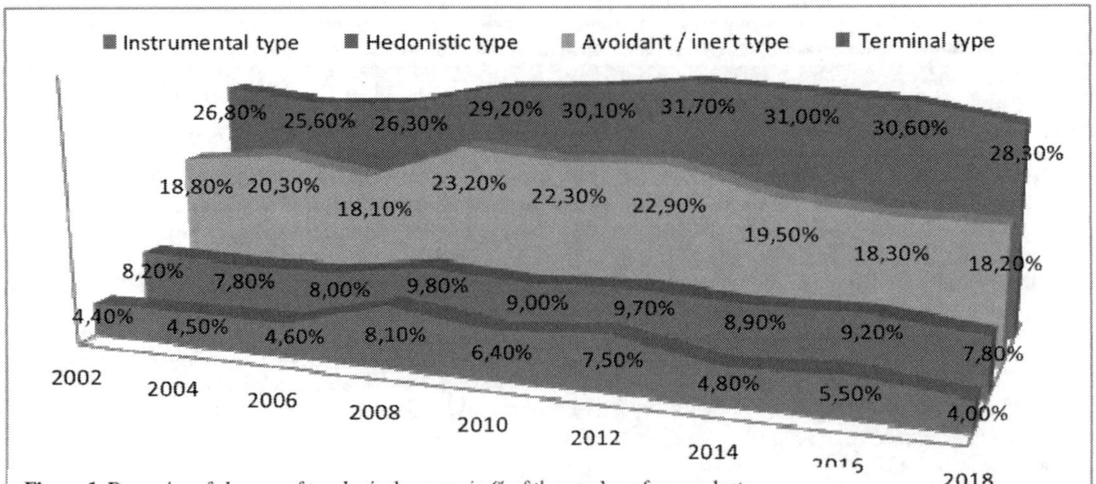

Figure 1. Dynamics of changes of typological groups, in % of the number of respondents.

groups which is constructed on the basis of prevailing social orientations to social transformations also 4 type - hedonistic, the basis of orientation for

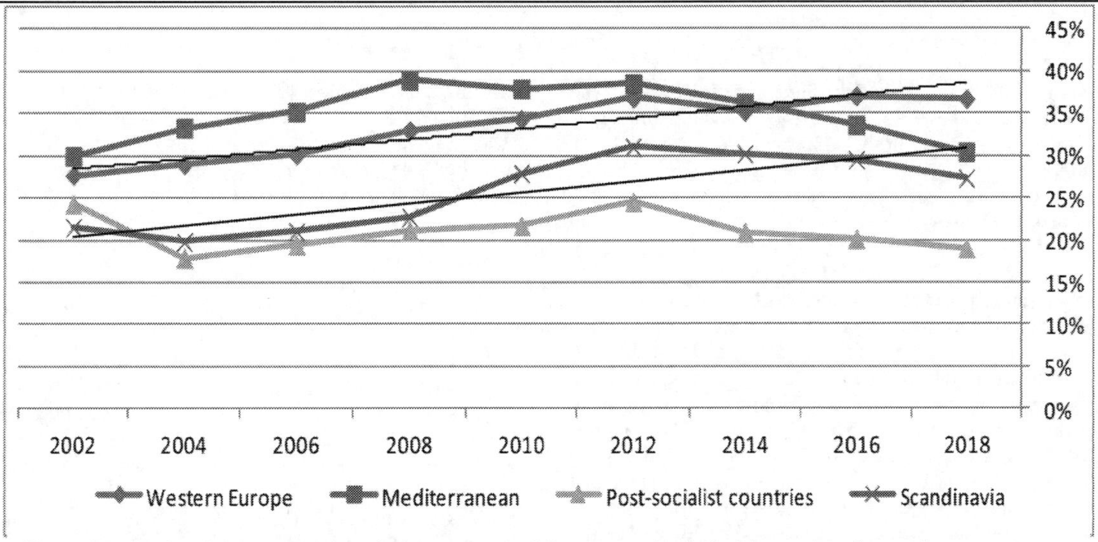

Figure 2. Dynamics of changes of the terminal type group in different countries, in% of the number of respondents.

which is the desire for pleasure and attention to the characteristics of the pastime.

The study involved 38 countries, the total sample consisted of more than 425,000 respondents. According to the analysis of data from nine waves of research (see Figure 1) from 2002 to 2018, it can be seen that the volume of typological groups remains practically unchanged. Fluctuations for the entire period do not exceed 5%. The most representative are the groups of the terminal type (on average 28.8% of the number of respondents) and the avoidant / inert type (20.2% on average). However, if we differentiate the survey data by country, the difference in the representativeness of typological groups will be clearly traced. The countries participating in the project were grouped:

Western Europe: Austria, Belgium, Germany, Ireland, Luxembourg, Netherlands, Portugal, Switzerland, United Kingdom;

Post-socialist countries: Bulgaria, Czechia, Estonia, Hungary, Lithuania, Latvia, Poland, Romania, Russian Federation, Serbia, Slovakia, Ukraine, Kosovo;

Scandinavia: Denmark, Finland, Iceland, Norway, Sweden.

Let us compare the dynamics of change in the representativeness of the terminal type group (see Figure 2) between the groups of countries for the period from 2002 to 2018. This typological group has a different number in different countries; the most representative is the terminal type in the countries of Western Europe and the Mediterranean (36.7% and 30.4% in 2018, respectively). There is also a clear trend in the growth of the number of the terminal type group in Western Europe and Scandinavia. In the Mediterranean countries, despite the peak periods (from 2006 to 2014), when the share of the terminal type was the largest among all countries, there was a further decline and

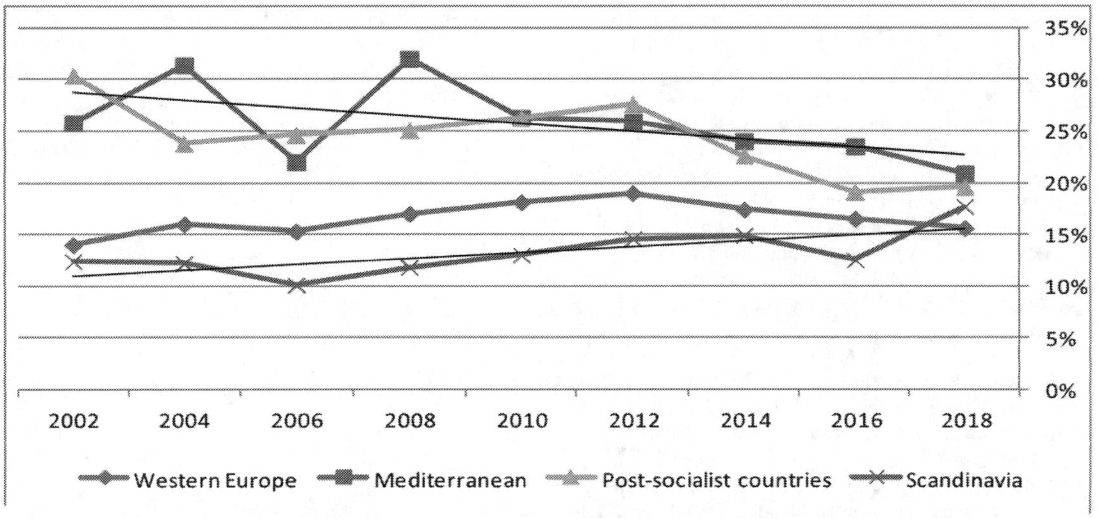

Figure 3. Dynamics of changes of the avoidant / inert type group in different countries, in% of the number of respondents.

Mediterranean: Albania, Cyprus, Spain, France, Greece, Croatia, Israel, Italy, Slovenia, Turkey;

return to the initial values in 2018. In post-socialist countries, the share of the terminal type is the smallest among all countries, while there is a downward trend since 2002. (24.3%) to 2018 (19%). A similar situation is observed in Russia from 18.7% in 2006 up to 16.3% in 2018.

Comparison of the dynamics of change in the representativeness of the avoidant / inert type group (see Figure 3) shows a different picture. This group of typology is most represented in the post-socialist countries and the Mediterranean countries. As the share of this group decreased from 30.4% and 25.8% in 2002 up to 19.6% and 20.9% in 2018 accordingly, as a result, the share of the avoidant / inert type group in 2018 became almost the same in all countries (the spread of values is in the range of 5%). However, in the countries of Western Europe and Scandinavia, there is an upward trend, and its further prolongation allows us to judge that in the future 6-8 years, the proportion of representatives of the avoidant / inert type in these countries will be the largest in comparison with the rest of the countries participating in the study.

4 Conclusion

In the information society, the potential of the cognitive power of all participants in social production is growing unprecedentedly, including the participation of subjects in social institutions of the superstructure, from politics to art, religion and morality. As a result, the possibility of overcoming dangerous crises internally regularized dangerous crises overcoming grows, which will receive an external manifestation with a great delay. This is due to the fact that systemic identity is different from individual identity ideologically leveled by political institutions in the interests of the civil world. Since the complexity potential is coupled with unity through value-rational intersubjectivity, then all the risks associated with the poverty line, environmental, civil, industrial crises can be fixed and overcome by the increasing complexity of a differentiated, internally integrated society. The main risk for the social sphere in any country is the loss of self-identity, which is much more dangerous than any contradictory diversity. It is the loss of self-identification that can explain the high prevalence of the avoidant / inert type among other personality types in all countries.

In our opinion, education as a social institution, built on the principle of differentiating the ability to work, communication and thinking, is the main source of the preservation of the objective tendency of society towards self-development. The former power of the USSR high-tech industries in the field of space, mechanical engineering, energy and other high-tech industries was the embodiment of such approach to education. The resource orientation prevailing in the world today is a consequence of the disorganization of creative thinking abilities due to the rupture of humanitarian relations and the redundancy of high technologies. Today, between the digital society, on the one hand, and resource orientation, on the other, it is necessary to create a support for the meaning of the activity of any social orientation, to prevent social entropy, the collapse of normal, institutionally traditionally and objectively verified conditions. Education, which has not free from the caste system in all previous epochs, today receives a new megatrend of restoration of the person semantic ego-identity lost in the process of technization.

References

1. V.V. Afanas'eva, N.S. Anisimov, *Questions of philosophy*, **8**, 28-41 (2015) [In Rus.]
2. D. Forrester, *World dynamics*. (M.: AST; SPb.: Terra Fantastica, 2003) [In Rus.]
3. J. Habermas, *Moral awareness and communicative action*. (SPb.: Nauka, 2000) [In Rus.]
4. M. Weber, *Selected works*. (Moscow: Progress, 1990) [In Rus.]
5. V.N. Stegnij, *PNRPU Bulletin. Socio-economic sciences*, **2**, 8-17 (2017) [In Rus.]
6. J. Schumpeter, *Economic development theory*. (Moscow: Progress, 1982) [In Rus.]
7. N.K. Okonskaja, *Philosophy and Culture of the Information Society*, 101-103 (2019) [In Rus.]
8. J. Ortega y Gasset. *Aesthetics. Philosophy of culture*. (M.: Iskusstvo, 1991) [In Rus.]
9. A.G. Aganbegjan, A.N. Klepach, B.N. Porfir'ev, M.N. Uzjakov, A.A. Shirov *Forecasting Problems*, **6(183)**, 18-26 (2020) [In Rus.]
10. N.K. Okonskaja, M.A. Ermakov, O.A. Reznik, *Philosophy and Humanities in the Information Society*, **4(14)**, 14-23 (2016) [In Rus.]
11. A.G. Aganbegjan, Population, **21(4)**, 4-13 (2018) [In Rus.]
12. M.A. Ermakov *Social orientations and representations of student youth about the information society*. (Perm', 2008) [In Rus.]
13. N.K. Okonskaja. *Intellectual property in the information age, sociogenesis and development prospect*. (Perm': PNIPU, 2018) [In Rus.]
14. Z.T. Golenkova, Ju.V. Goliusova, T.I. Gorina. Vestnik *RUDN Bulletin. Sociology*, **20(4)**, 821-836 (2020) [In Rus.]
15. Z.T. Golenkova, G.B. Kosharnaja, V.P. Kosharnyj. *Education Integration*, **22(2)**, 262-273 (2018) [In Rus.]

Organizational and economic aspects of innovation management in agro-industrial enterprises

Vladislav Anichin*, *Dmitry* Chugay, *Galina* Khudobina, and *Natalia* Yakovenko

Belgorod State Agrarian University named after V. Gorin, 1, Vavilova Street, Mayskiy, Belgorod region, 308503, Russia

Abstract. The economic aspects of innovation activity are determined by the fact that innovations serve to achieve the economic goals of the enterprise, act as a means of competition and require the use of economic resources. When organizing innovation activities, it is important to take into account macroeconomic factors and intra-company conditions. The authors identify three large-scale macroeconomic factors that have a significant impact on the chain of interrelated innovations in the activities of modern Russian agro-industrial enterprises. At present, the factor of creating institutions of the digital economy is beginning to take effect, in connection with which the third wave of organizational, technological and marketing innovations and the next growth of agro-industrial production in Russia are expected. Due to the different rates of economic concentration in the three areas of the agro-industrial complex, historically, unfair vertical competition has been and remains one of the limitations for innovation at agro-industrial enterprises. In the digital economy, information about the essential conditions of concluded transactions becomes available to antimonopoly authorities in real time, which allows them to quickly develop and apply regulatory actions in cases of violations of antimonopoly legislation or in cases of "failures" of the market. It is also possible to create institutions that support fair competition in the vertical and horizontal dimensions. The opportunities that open up actualize the improvement of the conditions for innovation activity at the intra-company level.

1 Introduction

The economic nature of innovation lies in the desire and possibility of obtaining a significant effect from various types of innovations initiated by entrepreneurs, managers and innovators. The intensity of innovation activity at the enterprise level is influenced by two constant factors: the competitive status of the firm and the availability of investment resources. The greatest intensity of innovation activity takes place in those organizations that are simultaneously in a highly competitive environment and have the resources to introduce innovations. Many agro-industrial enterprises belong to such organizations in modern Russia.

Competition encourages the search for new technological, organizational and marketing solutions, and investment resources allow them to be implemented. A balanced combination of motivations and opportunities is ideal. Of course, in addition to investment, opportunities for innovation are also determined by relevant information [1] and the availability of qualified and proactive personnel who are able to justify and implement effective innovations [3].

Macroeconomic conditions, including state and regional policies, play an important role in the activation of innovative activities of agro-industrial enterprises [4, 10]. At the micro level, the organizational and economic mechanism created and regulated by the management of agro-industrial enterprises is of crucial importance [6, 8, 9].

The link between macroeconomic factors and intra-firm conditions of innovation activity is the economic relations between related enterprises in the value chains, for example, between a beet-growing organization and a sugar factory, between a sugar factory and a trading enterprise. The development of these relations acts as a kind of business background that allows economic entities to implement external and internal prerequisites for innovation. Conversely, the low level of interaction, the lack of prospects for mutually beneficial cooperation can serve as a bottleneck that restricts innovation activity.

At the same time, you should not ignore the individual characteristics of people who develop and implement innovative solutions. Entrepreneurs, managers, and ordinary performers are not always guided by the motives of profit in their behavior. Moral and religious principles, cultural traditions, political sympathies, and individual development plans play an important role [5]. As J. Schumpeter noted on this occasion "First of all, there is a dream and the will to find a private kingdom ..., also a dynasty... Then there is the will to conquer: the impulse to fight, to prove oneself superior to others, to succeed for the sake, not of the fruits of success, but of success itself...Finally, there is the joy of creating, of getting things done, or simply of exercising one's energy and ingenuity" [7]. One way or

* Corresponding author: vladislavanichin@rambler.ru

another, innovation serves the interests of the majority of people. However, these same interests in a certain part of them may prevent the introduction of effective innovations or promote the introduction of innovations with low efficiency. Therefore, innovation management will always be an art of compromise between the possibility of obtaining current and future income, between the interests of the owners of the enterprise and the interests of management, employees and public interests.

2 Main findings of the study

2.1. Macroeconomic factors of innovation activity

In dynamics, innovation activity can be represented as a growing and branching graph, the vertices of which are certain innovations (including technological, organizational and marketing), and the edges are paired causal relationships, since any innovation inevitably generates the next one. So, at one time, the conveyor (technological innovation) caused organizational innovations in the automotive industry, which provided mass production of cars. Relatively affordable and numerous assembly line vehicles (product innovation) have catalysed new sales and service methods (marketing innovation) and new economic activities, including road construction, oil production, and refining. The rules of the road acted as an organizational innovation. The subsequent improvement of conveyor technology is largely due to the development of robotics...

Similar dynamic innovation graphs are being developed in other sectors of the economy, including the agro-industrial complex. So, after the introduction of single-sprouted varieties of sugar beet, precision seeding drills and the rejection of manual labor in the care of crops became relevant in beet production. As a result, labor productivity in the cultivation of sugar beet has increased more than 10 times.

The intensity of innovation activity of agro-industrial enterprises is greatly influenced by macroeconomic factors, some of which are macro-innovations of the organizational type.

As applied to modern Russia, the initial vertex of the first dynamic innovation graph was the introduction of the institutions of entrepreneurial activity and private ownership of the means of production (Figure 1).

This was followed by the privatization of most industrial and agricultural enterprises. New owners, as a rule, initiated technological innovations, mainly process innovations, and also began to actively introduce organizational innovations in equipping workplaces, managing production and coordinating the activities of related enterprises. Currently, large agro-industrial formations (mainly agricultural holdings) make a decisive contribution to ensuring food security in Russia. In the Belgorod region in 2019 agro-industrial formations produced 68.8% of the total volume of agricultural products.

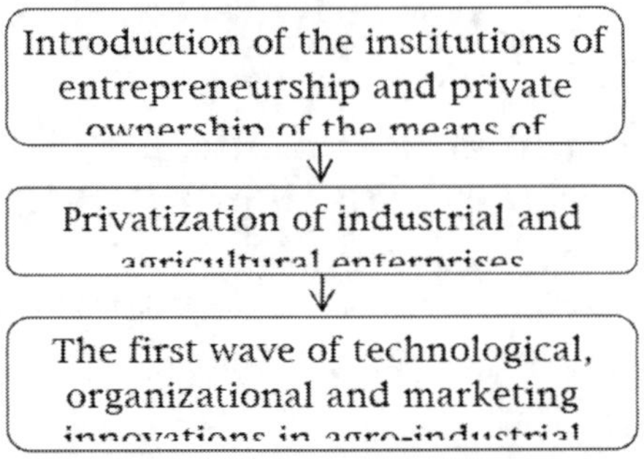

Fig. 1. The first dynamic innovation graph of the Russian agro-industrial complex

The regions of the Russian Federation, where the processes of agro-industrial integration have developed most intensively, are currently the leaders in terms of the volume of production of agri-food products in the country. For example, the Belgorod region, despite the fact that in 2019 it had only 1.79% of the sown area of Russia, produced 4.58% of gross agricultural output (the ratio of the share in gross production to the share in the total area of crops is 2.56). The region's share in the production of livestock and poultry for slaughter is 12.62%, and in the volume of net profit of agricultural organizations in Russia – 27.1%.

In recent years, an effective catalyst for innovation in the agro-industrial complex has been the embargo on food imports to Russia (forced organizational innovation in response to anti-Russian sanctions). Restrictions on the import of food products caused an increase in investment in the Russian agro-industrial complex and subsequent innovations of various types at the micro level (Figure 2).

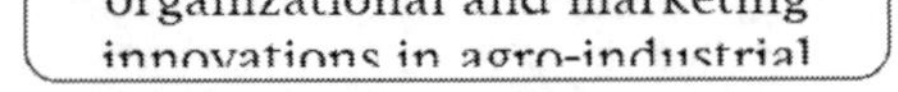

Fig. 2. The second dynamic innovation graph of the Russian agro-industrial complex

Any innovation sooner or later exhausts its potential and turns into a routine. Therefore, the next growth of agro-industrial production in Russia is expected from the creation of digital economy institutions (Figure 3).

Historically, unfair vertical competition has been and remains one of the constraints on innovation at agro-industrial enterprises. Unfair vertical competition occurs between related enterprises in value chains and is expressed either in overcharging if the seller has more market power, or in undercharging if the buyer has more power. The noted historical pattern is due to different rates of economic concentration in the areas of the agro-industrial complex (production of means of production, production of agricultural products, processing of agricultural products), as well as in the field of wholesale food trade. Producers of the means of production (such as seeds and fertilizers) and wholesale buyers of food have the greatest market power.

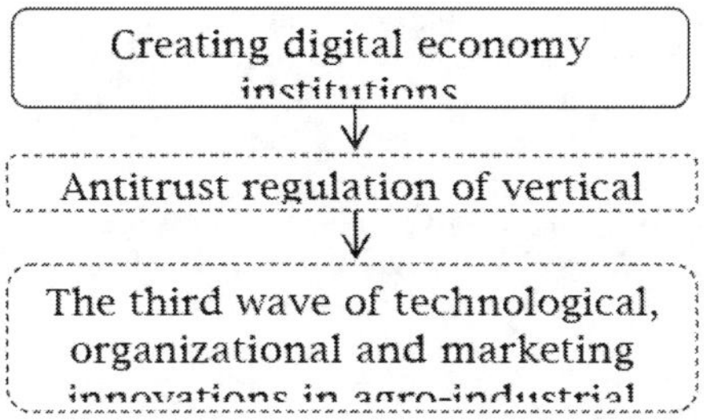

Fig. 3. The third (expected) dynamic innovation graph of the Russian agro-industrial complex

Inequality in market power is also evident in the interaction of organizationally independent agricultural and processing enterprises. The costs of unfair vertical competition here are largely offset by the creation of agro-industrial formations, in which the interaction of related enterprises in the value chains is regulated at the holding level.

One of the most important advantages of the digital economy is that information about the essential conditions of concluded transactions becomes available to antimonopoly authorities in real time, which allows them to quickly develop and apply regulatory actions in cases of violations of antimonopoly legislation or in cases of "failures" of the market. In addition, it becomes possible to create institutions that support fair competition in the vertical and horizontal dimensions.

Such an institution can be a pulling system of commodity movement [2], which ensures fair pricing, a combination of competition and centralized distribution of quotas for production and sales of products. It is quite appropriate in the beet-sugar, dairy-food and a number of other subcomplexes of the agro-industrial complex.

In the digital economy, it becomes possible to measure the market power of any economic entity that carries out official transactions. The market power of an enterprise should be evaluated comprehensively, taking into account its position in the market as a seller and buyer. To measure this phenomenon, the market power index is suitable, calculated using the formula:

$$Ip = S\,/\,FC \times C\,/\,FS$$

where S is the number of possible suppliers of raw materials or finished products for the i-th enterprise;

FC – number of buyers of raw materials or finished products from suppliers of the i-th enterprise (the total number of competitors of the i-th enterprise in the procurement market);

C – number of buyers of products from the i-th enterprise;

FS – number of sellers whose products are purchased by the buyers of the products of the i-th enterprise (the total number of competitors of the i-th enterprise in the sales market).

For example, the i-th bakery is offered to supply flour to twenty organizationally independent enterprises ($S = 20$). These same suppliers sell flour to ten organizationally independent bakeries (including the I-th bakery) and six other wholesale customers ($FC = 10 + 6 = 16$). The i-th bakery supplies its products to twelve organizationally independent retailers ($C = 12$). These retailers purchase bakery products from ten organizationally independent enterprises, including the i-th bakery ($FS = 10$).

Therefore, the market power index of the i-th bakery is equal to:

$$Ip = 20\,/\,16 \times 12\,/\,10 = 1.5.$$

For the i-th retailers (one of 12, having the opportunity to purchase 10 of bakery enterprises) that implements the bakery one thousand buyers ($C = 1000$), which can buy these products from five retailers ($FS = 5$), the index of market power is:

$$Ip = 10\,/\,12 \times 1000\,/\,5 = 166.7.$$

Such differences in the market power of related enterprises largely determine the unequal distribution of income between them, which generally creates problems for the implementation of innovation activities.

2.2 Intra-company conditions of innovation activity

In addition to macroeconomic factors, the intensity of innovation activity in agro-industrial enterprises is greatly influenced by the organizational and economic conditions formed at the micro level. Unlike macroeconomic factors, these conditions can be changed at the will and desire of the company's stakeholders.

Innovation activity in agro-industrial enterprises has two main aspects. The first is associated with the introduction of effective innovations at the initiative of the owners and management of the enterprise. The second-with the management of innovative activity of human resources available to the agro-industrial enterprise. The second aspect has two directions: 1) mass innovation creativity; 2) a single, but more focused and large-scale innovative proposal, which is, in fact, the initiation of an investment project (Figure 4).

Effective management of mass innovation activity is achieved by the implementation of the following organizational and economic measures: 1) formation and maintenance of the corporate database of innovation proposals; 2) development and application of the procedure for consideration of innovation proposals and material remuneration of innovation activity; 3) creation of a production and technical commission for the evaluation of innovation proposals and organization of its work.

The database of innovation proposals should include information about the authors of rationalization proposals, the content of rationalization proposals, the terms of their submission and consideration, information about the conclusion of the production and technical commission, a description of the results of implementation, the amount and nature of remuneration to the authors of rationalization proposals. The size and nature of the royalties should be commensurate with the beneficial effect of the implementation of the rational offer. The composition of the production and technical commission should be selected from specialists who are able to perform work on the assessment of the technical and economic efficiency of the innovation proposal.

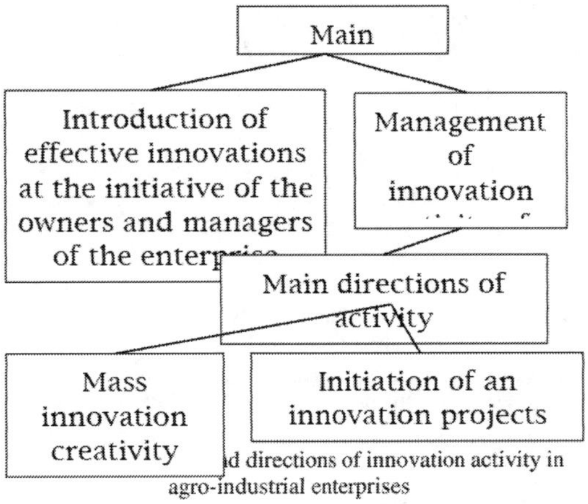

d directions of innovation activity in agro-industrial enterprises

Most agro-industrial enterprises grow dozens of crops and several types of farm animals. In addition to the main business units, many of them have industrial and auxiliary production facilities. For such multi-industry enterprises, there is always the problem of choosing the optimal combination of activities.

Quantitatively, the combination of industries can be expressed by the distribution of resources between activities, in particular, the structure of acreage. This is true for both an agricultural enterprise and an enterprise that combines the production of agricultural products and their processing.

Despite the fact that the rational composition of industries is one of the conditions for the efficiency of agro-industrial production, the variety of technological solutions in each of the industries creates the problem of choosing the product and technological structure of the enterprise.

This choice should ideally give an answer to the question of which crops, in what ratio, what technologies should be grown, how to use the resulting crop and livestock products, taking into account the available technologies of on-farm processing, in order to ensure the achievement of agribusiness goals in a particular enterprise.

The choice of an effective product and technological structure of the enterprise is possible if:

1) there is a political will of the management entity (owners and management of the agro-industrial enterprise) to take actions to improve the management object;

2) management entity has information about all economically significant costs and results of production and commercial activities in the controlled links of the value chain (starting from crop production units and ending with units for the production of consumer products and their marketing);

3) consistent target indicators have been formed that allow for the targeted selection of the product and technological structure of the enterprise;

4) optimization or heuristic methods are available that allow you to quickly get the desired parameters of the product and technological structure of the enterprise;

5) organizational and managerial measures for goal setting, delegation of authority and distribution of responsibility between managers and specialists have been implemented.

Together, these five conditions represent the organizational and economic mechanism of innovative activity of an agro-industrial enterprise.

3 Conclusion

The intensity and content of innovative activity of agro-industrial enterprises are determined by macroeconomic factors and intra-company conditions. Economic relations between related enterprises in value chains play an important connecting role. The differences in market power between producers of the means of production, agricultural, processing and commercial enterprises are so great that it significantly affects the level of prices and the distribution of income. The excess profits of some counterparties and the lack of funds of others equally, but for various reasons, reduce the intensity of innovation activity. Agro-industrial integration only partially solves this problem. Significantly greater opportunities are provided by the institutions of the digital economy. One of them can be a pulling system of commodity movement, which ensures fair pricing, a combination of competition and a centralized distribution of quotas for the production and sale of products.

When organizing innovation activities at the enterprise level, it is important to proceed from the fact that innovation is not only an implemented novelty, but first of all a useful novelty. The utility criteria are set by the owners and managers of the enterprise, based on their own economic and other goals. Therefore, the effectiveness of the organizational and economic mechanism of innovation activity will primarily depend on the quality of goal setting.

An equally important aspect is how to ensure the coherence and reliability of the innovation mechanism. Often, expensive technological innovations are promoted by manufacturers of agricultural machinery and industrial equipment. Dealers use various marketing techniques, including sales promotion. Under the pressure of such a proposal, the right decision is not always made by the management of the agro-industrial enterprise. Therefore, further research on the development of the "anti-corruption" component of the innovation mechanism will be relevant.

References

1. V. Alt, S. Isakova, E. Balushkina, Selection of crop technologies: approaches and methods applicable in information systems, *Vestnik of Kazan state agrarian university,* 1, 52-58, (2020)
2. V.L. Anichin, Regulation of vertical competition as the means of modernisation of the agro-industrial complex, *The problems of economy,* 1, 28-33 (2013)
3. V.L. Anichin, Yu.Yu. Vashcheykina, G.I. Khudobina, N.Yu. Yakovenko, Innovations in the personnel management of an agricultural enterprise, A*dvances in Economics, Business and Management Research,* **157**, 209-214 (2020)
4. K.A. Kalashnikov, A.P. Shindryaeva, The role of innovation in the development of agroindustrial complex, *Journal of international economic affairs,* 2, 1061-1070 (2019)
5. R. Kapinos, A. Dobrunova, O. Akupiyan, D. Kravchenko, Distinctive features of classification of factors determining the economic behavior of economic entities predominantly rural areas from the standpoint of the theory of long waves, *International Review* **(3-4)** 26-34, (2018)
6. S.A. Saninsky, K.A. Kalashnikov, Formation of the mechanism for the implementation of innovations in the production and economic activities of agricultural enterprises, *Problems of Contemporary Science and Practice. Vernadsky University,* 3, 79-87 (2019)
7. J.A. Schumpeter, *The Theory of Economic Development: An Inquiry into Profits, Capital, Credit, Interest and the Business Cycle,* (New Brunswick (U.S.A) and London (U.K.): Transaction Publishers, 2008)
8. F. Suboch, Study of the levers and mechanisms that form the potential for innovative development of enterprises of the agroindustrial complex, *Agricultural economy,* **8**, 3-23 (2019)
9. I. Yasnolob, Ya. Radionova, The organizational fundamentals of innovation development management of agro-industrial enterprises, *Economics, Management and Sustainability,* 2 (1), 60-66. (2017)
10. Yu. I. Zdorovec, O.V. Goncharenko, Characteristics and trends of development of agrarian production of the holdings in Belgorod region, *Scientific notes of the Russian academy of entrepreneurship,* **49**, 296-308 (2016)

The recurrent approach to the study of innovative economic development

Oksana Butorina[1*], *and Irina* Shishkina[2]

[1] Perm National Research Polytechnic University, Perm, 614990, Russian Federation,
[2] Perm State National Research Polytechnic University, Perm, 614990, Russian Federation,

Abstract. The recurrent approach is an integrated approach that combines existing studies of cyclic macroeconomic processes, taking into account the nature of their intercyclic and interphase dependencies. According to this approach, the innovative economy is a natural phase of depression (interphase transition from the neo-industrial economy to the information innovative economy) within the modern transformation cycle. Recognition of the significance of the depression phase allows us to assert that the innovative economy has a twofold nature; it intertwines the processes characteristic of the neo-industrial economy, and forms processes that ensure a "breakthrough" to the neo-economy. In other words, it is dependent on the neo-industrial processes, implying increased digitalization, and becomes the basis for the development of the neo-economy. Taking into account the objectivity of the twofold nature of the innovative economy, it can be argued that it is a special type of the economic system based on a continuous stream of innovations, constant technological improvement, production and export of high-tech products with high added value and technologies, determined by a combination of information, technological, investment, industrial, structural and social transformations.

1 Introduction

In modern world science, the entire set of macroeconomic processes is considered to be the basis for a new type of cycle. Kleinknecht and Van der Velde point out that modern economic development, due to its uniqueness based on direct and inverse relationships between the basic macroeconomic processes, cannot be limited to studies of traditional short-term, medium-term and long-term cycles. Glazyev, Ajvazov and Belikov focusing on the priority of objective technological transformations, associate the macroeconomic processes with the transition from the industrial to postindustrial economy [6].

On the basis of numerous studies of the current development of the domestic and world economic systems, the basic processes of the transition stage can be identified. These include

- formation, dissemination, commercialization of new knowledge embodied in material means and objects of labor, which transforms social production and ensures the accumulation and dissemination of information, innovation and technological progress [1];
- structural changes in production and in the economy as a whole, implying an increase in the share of the non-industrial sector in the value of the goods, based on the development of digital and IT industries;
- large-scale development of innovations in the production and non-production spheres of human life [1];
- increasing the efficiency of innovations based on the effective global and national markets for innovations, based on the market mechanism dependent on the aggregate action of the market laws: demand, supply, equilibrium price, competition, value and money circulation;
- transformation of labor resources from the mechanical "producing" process it becomes "creative" based on qualitative changes in the human capital through the commercialization and industrialization of knowledge and individual's ability to increase intellectual property [8];
- improving science intensity and technotronic potential of production technologies, which implies an increase in the share of R&D costs, constant renewal, transformation of means of production by implementing breakthrough technologies;
- search and implementation of new forms of investment support for the economic development, creation of the monetary and informational forms of capital. MIFC is investment resources in real assets through the digital and IT technologies. These may include crowdfunding, crowdlending, crowdinvesting and P2P lending. In a broad sense, the MIFC as an independent form of capital associated with the self-increasing cost of investments in the development of information and labor resources to create new high-tech products capable of independently increasing surplus value.

2 Materials and methods

The uniqueness of the current state is based on the

* Corresponding author: ok.butorina@yandex.ru

transformation of the industrial and post-industrial eras: the objectivity of reasons for the formation of a transformational crisis, the duration and amplitude of cyclical fluctuations, specification of processes, forming the uniqueness of the modern macroeconomic state, emphasizing direct and inverse relationships between these processes.

Expansion of the range of research areas of cyclic processes has actualized the search for new methodological approaches. Along with the traditional dialectical and systemic approaches, we consider it expedient to use a new methodological approach to the study of modern macroeconomic processes, taking into account the nature of their intercyclic and interphase dependencies.

3 Results and discussion

Due to the uniqueness of the modern macroeconomic cycle, the authors suggest using the recurrent approach, which can be considered as a synthesis of existing studies of cyclic processes. Its main provisions can be formulated as follows:

- in terms of the cyclical dynamics, recurrence is a general economic regularity, which is a multilevel, multifactorial, multi-criteria characteristic of specific dependencies of various cycles and their internal phases. It makes it possible to reveal causal relationships between the phases within each cycle and between the cycles that form the general macroeconomic pulsation;
- specification of the macroeconomic processes of the modern transition period from the industrial era to the post-industrial era can become the basis for identifying basic cycles that are in direct and inverse relationships with each other. The list of these cycles is determined by the researcher based on the purpose of the study, therefore it is not devoid of subjectivity and requires additional justification of the choice. These cycles are identified based on the processes that describe current transformational changes;
- the presence of a relationship between macroeconomic processes, which is formed by the essential characteristics of the cause-and-effect relationships between them and is defined as an intercyclical recurrence. When characterizing the relationships, it is necessary to take into account that each cycle is characterized by its own amplitude, time duration and a system of indicators. The cycles can be divided into several groups: 1) cycles of a providing nature (their phase state forms the basis of the states of the cycles interconnected with them, acting as factors of its dynamics), 2) the cycles that characterize the transformation of macroeconomic processes; 3) the cycles reflecting the effectiveness and / or completeness of the transformation at each stage;
- recognition of the existence of recurrent dependencies between the cycles can become the basis for the development of a comprehensive methodology for assessing the current state and subsequent medium-term and long-term forecasting of the cyclical dynamics of economic systems;
- the intercyclical and interphase recurrent dependencies can specify directions and methods of management of the cyclical processes in the economy, increasing the effectiveness of countercyclical policies in the regions and the country as a whole.

The recurrent approach has become the methodological basis for the study of the modern macroeconomic cycle.

The current macroeconomic cycle can be interpreted in two ways. Within the narrow approach, the modern macroeconomic cycle is a special unique type of cycle that has a transformational nature, within which the information cycle, as an independent type of cycle, can be considered as a cycle of an opportunistic nature, while the technological, innovation, investment cycles are production cycles, and the social and structural cycles are resultant.

In a broad sense, the modern macroeconomic cycle is a system of intra-phase transitions within the industrial and post-industrial mega-cycle, which are characterized by the uniqueness of transformation of the forms of economic development. The system of intra-phase transitions includes the transition from the industrial to neo-industrial economy (the phase of the crisis within the modern macroeconomic cycle), from the neo-industrial to information and innovation (depression phase) economy, from the innovation and information to neo-economy (exhilaration phase), from the neo-economy to the "knowledge economy" or post-industrial economy (recovery phase).

This interpretation can be interesting for identifying the nature of the innovative economy, as a natural phase of depression (interphase transition from the neo-industrial to the information-innovative economy) within the modern transformation cycle. Karl Marx considered the depression phase as the main phase of any cycle. It is in it that the last crisis processes fade away and new ones arise, ensuring the progressive development (transition to the phases of revitalization and recovery), and determining the general vector of the incipient new cycle. Recognition of the significance of the depression phase allowed us to assert that the innovative economy has a twofold nature; it intertwines the processes characteristic of the neo-industrial economy, and forms processes that ensure a "breakthrough" to the neo-economy. In other words, it is dependent on the completeness of neo-industrial processes, implying increased digitalization, but it also becomes the basis for the neo-economy.

Taking into account the objectivity of the twofold nature of the innovative economy, it can be argued that it is a special type of the economic system based on a continuous flow of innovations, constant technological improvement, production and export of high-tech products with high added value determined by a combination of information, technological, investment, industrial, structural and social transformations. Tables 1 and 2 present the characteristics of the innovative economy within the interphase transition from the neo-industrial economy and from the innovative economy to the neo-economy in terms of the intercyclical recurrence.

Table 1. Features of the innovative economy within the interphase transition from the neo-industrial economy in terms of the intercyclical recurrence

Transformational processes within the inter-cycle recurrence	Characteristics of transformation processes	Degree of transformational transition
Informatization processes	- commercialization of information; - socialization of information; - use of ICT in production, management, communication, entertainment [5] .;	- use of ICT in all spheres of the economy and human life
Technological transformation processes	- computerization as the basis for digitalization; - massive digitalization; - increasing the level of automation, robotization; - high technology technotronic industrialization;	- dominance of V - VI technological orders;
Innovation processes	- development of innovation; - development of borrowed innovations; - development of own innovations	- the continuity of the process of generating their own, not borrowed innovations [2].
Investment provision processes	- flourishing of venture capital; - electronic payment systems	- stable investment support for the innovation sector
Manufacturing processes	- creation of new products; - creation of new business models based on the capabilities of the digital economy [4];	- accelerating the pace of economic development; - high production efficiency based on cost reduction
Structural transformation processes	- development of e-commerce; - development of new industries within the ICT; - priority importance of "critical industries" capable of increasing the surplus value; - development of the knowledge industry [8].	- the formation of a new economy focused on knowledge derived from information [4].
Social transformation processes	- formation of new sources of income by increasing the share of intellectual property; - intellectualization of workers' labor; - large-scale capitalization of professional and intellectual abilities	- the created knowledge industry as a mechanism for building up human capital [7]; - an increase in the quality of life

The following conclusions can be drawn:

- at the present stage of the innovative economy formation and development, the priority is given to the processes of informatization. It based on the large-scale use of information and communication technologies in all spheres of the economy and people's life. At the same time, the conjunctural nature of informatization as a process should be particularly noted. In other words, it is basic for technological, industrial, structural, and social transformations.

Table 2. Features of the innovative economy within the interphase transition from the innovative economy to the neo-economy in terms of the intercyclical recurrence

Transformational processes within the framework of inter-cycle recurrence	Characteristics of the transformation process	Degree of transformational transition to the innovation and information economy
Informatization processes	- constancy of informatization and generation of knowledge and innovations	- formation of a stable dependence of productivity on information and management quality
Technological transformation processes	- dominance in all sectors of the ICT economy, transforming the material basis of the modern world	- technology for the dissemination of knowledge
Innovation processes	- innovation of production based on the mass transmission of knowledge and innovations	- development of the innovative potential of the national economy
Investment provision processes	- creation of new forms of capital (monetary and informational form of capital)	- accelerated development based on intangible assets and intangible components of the economy
Manufacturing processe s	- production of intelligent products	- intellectualizatio n of production
Structural transformation processes	- predominance of industries that create intellectual products	- self-development of intelligent technologies and industries that create intelligent products [7].
Social transformation processes	- intellectualizatio n of labor	- constancy of the increase in the value of the human factor of production as the only transformer of knowledge

Informatization defines the general vector of the cycles identified by us, connected with each other by recurrent dependencies within the modern macroeconomic cycle. Today the essence of the technological cycle is characterized by the knowledge-intensive technotronic industrialization of the economy, the essence of innovatioal cycle characterized by the continuous process of generating its own (national) innovations rather than borrowed, the essence of the investment cycle characterized by the innovation sector stable investment provision, the essence of the production cycle characterized by the essence of the structural cycle is the formation of knowledge industry new branches and their development within the boundaries of ICT, focused on the continuous increase of surplus value in the long term, the essence of the social cycle characterized by the formation of new sources of income based on increasing the share of intellectual property, as well as the increase of "traditional" income based on large - scale capitalization of professional and intellectual abilities. Based on this, we consider it possible to define this phase of the modern macroeconomic cycle as an innovation and information one;

- the above-mentioned essential processes can be considered as conditions for the transition to a new more macroeconomic cycle progressive phase. Move from the innovation and information economy to the neo-economy and / or the knowledge economy. At the same time, within the framework of this transformational transition, the process of intellectualization can be distinguished as a conjunctural one. It can determine the general direction of the dynamics of cyclic processes. Thus, the essence of the information cycle can be determined by the stability of the dependence of productivity on the use of information and the quality of management, the essence of technological cycle can be determined by the formation of technologies for the dissemination of knowledge, the essence of innovative cycle can be determined by the continuous increase in innovative potential, the essence of the investment cycle can be determined by the stable increase in the use of intangible assets and the intangible component of management as a whole, the essence of the production cycle can be determined by strengthening the intellectualization of production, the essence of the structural cycle can be determined by ensuring the continuous self-development of intelligent technologies and industries that create intelligent products, the essence of the social cycle can be determined by the constant increase in the value of the human factor of production as the only converter of knowledge.

Based on the presented features of the cyclical processes that determine the phase transformation transitions within the modern macroeconomic cycle, the concepts of both "innovation economy" and "knowledge economy" can be concretized.

The term "knowledge economy" was introduced in 1962 by F. Machlup characterizing the economic system of the post-industrial era. Modern economists interpret it in a similar way. L.M. Gokhberg believes that this is an economy that is based on the effective application of knowledge. G.B. Kleiner defines knowledge as a product with unique properties that are a factor of production [7]. Based on these interpretations, the processes of the information-innovative phase can be identified:

1) formation of a stable dependence of productivity and production efficiency on the use of the achievements of science and technology and the quality of information and management;

2) dominance of information activities as a zone of mutual interests of producers and consumers;

3) continuous transformation of the production process (from standardized to "customized");

4) globalization of the economy, as a result of which capital, production, management, markets, labor, information and technology can function beyond the national boundaries;

5) constancy of technological changes based on the accelerated development and use of intangible assets and the intangible environment of economic activity.

These processes are basic, they can be differentiated and / or unified based on the purpose of the study.

4 Conclusion

Thus, the recurrent approach is an integrated approach, which is a synthesis of available studies of cyclic processes, taking into account the nature of intercyclic and interphase dependencies. The innovative economy is a natural phase of depression (interphase transition from the neo-industrial economy to the information-innovative economy) within the modern transformation cycle. Recognition of the significance of the depression phase allowed us to assert that the innovative economy has a twofold nature; it intertwines the processes characteristic of the neo-industrial economy, and forms processes that ensure a "breakthrough" to the neo-economy. In other words, it is formed under the influence of completed neo-industrial processes, implying increased digitalization, but it also becomes the basis for the development of the neo-economy. Taking into account the objectivity of the twofold nature of the innovative economy, it can be argued that it is a special type of the economic system based on a continuous flow of innovations, constant technological improvement, production and export of high-tech products with high added value. At the same time, transition to the information-innovative economy is ensured by digitalization and innovation processes, and to the neo-economy - by intellectualization of production and economic processes in general.

References

1. I.A. Sushkova *Information security of regions* 4 (25) (2016)
2. J.S. Metcalfe, J. Foster, R. Ramlogan *Cambridge Journal of Economic*, **30(1)**, 7–32 (2006).
3. T. Romanova Digital economy in Russia. Retrieved from: http://fb.ru/article/334484/tsifrovaya– ekonomika – v – rossii
4. K.V. Ilyina *Izvestiya SPbGEU*, **3 (117)**, (2019)
5. R.M. Nureev, O.V. Karapayev *Journal of economic regulation*, **10 (2)**, (2019)
6. O.V. Butorina, E.A. Tretyakova *Bulletin of the Perm University. Ser. "Economy"*, **14 (2),** 289-312, (2019)

7. G.N. Stepanova, E.V. Krasilnikova *Bulletin of Moscow State University named after Ivan Fedorov*, 3, 88-93, (2016)
8. A. A. Volkova, V. A. Plotnikov, M. V. Rukinov, *Management consulting* **4**

Ecological innovations for regional sustainable development: an empirical study

Maksim Kurganov* and *Elena* Tretiakova

Perm State National Research University, 15, Bukireva St., Perm, 614990, Russian Federation

Abstract. The article is devoted to the study of sustainable regional development from the perspective of the value-driven approach and the assessment of the environmental innovations impact on the sustainability of the regional development. The methodological framework included the sustainable development concept, the main provisions of the regional economy, the rational use of natural resources principles and the value-driven management. The study determined the level of realization of the values and interests of the population, the business community and authorities in the economic, environmental and social spheres. It has been established that achieving sustainable development goals at the regional level requires reducing the environmental intensity of enterprises, reducing the volume of pollution of the atmosphere and water resources, as well as the volume of generated production and consumption waste. The regional environmental innovations implementation is considered the main driver of positive changes in the field of natural resource management and environmental protection. Using econometric modeling, an assessment was made of the impact of environmental innovations and factors of the regional environment on the level of realization of the sustainable development values in the context of key stakeholders of the territories and the degree of coordination of their interests, as well as on key indicators of the economic activity. The environmental innovations importance in the regional sustainable development is mathematically proved and the environmental innovations specificity in the context of improving regional policy is characterized.

1 Theoretical and methodological provisions of the research

1.1 Relevance of the research

Against the background of the aggravation of the economic, social and environmental problems of the regions in recent years, the concept of sustainable development (hereinafter the "SD") is gaining special relevance. The importance of the regional vector of research on the basic provisions of the concept of SD is due to the need to balance the implementation of economic, environmental and social imperatives in combination with the improvement of methods and approaches to the management of regional systems [1].

Despite the fact that the concept of SD covers the most pressing problems in the social, economic and environmental spheres, it does not consider the problems of coordinating the goals and interests of the population, the business community and government bodies interacting in the process of economic relations. This impedes the achievement of socio-ecological-economic balance in the development of territories. Coordination of the priorities and interests of these stakeholders is fundamentally necessary, since the rates of socio-economic development, natural resource potential, and the level of infrastructure development in the regions are different. This significantly affects the structure of economic relations between the stakeholders of the territories, the characteristics of their behavior and readiness to compromise.

Overcoming this barrier may be associated with the use of a value-based approach to management, which is acquiring an increasingly significant role in the complex of modern methods of managing socio-economic systems of various levels. Thus, identifying the goals and interests of stakeholders, determining the values received by stakeholders, assessing the degree of their coherence with each other and correlating them with the results of the activities of subjects of economic relations will take into account the high level of environmental uncertainty and determine key areas of managerial influence, reducing the likelihood of antagonistic behavior towards each other.

Let us emphasize the fact that the results of studies in the field of sustainability of regional development indicate a relatively low level of achievement of goals in the context of the ecological component. As a rule, the key reasons for this phenomenon are: a high level of environmental intensity of enterprises operating in the region; large volume of generated production and consumption waste; rejection of resource-saving technologies and switching to circular business models in favor of traditional and well-established approaches to conducting production activities (see [2–4]).

One of the solutions to these problems is to stimulate economic agents to create and use technological and social innovations aimed at protecting the environment.

* Corresponding author: kurganovma@gmail.com

The modern studies note that environmental innovations are a tool that ensures the natural resources management and creates environmental values and behavioral attitudes of key regional stakeholders in the context of the economy co- development [5, 6]. In the given context, it is interesting to study the role of environmental innovations in the SD of the region and the realization of the values and interests of the population, the business community and regional authorities.

Thus, the purpose of the study is to assess the environmental innovations impact on the level of the key regional stakeholders SD values realization.

1.2 Description of the methodological toolkit

To achieve the stated goal, a methodology was developed that includes three stages: (1) conducting an integral assessment of SD from the perspective of a value-driven approach; (2) identification and analysis of imbalances in the level of value realization; (3) analysis of the environmental innovations role in the regional SD and the values and interests of economic agents realization.

At the first stage of the study, a set of indicators was formed that characterize the interests of the population, the business community and the state in the economic, environmental and social spheres. When selecting indicators, the following principles were taken into account:

- the indicator substantively characterizes the manifestation of the stakeholder's value;
- the indicator allows assessing SD in dynamics;
- the indicator allows characterizing the SD level of a particular region;
- statistical information about the indicator should be available in open official sources.

The selection of indicators resulted in the formation of a statistical base that includes time series for 51 indicators for the period from 2010 to 2018.

For further integral assessment, all indicators were standardized and, on their basis, nine group indices were calculated by computing a multidimensional average value in the context of three types of stakeholders and three types of values, the value of each of which comprehensively characterizes the level of implementation of each type of value for each regional stakeholder.

The calculated group indices served as the basis for calculating integral indices using the arithmetic mean of three group indices (in the context of SD components and in the context of regional stakeholders), which enabled a comprehensive characterization of the implementation degree of the SD values.

At the second stage of the study, the coefficients of variation between the values of the integral indices were calculated. This procedure allowed identifying the disparities in the level of realization of values (the value of the coefficient is more than 0.33) and quantifying the degree of consistency of values and interests of SD.

At the final stage, a regression analysis was carried out, which allowed determining the strength of the relationship and influence of environmental innovations and other environmental factors of the regional environment on the level of SD of the region and the degree of realization of the interests of the population, authorities and the business community.

Thus, the proposed methodological toolkit is focused on a comprehensive study of the role of environmental innovations in the SD of the region, taking into account the goals, interests and values with which the stakeholders enter into economic relations.

2 Empirical research results

2.1 Assessment of the level of implementation of sustainable development values

The methodological toolkit was tested on the regions of the Volga and Ural federal districts. It should be noted that the Tyumen region and its constituent autonomous okrugs were excluded from the sample due to the lack of statistical indicators sufficient for empirical research.

Tables 1 and 2 contain information on the level of realization of economic and environmental values of the population, business and government (values of group indices), as well as integral indices characterizing the degree of achievement of SD goals for each of the analyzed components.

Noteworthy is the relatively low level of implementation of the environmental values of the business community and government bodies. According to the results of the study, the most negative impact on the level of implementation of the environmental values of a business is exerted by relatively high indicators of the volume of fresh water use for industrial purposes, the volume of emissions of air pollutants and the volume of polluted wastewater discharge. This indicates the need to improve the technological base and search for innovative solutions that reduce the negative impact of enterprises on the environment.

The low level of implementation of the environmental values of regional authorities is due to the relatively high volumes of industrial and consumer wastes generation per capita, as well as the relatively low share of current expenditures on environmental protection in the GRP. In some regions (for example, in the Udmurt Republic and the Republic of Mari El), relatively low volumes of reforestation were recorded, correlated with the volume of timber harvesting, and very small volumes of utilized production and consumption wastes.

The environmental values of the population are also characterized by a low level of implementation in most of the analyzed subjects, despite positive developments in the Republic of Mari El, Perm Territory, Nizhny Novgorod and Orenburg regions. During the study, it was found that these positive trends are due to a significant decrease in the proportion of air samples with an excess of the MPC by 2018 and an increase in the proportion of the population supplied by with drinking water that meets all health-based standards. The worst situation is observed in the Kurgan region.

Dwelling on the analysis of the economic values of the stakeholders of the territories, let us pay attention to the low level of the authorities' economic values realization. It is also characterized by minor fluctuations in most of the regions under consideration. The leaders in the level of population economic values realization are the Republic of Tatarstan, the Sverdlovsk, Nizhny Novgorod regions and the Perm Territory. At the same time, downward trends in the realization of the economic values of the population were noted in the Chuvash Republic and the Chelyabinsk region. The authors identify the main reason for this trend as the growth in the proportion of population with incomes below the subsistence minimum. The lowest level of realization of the economic values of the business community was identified in the Kurgan and Orenburg regions and the Republic of Mari El. During the analyzed period, the number of regions with a high level of realization of economic values decreased from five (in 2010) to two (in 2018).

With regard to the social values of the considered stakeholders, the authors established a high level of their implementation in all analyzed regions for each economic agent (values of group indices are more than 0.7). This trend can be caused by the establishment at the federal level of social standards and social norms to ensure a decent quality of life, the implementation of federal state programs, and control over their observance at the regional level.

Thus, the SD integral assessment results based on value-driven approach confirm the trends identified in other studies on the problem under study. The level of implementation of environmental values is one of the main barriers hindering balanced socio-ecological and economical regional development.

2.2 Analysis of the role of environmental innovations in sustainable development of the region: results of econometric modeling

Taking into account the regional SD integral assessment results from the standpoint of the value-driven approach, as well as the results of national and international studies, the authors propose to consider the possibility of increasing the regional SD level by influencing the environmental values of the considered stakeholders.

The introduction and implementation of environmental innovations in the region is considered as a key instrument of influence, which creates conditions for the transition of regional socio-economic systems to a new qualitative level. To solve this problem, a regression analysis is carried out in the study. Based on the existing scientific base in the studied problem area, the first prerequisite for conducting econometric modeling can be designated as follows: the level of implementation of environmental values is influenced by regional environmental factors that characterize the degree and scale of management impacts to improve the situation in a particular region. It is important to emphasize that one of the most significant characteristics of the regional environment factors is controllability.

Therefore, in this study the following is considered as factors of the regional environment:

- the share of organizations that carried out environmental innovations in the reporting year in the total number of organizations surveyed, %;
- the share of investments in fixed assets aimed at environmental protection and rational use of natural resources in GRP, %;
- the number of violations in the field of environmental management.

The second prerequisite for the construction of economic and mathematical models is that the particular stakeholder environmental values realization level depends on the degree of goals achievement and interests satisfaction of the other economic agents. In other words, the environmental values of the population, business community and government are interrelated.

In view of the above, the analytical presentation of econometric models takes the following form:

$$ECOL\_IND_{people} = f(IND_{ecol}, X_{region}) + \varepsilon, \quad (1)$$

$$ECOL\_IND_{business} = f(IND_{ecol}, X_{region}) + \varepsilon \quad (2)$$

$$ECOL\_IND_{government} = f(IND_{ecol}, X_{region}) + \varepsilon, \quad (3)$$

where:

$ECOL\_IND_{people}$ is group index characterizing the level of implementation of the environmental values of the population;

$ECOL\_IND_{business}$ is the group index characterizing the level of implementation of the environmental values of the business community;

$ECOL\_IND_{government}$ is the group index characterizing the level of implementation of the environmental values of the authorities;

IND_{ecol} is the vector of variables reflecting group indices for the ecological component of other stakeholders;

X_{region} is the vector of variables reflecting the state of the regional environment in terms of the ecological component;

ε is the regression error.

Table 1. Group and integral indices characterizing the level of implementation of economic values*

Region	ECONOMIC VALUES									Integral index (economic values)	
	Population			Business			Authorities				
	2010	2018	Δ**	2010	2018	Δ**	2010	2018	Δ**	2010	2018
Republic of Bashkortostan	0.65	0.62	↓	0.72	0.58	↓	0.53	0.40	↓	0.63	0.53
Mari El Republic	0.59	0.56	↓	0.49	0.47	→	0.43	0.42	→	0.50	0.48
The Republic of Mordovia	0.55	0.55	→	0.62	0.59	→	0.51	0.48	↓	0.56	0.53
Republic of Tatarstan	0.92	0.92	→	0.78	0.77	→	0.67	0.64	↓	0.78	0.77
Udmurtia	0.66	0.63	↓	0.55	0.56	↑	0.53	0.39	↓	0.58	0.52
Chuvash Republic	0.65	0.58	↓	0.53	0.58	↑	0.37	0.34	↓	0.51	0.49
Perm Territory	0.70	0.72	↓	0.55	0.55	↑	0.56	0.60	↑	0.60	0.62
Kirov region	0.63	0.60	↓	0.50	0.64	↑	0.42	0.40	↓	0.51	0.53
Nizhny Novgorod Region	0.73	0.78	↑	0.86	0.74	↓	0.63	0.58	↓	0.73	0.69
Orenburg region	0.55	0.56	↑	0.63	0.46	↓	0.58	0.49	↓	0.58	0.50
Penza region	0.59	0.60	↑	0.52	0.53	↑	0.40	0.41	→	0.50	0.51
Samara Region	0.69	0.69	→	0.80	0.61	↓	0.58	0.49	↓	0.68	0.59
Saratov region	0.62	0.58	↓	0.72	0.52	↓	0.44	0.41	↓	0.58	0.50
Ulyanovsk region	0.63	0.58	↓	0.63	0.60	→	0.43	0.50	↑	0.56	0.56
Kurgan region	0.54	0.49	↓	0.53	0.45	↓	0.43	0.34	↓	0.50	0.42
Sverdlovsk region	0.88	0.81	↓	0.60	0.53	↓	0.51	0.53	↑	0.64	0.61
Chelyabinsk region	0.70	0.62	↓	0.69	0.58	↓	0.54	0.50	↓	0.64	0.56

* the gray filling in the table indicates the values of the indices, which indicate a high level of realization of values (the values of the indices are 0.7 and higher); the underscore indicates the values of the indices that characterize the low level of realization of values (the values of the indices are below 0.5).

** direction of arrows characterizes the direction of the trend during 2010-2018:

↑ - ascending; ↓ - descending; → - horizontal

Table 2. Group and integral indices characterizing the level of implementation of environmental values *.

Region	ENVIRONMENTAL VALUES									Integral index (environmental values)	
	Population			Business			Authorities				
	2010	2018	Δ**	2010	2018	Δ**	2010	2018	Δ**	2010	2018
Republic of Bashkortostan	0.68	0.38	↓	0.37	0.33	↓	0.39	0.36	↓	0.46	0.36
Mari El Republic	0.56	0.93	↑	0.25	0.25	→	0.19	0.20	↑	0.30	0.36
The Republic of Mordovia	0.56	0.62	↑	0.37	0.42	↑	0.43	0.31	↓	0.45	0.43
Republic of Tatarstan	0.42	0.44	↑	0.52	0.43	↓	0.33	0.24	↓	0.42	0.36
Udmurtia	0.47	0.38	↓	0.29	0.24	↓	0.21	0.16	↑	0.30	0.25
Chuvash Republic	0.65	0.48	↓	0.44	0.38	→	0.33	0.25	↓	0.45	0.36
Perm Territory	0.36	0.63	↑	0.24	0.23	↓	0.45	0.40	↓	0.34	0.39
Kirov region	0.47	0.56	↑	0.23	0.22	↓	0.30	0.35	↑	0.32	0.35
Nizhny Novgorod Region	0.35	0.67	↑	0.61	0.62	→	0.31	0.28	↓	0.40	0.49
Orenburg region	0.52	0.96	↑	0.19	0.19	↑	0.58	0.58	↑	0.39	0.47
Penza region	0.57	0.61	↑	0.26	0.22	↓	0.22	0.22	↑	0.32	0.31
Samara Region	0.51	0.62	↑	0.34	0.31	↓	0.43	0.32	↓	0.42	0.40
Saratov region	0.41	0.38	→	0.55	0.37	↓	0.37	0.34	→	0.44	0.36
Ulyanovsk region	0.51	0.40	↓	0.30	0.29	→	0.41	0.26	↓	0.40	0.31
Kurgan region	0.27	0.30	↑	0.27	0.24	↓	0.41	0.27	↓	0.31	0.27
Sverdlovsk region	0.39	0.40	→	0.36	0.40	↑	0.67	0.55	↓	0.45	0.44
Chelyabinsk region	0.43	0.35	↓	0.30	0.26	↓	0.53	0.54	↑	0.41	0.37

* the gray filling in the table indicates the values of the indices, which indicate a high level of realization of values (the values of the indices are 0.7 and higher); the underscore indicates the values of the indices that characterize the low level of realization of values (the values of the indices are below 0.5).

** direction of arrows characterizes the direction of the trend during 2010-2018:

↑ - ascending; ↓ - descending; → - horizontal

Thus, the regression analysis is based on panel data on the factors of the regional environment and the level of implementation of environmental values by stakeholders for the period from 2010 to 2018. To ensure the possibility of comparing the simulation results and reducing the statistical space, the absolute and index variables were logarithmized.

Regression analysis was performed using the construction of panel data models with fixed effects. The simulation results are presented in Table 3.

Table 3. Results of econometric modeling

Independent variable	*l*_ECOL_PEOPLE_IND	*l*_ECOL_BUS_IND	*l*_ECOL_GOV_IND
*l*_ECOL_VIOLAT	–0.11 (0.03)	0.02 (0.02)	0.02 (0.02)
ECOL_INN	–0.02** (0.00)	–	0.02*** (0.02)
ECOL_INN (-1)	–	0.01* (0.00)	–
INV_EP	–0.09 (0.06)	0.01 (0.03)	0.04 (0.03)
*l*_ECOL_PEOPLE_IND	–	0.05 (0.03)	0.09 (0.06)
*l*_ECOL_BUS_IND	0.19 (0.12)	–	0.04 (0.11)
*l*_ECOL_GOV_IND	0.23 (0.15)	0.01 (0.04)	–
Const	–0.13* (0.31)	–1.27*** (0.15)	–1.16***
R^2	0.74	0.95	0.89
p-value for F-statistics	0.00	0.00	0.00
Number of observations	136	119	136

Based on the analysis performed, a number of conclusions can be drawn. First, the synergy effect in the context of the environmental values of the population, business community and government authorities was not found, which indicates the need to influence the factors of the regional environment.

Secondly, it was established that there is a relationship between the level of implementation of environmental values by key stakeholders in the region and the number of organizations that carry out environmental innovations (ECOL_INN). Consequently, the improvement of innovation policy in the field of environmental protection can become one of the drivers for increasing the level of sustainability of regional development.

Third, it was revealed that there is a direct relationship between the number of organizations implementing environmental innovations and the level of realization of environmental values by the business community (with a lag of one year) and authorities. In turn, the relationship of the specified independent variable with the population environmental values realization degree is inverse. This finding can be explained by the fact that business and government bodies are the direct subjects of the introduction of any kind of innovation, as well as the owners of investment resources for the development of the innovation base in the region. The population, in turn, does not have enough resources and tools to influence the process of launching environmental innovations. Thus, the results of regression analysis prove the existence of a problem associated with a decrease in the number of organizations implementing environmental innovations, which prevents the creation of environmentally friendly living conditions.

In order to more comprehensively assess the identified patterns, it is necessary to assess the impact of environmental innovations on the level of consistency of stakeholders' interests, GRP per capita and on the investment potential of the region. For this, three equations were built for each of the indicated dependent variables without combining them into a system of equations:

$$lALIG\_AGENTS = f\left(X_{region}\right) + \varepsilon, \quad (4)$$

$$lGRP = f\left(X_{region}\right) + \varepsilon, \quad (5)$$

$$INV\_POTEN = f\left(X_{region}\right) + \varepsilon, \quad (6)$$

where:

$lALIG\_AGENTS$ is the logarithmic value of the level of consistency of interests of stakeholders in the context of SD of the region;

$lGRP$ is the logarithmic value of GRP per capita;

INV_POTEN is the share of the investment potential of a particular region in the all-Russian potential in percent;

X_{region} is the vector of variables reflecting the state of the regional environment in terms of the ecological component;

ε is the regression error.

Table 4 contains the results of building regression models and criteria for their quality.

The results of the regression analysis presented in Table 4 indicate a significant impact of environmental innovations on the level of consistency of values of the interests of key stakeholders in the region and indicators of the economic performance of the territories as a whole.

In particular, attention is drawn to the inverse relationship between the level of GRP per capita and the number of organizations carrying out environmental innovations, along with the volume of investments in fixed assets aimed at protecting the environment (INV_EP). This pattern can be explained by the fact that increasing the volume of GRP on the existing technological base, which is harmful to the environment, is impossible without investment in the development of new energy-saving and green production technologies with a subsequent transition to environmentally friendly business models.

Table 4. Assessment of the influence of factors of the regional environment on key indicators of regional development

Independent variable	*l*_ALIG_AGENTS	*l*_GRP	INV_POTEN
*l*_ECOL_VIOLAT	–0.00 (0.03)	–0.04 (0.03)	0.01 (0.00)
ECOL_INN	–0.01** (0.00)	–0.07*** (0.00)	0.01** (0.02)
INV_EP	–0.02 (0.01)	–0.11** (0.04)	0.02 (0.02)
Const	–0.18** (0.3)	–13.01*** (0.2)	1.2*** (0.03)
R^2	0.81	0.84	0.9
p-value for F-statistics	0.00	0.00	0.00
Number of observations	136	136	136

At the same time, it was found that an increase in the number of environmental innovations in the region favors the regional investment potential development, and also reduces the disproportions in the level of implementation of the values of the population, the business community and government bodies.

Summing up, we can unequivocally speak about the significant role of environmental innovations in the regional SD. The results of the study proved that the use of green economy tools has a positive impact not only on the consistency of goals and interests of economic agents, but is also one of the key factors in attracting investment resources and improving the investment climate of territories.

3 Environmental innovation as a tool for increasing the sustainability of regional development

The key purpose of green innovation is to reduce the destructive impact on the environment and natural resources from economic activities in the region. In this case, an understanding of the main vectors of managerial impact and directions for improving regional policy plays a special role. The introduction and implementation of environmental innovations can cover various spheres of society, which is fully consistent with the goals of SD [5, 7]:

- building;
- energy;
- agriculture and reforestation;
- industry;
- tourism;
- transport infrastructure;
- waste treatment;
- water supply and water resources;
- climate change.

It should also be noted that the content of environmental innovations and the nature of their impact on the level of sustainable regional development depends on the type of innovations themselves. In this context, it is customary to distinguish the following: technologies for environmental protection (for example, technologies for reducing emissions of air pollutants), organizational eco-innovations (for example, the formation of "green" supply chains); innovative products and services aimed at improving the environmental conditions of life, as well as innovations in the field of improving regional ecosystems (for example, alternative methods of production) [8].

Since the expected benefits and effects from the introduction of environmental innovations in the region are characterized by both technological and non-technological changes, the choice of instruments for improving regional policy in the field of SD should take into account the existing resource base and existing barriers to the implementation of green innovations. In particular, [9]:

- technological barriers caused by the complexity of developing innovative products and technologies due to the lack of environmentally friendly materials;
- financial barriers associated with the high cost of research in the field of environmental technologies and the difficulty of forecasting costs;
- barriers associated with the lack of highly qualified personnel;
- barriers associated with a high degree of volatility in consumer preferences and a well-established environmental mentality;
- management barriers caused by the resistance of enterprises and the need for network and intersectoral interaction.

Overcoming these barriers, according to the results of the study, is possible when influencing the factors of the regional environment. To achieve the SD goals and increase the realization level of the population, business community and authorities values, it is necessary to ensure diversification of funding sources, to form a socio-ecological identity among all stakeholders in the region, to stimulate consumer and investment demand for environmental innovations created in the region with the help of institutional measures of the state policy [10].

Summarizing the results obtained, it is important to emphasize that despite the growth of socio-economic and environmental costs associated with the search, development and implementation of green innovations in the economic activities of subjects of economic relations, the long-term effects of the implementation of environmental innovations have a positive impact on the SD of territories. The study showed the importance and necessity of stimulating economic agents to create green innovations, which will ensure an increase in the level of coherence of interests and values of all stakeholders and a balance of goals in the economic, environmental and social spheres.

References

1. M.A. Kurganov, E.A.Tretyakova. *Journal of New Economy*. **4**, 104–130. (2020).
2. E. Holden, K. Linnerud, D. Banister. *Global Environmental Change* **26**, 130-139. (2014).
3. M. Howes, L. Wortley, R. Potts, A. Dedekorkut-Howes, S. Serrao-Neumann, J. Davidson, P. Nunn. *Sustainability (Switzerland)*. **9**. (2017).
4. M. L. Graymore, N. G. Sipe, R. E. Rickson. *Ecological Economics*. **3**, 362–372. (2008).
5. T. Tambovceva, M. Tereshina, V. Samarina. *Engineering for Rural Development*. 1832-1839. (2019).
6. J.H. Jo, T.W. Roh, S. Kim, Y. C. Youn, M. S. Park, K. J. Han, E. K. Jang. *Sustainability*. **7**, 16820-16835. (2015).
7. M. Park, R. Bleischwitz, K. Han, E. Jang E., J. Joo. *Comparing eco-innovation indices: ASEM eco-innovation index & eco-innovation scoreboard*. (ASEIC and University College London, 2016).
8. A. Arundel, R. Kemp. *Measuring eco-innovation*. (Maastricht: UM Merit, 2009).
9. N. Yu. Arefyeva. EurAsian Journal of BioSciences. The **12**, 263-269. (2018).
10. A. Sarkar, *European Journal of SD* **2** (2013).

Possibility of implementing a cluster based on the territorial public self-government that includes sports, retailers and consumer services

Arkadii Simdiankin[1], *Egor* Khludnev[1,*], *Pavel* Probin[1], and *Alexey* Yamalutdinov[2]

[1]Russian State University of Physical Culture, Sport, Youth and Tourism, Sirenevy Blvd. 4, Moscow, 105122, Russia
[2]Russian Foreign Trade Academy, 6A, Vorobiovskoe high-road, Moscow, 119285, Russia

Abstract. The economic crisis has had a negative impact on the clustering processes. Nevertheless, in the current conditions there is a prospect of changing the very ideology of cluster models. It is very likely that the high impact of the volatility factor in the business environment will determine a qualitatively new format for the implementation of cluster initiatives. In most cases, the population of a particular territory is not directly involved in the process of creating a cluster but is considered as a labour force capable of achieving the goals of clustering. However, it is impossible to say that cluster projects cannot be initiated on the principle of "bottom-up development" where the population performs are not only the functions of "personnel" but also determines the feasibility of forming a project. This ensures open an equal access to the resources of absolutely all cluster participants and the population acts as an initiator, coordinator, investor and source of labour resources. In this regard, we think that it is appropriate and possible to consider the prospect of creating cluster (and similar ones) projects by the example of territorial public self-governments.

1 Introduction

As noted in the literature, in many developed countries the sustainability of economic development is directly related to small and medium-sized businesses (SME). Our country is also aware of the importance of this sector of the economy. In order to develop it, a set of measures is being carried out aimed at improving the conditions for small and medium-sized businesses and promoting them [2].

These measures should lead to an increase in the number of SME employees from 19.2 to 25 million people in the period from 2018 to 2024. Besides the fraction of small and medium-sized businesses in the country's GDP should rise from 22.3 to 32.5% [3]. However, in 2019 the number of SME decreased by 2% (from 6,041,195 to 5,916,906) as well as the number of employees decreased by 3.5% (from 15,873,589 to 15,321,788) [4]. During 2020 the situation only worsened because of COVID-19 pandemic. However, it does not cancel other factors that cause the economic vulnerability of SME such as their inability to compete effectively with representatives of large businesses [5]. The above circumstances clearly indicate the need to find new ways to create favorable conditions for the development of small and medium-sized businesses in our country.

2 Review of literature

In view of this it is advisable to pay attention to the fact that the economic stability of SME is markedly increasing within clusters where logistics, personnel and other opportunities are becoming wider because of it's realized on the basis of common efforts [6].

The practical effectiveness of the cluster approach as a factor in the development of SME is repeatedly mentioned in the scientific literature where the positive experience of implementing cluster projects in different regions of the Russian Federation describes [7,8]. For example, in

* Corresponding author: hei14@mail.ru

the Tambov region the trade turnover of small and medium-sized enterprises reached over 160 billion rubles in 2013. This is equivalent to 43.8% of the total turnover of all organizations operating in this region [9].

This suggests the feasibility of conducting a policy on the territory of our country aimed at involving small and medium-sized businesses in the activities of clusters. As one of the possible vectors of such a policy the authors see the formation of clusters around territorial public self-governments (TPSG) - self-organizations of citizens at their place of residence for the independent implementation of their own initiatives on issues of local significance. The boundaries of the territory are established by the representative body of the municipal formation based on the suggestion of the population and can cover both a small territory (for example, the entrance of an apartment building) and a residential community or a locality that does not have the status of a municipal formation. Citizens living in the relevant territory form TPSG' management bodies that represent their interests interacting with local self-government authorities and can carry out economic activities aimed at social and household needs of residents.

Despite the fact that the above-mentioned economic activities can be carried out not only on a donation basis of the citizens themselves but also on the taxpayers dime the financial resources of the TPSG' management bodies are not enough as a rule.

3 Methods

It seems the establishment of symbiotic relations between TPSG and SME entities that conduct economic activities in the territory within the boundaries of which territorial public self-government is carried out can reduce the acuteness of this problem. We see the direction of their mutually beneficial cooperation in the implementation by the TPSG' management bodies of a common bonus loyalty scheme. In this condition SME when purchasing goods (works, services) by program participants (i.e., citizens) would accrue bonus points (de jure – the obligation to provide a discount) and would make deductions to the program organizer in the amount of this discount. For example, a citizen purchasing a product in a grocery connected to the loyalty program can spend bonus points from any other partner of the program – a sports club, a barber shop, a pharmacy, etc.

The tasks of the TPSG' management bodies will include the reimbursement of the discounts provided to the partners of the loyalty program with the deduction of their agency fee that can be used to achieve the goals for which the TPSG was created. Thus, citizens will understand that by buying goods or ordering works and services from the program partners they contribute to the improvement of the territory where they live.

Of course, this system can be effective only if the TPSG authorities make efforts to ensure that citizens know what the economic entities have joined the loyalty program. Therefore, it is advisable to post information about the program's partners on the Internet. Thus, people will be able to easily find sellers and performers which can be not only organizations but also individual entrepreneurs and self-employed persons [10].

The main problem in the implementation of the described concept is likely to be the technical side of its implementation. But it can be solved by creating – maybe with the support of the state – a common Internet portal where TPSG' management bodies could register as organizers of loyalty programs and SMEs and citizens could connect to the programs, they are interested in. It is worth considering the possibility of an automated payment system between the organizers and partners of the loyalty program as part of such an online platform. It also makes sense to develop an appropriate mobile app.

Local self-government bodies can also make a significant contribution to the implementation of the idea under consideration by help to inform the population about TPSG that implement the described loyalty programs as well as encouraging TPSG to independently attract financial resources.

From this perspective it is interesting that the experience of the Perm urban district where a system of competitions for socially significant projects is implemented in order to increase the activity of the population in solving local issues. To participate in the TPSG competition it is necessary not only to develop a project but also to independently attract at least 30% of the resources needed for its implementation. The winners of the competition receive the corresponding grants [11]. It seems that such a system makes it possible to use the local budget more effectively not only because the solution of local issues is partially carried out at the expense of private capital. But also because of its involvement the population in this process since often only the residents themselves know exactly how to solve certain current issues for them [12].

Let us pay attention to a number of features inherent in clusters formed on the basis of TPSG. First, this type of cluster can be called complex since they can include organizations of a wide variety of economic sectors. The main thing is their territorial location and the opportunity to be useful for the local community. It makes them possibility to carry out activities in a more favorable business environment than enterprises that cannot become part of an industry cluster. For example, this applies to sports clubs in the place of residence.

Second, the cores of the clusters under consideration (i.e., TPSG) have certain public functions. In particular, they have the right to submit to local self-government bodies mandatory draft municipal legal acts for consideration. This means that the TPSG can effectively act as a mediator in the dialogue between the cluster participants and local authorities including the issues related to improving the investment climate in the territory of the municipality.

Third, TPSG is created and functions thanks to the "bottom-up development" initiative that is in a natural way. It is very important taking into account the fact that effective clusters (as world experience shows) very rarely arise artificially [13].

Fourth, within a municipality there can be as many TPSG as the population needs since the legislation does not limit their maximum number. Since the municipalities are formed throughout the entire territory of our country, there is a possibility of formation of such a number of clusters around territorial public self-governments that the economy needs at both the meso- and macrolevels.

To analyze the possibility of clustering the authors propose to apply some operations of set theory in particular the intersection operation (denoted as $A \cap B$ and is a set of elements contained in both set A and set B). At the same time, both A and B is a set of elements that describe the relationship of the business cluster groups to the sources of funding, potential involvement and interest to be in the cluster. This approach allows us to identify the "points of contact" of the interests of various groups of the cluster being formed as well as to determine (with a high degree of confidence) the possibility or impossibility of creating a cluster at this stage and with these participants.

4 Results

We consider the "motivating" features of economically active organizations that encourage them to the clustering process. Based on the above-mentioned analysis of literature sources as well as actual data [4], [13-18] the authors identified the "interests" of TPSG and business clusters (BC) with industry (Table 1), sources of financing (Table 2) and the potential involvement of various groups potentially interested in the work of the cluster (Table 3).

Let's consider TPSG and BC as two sets and find their intersection (Table 1):

$$TPSG_1(0,1;0,1;0,3;0,1;0,3;0;0,1) \cap$$

$$BC_1(0,15;0,2;0,1;0,1;0,3;0,1;0,05) \quad (1)$$

The result of the intersection of these two sets is the following set:

$$SI_1(0,1;0,1;0,1;0,1;0,3;0;0,05) \quad (2)$$

and its maximum value is $\max(SI_1) = 0,3$.

Table 1. Intersection of sets TPSG and BC within an industry.

	TPSG	BC	Intersection
Industrial production	0,1	0,15	0,1
Food and pharmaceutical industries	0,1	0,2	0,1
Service	0,3	0,1	0,1
Farm sector	0,1	0,1	0,1
Distributive industries	0,3	0,3	*0,3*
Building activity	0	0,1	0
Residential and public landscaping	0,1	0,05	0,05
Sum	1	1	0,75

Based on the analysis of the SI_1-set it can be concluded that cooperation within the cluster between business and territorial public self-government is possible in the field of wholesale and retail trade. At the same time, the small value of maximum indicates the limited maneuver of the business represented in this field in the development of the cluster. First of all, this restriction is due to the small financial resources of this kind of business. Moreover, it is difficult to call such business as a start-up and therefore no one should expect an "explosive" growth in its activity as well as the same

"explosive" growth in attracting funds from potential investors and the growth in the cluster's revenue. This situation is likely to be accompanied by a rather slow attraction of budget funds from the TPSG which will be invested in the cluster. Of course, business is interested in additional financing and will be willing to consume additional budget funds attracted by the TPSG. This approach will determine the very existence of the cluster on the principle of "there is money - there is a cluster, there is no money - there is no cluster" since budget financing will be perceived by the business exclusively as an additional source of income. This interpretation will be a barrier to the development of the cluster.

Table 2. Intersection of sets TPSG and BC within the sources of financing.

	TPSG	BC	Intersection
State aid (programs, funds)	0,6	0,3	*0,3*
Foreign investors	0	0,1	0
Venture funds	0	0,1	0
Local Russian investors	0,1	0,2	0,1
Business angel investors	0	0,1	0
Banks	0,3	0,2	0,2
Sum	1	1	0,6

Let's perform the similar actions with the data in Table 2.

$$TPSG_2(0,6;0;0;0,1;0;0,3)\cap BC_2(0,3;0,1;0,1;0,2;0,1;0,2) \quad (3)$$

The result of the intersection of these two sets is the following set:

$$SI_2(0,3;0;0;0,1;0;0,2) \quad (4)$$

and its maximum value is $\max(SI_2) = 0,3$.

Thus, financial support for the possible formation of a cluster based on the interaction of TPSG-BC is again supported by budget funds with a focus on financing TPSG but not business.

In SI_2-set the second largest member of this set is "banks" (0,2). Banks' position will significantly affect the financing of the cluster at the local level. However, the member of the $TPSG_2$-set for banks is larger than the one for business - it indicates a higher attractiveness in the financing of TPSG by banks than for business. On the other hand, such financing can also be carried out not at the initiative of banks but by order of "governmental authorities" - as financing for a group that developed the local community.

Table 3. Intersection of sets TPSG and BC of various groups potentially interested in the work of cluster.

	TPSG	BC	Intersection
Industrial plants, raw material processing, service, farm sector	0,15	0,3	0,15
Small and medium-size enterprises	0,3	0,4	*0,3*
Individual entrepreneurs	0,3	0,1	0,1
Self-employed persons	0,15	0,1	0,1
Retirees	0,05	0,05	0,05
Unemployed population	0,05	0,05	0,05
Sum	1	1	0,75

Let's perform the similar actions with the data in Table 3.

$$TPSG_3(0,15;0,3;0,3;0,15;0,05;0,05)\cap BC_3(0,3;0,4;0,1;0,1;0,05;0,05) \quad (5)$$

The result of the intersection of these two sets is the following set:

$$SI_3(0,15;0,3;0,1;0,1;0,05;0,05) \quad (6)$$

and its maximum value is $\max(SI_3) = 0,3$.

It is clearly seen that small and medium-sized businesses (as a rule specializing in wholesale and retail trade) are most interested in clustering. At the same time, this type of business is almost equally represented by both TPSG (0.3) and business structures (0.4).

5 Conclusions

Considering the total values of all intersections we can come to the following conclusion: the similar “desire” of TPSG and BC to enter the cluster within the industry and the involvement of various groups potentially interested in the work of the cluster is not supported at the proper level by potential investors. This indicates some concern about the "sources of funding" in the success of this type of cluster. The proposed approach to assessing the formation of a cluster based on the compilation of sets that include various aspects of the cluster, sources of funding, groups of people interested in its work and direct participation in it as well as sectors of the national economy allows us to comprehensively consider the impact of the current situation on the clustering process and identify the main obstacles on this path.

References

1. N. Goridko., R. Nizhegorodtsev *The econ. of the reg.* 14, 30 (2018)
2. YU. Baldina *Bull. of the Udm. Uni. Econ. and Law ser.* 26, 7 (2016)
3. Presidium of the Presidential Council for Strategic Development and National Projects Protocol No. 16 *Passport of the national project «Small and Medium-sized Enterprises and Support for Individual Entrepreneurial Initiatives»* (2018)
4. Official website of the Federal Tax Service of Russia *Unified Register of Small and Medium-sized Businesses* Retrieved from: https://ofd.nalog.ru
5. M. Vyshegorodtsev *Mod. competit.* 5(17), 53 (2009)
6. G. Zelenkov *Innov. and invest.* 12, 212 (2017)
7. N. Solodilova, R. Malikov, K. Grishin. *Econ. Pol.* 13, 140 (2018)
8. N. Zubarevich. *J. of NEA*, 2 (46), 162 (2020)
9. M. Pakhomov, I. Narizhny, *D. Sechenov Soc.-econ. phenom. and proc.* 10, 73-80 (2015)
10. E. Kiseleva, *Fin.: th. and pr.* 24(5), 80 (2020)
11. YU. Aksyutina, *Mod. res. on soc. probl.* 11-2, 16-25 (2016)
12. S. Belev., V. Veterinarov, O. Suchkova. *HSE Econ. J.* 25(1), 30 (2021)
13. T. Tsatkhlanova, E. Erdnieva *Bull. of the Uni.* 16, 83-87 (2014).
14. *Federal State Statistics Service.* Retrieved from: https://rosstat.gov.ru/folder/10705
15. *Map of clusters in Russia.* Retrieved from: http://map.cluster.hse.ru/list
16. *Open data portal of the Russian Federation.* Retrieved from: https://data.gov.ru/
17. *Ministry of Industry and Trade of Russia. GISIP. Industrial clusters.* Retrieved from: https://gisp.gov.ru/gisip/#!en/clusters/
18. *SME Corporation.* Retrieved from: https://corpmsp.ru/upload/19.10.2020.pdf

Social aspects of innovation management in digital sector

Milana Kolesnichenko*, and *Olga* Burova

Perm National Research Polytechnic University, 29, Komsomolsky Prospekt, Perm, 614990, Russia

Abstract. Innovation management in the digital sector requires taking into account the institutional, socio-structural and socio-cultural context. The article discusses the thesis about unevenness of digitalization in different spheres of society and for different social groups. The actual implementation and use of innovations in everyday life depends not only on development of infrastructure and digital literacy, but also on emerging social norms, including associated with peculiarities of culture of society, belonging to the "artificial" or "natural" type. Artificial intelligence and digital innovations affect the field of health, human corporeality, bioethics, biosociality, etc. There are new sociological approaches which make it possible to study the management of innovations, for example, atypical sociology, augmented reality, concept of posthumanism, etc. An information policy is needed that allows representatives of social groups to perceive innovations more adequately.

1 Introduction

Innovations in the digital sector affect different spheres of life of modern peoples and are associated with the formation of new social norms and practices at the individual, group, national and global levels. In this regard, in order to manage innovations in the digital sphere, it is necessary to study their institutional, socio-structural and socio-cultural context.

Institutional goals and norms that stimulate and regulate the digitalization process in Russia are defined in the national program "Digital Economy of the Russian Federation" for the period of 2018-2024. The criterion for the digitalization of society (digital transformation) is the level of use of digital technologies in business, activities of government agencies and everyday practices [1].

The socio-structural context of the introduction of new digital technologies involves the identification of social groups that have more or less resources for development and use of these technologies.

Three levels of the digital divide in modern society can be distinguished: the infrastructure that provides access to information and communication technologies, the skills to use modern technologies, as well as life chances, defined as opportunities or advantages of using technologies [2]. Some social groups, such as the elderly, persons with low incomes and low levels of education, certain ethnic groups, turn out to be the least involved in the digitalization processes in all the indicated parameters.

2 Materials and methods

Digital literacy is not just the ability to use software or digital equipment (devices), it includes the more complex cognitive, emotional and social skills required to function effectively in a digital environment. As shown in a study conducted in Indonesia's telecommunications industry, digital literacy of company employees affects the effectiveness of introduced innovations: digital literacy significantly moderated the relationship between innovative work behavior and employee performance [3]. It is natural to assume that digital literacy of the population is also a resource for effective digitalization of society as a whole.

In Russia, an annual survey is being conducted to analyze the digital literacy of the population. At the same time, the European Commission DigComp 2.0 methodology is used, according to which the analysis of digital competencies is carried out according to 5 main parameters: information literacy, communication literacy, digital content creation, digital security and problem solving skills in the digital environment.

According to the results of a survey in January 2020 (a representative sample of the Russian population, N = 1300 respondents), the digital literacy index of Russians was 58 percentage points on a scale from 0 to 100. With the highest values obtained for the parameter "communicative literacy" (the ability of Russians to use various types of online services and electronic devices for communication on the network) it was 62 percentage points. The lowest values were obtained for the parameter "digital content creation" (human competence in creating and editing digital content, skills in working with copyright on the network) - 53 percentage points. Only 27% of Russians have a high level of digital

* Corresponding author: milana72000@mail.ru

literacy and key competencies of the digital economy (this share remained practically unchanged from 2018 to early 2020) [4].

The digital literacy of Russians is largely determined by the region of residence. The residents of the South and North Caucasian Federal Districts have the lowest rates of digital literacy, the highest are the residents of the Northwestern Federal District. Also, the level of digital literacy is higher in the largest settlements, and the lowest in rural areas. There are no significant differences in the values of digital literacy between men and women. To the greatest extent, the level of digital literacy is associated with age characteristics: the lowest values of digital literacy are among Russians over the age of 55, especially non-working pensioners. Professional activity affects the level of digital literacy, and this indicator is higher among working students as compared with other groups [4].

Differences in the level of digitalization depending on age are also noted in other studies: "different generations live in different virtual realities", i.e. "Analog" (mainly watching TV) and "digital" (active use of gadgets) [5, p. 49]. Representatives of the older generation themselves assess their digital competencies at a low level (see Figure 1): if among young people aged 18-24, 82% of respondents are satisfied with the existing level of digital competencies, and only 17% are not satisfied, then among those aged 60 years and older 42% believe that they do not have enough computer, laptop and telephone skills [6].

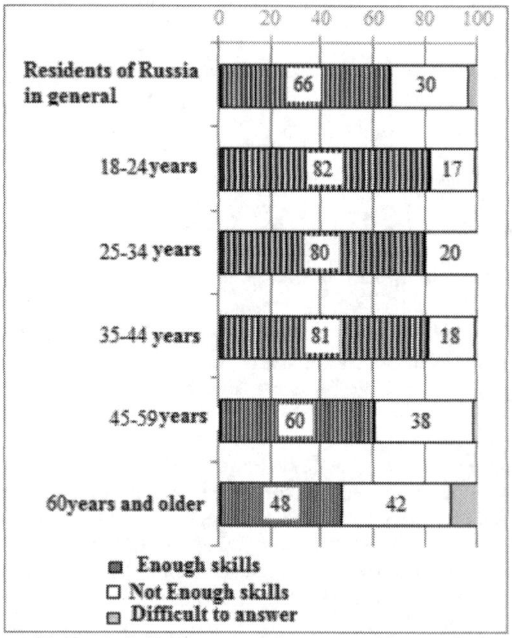

Figure 1. Self-assessment of computer, laptop, telephone skills in different age groups, in % (WCIOM survey data, N = 1600 respondents, April 2020 [6])

In a study by PwC (USA, 2017), groups of business managers, millennials, and consumers (N = 2500 respondents) were identified [7, p. 12]. 72% of business execs, 53% of millennials, 42% of consumers already have been using digital assistants. Moreover, half of managers noted that the free time allows them to immerse themselves in a deeper study of any area or in creativity. Over 40% of consumers believe AI will expand access to financial, medical, legal, and transportation services to those with lower incomes [7, p. 6]. 80% of consumers answered that it is more important for them to have access to legal advice than to keep the lawyer's job due to the use of robots.

According to the index of readiness for the future [8, p. 5, 30] in the "Technologies" section, we used such criteria as the number of mobile subscriptions per 100 inhabitants, the volume of venture capital, the level of innovation, the level of technological complexity of production. Russia ranks 12th (Table 1), the final index for all spheres was 0.38.

Table 1. Technologies Integrated index [8]

Country	Index
Japan	1,0
USA	0,94
Germany	0,89
Republic of Korea	0,85
United Kingdom	0,76
China	0,71
EU	0,55
France	0,52
Canada	0,51
Italy	0,43
Australia	0,42
Russia	0,36
Indonesia	0,15
India	0,15
South Africa	0,11
Saudi Arabia	0,11
Turkey	0,07
Mexico	0,07
Argentina	0,06
Brazil	0,00

3 Results and discussion

The socio-cultural context - the ideas, values, norms and symbolic values that have developed in society - requires special consideration in the process of digitalization of society. A number of sociological studies are devoted to the analysis of the perception of technological (digital) innovations in different cultures.

The data of the World Economic Forum (2019) and the model of national culture by G. Hofstede (2020) make it possible to single out such indicators of artificial sociality as the use of ICT (information and communication technologies) in the country and the digital skills of the active population [9, p. 52]. Such indicators of culture according to G. Hofstede, as Power Distance Index (PDI), Uncertainty Avoidance Index (UAI), Individualism versus Collectivism (IDV), Long Term Orientation versus Short Term Normative Orientation (LTO), (especially the last two), were important for assessing the potential of human-machine interaction. For example, short Power Distance is characteristic of Austria, Denmark, a large one - Russia, the Philippines, Individualism - the USA, Australia, Collectivism - Venezuela, Indonesia [9, p. 61]. Conditionally "artificial" cultures include countries with short Power Distance, values of Individualism, Acceptance of Uncertainty, Long-Term Orientation. Conditionally "natural" cultures include countries with long Power Distance, values of Collectivism, Avoidance of Uncertainty, Short-Term Orientation. Denmark was among the leaders of the "artificial" culture, and Venezuela was among the "natural" ones.

Sociology also examines attitudes towards specific types of innovation. In 2017, a study was conducted of the attitude of Russians (a voluntary response sample of active Internet users, N = 2314 respondents) to the introduction of robots in medicine, genetic modification, cyborgization and unmanned vehicles, as well as the influence of the information factor on social attitudes (obtaining additional information about the essence and value of technology). According to the data obtained, the overwhelming majority of respondents (95.5%) have a positive and rather positive attitude to the use of robotics in medicine and even express their readiness, in case of medical indications, to undergo surgery with a robotic surgeon (78.7%). 17.0% of respondents expressed their disinclination (unwillingness) to undergo surgery with a robotic surgeon. Every fourth respondent (26.1%) expressed a negative attitude towards the possibility of changing the human genome. The introduction of unmanned vehicles is positively or rather positively assessed by 89.4% of the respondents, negative - only 7.2% of the respondents. At the same time, they are completely unprepared or rather unprepared to use an unmanned vehicle themselves - 17.7% of respondents.

Thus, even among persons who are more familiar with new technologies, there is a negative or wary attitude towards the introduction of these technologies into their lives. The authors of the study conclude that it is necessary to systematically prepare society for the introduction of a new technology, to pursue an appropriate information policy, otherwise new technologies can become not only a source of new conveniences, but also a potential determinant of the growth of anxiety in society [10].

In 2018, an assessment was made of possible scenarios for using AI / robots in different spheres of life, according to representatives of different social groups (N = 211 respondents). In general, respondents in Russia are the most supportive of the transfer of functions from humans to artificial intelligence and robots in areas such as driving, disaster prevention and military activities, and the least supportive in areas such as childcare, personal assistant for making important decisions and creative activity [11]. Thus, robotization is largely supported by the public in those areas that are more formalized, algorithmic and not associated with interpersonal, emotionally colored relationships. Similar results were obtained in a survey in Japan in 2015, although there were also reduced assessments of the possibility of robotization (complete replacement of human labor) in the healthcare sector [12].

As of 2020, Russians are ready to use the achievements of AI, first of all, in obtaining public services, in solving everyday problems, leisure and healthcare [1, p. 30], which indicates that everyday consumer practices and the speed of settling related problems are important, as well as culture (rather, with a predominance of hedonistic values) and physicality, bordering on the medical sphere. So, among the main achievements of the XXI century, according to the respondents, bionic prostheses of arms and legs came out on top [1, p. 7].

Particularly acute are such dilemmas as human / subhuman (embryos, genetic "design"), living / nonliving, life extension / death (euthanasia, care for the elderly), physical / non-physical, etc. The unifying factor is ethical issues, which equally relate to both the AI and biomedical (bioethics, which overlap with artificial biosociality), as well as the development of telemedicine.

If we assume that the practice of caring for family members (medical and ethical) in the future will become the sphere of robots (in countries such as Japan, they have already begun), then the attitude of the population to special, sometimes taboo topics (euthanasia, IVF procedures, egg freezing, abortion, some diseases, disability, etc.) is needed to be revealed now. So, on the question of whether it is worth allowing euthanasia in the event of a serious illness that brings suffering, Russian society is split by half - 50% for, 40% against, and 10% found it difficult to answer (2020). To the question "Is it permissible to have a baby with IVF?" - "rather acceptable" was answered by 84% of the respondents [13]. Those who disagree have religious, ethical motives (it is better to take children from orphanages).

Bioethics also touches upon the issues of modern burial practices. There are very few such studies in the field of the sociology of death (death studies) in Russia. Sociologists from Voronezh city conducted a survey in 2016. In particular, the question was asked: Would you agree to be cremated after death? (N = 600 respondents, telephone interview). 40% of respondents answered no, 24% -yes. There are more women among those who answered yes. Among those who believe that this issue should be resolved by relatives, but they do not care, men predominate. Among the respondents with incomplete secondary education, nobody would agree with own cremation [14]. Therefore, a special segment of innovation management is making decisions on construction of crematoria as objects of more modern burial practices.

In the process of managing the implementation of innovations, the importance of the emotional component of social behavior increases. Taking into account the type of corporate culture, it is important to monitor subjective well-being, i.e. emotional assessment of satisfaction with life in general [15, p. 42]. According to researchers, subjective well-being is better in innovative companies as compared to ordinary companies (N = 260 people). The class of professions plays a role under the same conditions for the introduction of innovations: women over 55 and men over 60 (in a situation of raising the retirement age), ordinary organizations. Physicians turned out to be more ready to introduce innovations, to continue working for a five-year period than engineers [15, p. 48]. Perhaps this is due to the fact that Physicians regularly (every 5 years) confirm their qualifications.

Thus, sociological studies reveal the presence of contradictions in the perception of digitalization in the public mind. It is difficult for Russians to decide what AI is, how to make a choice of services and preferences, an emotional attitude is not formed, nevertheless, 1/3 of the respondents have an interest [1, p. 30].

According to the results of an all-Russian survey conducted in January 2020 (a representative sample of the Russian population), the stereotype about the negative impact of artificial intelligence on the labor market is rejected: 68% of the population are not afraid of replacing humans with AI technologies in their profession [16]. When answering an open-ended question (with no answer options) about why the spread of artificial intelligence technologies causes concern, Russians most often talk about a possibility of technical failures: problems in operation of equipment, machines out of control, their unreliability (in total, these problems were indicated by 31% from the number of those who expressed a negative attitude towards AI). Also, the negative attitude is due to the threat to the security of personal data: hacking and theft of personal data, violation of personal space (21%); replacement of live communication between people, personality degradation (16%), lack of understanding of new technologies or lack of trust in them without specifying the reasons (12%).

4 Conclusion

The COVID-19 pandemic and lockdown led to an accelerated and, for the most part, forced digitalization of all spheres of life (consumption, labor, education, leisure, communication, everyday life, etc.), which, on the one hand, increased the digital literacy, on the other hand, it has led to an increase in the level of technophobia and anxiety about the use of digital technologies to spy on humans and manipulate public opinion. In this regard, the implementation of the Digital Economy program, including the National Strategy for the Development of Artificial Intelligence, may require, according to T. Nestik, "significantly more efforts, first of all, a broader public discussion of digital initiatives" [1, p. 22]. Further implementation of digital technologies in the economy and social sphere requires not only the development of appropriate infrastructure, but also the willingness to participate in the digitalization of various social groups, which largely depends on how these groups perceive and evaluate new technologies.

The sociological study of innovation in the digital sector involves development of new theoretical (explanatory) models.

For a long time, social processes taking place in a virtual, digital environment have been explained in sociology in the postmodern paradigm. For example, it was suggested that the formation and spreading of Internet memes as a new cultural product occurs by analogy with simulacra according to the concept of J. Baudrillard. Now sociologists are increasingly talking about post-globalization, post-virtualization, post-humanization. What is the new modernity?

In the conditions of post-globalization and post-virtualization, the concept of augmented reality has emerged. The old modernist structures do not disappear, but are supplemented by new ones, forming a rich active life in urbanized enclaves. In the periphery, there is a "depleted" sociality ("depleted" modernity), poor in cultural and socio-political events [5, p. 51-52].

Artificial sociality is understood as participation of artificial intelligence agents in social interactions as active mediators or participants in these interactions [17, p. 43]. AI in a certain sense is inherent in humans when humans have to think algorithmically, schematically. The boundaries of natural and artificial sociality can shift. Studying the features of human-robot communication, researchers come to the conclusion that it is possible to teach robots to communicate socially, for example, by avoiding long pauses in conversation. And then it is possible to predict the expansion of the "natural (associative) sociality" of robots. Even R. Collins noted that storing information about previous conversations in the memory of a robot allows accumulating cultural capital [18, p. 589]. In the case of humans, the mechanism of communication appears that corresponds to an artificial or natural sociality (culture).

Despite the fact that many social structures and processes work in the same way as before, perhaps an atypical sociology which gets rid of any typification is needed to explain new phenomena [19, p. 10]. Probably, a new direction is also possible in the mainstream of posthumanism, which manifests itself through a rapid increase in the rate of social processes (transience), metamorphism (constant renewal), "desecuritization", i.e. lack of security and an abundance of risks up to total control using artificial technologies, etc. [20, p. 18-19].

Gadgets, virtual services, online learning, telecommuting, artificial intelligence technologies have already spread enormously. However, the problems of involving different social groups, classes of professions in digitalization, the social consequences of digital technologies and ways to regulate them remain an open question and require further study.

References

1. *SOCIODIGGER. Digitalization and artificial intelligence.* October **1(3)** (2020) retrieved from: https://wciom.ru/fileadmin/file/nauka/podborka/wciom_sociodigger_102020.pdf
2. D. E. Dobrinskaya, T. S. Martynenko *Monitoring of Public Opinion: Economic and Social Changes,* **5**, 100-119 (2019) https://doi.org/10.14515/monitoring.2019.5.06.
3. H. Santoso, S.B. Abdinagoro, M. Arief *International Journal of Technology,* **10(8)**, 1558-1566 (2019) DOI:10.14716/ijtech.v10i8.3432
4. Analytical Center NAFI. *Digital literacy of Russians: study 2020.* Retrieved from:

https://nafi.ru/analytics/tsifrovaya-gramotnost-rossiyan-issledovanie-2020/

5. D.V. Ivanov, *Sociological Research,* **5**, 44-55 (2020)
6. *Digital literacy and remote works in a pandemic: Joint analytical report by VTsIOM and Social Business Group.* May 15, (2020) Retrieved from: https://wciom.ru/analytical-reports/analiticheskii-doklad/czifrovaya-gramotnost-i-udalennaya-rabota-v-usloviyakh-pandemii
7. Bot.Me: A revolutionary partnership. How AI is pushing man and machine closer together (2017) Retrieved from: https://www.pwc.com/CISAI
8. Valdai. Future readiness index. Joint project of VTsIOM and Valdai Club (2019) Retrieved from: https://ru.valdaiclub.com/a/reports/indeks-gotovnosti-k-budushchemu-2019/
9. V. Komarova, M. Kolesnichenko, *Bulletin of social sciences,* **2 (31)**, 40-65 (2020)
10. E.N. Yadova, P.A. Levich, *Technologos,* **2**, 25-41 (2020) DOI: 10.15593/perm.kipf/2020.2.03
11. E.V. Seredkina, O.A. Burova, I.I. Bezukladnikov *Social and Humanitarian Knowledge,* **10**, 109-114 (2018)
12. Arisa Ema, Naonori Akiya, Hirotaka Osawa et al. *IEEE Technology and Society Magazine,* **4(35)**, 68-75 (2016) DOI: 10.1109/MTS.2016.2618719
13. *Health. Healthcare. Bioethics. SOCIODIGGER (extended selection).* November, **1(4),** (2020) Retrieved from: https://profi.wciom.ru/fileadmin/file/nauka/podborka/wciom_sociodigger_112020.pdf
14. *Sociologists have found out how Voronezh residents feel about cremation* Retrieved from: https://moe-online.ru/news/society/347707
15. I.S. Leonova, L.N. Zakharova *Bulletin of PNRPU. Socio-economic sciences,* **4**, 39-53 (2020)
16. *Artificial Intelligence: Threat or Opportunity? Analytical review of VTsIOM.* January 27, (2020) Retrieved from: https://wciom.ru/analytical-reviews/analiticheskii-obzor/iskusstvennyj-intellekt-ugroza-ili-vozmozhnost
17. A. V. Rezaev, N.D. Tregubova *Monitoring of public opinion: economic and social changes,* **6**, 35-47 (2019) https: //doi.org/10.14515/ monitoring.2019.6.03.
18. *Personality-oriented sociology.* (Moscow: Academic project) (2004)
19. A.V. Rezaev, V.S. Starikov, N.D. Tregubova *Sociological studies,* **2,** 3-12 (2020)
20. I.V. Katerny, *Monitoring of public opinion: Economic and Social Changes,* **6**, 13-34 (2019) https://doi.org/10.14515/monitoring.2019.6.02.

 https://doi.org/10.1051/shsconf/202111600046

Energy management as an innovative practice of public administration in the Leningrad Region

Nadeshda Kosmacheva[1,*], *Galina* Cherkasskaya [2], and *Irina* Dolgova[3]

[1] Pushkin Leningrad State University, 10 Petersburg Highway, 196605 St. Petersburg, Russian Federation
[2] Pushkin Leningrad State University, 10 Petersburg Highway, 196605 St. Petersburg, Russian Federation
[3] Bauman Moscow State Technical University, 5, 2-ya Baumanskaya, 105005 Moscow, Russian Federation

Abstract. The purpose of the research is to identify vector of sustainable industrial development in the context of limited resources, in particular, in the context of the efficient energy use as a decisive factor. An attempt has been made to determine the level of managerial and economic efficiency of energy sector management in the Leningrad Region of the Russian Federation through the introduction of energy management methods and measures. The energy sector economy basics and peculiarities have been considered in the course, including goal setting, planning and execution of energy management activities in the Leningrad region at the state, municipalities and corporations level. A certain readiness of the main state municipal and corporate levels management institutions to implement some energy management areas, methods and activities of the world standard aimed at achieving efficiency indicators set by the planning and legal documents of higher organizations, is set up. At the same time, the executive and consumer levels of activity demonstrate a certain lack of energy management importance understanding as a strategy and set of rules conduct in production and in everyday life. The level of management and economic efficiency of the energy sector management in the Leningrad Region has been determined as an average. And it was concluded on the basis of the conducted research that it is necessary to introduce a systematic approach to the solution of strategic issues of region sustainable development into management practice of the Leningrad Region, as well as to ensure a closer connection between energy, socio-economic and environmental projects.

1. Introduction

In the conditions of population growth and depletion of energy sources, the tasks of ensuring sustainable social and economic development of the territories cannot be solved by the regional system of power without the use of innovative solutions in the field of energy management in the energy sector.

Increasing the energy sources management, resources and consumption efficiency becomes one of the main trends in the theory and practice of public and corporate governance of socio-economic systems of various types and levels.

According to the World Bank's known data (for 2009), the implementation of the existing energy saving potential at that time would have allowed Russia to save up to 100-150 bil. RUB of federal and municipal budgets per year (25-40% of GDP (in 2009 prices)) [1].

However, it is important that the declared and/or implemented energy saving and energy procurement management policy in the regions themselves are consistent with the task of ensuring sustainable development of the territories and improving the life quality of the population.

However, the studies are pointing to certain problems in this area, both at the level of planning and the level of implementation of planned activities [2, 3].

Due to the significant differences in the location of administratively designated areas in the Russian Federation, their geographical and climatic characteristics, the location of productive forces, including production and distribution of energy, the research of local practice of applying strategic guidelines in the economics field of the energy sector and energy management at the level of specific regions are of scientific and practical interest.

The purpose of this article (research) is to determine the level of managerial and economic efficiency of energy sector management in the Leningrad region of the Russian Federation through the introduction of energy management methods and measures.

2. Materials and methods

The scientific and informational materials on the energy sector economics, theory and practice of energy management, including information about the work of the energy sector, corporations, state and municipal authorities and management of the Leningrad region of the Russian Federation, as well as documents of the legal framework of energy management of the Russian Federation and the UN, are used in the paper.

The modern methodology of socio-economic research, including system and process approaches,

* Corresponding author: n.kosmacheva@lengu.ru

methods of forecasting, comparison, formal-logical and statistical analysis, as well as tools of scientific analysis of documents and results of activities, are used in the paper.

3. The result

As is known, the Leningrad Region (a constituent entity of the Russian Federation, with 94.6 thousand square kilometers total area) is a highly urbanized territory. 19 cities of the region and 10 cities of district subordination are home to almost two-thirds of its population, and eight cities of the region (Vyborg, Gatchina, Tikhvin, Sosnovy Bor, Kirishi, Vsevolozhsk, Sertolovo, Murino) belong to the average category, in which the number of residents is over 50 thousand people.

As of the 1st of January, 2020, the Leningrad Region population was 1875.9 thousand people, including 615.6 thousand people in a rural areas.

In general, the region has an imbalance in the settlement system, uneven development of municipalities and sometimes an increase in depreciation of fixed assets of the municipal infrastructure.

In total, there are 205 municipal entities on the territory of the Oblast, but the highest public authorities of the Leningrad Oblast are currently located in St. Petersburg.

The region has a vast territory (500 km from west to east and 320 km from north to south), an external (state) border with two countries: Finland and Estonia, and an internal border with 5 subjects of the Russian Federation: the Republic of Karelia and regions: Vologda oblast, Novgorod oblast and Pskov oblast, as well as the federal city of St. Petersburg.

The Leningrad Region is one of the leaders in economic development in the North-Western Federal District of the Russian Federation.

The region's economy is based on industry, accounting for about a third of the GRP structure.

The following industries are represented in the industrial complex of the Leningrad Region: mechanical engineering, automotive industry, shipbuilding, chemical production, petrochemistry, timber processing, pulp and paper production, aluminum industry, construction materials industry, etc.

The key companies located in the Leningrad Region are Philip Morris Izhora CJSC, Novatek-Ust-Luga LLC, Kirishinefteorgsintez Production Association LLC, Tikhvin Freight Car Building Plant JSC and others [4].

The Program of the region development until 2030 provides the implementation of several large investment projects in the sphere of petrochemical, transport and energy industry [5, 6].

The main indicators of the regional economy, including the energy sector of the Leningrad Region, are presented in Table 1.

Table 1. Leningrad Region economy and energy sector key indicators.

Indicator	2013	2015	2017	2019
Gross Regional Product in the basic.prices, bil. RUB/bil. EUR	678.7/ 16.1	853.6/ 12.2	963.8/ 13.8	1100.0/ 12.9
- per capita, thous. RUB	386.2	480.3	534.6	610.0
Electricity consumption, bil. kWh	18.951	19.167	25.16 1	21.500
Maximum power consumption, MW	3129	2995	4511	7703
Balanced financial result (profit minus loss) in the economy, bil. RUB	96.0	109.6	161.9	144.2
Investments in the fixed assets, bil. RUB	253.6	225.9	337.7	419.1
Consumer price index (December vs. the previous year December), %	106.0	113.3	102.7	102.9

*Reference: EURO exchange rate for the 2013 – 2019 years (43 rub.; 70 rub.; 70 rub.)
The figures for 2019 are estimates.
Source: authors' calculations based on [5-7] and statistical data http://petrostat.gks.ru

The Leningrad Region produces about 4% of electricity and 7% of oil products from total Russian volumes.

At the same time, according to the data of the special GIS "Energy Efficiency" (expert portal on energy saving issues https://gisee.ru/regionsupport/), the energy consumption in the region was 7569157 toe in 2018, and the energy intensity was 214,30, which is lower than the average level. The fuel and energy balance of the Leningrad Region is represented by six types of primary resources (shale, peat, electricity, crude oil, natural gas and coal). The last three of these are imported into the region, and the first three are produced. Coal is imported from the Kuznetsk and Pechora coal basins, natural gas is supplied from the Komi Republic and northern regions of the Tyumen Region, and refineries from the cities of Kirishi, Ufa, Perm and Yaroslavl are the main suppliers of fuel oil. Thus, the supply of all natural gas and coal required to meet the regional energy generation needs, primarily thermal power, is carried out through imported procedures, while almost entire volume of oil products produced by "Kirishinefteorgsintez" is sent outside the region and outside the Russian Federation.

Electricity supply to consumers located in the Leningrad Region is carried out by an integrated power system (Saint-Petersburg and the Leningrad Region). It, being the largest, is a part of the United Power System of the North-West (IES of the North-West), along with the power systems of the Murmansk, Novgorod, Pskov Regions, Arkhangelsk and Kaliningrad Regions (since 2004) and the Nenets Autonomous District, as well as the power system of the Republic of Karelia and the Republic of Komi. At present, more than a dozen of power generating and distribution companies work in the Leningrad Region, including TGC-1 PJSC, Rosenergoatom Concern JSC, Rosseti Lenenergo PJSC, various CHP, and industrial power plants.

Leningrad NPP generates the main proportion of electricity in the region. The LNPP is developing progressively (work in the second power unit is being completed), so the Leningrad Region is a major supplier of electricity to the country's unified energy system, supplying not only St. Petersburg and the entire North-West, but also to the EU countries, primarily Finland. The structure of electricity consumption in the region is stable: half of it is consumed by the primary sector; about a tenth is a domestic consumption. The rest is distributed between the secondary and tertiary sectors, including industry and transport. The fuel and energy resources conversion sector remains of key importance in the use of the gross mass of primary energy and its equivalents, as the main consumers of fuel and energy resources of the Leningrad Region are the oil refining industry and energy production. In general, the balance of the electric power industry is steadily positive and has available excess capacity.

As it can be seen, the energy sector plays an important role in the economy of the Leningrad Region. The efficiency of economy and management in this sector largely determines the final efficiency of public administration in the region.

In order to improve the economic and managerial efficiency of the energy sector, the Oblast Government applies, using the regulatory framework and experience of the federal [8, 9] and world-class [10], methods and measures of energy management [11], implementing them at the state, municipal and corporate levels of public production and management, taking into account environmental problems [12].

The set of methods and measures of energy management can include demand management, energy consumption and load [13], and supply, as well as procurement management, energy production and energy consumption management, carried out at the state, municipal, corporate and individual levels.

According to the region data, this activity has some success [14]; and, moreover, the rating of the Leningrad Region at the country level is 28 (2015) and 5 (2016 - 2019). The results of management activities of municipalities' energy efficiency for 2016-2019 are determined by a special rating and selectively presented in Table 2.

Table 2. The results of municipalities' energy efficiency management activities for 2016-2019.

Region	Population for 2020	Municipal area rating based on 2015 data	Municipal area rating based on 2017 data	Municipal area rating based on 2019 data
Leningrad region, total	1875.9			
including the municipal neighborhoods:				
Boxitogorsky	48.6	7	16	12
Volosovsky	51.8	14	18	14
Volkhovsky	88.2	6	10	3
Vsevolozhsky	438.6	9	9	13
Vyborgsky	198.2	13	17	9
Gatchinsky	238.6	9	12	10
Kingiseppsky	74.9	16	11	5
Kirishsky	61.5	1	1	6
Kirovsky	106.0	12	4	7
Lodeynopolsky	28.0	10	5	1
Lomonosovsky	76.8	11	15	13
Luzhsky	70.8	3	3	4
Podporozhsky	27.7	9	8	8
Priozersky	60.4	2	2	7
Slantsevsky	42.3	5	14	3
Tikhvinsky	69.5	2	6	1
Tusnensky	126.9	8	13	2
urban district Sosnovoborsky	67.7	4	7	11

The rating of energy efficiency of the MD is determined by 13 criteria in accordance with the value of the consolidated energy efficiency index calculated for each MD (UD) of the LR based on the maximum number of points – 115. The main sources of information for the indicators of energy efficiency of LR MD (UD) ranking were the information systems (GIS "Energy Efficiency" module, GIS "Energy Efficiency", RGIS "Energy Efficiency"). These include the official information prepared by the administrations of LR MD (UD) within the framework of the information preparation on the Leningrad Region for inclusion to the state report on the state of energy saving and energy efficiency in the Russian Federation [15, 16]. As it can be seen, there is a multidirectional dynamics of the rating, but a long-term high rating of the Kirishi MD as a consequence or a reflection of the conscious corporate energy efficiency of the enterprises located in it, is observed. Conversely, there is also a consistently low rating of Volosovsky MOE as a consequence of the negative individual and entrepreneurial consumer behavior in its energy sector.

One of the important areas of energy management reflected in this rating is the availability of metering devices for all used energy resources, for both general and individual use. The equipping level with energy metering devices in the Leningrad Region in 2016 - 2019 is presented in Tables 3 and 4.

Table 3. Level of equipment with household energy metering devices in the Leningrad Region in 2016 – 2019.

Type of energy resource	Household metering devices % of equipment in			
	2016	2017	2018	2019
Cold water supply	38.84	44.90	47.30	50.42
Hot water supply	31.08	36.86	43.34	46.98
Thermal power	36.59	35.18	40.41	44.61
Electrical energy (without public area meters)	34.96	37.87	43.12	44.95
Gas	5.34	4.78	6.95	9.96

Table 4. Level of equipment with individual energy metering devices in the Leningrad Region in 2016 – 2019.

Type of energy resource	Individual metering devices % of equipment in			
	2016	2017	2018	2019
Cold water supply	72.42	74.24	77.96	84.54
Hot water supply	74.55	79.58	82.71	86.20
Thermal power	5.06	4.13	22.29	38.19
Electrical energy (without public area meters)	89.90	97.68	98.63	99.22
Gas	45.74	44.18	48.16	47.10

As it can be seen, the energy metering devices equipping level in the Leningrad Region is not sufficient, moreover, its dynamics for 2016-2019 is multidirectional, which indicates a certain problematicity of standard energy management measures implementation at the individual consumption level.

Indicators of equipment with energy meters in the Leningrad Region of buildings, structures, and structures of municipal property occupied by administrations of MD (UD) of the LR and settlements of MD of the LR, MI of the LR for municipal entities are ranging from 71% to 100%.

It should be noted that since 2003, the region has been implementing the 'Concept of Fuel and Energy Resources Development for Municipal Boilerhouses of the Leningrad Region until 2020'. It has been aimed at optimizing the fuel balance structure by means of more local fuels, gasification of heat supply facilities, reduction of fuel oil and coal consumption, and general improvement of fuel efficiency at municipal boilers of the Leningrad Region. So far, the concept is being implemented in a sluggish manner, as the Leningrad Region Government is only looking for huge investors to convert boiler houses to local fuels and to introduce mini-CHPs for combined heat and power generation in the Leningrad Region's communities. At the same time, existing players in the energy sector are successfully implementing their own investment programs. For example, in 2015, LOESK invested over 2.5 billion RUB in the Leningrad Region's power grid sector, and about 3.3 billion RUB in 2016. In 2017, LOESK's specialists connected about 2.5 thousand subscribers to the network with a total capacity of about 100 MW. Experts believe that the number of technical connections and the volume of capacity allocated by the company will grow year by year [3]. In general, energy efficiency improvements are implemented by the Government at the oblast and municipal levels, with a breakdown by sectors (energy, housing and public utilities, agriculture, transport, public sector), with the involvement of budgetary and extra-budgetary funds. Within the framework of the subprogram "Energy Saving and Energy Efficiency in the Leningrad Region" of the Leningrad Region state program "Ensuring the sustainable functioning and development of municipal and engineering infrastructure and energy efficiency of the Leningrad Region", a set of measures is implemented in the heat and water supply systems, in the housing stock of the region (500 million rubles annual financing).

In general, the measures, planned and undertaken by the Leningrad Region Government in 2018-2019, to improve the regional policy in the energy saving and energy efficiency field may include measures to increase the responsibility of Leningrad Region sectoral executive authorities for the implementation of energy saving and energy efficiency policies in the sectors of the Leningrad Region economy. This also includes recommendations to the Leningrad Region municipalities to adjust the municipal programs in the energy saving and energy efficiency field, including the activities of the "road maps" of the municipal districts (urban district) for the installation of energy metering devices and automated individual heat supply points with weather and hourly regulation, as well as the modernization of internal lighting systems in Leningrad Region municipal institutions. This aims to provide funds from the local budget for their realization and other equally significant events.

Thus, the study allowed one to determine the management level and economic efficiency of the energy sector management in the Leningrad Region as an average. It showed a certain readiness of the main management institutions at the state, municipal and corporate levels to implement some areas, methods and activities of energy management of the world standard, aimed at the achievement of performance indicators set by the planning and title documents of higher organizations. The executive and consumer levels of activity demonstrate a certain lack of understanding the of energy management importance as a strategy and a set of rules conduct in production and everyday life.

4. Discussion

It should be noted that although the energy sector economics is a rather topical area of scientific research in recent years, the problems of energy management from 2008 to the present time in our opinion are not an acutely relevant research subject for Russian scientists. Thus, there were less than 0.003/0.001% (about 900/300 works out of more than 28 million works) publications on the first/second theme at the end of 2020 in the Russian electronic library E-library (https://elibrary.ru/).

It is noteworthy that the discussions in this sphere are particularly acute [17, 18] and descriptions of specific experiments and methods rather prevail [19, 20, 21]. As our review of domestic publications has shown, the discussion on energy management issues remains exclusively at the specific level of management [22, 23, 24], responsible for planning and implementing energy efficiency plans, because both the strategic need and the operational cost of required activities are obvious, and the resource base is always limited. This statement applies equally to corporations in any field of economic activity and to the authorities at the industry, region and municipality level. Discussion is also possible in the field of methods for calculating the efficiency of specific energy management measures at the stage of achievements evaluation [20].

5. Organizational conclusion

Introduction of innovations in the energy management system in the region and achievement of energy management goals, including indicators of economic efficiency of the energy sector and energy efficiency of the regional economy in the Leningrad Region as a whole is currently carried out through compliance with the provisions and indicators of a special federal law, the Energy Strategy of Russia (2030), the development strategy of the Leningrad Region for the period up to 2030 and the implementation of the Leningrad Region State Program «Ensuring sustainable functioning and development of communal and engineering infrastructure and energy efficiency in the Leningrad Region» (in 2013).

In our opinion, sluggish implementation of the planned activities in the region is explained by the fact that, despite the certain readiness of the main management actors at the state, municipal and corporate levels to implement some areas, methods and activities of energy management of the world standard, aimed at achieving efficiency indicators set by the planning and legal documents of higher organizations, the executive and consumer levels of activity and consumption are demonstrating a certain lack of understanding of the energy management importance as a strategy and a set of rules conduct in production and everyday life.

In this regard, the regional administration should be concerned about the introduction of elements and even a system of knowledge on personal and industrial energy management into the educational practice (at different levels) [25, 26], starting from junior and secondary schools (the individual consumption level), professional schools (the collective and industrial consumption level) and ending with special programs of additional education. This is aimed at the mentality changing of existing managers, public and municipal employees, so that energy management becomes a natural part of management practice, as well as quality management, so that all decisions have been made taking into account the requirements of fundamental documents in this area, not only in Russia, but also at the international level.

The aforesaid allows one to conclude the urgent need for introduction of a more systematic approach to solve the issue of power efficiency at the state and municipal management level of the Leningrad region of the Russian Federation. Namely, increasing power efficiency of the regional social and economic system as the purpose of power management should be considered as a way to improve the system of public production, the environment quality and the life of the population in the region.

References

1. World Bank Group Report. (2009) Retrieved from: http://www.cenef.ru/file/FINAL_EE_report_rus.pdf
2. I.A. Bashmakov, V.I. Bashmakov, *report*. Moscow: CENEF (2012) Retrieved from: http://esouz.ru/ENERGOAUDIT/portal-energo_ru_sravnenie_e_ef_v_rossii_i_za_rubezhom.pdf
3. Energy of the Leningrad Region: vector of development, *Const. Techn. Org .J.* **03/44**,54-57 (2016)
4. *Issues of managing the socio-economic development of the Leningrad Region.* (St. Pb, PSU, 2018)
5. *Strategy of socio-economic development of the Leningrad Region until 2030* (St. Pb, 2017). Retrieved from: http://econ.lenobl.ru/work/planning/concept2030
6. *Action Plan for the Implementation of the Strategy of Social and Economic Development of the Leningrad Region until 2030* (St. Pb, 2017) Retrieved from: http://docs.cntd.ru/document/456095537.
7. *Forecast of socio-economic development of the Leningrad region for the period up to 2030.* (St. Pb, 2015) Retrieved from: http://econ.lenobl.ru/work/social/prognoz/prognoz2030
8. *On energy saving and energy efficiency and on amendments to certain legislative acts of the Russian Federation*: Fed.L. **261-FZ** (2009)
9. *Energy Strategy of Russia 2030, Min. of Ener. of the RF* Retrieved from: http:// minenergo.gov.ru/aboutminen/energostrategy/
10. UNIDO Centre for International Industrial Cooperation in the Russian Federation. An. Rep. (2017) Retrieved from:: http://www.unido.ru/upload/files/a/annual_report_2017_rus.pdf
11. Ensuring sustainable functioning and development of communal and engineering infrastructure and the energy efficiency improvement in the Leningrad region, *Resol. of the Gov. of the LR* **400** (2013)
12. N.M. Kosmacheva, G.V. Cherkasskaya, *Intern. SGEM.* **18** (5.3), 343-350 (2018)
13. L.D. Gitelman, B.E. Ratnikov, M.V.Kozhevnikov, *Reg. Econ. J.* **2**(34), 71-84 (2013)
14. *Regional report on the state of energy saving and energy efficiency improvements of the Leningrad*

Region at the end of 2019, St. Pub. Ins. of the LR, CESE LR, (2019). Retrieved from: http://www.lenoblces.ru/
15. *Rating of administrations of municipal districts (urban district) of the Leningrad region in the field of energy saving and energy efficiency improvements,* (2016). Retrieved from: http://www.lenoblces.ru/wp-content/uploads/2017/01/Rejting-e-nergoe-ffektivnosti-po-itogam-2016-goda.pdf
16. *Rating of administrations of municipal districts (urban district) of the Leningrad region in the field of energy saving and energy efficiency improvements,* (2017). Retrieved from:: http://www.lenoblces.ru/wp-content/uploads/2018/01/Rejting-e-nergoe-ffektivnosti-po-itogam-2017-goda.pdf
17. E.V. Lyapuntsova, Z.M. Kertbiev, *Econ. and Entrep. J.* **3**(56), 406-410 (2015)
18. A.A. Tupikina, *Bus. Educ. Law.* J. **2 (31)**, 219-223 (2015)
19. A.A. Lukinov, I.L.Pichugin, *Real Est: Econ. Men. J.* **1**, 73-77 (2013)
20. I.I. Kashapov, B.G. Ziganshin B.G, *Innov. in Agric.* J. **2** (23), 19-24 (2013)
21. Yu. E. Polyak, *Infor. Resour. of Rus. J.* **4** (134), 13-15 (2013)
22. M.N. Dudin, E.E.Frolova, Yu.A.Artemeva, V.V. Bezbakh, A.N. Kirsanov, *Intern. J. of Ener. Econ. and Pol.* **6. (4),** 714-720 (2016)
23. S.M. Voloshin, *Innov. in Agric.* **2** (12), 9-16 (2015)
24. Z.V. Bragina, E.A. Makhova, *Vest. of Kostr. St. Un. named after N.A. Nekras.* **17.** (5-6), 303-307 (2011)
25. M.S. Dmitrieva, I.V. Bogomolova, *Innov. in Inf. Tech., Mech. Eng. and Mot. Trans. ISP Conf. KS TU T.F. Gorbachev*,20-22 (2017)
26. G.V. Cherkasskaya, N.M. Kosmacheva, *Vocat. Educ. Scien. Innov. in the XXI Cen*t. (VIII St. Pb. Congr. NMRMU "Gorny", 84-89 (2014)

Mechanisms of knowledge-driven innovation: evidence from Russia

Mariia Molodchik[1,*], *Anatoliy* Molodchik[2], and *Irena* Esaulova[2]

[1] National Research University HSE, Perm campus, 38, Studencheskay str., Perm, 614070, Russia
[2] Perm National Research Polytechnic University, 29, Komsomolsky pr., Perm, 614990, Russia

Abstract. The paper aims to propose theoretical framework for knowledge-driven innovation and justify it empirically using database of 110 Russian companies. The study utilizes multidisciplinary approach and develops theoretical framework in the intersection of knowledge-based theory of a firm, the concept of learning organization, the concept of intellectual capital and the concept of self-development and self-organizational systems. The partial least square was applied for modelling the impact of the flexible structure, knowledge culture, transformational leadership on proactive employee behavior that in turn influences knowledge resources and organizational learning followed by company's innovation. The study confirmed the key role of proactive behavior in knowledge-driven innovation. Findings revealed the highest impact of transformational leadership on employee behavior and innovation, as well as positive and significant impact of knowledge culture and indirect effect of flexible structure.

1 Introduction

Innovation activities are one of the main sources of company success and sustainable development [1, 2]. Consequently, the search for innovation drivers attracts scholars' attention and has wide practical implications for business. For Russian business context such research is of particular importance because Russia has a catching up position in innovation activities. Namely, the level of innovation activities in 2019 in Russia was equal 9,1; while in developed countries it is higher than 60 points. Another example of innovation gap is the share of innovation products in sales, which was 5,3% in 2019 for Russian companies [3].

Knowledge-driven innovation is an emerging topic and allows at investigating sources of innovations that are connected with knowledge management. Most theoretical and empirical studies discussed the question of knowledge impact on innovation performance [2, 4], meanwhile the question of mechanisms, which intensify knowledge resources and push the processes of organizational learning, is still underdeveloped.

This study aims to propose a novel theoretical framework for knowledge-driven innovation of a company. It will contribute to innovation and knowledge management field with discovering mechanisms of knowledge transformation into innovation. The rest of the paper is organized as follows. Next section is devoted to theoretical background. Further the authors describe the methodology of the study. Results and conclusions finalize the paper.

2 Theoretical background

The theoretical framework for knowledge-driven innovation is elaborated in the intersection of the following theories: knowledge-based theory developed by Grant [5], organizational learning concept proposed by Argiris and Shon [6], intellectual capital concept developed by Stewart [7] and concept of self-development and self-organizational systems proposed by Molodchik et al. [8].

According to Oslo Manual an innovation is "a new or improved product or process (or combination thereof) that differs significantly from the unit's previous products or processes and that has been made available to potential users (product) or brought into use by the unit (process)" [9]. Scholars confirmed that knowledge resources are a base for innovation, but what mechanism states behind this link is still a black box and call for detailed investigation. Knowledge resources contain human, organizational, innovation and relational resources according to the concept of intellectual capital [7]. These resources are organized and managed within the processes of organizational learning. According to Garvin et al. [10] organizational learning is a spiral mechanism of creating and sharing knowledge among individuals, groups, and the entire organization. Concrete learning processes include experimentations, information collection and transfer, analysis, education and training.

The authors of this study suppose that organizational learning mediates the impact of knowledge endowment on innovation. Therefore, the *first proposition* of this study is as follows, the organizational learning has positive impact on company's innovation.

* Corresponding author: mmolodchik@hse.ru

The second question addressed in the study concerns the drivers of organizational learning and knowledge resources. One of the answers might be a vector of employee proactive behavior. The concept of proactive behavior was introduced in early 90th by Crant [11] and further developed by Parker et al. [12]. The proactive behavior includes the patterns of self-development, initiative and organizational commitment. Taking into account this concept the authors put forward the *second proposition* of the study: employees with proactive behavior intensify knowledge resources and participate in organizational learning processes. Altogether, proactive behavior, organizational learning and knowledge resources constitute knowledge capability of a company to drive its innovation.

Searching deeper in understanding the knowledge-driven innovation, the authors propose mechanisms that activate such proactive behavior. These organizational and motivational mechanisms include knowledge culture, transformational leadership and flexible structure. Knowledge culture is a specific type of culture that encourages knowledge creation, knowledge sharing, and knowledge application [10]. Transformational leadership is based on encouraging employees to accomplish extraordinary tasks [13]. Some studies found out that transformational leadership might facilitate individual learning and employee engagement. At the same time leadership is acknowledged as a shaping factor of the knowledge culture. Flexible structure determines the level of autonomy and decentralization within the company. According to [14] organizational structure is an antecedent for many KM practices and therefore, affects KM output. Meanwhile, organizational structure is the most understudied issue in KM topic. *Third proposition* of this study states that organizational and motivational mechanisms induce proactive employee behavior, which in turn intensifies knowledge resources and activates organizational learning.

Summarizing all propositions stated above the theoretical framework of knowledge-driven innovation chain can be elaborated (fig.1).

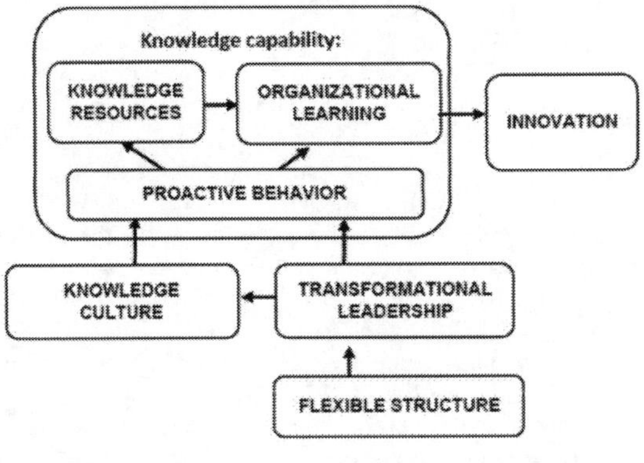

Fig. 1. Theoretical framework "Mechanisms of knowledge-driven innovation".

Elaborated theoretical framework considers flexible organizational structure as exogenous variable which is the premise for the chain of knowledge-driven innovation. Meanwhile, transformational leadership and knowledge culture have direct impact on proactive behavior. Moreover, following the study of Molodchik and Jardon [14] this framework proposes that transformational leadership has an impact on knowledge culture. To the best of our knowledge the presented framework is the first one giving the complex view on knowledge-driven innovation specifically for Russian business context.

3 Methodology and database

In order to justify empirically the elaborated theoretical framework, the authors performed quantitative study on the sample of 110 Russian companies. The importance of knowledge management and innovation promotion in Russian business context was underlined in several studies [15, 16]. The database was collected in 2019-2020 using google form shared with the help of "snowball" technique and questions with regard to company's innovation, knowledge resources, organizational learning, proactive behavior, knowledge culture, transformational leadership and organizational structure. The questionnaire contained statements with 5-point Likert scale answers. Table 1 represents metrics with loadings and latent variables of the study.

The companies are from different industries but mostly from manufacturing one, in particular 64% of the sample. 51% from the sample have R&D expenditures, and 21% has high technological level compared with foreign competitors, 48,5% of companies are certified according ISO. On average around 70% of employees have high education degree, at the same time only 43% upgraded their qualification during last year. Relational resources were estimated through long-term agreements, collaboration with educational institutions and export activities. According to the survey 47% of companies excel export activities, more than 60% have relationships with universities and other educational institutions, 53% have long-term agreements with customers. These indicators were used to estimate the endowment of knowledge resources. The index of knowledge resources was calculated using principal component analysis.

In order to estimate innovation activities, the respondents were asked about product, technological, marketing and management innovations in a company, whether they were increased or not during last three years. 5-point Likert scale was also applied to these questions.

For econometric methodology the authors used partial least square technique with the help of software smartPLS [17]. This method fits for exploratory studies, can be used for small samples and allows at investigating complex structural paths of the model. In particular, the authors implemented mediation analysis, in which an antecedent variable affects a mediating variable, which then affects a dependent variable. Therefore, a researcher can explain the process or mechanism by which one variable affects another.

Table 1. Loadings of latent variables

Metrics	Loadings
Organizational learning (CA=0,80; CR=0,86; AVE=0,56)	
Organization attracts, develops and retain talented employees	0,707
Organization experiments frequently, initiates new product development	0,765
Positive experience of one department is spread over the whole organization fast	0,717
Organization regularly conducts monitoring and compare itself with competitors and best practices of leading companies	0,757
Organization initiates joint projects with potential partners	0,794
Proactive behavior (CA=0,85; CR=0,89; AVE=0,57)	
Most employees are able to achieve the desired results with minimal control from the management	0,735
Employees are ready to upgrade their professional qualification during private time	0,831
Employees introduce initiatives on new product and services development	0,766
Employees experiment and initiate new ways of working with customers	0,808
Employees are ready to work for a long time in the enterprise	0,631
Employees voluntarily, even detriment personal interests, make additional efforts to achieve the desired results	0,760
Knowledge culture (CA=0,89; CR=0,92; AVE=0,70)	
Organization encourages strive for education	0,826
Organization encourages, values and rewards team work	0,875
Organization encourages their employees for initiative behavior, to elaborate new ideas for product, process and technological development	0,824
Organization encourages employees to define the task and to achieve the goals with high level of independency	0,870
There is a high level of trust between people in the company	0,782
Transformational leadership (CA=0,85; CR=0,89; AVE=0,69)	
Leaders inspire others to solve complex problems and to accomplish extraordinary results	0,793
Leaders take into account the opinion of other employees	0,821
Leaders delegate authority and decision-making for achieving strategic purposes of a company	0,834
Leaders are examples of entrepreneurial, responsible, innovative behaviour	0,867
Flexible structure (CA=0,85; CR=0,91; AVE=0,77)	
Organizational structure allows leadership behavior at all levels	0,876
Organizational structure allows one to create new structural units in short time period	0,877
Organizational structure allows one to delegate authority to employee, which decided to take responsibility for new project	0,883

4 Results

The model estimated in smartPLS package fits all indicators of reliability and validity. In table 1 one can see the values of Cronbach alpha (CA), complex reliability (CR) and average value extracted (AVE). The standardized root mean square residual of estimated model is SRMR=0.082, which reflects appropriate level for model acceptance. Exploratory power expressed with R^2 was higher than 15% for all dependent variables, that also reflects appropriate level for model acceptance. Moreover, the dependent variables "proactive behavior" and "innovation" demonstrated R^2 higher than 50% (see table 2). The significance of structural paths was estimated with the help of bootstrapping at alpha level equals 0,05. Figure 2 presents direct paths of estimated model.

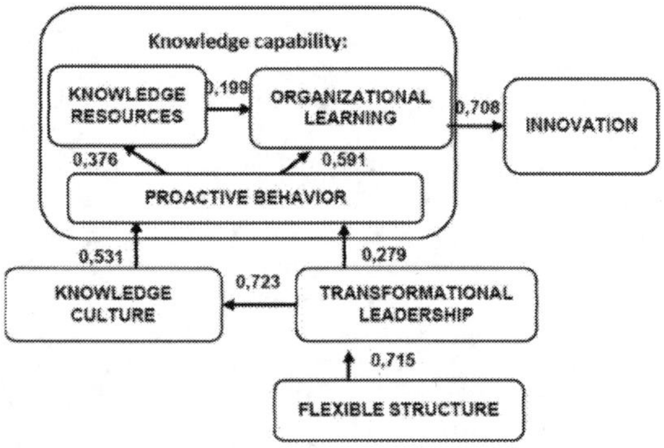

Fig. 2. Empirical results.

According to PLS modelling all three propositions put forward in the study were confirmed by empirical testing. Path modelling with bootstrapping procedure allowed to reveal all mediation effects and to estimate total effects of knowledge capability and organizational and motivational mechanisms on company's innovation. The study revealed full mediation effect of organizational learning on the link between proactive behavior and company's innovation. This empirical evidence underlines that for Russian companies organizational learning is a crucial capability for innovation performance.

Table 2 presents the final results of the study with regard to total effects of the model.

Table 2. Total effects of the model

	Element of the model	Proactive behavior R^2= 58,3%	Innovation R^2= 50,1%
Knowledge capability	Organizational learning		0,708
	Knowledge resources		0,140
	Proactive Behavior		0,472
Mechanisms	Knowledge Culture	0,531	0,250
	Transformational leadership	0,663	0,305
	Flexible Structure	0,473	0,215

As one can see from table 2, transformational leadership has the highest impact on proactive behavior (0,663); knowledge culture takes the second place and flexible structure the third one. All three mechanisms are significant and promote proactive patterns of employee behavior. Moreover, these mechanisms have indirect impact on innovation; it occurs through knowledge capability and confirms its mediation role in knowledge-driven innovation. Among three mechanisms, transformational leadership demonstrates the highest impact on innovation (0,305).

The study reveals that organizational learning has the highest impact on innovation (0,708), while the impact of knowledge resources is significant but weak (0,140). This empirical fact confirms that knowledge resources are not the only one antecedent of company's innovation. For successful innovation the organizational learning should be activated with the help of proactive behavior, which in turn can be induced through knowledge culture, transformational leadership and flexible structure.

5 Conclusions

This study elaborated the novel theoretical framework for knowledge-driven innovation of a company. This model underlines the key role of proactive employee behavior for successful transformation of knowledge into company's innovations. Moreover, the model proposes three mechanisms, such as knowledge culture, transformational leadership and flexible structure, which might induce proactive behavior. The elaborated framework for knowledge-driven innovation can be used by other scholars for further research and, therefore, contributes to knowledge and innovation management field.

Using the sample of Russian companies, the paper confirmed the crucial role of knowledge management for innovation in Russian business-context. The empirical evidence of positive significant impact of organizational learning on innovation confirmed the first proposition of elaborated theoretical framework of knowledge-driven innovation. Further, significant indirect effects of proactive behavior on company's innovation proved the second theoretical proposition of the model. Last but not least empirical conclusion obtained in this study supports the third proposition about the knowledge-driven innovation chain, which starts with three mechanisms conditioning proactive behavior and its impact on knowledge resources and organizational learning.

The study has practical implications expressed in empirical evidence of complementarity of knowledge culture, transformational leadership and flexible structure. For company it means that three mechanisms should be implemented or improved simultaneously for better impact on proactive behavior. The highest impact of transformational leadership determines this particular mechanism as a crucial for Russian companies.

Acknowlegements

The study has several limitations. Namely, the empirical evidence is restricted with Russian companies and the sample was collected using snowball technique, which does not produce a representative sample. At the same time, the study is based on the solid literature and provides interesting insights about antecedents of innovation in Russian business context. This knowledge is novel and gives other scholars new directions for future research.

References

1. S.Roper, J. Du, J. H. Love, *Res. Pol.* **37,** 6-7 (2008)
2. S.Cabrilo, ,S. Dahms, *J. Know. Man.* **22**, 3 (2018)
3. L. Gohberg, K. Ditkovskij, E. Evnevich *Science, Technology. Innovation* (NRU HSE, Moscow, 2021)
4. A.Kianto, J. Sáenz, , N. J. Aramburu, Bus. Res. **81** (2017)
5. R. Grant, *Strat. Man. J.* **17** (1996)
6. C. Argyris, D.A. Schön, *Organizational learning: a theory of action perspective* (Addison-Wesley, Reading, MA, 1978)
7. T.A. Stewart, *Intellectual Capital* (BDD Publ. G., New York, NY, 1997)
8. A. Molodchik, S. Komarov, I. Esaulova, D. Dimitrakiev, *Mechanisms and forms of self-development and self-organization* (PNRPU, Perm, 2019)
9. Oslo Manual 2018, 4th ed. Retrieved from: https://www.oecd.org/science/oslo-manual-2018-9789264304604-en.htm
10. D.A. Garvin, A.C. Edmondson, F. Gino, *HBR. March* (2008)
11. J.M. Crant *J. Man.* **26**, 3 (2000)
12. S.K. Parker, Bindl U.K., K. Strauss, J. Man. **36**, 4 (2010)
13. M.M. Pellegrini, F. Ciampi, G. Marzi, B. Orlando, *J. Know. Man.* 24, 6 (2020)
14. M. Mahmoudsalehi, M. K. Sahari *Learn. Org.* **19**, 6 (2012)
15. T. Gavrilova, A. Alsufyev, A. Pleshkova, *Meas. Bus. Exc.* **22**, 3 (2018)

16. M. Molodchik, C. Jardon, *Learn. Org.* **22**, 6 (2015)
17. C. M. Ringle, S. Wende, J.-M. Becker, 2015. *"SmartPLS 3." Boenningstedt: SmartPLS GmbH*, Retrieved from: http://www.smartpls.com

Development strategies for small and medium-sized enterprises with limited investment resources

Boris Gusakov [1], and *Anna* Sotnikova[2]

[1] Belarusian National Technical University, Minsk, Republic of Belarus
[2] Belarusian National Technical University, Minsk, Republic of Belarus

Abstract. The main source of investment resources for small and medium-sized enterprises is their own profit. The dynamic development of these enterprises is possible only with a strategic program aimed at implementing independent discrete measures. The following tasks are relevant: development of a set of program measures; optimization of measures; maintenance of the financial stability of the enterprise by adjusting development program indicators. The indicators are strategic guidelines for the development. The target functions of the effectiveness of strategic management are minimization of the period of implementation of the development program and the maximum asset growth rate. To assess the effectiveness of development, the following indicators developed by the authors can be used: the coefficient of innovation management, which characterizes the average geometric growth rate of assets attracted during the initial period; the index of innovation management, which characterizes the quality of management by comparing the average geometric growth rate of assets of the optimal and basic variants of the development program.

1 Introduction

Sustainable social and economic development of the country depends on small and medium-sized enterprises. This sector is quite mobile, easily adapts to frequently changing market conditions, and ensures socio-economic stability through the creation of new jobs. The National Strategy for Sustainable Socio-Economic Development of the Republic of Belarus assumes an increase in the contribution of small and medium-sized enterprises to GDP from 21.6% in 2019 to 50% in 2030.

The problem that hinders the development of small and medium-sized enterprises is the lack of investment resources. Small and medium-sized enterprises do not have liquid capital or it is not sufficient to obtain a bank loan. The absence of a state system for ensuring loans hinders entrepreneurial activities. The main source of investment for small and medium-sized enterprises in the start-up period is contributions of founders, and profit is used for development in the foreseeable future.

In the post-Soviet space, the traditional definition of the concept of profit is widely used: "Profit is the difference between income and costs". This definition helps to calculate profit, reveals the main properties and relationships of profit. This is a quasi or fake concept.

2 Materials and methods

The main properties of profit were defined by K. Marx: "a certain amount of value is put into circulation in order to extract a large amount of value from it." It is an excess surplus value" [1, p. 45]. Marx emphasized that profit is an integral part of the surplus value. In a letter to Engels of April 30, 1868, Marx wrote: "Profit is just another name or another category for the surplus value" [2, p. 145]. Marx indicated the possibility of using profit as part of the surplus value for the development of entrepreneurship. He argued "that accumulation, transformation of the surplus value into capital is a process of reproduction on an expanded scale, regardless of whether such expansion is expressed extensively, by building new factories in addition to the old ones, or intensively, by increasing the scale of production" [1, Ch. 22]. The provisions form the basis for the general and specific concepts of profit.

The general definition of profit reveals its main properties. It performs an indicative function. When developing the concept, it was taken into account that the owners can use assets for the production and provision of services, place financial assets on a deposit, or purchase shares of enterprises. In all cases, the owners will make a profit.

"Profit is an increase in the value of assets when they are used in entrepreneurial activities under the agreement."

Specific concepts of profit repeat its basic properties and show its connections based on the

* Corresponding author: b-99@yandex.ru

specifics of entrepreneurial activity. They perform an informational function and show directions of the drive for making and using profits.

Profit is an increase in the value of company's assets due to current, investment and financial activities (except for an increase in obligations to creditors and contributions of founders), directed to payments to the budget, accumulation, expanded reproduction, consumption and other needs permitted by law [3, p. 93].

The concept reveals the main properties and connections of profit with its sources. The concept is based on the provisions of Karl Marx's theory and requirements of the current legislation. When optimizing the value of profit received, the owners should focus on the possibility of choosing a rational direction for the use of assets. When rationalizing the use of profit, the criterion of allocative efficiency is used - the profit is directed to those objects or to encourage those members of society who provide additional profitability of the assets. Objects can be short-term or long-term assets, and members of society are employees and managers of the organization, as well as consumers of the products and services. The government provides tax incentives to encourage the development of the organization; owners and managers change the proportions of profit distribution when the economic situation changes, the financial stability of the organization and changes in the social needs.

3 Results and discussion

A set of factors reflecting the properties and relationships of profit are used in the development of a model for the strategic development of small and medium-sized enterprises. The model includes three parts: selection of a development strategy and proportions of distribution and use of profits; development and optimization of the development program; monitoring of parameters of the development program. The strategic development model is shown in Figure 1.

There are three variants of the strategy: stable development, accelerated development; survival. The choice depends on the potential of an enterprise. The strategic potential is assessed by an integral indicator calculated by eighteen standard technical and economic factors available in open statistical reporting [4]. Standard indicators are combined into three aggregates: financial and economic (asset liquidity); market (indicators characterizing the market share of the enterprise); labor and material and technical (quality of life and materialized labor). For each standard indicator, the indicator of prospects for the i-th factor is calculated, and the specific indicators of prospects are integrated with the organization's potential by formula (1):

$$I_{\text{CP}} = \frac{\sum D_i}{i} = \frac{\sum k_{ix}/k_{i\text{M}}}{i} \quad (1)$$

where I_{ds}– the organization's potential for choosing a development strategy;

Di– indicator of the organization's potential by the i-th factor;

i– number of potential's indicators.

k_{ix}– the i-th standard indicator;

$k_{i\text{M}}$ – the maximum value of the i-th standard indicator in the sample represented by enterprises with a development rate higher than the industry average.

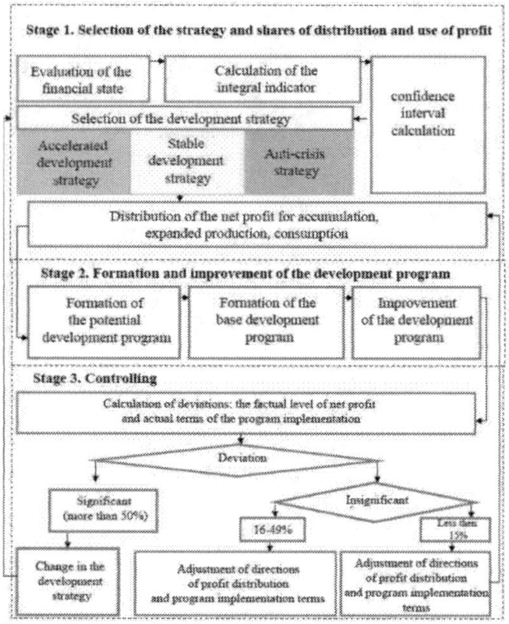

Figure 1. The model of strategic development

The integral indicator of the organization's potential is a positive value; its maximum value is equal to 1. The minimum indicator of 0.1 was obtained empirically from the analysis of calculations using five samples of small and medium-sized enterprises of various profiles.

The ranges of admissible values of the organization's potential for strategies of accelerated development, stable development, survival and their qualitative characteristics were determined.

The range of permissible values of the integral indicator for strategically stable development is defined as the average of the values of the integral indicator of a representative sample. It was revealed that small and medium-sized enterprises with a potential in the range of the confidence interval having a development rate close to the industry average, rarely experience financial difficulties, have a development rate close to the industry average. Accordingly, large values of the organization's potential correspond to the strategy of accelerated development, lower values of the indicator determine the strategy of survival. A qualitative analysis of specific indicators of small and medium-sized enterprises made it possible to identify their basic characteristics.

The accelerated development strategy is recommended for enterprises that have potential in the upper range, which corresponds to the presence of

https://doi.org/10.1051/shsconf/202111600048

significant own financial resources. They are able to attract loans. These enterprises are encouraged to grow their assets above the industry average. The sustainable development strategy is suitable for enterprises with limited investment resources, stable levels of production volumes and financial stability indicators close to the industry average. The survival strategy is recommended for enterprises with unsatisfactory indicators of financial stability.

When choosing a strategy for stable or accelerated development, the strategic development program for a small or medium-sized enterprise is developed. It involves the implementation of the first measure at the expense of own investment resources and subsequent independent measures. Financial resources of the depreciation fund reproduce the value of worn-out assets, and the profit increases enterprise's assets. The speed of implementation of the program depends on the ability of entrepreneurs to use project management methods.

The optimal development program is implemented at several stages.

At the first stage, it is necessary to develop a potential program based on the marketing research and analysis of possible development measures. Using the method of expert assessments, activities that cannot be implemented are eliminated from the potential program even in the strategic period of 5-10 years.

At the second stage, the basic program is developed. An individual business plan is developed for each measure.

At the third stage, priorities of the measures are determined. The first (highest) priority is technological. It determines the priority of measures required for the development of the enterprise. For example, when creating a roadside complex, the gas station has a technological priority.

The second priority is given to the measures with a short return on investment, allowing for the quick accumulation of financial resources for implementing subsequent measures of the development program. The third priority is given to the activities with a long return on investment. The fourth priority determines the need for the simultaneous completion of two or more activities. Infrastructure projects also have this priority. For example, a nuclear power plant is being launched in Belarus, and the power grid is limited in capacity, and potential consumers of electricity are not ready. Simultaneously with the commissioning of the nuclear power plant, it is necessary to ensure the demand for electricity.

At the fourth stage, the first optimization of the development program is carried out. Measures are arranged in a sequence according to their priority. At the same time, the activities united by the priority are arranged in a sequence that allows them to be completed simultaneously as a complex event with a long period of return on investment. The first optimization of the priority of measures of the development program has a significant drawback; it does not take into account the period of accumulation of financial resources for the next event. The first activity is funded at the starting point of implementation, since the initial financial resources for investment activities must be accumulated by the owners. Subsequent activities are funded using the net income from operating facilities and loans [5, 6, 7].

At the fifth stage, the second optimization of the development program is carried out. The method of combinatorics is used. It includes a sequential rearrangement of the order of implementation of the measures of the second and third priorities. Optimization requires several iterations. After each permutation, the average geometric growth rate of assets is calculated. The optimal variant of the strategic development corresponds to a sequence of activities that ensure the maximum asset growth rate. The growth rate of assets is the coefficient of innovation management based on technical innovations.

Technical innovation is a new or improved product or process created using modern scientific knowledge, with improved technical characteristics and increased added value.

In our case, a new management process ensures a reduction in the period of implementation of the development program and additional added value, since the income received earlier is of great value, it will provide additional income.

The coefficient of innovation management is determined by formula (2):

$$IMRR = \sqrt[T]{\frac{\text{Cash} + \sum_{i=0}^{n} P_i}{PVI}} - 1, \quad (2)$$

where *IMRR* is the innovation management ratio;

T is the term of implementation of the development program, years;

Cash is unused funds at the time of completion of the development program secured by the flow of unused net income from previously commissioned objects under the program during the implementation of the last measure, rubles;

∑ Pi is the market value of objects created, rubles;

N is the number of objects created, pcs;

PVI – the amount of the owner's discounted investment, rubles.

The dynamics of the value of assets is shown in Figure 2. It differs from the dynamics of the value of assets when implementing a business plan for one object. In our case, there are financial resources, three objects at by market value and cash accumulated during the implementation of the last third measure.

The market value of objects of the development program is determined for the date of its completion by the profitability method. The facilities will be put into operation in order of priority and, accordingly, have different periods of operation. Accordingly, the market value of each object should be determined individually. It consists of the residual value of property and Goodwill. Simplified Goodwill is capitalized economic profit calculated by formula (3):

$$\text{Pi} = \text{Coi} + \text{Pgi} = \text{Co} + \frac{\text{Pg}}{\frac{1}{t_g} + \text{Ek}}, \quad (3)$$

where Pi is the market value of the i-th object as of the date of completion of the development program;

Co is the residual value of the i-th object as of the date of completion of the development program;

Pg is the cost of goodwill;

Pg is the economic profit of the object (profit of goodwill);

t_g is the service life of Goodwill, years. The service life of Goodwill is assumed to be 5 years;

Eк is the relative value of the rate of payment for the loan.

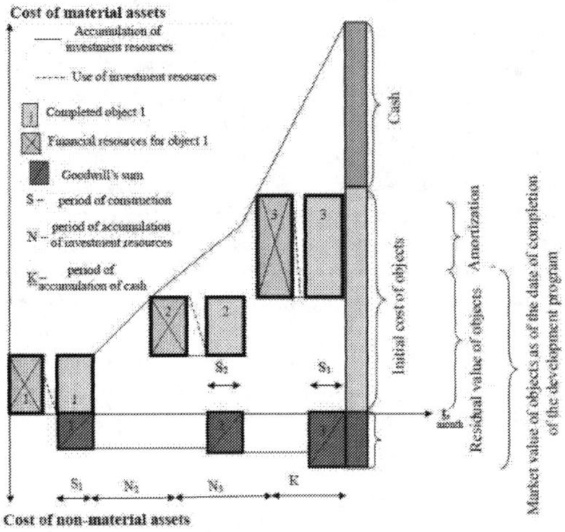

Figure 2. The scheme of implementation of the development program (three objects)

The effectiveness of the development program characterizes the general and specific index of innovation management.

The general index of innovation management shows how many times the growth rate of assets attracted for the development program is higher than the rate of loan payment. The program is effective if the index is more than 1. The index is calculated according by formula (4):

$$FL_g = \frac{IMRR}{E\kappa}, \quad (4)$$

where FL_g is the general index of innovation management;

IMRR is the innovation management coefficient;

Eк is the rate of payment for the loan in a decimal form.

The specific index of innovation management characterizes how many times the growth rate of assets in the used version of the development program is higher than in the version with the minimum coefficient of innovation management. It is calculated by formula (5):

$$FL_s = \frac{IMRRi}{IMRRmin}, \quad (5)$$

where FL_s is the specific index of innovation management;

$IMRR_i$ is the coefficient of innovation management of the i-th pti n f the devel pment pr gram;

$IMRR_{min-}$ is the minimum coefficient of innovation management in the development program.

When implementing the development program, after the completion of the next measure, monitoring measures should be implemented. The fulfillment of the strategic plan is checked in terms of time and profit. In case of deviation of the actual level of profit and the actual period of implementation of the next program measure, the following solutions are possible: use of a different development strategy, a differentiated approach to adjusting the profit. If the actual value of net profit decreased as a result of changes in the market conditions, the net profit does not change or decreases while maintaining the distribution proportions. If the actual amount of net profit has decreased as a result of changes in the internal environment, the profit is used in previously set proportions.

4 Conclusion

The development strategy developed for small and medium-sized enterprises is an important element of project management. The development and implementation of the strategic program allow us to consider long-term prospects for the development of the enterprise in a formalized form. In addition, it becomes possible to assess the organization's potential at the starting point of the strategy and control the quality of management by comparing current and long-term actual and planned indicators.

The specificity of the strategy for the development of small and medium-sized enterprises is self-financing from own net income. In this case, depreciation compensates for the cost of depreciation of assets, and profit allows you to increase the value of assets used to produce goods and services.

In the modern economy, part of the net profit should be used to encourage managers, workers, employees and customers. In these conditions, an important task is rational management of the size of incentive payments.

In contrast to large enterprises, in small and medium-sized enterprises, the process of development is discrete, it is necessary to choose the sequence of activities. The criterion for the optimal program is the maximum growth rate of assets and the minimum period of implementation of the development program.

The effectiveness of a development strategy is specific for two reasons. Firstly, investments are financed from the income received, and the period of implementation of the development program is optimized.

References

1. K. Marx, *Capital. Criticism of Political Economy. Volume 3. Political literature.* (Moscow: Politizdat, 1950)
2. K. Marx, F. Engels, *Collection, vol. 32* (Moscow: Political Literature Publishing House, 1955-1974)

https://doi.org/10.1051/shsconf/202111600048

3. B. Gusakov, A. Sotnikova, *Bulletin of University,* **12,** 10-14 (2018)
4. A. Tsarikayev, *Business in Law* **4**, 222-224, (2011)
5. G. Birman, S. Schmida, *Economic Analysis of Investment Projects* (Banks and Exchanges, YUPITI, 1997)
6. M. Bromwich, *Analysis of the economic efficiency of investments* (INFRA-M, 1996)
7. B. Gusakov, *Chief Economist*, **10**, 65-73 (2015)

Management of the investment and construction project cost under conditions of risk and uncertainty

Nadezhda Bobrova [1,*], *Andrey* Ivanov [2], *Denis* Kamenskikh [3], and *Lyubov* Plyusnina [4]

[1] Perm Research Polytechnic University, 29 Komsomolsky prospect, Perm, 614990, Russia
[2] Lukoil-Perm LLC, 62 Lenin st., Perm, 614068, Russia
[3] GKBUK "KCOP", 18A 25th October st., Perm, 614000, Russia
[4] Perm Research Polytechnic University, Komsomolsky prospect, Perm, 614990, Russia

Abstract. The solution to the problem of managing the cost of an investment and construction project under conditions of risk and uncertainty is presented in the issue. The external and internal factors of project risks are analyzed. The concept of uncertainty is characterized as a property of the external and internal environment of the project, which underlies the occurrence of risks. Additional prerequisites for the emergence of risks, both objective and subjective, are considered. An authors' model for managing the cost of an investment and construction project based on a process approach has been developed based on the above mentioned prerequisites. The content of the stages of the project cost management model has been determined. The possibility of managing the cost of the project through the analysis of budget spending has been investigated. In the course of the study, a comparison was made between the work schedule in accordance with the basic project plan and the actual work schedule with deviations from the project plan. Unfavorable deviations of the baseline indicators of the project were revealed, both in terms of work execution time and in the cost of work performed. It has been proved that the adverse impact of risk factors on the overall performance of the project, expressed by indicators of investment efficiency, is reduced as a result of the application of the process model for managing the cost of an investment and construction project.

1 Introduction

Management of the investment and construction project cost is a complex of actions. The deviation of the actual costs from the planned level occurs as a result of the influence of risk factors due to the uncertainty of the environment. The deviation of the actual costs from the planned level can be considered as a result of the influence of risk factors, the nature of which is due to such a property of the environment as uncertainty. "Uncertainty" is understood by the authors as environmental conditions with a low degree of predictability of events. Also an important prerequisite for the occurrence of risks, along with uncertainty, is the limited resources of the project.

Uncertainty in the external environment of the project is enhanced by the presence of uncertainty in the internal environment of the project caused by the increase in the number of participants involved in the construction project. Investors, customers, developers, contractors and subcontractors have to be named among the key participants in the construction project. Participants are represented by both legal entities and individuals, endowed with their own functions. At the same time, the functional diagram of the construction project implementation continues to change. The designated prerequisites determine the objective and subjective nature of the risks, which are based on factors of both a subjective nature and those that do not depend on the behavior of individual project participants. Among the risk factors external and internal (project) ones can be highlighted [1].

External risk factors are due to the uncertainty of the macroenvironment. They include legislative, political, economic and social factors. Significant factors of legislative risks include the adoption of laws related to the regulation of investment and construction activities, sources of attracting investments, tax payments, forms of placing orders for construction, updating regulatory documents [2]. Political risk factors are associated with government policy, as well as with the political structure of society, which can lead to a change in the general course of economic development and influence the adoption of laws that stimulate or hinder the development of the construction business. The main economic risk factors include the following ones: increased inflation, falling investment demand, deterioration of the financial and credit mechanism, changes in lending conditions and interest rates, late payment for work performed, delays in project approval, improper coordination between the customer and project implementers, changes in the contract documentation, incorrect insurance operations, damage to production assets, untimely deliveries of construction materials and equipment, changes in market prices for resources. Social risk factors include the human factor, which can

* Corresponding author: bobrova.nadejda@gmail.com

be the cause of possible failures of erected structures, accidents and fires during the entire construction period [3].

Internal risk factors are directly related to the implementation of the project at the construction site, they include organizational, technological, environmental and climatic ones. These risks are the higher, the lower the requirements for the qualifications of production personnel, the timing and quality of construction are. Organizational risks include irrational organization of the construction site, unclear planning of the production cycle, failure to meet the delivery schedules of material, technical and labor resources, lack of a quality control system for construction work and as-built documentation.

The most common technological risk factors that are associated with a lack of geological and geodetic surveys, inaccuracy of initial data and standards used in calculating production parameters and economic indicators, such as: completeness of the list and volume of work planned to be performed, the amount of material and technical resources, productive innovative technologies, effective building materials and structures, volume and structure of capital investments. Environmental risk factors can manifest themselves in the form of excessive noise and vibration effects on the territory of the construction site, unacceptable sedimentation of the foundation of the structure, flooding of the building area, the development of landslide processes and landslides, wind loads, precipitation in the form of rain and snow.

All the factors listed above interact and complement each other.

The impact of external factors that create the likelihood of adverse consequences caused by the impact of the macroenvironment is indirect and manifests itself through a change in the project parameters. Thus, the dominant factors in terms of influence on the cost of the project are internal risk factors for which the analysis is carried out, and decisions are made to clarify the technical, time and cost indicators of the project. .

2 Methods of research

Based on the process approach to project management, the authors have analyzed the implementation of the project in stages [4]:

1. Identification – revealing project indicators for which deviations from the planned parameters were observed. Responsible executives notify project managers of deviations using previously established written or electronic forms. The notifications record the nature of the deviations, give a brief assessment of the reasons and provide general data (date, location, type of work, etc.). In order to ensure the implementation of the change management process, notifications should be classified by source (for example, supply chain, production and technical department, estimate department, accounting department, etc.) and recorded in journals.

2. Analysis of deviations – determining the trend of the deviation is performed in order to assess whether the deviations are random or repeatable, and whether they are acceptable. A trend is a non-random deviation of the actual parameters of the project's performance from the planned indicators. In most cases, trends are the result of changes in market prices for material and labor resources. If a deviation is unacceptable, then an analysis of its content and probable causes is carried out, and then options for corrective actions are developed to eliminate these causes. Corrective actions are the solution of tasks to overcome deviations from the planned parameters of the project. Most often, an unfavorable deviation trend indicates that the project's performance in the future will also be unsatisfactory, so corrective actions are taken. At the same time, if deviations within the budget are the result of randomness and there are no total deviations in the analysis period, this does not mean that the project's performance will be satisfactory in the future. In other words, if the positive outcomes of the interim analysis are due to chance, then this also requires corrective action.

3. The processing of data on trends and random deviations (using logs) is carried out by the responsible executors (value engineers) and consists in the development of proposals for corrective actions and changes to the project baseline. The decision-making on these changes is made by the project management. As a result of the implementation of corrective actions, the project budget is updated.

The main goal of project cost management is the timely adjustment of its base plan to minimize the identified non-random and unfavorable deviations in cost and to ensure the effectiveness of investments [5].

3 Experimental part

The process model for managing the cost of an investment and construction project is shown in Fig. 1.

This model contains the following stages: planning, identification, performance evaluation, analysis, processing [6].

At the planning stage, the project budget and work schedule are drawn up. Below is the Schedule of work on the arrangement of the zero cycle of the industrial building construction at an estimated cost of 10 million rubles with the duration of the work - 43 days (Fig. 2).

In the course of the implementation of the project plan (before the start of work on the installation of the basement overlap), deviations in the fulfillment of the baseline indicators of the project were revealed. At the identification stage, deviations in the execution schedule for certain types of construction work presented in table 2 were determined.

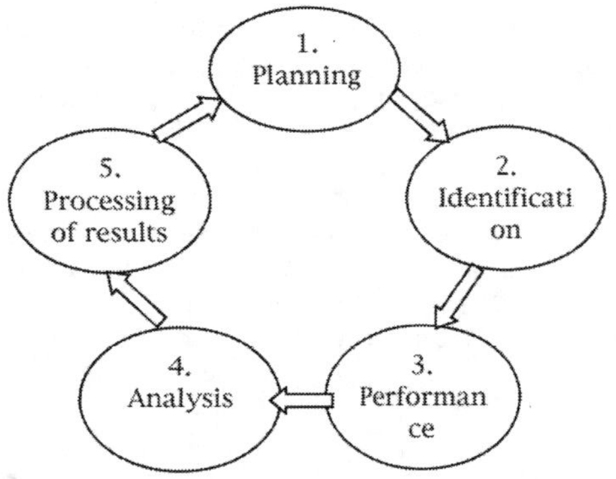

Fig. 1. The process model for managing the cost of an investment and construction project.

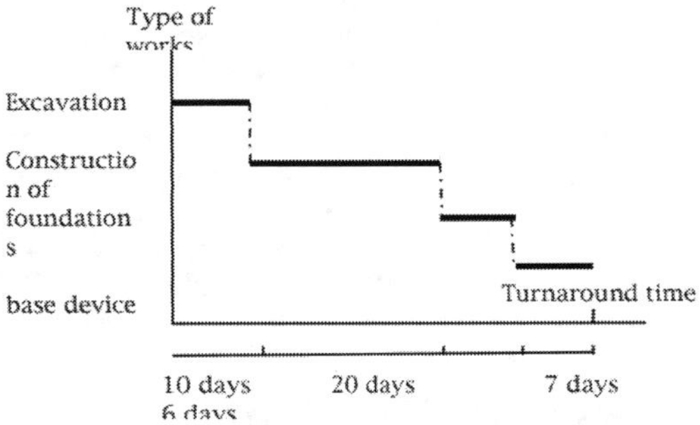

Fig. 2. Work schedule in accordance with the project baseline.

Also, at the planning stage, the list and the value of the baseline indicators of the project for the implementation of the zero cycle of work on the construction of an industrial building were determined (Table 1).

Table 1. Baseline project parameters

Types of work	Period of completion according to the schedule (days)	Cost of work according to the estimate (million rubles)
1. Excavation	10	2
2. Construction of foundations	20	4
3. Base device	7	1
4. Base overlap device	6	3
Total	43	10

Table 2. Fulfillment of the baseline schedule

Types of work	**Compliance with the plan**
1. Excavation	According to the plan
2. Construction of foundations	Deviation
3. Base device	Deviation
4. Base overlap device	Planned in accordance with the project schedule

Thus, it is planned to carry out work on the arrangement of the basement overlap in accordance with the updated project execution schedule (Fig. 3).

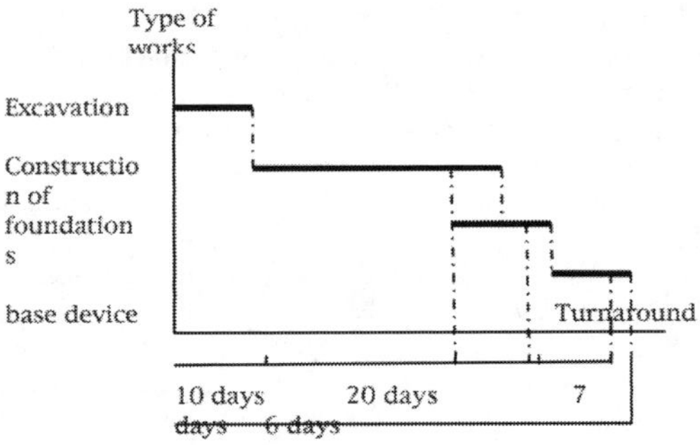

Fig. 3. Actual schedule with deviations from the project plan.

The effectiveness of the execution of an investment and construction project is determined by the level of compliance of the quantitative and qualitative indicators of the construction object with design solutions. Efficiency is expressed in comparing the actual indicator of construction costs and the value of planned costs to obtain the present state of the object.

$$Effectiveness = \frac{Result_{fact}}{Indicator_{planned}} \quad (1)$$

The effectiveness of project execution depends on the effectiveness of management and can be measured using three main parameters: time, quality and cost.

In general, the effectiveness of the project is assessed by comparing the actual schedule of its implementation with the budget. It is convenient to use the earned value method for this comparison. To use this method, the following conditions must be met:

- availability of the project schedule and budget;
- the use of the same accounting structure for both planned and actual construction costs (cost of resources);
- periodic fixation of indicators of the execution of the schedule and construction budget;
- constant quality control of the construction result.

Evaluation of the project execution effectiveness based on the assessment of the earned volume of work showed that with these deviations, the estimated cost of the zero cycle upon completion of the project will increase from 10 million rubles up to 11.5 million rubles. At the same time, the project execution time will increase - from 43 days to 45 days. The index of project implementation will be equal to 0.87. Thus, the probability of a decrease in the effectiveness of the project is stated [7].

At the analysis stage, we determine the reasons for deviations in the planned indicators of work on the construction of foundations and the basement, which may be associated with changes in the schedule of execution, the amount of work and cost. In a specific case, the study of these factors revealed the following reasons for deviations.

Failure to complete work on the construction of foundations on schedule and overstatement of their cost (by 0.8 million rubles) was due to the occurrence of additional unforeseen work associated with force majeure circumstances of a natural nature, which led to changes in the timing of the work. Table 3 shows the data of deviations in terms of work performance.

Table 3. Deviations in terms of the work schedule.

Types of work	Terms of work (days)		Deviation (days)	Type of deviation
	Plan	Fact		
Construction of foundations	20	23	+3	Unfavorable
Base arrangement	7	5	–2	Favorable
Total	27	28	+1	Unfavorable

Table 4 shows the data on the indicators of deviations in the cost of work.

Table 4. Indicators of deviation in the cost of work.

Types of work	Deviation in the cost of work (mln rub)	
	Unfavorable	Favorable
Construction of foundations	+0,8	– 0,4
Base arrangement	+0,2	no
Total	+1,0	–0,4

As a result of the analysis of deviations, it was revealed that the deviation of the planned time for performing work on the construction of foundations and the basement is unfavorable (an increase in the construction time by one day). The cumulative deviation of the planned cost of these types of work in the direction of increase amounted to 0.6 million rubles, which is also unfavorable. Thus, there is a general trend of unfavorable deviations from the baseline, which increases the likelihood of a decrease in project performance.

After determining the nature and causes of deviations, this information is processed and a corrective action plan is developed, which are applied at the next stage of the zero cycle work. Changes are made to the schedule of subsequent work, taking into account three identified deviation factors: unforeseen types of work, exceeding the construction time and an increase in the cost of the zero cycle of work.

3 Conclusion

1. Implementation of an investment and construction project is associated with the likelihood of the impact of various risk factors, the prerequisites of which are both the uncertainty of the external environment of the project and shortcomings in the organization of activities within the framework of the project itself.

2. Exposure to risk factors can cause adverse impacts on the cost of the project, which are of a recurring nature, which will subsequently negatively affect the effectiveness of the project.

3. The process model of project cost management makes it possible to identify deviations in the basic parameters of the project at the early stages of its implementation, to determine their nature and to update the budget and schedule of subsequent work, taking into account the negative trend of the identified risk factors.

4. Application of a process model for managing the cost of an investment and construction project reduces the adverse impact of risk factors on the overall performance of the project, expressed in terms of investment efficiency.

References

1. L. M. Plyusnina, L. V. Brezgina, N. M. Bobrova, *International Journal of Civil Engineering and Technology (IJCIET)* **8(10),** 1526–1535 (2017)
2. I.V. Elokhova, L.A. Nazarova, *Science of Krasnoyarsk* **8(4),** 49–68 (2019)
3. E.Yu. Kulikova, *Bulletin of Moscow State Mining University* **1**, 10–26 (2006)
4. Project Management Institute, *A guide to the project management body of knowledge* (PMBOK®Guide) (5th ed.). Newtown Square, PA: Project Management Institute (2013)
5. V. Popov, G. Ostapenko, *Proceedings of the 4th International Conference,* Prague, University of Economics in Prague (2016).
6. L.M. Plyusnina, L.V. Brezgina, V.A. Terentyev, *Scientific Review. Ser. 1. Economics and Law 1,* 177–183 (2013)
7. L.V. Brezgina, D.V. Kamenskikh, *Innovative development of self-regulatory construction organizations* **7,** 9–18 (2018)

Expenditures for innovations and foreign direct investments in Bulgaria - regional aspects, features and trends

Mariana Petrova*, and *Sergey* Radukanov

St. Cyril and St. Methodius University of Veliko Tarnovo, T. Tarnovsky 2, Veliko Tarnovo, 5003, Bulgaria

Abstract: The study examines the costs of innovation and foreign direct investment by district (NUTS 3), Planning regions (NUTS 2) and Regions (NUTS 1), according to the classification of territorial units for statistical purposes. The main trends in their development are outlined. The authors believe that foreign direct investment is closely linked to the cost of innovation. Of particular importance is the view that the intensity of R&D expenditure by statistical regions and districts almost coincides with foreign direct investment. The results of the analysis confirm the strong disproportion in recent years in the distribution of direct investment and R&D expenditures, which leads to the formation of a highly developed center - Sofia district (capital) and weak periphery - all other districts.

1. Introduction

In the context of European vector development and harmonious entry into the global economy, the Bulgarian economy has the task of ensuring international competitiveness, and the main channel for the transfer of innovation is foreign direct investment. The latter are the driver of economic growth and have an impact through the formation and accumulation of new capital and technological renewal.

Therefore, the study of research and development costs and foreign direct investment at the regional level is particularly **relevant**. The current development **focuses** on the costs of innovation in the context of foreign direct investment, and the **subject** - the dynamics of their development in the regional aspect.

The goal is to analyze the level and outline the trends on a regional scale. In order to achieve the formulated goal, it is necessary to solve the following tasks:

- analysis of the dynamics of foreign direct investments in the regional aspect;
- study of the dynamics of expenditures for research and development by statistical regions.

2. Theoretical background

The notion of innovation was put under analysis already in the middle of the previous century, J.A. Schumpeter defined innovation as [1]:

- introduction of a new product,
- introduction of a new production method (process innovation),
- opening of a new sales market,
- opening of a new supply market,
- introduction of a new organisation.

E.M. Rogers [2] defines innovation differently, relying on the assumption that it encompasses any concept, idea, relation, position or object that stands out among all previously existing due to its quality.

Peter F. Drucker formulated innovation as "the specific tool of entrepreneurs, the means by which they exploit change as an opportunity for a different business or a different service".

D. Cavagnoli defines innovation as the process of creative use of knowledge, transformation of knowledge owned by an organization or acquired from outside, to be used in new products, services or processes [3].

The opinions presented (positions) are just a few of the countless definitions formulated by a large group of scientists who have dealt with this issue. In this field we can distinguish part of the research of Pukala [4, 5], Pukala & Linde [6], Bacho et al. [7], Seitzhanov et al. [8], Adamišin et al. [9], Labunska et al. [10], Sushchenko & Trunina [11], whose multi-layered analysis of innovation has enriched and expanded knowledge in this area.

Scientists dealing with innovation in a global sense have developed a standard definition, which has been published by the EC under the title "The Oslo Manual" [12]. It represents a point of reference for the majority of studies on innovation. According to a definition contained therein, innovation is "the implementation of a new or significantly improved product (good or service), or process, a new marketing method, or a new organizational method in business practices, workplace organization or external relations". The problem of organizing efficient innovation activity invoked great scientific interests of Bulgarian and Latvian researchers, such as Komarova V., Selivanova-Fyodorova N., Ruza O., Kaźmierczyk J., Arbidane [13], Zahariev, Zveryakov, Prodanov, Zaharieva, Angelov, Zarkova,

Corresponding author: m.petrova@ts.uni-vt.bg

Petrova [14] Deneva & Grasis [15], Tepavicharova, Aleksejeva & Vazov [16], Tepavicharova & Zahars [17].

2. Materials and methods

At the first stage of the study, it seems appropriate to use methods of collecting, systematizing and analyzing literary sources.

Further, the official statistics of the National statistical institute of the Republic of Bulgaria should be used, since they will help characterize the current state of innovative activities in Bulgaria.

To identify differences in the development of the main elements of the national innovation system, it is advisable to use the method of comparison.

All results refer to relevant sources of information of the National statistical institute of the Republic of Bulgaria or are based on authors' own calculations.

3. Results and discussion

All the conclusions of the authors are based solely on the results of their own calculations, based on generally recognized ratings with a transparent methodology and a wide list of indicators ensuring their objectivity and published from National statistical institute of the Republic of Bulgaria.

The volume of foreign direct investment (FDI) attracted in the country is 21581579.6 thousand euros for 2014 and reaches a level of 24919589.3 thousand euros for 2018 (Fig. 1).

The data show that for the period from 2014 to 2018, foreign direct investment is growing. For the considered period in absolute terms the increase is by +3338009.7 thousand BGN, and in relative aspect it is 15.47%. The growth rate indicator compared to the previous period has the highest value for 2015 compared to 2014. - (+7, 33%).

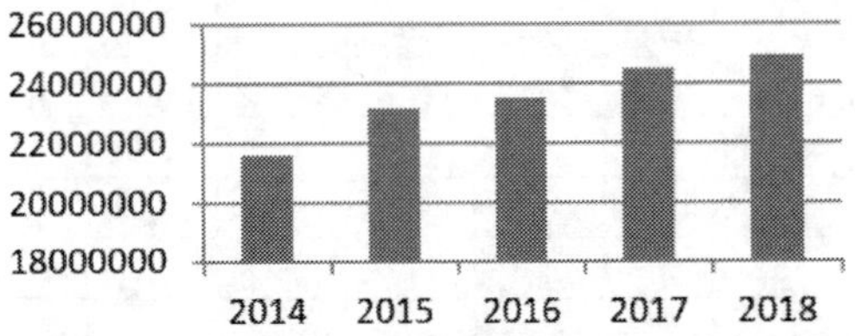

Fig 1. Foreign direct investment in Bulgaria for the period of 2014-2018 (in thousands of EUR)

For other periods it is characterized by lower values: 2016/2015 is + 1.49%, 2017/2016 - (+ 4.11%) and for 2018/2017 - (+ 1.82%). For the whole period there is a steady upward trend in foreign direct investment. The average annual growth rate is + 3.69%. The development of the gross domestic product in Bulgaria Mitkov [18].

For research and analysis of FDI at the regional level (see Table 1), the average value of foreign direct investment for the country is calculated by areas:

FDI (total for the country)/28 (districts count)

The obtained arithmetic average value shows the inflow of FDI of one conditional, statistically average area in Bulgaria.

Table 1. FDI in Bulgaria, by districts (NUTS 3) and average for the country for the period of 2014 – 2018 (in thous. of EUR)

Regions	**2014**	**2015**	**2016**	**2017**	**2018**
Vidin	51686	77842	71466	73043	80909
Vratsa	138052	85677	76903	108893	106294
Lovech	134000	147443	139150	114073	120982
Montana	34218	27355	36978	38191	35542
Pleven	306765	282479	187588	226473	125995
VelikoTarnovo	108176	121708	147375	134034	140662
Gabrovo	265205	284368	302616	326309	338201
Razgrad	114509	104855	131300	152460	175490
Rousse	385755	337023	367855	374815	429190
Silistra	31667	28803	32685	36013	28115
Varna	1679193	1760957	1785198	1946023	1869908
Dobrich	111566	202894	244206	270243	311763
Targovishte	198286	204744	227261	244815	280506
Shumen	104871	103320	103093	107798	117305
Bourgas	937227	1704392	1746015	1887318	2059421
Sliven	92702	90252	124199	139289	161505
Stara Zagora	969271	1023382	920971	905946	929285
Yambol	58091	79030	56527	50680	52899
Blagoevgrad	384666	446364	482987	474397	499327
Kyustendil	43524	42028	43720	42782	42946
Pernik	218999	178902	186558	204136	199062
Sofia district	1239078	1274063	1313943	1464101	1389012
Sofia city	11675227	12111790	12211562	12310741	12432131
Kardzhali	143951	176317	153124	256162	291633
Pazardzhik	472035	486357	480425	505856	516146
Plovdiv	1546360	1592531	1731002	1895015	1911939
Smolyan	75175	83064	83761	82920	90632,8
Haskovo	61324	105335	120400	102671	182793
National Total	21581579	23163273	23508865	24475196	24919589
National average	770771	827260	839602	874114	889985

Source: own calculations based on NSI (2018).

The value obtained is compared with foreign direct investment in specific areas for all years.

The study logically presents the comparison between the national average with the specific indicator by district.

 https://doi.org/10.1051/shsconf/202111600050

Table 2. Comparison of the FDI in Bulgaria by districts with the national average for the period of 2014 – 2018

Year	*Districts with FDI above the national average*	*Districts with FDI below the national average*
2014	*6 regions* - Sofiacity, Varna, Plovdiv, Sofia, Stara Zagora,Bourgas	*22 regions* – Pazardzhik, Rousse, Blagoevgrad, Pleven, Gabrovo, Pernik, Targovishte, Kardzhali, Vratsa, Lovech, Razgrad, Dobrich, VelikoTarnovo, Shumen, Sliven, Smolyan, Haskovo, Yambol, Vidin, Kyustendil, Montana, Silistra
2015	*6 regions* - Sofiacity,Varna, Bourgas, Plovdiv, Sofia, Stara Zagora	*22 regions* - Pazardzhik, Blagoevgrad, Rousse, Gabrovo, Pleven, Targovishte, Dobrich, Pernik, Kardzhali, Lovech, VelikoTarnovo, Haskovo, Razgrad, Shumen, Sliven, Vratsa, Smolyan, Yambol, Vidin, Kyustendil, Silistra, Montana
2016	*6 regions* - Sofiacity, Varna, Bourgas, Plovdiv, Sofia, Stara Zagora	*22 regions* – Blagoevgrad, Pazardzhik, Rousse, Gabrovo, Dobrich, Targovishte, Pleven, Pernik, Kardzhali, Veliko Tarnovo, Lovech, Razgrad, Sliven, Haskovo, Shumen, Smolyan, Vratsa, Vidin, Yambol, Kyustendil, Montana, Silistra
2017	*6 regions* - Sofia city, Varna, Plovdiv, Bourgas, Sofia, Stara Zagora	*22 regions* – Pazardzhik, Blagoevgrad, Rousse, Gabrovo, Dobrich, Kardzhali, Targovishte, Pleven, Pernik, Razgrad, Sliven, Veliko Tarnovo, Lovech, Vratsa, Shumen, Haskovo, Smolyan, Vidin, Yambol, Kyustendil, Montana, Silistra
2018	*6 regions* - Sofiacity, Bourgas, Plovdiv, Varna, Sofia, Stara Zagora	*22 regions*–Pazardzhik, Blagoevgrad, Rousse, Gabrovo, Dobrich, Kardzhali, Targovishte, Pernik, Haskovo, Razgrad, Sliven, Veliko Tarnovo, Pleven, Lovech, Shumen, Vratsa, Smolyan, Vidin, Yambol, Kyustendil, Montana, Silistra

Source: authors' calculation

The analysis for the period of 2014-2018 confirms that the number of territorial units in the two separate groups is unchanged - a separate group of six districts has foreign direct investment above the national average and for the rest (the same during the study period) twenty and two, the values are lower than the national average.

The leading place in attracting foreign direct investment is occupied by Sofia city (capital). According to the territorial-administrative division of Bulgaria, there are two districts with similar names, but it is important to distinguish them. One is Sofia city (capital), and the other is Sofia district The strong concentration is understandable, as in addition to the large population, it is the administrative center of Bulgaria and as such is an attractive place for localization of foreign companies and representative offices. For the considered period the dominant role is also played by the district of Varna, which for the last year gives way to the second place to the districts of Burgas and Plovdiv.

In the group of districts with foreign direct investment below the national average, the district of Pazardzhik has a dominant role. It is explained by the proximity to Plovdiv and the related productions in a number of areas. The districts of Blagoevgrad and Ruse are directly below the district of Pazardzhik, excluding the result in 2016. For the period of the study, the districts with the lowest values remain the same: Vidin, Yambol, Kyustendil, Montana and Silistra.

According to the Global Innovation Index [19], in 2020 Bulgaria ranks 37th, and for the previous year it ranked 40th. With reasonable optimism, we can determine a relatively positive trend and restore the 37th position from 2018.

In order to study innovations according to the classification of territorial units, the latter must be presented in the aggregate form - by NUTS 2 planning regions.

Table 3. Expenditures for research and development by NUTS 2 planning regions for the period of 2014 - 2018 (in thous. of BGN)

Planning regions	2014	2015	2016	2017	2018
Northwestern	11900	33265	29752	31285	37437
Northern Central	10661	37218	30654	30226	26612
Northeastern	33763	35566	51545	51626	49750
Southeastern	27459	36415	33892	39094	42137
Southwestern	550827	642876	522683	538651	603853
Southern Central	30219	65117	65748	69352	69115
National Total	664829	850457	734274	760234	828904

Note - the table is compiled according to the NSI [19]

For the researched 5-year period the expenses for innovations increased by 164075 thousand BGN or by 24.68%. The average annual growth rate is 6.71%. The Southwest region has the greatest weight - the costs for innovation are significant. Structurally, its share varies between 82% and 73%. It is understandable because it includes Sofia district and Sofia district (capital). The least represented are the regions Northwest and North Central.

The disproportion between the different regions stands out to a greater extent if the same data are considered, but in aggregate form - by zones, NUTS 1 level (see Table 4).

Table 4. Expenditures for research and development by NUTS 1 regions

Regions	2014	2015	2016	2017	2018
Northern and Eastern Bulgaria	83783	142464	145843	152231	155936
South-Western and South-Central Bulgaria	581046	707993	588431	608003	672968
National Total	664829	850457	734274	760234	828904

Note - the table is compiled according to the NSI [20]

The relative share of expenditures for research activities in Northern and Southeastern Bulgaria is insignificant - in 2017 it was 20.02%, when it registered a peak, and the lowest value was registered in 2014 with only 12.60%.

Tracking the relative share of expenditures for research and development by districts in relation to the gross domestic product by districts for the studied period is of interest and shows surprising results (Table 5).

Table 5. Expenditure for R&D related to GDP (by district), decremental order

	District	2014	District	2015	District	2016	District	2017	District	2018
1	Sofiacity	1.64%	Sofiacity	1.74%	Lovech	1.34%	Shumen	1.40%	Sofiacity	1.32%
2	Varna	0.50%	Lovech	1.38%	Sofiacity	1.32%	Sofiacity	1.28%	Lovech	1.26%
3	Pazardzhik	0.48%	Pazardzhik	0.80%	Shumen	0.96%	Lovech	1.08%	Shumen	1.13%
4	Lovech	0.43%	Pleven	0.76%	Pazardzhik	0.73%	Pazardzhik	0.84%	Pleven	0.81%
5	Stara Zagora	0.40%	Rousse	0.72%	Rousse	0.68%	Pleven	0.72%	Plovdiv	0.58%
6	Pleven	0.35%	Plovdiv	0.66%	Plovdiv	0.61%	Plovdiv	0.56%	Pazardzhik	0.57%
7	Plovdiv	0.33%	Yambol	0.53%	Varna	0.56%	Rousse	0.55%	Rousse	0.49%
8	Yambol	0.31%	Sofia	0.51%	Pleven	0.53%	Gabrovo	0.51%	Stara Zagora	0.45%
9	Rousse	0.28%	Varna	0.47%	Stara Zagora	0.35%	Varna	0.40%	Gabrovo	0.44%
10	Sofia	0.26%	Stara Zagora	0.44%	Gabrovo	0.34%	Stara Zagora	0.33%	Yambol	0.42%
11	Shumen	0.19%	Sliven	0.38%	Yambol	0.34%	Veliko Tarnovo	0.32%	Varna	0.36%
12	Bourgas	0.14%	Shumen	0.32%	Veliko Tarnovo	0.33%	Sofia	0.26%	Sliven	0.34%
13	Blagoevgrad	0.14%	Vratsa	0.24%	Sofia	0.31%	Dobrich	0.21%	Dobrich	0.30%
14	Veliko Tarnovo	0.12%	Veliko Tarnovo	0.22%	Kardzhali	0.31%	Blagoevgrad	0.17%	Kyustendil	0.28%
15	Gabrovo	0.10%	Bourgas	0.17%	Sliven	0.29%	Bourgas	0.16%	Sofia	0.25%
16	Kyustendil	0.08%	Blagoevgrad	0.16%	Blagoevgrad	0.21%	Montana	0.15%	Veliko Tarnovo	0.20%
17	Vratsa	0.04%	Smolyan	0.16%	Vidin	0.18%	Vratsa	0.12%	Blagoevgrad	0.16%
18	Sliven	0.04%	Montana	0.09%	Bourgas	0.18%	Haskovo	0.08%	Bourgas	0.15%
19	Haskovo	0.04%	Silistra	0.09%	Montana	0.16%	Vidin	0.07%	Montana	0.12%
20	Pernik	0.02%	Vidin	0.08%	Smolyan	0.13%	Targovishte	0.02%	Vratsa	0.11%
21	Vidin	n/a	Gabrovo	n/a	Vratsa	0.12%	Razgrad	n/a	Vidin	0.10%
22	Montana	n/a	Razgrad	n/a	Haskovo	0.04%	Silistra	n/a	Haskovo	0.10%
23	Razgrad	n/a	Dobrich	n/a	Razgrad	n/a	Sliven	n/a	Pernik	0.08%
24	Silistra	n/a	Targovishte	n/a	Silistra	n/a	Yambol	n/a	Targovishte	0.02%
25	Dobrich	n/a	Kyustendil	n/a	Dobrich	n/a	Kyustendil	n/a	Razgrad	n/a
26	Targovishte	n/a	Pernik	n/a	Targovishte	n/a	Pernik	n/a	Silistra	n/a
27	Kardzhali	n/a	Kardzhali	n/a	Kyustendil	n/a	Kardzhali	n/a	Kardzhali	n/a
28	Smolyan	n/a	Haskovo	n/a	Pernik	n/a	Smolyan	n/a	Smolyan	n/a

Source: authors' calculation

For certain years by districts, the NSI classifies data on research and development as confidential. In practice, the relative share of R&D expenditure in relation to GDP by district cannot be calculated due to a lack of official data.

It is indisputable and explainable that Sofia district (capital) dominates in all years with the highest relative share. Impressive are 2016 and 2017, when Sofia capital gave away the leading position to the districts of Lovech and Shumen, respectively. The latter are established as leaders for the period 2014-2018.

The districts of Vidin, Montana, Vratsa, Silistra and Targovishte stand out with an extremely low relative share for the studied period. This is not surprising; these areas are traditionally positioned in the last places and in a number of other areas - social, demographic and economic indicators.

4 Conclusions

The limitations of the scientific study that the study was conducted exclusively on statistical data.

The study allows us to make the following conclusions:

- there is an asymmetric distribution of foreign direct investment and research and development expenditure by districts and statistical regions;
- the intensity of research and development expenditures by statistical regions almost coincides with foreign direct investment. This means that there is a direct correlation between foreign direct investment and R&D expenditure;
- the largest amount of foreign direct investment and expenditures for research and development is the Southwest region, and with the lowest values are North Central and Northwest;
- the strong disproportion in the distribution of direct investments and the costs for research and development activity leads to the establishment in this respect of a highly developed center - Sofia district (capital) and weak periphery - all other districts.

References

1. J.A. Schumpeter. *The Theory of Economic Development. An Inquiry into Profits, Capital, Credit, Interest, and the Business Cycle.* (Cambridge, MAUS: Harvard University Press, 1934).
2. E.M. Rogers. *Diffusion of innovations* (Free Press, New York, 2003)
3. D. Cavagnoli, *Innovation: Management, Policy and Practice*, **13**, 4 (2011)
4. R. Pukala, *Economics and Management*, **8:3,** 1 (2016)
5. R. Pukala, MATEC Web Conf., **297,** 08002 (2019)
6. R Pukala, I. Linde. E3S *Conf.*, **174**, 04030 (2020)
7. R. Bacho, R. Pukala, S. Hlibko, N. Vnukova, P. Pola. *Marketing and Management of Innovations*, **3**, 297–307. (2019).
8. S. Seitzhanov, N. Kurmanov, M. Petrova, U. Aliyev, N. Aidargaliyeva, *Entrepreneurship and Sustainability Issues*, **7:4**, 2615-2629 (2020)
9. P. Adamišin, R. Vavrek, R. Pukala. *SGEM conf,* **3(5)**, 135-140.
10. Sv. Labunska, M. Petrova, O. Prokopishyna, *Economic Annals – XXI*, **165:5-6**, 3-18 (2017)
11. O. Sushchenko, I. Trunina, Actual Problems of Economics, **177:3,** 191-198 (2016)
12. *The measurement of scientific and technological activities.* Proposed guidelines for collecting and interpreting technological innovation data. OSLO Manual.
13. Komarova V., Selivanova-Fyodorova N., Ruza O., Kaźmierczyk J., Arbidane I. (2021). *Entrepreneurship and Sustainability Issues* **8(3):** 110-121.
14. Laktionova, O., Dobrovolskyi, O., Karpova, T. S., & Zahariev, A. *Management Theory and Studies for Rural Business and Infrastructure Development*, **41(1)**, 62-73. (2019).
15. A. Deneva, J. Grasis. *E3S Conf.*, **174,** 01026. (2020).
16. M. Tepavicharova, L. Aleksejeva, R. Vazov. *E3S Web of Conf.*, **174,** art. no. 04031. (2020).
17. M. Tepavicharova, L. Dikova, V. Zahars. *E3S Web of Conf.*, **105,** 04029, (2019)
18. M. Mitkov. *Social and Economic Analysis,* **10**(1), 34-44 (2018).
19. Global Innovation Index 2020: Who Will Finance Innovation? Retrieved from: ttps://www.globalinnovationindex.org/
20. *INFOSTAT information system* Retrieved from: https://infostat.nsi.bg/
21. *Global Innovation Index 2019 rankings,* Retrieved from:https://www.wipo.int/edocs/pubdocs/en/wipo_pub_gii_2019-intro4.pdf
22. *Global Innovation Index 2018 rankings*, Retrieved from:https://www.wipo.int/edocs/pubdocs/en/wipo_pub_gii_2018-intro5.pdf

Actual problems of project teams in the management of innovative projects

Yuri Belonogov, *Anna* Zuikina, *Anastasia* Kucheva, *Daria* Semenova*

[1] Perm National Research Polytechnic University, 29, Komsomolsky prospect, Perm, 614990, Russia

Abstract. Management of innovative projects is carried out under conditions of a high degree of uncertainty and risk. In the absence of competent management, an innovative project is either unclaimed by the target audience, or is closed due to overspending of resources. The purpose of this research is to identify the main problems in the management of innovative projects and to offer recommendations for improving the efficiency of management. The object of the research is the management of innovative projects. The subject of the research is the main problems of project teams in innovative projects. As part of the research, an expert interview was conducted with managers of innovative projects and representatives of project teams in various areas of project management. Based on the results of the interview, the key problems in the management of innovative projects were identified and recommendations for their elimination were proposed.

1 Introduction

In the modern world, which is characterized by instability, uncertainty, complexity and ambiguity (VUCA-world), almost every organization faces the need to implement the latest technologies, and, consequently, to manage innovative projects. Innovative projects, in the research, is a complex of interrelated management actions within the time and resource constraints aimed at creating an innovation – a fundamentally new product or service based on the advanced achievements of science and technology.

Traditional project management is often seen as a process for planning, scheduling, controlling. These practices may not work well when innovation is required.

The following features of innovative projects can be distinguished:

1) High risk of project failure. According to research, a large number of innovative projects fail, either due to overspending of resources, or are unclaimed by the target audience.
2) The need for the external and internal environment of the project to be ready for innovation.
3) A high degree of probability of changing the content of the project during its implementation.
4) The difficulty of calculating and planning time and resources at the project initiation stage.

In order to be in demand, an innovative project must:

1) Create a new solution to the problem, or a new product / service.

2) Meet the customer's expectations.

3) Reduce operating costs in the long term.

4) Increase customer/target audience satisfaction in the future.

However, a large number of innovative projects and startups, even in the conditions of timely financing and provision of resources, are unsuccessful.

The success of an innovative project largely depends on the project manager and team, their ability to make timely management decisions, the level of motivation and flexibility.

The purpose of the research is to identify the main problems in the management of innovative projects and to offer recommendations for improving the efficiency of management.

The research goals are to:

1) Conduct an expert interview with representatives of project teams in various segments of project management.
2) Based on the results of the expert interview, identify the key problems in the management of innovative projects.
3) Understand the differences between traditional and innovation project management.
3) Develop recommendations for improving the management of innovative projects.

2 Methods

At the moment, there are a number of studies in the field of innovative project management in Russian science. V. S. Tikhonov's works are devoted to the study of the issue of digital management of innovative projects [1], in particular, N. B. Kultin [2] considers the possibilities of using artificial intelligence in the management of

*Corresponding author: d.m.semenova@gmail.com

innovative projects. Machine learning systems can be used to solve the problems of the analysis of indicators of efficiency of investments, selection of suppliers and subcontractors, subject to the availability of sufficient amount of statistics necessary for the training of decision-making algorithm. Problems of evaluating the effectiveness of innovative projects are raised by A. A. Borisov and I. S. Chernat [3], I. I. Fedorova and G. S. Gabidinova [4]. The works of M. V. Korchagina and K. S. Mukhiddinov, O. A. [5] Suikova, E. V. Kudryashova [6] are devoted to risks in the field of management of innovative projects. O. V Kostenko and V. V. Schennikova have developed and substantiated a methodology for designing a financial risk management system for innovative projects [7]. The work of M. A. Romanenko is devoted to the peculiarities of competencies of the project teams of innovative projects [8].

The problems of development of concession project management mechanisms are raised by Pazdnikova, N. P., Pechenegina, T.A. Mingaleva, Z., Mitrofanova E.A [9, 10].

Features of quality management in innovative IT-Projects are raised by T. I. Vlasova [11]. The researcher comes to the conclusion that there are some methodological problems in the field of quality management of IT projects, which are associated with specifics of IT projects and with the fact that project quality management focuses on ensuring the quality of project's output product and quality work of the project, while not paying significant attention to the quality of IT project management process.

Solovieva's article, team management as a factor of increasing the realizability of innovative projects, presents a model, developed by the authors, for diagnosing a company's readiness to implement high-risk innovative projects [12].

A. V. Lomazov, V. A. Lomazov in research Fuzzy temporal periodization of innovative projects considered the problem of time management in the implementation of innovative production-economic and socio-economic projects [13].

However, the key problems of project teams in the management of innovative projects and the ways to solve them are not sufficiently covered in Russian research.

In the world science there exists an abundance of literature on innovation and project management. Kerzner H. [14] describes an experience of innovative project management in the work "Project management best practices: achieving global excellence".

Some researchers try to add structure to innovation by identifying categories of innovation according to elements such as complexity, life-cycle phases, levels of risk, strategic business importance, and information available (Garcia and Calantone; O'Connor and Rice) [15], [16].

In this study, the PMBoK 7 version of project management knowledge was used to define the key approaches to project management, the concept of a project team, and the project lifecycle: Waterfall and iterative [15].

A project team is a group of individuals that supports the project manager in the execution of project work to achieve project objectives [15].

The project life cycle is a set of phases through which a project passes from the initiating to the closing. An iterative project life cycle is a cycle, in which the scope of the project is determined at the beginning of the project life cycle, but the evaluating of time and cost is regularly adjusted. In the case of a predictive life cycle of the project, the scope, the schedule and cost of the project are determined at the initial phases, and any changes in the content require competent management [15].

To identify the main limitations and problems in the management of innovative projects, the study conducted an expert interview with 10 representatives of the following areas of project management: banking, information technology, transport planning, mining engineering, state and municipal administration. The respondents were the managers of innovation project teams, and 8 out of 10 respondents have successfully developed and implemented the product at the moment. A number of respondents had experience in developing their own innovative project (transport model, IT-projects), a number of respondents managed innovative projects of an enterprise. The respondents were asked the following questions:

- how many innovative projects did you participate in?
- what main problems did you face?
- what management decisions did you make to solve the problems?
- what is the difference between project management and innovative project management? The respondents had experience in managing 1-3 innovative projects in practice.

3 Results and Discussion

In the course of the expert interview, based on the answers of the respondents, the following groups of problems were identified:

1) A group of problems related to the planning and management of project schedule. Respondents noted the difficulty of evaluating the exact time of project implementation. The majority of respondents faced the problem of falling behind schedule due to the fact that a number of factors were not taken into account when planning. Also, the project teams had problems with providing the project with resources, since they were not calculated accurately at the planning stage. This problem may arise in the case of using the waterfall project lifecycle, when the stages of work on the project – from idea to implementation-go in strict sequence one after another, and after the planning stage is completed and the project plan is approved (already at the implementation stage), it is problematic to make changes to the plan. In innovative projects, since a product or service is being developed for the first time, it is quite difficult for experts to accurately assess the actual time

and cost of project implementation at the stage of project initiation and planning.

2) A group of problems related to the knowledge management of innovative projects: first, the team leaders had to find specialists with the appropriate knowledge and competencies to implement the project. ("It was difficult to find a team of like-minded people to implement the project and experts from whom you could learn", "The main difficulty is a large amount of data and information that needed to be collected and processed, such an experience in the Russian Federation was the first time"). Secondly, the respondents faced the difficulty of preserving and transferring knowledge after the project was closed. ("An important task was to retain unique specialists with unique project experience, since they are also in demand in other organizations", "One of the problems was the possible leakage of knowledge gained during the project").

3) A group of problems associated with bringing the product of an innovative project to the market to consumers. Experts stated that in some cases the product was not in demand, or there was a low motivation to use it. The external environment was not ready for the implementation of this innovation.

Most of the respondents noted the need for continuous improvement of the innovative product and its refinement after receiving feedback from the customer or the target audience.

The experts also identified the following problems: the lack of competitors, whose experience could be used in the development of the project, the need to find like-minded people and investors who are ready for high risks, the difficulty of forming a project team.

In the management of innovative projects in the field of state and municipal administration, the key problems differ from other areas of management because the level of project culture is not high.

It should be noted that project management has only recently begun to be implemented in the state and municipal authorities of Russia.

During an expert interview with representatives of the Ministry of Digital Development and Communications of one of the regions of Russia, the following problems were identified: the difficulty of combining project management tools with the current work of the Ministry. In many ways, the principles of project management contradict the basic regulations of the civil service.

Also, the use of Agile is not fixed in regulatory legal acts and is used more on personal initiative.

The state and municipal service of the Russian Federation does not have the opportunity to constantly make changes to the project budget and schedular, therefore the use of Agile is problematic.

A number of employees do not have the appropriate education and do not have experience in using the tools and principles of project management.

We can also state the low level of risk management culture in the state and municipal service.

The main recommendations for improving the quality of innovative projects management in the public service of the Russian Federation and the development of a project culture are staff training, development of project skills and a regulatory framework.

According to the results of the expert interview the main differences between traditional and innovative project management are:

- High degree of risk in the innovative project management. Innovative project management is more challenging than traditional project management practices.
- High impact on human factors in the innovative project management.
- The need for phases of research and implementation in the innovative project management.
- The need for contingency reserves in time to the project schedule for the testing of hypotheses and bug fixes.
- If in traditional project management a failure is perceived as a defeat, in an innovative project a failure is perceived as an experience.
- Innovative project managers may need a different skill set than traditional project managers.
- Innovative project management should include product management tools.

Based on the results of the interview, we can state the low efficiency of the Waterfall project life cycle when managing an innovative project. With this approach to project management, it is quite difficult to make changes to the project plan at the implementation stage. Therefore, most organizations currently use an iterative lifecycle, which allows the project schedule revision and budget, as well as making changes during each iteration.

Agile allow to respond faster to changes in the environment and customer requirements, so they are more effective in managing innovative projects.

In innovative projects, there are much more risks, since management is carried out in an area of high uncertainty, so the identification and assessment of risks should be carried out at each iteration, regularly updating the risk log.

Before initiating an innovative project, it is necessary to conduct a study of the problems and needs of the target audience of the project, using problem interviews. Work with the target audience of the innovation project should also be organized after the creation of the product, at the stage of its implementation. The project team must develop a product implementation plan with indicators of effectiveness. Respondents noted the need for active work on the implementation and popularization of the product, by presenting it at international exhibitions, conferences, publishing the results of the project in scientific journals, as well as presenting the product in social networks.

The management of innovative projects requires closer cooperation with the customer and stakeholders of the project, for timely correction of the statement of work, risk breakdown structure, risk matrix, WBS. Therefore, communication management becomes one of the priority areas in innovative projects. Using a communication plan will help reduce the number of risks.

When managing innovative projects, the importance of project knowledge management increases, since an innovative project is a set of management actions performed for the first time. The project manager must properly organize the collection and analysis of information about the existing knowledge of the team at the project initiation stage. Based on the analysis, the manager can timely draw conclusions about the lack of certain knowledge in the team and thereby reduce a number of risks.

Also, during the iterations, the manager must organize the collection, transfer and storage of new knowledge gained during the implementation of the project, ensure the creation of a knowledge base, and ensure the information security of the project. The innovation project manager should be focused on training the team throughout the entire project lifecycle.

Visualization tools of the project management process: dashboards and Kanban-boards also have a positive effect on management efficiency according to respondents. They allow you to track the progress of the project, see failures and lacks.

An important factor in the success of an innovative project, according to the respondents, is the high level of the project culture of the organization. Under the project culture, we understand a set of logically interrelated knowledge, skills and possessions based on generally recognized standards and principles in the field of project activity [16].

In addition, the project culture implies a high level of personal responsibility and motivation of the project team, the desire to learn, and also developed soft skills: trust building, communication, motivation, leadership, coaching.

Three respondents highlighted the importance of the leadership qualities of the team leader and the need for sufficiently broad powers to make managerial decisions.

More than half of the respondents noted the need to develop emotional intelligence among team members for successful work in an innovative project.

Emotional intelligence is the ability to perceive, evaluate and manage one's own emotions and those of others. Emotional intelligence affects the self-organization and motivation of the project team.

4 Conclusion

Project management, as a management methodology, currently provides a fairly large number of effective tools that allow you to successfully manage innovative projects and solve emerging problems in project teams. The use of management methods like Agile allows you to respond more quickly to changes in the internal and external environment of the project, make changes to the project plan, and focus on working with the customer.

Knowledge management, being a separate group of project management processes, also reduces the number of risks and problems of preserving and developing the team's knowledge during the management of an innovative project. In addition, the development of soft-skills in the team also has a positive effect on the probability of achieving the goal and creating a demanded product.

An important factor influencing the effectiveness of the management of an innovative project is the quality of the project team, its ability to respond flexibly to changes, the ability to learn, the ability to communicate with each other, its level of motivation and self-organization.

Developing the project team, its knowledge and experience, you can achieve high efficiency in the management of an innovative project.

References

1. V.S. Tikhonov, *Scientific and Technical Bulletin of SPbSPU. Economic sciences.* **12(1)** (2019)
2. N.B. Kultin *Innovations.* **12(254)** (2019)
3. A.A. Borisov I.S. *Scientific and Technical Bulletin of SPbSPU. Economic sciences.* **13(5)** (2020)
4. I.I. Fedorova, G.S. Gabidinova *Education and law.* **8.** (2019)
5. M.V. Korchagina, K.S. Mukhiddinov, R *Economy and ecology of territorial entities* **3(2)** (2019)
6. O.A. Suikova, E.V. Kudryashova, *Innovative development of professional education.* **1 (25).** (2020)
7. O.V. Kostenko, V.V. Schennikova *The Eurasian Scientific Journal.* **2(12).** (2020)
8. M.A. Romanenko*News of higher educational institutions. Volga region. Social sciences.* **4 (52).** (2019)
9. N.P. Pazdnikova, T.A. Pechenegina *International Journal of Economic Research,* **14(16)**, (2017)
10. Z Mingaleva, N. Pazdnikova, E.A. Mitrofanova, *Asian Social Science.* **10 (24).** (2014)
11. T.I. Vlasova *Bulletin of Moscow Region State University. Series: Economics,* **3**, (2020)
12. I.A. Soloveva, I.A. Mostovshchikova *Bulletin of the South Ural State University. Ser. Economics and Management,***13(4)**, (2019)
13. V.A. Lomazov, *Azimuth of Scientific Research: Economics and Administration.* **9**, **3(32)**, (2020)
14. H. Kerzner, P*roject management best practices; Achieving global excellence,* (4E. Hoboken, NJ, John Wiley and Sons; Chapter 19. 2018).
15. G.C. O'Connor, M.P. Rice, *Journal of Product Innovation Management.* **30**, (2013)
16. R. Garcia, R. Calantone, *Journal of Product Innovation Management* **19 (2)**, (2002)
17. Project Management Institute, a *Guide to the Project Management Body of Knowledge* (PMBOK Guide) (2018)
18. D.M. Semenova, *Modern city: power, management, economics,* **1**, (2018)

Small retailers in/out of the crisis in Germany – Re-inventing Retailing

Cornelia Scott*, and *Vanessa* Fehlig

Anhalt University of Applied Sciences, Department of Economics and Business Management, 06406 Bernburg (Saale), Germany

Abstract. The year 2020 required many changes in all areas of life as a result of the Covid-pandemic. Both privately and professionally, we had to deal with new challenges. The economy in Germany and especially the in-store retail sector was affected severely by the lock-downs. Small retail businesses are especially threatened because of their low financial reserves and many insolvencies are imminent. This is very dramatic because, on the one hand, many livelihoods are directly linked to these small retailers. On the other hand about 94% of the sectors are currently so called micro or small enterprises with less than 10 or 50 employees and an annual turnover under 2 or 10 million euros. These "local shops" are a characteristic of city centers and important for the economic stability and diversity of the country. To support these businesses, the state has introduced various direct financial aids and indirect ones such as short-time work or guarantees. Regardless of the difficulties associated with applying for and receiving aid, it is also the responsibility of retailers to use change management skills to develop their own solutions. Covid has now dramatically shown where unused potential exists, e.g. through the implementation of multi-channel or hybrid business models. As dramatic as the situation is, it should also be used as an opportunity to increase innovative skills and to take a purposeful approach to the transformative process.

1 The corporate landscape in Germany

The corporate structure in Germany shows a clear picture, as a total of 99.4% of companies are so called "KMUs" (Kleine und mittlere Unternehmen / small and medium-sized enterprises). Of these, micro-enterprises take up the largest share with 81.8%. The companies are differentiated on the basis of the number of employees and turnover. [1-3]

Table 1. KMU - Enterprise size categories in Germany

	Number of employees	Turnover
Micro-enterprises	< 10	Annual turnover or annual balance sheet total ≤ € 2 million
Small enterprises	< 50	Annual turnover or annual balance sheet total ≤ € 10 million
Medium enterprises	< 250	Annual turnover ≤ € 50 million o. annual balance sheet total ≤ € 43 million

The crisis induced by Corona affects the majority of the German economy, but micro-enterprises are particularly affected, especially in the retail sector. These represent the largest number of enterprises with about 79.6%. Despite the large number, micro-enterprises were already losing more and more market share before Corona, so that turnover fell to only about 15% of total turnover in the sector. Corona has now accelerated this process and threatens the existence of many retailers. These little "shops", which invite people to spend time strolling and browsing in the city centre, are particularly affected by the economic consequences of the crisis. Often there are no large financial reserves and the distribution channel is primarily focused on local, direct sales. As a result of the initially optional and later mandatory closures of shops as part of the lockdowns in the spring and winter of 2020, resulted in a high loss of demand, with the exception being supplies of daily goods. In the stationary retail sector, the decrease in sales since the second lockdown in December has been persistent at around 80%, so that 52% of businesses reported that the impact would be so severe that, without further assistance, they might have to quit their business in 2021. [4-7]

2 State aid measures for the retailer in Germany

With the intention of supporting businesses quickly in the crisis, the federal government has expanded and simplified existing measures and adopted additional programmes. One of the most popular instruments is financing by loans. Special support is available, for example, from the Kreditanstalt für Wiederaufbau (KFW) in the form of fast-track loans with federal guarantee. Approximately 40,000 fast-track loans have been applied for so far. [8]

https://doi.org/10.1051/shsconf/202111600052

Another measure that has been established in Germany for years is short-time work (Kurzarbeit). This measure is primarily intended to save employment when a low order situation results in low capacity utilisation and thus also in financial loss. Due to the effects of the Corona pandemic, special regulations were enacted for the receiving of short-time work payments. The current requirements are that at least 10% of the employees have a loss of work hours and related wages of more than 10% as well as that the overtime has been reduced. If these conditions are achieved, a company can apply for short-time work. The reference period has been extended to a maximum of 24 months. If the application is approved, the company pays the employees the wages earned plus the short-time allowance for the lost wages. Upon application, the short-time allowance will be refunded to the company for the previous month with retroactive effect. The short-time allowance amounts to 60% of the net final salary for the first three months or 67% for employees with children. The benefits increase in two stages to a maximum of 80% or 87% after seven months. In addition, social security contributions, minus unemployment insurance, are also reimbursed, thus additionally reducing the burden on the company. [9,10]

In 2020, 1,082,500 companies applied for short-time work across all sectors, which affected a total of 14,403,789 people. 57.7% of the companies made their application in April, i.e. in the period of the first lockdown. To be able to classify the dimensions, in 2019 there were only 16,420 applications for short-time work. Especially in small companies, as well as in the retail sector, this support was used a lot. Thus, 11% (116,637) of all applications came from the retail sector. The number of all employees subject to social insurance contributions in the retail sector who were on short-time work averaged approx. 6.7% in the period from May to December 2020. Around 74% of these work in micro-enterprises and around 94% in micro or small enterprises with fewer than 50 employees. [11-13]

In addition, there is special support that was introduced due to the pandemic. So called interim aid (Überbrückungshilfen), grants and emergency aid can be applied for or tax payments can be deferred interest-free for a limited period. In total, the federal government has made more than 350 billion euros available for all Corona relief measures and has pledged guarantees amounting to approximately 820 billion euros. [14]

The various aid offers are partly aimed at different target groups and have different eligibility criteria. Companies can apply for several subsidies, but if the subsidy periods overlap, they are offset against each other. At the beginning of the first lockdown, the first interim aid was introduced, which provides funding for June, July and August. The basis of eligibility for this is that the turnover in April and May 2020 must have fallen by at least 60% on average compared to the same months of the previous year. This assistance should be particularly targeted at small and medium-sized enterprises, which is why an exclusion limit of 750 million euros in annual turnover was introduced. The subsidy amount is 40%, 50% or 80% of the eligible fixed costs, but a maximum of €50,000 per month, and is calculated depending on the drop in turnover. In addition, there is a maximum amount of €5,000 for micro-enterprises. [15,16]

Subsequently, the second interim aid for the so called "funding months" September to December 2020 was decided with modified conditions. The prerequisite for receiving the aid is a decline in turnover of at least 50% in two consecutive months as well as an average decline of at least 30% in the period from April to August 2020 compared to the corresponding months of the previous year. All enterprises, regardless of sector and size, that meet the eligibility criteria can apply. The subsidy amount has been increased to 40%, 60% or 90% of the eligible fixed costs, but still amounts to a maximum of €50,000 per month and is calculated depending on the drop in turnover. [15,16]

In addition to the second lockdown, November and December allowances were decided. The eligibility criteria for these are that the company is affected by the closure ordinance. Companies in the retail sector that only had to close in mid-December are therefore not eligible. [15,16]

The third interim aid covers the funding months November 2020 to June 2021. The requirements were significantly lowered here. A company must show a drop in turnover of at least 30% in one month compared to the same month of the previous year. Once again, all relevant companies can apply for assistance, but they must not have exceeded an annual turnover of 750 million euros in 2020. The subsidy rates remain at 40%, 60% or 90% of eligible fixed costs, but the maximum amount has been increased from €50,000 per month to €1,500,000 per month compared to the previous bridging aid. [15,16]

With regard to interim aid, a special rule was adopted which allows retailers to declare depreciation on seasonal goods at 100% as fixed costs, provided they were affected by the closure order and made a deficit in 2020. Previously, only depreciation on fixed assets could be declared at 50%. [17]

Furthermore, there are grants in the form of emergency aid from the federal government for small businesses, which are based on the number of employees and the operating costs for three months, among other factors. Companies with up to five employees receive up to €9,000 and up to €15,000 for up to ten employees. These subsidies also do not have to be repaid by the companies. A total of €50 billion has been made available for this measure. [17,18]

As promising as the measures sounded at first, considerable difficulties have now become apparent in retrospect, especially in the disbursement of aid. By the end of 2020, only 8% of the funds available for the first and second interim aid had been approved and disbursed, amounting to only 623.4 million euros. One reason for this is the application requirements, which are perceived as very complicated. For example, companies must have the applications submitted by a tax advisor, accountant, auditor or lawyer. In addition, their capacities are also limited and already at full capacity. Another aspect that slows down the process is that economic aid has to be approved by the EU Commission. [16,19]

For the retail sector, there is also the problem that, despite closures, the companies are not entitled to the November and December aid and that an application for bridging aid III is only possible since 10 February 2021, which can lead to major liquidity problems. In order to counteract the late payments and to reduce the time until payment via the federal states, advance payments are now to be made by the federal government. [20,21]

Overall, the affected companies are rather dissatisfied with the measures. For example, 81% of the stationary retail businesses stated that the relief measures were not sufficient to secure their existence. However, in addition to the financial measures, there is also dissatisfaction with the approach and communication to the ongoing lockdown. Although the Robert Koch Institute classifies the risk of infection for retailers as low (if hygiene regulations such as shopping with a mask and a maximum of one person per $10m^2$ are observed), there is still no opening concept. Retail representatives are therefore requesting such a concept, e.g. with realistic gradual openings. It is important that the concept applies uniformly, regardless of the size of the company and the federal state, in order to avoid distortions of competition and uncertainties. [5,22,23]

3 The entrepreneurial spirit of retailers is in demand

Despite the measures decided by the federal government, there is considerable uncertainty among many retailers: are they really entitled to financial aid, how long will it take for the money to be paid out and will the amount even be enough? Therefore, many retailers were forced to develop their own solutions for the lockdown immediately.

Small retailers in particular face different difficulties from large retail chains due to their shop closures. Until the pandemic, many small shops had no or only a rudimentary online shop or delivery service and tended to rely on the stationary distribution channel, as they had for years. But of course, this was no longer possible overnight due to the lockdown. The entrepreneurial spirit of the retailers is required to find quick and useful solutions for this new situation.

A few practical examples are given below:

Retailer *Korbmayer* is located in the city center of Stuttgart and stocks exclusive women's and maternity fashion as well as a wide range of baby and children's clothing, prams, children's furniture and toys and gifts from selected brands. In the last 12 months, the company has been presented several times by the German press as a good example of how to adapt to the new challenge - lockdown. Before the pandemic, *Korbmayer* achieved about 98% of its turnover through stationary sales and had only a small online shop. But the young managing director Florian Henneka did not want to simply accept the perspective of short-time work for his employees with the onset of the lockdown and was also worried about losing customers permanently to online suppliers. He reacted quickly by setting up his online shop. The employees actively participated and quickly found their way into the new tasks: Photographing, online posting, logistics, online customer service. Already in March 2020, 1/5 of the usual turnover was achieved through the online shop and in April, about 200 orders per day were already being received and shipped. Henneka is convinced that "without the development of an online shop, he is not sure if the business would have made it into the 6th generation" [8]. Now the online shop is indispensable for the staff and customers. Other sales options, such as video calling, were also quickly added: Online consultations are arranged and customers are invited to a virtual tour of the shop, during which products are demonstrated. The products they like can then be picked up or delivered to their home to try them out. Online shopping events via Zoom/Webex/Microsoft teams are also planned and offer a convenient way for customers to see the products without having to leave home.

For companies without their own online store and no ability to set one up on short notice, there are alternatives such as the website *Yourlocal*, which enables small companies in Magdeburg to present themselves and their products online. On the website, which was set up in the first lockdown, there are various categories ranging from food, clothing and household goods to restaurants as well as sports and cultural offers, which provide a varied selection. Customers can then request the products and, depending on the shop, pick them up (Click & Collect) or have them delivered. Thus, businesses are relieved of having to deal with the technical challenges. In addition, there is the advantage that *Yourlocal* is free of charge and, as a project, has set itself the goal of supporting local retailers in the pandemic by providing a platform. [24]

4 Seeing the crisis as an opportunity - retailers should develop and be confident in themselves....

The lockdown has changed buying behaviour, perhaps in the long term, and retailers need to adapt to these changes. Retailers, especially small retailers, need to adapt to this customer behaviour and act instead of react if they want to survive in the market permanently. They must increasingly draw the customer's attention to their strengths and "bind" them to them. Creative, new sales channels (e.g. through online shops, online shopping, social media) must be developed further and a new, different "shopping experience" created. Small shops/stores often live from their flair, they have personality and are run by personalities - personal contacts and advice, a familiar and secure feeling is often what makes these retailers strong. This must be communicated effectively in the sales channels. In personal conversations with retailers in recent months, however, we have often also heard scepticism, especially among older business owners, about the implementation of the digitalisation strategy in sales and distribution. Many retailers attribute this scepticism to a lack of resources, especially now after the long lockdown. Here, retailers should rely on trusted networks and also new stakeholders and partners. Retailers are not alone in this situation and the environment (retail eco-system) appreciates the added value that retailers bring to the cities - our city centres, in small, medium and large cities - get a liveable atmosphere through the "small shops". It can also be "friends & family" who support the transformation - perhaps the children or grandchildren, the generation of "digital natives". For example, an older owner who has been running her business for over 50 years told me that her granddaughter, who is active on Instagram, helped her to digitally set up her traditional business to get over the lockdown. She was surprised and thrilled with the positive outcome. Already after the first day on "Insta", she was greeted in the morning the following day by a long line of young people in front of her shop, all wanting to buy her hats. Digitisation has not only saved your business through the lockdown period, but has also given you access to a new customer base.

But other groups (stakeholders) - institutes, associations, educational institutions e.g. chambers of industry and commerce (IHK) and universities of applied sciences (HAW) can also accompany this transformation. They have the expert knowledge and young, creative minds. It is in the interest of the citizens to preserve the retailers, because they are an enrichment for all of us and bring diversity and liveliness to our cities. It would be a great loss and very sad if these small, cosy shops were lost in the course of the lockdown.

5 Summary and outlook

The economic consequences of the lockdown affect small stationary retailers particularly hard. However, at around 80%, micro-enterprises with fewer than ten employees make up the largest share of the industry. It is precisely these small retailers that shape the picture in Germany's city centres. Due to the lockdown in spring 2020 and the still ongoing second lockdown since December, which led to sales slumps of around 80%, many are facing financial ruin. More than half of the stationary shops estimate that they will not survive the year 2021 without help.

The federal government has taken many measures to support retailers in this severe crisis; comprehensive direct financial aid payments, guarantees, tax aid as well as adjustments to the short-time work regulations have been initiated, but many retailers still complain that these measures are too bureaucratic and that the aid does not arrive at all or only late. The voices of retailers and their lobbyists are also getting louder and louder that a realistic strategy for an end to the lockdown and an opening of the shops is missing on the part of politics.

Looking for solutions from the state alone does not seem to be a good way forward. The entrepreneurial spirit of the retailer is needed. Some practical examples have shown how retailers have successfully dealt with the crisis through creativity and agility: the introduction/expansion of online shops, online shopping, social media channels (especially Instagram and WhatsApp), online calls, click and collect were quickly introduced successfully. The survival strategy in the crisis is digitalisation and customer loyalty.

The "shopping experience" has to be rethought and repositioned. Retailers, who often have limited resources, do not have to tackle this transition alone. Experts can be called in as advisors, such as chambers of industry and commerce (IHK) or universities of applied sciences (HAW) to support these processes and conversions. As Christine Lagarde said in an online interview with the Economist magazine a few weeks ago - it would be very sad if the small retailers disappeared from our cities because they couldn't withstand financially, because they bring diversity and joy and make cities liveable. I think she speaks for all of us. [25]

References:

1. Handelsverband Deutschland, *KMU-Definitionen*, Retrieved from: https://einzelhandel.de/index.php?option=com_content&view=article&id=974
2. Statistisches Bundesamt, *Kleine und mittlere Unternehmen (KMU)*, Retrieved from: https://www.destatis.de/DE/Themen/Branchen-Unternehmen/Unternehmen/Kleine-Unternehmen-Mittlere-Unternehmen/Glossar/kmu.html
3. Statistisches Bundesamt, *Verteilung der Unternehmen in Deutschland nach Unternehmensgröße im Jahr 2018*, Retrieved from: https://de.statista.com/statistik/daten/studie/731901/umfrage/verteilung-unternehmen-in-deutschland-nach-unternehmensgroesse/
4. Handelsverband Deutschland, *Auswirkungen der Coronavirus-Krise auf den Einzelhandel: Viele Handelsunternehmen leiden unter sinkender Kundennachfrage,* Retrieved from: https://einzelhandel.de/presse/aktuellemeldungen/12594-auswirkungen-der-coronavirus-krise-auf-den-einzelhandel-viele-handelsunternehmen-leiden-unter-sinkender-kundennachfrage
5. Handelsverband Deutschland, *Pressekonferenz des Einzelhandels,* Retrieved from: https://einzelhandel.de/hdepk
6. Statistisches Bundesamt, *Anzahl der KMU in Deutschland nach Wirtschaftszweigen im Jahr 2018,* Retrieved from: https://de.statista.com/statistik/daten/studie/731975/umfrage/anzahl-der-kmu-in-deutschland-nach-wirtschaftszweigen/
7. U. Bentele, *Zukunft Einzelhandel - Was kommt, wenn Corona geht?* Retrieved from: https://www.rbb-online.de/doku/x-z/zukunft-einzelhandel-was-kommt-wenn-corona-geht.html
8. SWR, *Wirtschaft im Corona-Schock - wie überstehen wir die Krise?* Retrieved from: https://www.ardmediathek.de/swr/video/betrifft/wirtschaft-im-corona-schock-wie-ueberstehen-wir-die-krise/swr-fernsehen/Y3JpZDovL3N3ci5kZS9hZXgvbzEyNDc3NDk/

9. Bundesagentur für Arbeit, *Kurzarbeitergeld*, Retrieved from: https://www.arbeitsagentur.de/m/corona-kurzarbeit/
10. Bundesministerium für Arbeit und Soziales, *Erleichtertes Kurzarbeitergeld*, Retrieved from: https://www.bmas.de/DE/Corona/erleichtertes-kurzarbeitergeld.html
11. Bundesagentur für Arbeit, *Angezeigte Kurzarbeit - Deutschland, West/Ost, Länder, Kreise und Agenturen für Arbeit (Zeitreihe Monatszahlen)*, Retrieved from: https://statistik.arbeitsagentur.de/SiteGlobals/Forms/Suche/Einzelheftsuche_Formular.html?nn=20726&topic_f=kurzarbeit
12. Handelsblatt, *Kurzarbeit sinkt leicht – Starker Anstieg im Handel*, Retrieved from: https://www.handelsblatt.com/politik/deutschland/ifo-institut-kurzarbeit-sinkt-leicht-starker-anstieg-im-handel/26766054.html?ticket=ST-2839684-jTrsCuJUFnMTM5LjbmVb-ap2
13. ifo Institut, *Tabelle: Kurzarbeit nach Branchen*, Retrieved from: https://www.ifo.de/themen/coronavirus
14. Bundesministerium der Finanzen, *Kampf gegen Corona: Größtes Hilfspaket in der Geschichte Deutschlands*, Retrieved from: https://www.bundesfinanzministerium.de/Content/DE/Standardartikel/Themen/Schlaglichter/Corona-Schutzschild/2020-03-13-Milliarden-Schutzschild-fuer-Deutschland.html
15. Bundesministerium für Wirtschaft und Energie, Bundesministerium der Finanzen, *FAQ zur „Corona-Überbrückungshilfe I für kleine und mittelständische Unternehmen"*, Retrieved from: https://www.ueberbrueckungshilfe-unternehmen.de/UBH/Redaktion/DE/FAQ/FAQs/faq-liste-01.html?nn=2323344
16. M. Greive, J. Hildebrand, D. Neuerer, and F. Specht, *Chaos bei den Corona-Hilfen: Das müssen Unternehmen jetzt wissen*, Retrieved from: https://www.handelsblatt.com/politik/deutschland/ueberbrueckungshilfe-chaos-bei-den-corona-hilfen-das-muessen-unternehmen-jetzt-wissen/26793762.html?ticket=ST-2651550-u7uG79D4unfmMBC0X5Ad-ap2
17. Bundesministerium der Finanzen, *Überbrückungshilfe vereinfacht und verbessert*, Retrieved from: https://www.bundesfinanzministerium.de/Content/DE/Standardartikel/Themen/Schlaglichter/Corona-Schutzschild/2021-01-19-ueberbrueckungshilfe-verbessert.html
18. Bundesministerium für Wirtschaft und Energie, *Eckpunkte „Corona-Soforthilfe für Kleinstunternehmen und Soloselbständige"*,Retrieved from: https://www.bmwi.de/Redaktion/DE/Downloads/E/eckpunkte-corona-soforthilfe.html
19. K. Köster, *Corona-Hilfen werden kaum abgerufen*, Retrieved from: https://www.stuttgarter-nachrichten.de/inhalt.coronavirus-corona-hilfen-werden-kaum-abgerufen.2e037953-2541-42e9-8284-89ab853f06fd.html
20. T. Hoppe, *Firmen erhalten mehr Spielraum bei Hilfen*, Retrieved from: https://www.handelsblatt.com/politik/deutschland/coronakrise-firmen-erhalten-mehr-spielraum-bei-hilfen/26888620.html?nlayer=Themen_11804704
21. M. Greive, J. Hildebrand, and K. Stratmann, *Späte Auszahlung: Wer ist schuld am Coronahilfen-Debakel?* Retrieved from: https://www.handelsblatt.com/politik/deutschland/unternehmensgelder-spaete-auszahlung-wer-ist-schuld-am-coronahilfen-debakel/26914424.html?nlayer=Themen_11804704
22. Handelsverband Deutschland, *Corona-Gipfel: Politik versagt - keine Perspektive für den Handel*, (2021, Feburary 11), Retrieved from: https://einzelhandel.de/presse/aktuellemeldungen/13181-corona-gipfel-politik-versagt-keine-perspektive-fuer-den-hande
23. RKI, *ControlCOVID Strategie und Handreichung zur Entwicklung von Stufenkonzepten bis Frühjahr 2021*, (2021, February 18), Retrieved from: https://www.rki.de/SiteGlobals/Forms/Suche/serviceSucheForm.html;jsessionid=B38FFA56AAA26699C2DA9FF7E1F7AA8B.internet061?nn=2725444&input_=2386720&gts=2725442_list%253DdateOfIssue_dt%252Bdesc&resourceId=2390936&submit.x=0&submit.y=0&searchEngineQueryStrin
24. YourLocal, *Dein Online-Marktplatz*, Retrieved from: https://magdeburg.store/start
25. C. Lagarde, *Editor's conversations: Christine Lagarde, president of the European Central Bank*, (2021, February 11), Retrieved from: https://economist-subsevents.zoom.us/rec/play/s8XPja9t4UzxPYZWd1W8dIIDLBvwkGqJNxl-8EcMAIunBJ1pvNh_YTiWmYa1QpCAYgw0Qv763rniURBe.6vUJQh64GCEDO-cV

The regional accelerator as a driver for the development of an innovative environment and technological entrepreneurship

Anna Ovchinnikova[1*], *Tatyana* Topoleva[1]

[1] Udmurt branch of Institute of Economics of Ural Branch of Russian Academy of Science, 4, Lomonosov st., Izhevsk, 426004, Russia

Abstract. Development of the innovative economy in Russia predetermines the need to adapt to new challenges and global economic trends. World experience demonstrates the effectiveness of the acceleration mechanism in supporting innovative projects, expanding the network interaction of the elements of innovative ecosystems, and accelerating the implementation of breakthrough ideas. The existing imbalance in the socio-economic development of the Russian regions, the differentiation of their resource potentials determine the search for institutional solutions that activate the innovation process. The conceptual and methodological basis for acceleration platforms and foreign and domestic experience of business accelerators are analyzed. The stages of change management in the context of acceleration logic are systematized. These are hackathon, foresight, acceleration, project development. The article substantiates the expediency of acceleration of innovations in the Udmurt Republic as an institutional element of the innovative environment that meets the specifics of the military-industrial complex, the declared diversification of defense industries and actualization of the request to increase the share of high-tech civilian products in the total volume. The mission and goals of the accelerator and the main program stages are described. The graphical model of a regional accelerator is presented. It integrates key elements of the innovation environment. The result is stimulation of the innovation process, support of the technological entrepreneurship, strong regional innovative competencies, qualitative growth of project activities, which will significantly increase the level of competitiveness of regional enterprises and the role of the region in the global economic system.

1 Introduction

In the context of the innovative economy, in order to achieve the technological leadership in the international arena, it is necessary to overcome the communication gap between the subjects of innovation activities, significantly increasing the level of interaction in the "science (innovators) - business - government" triad. The focus on structural changes in the national and regional economic systems actualizes the need to improve mechanisms, methods and tools for innovation, as well as to stimulate the motivation and competencies of innovators to generate and commercialize new ideas. Innovative activity is expressed in the reproduction of the intellectual component of an innovative product and in the ability to manage it qualitatively. In the institutional structure of successful commercialization of the results of innovative activities, one of the most effective elements are business accelerators - systems for supporting innovative projects at the initial stages. Acceleration programs carry out technological scouting, transform the level of participants, increase the adaptive abilities of companies, opening up objective opportunities for leapfrogging development and strengthening competitive advantages.

2 Materials and methods

The authors used the results of works describing the world and domestic experience of business accelerators. One of the first theoretical and methodological substantiation of the phenomenon of acceleration was provided by J.D. Christiansen in his work *Copying Y-Combinator: a framework for developing seed accelerator programs*, describing a business accelerator as a startup support program characterized by such features as the transparency of selection for participation, support, investment in exchange for a share in the charter capital of companies at the pre-sowing and sowing stages, project consulting, team work [1]. The issues of intensification of the technological transfer in the context of acceleration programs are described by K. Malek, N. Radojevich-Kelly, I. McCarthy, D. Hoffman, P. Miller, K. Bound, M. Szafranski [2-6]. The mechanism of acceleration of innovative projects has been analyzed by H. Etzkowitz, J. Pique, C. Scheel, and J. Berbegal-Mirabent [7-9]. The domestic vector of research is represented by the works by S. Makarov, A.A. Sytnik, M.A. Kamenskikh [10-12]. O.A. Zhdanova views business accelerators as an institution of the venture capital system, paying particular attention to the business model and marketing strategy [13]. Yu.V. Fedorov and

Corresponding author: tn-topoleva@mail.ru

A.V. Tsvetnykh define it as a form of integrated entrepreneurial activity aimed to support innovations and as an element of the regional innovation infrastructure that provides partner support for project commercialization [14]. M.R. Kozlov defines a business accelerator as a type of business incubator formed as a result of the focus of corporations on the selection of the most promising startups [15]. Characteristics of the regional acceleration process are presented by D.A. Statovsky who defines it as a tool or a set of principles in the arsenal of the subject of innovation (development institution) aimed at increasing the efficiency of the regional technological transfer. The hypothesis about the unification of the acceleration toolkit has been made forward in [16].

The authors used the general scientific methods, including: comparative and structural analysis, classification and typology, abstraction, generalization, and graphic visualization.

3 Results

At the national level, business accelerators perform a number of functions that facilitate innovation processes, including:

- increasing the systemic efficiency of development institutions at the federal and regional levels;
- involving the corporate sector in acceleration programs;
- promoting the development of social entrepreneurship [17].

A business accelerator is an organizational model for the accelerated development of startups through the investment, mentoring and expert support for early stage projects. Within the accelerator, structured practice-oriented programs are implemented. They allow participants to “grow” business models, improve professional skills, form a business strategy, obtain investments and increase economic performance.

In the mid-2000s, the concept of business acceleration (as opposed to business incubation) was proposed in order to promote projects. The first accelerator "Y-Combinator" was created in 2005 in the USA by P. Graham and D. Livingston. Until 2014, the investment conditions were as follows: $ 17,000 for a 7% stake and additional $ 80,000 converted into a stake, depending on the results of the next round of investments. Currently, seed investment is $ 150 thousand for a share of 7% [18]. The average business value of companies that implemented the Y-Combinator program exceeded $ 100 billion, which determines the image effect and positions its graduates as particularly attractive to the investors. According to the developers, the amount of seed funding is not a determining factor for acceleration participants, but it allows teams to concentrate efforts on project activities and cover the basic costs of startups. Among the most famous American business accelerators are Techstars, Plug and Play, 500 Startups, Alchemist, Techstarts, Angelpad, FoodX, and Coplex.

In Europe, the first business accelerator "Seedcamp" appeared in the UK in 2007. More than 380 startups have implemented acceleration programs on its platform, the investments of the world's leading investors have exceeded $ 5 billion [19]. Currently, over 50 business accelerators operate in Europe, including: Startupbootcamp, Lanzadera Accelerator, Future Fifty, Barclays, Entrepreneur First, Wilco, and others. Research on the startup support scale indicates the proportionality of the American and European innovation ecosystems [20]. Intensive growth of business accelerators is also observed in Asian countries, where their number is more than 100. According to various estimates, there are 500-700 business accelerators in the world; in 2019, the number of acceleration programs reached 2,191 thousand dollars; in Canada and Great Britain - 250 thousand dollars.

The acceleration logic provides for several stages of change management (Fig. 1): hackathon, foresight, acceleration and post-acceleration. The hackathon is an element of crowdsourcing of ideas and the starting point for successful projects. The most important components are interaction with mentors and technical specialists, new knowledge about technologies, taking into account tasks of innovators. Within the foresight sessions, the market needs are actualized, strategies for the participants of the innovation process are developed. At this stage, the circle of experts, interested investors and industrial partners is determined. The pre-acceleration stage is an additional service for the commercialization of projects and business plans. It is useful for teams that have not yet decided on a direction of movement and need some refinement of ideas. Pre-acceleration is aimed at popularizing innovative activities; a wide range of participants can be involved in this process (creative youth, students, young researchers, technology teams, representatives of science-intensive companies).

The acceleration program is implemented directly, including the preparatory and active stages, as a result of which the best accelerator projects are selected 3-8 months later. At the post-acceleration stage, expert consulting, project promotion and business scaling are accompanied by personal facilitators.

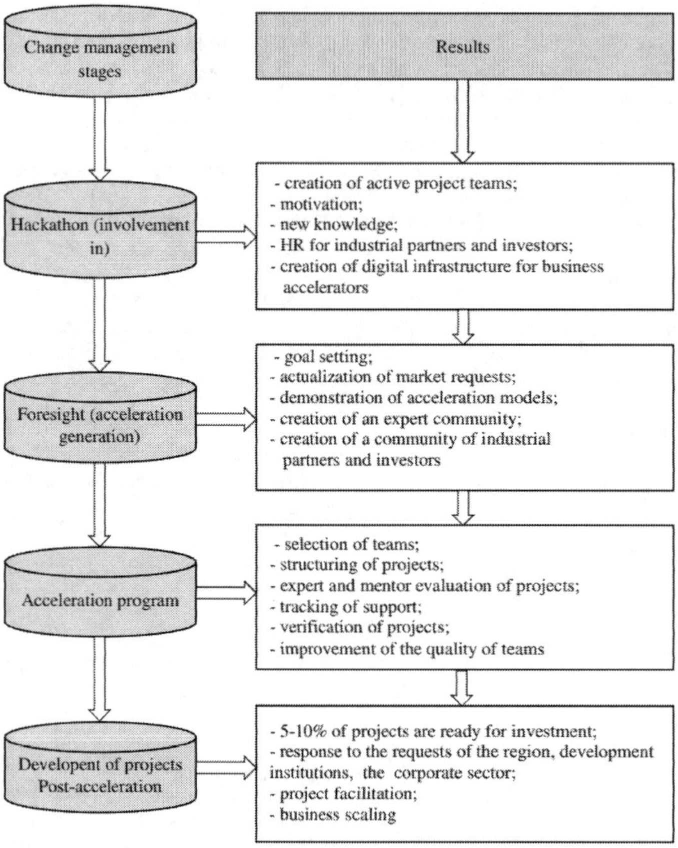

Figure 1. Stages of change management in acceleration logic.

The specific feature of the Russian innovation ecosystem is a predominance of the government component (grant programs, fund support for startups in investment rounds "A" and "B"). A significant share of the infrastructure support is entrusted to federal development institutions. Business accelerators were created later than in Western countries - in 2009; at the initial stage, IT projects were promoted. Currently, there are the following types of accelerators: corporate accelerators, private accelerators, accelerators with public capital, university accelerators and regional accelerators.

The largest Russian accelerators are GenerationS, Pulsar Venture, Accelerator of the Internet Initiatives Development Fund (IIDF), Skolkovo. In 2013-2020, more than 16 thousand startups took part in acceleration programs by "GenerationS", more than 10 thousand took part in acceleration programs by "IIDF", more than 2 thousand took part in acceleration programs by "Skolkovo". "GenerationS" is an accelerator with public capital that does not invest in startups, but facilitates communications with the corporate sector, with an impressive prize pool of industrial partners. The "IIDF Accelerator" operates at the regional level; up to 50% of the total number of transactions are carried out with regional startups in IT, retail, financial technologies. Interests of the accelerator in the share of companies vary within 3-7%; when the KPI is reached, investments in technology startups increase significantly; for example, round "A" involves financing in the amount of over 300 million rubles. [21].

The leading Russian universities have also initiated acceleration programs: Captains of Russia at the REU n.a. G.V. Plekhanov, IT Accelerator at the Ural Federal University, Ideal Machine at St. Petersburg National Research University of Information Technologies, Mechanics and Optics, etc. In 2020, the acceleration program of the Higher School of Economics ranked first in the world ranking of university accelerators *UBI Global* [22].

The corporate sector develops its own acceleration programs within a given specialization and tasks related to the growth of companies' capitalization achieved through the acquisition of new technologies that allow for occupying adjacent market niches, as well as for optimizing business processes, reducing costs, promoting internal projects, attracting new clients. Among the corporate accelerators are "Accelerator of Sberbank", "KAMAZ-digital", "Accelerator of NCO Nornickel", "MTS Startup Hub", "Severstal SteelTech

Accelerator", "Goznak Startup Lab". In 2020, 14 new corporate accelerators were launched, more than 30% of acceleration programs were ordered by the banking sector and food retailers (Fig. 2).

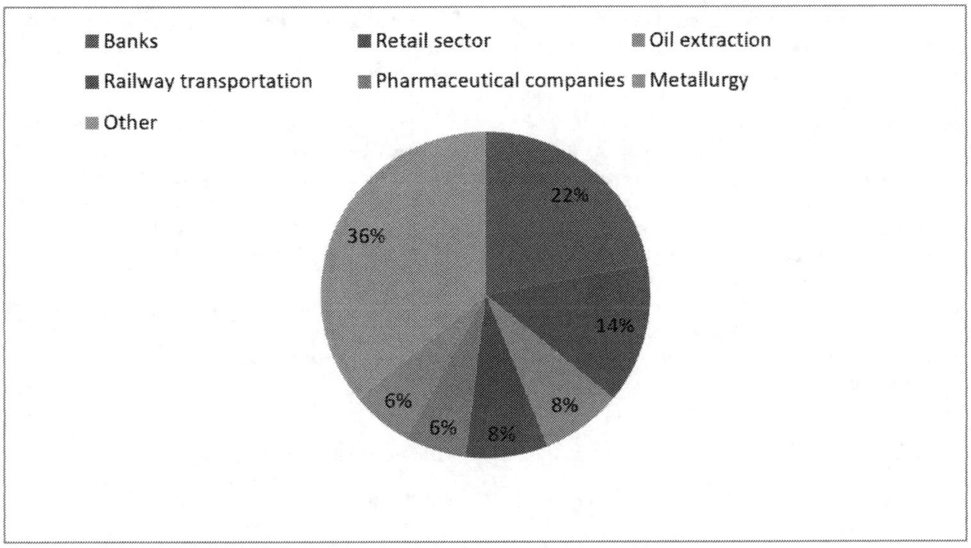

Figure 2. Distribution of corporate acceleration programs by industry in 2020.

In the context of geopolitical tensions, since 2019, the West has a negative attitude to the Russian capital. Startups with a high share of Russian transactions have been closed; in 2020, this problem spread to funds working in conjunction with business accelerators and the investment sector.

In 2020, the Covid-19 pandemic had a negative impact on all sectors of the economy, including the innovation sector. According to the Dsight study "Venture Russia - 2020", some of the acceleration programs were suspended [23]. Most business accelerators have changed the format of their activities, moving to the online space. Analysis of innovative solutions revealed a demand for technologies that are more likely to be relevant in the long term - Digital twins, IoT, BigData, Industrial Tech, robotization, biometrics, Hyperautomation, AR / VR, Healthcare. In 2018, 107 transactions of Russian accelerators with technology startups were concluded in the amount of 466.8 million rubles; in 2020, the downward trend remained. 16 contracts were concluded [24].

Summarizing the results of the activities of business accelerators, it is possible to formulate the main advantages characteristic of this organizational model:

• Building the potential of innovative entrepreneurship through the mentoring support.

The main stages of training involve the participation of mentors - qualified consultants accompanying the projects. According to the founder of LinkedIn R. Hoffman: "Initiating a startup is like throwing yourself off a cliff and trying to assemble an airplane during the flight down. A mentor is a person who has already assembled an airplane and will help others to do it" [25]. The largest business accelerators define mentoring as a key competence when working with startups. The training block includes the improvement of entrepreneurial skills in the following areas: business planning, legal and accounting, management, financing, work with personnel and clients, negotiation. The educational context of acceleration is of great public importance.

• Decrease in transaction costs, access to resource bases.

Since within the acceleration programs, the interaction of participants with representatives of the expert community, industrial partners, and potential consumers is intensifying, the information asymmetry is leveling: startups immediately "connect" to reliable resources, bypassing the stages of search and verification. On the other hand, resource suppliers can focus on the most promising projects. Thus, transaction costs are reduced on both the demand and supply sides.

• Effective use of investments from the public and corporate sectors.

By screening out unproductive ideas, business accelerators help minimize government expenditures intended to support unpromising venture capital projects. Likewise, entrepreneurs (investors) and innovators save time and resources through the rapid market validation which can help to select really valuable ideas, improve or abandon them at no cost. In addition, in recent years, the level of understanding of market mechanisms has significantly increased among the investors.

An effective approach to innovation management is largely due to the knowledge of the laws of the innovation cycle - a process of logically interrelated stages, including:

- creating a business idea, R&D;
- prototyping an innovative product;
- seed financing, prototype improvement;
- investing in an innovative project;

- entering the market;
- increasing sales volumes;
- achieving sales stabilization;
- bifurcation (modernization or assortment changes).

In the infrastructure system intended for supporting innovation, business accelerators are usually involved in the first four stages of the innovation cycle, which is an undoubted advantage for innovators in the face of a lack of knowledge, experience in the investment market and start-up resources for implementing technological ideas.

The uneven development of regional economic systems reflects the lag in innovation trends both in institutional and infrastructural aspects. Currently, 52% of regional accelerators are located in Moscow and Moscow region, 5.8% - in the Republic of Tatarstan, 5% - in St. Petersburg and Leningrad region, 4% - in Sverdlovsk region. Of the total number of accelerators, 5% were created under the program of the Presidential Grants Fund. The most successful regional business accelerators are TechStarter (Voronezh), the Regional Acceleration Program of the Republic of Bashkortostan (Ufa), A-Start (Novosibirsk), SturtupSamara (Samara), Big Intelligence (Perm). In more than 50 regions of the Russian Federation, there is no this element of the innovation environment [26].

In the context of the underdeveloped domestic venture capital market, most regional accelerators do not finance innovative startups. They set a fixed fee for acceleration services, thereby minimizing market risks. In the structure of investment in innovative startups, the share of accelerators does not exceed 2% (Fig. 2) [27]. There is a trend towards an outflow of innovative ideas from the regions to the federal sites, or to the regions with a developed innovation infrastructure. This aggravates regional disproportions, affecting the innovative potential of the territories.

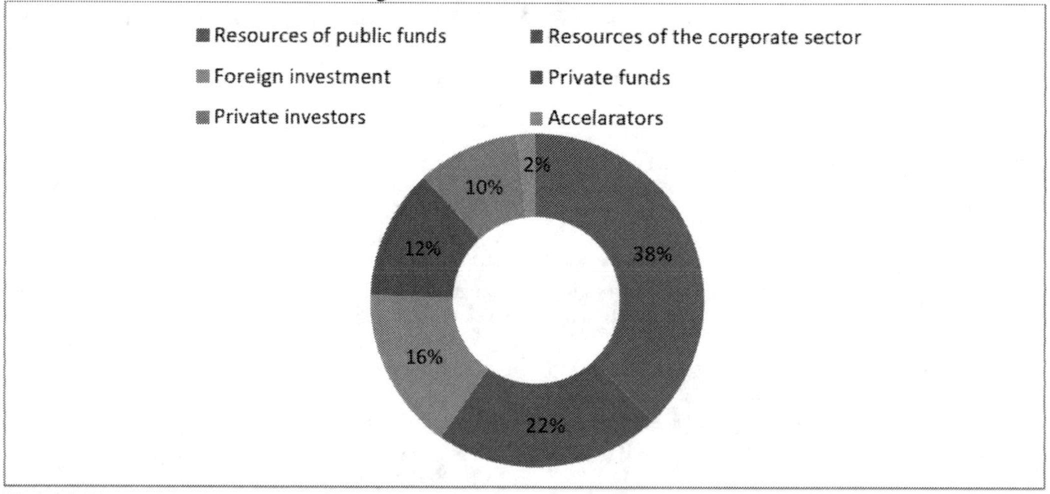

Figure 3. The structure of investments in innovative start-ups in Russia.

Currently, in the Udmurt Republic, with the assistance of the UR Development Corporation and the UR Export Support Center, two acceleration programs are being implemented. The export accelerator has trained five teams in the following tracks: IT, mechanical engineering, agriculture and food industry. The result was contracts concluded with foreign companies from the USA, Spain, Bulgaria, Venezuela, Kazakhstan, Belarus, and Moldova. The winners received 500 thousand rubles to conduct marketing research, create exhibition stands and promote projects online. Interaction with teams continues in a post-acceleration format.

In order to develop a model of the regional accelerator of innovations for the Udmurt Republic, it is necessary to analyze features of the innovative development of the region. Currently, the republic is forming an innovation ecosystem [28-32]. As of the beginning of 2020, 30 organizations are involved in R&D, of which 20% belong to public organizations, 33.3% are companies, and 36.6% are higher education organizations. The region has a high level of scientific and technical potential, the fundamental block of research is coordinated by the Udmurt Federal Research Center of the Ural Branch of the Russian Academy of Sciences, which includes: agricultural and humanitarian subdivisions. Studies of the socio-economic problems are carried out by the Udmurt branch of the Institute of Economics of the Ural Branch of the Russian Academy of Sciences.

Over the past five years, there has been a positive trend in the number of personnel engaged in research and development which has increased by 19% (2036 people). The number of production technologies used increased by 33% and amounted to 6642 units. In 2019, the innovative activity of organizations reached the level of 10.6%, which is 3.5% higher than in the previous period. At the same time, a number of critical trends have been observed: a decline in the cost of technological innovation of organizations - 5156.1 million rubles. (a 55% drop by 2018) and a drop in the number of advanced technological developments from 21 units in 2017 up to 3 units in 2019, as well as a decrease in the volume of manufactured innovative goods - 617043.6 million rubles. (a decrease by 2.6%). In the structure of current expenditures on research and

https://doi.org/10.1051/shsconf/202111600053

development, the negative situation is developing (a decline was 43.7% compared to the previous period, and in general there was a rollback to the level of the beginning of the 2010s). In the area of "fundamental research", on the contrary, there is an increase in costs by 14.6% compared to the previous period and by 47.2% over the last 5 years [33].

In the current conditions, in order to activate the processes of scientific and technological development, it is necessary to strengthen the regional innovation environment by creating a regional accelerator of innovations (Fig. 3). The accelerator should be modified taking into account the specifics of the Udmurt Republic, as a center for the production of weapons and military equipment, where 12 largest Russian military enterprises are located (Kalashnikov, IEMZ Kupol ", Izhevsk Motorcycle Plant" Axion-Holding ", Izhevsk Radio Plant, Votkinsk Plant, Chepetsk Mechanical Plant, Sarapul Electric Generator Plant, etc.).

The mission of the regional innovation accelerator is to develop an innovative environment and technological entrepreneurship in the region, to diversify the military industry.

The goals of acceleration are as follows:

- development of residents' skills and competencies for the effective implementation of entrepreneurial activities in the field of science-intensive technologies;
- experience in the investment market;
- mastering new methods and tools for doing business;
- multiple growth of sales in the Russian and foreign markets;
- improving the quality of innovative projects;
- financing startups;
- popularization of technological entrepreneurship and innovation in the region;
- ensuring the global technological leadership of regional companies.

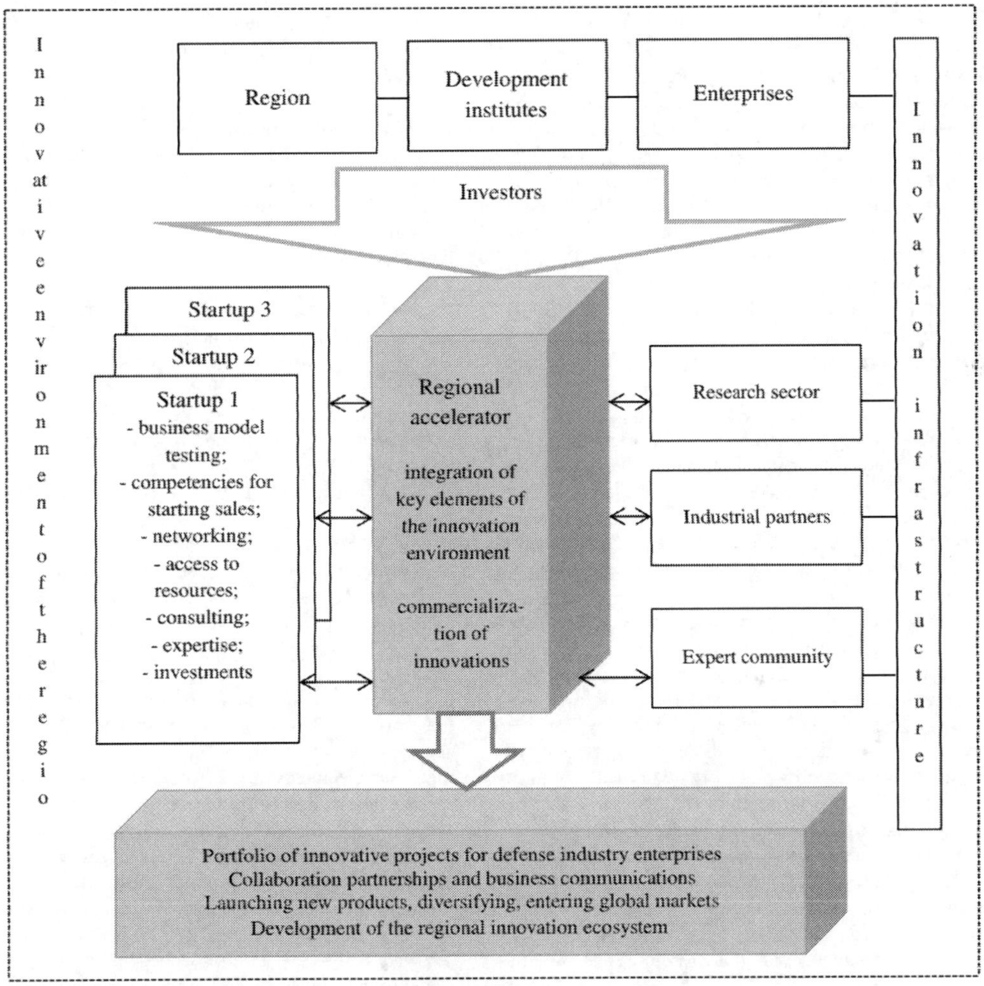

Figure 4. The regional accelerator of innovation in the Udmurt Republic.

The military enterprises employ more than 40% of the total number of workers in the region. The creation of a regional accelerator focused on the regional defense industry is in line with the federal agenda, according to which, since 2017, Russia has been implementing diversification measures in the defense industry by increasing the output of civilian and dual-use products, increasing exports of high-tech products and implementing an import substitution program. By 2030, the share of exports of high-tech civilian products in the total volume of military products should be 50%. In 2020, the output of the republic's military enterprises amounted to 148 billion rubles. The leading exporters of are military enterprises (60% of total exports at the end of 2019), while the pace of diversification lags behind the declared ones. The best performance has been demonstrated by the Axion-Holding Izhevsk Motorcycle Plant, where the volume of civilian products reached 35-40%. Other enterprises are experiencing organizational and technological difficulties. The priority market niches for the defense industry are medicine, radio electronics, power engineering, etc. In terms of the competencies, the capacity of these markets is 1.2 trillion rubles.

Under these conditions, the regional accelerator can become a catalyst for innovation processes, and innovative projects will be funded. Given the growing interest in the "open innovation" paradigm, the regional accelerator can become a platform for implementing the mechanism of "open innovations" - defense industry technologies without the stamp of secrecy, as well as civil technologies with potential dual-uses. Thus, it becomes possible to link the actual needs of the defense industry with the needs of the commercial technology market in the accelerator space.

Since leading business accelerators are interested in projects at the prototyping stage, the value of a regional accelerator can increase in the context of supporting projects at the earliest stages (the "pre-seed" investment round), involving fundamental, exploratory and applied research. At this stage, costs are lower than at subsequent stages of the innovation cycle, which is important in the conditions of investment "vigilance" of the corporate sector and makes it possible for promising ideas to be checked. In world practice, pre-seed investment is the most dynamically growing segment; investors assess the hypotheses of potentially successful companies.

The investment fund of the regional accelerator is formed at the expense of the corporate sector, which is interested in innovations. Regional budget funds can also be attracted; the problem is caused by an unfavorable trend of the budget deficit. Nevertheless, the region has opportunities to facilitate activities of the accelerator in terms of expert, mentoring and organizational support. As a result, the integrative solution of businesses and the government will contribute to the formation of a regional innovation ecosystem. The funds will be used to support the best projects that have passed through the acceleration stage, coworking services, expertise and educational services within the developed program.

The implementation of the acceleration process involves two stages:

Stage 1. Development of the Acceleration Program (AP):

- development of the application form, selection criteria, rules and conditions for the AP;
- development of the site for the AP;
- attraction of participants;
- selection of applications and assessment of AP participants (assessment interviews);
- preparation of companies for implementing the AP (self-diagnostics of companies, work with trackers);
- selection of experts;
- information support of the AP;
- bootcamps (practice-oriented and interactive group sessions with the involvement of mentors and experts, which ensure the involvement of AP participants in the process, form practical skills, abilities and work out design solutions).

The level of expertise and mentoring is crucial. Mentors should be high-skilled specialists. It is difficult to overestimate the amount of support for teams; in a few months, startups travel a path that would otherwise take years. The accelerator methodology provides for an interactive, flexible approach with experimental and competitive components when creating (testing) business models and technologies.

Stage 2. Implementation of the Acceleration Program:

- application of the methods and tools obtained during the bootcamps in the existing business;
- informational coverage of AP events by federal and regional media and Internet resources;
- AP information and consulting block (expert consulting, export tracks, seminars, master classes, business games, case studies);
- practical block of the AP (work accompanied by trackers - mutual exchange of ideas, recommendations and results in a group format);
- expert evaluation of projects (representatives of federal and regional development institutions, the investment sector;
- revision of projects according to experts' comments;
- holding a demo day - the final presentation of AP projects;
- selection of t winners;
- presentation of certificates to AP finalists.

According to experts, in 30% of startups, as a result of acceleration, the business model undergoes dramatic changes, the so-called "pivot" occurs. Another 30% of startups update their metrics. The life cycle of a startup is 18 months, and if there are no attracted investments, the project becomes unprofitable. That is why accelerators are so important for innovators - analysis of ideas, development of technologies and business models, identification and elimination of shortcomings - all processes are much faster and more efficient.

Thus, the AP helps to attract investment, conclude contracts, and receive support for further promotion of projects. An analysis of startups that have successfully gone through the acceleration stage shows that within 1.5 years 30% of startups raise an investment round, scale up their business and expand partner programs.

4 Conclusion

1. The development of Russian business accelerators is in line with global trends: increased specialization, cooperation with corporations, interest of program residents in international markets, new acceleration formats.

2. The crisis contributed to a significant decrease in the volume of investments of business accelerators in new projects and in the number of concluded deals with startups. At the same time, in 2020 the number of acceleration programs increased due to the active involvement of the corporate sector.

3. The regional level of acceleration is not optimal, 65% of the regions do not use the acceleration mechanism. The regional authorities should create business accelerators to stimulate the innovation processes.

4. Most of the existing regional accelerators enjoy the support of federal development institutions, regional authorities and business representatives. The role of acceleration in the regional intellectual property market, as well as its consistent integration into the international market system, should be emphasized.

5. Regional accelerators are increasingly focusing on support mechanisms and longer interaction with teams. Industrial partners show interest in pilot implementations and synergies; they want to investment taking into account the possibility of integrating new solutions into the existing production systems.

6. The model of a regional accelerator of innovations developed for the Udmurt Republic corresponds to the sectoral profile of the region, characterized by a high concentration of military enterprises. The most effective format for the model is a public-private partnership that takes into account interests of the region, innovators, military enterprises and other representatives of the corporate sector.

References

1. J.D. Christiansen, *Copying Y-Combinator: a framework for developing Seed Accelerator programmes*, (Individual Project Judge business School & Jesus College, Cambridge University, Cambridge, 2009)
2. K. Malek, E. Maine, I. McCarthy, *Journal of Engineering and Technology Management*, **32,** 26-39 (2014)
3. D. Hoffman, N. Radojevich-Kelly, *Small Business Institute*, **8 (2),** 54-70 (2012)
4. P. Miller, K. Bound, *The startup factories: the rise of accelerator programmes to support new technology ventures* (NESTA, London, 2011)
5. M. Szafranski, *Business acceleration* **21** (2018)
6. H. Etzkowitz, C. Zhou, *The triple helix: university-industry-government innovation and enrepreneurship*, (Routledge, 2017)
7. H. Etzkowitz, *Technol Forecast SOC*, **123**, 122-129 (2017)
8. J. Pique, J. Berbegal-Mirabent, H. Etzkowitz, *Triple Helix*, **1 (aop)**, 1-45 (2020)
9. C. Scheel, L. Pineda, *Innovacities: Impacto de los sistemas regionales de innovación en las estrategias competitivas de las ciudades*, (Universidad Jorge Tadeo Lozano, Bogota, 2015)
10. S. Makarov. E. Ugnich, *Foresight-Russia,* **9** (1), 56-76 (2015) DOI: 10.17323/1995-459X.2015.1.56.67
11. A.A. Sytnik, *Bulletin of Saratov State Social and Economic University,* **5 (59),** 51-54 (2015)
12. M.A. Kamenskikh, *Regional Economics: Theory and Practice,* **9 (456)**, 1725-1734 (2018)
13. O.A. Zhdanova, *Theory and Practice of Social Development,* **12**, 76-78, (2016)
14. Yu.V. Fedorova. A.V. Tsvetnykh, *Actual problems of aviation and astronautics,* **2(9)**, 348-349, (2013)
15. M.R. Kozlov, *Creative economy,* **8 (80)**, 74-82 (2013)
16. D.A. Statovsky, *Innovation,* **4(198),** 49-52 (2015)
17. V.V. Danshina, *Bulletin of Belgorod University of Cooperation, Economics and Law,* **5**, 79-87 (2018) DOI:10.21295/2223-5639-2018-5-79-89
18. Official website «Y-Combinator» Retrieved from: https://www.ycombinator.com/
19. Official website «Seedcamp». Retrieved from: https://seedcamp.com/
20. Yu. Yu. Pavlova, *Economics and Business: Theory and Practice,* **2-2 (60),** 71-78 (2020) DOI: 10.24411/2411-0450-2020-10128
21. Official website of the IIDF Accelerator. Retrieved from: https://accelerator.iidf.ru/
22. The HSE Business Incubator has become the best university accelerator in the world. Retrieved from: https://www.hse.ru/news/edu/316184838.html
23. Review of corporate accelerators 2020. Retrieved from: https://vc.ru/u/495032-sergey-kancerov/224726-obzor-korporativnyh-akseleratorov-2020-goda
24. Venture capital market in Russia in 2020. Retrieved from: https://incrussia.ru/understand/vc-2020/
25. Who are mentors and why are they needed? Retrieved from: http://www.lookatme.ru/mag/how-to/inspiration-howitworks/213055-faq-mentors

26. Yu.N. Alexandrin, Yu.V. Gorbova, *Economics: Theory and Practice*, **3 (59)**, 40-50 (2020)
27. Who and how invested in startups in 2019. Retrieved from: https://incrussia.ru/understand/vc-2019/
28. A.V. Ovchinnikova, *Bulletin of the Udmurt University. Series Economics and Law*, **2**, 61-70 (2014)
29. T.N. Topoleva, *Vestnik of the Plekhanov Russian University of Economics*, **17 3** (111), 49-60 (2020) DOI: 10.21686/2413-2829-2020-3-49-60
30. V.V. Matveev, *Bulletin of the Udmurt University. Series Economics and Law*, **1** (31), 21-29 (2021) DOI: 10.35634/2412-9593-2021-31-1-21-29
31. I.I. Rakhmeeva, A.N. Lysenko, *Bulletin of PNIPU*, **4**, 152- 166 (2020) DOI: 10.15593/2224-9354/2020.4.13
32. V.I. Nekrasov, A.I. Sutygina, P.F. Sutygin et al., *Regional economy under WTO conditions*, (Russian Academy of Sciences Ural Branch Institute of Economics Udmurt Branch, Yekaterinburg , 2013)
33. Science and innovation: Federal State Statistics Service for UR. Retrieved from: https://udmstat.gks.ru/folder/51964

Innovation management in high school visual communications: from heraldry to animation

Svetlana Shlyakhova, *Yulia* Lektorova*, and *Natalia* Bolshakova

PNRPU, 29, Komsomolsky prospect, Perm, 614990, Russia

Abstract. The article considers problems of innovative practices of building visual communication in higher education. In the theoretical part of the article, the authors' model of university brand identity and a model of university logo identity are proposed. The research part presents the initial analysis results of the graphic design of university logos and the data on 80 logos comparative analysis, performed according to the color, font, symbol, shape criteria. Based on the results of the study, comparative analysis of the identified trends in formation of logos of Russian and foreign universities was conducted.

1 Introduction

The key task of higher school modern visual communications remains search for tools and solutions that ensure recognition and loyalty among both external and internal audiences, regardless of the university's position in the international ranking, its history and location. The arsenal of possible directions of visual communication in modern conditions includes mechanisms of brand innovation, among which there should be considered methods of developing and restyling corporate identity (logos), aimed at creating and maintaining a holistic image of higher education in general public mind.

Foreign researchers understand brand innovation as the degree to which consumers perceive brands as capable of offering new and useful solutions to their needs [1,2]. From our point of view, this approach, which implies, first of all, the degree of innovativeness of the product it offers, with which the consumer associates it with, cannot be exhaustive. Brand innovation should also be understood as the degree of innovation of the very nature of brand communication with consumers, including visual communication.

Today, visual communication is one of the main components of modern mass media, transforming any content into visual codes. During the period of active development of electronic digital technologies, visual communications resemble the process of transferring information "through visual language (images, signs, pictures, typography, infographics, etc.), on the one hand, and visual perception (organs of vision, psychology of perception, etc.) on the other hand" [3].

If we proceed from the statement of A. Berger that the purpose of any visual communication is to create an image [4], and the image remains the main acting agent of influence on consciousness of the target audience, we can assume that innovations in visual communication in the academic environment and beyond it include modern trends in graphic design of logos.

In the language of professionals, graphic design is a visual display of any information about a specific object by applying an image to a medium using artistic techniques. Targeted at the mass consumer, such art and design activities include various illustrations, fonts, lettering, design of packaging, buildings, classrooms, laboratories and much more.

Taking into account the behavior of the main target audiences of corporate communications (both Russian and foreign), higher schools are testing unique corporate styles that identify originality of each educational unit in the digital media space (on websites, in social networks) and in the context of off-line contacts (on letterheads, souvenirs, advertising modules).

In visual representation of universities today, there is a clear movement from the coat of arms (the insignia of its owner, drawn up according to the rules of heraldry, must be on the heraldic shield) and the emblem (concepts, ideas, etc. in the form of a symbolic image) to the logo (trademark, graphic image, embodying and representing the brand) of the university.

If coats of arms and emblems tend to be academic, traditional, have long history, then logos are associated with dynamism, modernity, branding, marketing and advertising. It is also obvious that there is a clash of ideas about the university as an educational and scientific institution or as a business corporation [5].

In real practice of visual communication, the university image is not terminologically defined: the terms *coat of arms, emblem, symbol* and *logo* are often used as synonyms. Often the university coat of arms, freed from some details, acts as a logo.

The logo of any organization is a reflection of its essence, which it individualizes. "It is equally important for a logo to be recognizable and readable ... as a rule,

* Corresponding author: Lektorova2015@yandex.ru

people can only catch a glimpse of the logo, and often from a great distance. In such a situation, the main thing is recognition, not ease of reading " [6]. On this basis, most researchers focus on logo design that integrates recognizable characters, shapes, fonts, and colors.

The entire set of basic components of visual communication is included in the higher school brand identity system, along with functional advantages, emotional benefits, consumer values, the uniqueness of the university and of its educational service and the higher school mission.

Based on the existing models of brand identity [7], we propose a model of university brand identity, taking the pyramid model as a basis:

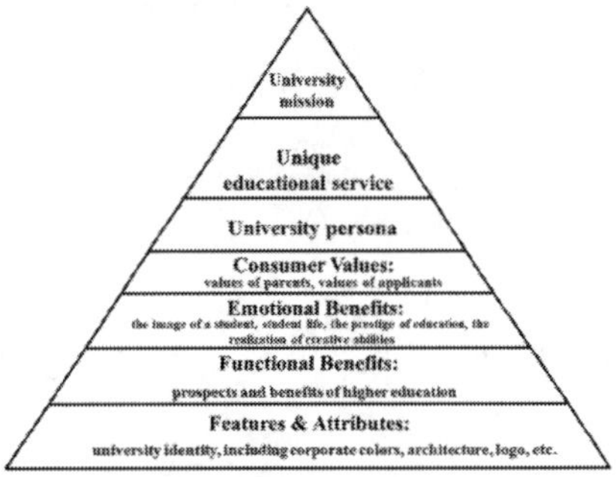

Figure 1. University brand identity model.

In this model, the logo as an element of attributes (identity, i.e. visual communication) "cements" and acts as the "foundation" of the higher school identity.

Identity of the university brand logo is built from three aspects: individuality, positioning, essence. Such aspects as individuality and positioning of the university brand logo have specific features.

Individuality of a university brand includes three elements: visual components (color, symbol, font), modernity, and memorability.

Positioning of the university brand includes reflection of three characteristic features in the logo: the main brand values (history and specificity of the university), territory (location of the university), mission, and the university motto.

Effective communication of the essence of the brand to the consumer is a logical combination of correct reflection of the characteristic features of positioning and brand personality. Based on the above, below is a model of university logo identity:

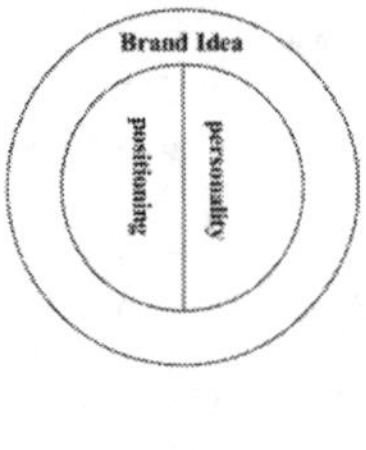

Figure 2. Brand identity model of the higher school logo

A logo as a visual metaphor is a sign system consisting of visual and semantic categories. The visual component is a sign with specific structural elements (font, symbol, shape) that form the external shape of the logo.

The semantic component is a culture sign that conveys meaningful symbols and senses to target audiences. In the era of networked communication, visual communication in general and graphic logo design in particular must be adapted to electronic multimedia as well.

2 Methods

This study attempts to identify trends in the design and architecture of logos of Russian and foreign universities.

The research material was selected according to the rating of the university on the official websites of the universities. The logos of included in the Top-20 and not included in the top rankings of universities were analyzed. A total of 80 university logos were analyzed: 40 Russian and 40 foreign ones:

Figure 3. Logos of Russian universities

Figure 4. Logos of foreign universities

Analysis and assessment of the brand identity of the logos of Russian and foreign universities are carried out according to the following method:

Initial analysis of logos according to the following criteria: logo evolution; verbal part, language; non-verbal part. When analyzing the non-verbal part, the following were taken into account: symbolism of the subject-material environment, emblems, color, geographic referencing.

Comparative analysis of logos of Russian and foreign universities according to the following criteria: quantity and quality of colors; adaptability of color for different media; font type; symbols (national, heraldic, industry, educational, natural, operational); the form; personality and modernity; availability of a brand / guidebook; brand positioning and values (year of university creation, continuity in brand evolution, reference to historically significant symbols, territory, training profiles).

Comparative analysis of the identified trends in the formation of logos of Russian and foreign universities, assessment of their brand identity.

Research time: 2019-2020

3 Results

The volume of the publication does not allow posting full analysis of the logos under study, so we will present some fragments of the initial logos analysis.

An example of the initial analysis of a foreign university logo (Harvard University):

Figure 5. Evolution of the graphic design of the Harvard University logo (old and new university logos)

The analysis revealed the following results:

* logo evolution: yes

* verbal part, language: the word *university* (at present); in Latin *truth*

* non-verbal part:

- symbolism of the material environment, emblems: the symbol is the coat of arms - a shield with three open books, on the pages of which *the truth* is written; in the past, a wreath symbolizing knowledge and excellence. The signs of the shield and the open book symbolize strong and powerful education provided by the university;

- color: the crimson color is associated with love of freedom, it adjusts to a free atmosphere;

- geo-referencing: the name of the city of Cambridge, USA.

An example of the initial analysis of the logo of a Russian university (Tomsk Polytechnic University):

Figure 6. Evolution of the graphic design of the Tomsk Polytechnic University logo (old and new university logos)

The analysis revealed the following results:

*logo evolution: yes, the trend towards simplification

*verbal part, language: abbreviation and the university name in Russian

*non-verbal part:

- symbolism of the subject-matter environment, emblem: consists of blocks forming capital letters T and P (Tomsk Polytechnic) or T and U (Tomsk University); symbolizes multidimensionality, interconnection, harmonious coexistence of individual elements of the university: students, employees, graduates, structural units, departments, laboratories, centers, institutes. The square consists of six rectangular parts, which correspond to the six main directions of scientific development of this university. Continuity is only in abbreviation and green, which is the main color in all previous versions of the university logo.

- color: the green color of the letter "T" (Tomsk) is traditional for the university. This is the color of the natural environment from which man emerged. The black color of the letter "P" (Polytechnic) is technological, symbolizing modern engineering solutions. The combination of colors used in the logo indicates the basis of the university's scientific and educational activities: solving global problems of mankind through development of resource-efficient technologies.

- geo-referencing: the name of the city of Tomsk, Russia.

All the logos of Russian and foreign universities under study were analyzed and systematized in a similar way.

There are obviously two main styles in the university logos:

* heraldic (shields, ribbons, mottos in Latin, fittings, laurel crown, wide color palette, tightness of the visual range, etc.): the desire of the university to emphasize duration of its existence, prestige, fundamentality, tradition, etc.;

* minimalist (conciseness, simplicity, clarity, limited color palette, contrast, a lot of empty space, etc.): the desire of the university to emphasize modernity, dynamism, innovation, aspiration for the future, etc.

In the logos of foreign universities, striving for minimalism slightly dominates, in the logos of domestic universities - heraldic elements.

The main means of visual communication are the basic components of a logo graphic design - color, font, symbol.

Color. The palette of university logos used is represented by a variety of color and compositional solutions.

According to designers, it is color that creates visual sensation and emotional impact in visual communication. The choice of the color palette should take into account coloristic preferences of the target audience and correlate with objective characteristics of the identified object.

Analysis of such a graphic component as color showed that the top foreign universities mostly have two types of logos - tricolor and bicolor (40%), while the other universities are dominated by tricolor logos (33%).

Top Russian universities most often have logos consisting of two colors (54%), while other universities are dominated by one-color logos (40%).

The adaptability of the main color is insufficient for all universities: in 80% of foreign universities, the logo color is not adapted for various media. In 52% of Russian top universities the logo color is not adapted for media, in the rest - 60% of the logos.

The most common primary colors in foreign universities are blue, burgundy / red, black and white; the Russian universities have blue and white.

Analysis of the primary colors of the logos did not show significant differences between the top and other universities in Russia and abroad. There are no significant differences in the design of logos between Russian and foreign universities:

* dominant colors in university logos - blue, white (as the main color of the field), burgundy / red; foreign universities often have black;

* poor adaptation of logos to different media;

* prevalence of one- and two-color logos; in the logos of foreign universities, preference is often given to emblems in which there are more than three colors.

The dominance of blue, white and burgundy / red in the logos of universities can be associated with tinctures in heraldry (the name of paints and coatings of heraldic bodies).

Today, many universities use color coding for faculties and departments, assigning each its own color and an additional symbol (icon). This approach solves the problem of self-identification of university departments.

Font. Of particular importance is the font, which is considered the basis of graphic design, representing a certain style of characters, their general image and size. Fonts are conventionally divided into four main groups: serif fonts (antiqua), sans serif fonts (sans serifs), decorative fonts, and handwritten fonts. For each font, there are several types of styles, features of their image: normal, italic, bold and bold italic.

The variety of fonts used in university logos is an additional tool for visualizing information. When choosing an appropriate font, designers are guided by the basic requirements: harmony of proportions; simplicity and clarity; clarity and proportionality to ensure readability; "associativity" (grace - massiveness, rigor - liveliness, etc.); the purpose of the font (font readability, relevance) [8].

Analysis of the font showed that the most frequent font type for the best foreign universities is sans serif (62%), for other universities - serif in the same proportion (62%). The top Russian universities are dominated by serif (70%), the rest have sans serif (44%) and serif (45%) in almost equal proportions.

The typeface gave almost the same results for all universities: the "normal" typeface dominates in the logos of foreign (60-80%) and domestic (70-90%) universities.

Analysis of the logotype did not show significant differences between the top and other foreign and Russian universities. There are no significant differences in font between Russian and foreign universities:

* dominant fonts in university logos – serif and sans serif; at the top foreign universities sans serif prevails (62%), at the top Russian universities - serif (70%);

* "Normal" typeface dominates in the logos of foreign (60-80%) and domestic (70-90%) universities.

Symbolism. At the first stage, the analysis of visual semiotics in university logos was carried out on the basis of the classification of P.V. Demin, who identifies national, heraldic, natural, sectoral, educational, operational (figures, rectangles, ovals, triangles, lines, points and stars, etc.) symbolism [9]. We will also include state symbols as national symbols.

The analysis showed that the TOP 20 foreign universities are dominated by heraldic and operational (26% each) symbols, but taking into account the fact that operational symbols are additional graphic elements of the logo, we can say that the dominant symbols of the best foreign universities are heraldic. Other foreign universities are dominated by industry (33%) and educational (27%) symbols.

Operational (27%), heraldic and national-state (23% each) symbols dominate in the logos of Russian universities. The rest of the universities have national-state (23%) and industry (19%) types of symbols.

Heraldic symbols are popular with top universities because the TOP-20 is occupied by the oldest universities in the world, and their logo since its foundation is usually a coat of arms, modernized to match the logo.

Figure 7. Logo and coat of arms of Moscow State University, St. Petersburg State University

The logo, which has a five-hundred-year history (University of Aberdeen), consists of four coats of arms: the coat of arms of Old Aberdeen (seat of King's College), the coat of arms of Bishop Elphinstone (founder of King's College), the coat of arms of Keith, the Earl of Marischal (founder of Marischal College), coat of arms of the city of Aberdeen.

On the other hand, the reference of university emblems to heraldic semiotics is at the same time territorial symbolism, since logos and coats of arms of universities often include elements of the coats of arms of cities and states, including mottos in Latin. Thus, the coat of arms of the University of Nairobi includes elements of both the coat of arms of Kenya and the coat of arms of the capital of Kenya - Nairobi:

Figure 8. Coat of arms of the city of Nairobi - the capital of Kenya, Coat of arms of the University of Nairobi, Coat of arms of Kenya

In modern versions of university logos, there is no heraldic reference to the territory, but the territorial reference remains in the name of the university (an indication of the city, country, region). This is typical of the overwhelming majority of the names of universities.

For some Russian universities, state symbols prevail in logos, with the help of which universities indicate the status of a Russian and a state university, which in Russia is a sign of reliability of the university and quality of education. State symbols are also typical of universities related to state structures (Ministry of Foreign Affairs, Ministry of Emergency Situations, Ministry of Internal Affairs, government, etc.).

National (ethnic) symbols are far less common in logos. In our material, ethnic symbols in the form of Zilant are present in the logo of the Kazan (Volga Region) Federal University. Zilant - a mythological creature in Tatar folklore - comes from the polymorphic monsters of Central Asia, including the ornithomorphic features of the Semurg (king of birds) and the dog-headed griffin. The dog-headed griffin goes back to the ancient Iranian Senmurv, which combines the features of a serpent, eagle, dog and lion.

Industry symbols are relevant mainly for the logos of specialized universities (medical, agricultural, architectural, etc.).

The study showed that architectural symbols (the main building of the university in the logo) are most typical for domestic universities (about 30-40%). In the logos of foreign universities, the image of the main university building is rare (10% for top universities) or is completely absent.

The indication of the foundation year of the university is most typical for Russian universities, however, there is also a tendency towards minimalism: in new versions of logos, the year of foundation disappears (NSU, KFU, SPbPU Peter the Great, TPU, NSTU, Oregon State University, Stanford University, University of Illinois, University of Florida, Johns Hopkins University, Massachusetts Institute of Technology, etc.).

Analysis of the logo symbols did not demonstrate significant differences between top and other foreign and Russian universities. There are no significant differences in symbolism between Russian and foreign universities:

* dominated by the operational, heraldic, national-state and industry type of symbolism;

* symbol of a university building is most typical of Russian universities;

* indication of the university foundation year is typical for Russian universities, however, there is also a tendency towards minimalism: in new versions of logos, the year of foundation disappears.

The predominance of national and architectural symbols reflects the identity of the university brand on the basis of territory (location), industry symbols - identity in the direction of research, heraldic - in terms of the duration of the university history, its prestige, fundamental nature, tradition, etc.

The shape. Analysis of the shape of the logos showed that the predominant general shape of the logo in the best foreign universities is the form of the heraldic shield (50%), while the other universities are dominated by a square (30%) and a rectangle (20%). Russian universities are also dominated by the form of a heraldic shield (38-42%), a circle (41%) and a rectangle (31%).

Analysis of the logo symbols did not show significant differences between top and other foreign and Russian universities. There are no significant differences in symbolism between Russian and foreign universities:

* the shape of the heraldic shield dominates (38-50%), the rest of the universities have a rectangle (20-31%);

* for foreign universities the frequency form is a square (30%), for Russian universities - a circle (41%).

Symbolically, a rectangle is the most rational, reliable and correct of all geometric shapes, with the help of which a person has adapted the space for use in everyday life (house, room, table, bed). The square implies hard-to-hold dominance born of an abstract desire for power, while the circle excludes earthly associations due to heavenly symbolism.

Predominance of the heraldic form reflects the identity of the university brand in terms of the duration of the university history, prestige, fundamental nature, tradition, etc.

Thus, the analysis of the university logos showed that there is no significant difference between top universities and universities that are not included in the Top-20, both in Russia and abroad. It should be noted that foreign universities have a more developed branding of products (T-shirts, baseball caps, etc.) with university symbols. In Russia, this advantage is not used enough, while abroad they have already made a separate product niche from branded products of universities.

Most universities do not yet use logos and corporate identity to position themselves in the higher education market. We can only talk about individual initiatives, where the process of using the visual brand identity has become an integral part of the strategic development of the university.

Evolution of the high school logo. The practice of global brands and branding of transnational companies shows that the heraldic (shields, ribbons, mottos in Latin, fittings, a laurel crown, a wide color palette, a tight visual range, etc.) logo style is actively transforming towards minimalism (brevity, simplicity, clarity, limited color palette, contrast, minimum of elements, a lot of "air", etc.).

Comprehensive analysis of the logos shows that more than 50% of the logos of foreign universities and more than 40% of Russian universities are designed in the minimalist style or tend to it. Universities (even those with a long history) strive for a logo that is associated with modernity, dynamism, and innovation. Many universities, trying to preserve the traditional (heraldic) appearance of the logo (emblem), nevertheless try to free it from "unnecessary" elements. Today it is a universal trend in the design of logos for Russian and foreign universities.

IT technologies have given rise to a new trend in creating university logos - generative, "animated" logos: movements and / or changes in the shape of logos (morphing), which are constantly changing, generate a new shape based on a given algorithm of the brand language. *Morphing* is a visual effect that gives the impression of a smooth transformation of one object into another.

Generative logos are relevant in universities related to design and computer graphics. Examples of such logos are OCAD University logos (Toronto, Canada, university foundation year: 1876) and Design Academy Eindhoven (Eindhoven, Netherlands, academy foundation year: 1947).

OCAD University logo is based on black and white pixel "windows" - modular frames that are filled with variable visual content. The visual content options are the work of students and designers from OCAD University. Every year, the best graduates of the university, within the framework of the basic "window" structure, basing on their work, develop a new set of logos. In fact, this logo format is an archive of the university's brand identities.

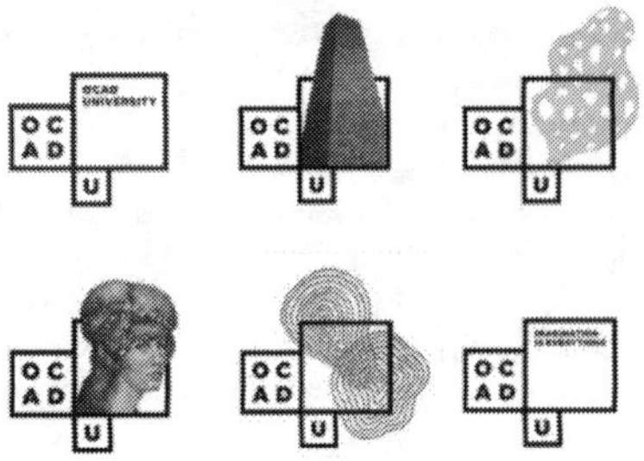

Figure 9. OCAD University and Design Academy Eindhoven logos

The design Academy Eindhoven logo is based on three stripe windows (symbolized by the letter E - Eindhoven), the content and type of font of which is constantly changing. The invariant elements of the logo filling are the words Design Academy Eindhoven, the options are any phrase or combination of the words Kiss the Future, Dare to Dream, Nothing is True. Innovation, mobility, variability, dynamism of the Design Academy Eindhoven logo is beyond doubt, but its perception is somewhat difficult ("flickering" in the eyes due to the rapid change of words).

Thus, the following trends can be identified in the evolution of Russian and foreign university logos:

* change of heraldic tradition (shields, ribbons, laurel crown, wide color palette, tight visual range, etc.) in logos for minimalism (brevity, accuracy, simplicity of geometric shapes, abundance of empty space, neutral or limited and striving for monochrome colors, which provides good contrast, emphasizes significant elements, etc.);

* simplification of logos in a heraldic style (exemption from individual elements);

* color or symbolic coding of university logos within the invariant base logo;

* formation of a trend for generative, "animated", "live" logos that are constantly changing, generate a new form based on a given algorithm of the brand language.

When building relationships with various target audiences of higher education, graphic logo design acts as a tool that conceptualizes visual content and constructs an image in the landscape of visual communication.

References

1. A. Eisingerich, G. Rubera, *J. Int. Mark.*, **18(2)**, 64-79 (2010)
2. B. Hetet, C.-L. Ackermann, J.-P. Mathieu, *J. Prod. Brand. Manag.*, **29(5)**, 569-581. (2019)
3. P. Rodkin, *What is visual communication?* Retrieved from: www.prdesign.ru/text/2010/visualcommunication

4. A. Berger, *Seeing is believing. Introduction to Visual Communication* (Publishing House Williams, 2005).
5. Bullard D.B. *Academic Capitalism in the Social Sciences: Faculty Responses to the Entrepreneurial University:* Ph.D. diss. (University of South Florida, 2007.)
6. B. Vaneken, *Brand Aid* (Peter, 2005).
7. V. Domnin, *Branding: New Technologies in Russia. 2nd ed.* (Peter, 2004).
8. A. Buzinova, *Mediascope,* **4** (2013).
9. P. Demin *University Management: Practice and Analysis,* **22(6)** (2018).

Managing budgetary openness of public services

Elena Razinova*

[1] Perm National Research Polytechnic University, Komsomolsky prospekt, 29, Perm, 614990, Russia

Abstract. The analysis of the aspects of information transparency of public finance in terms of public services (lists of public services, government assignments and information on subsidies for financial support of government assignments, cost standards for the provision of public services) at the federal and regional levels. Regulatory legal acts of the constituent entities of the Russian Federation, official websites of executive authorities and specialized federal resources for publishing budget data are considered. In accordance with the identified contradictions, the information objects published at the federal and regional levels were ranked according to the level of budgetary openness, and areas of improvement associated with increasing the level of accessibility and transparency of public finance data were identified.

1 Introduction

According to the recently published materials of the Federal Service for Supervision of Communications, Information Technology and Mass Media of the Russian Federation, the restriction of the right to receive information is at the top of the pyramid of digital threats [1]. Representatives of the expert community call big data "new gold" [2] and "new oil", which actualizes the value of information accessibility in the modern world, and increasing openness and transparency is one of the global trends and prospects for the development of the public sphere.

Proactive mode of public access to official, statistical, budgetary and other information administered by public authorities that meets the criteria of relevance and cognitive accessibility, interaction between the state and public institutions, citizen participation in decision-making ensure the implementation of the concept of openness of the state [3].

At present, special importance is attached to budgetary openness, the high level of which scientists associate with positive economic effects: a decrease in the level of corruption, an increase in the level of public trust and an intensification of an open dialogue [4]. Despite the obvious successes of the Russian Federation in this area (in the Open Budget survey of the International Budget Partnership for 2019 [5], Russia took 14th place out of 117 countries, moving up two positions compared to the previous version of the ranking), researchers highlight contradictions and areas of improvement. It is associated with an increase in the degree of detail of information on budgetary indicators, including about state and municipal institutions, including information about the services they provide, parameters of compliance with requirements for the quality of services and the principles and conditions of their financial support [6].

2 Methods

At the moment, the general concept of "openness" in the aspect of the analysis of public institutions is debatable, in the scientific and practical environment there is no uniform approach to its definition, while the concepts of an applied and instrumental nature are most often used. So, according to experts of the Accounts Chamber of the Russian Federation, the directions of the state's openness are open dialogue (the availability of mechanisms for public participation), open data (the availability of tools for proactively obtaining the necessary and relevant information, technical and functional accessibility) and the openness of information - the availability of regulatory and communication tools, allowing citizens and organizations to receive data supervised by state and local authorities and budgetary organizations [7].

Openness of the budget as a direction of research has a broader theoretical and practical basis. So, in terms of research on foreign experience that determines the vector of development of Russian budgetary openness, the works of N.A. Guz [8], N.V. Golovanova [9] and others. Analysis of domestic practice, identification of problems and areas of improvement is carried out in the studies by A.M. Lavrova, O. I. Timofeeva, V.V. Klimanov and A.A. Mikhailova [6] and others.

The principle of transparency (openness) of the budget since 2007 is enshrined in Art. 36 of the Budget Code of the Russian Federation and presupposes the obligatory publication in the media of approved budgets and reports on their implementation, completeness of information on the progress of budget execution, as well as the availability of other

* Corresponding author: evfefilova@gmail.com

 https://doi.org/10.1051/shsconf/202111600055

information on budgets by decision of legislative (representative) government bodies; mandatory openness to society and the media of draft budgets, procedures for considering and making decisions on draft budgets; providing access to information posted on the information and telecommunications network "Internet" on a single portal of the budgetary system of the Russian Federation; stability and (or) continuity of the budget classification of the Russian Federation, budget indicators. These indicators, as a rule, are analyzed and evaluated in the framework of world ratings and reviews.

Speaking broadly about budgetary transparency, we should mention the relevance of publishing the expanded and detailed set of data on public finance, in particular, on financing state and municipal institutions, including quantity, scope, cost of services provided (works performed), and specifying the applicable standards in official documents [4]. Thus, we speak about information transparency within the context of different structural elements of public finance, which we identify as budgetary transparency within our study. We should mention that we could rate highly such information in terms of social significance as it is aimed at understanding social processes, including the process of financing social organization directly connected to the provision of decent standard of living for citizens.

Researchers distinguish the following approaches to the exercise of the right for access to information provided by government authorities or organizations: the active approach (individual search of information in various sources) and the proactive approach when citizens consume information whether they have or do not have the purpose to do so (through mass media, social media, etc.) [10]. From our point of view, now there is the third approach which combines the features of the above-mentioned approaches: information is published in the public domain (on special resources or websites of executive authorities) in a maximum available form, which makes it convenient to process and analyze it. As an example demonstrating the synthesis of the described approaches, we can consider the Budget for Citizens project—a document that provides publishing the budget and budget implementation reports in a maximum available form. This document is drawn up to familiarize users interested with the tasks and priority areas of the budgetary policy, major conditions of budget formation and implementation, major budget parameters; since 2013, it shall be published annually at all levels of governmental units: federal, regional, and municipal. At the same time, the structure of data published within the project generally complies with the requirements of international ratings (in particular, the International Budget Partnership rating) and excludes the information on the finance of public organizations.

We should note that within the study the analysis object includes documents (regulations) and public information on government services in terms of the active approach (documents, presentations, open data, interactive websites) in the context of activities of governmentally, municipally and privately owned institutions which provide satisfaction of material and spiritual needs of citizens. We used the methods of content and document analysis: to determine the level of budgetary transparency of government services, we analyzed the regulations of the supreme executive bodies of the constituent entities of the Russian Federation, including the procedure for preparing the government assignment for government services (works) and the financial security of the government assignment issued in accordance with Article 69.2 of the Budgetary Code of the Russian Federation, and public information published on the Single Portal for the Budget System of the Russian Federation (http://budget.gov.ru), the Official website for information on government and municipal institutions (https://bus.gov.ru, the website of GMU), official websites of executive authorities of the federal and regional levels.

3 Results and discussion

In accordance with Article 69.2 of the Budget Code of the Russian Federation, the composition of public services and works included in the state assignment - the basis for financial support of state organizations, is accumulated within the framework of the all-Russian basic (sectoral) lists (classifiers) of state and municipal services provided to individuals (can be used by institutions of all levels publicly - legal entities), federal and regional lists (classifiers) of public services that are not included in the all-Russian basic (sectoral) lists (classifiers) of state and municipal services provided to individuals, and works, the provision and implementation of which is stipulated by the regulatory legal acts of the Russian Federation (subjects of the Russian Federation). Up-to-date information on all-Russian, federal and regional lists is published on the Single Portal for the Budget System of the Russian Federation, including in the open data format.

Government assignments for government services (at the federal and regional levels) are placed both on the Single Portal for the Budget System of the Russian Federation and the website of government and municipal institutions. This information object has a relatively high level of transparency, where information is published both as an original document and as data adapted for an unlimited set of people. According to the website of government and municipal institutions, in 2021 there have been published 93.41% of government assignments of federal budgetary and autonomous institutions (government assignments may not be drawn up in respect of public institutions, that is why they are excluded from the fetch), 87.65% of government assignments of institutions in the constituent entities of the Russian Federation, and 85.58% of municipal institutions' assignments (placed by such institutions). Note that a high level of data openness has been observed for a significant period of time, which is demonstrated by retrospective indicators (Fig.1) at the federal level. At the same time, the level of information placement by autonomous institutions is significantly lower than the corresponding indicator for budgetary institutions, however, the bottom line is consistently over 90%.

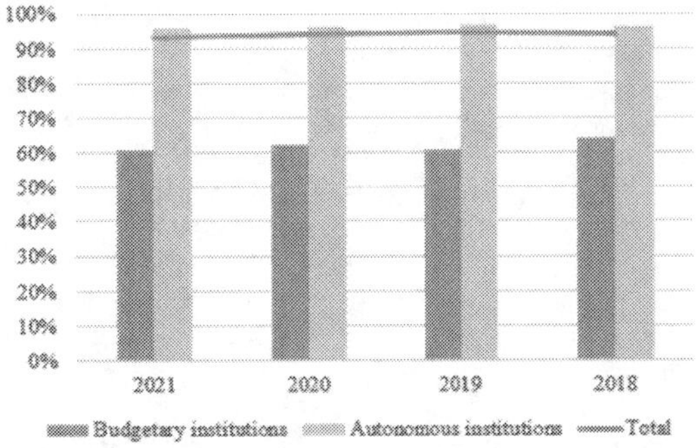

Figure 1. Dynamics of publications of state assignments, federal level

But there are some regional differences: the majority of regions (64%, or 54 constituent entities of the Russian Federation including St. Petersburg, the Orenburg Region, the Perm Territory, etc.) are characterized by a high transparency level for information on government assignments (they place more than 90% of approved documents); institutions from 29% of regions (25 constituent entities of the Russian Federation including the Arkhangelsk Region, the Irkutsk Region, Moscow, the Nizhny Novgorod Region, etc.) place data on assignments for services (works) partially—they publish 60–90% of documents; less than 10% of regions (the Kurgan Region, the Republic of Tatarstan, etc.) demonstrate an insufficient transparency level (they publish less than 60% of approved documents).

It is noteworthy that the analysis of publications of state assignments of regional autonomous and budgetary institutions in retrospect (Table 1) reflects a decrease in publication activity and, as a consequence, budgetary openness. This situation may be due to some weakening of attention to the issues of openness at the regional level, insufficient interest of the inspectors and the expert community to the Official website for information on government and municipal institutions. At the same time, this is the only open area where you can get comprehensive information about state and municipal institutions.

Table 1. Transparency level of regional government assignments

Year	Share of published documents		
	>90%	**60-90%**	**<60%**
2021	54 regions	25 regions	6 regions
2020	65 regions	20 regions	-
2019	67 regions	18 regions	-
2018	72 regions	13 regions	-
2017	71 regions	14 regions	1 region

Overall, data on government (municipal) assignments can be characterized as objects with a high transparency level: they established legal principles and rules of placement, implemented special-purpose platforms for publication within which the information on services (works) is provided in available and clear form.

The Official website for information on government and municipal institutions, based on an internal methodology, maintains its own rating of information transparency of institutions in the context of the constituent entities of the Russian Federation. The city of Sevastopol has the highest rating at the moment in 2021. Another 10 regions (Lipetsk region, Penza region, etc.) receive estimates from 0.8 to 0.9 (maximum value 1). The value is lower than 0.4 in 7 regions (Novosibirsk region, St. Petersburg, etc.). It should be noted that this rating characterizes the information openness of institutions as a whole, taking into account the placement of data on a number of parameters, including the publication of government assignments, financial statements, information about property, etc.

At the same time, the norms of the Government Decree of 06/26/2015 for federal executive bodies engaged in the development of policy in the established area of activity (they maintain the above-mentioned all-Russian lists, determine the principles of financing the industry), the obligation is established to publish information on the Single Portal for the Budget System of the Russian Federation (until 2020 - on the Official website for information on government and municipal institutions) on the values of the basic cost standards and industry coefficients to the basic cost standards, which form the basis for the formation of the volume of financial support for public institutions. It should be noted that the section for presenting the relevant data provides the possibility of placing any documents of a budgetary nature and has a low level of functional openness: the published documents are not subject to systematization and are partially identifiable before they are opened. In addition, information on the values of the basic cost standards and industry coefficients to the basic cost standards at the federal level was published in relation to 63% (17 out of 27) spheres of activity, and the assessment of the relevance of documents and the degree of their applicability in the current budget cycle seems to be limited due to the virtual lack of appropriate identifiers.

Also, in accordance with the Procedure for maintaining the register of agreements (contracts) on the provision of subsidies, budget investments, interbudgetary transfers (order of the Ministry of Finance of the Russian Federation dated July 30, 2020 No. 153n), the corresponding register of subsidies, including data on subsidies for financial support for the implementation of state assignment, are subject to publication on the Single Portal for the Budget System of the Russian Federation. For several months, this section of the system (including posted open data) has been unavailable. In addition, amendments to the Procedure have been prepared that significantly reduce the list of published information: information about recipients, the intended purpose of the subsidy, targets for the results of its use, the transfer schedule and other information will no longer be public. Thus, according to a number of parameters, at the federal level, it is advisable to increase the level of budgetary openness of public services, which is due to the formal nature of posting information on standard costs and subsidies for financial support of the state assignment.

Regional procedures for the formation of a state assignment for the provision of public services (performance of work) and financial support of the state assignment, as a rule, determine the need to publish information on the standard costs of the provision of public services (basic cost standards, sectoral adjusting coefficients). At the same time, in 36 constituent entities of the Russian Federation (Arkhangelsk Oblast, Vologda Oblast, Altai Republic, Moscow Oblast, etc.), such a duty is not fixed (Fig. 2), which means there is an actual lack of open information about public finance in terms of services.

The most common ways of publishing the values of basic cost standards specified in regional orders are publication on the Official website for information on government and municipal institutions (20 constituent entities of the Russian Federation: Lipetsk Region, Pskov Region etc.) or posting on the official resources of the executive authorities of constituent entities of the Russian Federation (24 constituent entities of the Russian Federation: Perm Territory, Tver Region, Sevastopol, etc.).

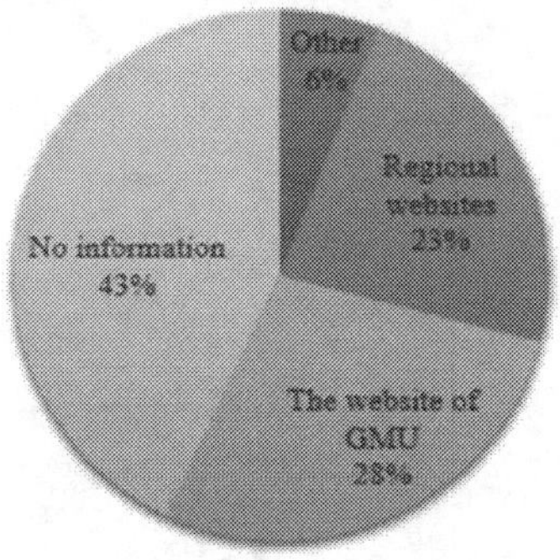

Figure 2. Baseline publication information

It should be noted that in accordance with the Order of the Ministry of Finance of the Russian Federation of July 21, 2011 No. 86n "On approval of the procedure for providing information by a state (municipal) institution, posting it on the official website on the Internet and maintaining the specified site", information on standard costs is not included in list of data published on the site. And according to the technical documentation, the platform does not provide the corresponding functionality. Thus, the implementation of the norm on the publication of standards on the Official website for information on government and municipal institutions, enshrined in the procedures, is not possible and is formal in nature; this information remains unavailable for public consideration. We note that in a number of cases (for example, in the order of the Krasnodar Territory), the wording of the norm on the publication of standards for the cost of providing services provides for such publication only if there is an appropriate technical capability on the official website for information on government and municipal institutions.

At the same time, the content analysis of the official resources of the authorities showed that the obligation to publish the standards for the provision of services on the official websites of regional authorities in most cases (92%, 22 constituent entities of the Russian Federation) is observed, but often the relevant information, as in the case with federal regulations, is not systematized, the ability to search and identify the validity period of documents is limited, which makes it difficult to analyze them by interested parties.

Other options for publishing standards, provided for by regional orders (5 constituent entities of the Russian Federation), combine the provisions on the placement of documents on the Single Portal for the Budget System of the Russian Federation (such functionality is provided) and wording that does not allow one to unambiguously determine the publication procedure and the corresponding resource (for example, "in the prescribed manner on the official website" - Vladimir region, "on the official website" - Tambov region) [11].

Analysis of documents and official information resources makes it possible to group the categories of information according to the level of openness (Table 1). A high level of budgetary openness is possessed by data on all Russian, federal and regional lists of services, as well as state assignments for the provision of public services (Table 2). Partially open one is information about the standards for the provision of public services (not published by all subjects of data provision, low level of functional accessibility) and a subsidy for financial support for the fulfillment of a state task at the federal level. However, there is no public information on subsidies at the regional level and is not available for public review [12].

Table 2. Level of budgetary openness

Categories of information	Federal level	Regional level
Lists of services	High level of openness	High level of openness
Government assignments	High level of openness	High level of openness
Service cost rates	Partial openness	Partial openness
Subsidies	Partial openness	No open information available

Thus, it is advisable to take actions to increase the level of openness of information on the financial support of public services at both the federal and regional levels, including ensuring the timely publication of relevant documents in machine-readable form and the functional availability of information resources.

4 Conclusion

Under the conditions of increasing relevance of issues on increasing the transparency level of government institutions, it is necessary and reasonable to extend the scope of open information on public finance because such information directly demonstrates the target nature and areas of the government budget disbursement.

We should note that the transparency of financial information in the Russian Federation is at an extremely high level in comparison with other areas. For example, it is easier and faster to make a retrospective list of contracts concluded by an educational institution than to find statistical information on mean grades for final examinations of its graduates. The positions in international ratings also confirm this statement.

At the same time, we think we should focus on the transparency of information interesting not only for subject specialists and experts but also for a wider set of people. Such data may include information on providing and financing government (municipal) services as an activity which can be seen and evaluated by citizens in their daily life. The exercise of the right for free public access to the detailed information on the structure of budget expenses enables people to participate in social life, increase community commitment and confidence in government and municipal authorities.

References

1. *Official site of The Federal Service for Supervision of Communications, Information Technology, and Mass Media, The digital threat pyramid* Retrieved from: https://rkn.gov.ru/news/rsoc/news73380.htm
2. *Data is the new gold*, Retrieved from: https://www2.deloitte.com/global/en/pages/real-estate/articles/future-real-estate-data-new-gold.html
3. *OECD Policy Brief 2005* Retrieved from: http://www.oecd.org/dataoecd/1/35/34455306.pdf
4. V.V. Klimanov, A.A. Michailova, *Finance,* **7** (2014)
5. *Open Budget Servey* Retrieved from: https://www.internationalbudget.org/open-budget-survey/open-budget-survey-2019-0
6. V.V. Klimanov, A.A. Mikhailova, *Working Papers* **041922,** (2019).
7. *Openness of the state in Russia* Retrieved from: https://cpur.ru/research_pdf/open-gov-2019.pdf
8. N.A.Guz, *Economy. Taxes. Right,* **6** (2016)
9. N.V. Golovanova, *Financial journal,* **3** (2015).
10. S.E. Gasumova, *Sociology of authority,* **6** (2011).
11. O.V. Butorina, N.P. Pazdnikova, Y.V. Karpovich, *Espacios,* **39(27)** (2018).
12. Z. Mingaleva, N. Pazdnikova, E.A. Mitrofanova, *Asian Social Science*, **10** (2014).

The impact of leadership style on ethical decision-making, the moderating role of strategic thinking: a study among Malaysian managers

Vishalni Balasubramaniam[1], *Yashar* Salamzadeh[1*], *Christopher* Richardson[1], and *Andrey* Plakhin[2]

[1]Graduate school of business, Universiti Sains Malaysia, 11800, Pulau Pinang, Malaysia
[2]Ural state university of economics, management and entrepreneurship department, Ulitsa 8 Marta, 62/45, Yekaterinburg, Sverdlovsk Oblast, 620144, Russia

Abstract. This study develops and empirically tests a multilevel model of cross-level interactions between leadership style and ethical decision-making at the individual level. The study also examines the moderating role of strategic thinking support in the proposed cross-level relationships. Data were collected from 389 managers of electrical and electronic (E&E) companies in Penang. These data were evaluated using a partial least square-structural equation demonstrating via SmartPLS. Results confirmed the relationship between leadership style and ethical decision-making. Additionally, strategic thinking was found to perform the moderating role between leadership style and ethical decision-making. This research contributes to the literature on leadership styles and ethical decision-making. In recent times, no such study in (E&E) manufacturing context has been reported in Malaysia. Also using strategic thinking as a moderator is another novelty of this research.

1 Introduction

The concept of leadership is one of the most studied organizational phenomena by researchers and is identified as the pivot of the evolution of the organizations [1]. Therefore, the organizations need the right leadership style to make an appropriate decision about their business in order to survive in this competitive environment.

Leadership is the capability to inspire others to perform to accomplish the objectives set [2]. Building the style of leadership is the way of encouraging people, providing path to accomplish the wanted objective [3]. According to Kumar et al. (2019), a leadership style may either inspire or depress workers which, in effect, contribute to higher or lower rates of efficiency [4]. Successful leaders also follow styles that deliver employees with idea and empower them to think creatively [5]. Various forms of leadership exist, including transformational and transactional leadership. Transformational leaders concentrate on the creation of human resources which can transform an organization into creative [6].

Decisions are taken at various stages of the organization. Those who reach a higher level of moral reasoning are more likely to exhibit greater consistency with higher levels of moral action, ethical decision making depends upon personality differences. It is also explained that other factors may also be influential in ethical decision making that include self-concept, life experiences, together with wider personality differences that include emotions. Each person has one's own views, thoughts and understanding.

Strategic thinking is a method of analysis performed by individuals to innovatively reinterpret the roles and behaviors most frequently held in relation to the strategy of the organization. This style of thought is also called out-of-the-box thought. Gross (2015), identified that 'a way to solve a strategic complication that blends logical and convergent tactic with innovative and different thought' also can be defined as strategic thinking [7]. According to Pisapia and Glick-Cuenot (2010), strategic thinking is the opposite of linear thinking or preserving the position quo, and therefore this method of thinking involves the individuals to use reflection, system thinking, and reframing expertise [8].

Being an important contributor to Malaysia's economic, Malaysian E&E industry still faces significant challenges in maintaining growth with growing competition from Taiwan, Singapore, China and other Asian countries. The E&E's share of Malaysia's exports has gradually declined from 59 percent in 2010 to 41 percent in 2017.

Previous research results showed there are positive and negative significant occurred among the leaders. However, the research found that there is systematic leniency bias when respondents rated subordinates. According to Graham, Ziegert, and Capitano (2015), inspirational and charismatic transformational leaders have resulted in higher rates of unethical pro-organizational behavior (UPB) than transactional leaders in their research [9]. Their findings expand the

* Corresponding author: yashar@usm.my

interpretation of UPB, include theoretical frameworks to clarify why this activity happens, and add to the philosophy of leadership and work on ethical decision-making.

Several leaderships and ethics researchers argue that the cycle of transformational leadership is centered on a different collection of ethical principles relative to transactional leadership. Transformational leadership describes a large amount of change in the formal code of ethics, and transactional leadership describes a considerable increase in diversity preparation when the size of company and values are controlled. Additionally, Kumar et al., (2019), said the transformational leaders has a huge influence than laissez-faire and transactional leaders on decisions, in banking industry [4]. However, there are very little empirical research done about the impact of leadership styles on ethical decision making. Thus, this paper will investigate the impact of leadership style on ethical decision making with moderating role of strategic thinking. The following research objectives have been developed to achieve this research aim. In short, this study aims:

i. To test the impact of transformational leadership style on ethical decision making.

ii. To test the impact of transactional leadership style on ethical decision making.

iii. To test the impact of transformational leadership style moderating role of strategic thinking.

iv. To test the impact of transactional leadership style moderating role of strategic thinking.

2 Literature Review

2.1 Ethical Decision-Making

The challenge that researchers have is that ethical decision making is not a term that can be simply defined, as it appears to include multiple influences such as individual difference, that include personality and environmental factors, that can affect ethical/ unethical behavior. There appears to be a lack of agreement as to a single agreed definition of ethical judgment within business ethics literature. This is supported by Onyebuchi (2011) who advises that there is no universal consensus, and the definition of ethics is broad. As claimed by researchers in certain cases, ethical decisions are simple enough for people to be able to determine what to do and for business executives to behave appropriately [10]. Ethical decision-making is a problem when exceptions are made and there is confusion regarding ethical responsibilities.

A hermeneutical viewpoint has been added by Keller et al, (2007) to strengthen the previous ethical model, which include egoism, utilitarianism, religion and deontology [11]. The consequential theory is one of the perspectives of normative ethics, which embraces egoism and utilitarianism as two main components of ethical decision making [12,13].

Utilitarianism is one of the moral sources which is rooted in the desire to produce an optimum outcome. Therefore, the greatest value is the output of the ethical decision-making process. A given behavior directly serves the desires of a person as it contributes something to the total amount of satisfaction for him or her or decreases the overall level of discomfort for him or her.

Moreover, in this model Keller et al, (2007) found that the matters that brings only positive for the decision-maker must take into consideration [11]. This implies that while a decision is conflicting to the needs of others, it is right if it serves the benefit of the decision-maker. Furthermore, Rodgers et al. (2009) illustrated egoism as it is focused on persons behaving in accord per their own interests [14]. This has been also explained that egoism is the wish of maximizing one's own interest.

Vivier (2013) theorized Deontological ethics as an ethic that refers to moral roles, obligations, responsibilities and values, irrespective of the implications of an action [15]. According to Modarres & Rafiee, (2011) this concept hence concentrates on the action and also the outcome of the respective action [16]. Deontology is the desire to do the absolute right thing regardless about action's outcome and whose interests are affected by doing the right thing [17]. The deontological ethics suggests that the ethical approach is one that includes acts which have an underlying moral meaning. As rules, what can and cannot be achieved occurs.

A different use of the term "moral" is the opposite of the term "amoral." The term "moral" has been defined as there is something within the subject of "right and wrong", whereas "amoral" term means something outside that subject. When subordinates perceive a lack of integrity and ethical intentions in their leadership, they are more likely to engage in unethical behaviors themselves [18,19]. A mitigating factor in the increase in unethical behavior is an individual's ethical beliefs. One study found that those individuals who had a strong belief in universal moral rules were less likely to behave unethically [20].

2.2 Leadership Style

Leadership is discussed widely in the management literature as a soft skill and individual knowledge [21,22]. Leadership also can be defined as the power which can persuade others to perform to accomplish the objectives set [2]. Leadership is the capacity to inspire others to collaborate for common objectives [23]. Leadership is the process of directing and steering a group in one direction. Leadership style is a critically important characteristic of managers. It is described that leadership style is a way used by a leader to influence the behaviors of followers. The framework of transactional and transformational leadership has provided significant support for leadership style research. The methods of transactional and transformational leadership are the most critical and commonly used for leadership studies since 1990. Burns (1978) firstly developed the both leadership style then Avolio and Bass (1991) expanded the styles [24,25]. Transformational leaders satisfy higher needs by engaging the whole person and identify potential in order to motivate followers to accomplish a goal [26], whereas leaders who employ the exchange system, in which they motivate followers through rewards, and punishments for something in return, are considered transactional leaders [26]. It is clear that each of leadership styles has their own strengths and weaknesses.

Transformational leaders are constructive, encourage and empower followers to accomplish higher order goals by shaping the values, convictions, behaviors and actions of supporters [6,27]. Transformational leadership style values employees' ideal and values, motivates them to put organization interest on the first place, and drives them to achieve the state of best self [25]. Additionally, transformational leadership discusses the fundamental processes that affect various outcomes. Furthermore, the leaders who following transformational leadership style inspire their subordinates knowledgeably, creatively and deliver a clear idea within their organizations. Transformational leaders also concentrate on the creation of human resources which can assist to transform the organization into a creative organization [28].

Transactional leadership style is a leadership model using various compensation to exchange outcomes needed. Transactional leadership style is a task-oriented and interactive leader manner. Transactional leadership style helps organizations achieve their goals by making job reward and remuneration clear [25]. The transactional leadership style has been categorized through leader-follower conversation. Transactional leadership style involves a sort of a carrot and a stick approach or a combination of a reward and punishment. According to Yahaya and Ebrahim (2016) leaders who employ the exchange system, in which they motivate followers through rewards, and punishments for something in return, are also considered as a transactional leadership style [26]. Transactional leadership supports the logical and money-oriented approach by explaining the purpose of distributing the incentives for objective achievement [29,30].

2.3 Strategic Thinking

Baumgartner and Korhonen (2010) added that strategic thinking is contributing to sustainable organizational growth [31]. In addition, Low & Ang (2012) has defined strategic thinking that an informal process of thinking regulation can help itself or the organization to look forward to a long-term goal [32]. This has been more explained by Gross (2015), "a way to solve a strategic complication that blends logical and convergent tactic with innovative and different thought" [7].

Current researchers stress the importance of three primary cognitive mechanisms, including systems thinking, reframing, and reflection as performance drivers for corporate leaders in coping with situational uncertainty [33]. Furthermore, Pisapia and Glick-Cuenot (2010) explained more about strategic thinking as it is an opposite of linear thinking [8]. Leaders struggle to achieve the goal in complicated situation, because they are depending on linear thinking [33].

Such three mechanisms assist leaders in interpreting the problem over the reframing process; build practical theories to direct behavior over the reflecting process; and using system thinking in a method of holistic [33]. Modern researchers highlight that the whole in system thinking is main then the parts are minor [33]. Reframing is described as a reasoning method to gather and organize the information collection to describe the situational realities. Reflecting is defined as the capacity to interpret information and adapt it according to the specific requirements [33] through practice.

3 Hypothesis Development

3.1 Leadership Style and Ethical Decision-Making

The purpose of this correlational study was to determine if an impact exists between leadership styles and ethical decision-making as perceived by managers in E&E organizations. Leadership has been one of the most researched and still being perplexing areas of inquiry in business [3]. By valuing the ethic of justice, leaders can contribute to management policies and rules. There are many researchers identified the impact of leadership style on ethical decision making. Eberlin and Tatum (2008) described said that transformational leaders were valued higher on social justice, while transactional leaders were valued high on structural justice [34]. This proved that ethical decision making playing an important role in an organization for better outcomes. Transactional leadership is associated with the computational belief between transformational leadership, which is associated with a lower grade of non-ethical behavior [35].

Burns (1978) proposed that transformational leaders appeal to higher ideals such as justice and equality and that "both leader and follower are raised to higher levels of motivation and morality" [25]. Bass and Bass (1985), however, in his initial iteration of transformational leadership argued that depending upon their motivation, transformational leaders could be moral or immoral [36]. Burns (1978) clearly believed that transformational leadership was moral leadership [25]. In accordance with Burns' declaration, "The effect of transforming leadership is a bond of common stimulation and upliftment". Priestino's (2014) studies examined transformational leadership and ethical action beliefs to assess if there is a relationship [37]. The study initiates a strong link between perceived ethical behavior and transactional and transformational leadership. In short, leadership has been linked to a person's capability, ability, and degree of influence to move people in a direction, to make decisions, and to do things that they would not normally have undertaken. leaders following transformational style use an extensive variety of data sources to make corporate decisions meanwhile the transactional leaders are more likely to limit the details they use in making decisions. Based on previous studies, this research introduced leadership styles as a mediator between crisis traits and decision-making styles [38,39].

H1: There is a significant relationship between transformational leadership and ethical decision-making.

H2: There is a significant relationship between transactional leadership and ethical decision-making.

3.2 Leadership Style and Ethical-Decision Making, Moderating Role of Strategic Thinking

Strategic thinking has been studied in both management and psychological literature, with management literature concentrating on the mechanism of strategic thinking and strategic decision-making, while psychological literature focuses on factors influencing strategic thinking and decision-making. Kaplan et al. (2014) pioneered the idea of a strategy map that "provides a graphic representation of the vital goals of a company" in order to support the strategic thinker's "system thinking" [40]. Strong strategic skills are correlated with strong organizational success. Strategic thinking is concluded as an important for current and future business strategies to lead organizations [41].

The emphasis of the transactional leaders is on established performance objectives, mission commitment, job priorities and task-related skills. This form of leadership is underpinned by a clear interaction between each person concerned on the basis of mutually agreeable promises. O'Regan et al. (2004) quantitatively evaluated the style of leadership, strategy, and performance achievement [42]. They observed that transactional leadership was strongly associated with short-term success and had an internal focus. Love and Roper (2015) indicated that meanwhile the transactional style became associated with management, smaller organizations saw the advantages of transactional style managers because they can track and monitor subordinates' output [43]. Norzailan, et al. (2016) suggested a range of strategic thinking skills that are necessary for successfully participate, reasoning, and to deliver critical judgement [44].

H3: Strategic thinking moderates' relationship between transformational leadership and ethical decision-making.

H4: Strategic thinking moderates' relationship between transactional leadership and ethical decision-making.

The literature was used as the basis for the creation of a conceptual structure for this analysis, as shown in Figure 1.

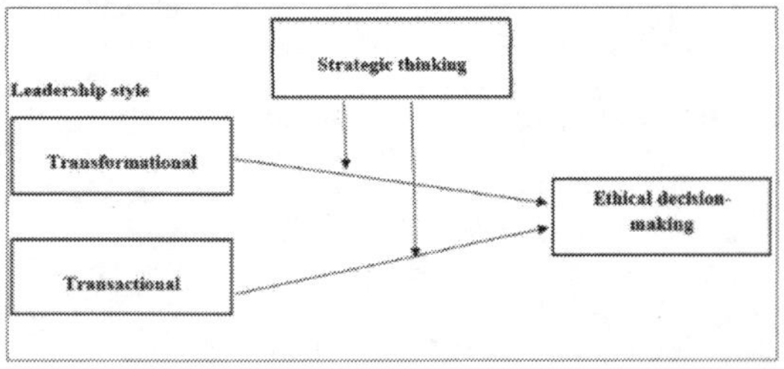

Fig. 1. Conceptual framework

4 Methods

4.1 Research Design

This study focusses on managers from electrical and electronics (E&E) companies located in Penang. The mentioned sampling frame was selected because the E&E industry is the largest single contributor to the manufacturing sector, accounting for 26.1% of total manufacturing output. Measured by share of value added by state, in 2018 Penang made the highest contribution at 32.5% of total value added in the E&E sector. In short, Malaysia's rapid industrialization and high ranking among the world's leading trading nations have contributed largely to this E&E industry. The sampling frame of this study is gained from the Federation of Malaysia Manufacturers (FMM) directory 2018. Based on the FMM, there are total of 2,787 manufacturing firms in Malaysia and total of 39 E&E in Penang. Total Employment in E&E Industry is approximately 100,000. The sample of 389 Penang E&E managers are selected in this research.

A review of leadership studies shows that almost every research method has been applied to study leadership, which include observation, experiments, surveys, interviews, and content analysis. Similarly, studies of ethical leadership, have applied both quantitative and qualitative methods. In this study, the quantitative methodology was chosen. To accomplish the desired research objectives, an online survey form has been shared to the managers through WhatsApp and emails. 389 managers answered the online questionnaire and there are 55 questions included demographic. Respondents were required to answer all the questions mandatorily.

4.2 Measurement instruments

A structured questionnaire has been used in this study. Multi-item scales were used to measure leadership style, ethical decision-making and strategic thinking. A questionnaire was adapted from previous research and updated to suit the framework of this research. The survey questionnaires are divided into four parts Including (1) Transformational leadership with 10 items adapted from Avolio and Bass's (2004) multi factor leadership questionnaire [45], (2) Transactional leadership with 5 items adapted from the same source as previous section, (3) Ethical decision-making adapted from Keller et al. (2007) and Epstein and Spalding (1993) [11,46] and finally (4) Strategic thinking with three sub-dimension as system thinking, reframing and reflecting with total 23 items adapted from Pisapia et al. (2005; 2010) [8,33].

5 Findings

5.1 Data Analysis

Common method variance needs to be examined, when data have been obtained from a single source. The results indicate that 9 of the items explains 63. 36 per cent and the first element clarified 36.94 per cent, which is less than 50 per cent. The Partial Least Squares technique by the SmartPLS 3.0 software [47] has been applied to analyse this study model. This software was used to test the measurement model (validity and reliability) and structural model (testing the relationship among variables).

5.2 Measurement Model

In this study, the latent variable of ethical decision-making was calculated as a second-order, reflective construct where first-order constructs leadership style (transformational leadership and transactional leadership) keep the reflective measurements corresponding to the reflective form. On the basis of literature, higher order is used to limit the number of interactions (and at the same time the number of theories to be tested) in the structural model; and to make the PLS path model more parsimonious and simpler to comprehend.

The validity of convergence and discrimination were used to study the measurement model. The average variance extract (AVE), factor loading, and composite reliability need to be considered when assessing convergent validity. The test shows that all the items loading and AVEs values are higher than 0.5, and CR was above 0.8 (Table 1).

This study used HTMT technique in assessing discriminate validity. As shown below in Table 2, all the values are lower than the required threshold of HTMT 0.90, this indicates that discriminate validity has been ascertained.

Table 1. Measurement table

	Items	Loadings	AVE	CR	Rho_A
Transformational Leadership	A1	0.948	0.972	0.994	0.997
	A2	0.945			
	A3	0.974			
	A4	0.989			
	A5	0.972			
	A6	0.956			
	A7	0.987			
	A8	0.986			
	A9	0.950			
	A10	0.995			
Transactional Leadership	B1	0.901	0.855	0.983	0.994
	B2	0.895			
	B3	0.881			
	B4	0.895			
	B5	0.896			
Ethical Decision Making	C1	0.983	0.878	0.728	0.999
	C2	0.942			
	C3	0.792			
	C4	0.979			
	C5	0.978			
	C6	0.905			
	C7	0.980			
	C8	0.919			
	C9	0.894			
	C10	0.981			
Strategic Thinking	D1	0.524	0.513	0.825	0.878
	D2	0.580			
	D3	0.557			
	D4	0.573			
	D5	0.646			
	D6	0.510			
	D7	0.629			
	D8	0.596			
	D9	0.523			
	D10	0.561			
	E1	0.850			
	E2	0.889			
	E3	0.831			
	E4	0.879			
	E5	0.900			
	E6	0.880			
	E7	0.882			
	F1	0.591			
	F2	0.511			
	F3	0.567			
	F4	0.510			
	F5	0.584			
	F6	0.562			

5.3 Structural model

It is advisable to ensure that there is no lateral collinearity issue in the structural model, before evaluating the structural model. This study refers to the Variance Inflation Factor (VIF) to assess multicollinearity issue, since multiple predictors are tested. Based on the result shown in table 3, all the VIF values are lower than 5.

There were 4 direct hypotheses developed between the constructs, in this study. To examine the significance level, t-statistics for all paths are generated using SmartPLS 3.0 bootstrapping function. Based on the assessment of the path coefficient, four relationships were found to have t-value 3.471, thus significant at 0.01 level of significance. Specifically, the predictors of transformational leadership ($b = 1.393$, $p < 0.01$), and transactional leadership (b = 1.336, p < 0.01), are positively related to leadership style, which explains 65.4 per cent of ethical decision-making. Thus, H1 and H2 are supported.

Hypotheses H3 and H4 focused on the testing of the moderating effect of strategic thinking between leadership style and ethical decision-making. Therefore, assessment of moderation effect for Structural Equation Model (SEM) with the bootstrapping test was conducted. Based on the earlier examination, it was found that the two hypotheses (H3 and H4) moderated the relationship, and these hypotheses fulfilled the conditions required to establish the moderation relationship proven through their significant indirect effects where t-value 3.471, thus significant at 0.01 level of significance.

Table 2. Discriminant Validity (HTMT)

	Ethical Decision-Making	Strategic Thinking	Transactional Leadership	Transformational Leadership
Ethical Decision-Making				
Strategic Thinking	0.530			
Transactional Leadership	0.450	0.660		
Transformational Leadership	0.465	0.717	0.883	

To measure the moderating effect of strategic thinking, an interaction, as an extra latent variable has been added to the model. This interaction describes how the outcome of leadership style on ethical decision-making changes when the moderating variable is increased by one standard deviation. Consequently, the positive effect of leadership style on ethical decision-making is stronger when strategic thinking is high.

In addition, the effect size (f^2) is evaluated. The Cohen's recommendation is used for computing the size of outcome. The standards of 0.02, 0.15 and 0.35 characterize small, medium, and large outcome correspondingly. In table 3, we observed transformational leadership (0.135) and transactional leadership (0.087) has small effect size in producing R^2 for ethical decision-making.

Blindfolding procedure was applied to study the predictive relevance of the model. If Q^2 value is larger than 0, the model has predictive relevance for a certain endogenous construct. The transformational leadership (0.120) and transactional leadership (0.089) are more than 0, indicating that the model has sufficient predictive relevance. Table 4 shows the indirect hypotheses testing results.

Table 3. Direct Hypotheses testing

Hypothesis	Relationship	Std Beta	Std error	|t-value|^	Decision	f2	q2	95% CI LL	95% CI UL	VIF
H1	Transformational Leadership -> Ethical Decision-Making	1.393	0.369	3.946**	Supported	0.135	0.120	0.889	2.002	4.369
H2	Transactional Leadership -> Ethical Decision-Making	1.336	0.344	3.875**	Supported	0.086	0.089	0.827	1.937	4.759

Table 4. Indirect Hypotheses testing

Hypothesis	Relationship	Std Beta	Std error	|t-value|^	Decision	95% CI LL	95% CI UL
H3	TF-ST-EDM -> Ethical Decision-Making	1.426	0.417	3.471**	Supported	0.768	2.161
H4	TS-ST-EDM -> Ethical Decision-Making	1.932	0.45	4.297**	Supported	1.181	2.735

6 Discussion

The purpose of this study was to explore the association of transformational and transactional leadership styles with various ethical decision-making among Penang E&E managers. We hypothesized that transformational leaders significantly use ethical decision-making and we proved where the regression analyses supported the hypotheses. This research explains the managers of Penang E&E companies are following transformational leadership style in leading their team. As a result, the managers those following transformational leadership style are attentive to the needs of their employees, focuses on helping them to meet their potential, and consider ethical consequences while making a decision. The transformational managers are also being more integrity to their employees which accordance with prior works.

The research has showed that the style of transactional leadership has a significant relationship with ethical decision-making as per mentioned in the literature. The positive relationship of transactional leadership style with ethical decision-making is shown a lesser degree than the relationship between transformational leadership style and ethical decision-making. Transactional leadership is based on a give and take between leader and follower that includes reward and discipline. The Penang E&E company managers those applied transactional leadership style in their job responsibility tend to make their own ethical decision even they consider as reasonable and money-oriented attitudes toward leader and employees [30]. The managers make a clear expectation regarding the given task towards to their team and work out agreements with their team.

Moreover, the moderator role of strategic thinking also given a positive impact on both leadership style and ethical decision making. Transformational leaders tend to follow the system thinking and reflecting criteria from strategic thinking variable. The reframing criteria has been highly rated by the managers who following transactional leadership style. This study proven that the moderating role of strategic thinking strongly supporting the managers to make an ethical decision-making. The strong ties between transformational, transactional and strategic thinking indicate that, since strategic thinking is therefore linked to leadership types, leaders should support and develop strategic thinking habits. Strategic thinkers can deal with strategy variations because they think about why, what, how and when in each case, consider capital, and are effective at scheduling to find the right response [48].

To conclude it can be stated that decisions are highly situational, and leaders need to apply their wisdom in making such ethical decisions. The study found that there is slightly difference among the rates of transactional and transformational style of leadership where the managers in Penang E&E following more transformational leaderships style than transactional. Understanding the relationship between leadership style and ethical decision-making can be important for E&E manufacturing companies to avoid bias and increase effectiveness.

6.1 Implications

The present study outlines various theoretical implications for E&E managers by identifying leadership styles that foster positive result in ethical decision-making. Precisely, outcomes of this study indicate that transformational and transactional leadership style have a positive impact on ethical decision-making. Ethical decision-making is important to managers, which eventually translate into organizational success. As established in the previous research, this research also proven that ethical decision-making helps the transformational managers to avoid unethical behaviour. Additionally, this research has identified that there is a strong relation between transformational leadership style and leader integrity which has been explained earlier. As an evidence we can evidently see that E&E managers listen to their team concerns and consider moral and ethical consequences while applying a decision. On the other hand, the managers following transactional leadership style, feel the need to make immediate choices thus limit the details they use in making decisions. Penang E&E company managers tend to make their own ethical decision even they consider as reasonable and money-oriented attitudes toward leader and employees. As shown in this study the manager with strategic thinking is determined as a significant for ethical decision-making which leads to organization success.

Additionally, the present study also provided practical contribution. By empathetic the impact and differences between both leadership styles on ethical decision-making, managers will get an opportunity to know the accurate leadership style that they are currently following. Furthermore, managers also will realize and understand the importance of ethical decision-making, which can give an impact by their own leadership style. Organizations could give considerable weight for moderating role of strategic thinking in the process of making ethical decisions. This study positively will help the organizations in hiring or positioning the managers into specific team based on their leadership style. This method may ease the management to avoid giving a negative impact or conflict to occur among the team, due to the ethical decision making by the managers.

In conclusion, the rather incidental findings regarding the impact of leadership styles and ethical decision-making are fascinating. They lend credence to the concept of leadership and may serve as an additional support to its concurrent validity. Regarding leadership and ethical decision-making more studies are needed. A good idea might be narrowing the study by limiting the population of participants and demographic in the study. It is logical to assume that being aware of one's ethical decision-making style, as well as leadership styles, may help to focus on achieving the organization objectives, develops necessary skills, and deals in a better way with a given situation such as problem solving, motivating, and interacting with others in organizations.

7 Conclusion

Though insightful, this study's findings need to be considered in light of certain limitations. The data of this study were only collected from the E&E sector based in Penang since it is the most important state in terms of Malaysia's E&E industry according to reports. Given the variations among different sectors and regions, the empirical validity of the current study may not be so solid, and we propose that future scholars test our study with more evidence. The demographic outcome of this current study does not reflect our nation's population pattern where the Malays respondents were represented as the smallest group of respondents. Future studies can choose different geographical or industrial which can show the accurate demographical based of ethnicity of our nation to test the main impact.

Furthermore, this study finds strategic thinking moderating role in the relationship between leadership style (transformational and transactional) and ethical decision-making. However, leaders might follow different leaderships styles which can also impact the ethical decision-making. Therefore, a similar study could be developed by future researchers using different leadership styles such as authentic leadership style and servant leadership style in different sectors with different employee's positions. Even the future research can use different moderating role such strategic vision and strategic mindset to find the more stronger moderating role between leadership style and ethical decision-making.

In spite of the shortcomings of this article, a number of useful contributions to the discourse on leadership style and ethical decision-making are nevertheless provided. The relevance of these problems needs to be more addressed in future studies.

References

1. L. Cappelletti, S. Khalla, F. Noguera, A. Scouarnec, and C. V. Fourboul, *Manag. Avenir* **263** (2010)
2. Z. M. Elqadri, Int. Educ. Stud. **8**, 183 (2015)
3. S. A. Z. Kazmi, M. Naarananoja, and J. K. Wartsila, *Procedia-Social Behav. Sci.* **229**, 387 (2016)
4. M. Kumar, S. Pandey, V. Sinha, and A. P. Srivastava, **J. Dev. Entrep**. **24**, 1950025 (2019)
5. D. O. Almutairi, *Int. J. Bus. Manag.* **11**, 231 (2016)
6. J. Giddens, *J. Prof. Nurs.* **34**, 117 (2018)
7. R. Gross, *Glob. Bus. Econ.* (2015)
8. J. Pisapia and S. Glick-Cuenot, *Am. Inst. High. Educ.* **3 (1),** 179-188, (2010)
9. K. A. Graham, J. C. Ziegert, and J. Capitano, J. *Bus. Ethics* **126**, 423 (2015)
10. V. N. Onyebuchi, *Int. J. Bus. Soc. Sci.* **2**, (2011)
11. A. C. Keller, K. T. Smith, and L. M. Smith, *Crit. Perspect. Account.* **18**, 299 (2007)
12. O. C. Ferrell and L. G. Gresham, *J. Mark.* **49**, 87 (1985)
13. S. M. Goodarzi, Y. Salamzadeh, and A. Salamzadeh, in *Compet. Emerg. Mark.* (Springer, 2018), pp. 503–539
14. W. Rodgers, A. Guiral, and J. A. Gonzalo, J. *Bus. Ethics* **86**, 347 (2009)
15. E. Vivier, *African J. Bus. Ethics* **7**, (2013)
16. A. Modarres and A. Rafiee, *Soc. Responsib. J.* (2011)
17. K. D. Martin and J. B. Cullen, *J. Bus. Ethics* **69**, 175 (2006)
18. Y. N. Ng and Y. Salamzadeh, *Int. J. Bus. Sci. Appl. Manag.* **15**, (2020)
19. D. K. Peterson, *J. Bus. Ethics* **49**, 371 (2004)
20. M. Nejati, Y. Salamzadeh, and C. K. Loke, *Soc. Responsib. J.* (2019)
21. C. Kimble, J. B. de Vasconcelos, and Á. Rocha, *Inf. Syst. Front.* **18**, 1119 (2016)
22. J. B. de Vasconcelos, C. Kimble, and Á. Rocha, *Inf. Syst. Front.* **18**, 1035 (2016)
23. P. G. Northouse, *Leadership: Theory and Practice* (Sage Publications, 2021)
24. B. M. Bass and B. J. Avolio, *Res. Organ. Chang. Dev.* **4**, 231 (1990)
25. J. M. Burns, *Leadership* (New York: Harper & Row, 1978)
26. R. Yahaya and F. Ebrahim, *J. Manag. Dev.* (2016)
27. Y. Salamzadeh, M. Kianmanesh, and T. Vahidi, *Int. J. Sustain. Strateg. Manag.* **7**, 186 (2019)
28. M. K. Copeland, *J. Leadership, Account. Ethics* **13**, 79 (2016)
29. F. Bushra, U. Ahmad, and A. Naveed, *Int. J. Bus. Soc. Sci.* **2**, (2011)
30. M. L. Khan, N. Langove, F. A. Shah, and M. U. Javid, *Glob. Bus. Manag. Res.* **7**, (2015)
31. R. J. Baumgartner and J. Korhonen, *Strategic Thinking for Sustainable Development* (Wiley Online Library, 2010)
32. K. C. P. Low and S.-L. Ang, *Educ. Res.* **3**, 502 (2012)
33. J. Pisapia, D. Reyes-Guerra, and E. Coukos-Semmel, *Leadersh. Rev.* **5**, 41 (2005)
34. R. J. Eberlin and B. C. Tatum, *Manag. Decis.* (2008)
35. L. T. Tuan, *Soc. Responsib. J.* (2012)
36. B. M. Bass and M. Bass Bernard, *Leadership and Performance beyond Expectations* (Free press New York, 1985)
37. C. A. Priestino, *Leadership Styles and Ethical Behavior: An Examination of Subordinate Perceptions,* (Capella University, 2014)
38. Y. Hur, P. T. Van Den Berg, and C. P. M. Wilderom, *Leadersh. Q.* **22**, 591 (2011)
39. S. Geiger, P. Guenzi, N. Panagopoulos, and S. Dimitriadis, *Eur. J. Mark.* (2009)
40. S. Kaplan, J. Cortina, G. Ruark, K. LaPort, and V. Nicolaides, *Leadersh. Q.* **25**, 563 (2014)
41. E. F. Goldman and A. Casey, J. *Leadersh. Organ. Stud.* **17**, 119 (2010)
42. N. O'Regan, A. Ghobadian, and M. Sims, *J. Small Bus. Strateg.* **15**, 45 (2004)
43. J. H. Love and S. Roper, *Int. Small Bus. J.* **33**, 28 (2015)
44. Z. Norzailan, S. M. Yusof, and R. Othman, *J. Adv. Manag. Sci.* **4**, (2016)
45. B. J. Avolio and B. M. Bass, *Mind Gard. Inc. Menlo Park.* CA (2004)
46. M. J. Epstein and A. D. Spalding, *The Accountant's Guide to Legal Liability and Ethics* (1993)
47. C. Ringle, D. Da Silva, and D. Bido, Bido, D., Da Silva, D., Ringle, C.(2014). *Struct. Equ. Model. with Smartpls. Brazilian J. Mark.* **13**, (2015)
48. O. Szekely, *Transitional Justice Arab Spring* 94 (2015)

Banking front-line personnel assessment by fuzzy logic approaches

Vladimir Koretskiy[1,*], *Marina* Degtiareva-Galiakhmetova[1], and *Evgeniy* Kostitsyn[2]

[1]Kalashnikov Izhevsk State Technical University, Institute of Digital Economy, 7, bld. 1, Studentcheskaya str., Izhevsk, 426069, Russia
[2]PJS Company BystroBank, 18, Krasnogeroiska str., Izhevsk, 426008, Russia

Abstract. The article is dedicated to the qualitative assessment of banking personnel and the interpretation of the results with a developed fuzzy logic expert system. The authors proposed to evaluate human resources based on the Company Loyalty, Customer Service Quality and Intra-Corporate Communication which are linguistic terms for personnel to be assessed. To interpret the results of the received staff assessment, a fuzzy expert system was developed which enables the Business Efficiency of Personnel to be estimated. The expert system was tested at the front-line office of the regional bank. Regression and correlation analysis revealed high correlations between the Business Efficiency of Personnel and the quantitative results of employees. The practical relevance of the research is conditioned by the growing need to assess credit managers during the trial period or at the introduction of new products, while quantitative indicators are absent. The methods used for research comprise survey, ranking of the factors and factor analysis.

1 Introduction

The effectiveness of personnel management becomes an essential factor in ensuring the competitiveness of any economic entity, including in the banking sector. For a few medium-sized, regional banks in Russia, this factor is becoming a key factor of surviving in their competition with transnational bank chains. At the same time, it should be noted that the quantitative criteria of staff assessment do not always sufficiently represent the value of the employee to the company. In particular, they failed to reflect the risks associated with, for example, lending, since they generally account for only the number of contracts gained by a bank clerk and their cost.

Staff assessment, as a decision-making process, involves determining the level of compliance of individual indicators of an employee with the requirements imposed on him by the employer. Besides, the staff assessment, especially for front line staff and managers, should identify both qualitative and individual characteristics of them, and the amenability to the team playing, flexibility to new demands and conditions, personalities and customers' challenges. Labour activity of the credit manager takes place in a volatile environment.

As noted in the references relevant to the subject, in the vast majority of banks there is no formalized mechanism for staff assessment and that is one of the main reasons for the lack of efficiency of performance of staff members [1]. As a rule, this has the following negative results:

- increased turnover of staff;
- vast of vacant jobs in the bank's subdivisions, unfilled for a long period;
- labour dispute;
- complaints from consumers;
- losses of efficiency of the bank's employees;
- loss of competitiveness.

Most banking organizations face challenges in institute procedures for staff to be assessed including the following ones: low level of the amenability to work ethic, subjectivism or excessive formalism in the interpretation of results, lack of follow-up measures, in particular, the motivation means based on data fetched while the evaluation of personnel was carrying out.

All staff assessment criteria, regardless of economic sectors and areas of activity of the entity, can be divided into three broad groups. They are:

- results of individual actions;
- behavioural characteristics;
- merits of the employee.

Whereas the first group usually includes quantitative indicators of the staff performance for a period, the second and third are subjective, evaluative or qualitative. At the same time, their relevance to the staff assessment is not lower than quantitative indicators weight, as they determine the staff morale, customer satisfaction, delayed risks, readiness team player to alter or take up a new project. In particular, as pointed out by some authors [1, 2] the following factors influence final results:

- biological (sex, age, mental and physical abilities);
- social and economic (state of the economy, standard of living, social protection);
- technical and organizational (challenges, conditions and complexity of work);

* Corresponding author: vpk1973@yandex.ru

https://doi.org/10.1051/shsconf/202111600057

- social and psychological (attitude to work, the psychological climate in the team);
- market (competition, inflation).

These all factors should be allowed for a comprehensive assessment of staff performance to be carried out and it increases the credibility objectivity and reliability of the findings of an evaluation of personnel activity point out these authors.

On assessing personnel to interpret qualitative indicators the approaches of fuzzy logic proved their effectiveness. Currently, fuzzy logic is used in the automotive, aerospace and transport industry, robotics, risk evaluation, management, including personnel assessment in the financial sector [3]. The way of fuzzy logic implementation for personnel management is shown in numerous works [4, 5, 6].

The fuzzy analytical approach for study competencies in employee selection was developed by Akyol &Guler [7]. They suggested a fuzzy expert system to select project managers by comparing the competencies of applicants.

There are some works on measurement analytical competencies of employees by fuzzy logic being applied [7]. For example, Kalvakolanu & Madhavaiah focused on the capability of the staff to demonstrate the analytical competencies.

Competency assessment methods proposed by Holt et al. for personnel to be assessed [8]. Jevscek stated that the assessment competencies of professionals fit the fuzzy logic approach [9]. Analytical abilities and the performance of HR professionals and appraisal using fuzzy evaluation methodology are relevant to this research [10, 11].

Thus, it can be concluded that both quantitative and qualitative indicators should be taken into account in assessing staff. At present, it is necessary to solve the problem of a comprehensive assessment of the personnel, i.e., their professional, personal and business qualities, which are necessary to be assessed as the effectiveness of the employees. At the same time, it should be borne in mind that efficiency means performance, i.e., the ratio of achieved result to planned ones – state some researchers [1].

2 Methodology

Among the factors relevant to personnel assessment the essential are the following ones:

- corporate culture (strategic goals, values, regional cultural features, management style, system and communication features);
- regulatory framework (legislative acts corporate standards and rules);
- number and quality of staff;
- staff turnover rate;
- seasonality of business (allowing for peak load or downtime) [2].

For example, the indicators related to the first factor, corporate culture, are qualitative. As a result, due to erroneous management decisions amenable to personal sympathies and dislikes of HR managers, serious risks for the management of human resources are rising. It should be noted that risks in the credit sector can be assessed on fuzzy logic bases [12].

The problem is aggravated by the fact that the results of actions of an individual employee performed under circumstances of odds financial risks and uncertainty are extremely unpredictable and hardly could be estimated both qualitatively or quantitatively. This issue is particularly common in the group assessment of staff or in cases where the contribution of a particular employee cannot be distinguished from the teamwork results (Business Game Laboratory).

It should be added that the assessment of the personnel is a quite challenging process both for the personal probation and adaptation because of lack of quantitative data.

For quantitative and qualitative parameters of evaluation indicators to be determined appropriately, it is extremely recommended:

- to avoid amenability to any single indicator or group of ones for fear that other equally important aspects of employee activity could be left beyond observation;
- to determine a sufficient number of indicators to maximize the performance of the staff and not to be overloaded with data.

As factors of assessment of loan officers, based observations, surveys of experts in the banking sector and factor analysis were chosen the following:

- Behaviour Towards Colleagues and Costumers.
- Customer Service Quality.
- Company Loyalty.

As an output variable defined the following one:

- Efficiency of the Manager.

It should be noted that indirect or qualitative assessments characterize the performance of an employee as one is the flawless performer of job duties and functions. For their assessment can be used purposefully developed scorecard to measure their compliance to the well-designed perfect employee.

Hence, the meaningful interpretation of the fuzzy model includes the selection and specification of three input and one output variable of the fuzzy output system. Table 1 presents the input and output variables of the regional banking sector staff assessment system.

Table 1. Input and output variables of the staff assessment system for banking

Variable Symbol	Variable Name	Term-set	Input Options
x_1	Behavior Towards Colleagues and Costumers;	low, medium, high	0 — 10 points
x_2	Customer Service Quality		
x_3	Company Loyalty	positive, neutral, negative	0 — 10 points
y	Business Efficiency	low, medium, high	0 — 1 point

The variables and the indicator, the input options were identified by surveys and observation.

Fuzzy Expert System was implemented in MATLAB with the Fuzzy Logic Toolbox extension package containing the following tools:

- Fuzzy Inference System Editor (FIS Editor);
- Membership Function Editor;
- Rule Editor;
- Rule View.

3 Results

The membership functions for the corresponding variables, allowing for variables, measured in points in the range of numbers from 0 to 10, are normalized to a single segment. Their type was determined based on surveys of experts. Figures 1-3 present the membership functions of corresponding variables - x_1, x_2, x_3. Figure 4 shows the membership function of the output variable y (Manager Business Efficiency).

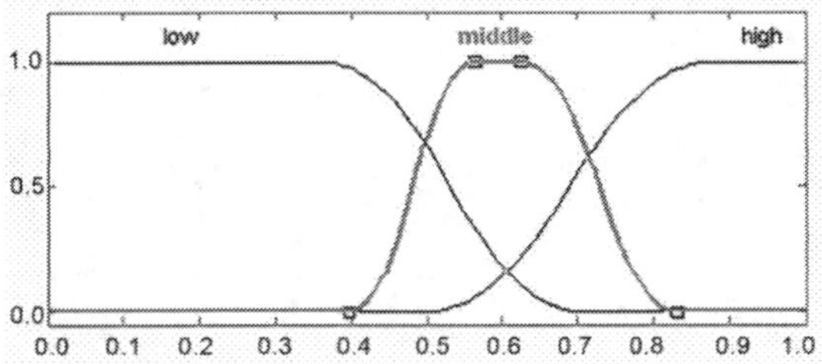

Fig. 1. The function of membership of variable x_1= Behaviour towards Colleagues and Costumers (Three linguistic terms: low, middle, high).

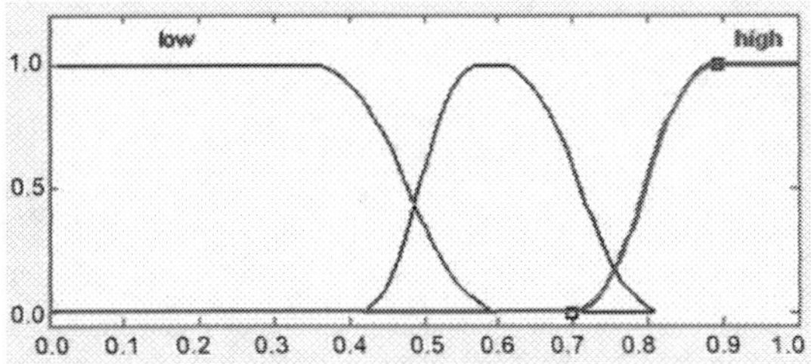

Fig. 2. The membership function for x_2=Customer Service Quality (Three linguistic terms: low, middle, high).

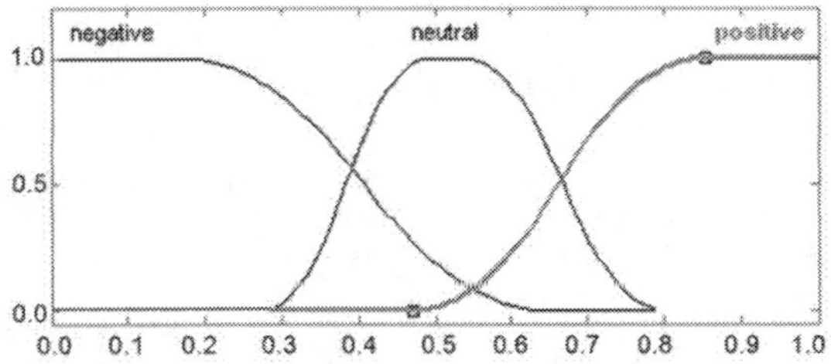

Fig. 3. Membership function for variable x_3=Company Loyalty (Three linguistic terms: negative, neutral, positive).

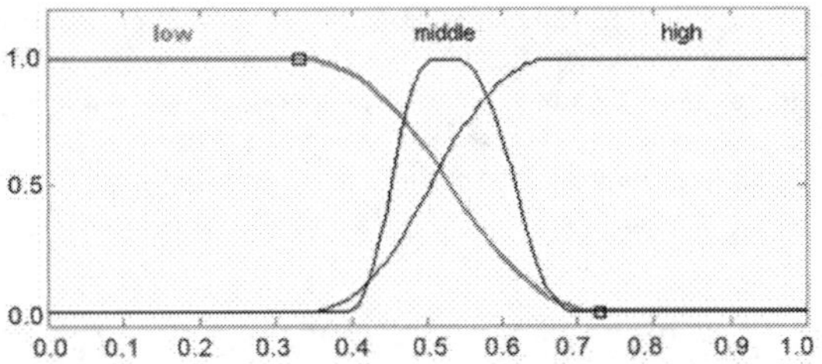

Fig. 4. The membership function of the output variable y=Business Efficiency of the Manager. (Three linguistic terms: low, middle, high).

Based on expert opinions and personal observations, the rules for the fuzzy expert system are set forth like following:

1) IF x_1 = Low AND x_2 = Middle AND x_3 = High then y = Middle;

2) IF x_1 = Middle AND x_2 = High AND x_3 = High, y = High, etc.

Fig. 5 presents a visualization of the developed fuzzy expert system, indicating the connection between the input variables Company Loyalty, Quality of Service, Behaviour towards Colleagues and Costumers and the output variable Business Efficiency of the Manager.

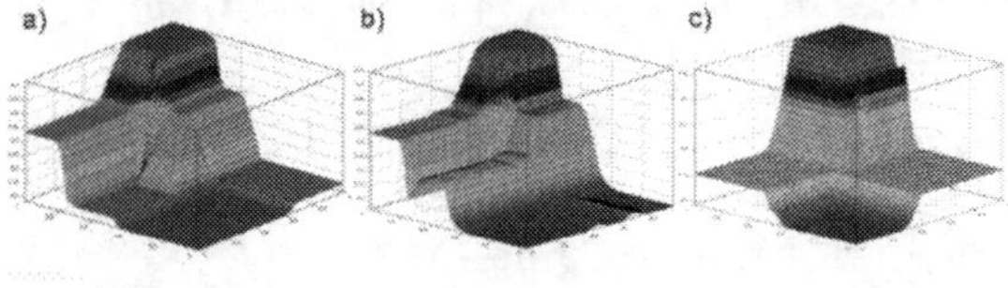

Fig. 5. The Fuzzy Logic Expert System:
a) dependence of output variable y= Business Efficiency of Manager on input variables x_1=Company Loyalty and x_2= Customer Service Quality;
b) dependence of output variable y on input variables x_1=Company Loyalty and x_2=Customer Service Quality;
c) Dependence of output variable y on input variables x_1=Customer Service Quality and x_2=Behaviour towards Colleagues and Costumers.

Table 2 presents qualitative parameters of the bank staff assessment system fetched from the survey and through the fuzzy expert system. The last column presented quantitative indicators of managers' work which is the number of loan contracts for a given period.

Table 2. Qualitative parameters of the staff assessment system

Staff	x_1	x_2	x_3	y	Number of credit agreements (pieces)
Manager 1	0.8	0.5	0.7	0.563	17
Manager 2	0.6	0.8	0.8	0.729	19
Manager 3	0.5	0.4	0.4	0.29	11
Manager 4	0.7	0.9	0.8	0.738	20
Manager 5	0.4	0.3	0.6	0.271	10
...					
Manager 100	0.7	0.8	0.3	0.56	14

Considering the variable Business Efficiency of the Manager as an explanatory variable, denoted as X, and N as Number of Concluded Loan Contract as an explained variable, and based on the scatter plot, it is possible to put forward a hypothesis that the relationship between all possible values of X and Y of the population is linear:

$$N = B1 \cdot X + B0 + err.$$

The regression equation, based on the data presented in Table 2, is the following:

$$N = 18.4815 \cdot X + 5.782.$$

The sample linear correlation coefficient is 0.888, which defines strong bidirectional causation between the Y and the X factor. The Student's criterion confirms the statistical significance of the regression coefficient.

The Fisher criterion for this model is met. Since the actual value is $F > F_{tab}$, the determining factor is statistically significant and the estimate of the regression equation is statistically reliable.

As a result, the following may be noted: the dependence of Y on X is studied. At this stage, a pair linear regression was selected. Its parameters are estimated by the least square method. The statistical significance of the equation is verified by the Fisher criterion. It is established that 78.91% of the total variability Y is explained by the change of X.

It is also established that model parameters are statistically significant. An economic interpretation of model parameters is possible. An increase in X by 1 unit leads to an increase in Y on average by 18.481 units. The obtained estimates of the regression equation can be used to predict the effectiveness of the credit manager based on his qualitative estimates with no quantitative parameters.

4 Conclusions

In the research carried out, we showed that qualitative indicators are indispensable to assess the work of specialists and employees in the banking sector.
It is proposed to avail the fuzzy logic opportunities to interpret qualitative indicators of labour efficiency of the banking sector's employees.

The paper presents a fuzzy expert system that enables the credit personnel of banking institutions to be assessed on the basis of qualitative data and the results of the evaluation of the effectiveness of credit managers.

The developed fuzzy expert system can be used to assess the banking personnel, and in particular, the effectiveness of credit managers, especially in the recruitment of personnel; implementation of new projects is going on without any available quantitative indicators. This makes the approach effective for managerial decision making for staff hiring and promotion to be made in the banking and financial sector to minimize risks and increase efficiency.

References

1. N. Shportiuk, M. Sabadin, *Int. Sci. J. Internauka.* **15 (37)** (2017).
2. Y. Odegov, M. Kulapov, O. Rakauskiene, P. Karasev, N. Fomenko, *Mon. J. Econ.* **17 (1)** (2021).
3. I. Hatzilygeroudis, J. Prentzas, J. (2014). *Sci. J. St. Comp. Int.* **514** (2014).
4. A. Hernández, D. Hidalgo, *Op. J. Bus. Mng.* E **8, 6** (2020).
5. A. Ibragimov, L. Ibragimova, M. Karavayeva, *Vestnik S. St. A. Un. E* **1 (32)** (2012)
6. V. Koretskiy, L. Józsa, I. Mardanova, M., Galiakhmetova, *JEECAR,* **6 (1)**, (2019).
7. S. Kalvakolanu, C. Madhavaiah. J. *Adv. R. Dyn. Control. Sys.* **11 (6)** (2019).
8. J. Holt, S. Perry, *A Pragmatic Guide to Competency: Tools, Frameworks and Assessment* (BCS Learning & Development Limited, 2011)
9. M. Jevscek, *RUO.* **5 (2)** (2016).
10. D. Kryscynski, C. Reeves, R. Stice-Lusvardi, M., Ulrich, G. Russell, *HRM,* **57 (3)**, (2018).
11. N. Macwan, S. Srinivas. *IJ End. I. Technol.* **3** (2013)
12. N. Hasan, H. Elghareeb, F. Farahat, A. *Abo Elfotouh, Int. J. F. L. and Int. Sys.* **21 (1)** (2021).

HR branding for the forest industry: innovative practices

Andrey Prudnikov*, *Yulia* Lektorova, and *Svetlana* Dmitrieva

[1] Perm National Research Polytechnic University, Department of Foreign Languages and Public Relations, 13 Professor Pozdeev st., Perm, Perm Krai 614013, Russia

Abstract. The study examines innovative HR branding strategies for the forest industry. The need for innovation is driven by the growing influence of negative trends that complicate the reproduction of human resources within the framework of timber industry. These alarming trends are largely characteristic of both markets economically developed and developing countries. The paper presents a quantitative and qualitative analysis of HR innovations in communication policies of Russian and foreign companies. The benchmarking analysis of innovative HR branding practices enables us to build the best communication model for companies working in the forest sector of the economy.

1 Introduction

Economic globalization contributes to the growing strategic interest of foreign and domestic companies in the development of their HR brands. Researchers describe the concept of an HR brand as "a combination of employer's economic, psychological and emotional advantages in the market", as "a system of images, and ideas about an organization as an employer conveyed to external and internal publics", and as "a set of targeted measures to create a positive image of an employer and to attract high-quality job seekers" [1,2,3].

Ultimately, benefits, images, and events are focused on the creation of a favorable and manageable image of an industrial company as an employer for current and potential employees. Potential employees may include direct job seekers, professionals from other companies, and students and graduates from secondary and higher educational institutions. A positive company's image as an employer builds the confidence and trust of employees and job seekers. The combination of these features allows an employer to recruit and retain the top job candidates and employees at the lowest cost, which contributes to the effective strategic development and growth of an organization.

Building an HR brand is the key aim of industrial enterprises. A positive experience has been gained by companies in the industrial sector including the forest industry.

The forest industry employs around 13.7 million workers worldwide that account for 0.4% of the total workforce. The top ten countries in terms of the number of employees in the sector (including three subsectors: logging, woodworking, and pulp and paper industry) include China, the USA, Brazil, Russia, India, Japan, Germany, Indonesia, Italy, and Malaysia. Employment trends in the forest industry vary across regions in the world. Thus, in recent years, the employment rate has declined in Europe and North America but has increased in Latin America and Asia. The shift in the labor market is due to automation, the abundance of cheap labor, and measures to stimulate the development of the forestry sector and attract investments.

Jobs in the forest industry can provide job seekers with a competitive income and career prospects. However, it can vary greatly. For example, studies conducted in the United States show that average wages in the sector are close to or higher than the national average. High wages in the timber industry are recorded in North America and Europe, while low wages are typical for Asia and CIS countries.

Researchers identify the following essential characteristics of the world forestry labor market:

• gender imbalance: women make up about 20% of the forestry workforce [4]. The presence of women in the industry varies from country to country and from subsector to subsector: high employment rates are determined in the paper industry, while a low proportion of the female labor force is recorded in forestry. Given the seasonality of forestry activities, their participation is usually informal and low-paid [5];

• a high level of informal employment due to illegal deforestation;

• a high degree of injury risks at work;

• personnel aging: this trend is associated with new social risks, such as changes in the pension scheme and other risks in social security systems [6];

• low level of interest in the profession among young people.

HeadHunter, an Internet recruitment agency, analyzed more than 3,840 vacancies in the forest, timber, and woodworking industry posted on hh.ru by employers

* Corresponding author: prudnikow@gmail.com

from different regions of Russia in 2019. The analysis identified the following patterns for the Russian labor market:

• Job market shrinking. On the whole in Russia, the number of job offers is steadily decreasing: vacancies amounted to 65% in the 3rd quarter of 2018, and 14% in the 4th quarter of 2019. Two capitals Moscow and St. Petersburg offer the largest number of open vacancies, 10.6%, and 10.1% respectively. Irkutsk Region (7%), Moscow Region (6.1%), and Perm Krai (4.8%) are the leaders in labor demand.

• Increased competition for personnel. On average across regions, there are 3 CVs per job offer among forestry specialists (Moscow has 5), which is several times lower than the average level of competition among job applicants.

• Growth of wages in the industry. From the 1st to the 4th quarter of 2019, there was a noticeable increase in the average wage in the Russian timber industry: from 45,000 rubles to 45,977 rubles per month. The growth in St. Petersburg was higher: from 64,974 rubles to 67,272 rubles per month.

• Age shifts and gender imbalance. The proportion of male to female CVs is 86% male to 14% female; every fifth CV is from an applicant aged 46-55. Employees aged 50 and over make up 35% of the total number of employees at forestry enterprises. As a result, the annual demand for young personnel in Russia is 3 to 15 thousand people; over 7 thousand employees have to be trained and retrained [7].

Russian experts observe low interest in working in the forest industry among young people due to several reasons: lack of material incentives, lack of high-quality knowledge and skills, since the requirements for the level of training are constantly changing and the decline in the prestige of working professions [8]. Another reason is that the forestry sector does not appeal to prospective employees due to its closeness and lack of information about the industry [9]. Besides, there is a widespread bias about the conservatism of the industry and the unattractive image of employers among young people.

The current labor market in Russian and abroad confront with similar problems. The development of an effective HR branding strategy in the forestry industry is the way to successfully address them.

Backhaus and Tikoo have developed a theoretical model that describes HR brand as building the relationship between employer branding, corporate culture, and employee loyalty [10].

The researchers claim that employer branding holds two main intangible assets: HR brand associations and brand loyalty. Associations with employer branding create an employer image that enhances its attractiveness for job seekers. HR brand has an impact on the organizational culture and identity of a company as an employer, and increasingly contributes to the increase in employee loyalty to employer branding. All these have a positive impact on employee's productivity.

While researchers discuss a variety of algorithms and stages of building an effective HR brand, for example, they distinguish from several to dozens of stages, all of them describe HR branding as a three-step process.

First, a company develops value for employees, in other words, it offers "Employment Value Proposition (EVP)". EVP is a reflection of who the company is and what impression it creates in the market. This is the company's proposal that is addressed to current and potential employees.

The second step is to market the value proposition to external audiences, i.e., potential employees, recruitment agencies, job counselors, and others.

Following the value proposition, the third aspect of employer branding is internal branding. It incorporates the company's declared position as an employer in the internal culture. It leads to the development of employee's loyalty and commitment to the company.

Building effective HR branding for the forest industry is an important and urgent task. The industry's current global labor market trends confirm its significance. Thus, the analytical reports of the Food and Agriculture Organization of the United Nations (FAO) have revealed that even if the demand for forest products grows in the world; even if the trend is accompanied by the expansion of production facilities at enterprises, it will not result in a proportional increase of product supply in the market due to the acute shortage of workforce in the production process. This is especially true for the markets of well-developed Western European countries. Due to the hazardous and dangerous conditions of work in the forestry sector, employment is becoming less and less attractive for young people. As a result, forestry companies from such countries as Finland, France, and Sweden repeatedly complain about the lack of personnel in the industry [11].

The problems under consideration also have their downside. In the case of increasing immigration to alleviate labor shortage, it may lead to the degradation of the forest industry on the whole. Immigrants are less demanding for safe and attractive working conditions and adequate compensation, which may result in the substantial growth of environmental costs for the entire world community. One of the conditions of sustainable development of timber production is high-skilled personnel. A well-known fact is that in developing countries forestry work is traditionally carried out by predominantly low-skilled employees who work in hazardous conditions with a risk to health and life. This entails excessive waste in the course of production activities, significant damage to the environment, a high rate of accidents, and, consequently, high staff turnover. [12]. However, the timber enterprises in Western countries may experience the same problems, since they often encounter the urgency of filling job vacancies by hiring yesterday's immigrants who agree to work for a small remuneration due to their low qualifications. One of the ways to tackle the issue is to enhance the efficiency of communication between enterprises-employers and personnel, identify employees' problems and concerns and search for ways to solve them. In other words, building a powerful corporate HR brand.

2 Methods

The researchers carried out the benchmarking analysis of the existing mechanisms for attracting and retaining personnel in the forest industry. It revealed the best-implemented HR branding practices of the largest enterprises in the industry. The goal of the analysis was to develop an HR brand for the forestry sector. The following companies were included in the analysis:

• Stora Enso, the largest Finnish-Swedish forestry company;
• International Paper, an international paper manufacturer;
• UPM-Kymmene Oyj, a Finnish forest industry company,
• Sveza Group of Companies, a Russian plywood manufacturer;
• West Fraser Timber Co. Ltd, a Canadian forestry company.

The data from Perm Pulp and Paper Plant provided insight into Perm Krai's experience in HR branding. The region was selected due to objective reasons. Perm Krai is a leading forest industry region in the Volga Federal District. Rosstat, Federal State Statistic Service, counted 513 forestry enterprises in the region at the end of 2018. According to the Service, the average number of workers employed in the timber industry in Perm Krai accounted for 19.4 thousand people in January-February of 2020.

Since Perm Pulp and Paper Plant is concerned about attracting young professionals to work at the company, it was selected for the research. The company is a member of ongoing cooperation with Perm educational institutions within the vocational program framework "School - secondary vocational school - universities - enterprise".

The researchers conducted two opinion polls. The first survey was aimed at assessing the degree of Perm Pulp and Paper Plant's recognition among the residents of the Krai. The total sample included 207 respondents: 59.4% male and 40.6% female. The age-related composition of the respondents broke down to 6.8% aged 16 to 18 years old, 8.7% from 18 to 20; 64.3% from 20 to 22; and 20.3% over 22. The survey was conducted on the Google Forms platform for 20 days (4-24) in May 2020.

The second survey was conducted to analyze the effectiveness of the Bumfest HR event implemented by the company. The total sample consisted of 87 people (direct participants): 65.5% male and 34.5% female. The age-related composition of the respondents included 85.1% aged 20 to 25; 9.2% under the age of 20 and 5.7% aged 25 to 35. The survey was conducted on the Google Forms platform for 7 days (16-22) in December 2019.

Perm Pulp and Paper Plan held the Boomfest Forum as part of its HR brand strategy to create the image of the engineering profession that young people would appreciate. The forum engaged college and university students in a game where they had to develop a unique product made of cardboard. Besides, the forum included four educational blocks: a discussion "10 myths about the paper", two masterclasses "Creative Thinking" and "Over-professional skills", and a lecture "Industry 4.0".

3 Results

Benchmarking analysis, the study of the benchmark achievements of competing companies is an effective tool for using third-party experience to optimize a company's business processes [13]. In this research, the benchmarking analysis has enabled us to compare innovative practices and areas of HR branding in the forestry industry:

1) Openness of the production process. Stora Enso, the Finnish-Swedish forestry, implemented the practice to attract young professionals. Together with the employment agency Maximum, the company announced the "Global Trainee Program". The program offered potential candidates the opportunity to learn the specifics of the production process and internal communication by reading interns' blogs and private chats. Personal text messages and live video footage from the production sites reduced the level of anxiety among potential employees and formed a complete picture of ongoing events in the production process.

2) Labor protection and safety. Stora Enso presented the Safety Week program. As part of the program, Enocell and Uimaharju Sawmill Branches in Finland organized classes at local schools. The Langerbrugge Mill Belgium subsidiary focused on road safety by providing employees with VR technology to test their driving skills with a drunk simulator. Wood Supply Sweden launched new safety signs and developed communication materials about the most serious risks in the workplace. The Murrow sawmills in Poland and Nebolchi in Russia organized a drawing competition for preschool children. Nebolchi also focused on environmental safety.

3) Inclusiveness. The International Paper, an international paper producer, implemented programs that ensured equal accessibility conditions for employees with special needs. The company established a Global Diversity & Inclusion Council focused on a multicultural strategy in the industrial sector.

4) Two-way communication between a manager and a subordinate. The Finnish company UPM developed additional communication means aimed at improving the interaction between the company's management and subordinates. For example, a production manager at the Oklahoma sawmill explained what his job included, what he did on a typical working day, and what he considered the best part of his job at UPM. A team leader at Cross SBU Supply, Magazine in Germany,

highlighted diversity, career opportunities, and multiculturalism as the best aspects of working for UPM. He believed that the ever-changing market conditions in the forestry sector make the job challenging. Barrier-free communication minimizes manager-subordinate conflicts and strengthens corporate culture in terms of monitoring employee expectations.

5) Adaptation of new employees. To simplify the adaptation period for new employees, Sveza, the largest plywood producer in Russia, introduced Dasha chatbot, a virtual assistant on the Teams platform, and VKontakte social network. The online assistant accompanies new employees throughout the entire adaptation period (3 months), provides the necessary information about the company, notifies them about scheduled meetings, and answers all questions. Today, it has become a global trend to use social media to facilitate the adaptation of "newcomers" to the company [14].

6) Taking care of retired employees. The Canadian forest company West Fraser Timber Co. Ltd. provides all employees with a stock purchase plan and helps save money for retirement by making contributions to a defined benefit plan. In 2020, West Fraser Timber Co. Ltd. was included in the list of Canada's Top 100 Employers.

The analysis of HR branding practices in Perm Krai forest industry indicated the following features. The conducted sociological survey indicated that Perm Pulp and Paper Plant was not included in the top list of Perm residents' preferences about a potential place of work. The three most demanded job places presented oil and gas companies Gazprom Mezhregiongaz Perm (39%) and LUKOIL-PERM (35%), and PROMOBOT (26%), a robotics company. Though Perm Pulp and Paper Plant recognition amounted to 47%, only 7% of respondents considered the enterprise as a potential place of work.

Residents of Perm ranged the order of job selection criteria as follows:

- salary (20%);
- company' reputation in the labor market (20%);
- prestige of working for the company (16%);
- social security (16%);
- corporate training (16%).

The "Boomfest", a role-based forum, can be considered a relatively successful practice of Perm Pulp and Paper Plant HR branding: 59% of respondents gave the organizers the maximum rate for the event, 77.3% of participants said that they gained new experience and 52.6% of respondents reflected a positive change in their attitude towards the company after the event. However, the aim of achieving the company's maximum recognition and prestige as an employer was not accomplished. Even though, the image of an employer is one of the most significant factors for students as prospective job seekers [15].

4 Conclusion

The identified problem of respondents' unwillingness to apply for a job in the regional forest industry due to its low prestige value requires complex and long-term solution mechanisms. Some of them could be borrowed from a bank of ideas implemented by the largest Russian and foreign forestry companies. Some HR branding know-hows require significant investment, while others are comparatively less expensive but no less efficient. Thus, there is the need to develop an innovative HR brand strategy for the regional forestry that could, on the one hand, meet the demands of potential job seekers, primarily from local communities, and on the other hand, satisfy the demands of business that is interested in cost, time and labor savings to implement innovations.

References

1. T. Ambler, S. Barrow, *J. Brand. Manag.* **4** (1996)
2. T. Arkova, *Bulletin of the Astrakhan State Technical University. Series: Economics.* **1** (2011)
3. 3.D. Martasov, L. Danchenok, *Business Imperatives* (Kemerovo Institute of Plekhanov Russian University of Economics, Kemerovo, 2017)
4. K. Fernholz *Trends in green jobs in the forest sector in the UNECE region* (United Nations Publications, Geneva, 2019)
5. P. Katila, C. Pierce Colfer, W. De Jong, G. Galloway, P. Pacheco & G. Winkel *Sustainable Development Goals: Their Impacts on Forests and People* (Cambridge University Press, Cambridge, 2019)
6. C. Ackerknecht, *Unasylva.* **61** (2010)
7. A. Veselova, E. Kerzina, *Actual problems of the humanities and natural sciences.* **9** (2014)
8. A. Sumchenko, I. Zinovieva, *Modern high technologies.* **10** (2013)
9. K. Backhaus, S. Tikoo, *Career Dev. Int.* **9** (2004)
10. G. Sukhanov, A. Kostylev, *Forestry Journal.* **2** (2016)
11. P. Blombäg, P. Poschen, M. Lövgren *Employment Trends and Prospects in the European Forest Sector* (United Nations Publications, Geneva, 2003)
12. B. Strehlke, *Commonw. For. Rev. E* 75, 3 (1996)
13. V. Revyakina, O. Starodubtseva, *Infrastructure sectors of the economy: problems and development prospects.* **9** (2015)
14. Z. Yuan, N. Li, Y. Mai, Z. Ye, J (J) Yu, *Hum Resour Manage.* **59** (2020)

15. J.H. Wayne, W.J. Casper, *Hum. Resour. Manage.* **51** (2012)

https://doi.org/10.1051/shsconf/202111600059

Innovative potential of enterprise from a differentiated approach standpoint based on the concept of innovations' diffusion

Vladimir Lazarev*, *Anna* Churashkina, and *Tatiana* Shulenkova

Ulyanovsk State Technical University, Ulyanovsk, Russia

Abstract: The article contains the key results of the study of the essence of the innovative potential of an enterprise with solving the problems of strategic management of modern enterprises operating in the context of the innovation of the economy. As a vector for improving the theoretical and methodological foundations for solving problems in the innovation sphere, a differentiated approach to understanding the essence of the innovative potential of an enterprise is proposed. The possibility and expediency of differentiation based on types (strategies) of innovative behavior, identified based on the concept of diffusion of innovations, have been substantiated. The article reflects the main provisions of the author's idea of the essence of the innovative potential of an enterprise within the framework of a given approach. The significance of the research results for solving theoretical, methodological, and applied problems is concretized.

1 Introduction

In modern economic realities, the external environment is characterized by a high rate of change and a high level of uncertainty. This circumstance leads to an increase in the role of innovative transformations in ensuring socio-economic growth. Today there are conditions for the emergence of large-scale technical changes, forming a new technological base of economic systems. At the same time, researchers and analysts note that the importance of innovations in modern socio-economic development will continue to increase, and many traditional growth factors will demonstrate a downward trend in importance. These processes are viewed as a "long-term systemic challenge" facing the Russian economy [1].

The response is the transition to an innovative economic growth model as a priority of state policy, which is directly reflected in the focus of national projects, programs, and initiatives. In particular, 2021 has been declared the year of science and technology in Russia. This macroeconomic policy vector is implemented by creating an environment that encourages market participants to innovative activity. Therefore, this vector of macroeconomic policy predetermines the growth in the value of innovations to ensure the long-term competitiveness of Russian enterprises: over time, an increasing number of market entities will be interested in the implementation of innovative transformations, which will face increased competition on an innovative basis.

It should be emphasized that the dynamics of strengthening innovation activity are significant for the formation and support of the competitiveness of Russian enterprises. This process is consistent with global trends: "knowledge and innovation are becoming the most important factors in the long-term advantages of firms in the new economy" [2]. In addition, it is appropriate to note the importance of innovations not only for ensuring the leadership positions of enterprises but also for their "survival": an increase in the concentration of innovative activity initiates adaptation processes, which are also based on innovations. In other words, the presence of an innovative aspect in the activities of enterprises is gradually becoming imperative.

The shift in priorities of the competitive advantages formation of enterprises is associated with an adjustment of the focus of strategic management. One of the key tasks of strategic management is to ensure competitiveness in the long term. Thus, an extensive research area arises, associated with searching for ways to solve traditional problems based on innovation. A vast research area arises, associated with searching for ways to solve traditional problems based on innovation. This area is heterogeneous and represents a collection of overlapping subsets of research, one of which focuses on the formation, assessment, and implementation of innovative potential as the basis for the innovative development of enterprises. To solve the indicated applied problems, understanding the innovative potential of an enterprise as an economical category plays a critical role. The paper presents the results of studying the essence of the innovative potential of an enterprise from the perspective of strategic management.

* Corresponding author: v.lazarev@ulstu.ru

2 Approaches to considering the essence of the innovative potential of enterprises in economic science

Analysis of specialized literature allows concluding that there is no universal approach to understanding the innovative potential of an enterprise. Traditionally, there are at least two coexisting approaches – the result-based approach and the content-based approach. The resulting approach focuses on reflecting the implementation of the innovative potential of enterprises by fixating the existing results of innovation. The meaningful approach is focused on disclosing the essence of innovation potential through the composition of the resources necessary for the implementation of innovative activities or through a set of components of a different nature (competencies, resources, mechanisms) that determine the ability to achieve innovative goals. Note that the approach that is focused on disclosing the essence of innovation potential through the composition of resources is a special case of a meaningful approach, which, due to its prevalence in classifications, is often singled out as an independent one [3, 4]. It is also appropriate to note the presence of attempts to link the effective and meaningful approaches, introducing the resulting component into the structure of the innovative potential of the enterprise. Their long-term coexistence indirectly confirms the viability of these approaches in science and practice.

This viability is explained by the multifaceted nature of the innovative potential as an economic category that predetermines many difficulties in the unification process. At the same time, there is no consensus of opinion even within the framework of a single approach. The lists of criteria for assessment vary, which reflects the variety of analysis methods. Also, the lists of components that require consideration vary. This fact signals already about the discrepancies in the essential understanding of the category under consideration. However, there is a general tendency towards universalization, despite the discrepancies in understanding the innovative potential essence of an enterprise, including its content. This trend is embodied in the fact that a significant number of attempts are being made to characterize a single structure of innovative potential, which can be used as the basis for selecting criteria for the subsequent quantitative and qualitative characteristics of its state. The potentials included in the composition act as structural components or a more complex arrangement can be used.

For example, in [5], it is proposed to consider the intellectual, marketing, organizational and managerial, financial, production, and technical and research potential. Moreover, in work [4], three blocks are distinguished in the structure of the innovative potential of enterprises – the resultant, resource, and management blocks. Each block is represented by several components, including organizational, financial, personnel, material, technical, and other components. After the presented concretization of the essence of innovative potential by referring to its structure, the transition to the system of indicators characterizing each component is carried out.

An integral indicator (index) is also calculated in some cases, reflecting the general state of the enterprise's innovative capabilities. The assessment results are the basis for making informed management decisions in terms of the formation and implementation of innovative potential, management of enterprise development based on innovations, and a wide range of related issues. Because of this fact, a significant part of the methods implies the interpretation of the results in the "high" / "low" categories (with varying degrees of detail).

The considered sequence schematically reflects the typical logic of actions when researching to improve the theoretical and methodological foundations for assessing the innovative potential of enterprises and/or solving issues that are naturally associated with this process. The above scheme allows tracing the close relationship between the tasks "defining the essence of the innovative potential of enterprises" – "assessing the innovative potential of enterprises" – "making management decisions in the designated area." In other words, it can be stated that the level of efficiency of solving a specific problem significantly affects the quality of solving problems that follow it in the chain. Therefore, it is advisable to look for the application of efforts for the comprehensive development of theoretical and methodological foundations both in the technology for solving each problem and in the fullest consideration of existing connections.

So, for example, in the development of assessment methods, researchers are guided by the construction of a system of criteria that will most fully characterize the readiness of an enterprise to solve problems based on innovation. At the same time, researchers pay attention to the issues of data availability for analysis, the simplicity of the calculations, the objectivity of the estimates obtained, the visibility of the presentation of the results, and many other criteria that seem to be the most significant to the developers. Assessment techniques are associated with tasks for the solution of which their results can be used. These methodologies are reflected in requirements such as the ability to conduct a comparative assessment and/or reflect separately the available potential and the level of its use. Practical examples are the introduction of integral indicators for individual components along with a general integral indicator of the innovative potential of enterprises. This technique allows obtaining detailed results, which is shown in [4].

The connection between applied problems and the concretization of the innovative potential of enterprises through the structure seems to be least used at the moment. The issue of forming a general idea of the structure of innovative potential is traditionally solved based on identifying groups of factors that affect the ability of an enterprise to carry out innovative activities and the level of success of this activity. At the same time, the main ways to improve this stage of the study mainly lie in taking into account advanced trends (such as globalization and digitalization). Trends are transforming the operating conditions of enterprises and the success factors in the innovation sphere and taking into account the specifics of industries. However, the result of this phase is a unified (unified) structure of innovation potential, according to which it is proposed to build a further assessment. In other words, the development of assessment tools demonstrates sufficient flexibility regarding the goals and objectives, the solution of which will be based on the results obtained. However, the solution to the issue of concretizing the structure of the innovative potential of enterprises is actually happening separately from the target settings. Namely, this structure affects methodological developments, while the number of tasks for which an assessment of the innovative potential of enterprises is required in the context of the economy's innovation is growing. These tasks are heterogeneous. There is a tendency to expand the boundaries of what is commonly referred to as innovation and innovation. This circumstance makes it possible to call into question the possibility of developing a unified idea of the structure of the innovative potential of enterprises, which makes it possible to solve all groups of problems most effectively.

Considering all of the above, it seems appropriate to consider the possibility and analyze the feasibility of implementing a differentiated approach to considering the essence of the innovative potential of an enterprise as a vector for improving the theoretical and methodological foundations for solving innovative problems. Due to the strategic orientation of the study, it is of interest to consider the variety of strategic alternatives available to enterprises. Therefore, the type (strategy) of innovative behavior is chosen as the basis for differentiation. There is no universally accepted classification of innovative strategies that unambiguously reflects the idea of alternatives in the area under study. Therefore, the specification of the list of types of innovative behavior was carried out based on the concept of diffusion of innovations.

3 The process of diffusion of innovations as the basis for identifying types and strategies of innovative behavior

The diffusion of innovation has many patterns, the totality of which allows the division of enterprises into several groups, depending on the form of participation in the designated process.

J. Schumpeter noted that the innovation, which was developed and applied by one enterprise, over time, "mastered" by others. Moreover, inclusion in the development process occurs at different speeds. The number of followers grows over time. This fact is explained by the gradual simplification of innovation implementation activities. The most challenging task is faced by the first in the chain; they are forced to overcome the most significant difficulties. At the same time, their success facilitates the development of innovation for the following processes: firstly, by the very demonstration of the possibility of achieving success, and secondly, by developing algorithms for solving part of the problems. In addition, some tasks lose relevance as the innovation spreads, simplifying the activities of followers. A similar trend applies to later groups of followers, which explains the gradual increase in their number [6]. The presented pattern reflects a key feature of the diffusion process, covering the period of time from the emergence of innovation to the moment when it ceases to be such, becoming a routine.

A sufficient basis for dividing enterprises into groups is the difference in the time of the entry of enterprises into the diffusion process and the connection between the moment of entry and the nature of the solved tasks. In other words, we can state that it is possible to classify types of innovative behavior based on the role of the enterprise in the diffusion process.

Let us turn to the expediency of such a division. It is pertinent to emphasize that the dynamics of a gradual decrease in the complexity of tasks solved by followers seems to be significant for the essential characteristics of each group of enterprises. These dynamics can serve as a basis for determining the factors and conditions that are necessary and sufficient for the implementation of a specific role.

There is interest in the relationship between diffusion and the benefits a company receives by participating in it. The prospect of innovative strategies lies in the fact that the initiation of innovation if accepted by the market, leads to the formation of competitive advantages. Competitive advantage tends to wane as innovation spreads. They completely lose their strength by the time innovation becomes routine. In turn, the need for market acceptance of innovation is a prerequisite for obtaining a competitive advantage on its basis.

The market selects viable innovations in respect of which the diffusion process is carried out in the future. This circumstance causes high risks for the "pioneers," and their decrease as innovation becomes routine. Both trends are capable of making a significant contribution to forming a comprehensive understanding of the types of innovative behavior identified based on the concept of diffusion. In addition, the use of the concept of diffusion of innovations as a theoretical basis for identifying types of innovative behavior has several more significant advantages:

- the concept of diffusion of innovations is sufficiently developed (there is a significant amount of research that reveals various aspects of this phenomenon); this fact determines the possibility of a comprehensive approach to identifying the characteristics of enterprises due to their role in diffusion, that is, we are talking about the possibility of ensuring a high level of objectivity of the results obtained based on the considered theoretical base;
- the mechanism of the diffusion process is characterized by a sufficiently high degree of uniformity for different types of innovations, which provides the breadth of coverage necessary for solving theoretical and applied problems, i.e., forms a relatively broad area of application of the results obtained.

The combination of the above characteristics makes it possible to state that the classification of types of innovative behavior based on the concept of diffusion of innovations is justified since it forms a theoretical basis that meets the goals and objectives of the research.

In the specialized educational and scientific literature [7, 8], it is customary to distinguish four groups of enterprises concerning their role in diffusion: innovators, early recipients, early majority, and lagging behind. The innovators are enterprises that act as a source of progress; their key area of competence is the generation of scientific and technical knowledge and the creation of innovations. Early adopters, also referred to as "pioneers," are focused on maximizing the speed with which an innovation can be adopted and brought to market. The enterprises of the early majority carry out the dissemination of the already mastered innovation based on the experience of the pioneer enterprises. Lagging enterprises are introducing "innovation" at a time when it has actually passed into the category of routine. It is important to understand that the name "lagging behind" characterizes the lag to mastering innovation, not the efficiency of the enterprise as a whole. The nature of introducing innovations into production provides the lagging enterprises with the possibility of functioning but not obtaining significant competitive advantages.

The theory of innovations defines all groups of subjects, except the first, as imitators. The reason is that the activity of enterprises does not ensure the emergence of innovation but only its diffusion. However, the most effective is a different type of innovative behavior from a strategic standpoint. Innovators and early adopters are focused on gaining a competitive advantage through the introduction and improvement of innovation. The early majority and lagging enterprises are characterized by an orientation towards operating in a new area, based on their existing competitive advantages of production, marketing, or other nature.

Thus, based on the concept of diffusion of innovation, a list of strategic alternatives has been formed, namely four types of innovative behavior that enterprises can implement.

4 Features of the innovative potential of enterprises, considering their role in the diffusion of innovations

The above classification of types of innovative behavior allows introducing differentiation in terms of the essential understanding of the innovative potential based on the requirements for it, set by the role of the enterprise in the diffusion process. In this context, the process of theoretical comprehension of the essence of the innovative potential of an enterprise takes the following form:

- identification of the specifics of the tasks solved by each of the four considered groups of enterprises in the diffusion process;

- determination of the characteristics of the enterprise, which determine the possibility of solving these problems and, as a consequence, the possibility of achieving an innovative goal (implementing a particular type of innovative behavior);

- establishing relationships and determining the degree of importance of the components of the innovative potential (identified at the previous stage) for the implementation of the chosen innovative path.

As a result of the implementation of a differentiated approach, the specificity of each type of innovative behavior is reflected in the composition and structure of the innovative potential of enterprises, determining their features.

At the first stage, an idea is formed about the structure of each enterprise's innovative potential. Let us characterize the basic goals and objectives of their activities and form the components of the first level of detail, in general terms, reflecting the company's readiness to implement the enlarged groups of tasks necessary to achieve the goal.

Objectively speaking, innovators are not focused on participation in the diffusion of innovation. Still, it results from their activities, which allows characterizing their role in diffusion as an initiator. This category of enterprises solves unique problems. The uniqueness of the tasks determines the impossibility of building activities based on the reproduction of existing samples due to their absence. Therefore, the first component in the enterprises' innovation potential should reflect the degree of readiness of the enterprise to generate ideas.

Accordingly, the second component in the structure of the innovative potential should cover the characteristics of the enterprise, which determine its readiness to implement the generated ideas. Pioneers, innovators need to solve all groups of problems to develop and implement innovation, having carried out a complete innovation cycle.

Key characteristics of early adopters distinguish them from other types of innovative behavior. Early recipients are focused on the diffusion, development, and improvement of innovation by implementing their innovation activities. They are also focused on gaining competitive advantages through the fastest possible inclusion in the diffusion process. Based on this, the most significant components of the innovative potential of this group of enterprises are:

- a block of characteristics that give an idea of the level of susceptibility to innovations. This group of enterprises must ensure both a high speed of obtaining information and the maximum coverage of the areas in which monitoring is carried out since unsatisfactory indicators for any of the indicated parameters actually exclude the possibility of playing this innovative role. Lag almost always results in a loss of competitive advantage.

- a block of characteristics reflecting the potential for improving innovation. For the diffusion phase in which early adopters operate, no consumer preference for innovation has yet been identified. This circumstance determines the high level of importance of improving and adapting innovation to consumers' needs to ensure competitive advantages.

The block of characteristics that reveals the readiness to spread the innovation seems to be less significant in the context of the enterprise profile.

In contrast, early majority enterprises have a primary goal of spreading innovation, as the priority is to capture a significant share of a new market. The early majority enterprises use their competitive advantages, namely, access to resources, brand presence, developed distribution network, and other resources. This circumstance determines the weight of this component in the structure of innovation potential. At the same time, the very possibility of being included in the diffusion process is coupled with the ability to recognize innovations at the stage when diffusion will bring the maximum benefit.

Therefore, the enterprise must have a certain level of susceptibility. However, the requirements for the speed of reaction and the range of areas in which monitoring is necessary are significantly lower. The time factor as a whole is not so critical for this category of enterprises since, up to certain limits, the time lag can be partially compensated for by the superiority in the possibilities of solving the problems of the dissemination of innovation.

However, at the same time, an unsatisfactory reaction rate (below the threshold level for this category) will deprive the company of the opportunity to engage in diffusion in time, "throwing it" into the category of lagging behind. The level of involvement of the early majority enterprises in the tasks of improving innovation is limited to the implementation of improvements as needed. This process predetermines both the presence in the innovation potential of a component reflecting the readiness to implement this type of activity and a significantly lower level of its significance and requirements. Thus, it can be noted that the constituent elements of the innovative potential of the enterprises of the early majority coincide with the elements highlighted in the composition of the innovative potential of early recipients. However, the ratio of their significance and the requirements for them are fundamentally different.

The activity of the "lagging behind" is focused on covering mass demand and is characterized by low innovation activity. Therefore, the requirements for the level of innovation potential are minimal. In fact, it is enough for this group of enterprises to have the ability to overcome entry barriers and the possibilities of internal adaptation of activities for the implementation of "innovation."

Based on the above, we can conclude that the innovative potentials of innovators and imitators of different orders are different from each other. Each role requires a unique set of characteristics that an enterprise must possess, which is reflected in the unique structure of innovation potential.

Figure 1 schematically shows the structure of the innovation potential of enterprises for each type of innovative behavior in the first approximation and with the minimum possible detailing to obtain meaningful results. In the structure of the innovative potential, enlarged blocks of its characteristics (components) are highlighted, directly correlated with the key tasks, the effective solution of which is necessary for the successful implementation of each type of innovative behavior. The size of the segments reflects the degree of importance of each enlarged block of components for the possibility of achieving the goals and objectives set by the enterprise.

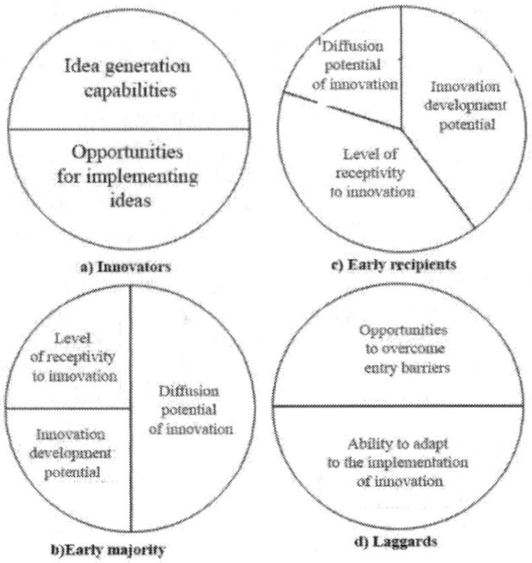

Figure 1. Visual representation of the features of the innovative potential of enterprises, depending on the implemented (planned for implementation) innovative role, compiled by the authors.

Despite the low level of detail, the model reflects a fundamental difference in the structure of capabilities that enterprises need to fulfill each of the roles, which predetermines the importance of further differentiation. This vision causes a shift in the focus of attention of researchers and decision-makers – from the search for a universal set of characteristics, the improvement of which lays the vector for increasing the innovative potential of the enterprise, to the analysis of combinations that form the growth potential within the framework of individual strategic roles. At this stage, it is also appropriate to focus on the fact that the solution of all designated groups of tasks is mandatory. This process leads to the requirement for a balanced structure as a determining criterion for the ability to play a particularly strategic role. In other words, if one or several components do not reach the minimum required level of development, then the potential as a whole cannot be considered satisfactory. Potential deficiencies are recognized even if other elements show superior development. This circumstance must be taken into account during the assessment, regardless of the methods used. The next level of detailing of the proposed model can be achieved by specifying the composition of each component while maintaining the logic of matching goals (subgoals) and enterprise characteristics that affect their attainability.

Thus, the component "the ability to generate ideas" can be concretized through the characteristics of the performance indicators of the R&D service, the assessment of the personnel composition of the enterprise and the level of openness inherent in the organizational culture, the analysis of the process of internal circulation of ideas and initiatives. Also, its composition may reflect the presence of its developments, patents (the resulting component). This fact demonstrates the flexibility of the model in terms of implementing basic approaches to understanding the essence of innovation potential. The relative simplicity of accounting for advanced concepts is also important to ensure the adequacy of the model to current conditions. Thus, what was listed earlier characterizes the emergence of ideas within the framework of the "closed innovation" paradigm, according to which this process is exclusively internal. It is possible to consider the "open innovation" paradigm, which allows the attraction of external sources of ideas – through the introduction of a component reflecting the level of interaction with external innovation structures.

At the same time, model varieties with different levels of detail are of practical importance. In particular, even with insignificant detailing of the model in the first approximation, it can be used for making managerial decisions in the role of a vector that sets the direction of analysis. This granularity is comparable to the level of detail of the Idea Generation component presented above. We use standard strategic management tools such as SNW analysis, focusing on decision-makers components and requirements. On this basis, it is possible to form a contour idea of the strengths and weaknesses of the enterprise and the gravitation of the innovative potential to the structure that best corresponds to a particular innovative role. This approach is entirely appropriate for express analysis for the initial diagnosis of the enterprise's innovative potential.

At the same time, the potential for further detailing the model is significant: an increase in the number of levels of analysis increases concreteness, providing the possibility of a gradual transition to a system of qualitative and quantitative indicators.

5 Conclusions

Summing up, let us concretize the advantages of a differentiated approach to considering the essence of innovation potential based on the concept of innovation diffusion:

- for theoretical developments: the proposed approach, due to the introduction of an appropriate target setting, allows at the initial stage of comprehension to take into account the multiplicity of strategic alternatives to innovative behavior, contributing to the convergence of theoretical concepts with the reality they represent, which has a positive effect on the applicability of the results obtained.

- for methodological developments: the proposed approach makes it possible to trace the motives for including specific components in the innovative potential by reflecting their influence on the possibility of implementing innovative strategies. In this regard, the results of estimates, the calculation of which will be based on the proposed theoretical model, will receive an additional layer of qualitative information essential for solving many applied problems.

- to improve the management process: the proposed approach contributes to the transformation of ideas about the process of formation and development of innovative potential, clearly demonstrating the limitations of the improvement strategy based solely on the characteristics of the level of development of individual indicators in isolation from the assessment of the structure of potential as a whole. So, going into innovators through improvement within the framework of the imitator's structure is not possible, and the development of individual characteristics of an innovator in the absence of a desire to fulfill this role and/or the absence of comprehensive development of the required structure seems unjustified.

Separately, it is appropriate to designate the traceable connection between the innovation diffusion process and the life cycle of industries (sub-sectors or areas of activity) [9], which suggests that there is a "request" for the implementation of a particular type of innovative behavior at a given moment in time. Because of this, other advantages can be indicated. A differentiated approach to studying the essence of the innovation potential of enterprises is possible, taking into account the concept of diffusion of innovation.

A differentiated approach deliberately ensures the comparability of the results of assessing external and internal factors when making management decisions regarding the implementation of innovative potential. A differentiated approach based on other features seems to be feasible. However, the composition of features and the feasibility of differentiation require additional research in each specific case.

References

1. *The concept of long-term socio-economic development of the Russian Federation for the period up to 2020.* Retrieved from: http://government.ru/info/6217/
2. V.S. Katkalo *Evolution of the theory of strategic management* (2011).
3. I.V. Antonenko *Problems of the modern economy*, **2,** 33–37 (2010)
4. S.V. Terebova *Scientific works: Institute for Economic Forecasting RAS*, **15,** 336–354 (2017)
5. O. I. Imaykina *Izvestiya of higher educational institutions. Volga region. Social Sciences,* **3**, 211–223 (2014)
6. J.A Schumpeter *Theory of Economic Development,* (M.: Eksmo, 2008).
7. S. D. Ilyenkova *Innovation Management* (M: UNITY-DANA, 2012).
8. V.V. Manuilenko, A.A. Mishchenko *Finance and Credit,* **41**, 21-31 (2012)
9. V.N. Lazarev, T.A. Shulenkova *Organizational, economic and technological problems of modernizing the Russian economy: collection of articles of the VII International Scientific and Practical Conference*, pp. 3-12 (2017).

SHS Web of Conferences **116**, 00060 (2021) https://doi.org/10.1051/shsconf/202111600060
ICSR 2021

Digital educational ecosystem as a tool for the intellectual capital development

Alzhanat Suleimankadieva[1,2], *Maxim* Petrov[1,*], and *Andrey* Kuznetsov[3]

[1]Saint-Petersburg State University of Economics (SPbSUE), 21, Sadovaya st., Saint-Petersburg, 191023, Russia
[2]Saint-Petersburg Electrotechnical University («LETI») (SPbETU), 5, Professora Popova st., Saint-Petersburg, 197022, Russia
[3]Udmurt Branch of the Institute of Economics, Ural Branch of the Russian Academy of Sciences, 4, Lomonosova st., Izhevsk, Udmurt Republic, 426000, Russia

Abstract. The analysis of the definitive aspect of the digital ecosystem concept is carried out, a generalized definition of this concept is given, taking into account the specific properties of its existence, which makes it possible to go beyond the IT-contextual understanding. The article provides a comparative analysis of existing approaches to assessing the digital ecosystem of education and forms the author's vision, which expands the content of the assessment mechanism, pays special attention to the impact of the digital education ecosystem on the intellectual capital development. The authors' attention is focused on studying the characteristics of the digital education ecosystem and analyzing the model for the intellectual capital development in the context of the transition of the world and Russian economies to a new technological level. Based on the analysis of empirical data, it is concluded that modern society requires a new structure (construction) and a completely different content of education. Considering the development trends of the world educational space and the characteristics of the modern intellectual economy, the necessity to develop a new model of education as a basic basis for the intellectual capital development is substantiated.

1 Introduction

The concept of "ecosystem of education" appeared at the beginning of the XXI century and immediately became the subject of public and scientific discussions, despite the fact that in biology the term "ecological system" has existed in scientific discourse for more than eighty years [1]. In biological science, an ecosystem or ecological system (biogeocenosis) is understood as a system consisting of a community of living organisms, their habitat, a system of connections that exchange substances and energy between them [2]. In other words, an ecosystem, as a more general concept, can be represented as a certain set of interrelated and interdependent elements that have general system properties: integrity, emergence, interdependence (connectedness), stability, etc.

Over a period of time spanning more than eight decades, this concept has received sustainable development, both from a theoretical and a practical points of view, confidently and actively being introduced into other branches of scientific knowledge, the field of social and business relations. In the last three decades alone, the concept of an ecosystem has become widespread in such spheres of human activity as: business, economics, finance, banking, manufacturing, education, etc. In this regard, new concepts and terms have appeared, such as: business ecosystem, financial ecosystem; banking ecosystem; an ecosystem for the development of innovations in education; innovation and production ecosystem; digital ecosystem; digital intelligent ecosystem; the ecosystem of the digital economy, the ecosystem of the digital society and the information (digital) ecosystem of the advocacy [3-10], etc. We believe that further analysis of the definitive aspect will lead to a wide variety of types of sectoral ecosystems.

At present, it is difficult to find a scientist-economist who would not admit that the main factor of economic growth and development of modern society is intellectual capital. And only those countries that actively reproduce new knowledge, whose economies are characterized by growing intellectual intensity, will be able to maintain leadership positions in the world economy. Today there is no such area of activity that would not undergo significant changes caused by the active development of technologies of the Fourth Industrial Revolution [11]. These changes are accelerating exponentially. According to K. Schwab and N. Davis [12], the exponential development of society takes place in the second phase of the industrial revolution. The technological changes associated with it form a new system of values that is constantly changing and transforms the existing culture and way of life of

* Corresponding author: petrov_maxim@rambler.ru

society, the education system through which new knowledge, skills and abilities are transmitted. All this predetermined the relevance of replacing the traditional model of human capital development, based on the transfer of a standard set of highly specialized skills necessary for an employee to perform a separate function and / or a limited group of functions, with a new model focused on the development of flexible skills, and containing the idea of inter- and transdisciplinary learning approaches. Thus, the purpose of this article is to analyze the definitive aspect of the concepts of "digital ecosystem", "digital educational ecosystem" and the formation of a new view (approach) to assess the impact of the digital educational ecosystem on the development of intellectual capital.

2 Materials and Methods

The article uses scientific and methodological approaches: 1) theoretical and methodological: content analysis of sources and materials of the Internet network on the problem (study and generalization) of the research; comparative analysis; system analysis and systematization of methods and approaches to assessing the digital ecosystem, including assessing the digital ecosystem of education; structural and functional analysis of the components of the digital education ecosystem that affect intellectual capital; 2) empirical: questionnaire survey (expert assessment); method of statistical processing of data obtained in the course of expert assessment.

The development of the problem under study was carried out on the basis of the study of regulatory documents, monographic studies of Russian and foreign authors, materials of the Internet network, periodicals devoted to the ecosystem approach to the formation of the educational environment as an element (subsystem) of the modern global digital space, which is a «mix» socio-economic and technical (digital environment) system of the public world order. Extrapolation of the research results made it possible to determine the prospects for the development of intellectual capital in the context of technological singularity and sustainable formation of educational ecosystems.

3 Results and Discussion

In accordance with the goal set in this article, we are interested in the digital ecosystem in relation to the educational environment, since it has experienced the greatest transformation processes over the past twenty years. Before proceeding to a descriptive analysis of the concept of a digital educational ecosystem, in our opinion, it is advisable to pay attention to a review of the definitions of the concept of "digital ecosystem". Despite the variety of existing definitions of this concept, it can be concluded that there is no unambiguous definition of this term in the literature. However, a common attribute for the formation of digital ecosystems is the use of digital technologies, and, in particular, access to the Internet. For example, in our country, the proportion of organizations using the Internet in their economic activities is steadily growing and it is shown in Figure 1.

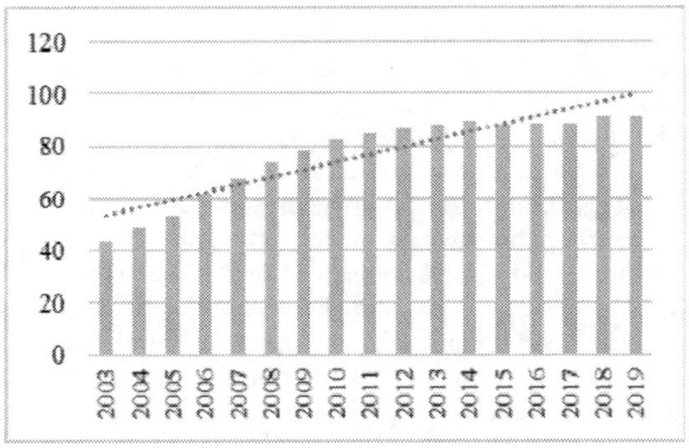

Source: Federal State Statistic Service (Rosstat)

Fig. 1. Share of organizations using the Internet in the Russian Federation, %

For example, for the purposes of modeling the process of implementation and development of information and communication technologies in 2002, European researchers and practitioners F. Nachira, P. Dini and A. Nikolay proposed the theory of a digital business ecosystem [13]. According to the authors, the digital ecosystem is characterized as a distributed, adaptive, open socio-technical system with the properties of self-organization, scalability, and sustainability. This definition notes that the basis for digital ecosystem models is knowledge about natural ecosystems, including in the field of competition and cooperation between various objects (primarily between enterprises and companies).

Benjamin Talin defines the digital ecosystem as a system focused on creating added value for customers (consumers) by optimizing data and workflows from various internal departments, tools, systems, customers, suppliers and partners. The main characteristics of such an ecosystem are: customer focus (for example, Amazon at first was clearly focused on the customer without an emphasis on monetization, but after a while, it focused on creating value); the ability to collect additional information (data) about customers, transactions, processes, which makes it possible for a global company (ecosystem) to make a more refined offer to its customers; automation of processes, which allows to significantly reduce the costs and prices of products and services; globalization (globality), which allows the ecosystem to scale its offerings, go beyond the terrain, region and country; dynamism, implying a quick response to changes in the environment and adaptation to them, the propensity of business intelligence to make quick decisions [14].

On the other hand, V. Artemiev (CEO of the digital ecosystem All.me) notes that the digital ecosystem can be understood not only as an information technology infrastructure that works on the principle of "win-win" everything, as an infrastructure that can harmoniously and profitably fit into the internal environment of the consumer's economy [15]. Thus, it is possible to generalize the content of the concept of "digital ecosystem", without being limited to its essence, defined in relation to the IT-sphere, and give a more comprehensive definition that characterizes it as a socio-technical system, which has, first of all, system-wide properties (integrity, emergence , interconnectedness and sustainability), and functioning in the context of information and communication technologies, digital technology platforms that allow ecosystem participants to carry out mutually beneficial accumulation, exchange of data (including information and knowledge) in a structured form and management of business processes. This ecosystem works on the basis of the implementation of additional principles: a) algorithmization of the relationships of the participants; b) "win-win" (mutual benefit); c) a unified information and communication infrastructure, which is a digital network structure of participants (agents, actors) and system elements; d) minimization of costs (both material and time) associated with the interaction of ecosystem participants; e) scalability, that is, the ability of the ecosystem to work efficiently and smoothly under conditions of increasing load (for example, with the addition of resources, an increase in the number of network participants, an increase in the volume of work); f) synergistic effect obtained as a result of optimization of business processes (including information exchange and control); g) creating additional value for the ecosystem through a competency model based on the concept of continuous learning (self-development) of the system.

With regard to the educational ecosystem (including the digital educational ecosystem), there are also different views on the acceptability of the ecosystem approach in terms of its formation and content. Thus, some authors (A. Tsvetkova, M. Gustafson) believe that the ecosystem approach, in principle, is not acceptable for the purposes of analyzing the education system [16]. Other authors are actively generalizing and developing the concept of the digital ecosystem in relation to the educational environment. In particular, D.O. Koroleva, T.O. Naushirvanov consider it as an ecosystem for the development of educational innovations [6].

A.M. Kondakov singles out the concept of "digital education ecosystem" and associates its emergence with the intensive growth of the digital economy, which substantiated the feasibility of developing non-specific competencies. This author defines the digital educational ecosystem as a socio-cultural environment in which the process of personalized teaching of students (learners) is based on the analysis of their requests and needs in the interaction of all participants in educational relations using variable, adaptive digital content [17].

According to the autonomous non-profit organization NTI University "20.35", by 2025 the main characteristics of the digital educational environment (ecosystem) should be the following: 1) management of the education system based on Big Data; 2) launching flexible, practice-oriented educational formats; 3) creation and development of methods of transfer and perception of knowledge about technologies "first hand", from technological leaders and experts of Russian and world levels; 4) the formation of a networked educational environment (ecosystem), where there is no border between universities, school and business, but at the same time there is a system of knowledge exchange, which is freely used by both those who teach and those who study; 5) the creation and use of digital tools, with the help of which a person will be able to determine his ability to create relevant educational goals and implement a step-by-step scenario for their achievement [18].

Thus, based on the above definitions of the concept of "digital ecosystem" and "digital educational ecosystem", in our opinion, it is possible to substantiate the main reasons for the emergence and intensive development of the digital ecosystem of education, which will make it possible to give a more precise definition of this term, taking into account its impact on growth and development of intellectual capital. Thus, the analysis of the previously reviewed works allows the authors of this study to note that the main reasons for the creation and development of the ecosystem approach to the educational environment are as follows.

First, the accelerated rate of obsolescence of knowledge has led to an increase in the rate of lag of the education system from the requirements of the real sector (primarily the labor market). In the era of Industry 4.0, the rate of knowledge update is two years. A student who enrolled in a bachelor's program today, after four years (that is, at the stage of completion of the educational process), receives knowledge that has twice become obsolete. The traditional higher education system does not keep pace with the accelerated pace of knowledge change. And this situation testifies to a crisis in the education system (including in the higher education system).

Second, in turn, the crisis of the modern system of (higher) education, associated with the fact that graduates of vocational educational institutions (primarily universities) do not have the necessary professional skills and remain unclaimed in the labor market, justified the expediency of searching for such technologies and approaches to the learning process, transfer of necessary and timely knowledge, which could significantly reduce the "circulation" of knowledge, that is, make the life cycle of the triad "Creation of knowledge - Learning (processing, analysis and transfer of knowledge) - Use of knowledge" very short. This situation requires a new (digital) infrastructure of the education system (including higher education).

Third, the rapid development of the ecosystem approach in relation to the education system is, on the one hand, a reaction to the general digitalization of the economy, and on the other, a consequence of the impact of the COVID-19 pandemic, which accelerated the transformation process at all levels of the educational process and, in fact, accelerated the process of technological singularity in the development of the world economic economy.

Fourth, the digital ecosystem of education can realize itself as a network structure developing on a single technological platform, where each of the participants (agents, actors) has access to common "ecosystem" resources, which he does not have enough or does not have at all.

Fifth, the digital educational environment (ecosystem) is becoming more and more innovative and project-based, which can be viewed as a synergistic effect of the network structure (the triad "Science-Education-Real sector (business environment)") on a single digital platform.

Sixth, the technological revolution in the educational process is actively and intensively converting all communications of the educational system into digital ones. An example is the intense growth of the EdTech segment (Fig. 2). Thus, the forecast of the annual growth rates of this segment in Russia in the next 5 years is about 20-25% [19].

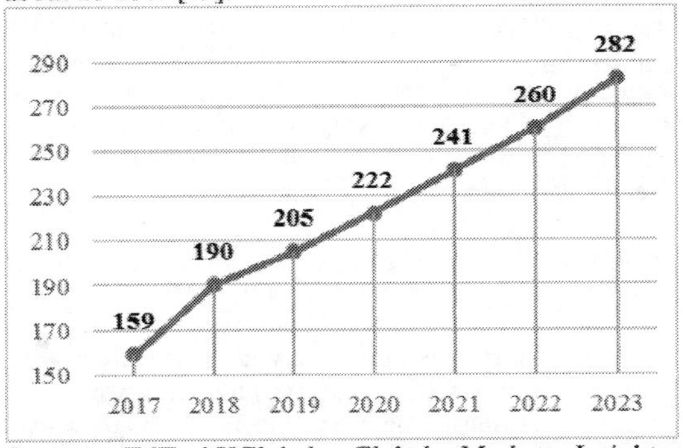

Source: EdTechXGlobal, Global Market Insights, Education International

Fig. 2. Market size of online education, USD billion.

In accordance with the national project of the Russian Federation "Education 2019-2024", the Federal project "Digital educational environment" is focused on the creation and implementation of a digital educational environment in educational institutions, as well as ensuring the implementation of digital transformation of the education system. Within the framework of this project, work is underway to equip organizations with modern equipment and develop digital services and content for educational activities, for the implementation of which a budget of 73.4 billion rubles has been allocated [20].

The main directions of the National Program "Digital Economy" are six federal projects (Table 1) [21].

Thus, a comparative analysis of definitions, characteristics, as well as the main directions of development of the modern education system allows us to note that its main and specific characteristics is the inclusion of digital technologies and platforms in all its areas of activity, that is, new forms and methods of generating, accumulating, storing, analysis, processing, contextualization and control of knowledge, which significantly reduce material and time costs, make knowledge more accessible (cheap) and easily updated. Consequently, the educational ecosystem must be understood not just as a platform on which the process of transmitting knowledge to students takes place, but the processes of generating, analyzing and processing knowledge would be taken outside the educational system. In the modern context, the digital educational ecosystem must be understood as a set of closely and intensively interacting stakeholders (agents, actors), which: a) includes an educational platform, teachers (developers and lecturers), students (students), educational and methodological materials, employers (business - environment), creators of new knowledge (research organizations, laboratories, etc.); b) organizes and carries out its activities either with the participation of only digital technologies, or on the basis of their preferential use; c) accelerates the process of generating intellectual capital and significantly increases the market capitalization of both the education system itself and the digital network structure.

In the methodology of intellectual capital management, there is a wide variety of approaches and methods for assessing intellectual capital, which can be conditionally grouped as follows: 1) quantitative approaches and methods; 2) qualitative and 3) structural approach [22]. The most widespread and actively used approach to assessing intellectual capital is a structural one, which allows a comprehensive assessment of intellectual capital and has a number of advantages, but this approach has its drawbacks. In general, the approach is attractive from the point of view of the possibility of aggregating various indicators, but the assessment mechanism does not allow focusing on the impact of the digital environment (digital component) on intellectual capital, although this approach "covertly" involves digital technologies as part of the information type of capital, which is part of the entire complex index of intellectual capital (along with human, organizational and client types of capital). It should be noted that the structural approach focuses on the building blocks and factors of a composite measure of intellectual capital, but not on technology.

At the same time, in new studies, various methods and approaches to assessing the digital educational space appear, but mainly they relate to the assessment of either the digital economy or the digital innovative educational ecosystem, but do not relate to the impact of the educational ecosystem on the growth of its intellectual capital, the growth of market value. educational ecosystem. In particular, in [6] D.O. Korolev and T.O. Naushirvanov analyze the availability and quality of digital education for the development of innovative educational projects. The assessment is carried out for six types of infrastructure, such as: 1) development infrastructure; 2) legal infrastructure; 3) telecommunication; 4) physical; 5) financial; 6) infrastructure of human resources (personnel). All types of infrastructure are assessed according to two criteria: "availability" and "quality" of digital education. As a result of the study, the authors revealed that the vast majority of innovators (who took part in the study as respondents), according to both criteria ("accessibility" and "quality"), the highest level of digital education in the communication infrastructure (93 and 63% of participants, respectively). The lowest score was found on the criterion "affordability" for financial infrastructure (only 25% of respondents gave a positive assessment), but according to the criterion "quality", the level of digital education in legal and physical types of infrastructure turned out to be the lowest (only 21 and

28% of respondents, respectively, noted a high level of education). The sample included representatives of innovative structures from 56 regions of Russia, of which 60% of respondents were residents of large cities with a population of more than 500 thousand. human.

Table 1. Federal projects of the Russian Federation "Digital Economy" national program

Main directions	Project content
Personnel for the digital economy	Improving the education system in order to provide the digital economy with competent personnel. Transformation of the labor market, which should be based on the requirements of the digital economy. Creation of a motivation system for the development of the necessary competencies and the participation of personnel in the development of the digital economy of Russia
Information infrastructure	Creation of the necessary telecommunication infrastructure and conditions to achieve the following tasks: provision of broadband Internet access to socially significant objects, as well as effective and safe use of online services by them; ensuring the possibility of broadband access to the Internet for households in sparsely populated, remote and hard-to-reach settlements; implementation of modern communication standards 5G / IMT-2020
Information security	Achieving a state of protection of the individual, society and the state from internal and external information threats, which ensures the implementation of constitutional rights and freedoms of man and citizen, a decent quality and standard of living of citizens, sovereignty and sustainable socio-economic development of the Russian Federation
Digital technologies	Development of promising high-tech areas, support of domestic companies-leaders of the information technology market, as well as stimulating demand for their solutions, creating conditions for the development of start-ups that develop solutions in the field of information technology in order to ensure national security, technological independence and competitiveness at the global level
Legal regulation	Formation of a new regulatory environment that provides a favorable legal regime for the emergence and development of modern technologies, as well as for the implementation of economic activities related to their use
Digital public administration	Implementation of digital technologies and platform solutions in the areas of public administration and provision of public services, including in the interests of the population and small and medium-sized businesses, including individual entrepreneurs

Undoubtedly, such an assessment allows one to get a certain picture of the digital educational space in terms of the availability of knowledge and the quality of its transmission (training), but in order to understand which elements of the digital educational ecosystem have the greatest impact on intellectual capital, other emphasis is needed in the study.

First of all, it is necessary to substantiate whether digital technologies affect intellectual capital, although it is obvious to any user that they have an unambiguous effect? But, on the other hand, it is advisable to highlight the main factors of the digital educational space, which are the main drivers of intellectual capital.

The authors of this work carried out an expert analysis of the elements of the digital infrastructure of the educational ecosystem that affect its intellectual capital and obtained the following assessment results, which are shown in Table 2. The analysis was carried out according to the following criteria:

1. Digital technology platform allows us to implement the following opportunities for digital education: accumulation, storage, processing and analysis, broadcast (training) and control of knowledge (placement and control of work performed, etc.); integration of all processes and types of educational activities (including lectures and practical classes, assessment work (for example, the organization of electronic testing in the process of monitoring students' knowledge, etc.); online organization of round tables, guest lectures, master classes, consultations, video conferences on various technological platforms and systems (for example, ZOOM, BigBlueButton (BBB), Microsoft Teams, Skype, WhatsApp, etc.); continuous replenishment of unlimited educational and methodological literature, including previously recorded lectures (videos, open online MOOC courses); access to library resources and databases, to third-party information resources and sources.

2. Educational and methodological complex of the discipline in digital infrastructure. There is an opportunity for the creation and unlimited improvement of the educational and methodological complex of the discipline in the digital environment. The author-developer has the ability to create online courses on a digital platform and continuously supplement their teaching materials, which ultimately contributes to the intensive development of approaches, theories (concepts), replenishment of the base of developed problems, practical examples for analysis and cases (situations) for solving ... This form of material presentation simplifies the process of translating the knowledge of students, reduces the costs associated with the personalization of training.

3. Stakeholders of the information network structure of the (digital educational) ecosystem. It is possible to integrate scientific organizations that develop (create) knowledge, the education system (including the higher education system), which processes, analyzes information and broadcasts knowledge to students, and the real sector (business environment), which is actively used (applied) knowledge gained by students in practice. As a result of such interaction of stakeholders on the digital educational platform, a synergistic effect is achieved.

4. Innovative projects in the digital education ecosystem. The digital educational ecosystem allows both employers and educational organizations (primarily teachers and students) to unite on a single digital educational platform and helps to implement a huge number of innovative projects offered by employers and developers through the efforts of the teaching and student environment. This approach significantly speeds up the process of commercialization of innovative projects.

5. Learners (students, listeners) in the digital educational ecosystem have: a) access to educational materials, library resources; b) the opportunity from anywhere in the world to attend classes, participate in online events, discussions, master classes, guest lectures, consultations, scientific conferences, competitions.

6. Teachers (course and module developers) on the educational platform have: a) access to information technologies and can place their developments and educational materials there (manuals, work programs, assignments, tests, regulatory documents, etc.); b) the opportunity to improve their qualifications using the online platform, take part in scientific conferences, symposia, congresses, etc.

Students and teachers of St. Petersburg State University of Economics (SPbSEU) participated in the survey as respondents. Column 2 of this table presents a point estimate of the degree of influence of each element (factor) of the digital education ecosystem on its intellectual capital, which is built in such a way that the total (final score) of the combined influence of all factors is 100 points. The assessment involves the most significant assessment criteria that have the same weighting coefficients.

Table 2. Influence of the educational ecosystem digital infrastructure elements on intellectual capital (by the example of SPbSEU)

Elements of the digital infrastructure of the educational ecosystem	Scoring the impact of an element on the intellectual capital
1	2
Digital technology platform	30
Educational and methodological complex of the discipline in digital infrastructure	15
Stakeholders of the information network structure of the (digital educational) ecosystem	15
Innovative projects in the digital education ecosystem	15
Learners (students, listeners) in the digital educational ecosystem	10
Teachers (course and module developers) on the educational platform	15

From Table 2, it is obvious that in relation to the digital educational environment, a digital (technological) platform plays a special role, providing unlimited opportunities for the integrated work of all stakeholders - participants in the digital network structure, as a result of whose joint work on a single platform, a synergistic effect is created that significantly increases the cost of an intellectual capital.

The most influential element (factor) of the digital educational ecosystem is the digital technological (educational) platform, the significance of which in the system of factors affecting intellectual capital is 30 points, and the least significant factor was the "learners" factor (10 points). The absence of at least one of the elements violates the principles of integrity and emergence. The factors noted in Table 2, according to the respondents, are the most significant in the process of ensuring the growth of the intellectual capital of the digital educational ecosystem. This is facilitated by the specificity of digital technologies that perform the functions: accumulation, storage, analysis, processing and transmission of knowledge to students.

The digital platform has a number of features: a) it breaks down geographic boundaries and limits the number of listeners; b) significantly reduces financial costs and eliminates the loss of time associated with the physical movement of both the teacher and the student (listener); c) allows the educational system to provide personalized educational services (for example, it makes it possible to choose a piece of program or discipline for study); d) provides students with unlimited access to library resources, sources and databases; e) "frees" the teacher from the routine process of continuous lecturing and reduces the workload, since there is a possibility of a one-time recording of a lecture and its replication using information and communication technologies. The lecturer has free time for other types of scientific and pedagogical activities.

4 Conclusion

The digital educational ecosystem should be formed on the basis of trends in the development of the world educational space and taking into account the characteristics of the modern intellectual economy. This process should proceed in the following directions:

1) the creation of a personalized educational environment (human - IT), the creation of unique (personal) educational technologies and trajectories.

2) the development of global educational platforms (the so-called "universities per million students").

3) the formation of a collective educational environment (person-person), communities of practice that provide support or mentoring.

4) creation of a local educational environment, which is a local ecosystem of lifelong education.

5) creation of an environment for self-education of students at any time and in any place.

The main principles of the digital education ecosystem should be:

1) consistency and complexity, designed to ensure interdisciplinarity and multidisciplinarity in teaching. The approach to teaching involves the development of disciplines that are different in their content, that is, from different angles. With this approach, each participating discipline does not lose its identity, but on the contrary

complements the others; the ability to build connections between phenomena is formed to manage all types of resources in a complex manner.

2) holism, which underlies the transdisciplinary approach to learning. This approach presupposes learning "through", "across" the boundaries of many disciplines, when at the junction of sciences new sciences and disciplines appear, new professions appear that did not exist before. Ecosystem education built on digital technologies is focused on the development of emotional intelligence, first of all, communication skills, the ability to work in a team, the ability to listen to others, analyze the situation and make a compromise.

3) rationalism, implying learning based on the development of cognitive abilities, the ability to perceive and analyze a large amount of information, the ability to work with a large amount of content, skills in solving complex problems.

4) pragmatism, focused on the development of the ability to assess possible and / or missed benefits, including economic, social, technological, etc. The pragmatic approach is focused on the development of skills to design, evaluate results, and adequately assess their capabilities.

The exponential development of the modern intellectual economy (that is, the intensive growth of the intellectual capital of the modern economy) depends on the intellectual intensity of the educational environment, its readiness to use knowledge-intensive techniques and technologies in teaching. This situation obliges modern society to move on to building a new model of education, which relies on the use of digital technology platforms that provide unlimited access to information sources and digital content and contribute to the intensive development of intellectual capital as the main resource for economic development. In conclusion, it can be noted that the new view substantiated the objective expediency of revising the structure and content of the traditional education system and the inevitability of its consistent and systemic digitalization, built on the principles of inter- and trans-disciplinary approaches to learning.

References

1. A.G. Tansley, *J. Ecology.* **16(3)**, 284 (1935)
2. *Ecosystem*, Retrieved from: https://ru.wikipedia.org/wiki/
3. J. Moor, J. *Harvard Business Review.* May-June. 75 (1993)
4. A. Katrich, *Digital financial ecosystems* Retrieved from: https://asros.ru/upload/iblock/4c5/19421_katrich.pdf
5. R. Isaev, *Banking ecosystems will capture many industries in the economy in the coming years* Retrieved from: https://www.businessstudio.ru/articles/article/bankovskie_ekosistemy_zakhvatyat_mnogie_otrasli_isaev/
6. D.O. Koroleva, T.O. Naushirvanov, *J. Facts of Education.* **6(31)**, (2020)
7. O.S. Kolesnikova, I.N. Molchanov, S.Y. Simonov, *J. Tekhnologiya textile industry.* **4(370)**, (2017)
8. V.V. Ivanova, *Current problems of economics and management* 190-193 (2017)
9. S.G. Grigoriev, R.A. Sabitov, G.S. Smirnova, Sh.R. Sabitov, *Computer science and Education* **5**, 15-23, (2020) https://doi.org/10.32517/0234-0453-2020-35-5-15-23
10. S.N. Gavrilov, S.I. Volodina, *J. Actual problems of Russian law.* **6(103)**, 156-166 (2019)
11. N. Zakharov, I. Britvina, A. Kuznetsov, *The impact of industry 4.0 on job creation*, 457-462, (2020)
12. K. Schwab, N. Davis, *Technologies of the Fourth Industrial Revolution,* **320**, (2018)
13. F. Nachira, P. Dini, *Network of Digital Business Ecosystems for Europe: Roots, Processes and Perspectives.* (2007)
14. B. Talin, *What is a digital ecosystem?* Retrieved from: https://morethandigital.info/ru/chto-takoye-tzifrovaya-ekosistyema-ponimaniye-naibolyeye-viguodnoy-biznyes-modyeli/
15. Markelov M., *Digital ecosystem: what it is, features of its perception and typical mistakes* URL: https://vc.ru/future/70143-cifrovaya-ekosistema-chto-eto-takoe-osobennosti-ee-vospriyatiya-i-tipichnye-oshibki
16. A. Tsvetkova, M. Gustafsson, *J. Clean. Prod.*, 246-254, (2012)
17. M.A. Kondakov, *Ecosystem of digital education* Retrieved from: https://firo.ranepa.ru/files/docs/cifr_didactika/plenar/kondakov_plenar.pdf
18. About the university 20.35. URL: https://2035.university/about/
19. Pandemic of digital education. URL: https://vc.ru/education/153144-pandemiya-cifrovogo-obrazovaniya
20. National project "Education". URL: https://edu.gov.ru/national-project/projects/cos/
21. Ministry of Digital Development, Communications and Mass Media of the Russian Federation "Digital Economy of the Russian Federation". URL: https://digital.gov.ru/ru/
22. A.E. Suleimankadieva, E.A. Tkachenko, M.A. Petrov, O.Y. Syrovatskaya, R.V. Klyarovskaya *ICICKM 2020.* 346–353, (2020) https://doi.org/10.34190/IKM.20.061

Multilayered balanced indicators system for evaluating the results of innovation activities within the framework of cooperation

Elena Kirillova*, and *Nataly* Uvarova

Smolensk Branch of the National Research University «Moscow Power Engineering Institute», 1,Energy passage, Smolensk, 214013 , Russia

Abstract. A modified system of indicators for assessing the innovation activities effectiveness in the region based on data on patent and licensing activities in the framework of scientific and industrial cooperation is described. Also in the article we presented a general mechanism for innovation process effectiveness managing of such cooperative interaction. They are intended to be used as tools to support decision-making, allowing one to coordinate the cooperative indicators management. The proposed system of indicators provides transfer and decomposition of goals for planning operational activities by levels and actors of the innovation process in the cooperative formation, as well as monitoring their achievement. Taking into account the peculiarities of the innovation environment, the proposed balanced indicators system (hereinafter referred to as the BSC) focuses on non-financial logistics performance indicators and patent`s information. The structure of the BSC assumes a pyramid-layer structure, the number of elements of each layer is determined by the structural decomposition of the cooperative interaction corresponding level. The article describes stages of forming such BSC-maps process. As a tool for improving the balance of strategic maps for generating management decisions, a transition to "horizontal dynamics" is proposed, according to which at each level of logistics management, among all goals there is the greatest deviation from the target values, i.e., from the target values achieved to the least extent. These goals are the "target limitations" of the enterprise effective operation and they take greatest importance for further analysis of critical situations. For this purpose, a procedure for managing innovation environment by deviations is proposed.

1 Introduction

In the modern literature, a lot of attention is paid to the analysis and evaluation of the innovative processes' implementation at the regional level. There are a large number of different approaches, models and methods of its assessment. Such close attention to the analysis and evaluation of innovation activity results on the part of Russian and foreign scientists is due to its essential importance for making operational and, above all, strategic decisions. In addition, technological progress, the growth of dynamism and environmental changes uncertainty, with the increasing globality of competition, significantly shorten the life cycles of products and services and forcing production and economic entities to quickly implement innovative technologies in their work experience. The complexity of technological processes makes innovations more expensive and risky, which emphasizes the importance and significant contribution of the evaluation and analysis stage in the innovative management. The need for a clear, transparent and logical system for analysing and evaluating the results of such activities is also due to the significant role of regional authorities in their implementation, the significant contribution of budget financing and co-financing, which involve strict reporting and ensuring flexibility in the allocation of resources to budgets of various levels. The size of Russian territory and the presence of industrial locations specialization separated by large distances lead to an increase in logistics costs when implementing the processes of material and technical support for the production and sale of finished products. This circumstance determines the feasibility of forming closed supply chains localized in one territory, as well as an integrated innovation process [1]. In addition, the climatic and resource conditions in each of the Russian regions differ quite significantly, which determines the need to implement an approach to concentrate efforts to catalyse integration processes within their localization territories and extract maximum benefits for them and the region.

It should also be noted that the crisis consequences caused by measures to prevent the spread of coronavirus had a significant impact on national innovation systems, which caused a reduction in budgets for innovative development programs and even more clearly raised the question of tools for assessing innovation activities effectiveness for making management decisions in this area. On the other hand, it is noted in [2] that these circumstances only further emphasized the need for

* Corresponding author: kirillova.el.al@yandex.ru

innovative and information development to ensure the sustainable development of society and world economies that were not ready for global challenges.

However, most of the existing methods and tools for assessing innovation activity results practically do not take into account such aspects as inter-organizational interaction and the synergistic effects generated by such cooperative formations. At the same time, these formations become an integral element of modern production and economic systems and have a significant potential for innovation development [3,4]. The level of such cooperation is most likely not taken into account for the existing objective problems in its analysis. In this case, it is assumed to analyse a significant number of qualitative characteristics, subjective and relative variables that have a probabilistic nature of implementation. This work is devoted to the development of tools for analysing innovation activity results in the region, which differ from others by using a multi-layer balanced indicators system for evaluating scientific and industrial cooperation based on data on patent and licensing activities.

2 Literature review and analysis

Among the statistical and information-analytical materials that reflect quantitative assessments of new forms strategic alliances, various partnerships and other cooperative relations implementation, we can distinguish the HSE collections. The indicators presented in them make it possible to identify and establish the existence of cooperative relations, research and development organization cooperation with other organizations, regardless of whether the organization is the customer or the performer of the relevant works and services. In particular, the following groups of indicators are presented: the underdevelopment of cooperative ties, technological partnership, the number of joint projects and in the context of innovations types, cooperative ties types, as well as by type of economic activity. It is worth noting that the methodological materials for the development of innovative development programs for Russian organizations and enterprises (approved by the decision of the Interdepartmental Working Group on the Implementation of Innovation Development Priorities of the Presidium of the Presidential Council for Russia Economic Modernization and Innovative Development (No. 1 of February 27, 2018)) include a section on assessing cooperation. In particular, this is interaction development analysis with third-party organizations, the implementation of "open innovation" principles, the development of partnership in education and science fields, the level of interaction with the subjects of innovation infrastructure in general, as well as the development of foreign economic activity and international cooperation in the innovation sphere. In [5], a comparative characteristic of the indicators of the analysis of innovative development, proposed by the European Commission, is given. Among all the statistical indicators that are most widely used in domestic and foreign practice and characterize business entities innovation activity and competitiveness, the following groups can be distinguished: cost, time and structural [6,7,8,9]. Modern scientists also assess the effectiveness of innovative development, including financial development, at the microeconomic level, from the point of view new technologies development of by industrial enterprises, the introduction of new technologies into economic circulation, the functioning of research and development departments of innovative business entities, etc. On the other hand, there is no unified methodology for assessing the effectiveness of innovation financing at the regional level, and the actual assessment of innovation financing results effectiveness of investment is based on a system of monitoring indicators that characterize the state of Russia's scientific potential.

The assessment of cooperation at the regional level is also reflected in the works of foreign and Russian scientists. So, Kuznetsova E. P. in [10] describes an approach to assessing the development of industrial cooperation in the region, based on the calculation of an integral indicator that reflects the "scale" of cooperation between corporate structures, small and medium-sized businesses and the "cost-effectiveness" of their interaction. The author defines the stages of assessing industrial cooperation in the region, and, in particular, at the first stage, identifies three groups of indicators:

- indicators that characterize change in the cost parameters of production activities in the region;
- indicators that characterize "efficiency" of production cooperation;
- indicators that characterize "scale" of production cooperation.

Aleshin A.V. [11] as the main areas of inter-organization interaction evaluation offers an assessment of the cooperation effectiveness both for individual business entities and for the regional economy as a whole, as well as a methodology for assessing the level of partnership relations development in regional business environment. A significant contribution to the theory of inter-organizational interaction evaluation is taking into account the fact of participation in the innovation process of various scales subjects. However, the author considered only business subjects and this assessment is not applicable for research organizations, educational institutions, government agencies, etc., which have mainly a different range of interests. Fedosova T. V. and Morozova T. V. [12] investigated the form of interaction "university-enterprise", the proposed assessment methodology identifies three vectors describing the state of the process according to three corresponding ones, namely: the university development vector, enterprise development vector, mutual development vector. Their spatial intersections give an idea of the level of integral interaction. The advantage of this method is the ease of visual representation, but there is no direction of interaction with government agencies. Ivashchenko N. S. [13] offers an assessment of the partnership innovation process subjects relations depending on the stage of their life cycle and defines the following groups of criteria, evaluated on a 4-point scale: the level of partnerships organization (the nature of goals and objectives, motivation, etc.), the level of their development (reliability, dynamics, trust) and the level of such partnerships effectiveness (depth, durability, etc.). Measuring the effectiveness of network forms of organization is a fairly new direction in research [14, 15]. In [16], network modeling is used for evaluating interoperable networks.

Tsiteladze D. D. [17] with regard to subjects' interaction features in the innovation process notes that it is largely determined by the nature and direction of transaction costs, while interaction is considered from the side of transaction costs of various subjects: business, state, population, expert community, financial institutions, etc. Interaction of subjects within clusters is reflected in the work of Stepnov I. M., Kovalchuk Yu. A., Gorchakova E. A. [18]. The authors describe an original model for evaluating the effectiveness of intra-cluster interaction, which includes a specially developed criterion "the degree of cluster leverage". Simachev Yu. V., Kuzyk M. G., Feigina V. V. [19] note that large research organizations are more often involved in interaction with business for R&D, and non-state research organizations are more often involved in one-time cooperation with business entities. An important common factor in the presence of integration interaction, as well as further plans for this interaction, as the authors note, is a relatively high scientific potential. Quite often, coefficients and indicators groups that reflect patent activity are included in the methods for evaluating innovation activities results and determining the degree of security of an enterprise with intangible intellectual resources. Most enterprises in the annual reports on innovation activities provide results on such indicators as: the number of RID registration applications, the number of registered certificates, patents for an invention, utility model, and so on. Patents and computer programs registration certificates secure the right to the result of intellectual activity and thus are the final research stage of the innovation life cycle. At the same time, the uniqueness of patent data lies in the fact that they characterize both the effectiveness of research and development and the potential for innovation, i.e. they are a kind of bridge between science and production. Patents not only perform the function of legal protection, but also are a unique source of technological information, since the information contained in patents is usually not presented anywhere else. In addition, patenting, usually several years ahead of scientific and technological achievements introduction, allows to take into account the possibility of technological innovations in advance. It is also necessary to note the strengthening of trends towards "open" borders between the world's economies, and the practice of open innovation is becoming increasingly popular. And the more such borders are open, the more acute the issue of patent protection becomes. For analytical purposes, absolute and relative indicators of patent activity are used, the information for the calculation of which is publicly available and has an objective quantitative assessment, which can serve as an objective basis for evaluating the innovative activity results of cooperative interaction at the regional level.

Thus, the presented analysis allows us to say that the currently existing methods and tools for innovation activities effectiveness assessing in the region have a number of limitations, which makes it difficult to apply them for a comprehensive analysis of modern integrated entities characterized by multidirectional interests of participants and high dynamism of changes.

3 Results and Discussion

Building an effective system of economic entities cooperative interaction in the innovative processes implementation at the regional level is impossible without evaluating and analysing the effectiveness of interaction processes, taking into account the interests of all its participants. The practice of network organization forms strategic management shows that to solve this problem in the planning and control of the organization's activities, it is advisable to use a balanced system of indicators (BSC), proposed in the works of Norton D. and Kaplan R. and developed in the works of Vetter M., Leonard D. [20]. The BSC for cooperative formations acts can be used as a tool to support management decision-making, allowing for coordinated management of indicators that characterize the activities of both individual entities and the group as a whole.

At the same time, the existing methodological apparatus for the construction and application of the BSC does not fully take into account the balance of interests of all participants in the innovation regional process in various functional areas and at different levels of management. At the same time, there is also a shortage of mathematical expressions that allows one to control a balance degree of both individual indicators and groups of BSC indicators, taking into account their importance. In this regard, as a basis for managing inter-organizational interaction in the process of implementing innovation activities, a multi-level modified BSC and its processing process are proposed, which provides for the transfer and decomposition of goals by levels and actors of the innovation process in the integration formation for planning operational activities, as well as monitoring their achievement.

The structure of the proposed BSC has a pyramid-layer structure, the number of each layer elements is determined by the structural decomposition of the cooperative interaction corresponding level. In order to build the structure of the intra-company base level of such BSC, taking into account the features of innovative processes, it is proposed to focus on non-financial logistics performance indicators.

The structure of the proposed matrix-hierarchical BSC is shown in Table 1.

Table 1. The structure of the proposed BSC.

BSC Aspect / Planning level	Components							
	1. Internal			2. Client Side	3. Innovative	4. Financial		
1. Interregional level	X^1_{11}	…	$X^1_{N^1_1,1}$	⋮	⋮	X^4_{11}	…	$X^4_{N^1_4,1}$
2. The level of the regional		⋮		⋮	⋮			⋮
3. Environments business		⋮		⋮	⋮			⋮
4. Intra-company level	X^1_{13}	…	$X^1_{N_3,1}$	⋮	⋮	X^4_{13}	…	$X^4_{N^3_4,1}$

It is possible to distinguish the following indicators that are part of BSC, taking into account the specifics of the cooperative innovative processes' implementation in the region. The system for evaluating innovation activity results in the region should be based on well-known indicators that characterize the main results of the functioning of enterprises and organizations. For example, it is proposed to include such indicators based on patent and licensing information as: coefficient for assessing the involvement of universities and research organizations in the patent and licensing activities of the industry, the indicator for assessing long-term interaction, the number of registered RID created within the framework of research contracts between an industrial enterprise and a university or research institute. It is also proposed to analyze the number of applications submitted for registration of RID, patents obtained and publications of a university or research institute, in publications included in the list of the Higher Attestation Commission and foreign bases, in the subject area of an industrial enterprise according to the IPC classifier. The calculation and analysis of these indicators will allow us to determine the presence, nature and level of scientific and industrial cooperation, which will thereby make it possible to significantly supplement the assessment of its innovative potential. To characterize the innovation component at the strategic planning level, it is proposed to use such indicators as the predicted ratio of the degree of applied novelty technologies in relation to their global counterparts, indicators of the long-term innovative projects' implementation, indicators of the energy and resource saving strategy implementation, etc.

The process of forming BSC is proposed to be divided into the following stages: determination of control actions (within the framework of local innovation cycles); determination of external and internal innovation environment states, as well as the general economic state and development trajectory of the region; determination of diagnostic parameters that characterize the external manifestations of the innovation system and the effectiveness of its functioning; determination of structural parameters that characterize the internal state of the innovation system; establishing the relationship between the structural and diagnostic parameters of the system; determining the vertical structure of BSC.

As a tool for improving BSC balance generating management decisions, a transition to "horizontal dynamics" is proposed, according to which at each level of logistics management, among all the goals, those that, as a result of the analysis of their indicators, have the greatest deviation from the target values, i.e., have been achieved to the least extent: these goals are the "target limitations" of enterprise effective activity and are of the greatest importance for further analysis of critical places.

To do this, we propose a procedure for managing inter-organizational interaction in the implementation of the innovation process in the region by deviations, consisting of the following steps: description of the desired state of cooperation innovation potential, i.e., the establishment of those qualitative and quantitative requirements for the state of the potential in all parameters that ensure the achievement of a specific goal and the implementation of the strategy; step-by-step establishment of the actual state of complex innovation environment in all parameters and the construction of the BSC; analysis of the discrepancy between the normative, target and actual values of the parameters, while for each BSC is proposed to calculate the imbalance separately using the formula of the form:

$$disb_m^n = \sum_{k=1}^{n} \left(\prod_{l=1}^{m} \Delta i_l\right) \cdot k_k^{norm} \tag{1}$$

where Δi – local imbalances (normalized differences) between identical indicators in different categories of BSC maps (in the simplest case for parametric indicators between planned and actual, planned and normative, and actual and normative indicator i), knorm – weight coefficient of group normalization for indicators different categories, m – the number of layers of the global BSC map, n – the number of management indicators groups.

This is followed by determining the total imbalance for individual contours, identifying the relationships between the unbalanced parameters and identifying possible causes of the discrepancy; defining a structured action plan for managing deviations.

The methodology for applying this mechanism includes the following main steps.

a) Using the results of external and internal cooperation environment analysis, strategic goals and corporate strategy of the participant are developed. The first line of the BSC is filled in.
b) The BSC is being built. The degrees of importance (weight coefficients of group rationing) of the BSC indicators for different planning levels are determined. A preliminary control of imbalance values is carried out.
c) Development of appropriate detailed strategic, medium-term and operational plans aimed at achieving the targets included in the BSC.
In the event of an imbalance, we adjust the medium-term and operational plans according to the procedure proposed in paragraph b.
This methodology allows us to take into account the degree of importance of individual indicators that are part of the proposed matrix-hierarchical BSC for different groups of subjects when implementing corporate governance procedures for regional innovation environment.
The evaluation of innovation activity effectiveness of individual subjects can be carried out in the form of calculation and comparison of the obtained indicators with the established basic values. Depending on the current state of the region, the results achieved to support and develop its innovation component, the basic values can be indicators for the last or other reporting year, the average values for the country or the federal district, or the corresponding indicators of other regions. At the same time, data from the survey of the innovation infrastructure of other subjects' activity or foreign experience can be used as threshold criteria. These coefficients, calculated for subterritorial entities within the region (districts, cities, municipal units) on a territorial basis, also lay the basis for comparative comparisons.
When calculating each of the above-mentioned indicators in the framework of comparative procedures or calculating dynamic series in a certain time interval, the result is a significant amount of quantitative economic parameters estimates of innovation financing in the region. Such a comprehensive analysis and the formation of evaluation indicators significant number on its basis make it possible to obtain, on the one hand, a multidimensional tool for evaluating the results of innovation activities, but, on the other hand, it makes it difficult to make decisions on determining the final dynamics of processes due to the significant amount of data being compared. Also the relevant one is the issue of comparative efficiency assessment based on territorial differences in the structure of the organization of the support infrastructure (economic and geographical analysis). It is based on a comparative analysis of innovative entrepreneurship development levels in regions/territorial entities with similar characteristics of socio-economic development, but different configurations of support infrastructure organizations.

4 Conclusion

Thus, calculation and analysis of indicators reflecting innovation activity results in the region according to the described methodology will allow us to determine the presence, nature and level of scientific and industrial cooperation, which will thereby make it possible to significantly supplement the assessment of its innovation potential. The breadth of the set of hierarchically arranged indicators delineating the zones, coupled with the possibility of transforming the described system of indicators for specific tasks, makes it possible to widely use integral indicators not only in the process of evaluating innovative potentials, but also in determining the key indicators of individual subjects functioning in the scientific and educational space. The use of such a tool will reduce the time for making decisions on the selection and implementation of innovative projects, reduce the level of uncertainty in the analysis and evaluation of scientific and industrial cooperation processes by using quantitative data available in the public domain and a more visual and simple system. And will also make it possible to predictively identify imbalances in the resulting economic indicators of innovation activities functioning in the region.

Acknowledgments

The research was carried out with the financial support of the «Council for grants of the President of the Russian Federation for state support of young Russian scientists – candidates of science» according to the project MK-4087.2021.2.

References

1. A. Zainchkovsky, V. Borisov, Smart Innovation, Systems and Technologies **172** (2020)
2. *Science, Technology and Innovation Outlook 2021: Times of Crisis and Opportunity* (OECD Publishing: Paris, 2021).
3. Y.K. Lew, J.-Y. Park, *Sustainable Development* **29**,2 (2021)
4. S. Liu, X. Zhang, W. Chen, W. Zhang *10th International Conference on Educational and Information Technology*. **9375561,** 185–189. (2021)
5. J. Janger, T. Schubert, P. Andries, C. Rammer, *Research Policy* **46**,1 (2017)
6. M. Ivanova, T. Selentyeva, *Proceedings of the European Conference on Innovation and Entrepreneurship*. **1,** 417–423, (2019)
7. R.P. Mamede, *Portuguese Journal of Social Science* **16**,3 (2017)
8. S. He, S. MacNeill, J. Wang, *European Planning Studies* **22**, 9 (2014)
9. M. Dziallas, K. Blind, Technovation **80–81** (2019)
10. E.P. Kuznetsova, *Ekonomika i biznes: teoriya i praktika* **12-2**,58 (2019)

11. V.A. Aleshin, *Terra Economicus* **9**, 4 (2011)
12. T.V. Fedosova, T.V. Morozova, *Planning and provision of training for industrial and economic complex of the region* **1** (2017)
13. N. Ivashchenko, *International research journal* **12**, 9 (2019)
14. P.S. Ferreira, A.H. Shamsuzzolta, C. Toscano, P. Cunha, *International Journal of Productivity and Performance Management* **61**,6 (2012)
15. S. Pekkola, *Measuring Business Excellence* **17**,1 (2013)
16. P. Graça, L.M. Camarinha-Matos, IFIP *Advances in Information and Communication Technology* **598** (2020)
17. D. D. Tsiteladze *Forms and methods of interaction of subjects of innovative activity in regional innovation systems:* PhD Dissertation (Moscow, 2015).
18. I.M. Stepnov, Yu.A. Kovalchuk, E.A. Gorchakova, *Forecasting problems* **3** (2019).
19. Yu.V. Simachev, M.G. Kuzyk, V.V. Feigina *Interaction of business and scientific organizations in the innovation sphere: the Russian experience of stimulating cooperation* (Mosocw, 2015)
20. N.-G. Olve, J. Roy, M. Wetter *Performance Drivers: A Practical Guide to Using the Balanced Scorecard* (UK: Wiley, 2001)

Formation of conceptual foundations for modeling the organizational structures of agricultural entities based on multilevel cooperation

Sergey Pestrikov[1,2*], *Dmitry* Maltsev[1,2], and *Elena* Gureeva[2]

Perm State Agrarian and Technological University named after academician D.N. Pryanishnikova, 614990, Perm, Russia
Perm National Research Polytechnic University, 614990, Perm, Russia

Abstract. The article presents an analysis of the organizational structures of cooperatives in the agricultural industry, offers conceptual approaches to modeling the organizational structures of agricultural formations in Russia, taking into account the specifics of the industry, using modern and innovative strategies for organizational changes. The creation of flexible multilevel systems of cooperation is substantiated, allowing, on the one hand, to consider the interests of both members of a single cooperative and a group of cooperatives that synergistically consolidate the available and attracted resources. An assessment of the importance of these principles for creating an effective cooperative structure is based on analyzing a survey of peasant farms in the Perm Territory. A promising organizational structure of multilevel cooperation has been developed.

1 Introduction

Topical issues of organizational excellence affect many sectors of the national economy [1]. However, as a rule, the most inert processes do not reflect the dynamics of rapid changes in the economy and proceed during the formation of complex organizational structures. For example, such complex structures include cooperatives, clusters, holdings. That is why in recent years, researchers have been looking at issues of analyzing the quality, adequacy, levels, and dynamics of development of organizational structures, the use of structures of a new type. These structures offer "soft" forms, on the one hand, and ensuring the manageability of organizational structures, on the other hand [2].

The system of agricultural cooperation in Russia began to revive again in the early 1990s after the adoption of the federal law "On Agricultural Cooperation." However, slipping in this direction is visible to the naked eye. It is evident to everyone that the role of cooperative formations in the country's economy should grow. Thus, the number of successful cooperatives in the western and eastern economies is significant (Valio, Mondragon, HKScan).

In Russia, the revival of agricultural cooperation is undoubtedly a priority task of the state since it guarantees the country's food security. Russian government realized the priority of the development of small forms of farming and rural cooperation. The state has developed and approved a Resolution of the Government of the Russian Federation, which provides funds for the creation of a support system for farmers and the development of rural cooperation in the period from 2019 to 2024 in the amount of 37.368 billion rubles [3].

There are examples of the successful operation of cooperative structures in the Lipetsk region, Tatarstan, and the Tyumen region in Russia. Despite the existing examples of the successful operation of cooperative structures in the country, the overall result of their activities on a national scale is indisputably low. In parallel with the cooperative policy in the country's agriculture, the creation and development of cluster associations are observed as one of the forms contributing to the synergistic manifestations of complex organizational services. However, there is still no broad clustering of all spheres of the agro-industrial complex, and little attention is paid to their interaction at the regional level. An extensive experience of cluster activity has been accumulated in specific industries.

However, the transfer and adaptation of this experience, taking into account the specifics of the agro-industrial complex, is not carried out. In addition, the investment attractiveness of such activities remains extremely low, both due to bureaucratic procedures and due to the lack of innovation and qualified personnel [4].

One of the goals of modeling organizational structures is to predict the leading economic indicators of a cluster using regression analysis methods [5]. Efficiency assessment is shown on the example of the analysis of the activities of agricultural enterprises in the Lipetsk region [5]. In work [4], a model of a livestock cluster in meat and dairy and food complexes is proposed on the example of the Saratov region. The model assumes the presence of a cluster core, which will

* Corresponding author: pestrikovsa@mail.ru

directly deal with animal husbandry, under the supervision of the executive authorities.

On the one hand, this work is joint work with the regional duma on organizing personnel training, conducting research, and attracting funding. On the other hand, this work is interaction with service companies in the field of legal support, transport companies, suppliers of machinery and equipment, feed. It is predicted that the introduction of the cluster will presumably reduce the costs of commodity circulation of products by 29.8% for meat and meat products and by 26% for milk and dairy products.

Obviously, this model is significantly enlarged and does not consider many factors: technical, technological, weather and climatic, socio-demographic, and other factors. A detailed analysis of this problem can be found in [6]. It is indicated that the organizational and legal forms of enterprises that are part of the cluster can significantly impact financial performance. Thus, according to the results of studies carried out in the EU, using the example of the Slovak Republic [7], the following was established: commercial organizations have more assets, higher creditworthiness, and investment attractiveness (by about 20%) compared to cooperatives. In this case, it is also necessary to consider the level of government support in different countries. For example, since 2019, the emphasis has shifted to supporting agricultural cooperatives in Russia. The introduction of joint-stock cooperation as an innovation in the management system of agricultural enterprises is also seen as promising [8].

Transport logistics cause significant difficulties, which is due both to the large distances between manufacturers and consumers and the fact that the products are perishable and require special transportation conditions. In [9], based on the analysis of the activities of 36 transport companies in China, a strategy for delivering agro-industrial products to consumers is proposed.

Similar difficulties are typical for the countries of Latin America. Thus, in [10], using the Crespo region (Argentina) example, a scheme of interaction between enterprises of the agro-industrial complex is presented, and a centralized model of interaction of all stakeholders through a regional agricultural cooperative is proposed.

These shortcomings are due to a poorly developed, ineffective organizational structure. Therefore, we need new approaches to their formation, taking into account the totality of all significant factors. Important questions of the methodology for modeling the organizational structures of agricultural formations remain not disclosed and cannot be used in practice accordingly.

The purpose of this article is to propose conceptual approaches to modeling the organizational structures of agricultural formations in Russia, taking into account the specifics of the industry, using modern and innovative strategies for organizational change.

2 Materials and methods

Conceptual problems of modeling organizational structures are in the area of close attention of scientists and specialists. The reason is the rapid changes in external and internal factors. External factors are socio-economic shocks in the form of the spread of coronavirus infection, which took the lives of millions of people and put the stability and resilience of many economies on the line. However, the reason is also in internal factors that can either accelerate the development of complex organizations or negate the implementation of opportunities associated with the synergy effect of improving structural units.

A feature of the current state of the agricultural industry in Russia, as a complex socio-economic system, is the need to adjust the industry structure based on innovative processes that allow actively building effective organizational structures. Today, in the agro-industrial complex of the Russian Federation, processes are taking place aimed at the development of cooperation, which is supported both "from below" and "from above." However, the dominant is the latter option.

So, J. Schumpeter singled out entrepreneurship as the main driving force of the economy. The entrepreneur relies on a general framework set by state institutions. An entrepreneur is an innovator who identifies opportunities for further business development. In this process, the entrepreneur, according to J. Schumpeter, implements the function of "new combinations of production factors" [11]. Among these functions, organizational and economic innovation is through the institution of economic activity of entrepreneurship, including through associations of entrepreneurs. The importance of this organizational and economic innovation is emphasized [12]. This fact means that entrepreneurs can determine the possibility and necessity of forming unifying structures to realize common goals. Thus, cooperatives can become the most effective instrument, functioning on the principles of initiative and independence. The creation of an appropriate business environment becomes a prerequisite.

It should be noted that such complex organizational structures (when they are formed "from below") often do not have effective management. There are not enough financial resources, managerial potential, complicated internal processes of interaction between the subjects of the structure). In the second variant of the development of cooperation, namely with direct support at the level of Heads of regions, line ministries, the organizational structure of agricultural formations takes on a more structurally consistent look, but, in our opinion, with elements of standard administrative management.

Stop factors restrain the development of agricultural cooperation. Stop factors are the insufficient level of the necessary competencies of both state bodies and agricultural entities, the lack of developed cooperative infrastructure and the small number of successful cooperative projects, and the difficulty of maintaining regular activities in monopolizing the agricultural market. This factor is also retail and, in many regions, the dominant presence of prominent market players. For example, in the Perm Territory, the UKMP group of companies has about 75% of the share of raw milk processing. In addition, socio-psychological factors associated with distrust of repeated forms of cooperation in Russia also negatively impact. The existing forms of cooperation have not brought significant positive results for the cooperation and industry members as a whole.

Consequently, a new methodology is needed for the formation of organizational structures of agricultural formations. One way to resolve the issue is to create flexible multi-level cooperation systems, allowing to consider the interests of both members of a single cooperative and a group of cooperatives that synergistically consolidate the available and attracted resources. The experience of the successful operation of the Lipetsk region is known [13].

3 Results and discussion

Thus, the organizational structure should be flexible and adaptable. Not only self-organized but striving for self-development. Thus, the Perm School of Innovation Management is based on the following position: A.V. Molodchik, S.V. Komarov, N.B. Akatov and other researchers. In this methodology, the pivot is the 2C-system based on self-development processes.

Let us highlight the key approaches in the formation of 2C systems.

1. People's behavior is focused on the priority of their internal motivation through the processes of self-development and self-organization.

2. Through the mechanisms of the tension theory, the organization seeks to innovate, gaining some competitive advantage by reducing the potential difference between the desired result and the tense goals.

3. Transition to a self-developing business model based on the principles of organizational excellence.

4. Leadership is a key factor for the formation of competencies in 2C systems [14].

We are in solidarity with the positions of those authors who reveal the development of organizational forms through their multiplicity. There is an urgent need to have numerous effective organizational forms and consider the social nature of their activities [15]. At present, the issue of forming the development of such structures has not been sufficiently worked out.

The state policy in the field of cooperation in agriculture has not been fully implemented. Related bodies for the development of cooperation have been created in each region. Organizational innovations, in our opinion, do not imply a rigid structure and stereotyped, and a new organizational structure should be built on the following fundamental principles:

1. self-organization and self-development,
2. much variance,
3. flexibility and adaptability,
4. initiative and independence,
5. dominant leadership and trust,
6. clarity and transparency of the goals,
7. clustering 8. adequacy of government support,
9. competence of state bodies.

An assessment of the importance of these principles for creating an effective cooperative structure is based on a survey of 24 peasant farms in the Perm Territory. Each of the nine principles was assessed on a five-point scale (1 – not significant, 5 – very significant). Figure 1 shows the results of this study.

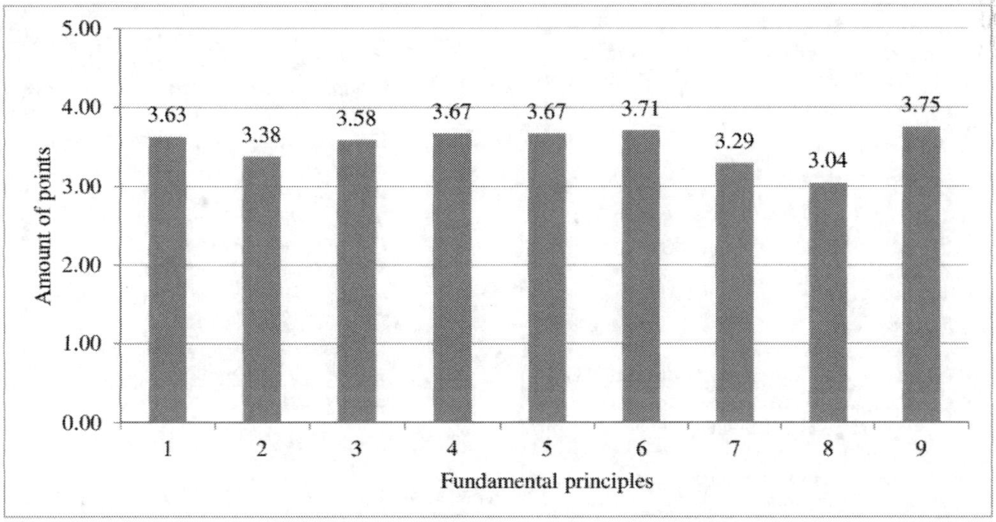

Figure 1. Results of the assessment of the fundamental principles of peasant farming: 1 – self-organization and self-development; 2 – much variation; 3 – flexibility and adaptability; 4 – initiative and independence; 5 – dominant leadership and trust; 6 – clarity and transparency of the goals; 7 – clustering; 8 – sufficiency of state support; 9 – competence of state bodies.

Based on the above principles of building an organizational structure, the following promising cooperation model is proposed (Fig. 2).

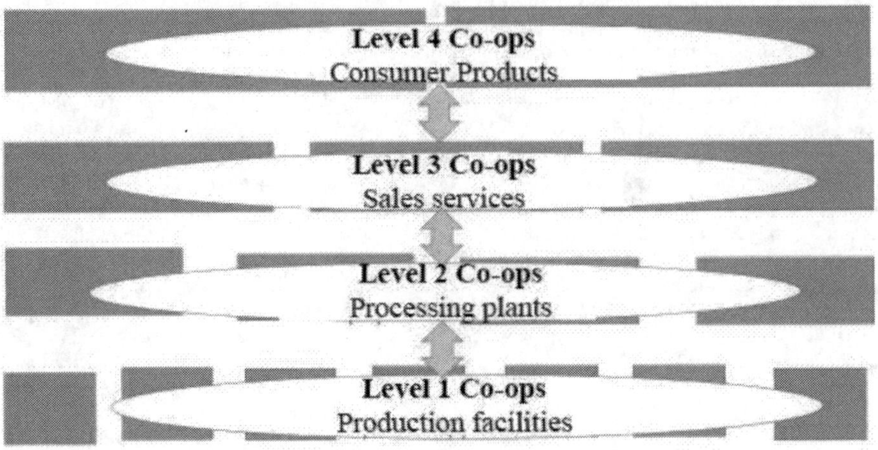

Figure 2.The perspective organizational structure of multilevel cooperation.

4 Conclusion

A feature of the current state of such a complex socio-economic system as the agricultural industry in Russia is the need to adjust the industry structure based on innovative processes that actively build effective organizational structures. Today in the agro-industrial complex of the Russian Federation, processes are taking place aimed at the development of cooperation, which is supported both "from below" and "from above," but the dominant is the latter option.

It is necessary to form a new conceptual approach to the criteria for forming complex organizational structures and measuring the results of the formation of cooperative relationships of self-sustaining structures of cooperative cluster formations

References

1. S.A. Pestrikov, S.Yu.Tyulkina, *Schumpeterianreading: Proceedings of the 3rd International Scientific and Practical Conference,* (Perm, 2013)
2. E.S. Mishchenko, *Organizational management structures* (Tambov, 2011)
3. Presidium of the Council under the President of the Russian Federation for Strategic Development and National Projects, N 16 *Passport of the national project "Small and Medium Enterprises and Support for Individual Entrepreneurial Initiatives"* (2018)
4. E. A. Aleshina, E. I. Anisimova, D. V. Serdobintsev, *IOP Conf. Ser.: Earth Environ. Sci.* **459** 062018 (2020).
5. N. Y. Timofeeva *International Journal of Engineering and Technology,* **7**, (2018).
6. O. Isaeva, M. Kabanenko, A. Chistyakov, L. Dubrova, N. Filin, *E3S Web of Conferences.* **175.** 13022. (2020). 10.1051/e3sconf/202017513022.
7. I. KravčákováVozárová, R. Kotulič, R. Vavrek, *Applied Economics,* **51:56**, 5999-6008, (2019) DOI: 10.1080/00036846.2019.1645285
8. L. Xiao, *International Journal of Security and Its Applications* **10 (3)** pp.333-344 (2016)
9. C. Han, A. Pervez, J. Wu, X. Shen, D. Zhang, *Sustainability* **12(16),** 6547; (2020) https://doi.org/10.3390/su12166547
10. R. SolenoWilches, *Cambiosestructurales e institucionales delcooperativismo agrícola entrerrianoen la postconvertibilidad.* (2017).
11. J. Schumpeter *The theory of economic development: a study of entrepreneurial profit, capital, credit, interest and the conjuncture cycle.* (1982)
12. E.V. Antipina *The innovative concept of entrepreneurship by J. Schumpeter* (2014).
13. R.G. Yanbykh, A.V. Morozov, G.I. Yavkina *On the Development of Agricultural Consumer Cooperation in Russia* (2014)
14. S.V. Komarov *From self-organization to self-development: a change in the management paradigm.* (Yekaterinburg, 2013)
15. L.V. Obedkova *Improving the management system of cooperative and structures in modern Russia* (2011).

 https://doi.org/10.1051/shsconf/202111600063

Analytical grouping of accounting for wages in health care institutions

Elena Golova[1,*], *Irina* Baranova[1], and *Marina* Gapon[1]

[1] Omsk State Agrarian University named after P.A. Stolypin, 8E, Fizkulturnaya street, 203, Omsk, Russia

Abstract. An important factor in the uninterrupted work of healthcare is a modern and decent salary for medical personnel. This issue has become especially acute in coronavirus infection when all the problems that have accumulated in the field of healthcare have become especially urgent. There was a situation of a shortage of medical personnel and the need for round-the-clock work, which led to the issue of incentive and compensation payments. In this regard, accounting is one of the innovative links that ensure uninterrupted payments to employees of medical institutions. All facts of the economic activity of the institution must be documented. However, the existing set of documentation is not always able to ensure the prompt provision of information on payroll and incentive payments, even in the context of accounting automation. The article contains research of the workflow organization in terms of remuneration on the example of one of the medical institutions in the Omsk region. The study revealed the need to improve the accounting area as one of the ways of innovative development. The article proposes a new analytical accounting register, which allows presenting information on charges to medical personnel in the context of personnel categories and comparing analytical and synthetic accounting data in terms of the amounts of accrued wages. It is proposed to integrate the proposed document into the Sail Budget 7 program, the Payroll module, which will make the calculation process more convenient, less labor-intensive, and will allow the data transfer process to be carried out automatically.

1 Introduction

The health care system is one of the basic elements of the quality of life of the population of any country and an area open to innovation. It is healthcare that ensures life expectancy and dignified self-awareness of a person. A comfortable standard of living is not possible in this area without a well-applied system. Health care is going through tough times right now. Nevertheless, according to the Decree of the President of the Russian Federation of 06.06. No. 254 "On the strategy for the development of healthcare in the Russian Federation for the period up to 2025" listed the threats that may arise in the development of this area. Considering the importance and significance of this system for the population of Russia, the state should always be ready for the manifestation of various dangerous epidemiological situations. The provision of hospitals with personnel plays an important role in this. Reducing the cost of financing the health sector led to the fact that staff shortages came to the fore during the pandemic, which showed the urgency of this issue and the need to address it. The staffing issues are especially relevant today, primarily due to the low level of wages, which served to make various decisions at all levels of government to increase the average wage and optimize the structure of the wage fund and the level of part-time employment [1]. In this regard, accounting is the source of information for the analysis of medical personnel. It is obvious that labor costs in the structure of total expenditures of budgetary health care institutions occupy the largest share. Consequently, the section for accounting for payroll calculations is the most voluminous, labor-intensive, and responsible. In addition, this section interacts with other services of any institution: the planning and economic department, the personnel department [2].

Currently, much attention is paid to the study of accounting in health care. This situation is understandable: the growth and spread of coronavirus infection made us look at the financing of this area differently. The study of the peculiarities of accounting in non-profit organizations in domestic practice was carried out by such scientists as: E.A. Ananieva, M.V. Belousova, A.S. Bakaev, E.R. Bakhankova, N.P. Balandina, R. Busse, A. R. Gazizova, N. A. Koshevarova, G.V. Kuzmin, I.N. Lozhnikov, A.S. Morozov, S.I. Oparina, V.I. Radachinsky, S.N. Starostin, V.V. Semenikhin, I.A. Feldman, K. A. Fedotova, O. V. Flek and others. Despite deep research in the field of accounting in budgetary health care institutions, it should be noted that the following issues still remain unaddressed: the validity of government funding, the reflection in the accounting system of cash flows from several budgets simultaneously, as well as the formation of a system of accounts and reporting in hospitals, the pricing of medical services, the lack of attribution of various kinds of costs to the cost of services and other problems [3].

In the context of the spread of coronavirus infection, financing of medical institutions came to the fore. The

* Corresponding author:ee.golova@omgau.org

source of information, in this case, is often accounting. It serves as a link between the authorities and the health care system. Therefore, well-organized and well-established accounting is of paramount importance. An important role in providing the information is assigned to the organization of document circulation. Fundamental in accounting is primary accounting and document flow, chart of accounts, accounting forms, and reporting forms [4]. However, recently there has been a need to form the forms of primary and consolidated documentation in a format that would make it possible to present information in the following way. It is necessary to provide generalized information about the accrued payments by medical workers in the context of various categories of workers. Such information could serve as a basis for the formation of operational, statistical reporting of health care institutions. In this regard, we believe that it is necessary to develop a system of an analytical grouping of the payroll of medical workers [5].

The subject of study is a set of theoretical and practical aspects related to document management and the creation of a perfect system of an analytical grouping of the wage fund. Work in these areas will make it possible to increase the transparency and efficiency of charges to medical workers in the context of personnel groups, types of charges, and sources of funding. The study's practical significance lies in improving the individual stages of the workflow for organizing the accounting of payments for labor in budgetary healthcare institutions. The object of the research was the activities of budgetary healthcare institutions of the Omsk region.

The purpose of this study is to develop a system for analytical grouping of the accrued wages fund, which makes it possible to evaluate it in the context of personnel groups, types of charges, and funding sources for subsequent analysis and compilation of operational, statistical reporting of health care institutions.

2 Materials and Methods

The methodological basis of the study is the general scientific principles of the study, suggesting an integrated and systematic approach. The economic substantiation of the main conclusions and recommendations was carried out based on general scientific techniques and methods, such as analysis and synthesis, induction and deduction, comparison, generalization. To process the initial information, the standard software package "Microsoft Word," "Microsoft Excel" was used, as well as the automated accounting system "Software package" PARUS – Budget 7" in terms of the modules "Salary" and "Accounting" [6].

3 Results and discussion

Forms of primary accounting documents for budgetary institutions, accounting registers, as well as guidelines for their application, approved by Order of the Ministry of Finance of Russia dated March 30, 2015, No. 52n "On approval of forms of primary accounting documents and accounting registers used by state authorities (state bodies), local self-government bodies, governing bodies of state extra-budgetary funds, state (municipal) institutions, and Guidelines for their application." Documentation in medical institutions must be properly maintained following the above requirements, which are indicated in Order No. 52. However, we believe that important factors are compliance with all regulatory requirements and the correct documentation of primary and consolidated accounting documents. The reason is that wages are calculated based on their basis, the timeliness and accuracy of settlements with employees, and the state regarding taxes and insurance premiums [7]. To substantiate the purpose of the study, let us consider the procedure for calculating wages in one of the medical institutions – the budgetary health care institution of the Omsk region "City Children's Clinical Hospital No. 2" (after this – the Institution). The collective agreement establishes the current remuneration system in the Institution, the Regulations on the remuneration of employees of the Institution, approved by order of the chief physician of 03/18/2016, No. 108, adopted following the current legislation, other applicable regulatory legal acts. Following the Regulation, the remuneration system in the Institution consists of two parts, one of which is basic and involves the inclusion of salaries according to the position, various kinds of incentives, as well as compensation payments, and other payments. In turn, official salaries are a fixed amount and are spelled out in the Regulations, which is established according to qualification groups by position. Institutions can establish the incentive part and compensation payments in absolute or relative terms; however, this must be indicated in the Regulation. The entire wage fund in institutions is divided into the basic fund and the additional one. The primary salary fund of medical personnel consists of salaries, compensation, and incentive payments, which are assigned according to the intensity of work of the employee of the Institution involved in the organization and implementation of entrepreneurial activity and an increase for the presence of a qualification category.

The distribution of the additional fund is carried out based on an expert assessment of the work of medical and pharmaceutical personnel prepared by a commission that appoints such payments to employees [8]. The main analytical form required for analyzing wage costs, including distributing the fund of incentive payments, is a chess sheet for individuals on accrued wages. There are more than 500 lines and 99 columns. Its processing is carried out by the planning and economic department by hand twice a month (Fig. 1). Considering the organization of work on the interaction of settlements in the Institution, it is obvious that this information has to be regrouped manually, which requires labor time. In addition, the likelihood of errors in the preparation of reports and operational information for the manager increases. Consequently, we have determined the purpose of the study, which requires improvement in the accounting of labor costs – the organization of analytical accounting, capable of ensuring the systematization and grouping of information in a convenient and understandable form for the organization of management accounting.

Sheet for accruals
Period: November 2019

Formulated for all personal accounts with accruals
Calculation of salaries; Calculation of salaries of state employees, FSS benefits, young professionals, lump sum payments, other benefits, secondary job calculations

Formed 12.05.2021 13-06-12

Personal account	SNILS	Position	Subdivision	Norm	Actually worked out	Accruals: remuneration	Secondary job	Scheduled duty	payments for harmfulness	Secondary job payments for harmfulness	Scheduled duty harmfulness	Total
1	2	3	4	5	6	7	8	9	10	11	12	98
				156	156	23400			936			83430
				144	144	11400			456			35902
				156	156	12000			480			27513
				156	78	15000	6000		480	240		7176

Figure. 1. Fragment of the chess payroll for November 2019

The accounting policy was approved for the purposes of accounting in the Institution by order of the head of the Institution dated December 29, 2017, № 203. According to the provisions of the accounting policy, the head is responsible for the organization of accounting, compliance with the law when performing business operations, storage of accounting documents in the Institution. At the same time, the accounting department fulfills the obligation to maintain accounting records. Also, the accounting policy established a workflow schedule, a fragment relating to the investigated accounting area. Expenses are accounted for using the codes of the budget classification approved by Order № 65n. An account working chart is approved in the appendix to the even policy, in which the accounts are included; however, they do not have an analytical code. Accounting in the Institution is carried out using the program "Sail 7 – Salary" [9]. Analytical accounting systems are based on the use of personal accounts of employees with whom the Institution has entered into an employment contract or civil law contracts. All information is entered and systematized based on primary documentation for payroll accounting. This information is transferred on a cumulative basis in the Journal of calculations for wages, salaries, and scholarships № 6. This Journal is kept separately for each source of financial support of activities and is automated. This operation is carried out by exporting the generated postings of their complex for wages to another automated complex, "Parus Budget-8 Accounting." However, in our opinion, the organization of analytical accounting of payments for labor in the Institution requires a separate consideration [10].

The accounting policy of the Institution enshrines the provision on maintaining analytical accounting of these expenses in the personal accounts of employees with whom labor contracts have been concluded and other individuals with whom civil law contracts have been concluded. At the same time, this provision does not specify the analytical accounting register, regulated by paragraph 257 of Order No. 157n. It follows from this order that the analytical accounting of payments for wages and scholarships is kept in the Journal of calculations for wages, salaries, and scholarships in the manner established by the Institution as part of the formation of accounting policies. In fact, analytics on accrued wages for subsequent analysis and various reports are carried out by the Institution in the chess register. This statement is quite lengthy and requires additional processing before using the data on the accrued wages of the primary person of the medical Institution. In modern automation, we consider these processes unacceptable in an institution because they require additional labor time for employees [11].

4 Conclusion

In order to improve the workflow process for calculating the salary of medical personnel, we propose to consider the possibility of detailing the analytical code of the accounting account in order to separate the accrued amounts of salary in the context of categories of the primary personnel of medical workers (Fig. 2).

№	Analytical code of a synthetic invoice	Code of the type of financial support for the activity	Synthetic invoice	Name of the type of financial security (activity)	Note
Applied analytical accounting account					
1	9013900000000110	7	30211	Funds for compulsory health insurance	It is applied by the BUSOO in accordance with the approved FHD Plan
Suggested analytical accounting account					
1	09011390000000Б1 11	7	30211	Funds for compulsory health insurance	It is used for grouping the amounts of payroll payments by the category of personnel "Doctors"
2	09011390000000С1 12	7	30211	Funds for compulsory health insurance	It is used for grouping the amounts of salary accruals by the category of personnel "Secondary medical personnel"
3	09011390000000Р1 13	7	30211	Funds for compulsory health insurance	It is used for grouping the amounts of salary accruals by the category of personnel "Administrative and managerial personnel"
4	09011390000000П1 14	7	30211	Funds for compulsory health insurance	It is used for grouping the amounts of payroll payments by the category of personnel "Other personnel"

Figure 2. Author's version of the working chart of accounts using analytics in the context of medical personnel

The proposed detailing will allow analytical accounting registers to form in a generalized form without specifying a specific person. The purpose of the analytical accounting registers will be efficiency in the preparation of analytical information and the timeliness and accuracy of the provision of monitoring, reports, and other operations [12].

The introduction of a detailed account of analytical accounting will allow comparing the data of analytical and synthetic accounting in terms of the amounts of accrued wages in the context of categories of employees of the main personnel. At the same time, there is a risk associated with the impossibility of implementing these proposals. This process is due to the centralization of accounting organization by the Ministry of Health of the Omsk region. In order to minimize this risk, it is proposed to develop an analytical accounting register form based on the available capabilities of the Parus Budget 7 software product, payroll module. In the "Personal account" tab, in our opinion, the detailing "Personnel category" should be added: doctors and pharmacists; nursing staff; administrative and managerial; other personnel (Fig. 3).

Personal account	SNILS	Post	Category	Division	Personnel category
1	2	3	3	4	5
					Doctors and pharmacists
					Average medical staff
					Administrative and managerial personnel
					Other personnel

Figure 3. Author's fragment of the "Personal account" tab

In continuing to improve the workflow in terms of remuneration, we consider it necessary to introduce into the work of the program the report form "Summary of accrued wages by categories of workers" (Fig. 4).

Institution: BUZ NGO "City Children's Clinical Hospital No. 2 named after V. P. Bisyarina"
Structural division: Ministry of Health of the Omsk Region
Name of the founder
Name of the budget
Unit of measurement

Summary of accrued wages by category of employees for the period " " ______ 20___

Name of payments		Administrative and managerial staff	Doctors	Secondary medical personnel	Other medical personnel	Total
Salaries						
Compensation payments						
Payments to employees for harmful working conditions						
Premium for working with state secret information						
Payments for working in conditions that deviate from normal conditions	Combining professions					
	Increasing the scope of work					
	Performing the duties of a temporarily absent employee					
	Overtime work					
	Working at night					
	Working on weekends and holidays					
	Working in other conditions that deviate from normal conditions					
Payments based on the district coefficient						
Incentive payments						
Intensity and high results	Labor intensity allowance					
	Award for the performance of particularly important and responsible work					
	Award for high performance					
Quality of work performed	Premium for the presence of a qualifying category					
	Premium for the performance of employee performance indicators					
Bonus payments based on the results of the period						
Basic salary total						
Other payments						
Total						

Figure 4. The author's form of the analytical accounting register "Summary of accrued wages in the context of employee categories."

Thus, the need to improve the organization of analytical accounting of these calculations was identified. A system of an analytical grouping of the accrued wages fund "Summary of accrued wages by categories of employees for the period" was proposed for implementation, the implementation of which is possible in two ways:

- by detailing the analytical code of account 302 11 "Payroll calculations," – a set of the author's working chart of accounts for this account has been developed;
- by inroducing additional detailing "Personnel category" in the "Personal account" tab "Software package" PARUS – Budget 7 "module on wages.

References

1. A.G. Basova, *Problems of Science.* **4 (17)**. 53-59. (2017)
2. O. P. Goleva, Z. B. Tasova, O.B. Kun, E. S. Tkachenko, A.V. Sabaev *Health Economics: A Study Guide.* (Omsk: Publishing house of Omsk State Medical University, 2019)
3. E. E. Golova, I.V. Baranova, Yu. G. Balandaeva. *Proceedings of the International Conference on Policies and Economics Measures for Agricultural Development Advances in Economics, Business and Management Research.* Pp. 30-35. (2020)

4. E. Golova, D. Baetova, O. Zaitseva, Y. Novikov *The Fifth Technological Order: Prospects for the Development and Modernization of the Russian Agro-Industrial Sector.* Pp. 103-108. (2019)
5. E.S. Dashkova *VSU Bulletin. Series: Economics and Management*, **2**, 67–74. (2018)
6. F. N. Kadyrov *Economic methods for assessing the effectiveness of medical institutions.* (Moscow: Healthcare Manager, 2019)
7. E. Kalinichenko *Top 10 changes from 2021 for a public sector accountant.* Retrieved from: https://school.kontur.ru/publications/1972 (2021)
8. R. Mukhamedzhanov, Increase of salaries for doctors in 2021 in Russia by region. Retrieved from: https://www.9111.ru/questions/7777777771223773/ (2021)
9. V. V. Sidavskaya *Scientific and practical electronic journal Alley of Science,* **16**, 62–65. (2017)
10. V. N. Filatov, *Bulletin of the North-Western State Medical University named after I.I. Mechnikov,* **1,** 87–97 (2017)
11. N. A, Sharapova *Accounting in budgetary and non-profit organizations*, **19,** 9–20 (2017)
12. O.V. Shumakova, M.N. Gapon, O.A. Blinov, V.Y. Epanchintsev, Y.I. Novikov *Entrepreneurship and Sustainability Issues,* **5(4)**, 736-747, (2018)

SHS Web of Conferences **116**, 00064 (2021) https://doi.org/10.1051/shsconf/202111600064
ICSR 2021

On methodological individualism, holism and management in the light "Tectology" by A.A. Bogdanov

Alexey Antonov*

Russia, Perm National Research Polytechnic University, 614990, Perm, Komsomolsky Avenue, 29

Abstract. The consideration for the improvement of the efficiency of social economy made researchers look for names, views and methods which were not fully appreciated. This includes the general organizational science which occupies a prominent position by A.A. Bogdanov. There is no doubt that the ideas of A.A. Bogdanov are reflected in systems theory, cybernetics, medicine and philosophy of technology, aesthetics, linguistics and other disciplines. However they had little influence on economy, despite the fact that A.A. Bogdanov was a recognized expert in this area. This paper tries to fill this gap to some extent, and show that the ideas of Tectology are very relevant in modern economic discussions, including in the debate between methodological individualism and methodological holism. A.A. Bogdanov did not set out to study the organization of relations between people in the production process. This is the purpose of modern management. However, his approach to the study of the relationship between an individual and a team, the analysis of the conditions for the appearance of emergent effects in industrial cooperation deserve to be considered in economic theory and in management practice. Moreover, this is correlates to the modern trend to discover and study and, ultimately, use intangible factors of increasing the productivity of social economy in favor of man. In addition, the organizational views of the works of A.A. Bogdanov allow taking a fresh look at the real relationship between methodological holism and methodological individualism both in the structure of management and economy in general.

1 Introduction

There are several factors of productivity in history of economics. The classics, including Karl Marx, considered only labor productive. Jean-Baptiste Say expanded this list and included land and capital. Pierre-Joseph Proudhon believed that "... neither land, nor labor, nor capital is unproductive", [1, p. 118] and the surplus product is formed only through the joint use of these factors. Finally, in 1890 A. Marshall added a new factor of productivity in the economy – organization to the above-mentioned [2, p. 208]. This can be considered the beginning of the theoretical understanding of management.

At the same time, A. Marshall laid the basis of neoclassical theory in economics with its central concept of an economic person making rational choices. Since then, the principle of methodological individualism has become dominant in neoclassical mainstream.

Methodological individualism initially played the same role in management as well. For example F. Taylor noted: "... when working together in a group, each worker becomes less active than when his personal pride is hurt; when working in a group, the productivity of workers invariably decreases to the level of the most sluggish people in this group…" [3, p. 59]. Also the use of the conveyor was by G. Ford also aimed to increase the personal productivity of workers.

However by the twenties of the twentieth century, it became clear that the possibilities for the increase in the intensity of personal labor are not unlimited. As a result, a new source of labor productivity was discovered - the organization of relations between working people.

Within the framework of the program of scientific organization of labor, one of the leading specialists of the People's Commissariat of the Workers' and Peasants' Inspection of the USSR N.A. Vitke formulated the concept of "human relations". It is possible to say that already in its depths the principle of methodological holism was used (from ancient Greek ὅλος "whole"). This, in particular, is evidenced by the fact that N.A. Vitke actively used concepts such as "collective labor activity", "human factor of production" etc. However, under the conditions of tough party leadership, the new science of human resources management then seemed superfluous and even suspicious.

It is not surprising that new ideas eventually penetrated into the United States. As a result, in global community, the American researcher E. Mayo became the author of "management of human relations". After that, the ideas of collective labor activity clearly manifested themselves in the practice of Japanese management.

* Corresponding author: akvizit@yandex.ru

2 Problem Statement

How a certain surplus in relations between people is born suddenly due to good organization? If we turn to nature, then, of course, we will find how this happens. As soon as we streamline the internal bonds, we will not get graphite, but diamond from the same carbon atoms.

Methodological individualism considers people as ordinary atoms, but atoms have no will, so they can only be rearranged externally. It is not surprising that this is the path that the American type of management follows today. However just as in the real transformation of graphite into diamond, high pressure is required, so the same pressure is required in the external restructuring of a group. As a result, American management got used to formal, almost cool attitude towards people. The phrase "Nothing personal" is just from the vocabulary of American managers.

At the same time, the American type of management is just the application of neoclassical theory in practice. From the point of view of methodological individualism, "a complex is made up of the simple" (highlighted in bold italics by M.V. Popov – A.A.) [4, p. 7]. Whereas, according the opinion of methodological holists, a complex is always something more than a simple sum.

That is why the issue of the fundamental nature of the principle of methodological individualism in management is the issue of the correctness of neoclassical theory, or at least its completeness.

3 Research Questions

Of all the variety of aspects of management, in this paper we turned to only one of its aspects – the consideration of the manifestation of the principles of methodological individualism and holism in it in the light of "Tectology" by A.A. Bogdanov.

4 Purpose of the Study

The purpose of this paper is to try to show the extent to which the practice of modern management confirms the fundamental principle of methodological individualism in economics.

5 Research Methods

The main methods used in this research are historical-comparative and logical.

6 Findings

Of course, people are not atoms. They are conscious and it seems that this makes them much easier to organize. It is exactly because people are not atoms they turn out to be very different, even when they are doing the same thing in fact, as is evident from the parable of Chartres Cathedral.

Three people pushing carts with stones were asked what they were doing. The first man replied: "I move the damn stones". The second man said that he was making money for his family and only the third man answered: "I am building Chartres Cathedral". As we can see, the goals of people participating in joint work almost never coincide in fact.

It follows that joint work must always be managed in order to achieve unity of purpose. As A.A. Bogdanov noted "to organize means to group people for some purpose, coordinate and regulate their actions in the spirit of *expedient unity*" (italics by A.A. Bogdanov – A.A.) [5, p. 69].

However, A.A. Bogdanov had little attention of his fellow economists. First of all, "... there is a shortage of scientific works, where the connection of the general organizational science with management itself is considered" [6, p. 155].

It is A.A. Bogdanov discovered that, unlike atoms, which have no will, the internal orientation of people (or their intentionality) can be different. This means that the actions of people themselves must be described by direction (that is, directed) and not scalar values.

In "Tectology" A.A. Bogdanov studied a clear paradox: two or three people who worked together could do more or less than double or triple work. As A.A. Bogdanov "both cases entirely depended on the method of combining these forces. In the first case, the statement that the whole turned out to be *practically larger* (italics by A.A. Bogdanov – A.A.) was quite legitimate than the simple sum of its parts, in the second case it was practically less than it" [5, p. 114].

From the point of view of everyday consciousness, "here we get some kind of creation out of nothing". Meanwhile, the emergence effect (most often it is expressed mathematically: $2 + 2 > 4$) can be easily explained if we pay attention to the fact that any of the employees of a team has his own "activity" as A.A. Bogdanov mentioned. Moreover, in terms of its direction, it practically never coincides with the activity of other workers, or it may even meet their "resistance" (that is, oppositely directed activity). The less resistance the activity of workers will meet, the greater will be their total "analytical sum". This is what determines the "surplus" of jointly organized labor.

Of course, A.A. Bogdanov broke new ground in the creation of organizational science and formulated its laws only in the most general form. That is why he also considered the internal orientations of workers. However, when solving economic problems, it seems that this can be abandoned. Indeed, during the production of goods, it does not matter whether someone hesitates or keeps complete calm. If there is no external activity, there will be no produced goods either. It turns out that it makes no sense to talk about the "analytical sums" of "activities", taking into account the internal ones, but it is necessary to talk about the vector addition of externally expressed forces and their resultant.

Nevertheless, A.A. Bogdanov clearly shows that vectoriality and expediency are identical only in the work of the atomic individual. In a joint or "holistic" work, the addition of vector values can give not necessarily a positive, but even a negative result. In fact, in this regard we have the anatomy of the principle of methodological holism. However, this discovery by A.A. Bogdanov did not find understanding among his fellow Marxists, although, it seemed to be that K. Marx agreed that "... the social productive force of labor arose from the cooperation itself" [7, p. 341].

The point is that cooperation was known long before the appearance of capitalism. K. Marx considered it as nothing more than an important and necessary prerequisite. All the advantages that cooperation gave were considered as something like free water or air for the first mills. K. Marx had to discover the law of self-movement of capitalist society itself.

However, it is most important that K. Marx was a methodological individualist in contrast to popular belief and not a holist in theory. Everywhere he viewed capitalist society as atomistic: "In the form of society now under consideration, - wrote K. Marx, - behavior of men in the social process of production is purely atomic" [7, p. 102-103]. This is why he focused on personal "labor" in order to explain the emergence of surplus value and not on the social productive power of labor, born of capitalist cooperation. At the same time, K. Marx for the complete view noted that "... the capitalist form of cooperation from the very beginning presupposed a free hired worker selling his labor power to capital" [7, p. 346].

Pierre-Joseph Proudhon was a methodological holist. He believed that the capitalist was not in hurry to pay for "... the enormous power that arose due to the cooperation of workers, simultaneity and harmony of their efforts. Within a few hours, two hundred grenadiers erected the Luxor obelisk. Is it possible to assume that a person within 200 hours could also do this? After all, according to the calculations of the capitalists the remuneration of these 200 grenadiers and this person should be the equal" [1, p. 86]. Proudhon indicated the very essence of capitalist cooperation, arguing that "even if you paid for all the individual forces, then you did not pay for the collective forces ..." [1, p. 87].

Nevertheless, according to the analysis by A.A. Bogdanov, the methodological individualist K. Marx and the methodological holist P. Zh. Proudhon were clearly mistaken, believing that the "social productive force of labor", being the resultant of the forces of all workers, arises in cooperation by itself, and even in the form in which the capitalist needs it. For example, as A.F. Sayfutdinova notes "... in practice, it is not uncommon for an accidental synergy of employees to occur, which most often leads to a negative synergy effect at an enterprise" [8, p. 134].

For this reason, in the early stages of the development of capitalism, the bourgeois had to not only invent the necessary form of cooperation, which led to the emergence of the social productive force of labor, but also manually maintain its functioning. Only later this function was transferred to professional managers. And today, the forms of cooperation that ensure the emergence of the social productive force of labor, as ready-made business models, are already freely sold and leased like ordinary goods, for example, in the form of franchising.

However, it is necessary to admit that unlike P. Zh. Proudhon, Marx was mistaken twice in this case. Indeed, the fact that he replaced labor with "labor force", in fact changed nothing. In "Capital", K. Marx assured readers that an ordinary deal between a capitalist and a worker was completely equivalent. Then it turned out that, being a monopoly in terms of the means of production, a capitalist simply forced a worker to agree to his terms.

In this case, what remains of the equivalence of exchange?! After all, any monopoly, in fact, denies the laws of the market. This means that here we have either the exchange of equivalents in the absence of exploitation, or, conversely, the operation without the exchange of equivalents. Karl Marx clearly failed to combine these two extremes.

K. Marx called labor force as "the ability to work" [7, p. 178]. At the same time, according to his own words, "the ability to work does not mean work, just as the ability to digest food does not yet coincide with the actual digestion of food" [7, p. 184]. It seems that we have the right to agree with K. Marx here. After all, as a rule, a worker can hardly know what he has to do with a capitalist. That is why he offers his pure "ability to work" on the market. A capitalist does not buy the same thing from him and it is another matter. "Future labor" or "future commodity" is the precise name of what a capitalist buys from a worker. In fact, this is not an employment contract at all, but a futures transaction.

As K. Marx wrote: "In all countries with a capitalist mode of production, labor is paid only after it has already functioned during the period established by the contract..." [7, p. 185]. This means that in reality a capitalist did not buy Marx's "labor power" from a worker, but simply labor that delayed for a while. In the end, even K. Marx was forced to admit that the "use value" that a worker delivered to a capitalist was not actually labor power, but its function, certain useful labor, the labor of a tailor, shoemaker, spinner, etc." [7, p. 551]. Thus, in fact, "...labor was bought, although this was mediated by the exchange for labor, and not directly for work ..." [9, p. 62].

It turns out that the substitution of "labor" for "labor power", which the Marxists declared as a solution to the problem of surplus value, in fact, is not such a solution. The only solution would be such a case when a capitalist have surplus value after he paid all the labor time of all workers at market prices. "These are the conditions of the problem. *Hic Rhodus, hic salta!*" [7, p. 177], as K. Marx himself expressed it in such cases.

It is curious that in practice the capitalists have solved this task long ago. The same cooperation, which is holistic in nature, helped them in it. That is why Proudhon accused the capitalists of the fact that having paid for all "individual forces", they did not pay for "collective forces". On the other hand, there is no need to blame the capitalists for exploitation. After all, they can pay workers the market equivalent sought by K. Marx. To do this, a capitalist only needs to find a suitable form of cooperation of workers with the means of production and with each other and calculate its organizational consequences. To do this, it is not at all necessary to be engaged only in production. In any case, capitalism began with its commercial form both historically and logically.

But, of course, no one will fully pay the workers for their "individual strength", being content only with "collective strength" as long as it is possible to obtain surplus value in the usual way. The surplus of labor generated by management becomes in demand only in conditions of developed competition.

K. Marx ignored these theoretical possibilities because he was a methodological individualist. In all cases, K. Marx considered collective labor as the usual sum of expended personal forces "... from the point of view of the production of value, it is completely indifferent whether 1,200 workers each produce separately, or whether they are united together under the command of the same capital" [7, p. 333-334].

However, according to "Tectology" by A.A. Bogdanov "the workers do not behave like methodological atoms. They all have different degrees of directed activity and as machines are always fully involved in the labor process and always only partly in the process of value formation", [7, p. 399] and workers united by joint labor all participate in the labor process, but not all of them participate in the formation of the social productive force of labor.

In the first volume of Capital, K. Marx considered the organic structure of capital (c + v) as an average [7, p. 626-627]. At the same time, K. Marx always studied variable capital (v) as a whole value. This means that in fact, K. Marx always considered the variable capital also as averaged one. Indeed, as it was revealed in "Tectology", behind V = 1 the ratio of 50: 50 of Bogdanov's activities and resistances was hidden.

However, in real life, only 40 percent of the staff of a company can actively strive for overall success, while in other firms this proportion may be 35/65, 20/80, 60/40, etc. While capitalism was experiencing a phase of extensive development, the difference in the internal structure of variable capital simply could not be noticeable. However, as soon as competition intensified, managers had to pay attention to those human reserves that they have in their teams.

In this regard it became clear that in order to turn the average statistical team - 50/50 (and even more 40/60 or 30/70) into a monolith aimed at success we need a great organizational work on the exact setting of tasks, distribution of powers, regulation of the responsibilities of each employee, etc. At the same time, the "organizational" increase in labor occurs not in additional, but in the same working time. And here we are not talking about a change in the quality of labor or its intensity, but only about the quality of the organization of external forms of the same labor that forms the value.

This organizational source of surplus labor will be used in the future in the practice of management. In other words, management has discovered in practice those integral truths that were already theoretically discovered in A. A. Bogdanov's "Tectology".

7 Conclusion

Management clearly showed the limitations of only one principle of methodological individualism. In any case, in Bogdanov's Tectology it is obvious that the profit that managers get from the organization of relations between people nevertheless is entirely holistic in nature although it is based on the activity of individual persons.

This means that the debate between methodological holists and methodological individualists is essentially not productive, because "... two ontological theories can disagree, not because they assert inconsistent claims about the world, but because they are partial representations of aspects of the world, and both perspectives may potentially be valuable for theory and partially true; then the fact that two ontological theories are different and highlight different aspects of the world does not entail that one of the theories must be false." [10, p. 21].

References

1. P.Zh. Proudhon What is property? Or Study on the Principle of Law and Power; Poverty as an Economic Principle; Pornocracy, or Women at the Present Time. (Moscow: Republic, 1998)
2. Marshall Principles of Political Economy. (Moscow: Progress Publishing house, 1983)
3. F. Taylor Scientific foundations of the organization of industrial enterprises. (SPB, 1912)
4. M.V. Popov Social dialectics. (SPB, Polytechnic University publishing house, 2014)
5. A.A. Bogdanov (Malinovsky) Tectology: A General Organizational Science. Book 1.(Moscow: Economics, 1989)
6. E A. Kuznetsov EKON.: real. hour. 3 (19) (2015)
7. K. Marx. Capital. Criticism of Political Economy. Volume I. (Moscow: Politizdat, 1960)
8. A.B. Sayfutdinova. Actual issues of economy, 3 (2011)
9. K. Marx. The theory of surplus value (IV volume of "Capital"). Part one (chapters I - VII). (Moscow: Politizdat, 1962)
10. D. Little. Philosophy of the Social Science, 51 (1) (2020)

 https://doi.org/10.1051/shsconf/202111600065

Conceptual approach to depreciation potential

Mariya Shchukina[1*]

[1] Perm National Research Polytechnic University, Department of Economics and Finance, 614000, Komsomolsky avenue 29, Perm, Russia

Abstract. The paper considers the category of "economic potential", reveals the approaches of researchers to the understanding of economic potential. Investment potential is one of the components of economic potential. The presented indicators (factors) reflecting the investment potential do not reflect the sources of investment in fixed assets. In this regard, the author proposes to introduce the concept of "depreciation potential". The content of the depreciation potential is clarified. The need for the importance of the use of depreciation charges as a source of funding is substantiated. The author proposes to single out the real and nominal depreciation fund, the ratio of which allows calculating the coefficient (index) of the use of the depreciation fund, which is one of the indicators to assess the depreciation potential. The proposed coefficient is calculated on the basis of statistical data from the Perm Territory. The obtained values indicate the intended use of the depreciation fund of enterprises of the Perm Territory.

1 Introduction

The risks of economic activity, crisis phenomena and unfavorable epidemiological situation negatively affects the functioning of any socio-economic system: an enterprise, a region, a country. One of the ways to improve the efficiency of the system at various levels is the mobilization of internal reserves and the rational use of available resources. Competent depreciation policy, contributing to the intended use of depreciation deductions, is one of the mechanisms leading to the increase in the competitiveness of an economic entity by increasing the technical level of the applied fixed assets.

When investment financing in fixed assets, economic entities use both their own funds and borrowed funds. Net profit and depreciation charges are considered as own sources of financing. Loans from banking sector and other organizations, financial resources from the budget, foreign investments, funds of state extra-budgetary funds are considered as attracted funds.

Mazurina T.Yu. notes that in the period from 2000 to 2009 there was a tendency to the increase in the share of borrowed funds, which reflected the instability of financial situation [1]. However, as statistics show, over the past decade, the opposite situation has been observed. Today there is an increase in the proportion which is attributed to own sources when financing fixed assets. Table 1 reflects the structure of sources of financing investments in fixed assets [2,3].

According to the data in the table, it can be seen that there is a steady trend towards the increase in the share of own funds over the studied period. Average share in the period from 2008 to 2013 is 41%, and in the period from 2014 to 2019 - 51%. A 10% reduction in the average share of borrowed funds indicates the increase in the financial stability of organizations.

Table 1. Structure of sources for investment financing in fixed assets

Source share, %	2008	2009	2010	2011	2012	2013	2014	2015	2016	2017	2018	2019
Own	39,5	37,1	41	41,9	44,5	45,2	45,7	50,2	51	51,3	53,1	55,1
Attracted	60,5	62,9	59	58,1	55,5	54,8	54,3	49,8	49	48,7	46,9	44,9

Table 2. Share of unprofitable organizations from the total number

Item	2008	2009	2010	2011	2012	2013	2014	2015	2016	2017	2018	2019
Share of unprofitable organizations from the total number,%	28	32	29,9	30	29,1	31	33	32,6	29,5	31,9	33,1	32,5

* Corresponding author: mvkrasn@mail.ru

However, there is the increase in the share of unprofitable enterprises. The presence of a high share of unprofitable organizations from the total number of enterprises and its insignificant growth over the past three years (up to 32% on average) indicates that the investment opportunities of organizations are limited by depreciation deductions.

The statistics presented in Table 2 indicate the importance of the use of depreciation deductions as a source of financing [3].

The only own source of loss-making enterprises to finance investments in fixed assets is depreciation charges, which makes the problem of the intended use of depreciation charges actual. Depreciation charges form a depreciation fund that should be used for simple or extended reproduction of fixed assets. The level of investment in fixed assets determines the potential of an economic entity.

2 Materials and methods

The following types of potential are described in literature: "economic", "financial", "investment", "property", "innovative", "socio-economic", "digital", "network", "personnel", "market" and "industrial". Economic potential is a fairly broad concept. There are many studies devoted to the assessment of economic potential, including methods for the assessment of potential. Over the past year, the greatest interest of researchers studying the economic potential has been provoked by the issues of measuring and modeling the economic potential of a region, city [4-6], the economic potential of individual enterprises and industries [7-8] and the development of the economic potential of municipal units [9].

Economic potential integrates several private potentials that are functional. Researchers determine the structure of economic potential in different ways, highlighting many components that form economic potential. For example, Bakhtin M.N., Petrykina I.N., Pleshakov S.S. distinguish the following partial components of economic potential: demographic, natural resource, financial, labor, production, consumer, infrastructural and institutional. Table 3 presents the systematization of existing opinions about private potentials included in economic potential.

According to Table 3, we can see that there is no single approach to the allocation of private potentials. In addition, if we consider the category of «socio-economic potential », for example, A.Yu. Vavilova, A.V. Letchikov, M.A. Sokolova identify a larger list of components of socio-economic potential: production, budget, financial, investment, labor, energy, agricultural, consumer, foreign trade and others [10]. The private potential associated with investment activities is investment potential. The study of the literature related to investment potential allows stating that there are different interpretations of this term. There is no single methodology to calculate indicators that determine the investment potential.

Table 3. Approaches to structuring economic potential into private (functional) potentials

Authors (author) of the approach	List of private (functional) potentials
Kharina V.S.	natural resource, investment, labor, financial, innovative
Mukhina E.G.	industrial natural, geopolitical, demographic (labor), potential of the social sphere
Bakhtin M.N., Petrykina I.N., Pleshakov S.S.	demographic, natural resource, financial, labor, production, consumer, infrastructural, institutional
Kovalev V.V.	property, financial
Krivoruchko O. N.	labor, investment, natural resource, innovative
Gunina I.A.	personnel (labor), financial, production, innovative, informational
Efendieva A.A., Temrokova A.Kh.	investment, financial, organizational, innovative

According to Vavilova A.Yu., Letchikov A.V., Sokolova M.A., the investment potential is measured by such indicators as investments in fixed assets, investments in fixed assets with the participation of foreign capital and the volume of investments from foreign investors [10]. All of these indicators are calculated per capita. Grishina I.V., Polynev A.O. proposed a method to assess the socio-economic situation of the regions of the Russian Federation, considering 16 indicators grouped into four blocks [11]. The investment and financial block includes such indicators as the balanced financial result per employee, the amount of investment in fixed assets, the share of profitable enterprises and organizations and the amount of payable accounts.

We see that both «economic potential» and «socio-economic potential» include investment potential, but there is a substitution of investment potential for investment and financial one. The existing methods for the assessment of the indicators of investment potential are not related to sources of financing for investment activities and do not reveal the sources of investment in fixed assets. In a number of existing interpretations of the concept of «investment potential» we are talking about the resources of an individual economic entity, which are directed to investment needs in order to ensure economic growth and increase efficiency. However, the source of these resources is not revealed: whether they used their own or borrowed sources.

3 Results and discussion

It is proposed to introduce the concept of «depreciation potential ». Under the depreciation potential we mean the ability of an economic entity to create the increase in the value of fixed assets at the expense of its own internal source - the depreciation fund. Figure 1 shows the relationship between economic, investment and depreciation potential.

The depreciation potential is considered as a private (functional) potential in relation to the investment potential, which is a component of economic potential. The depreciation potential can be considered in the context of three levels: enterprise, region and country. Depreciation potential includes two components: resources and directions of their use. Let us consider the process of formation of resources, formed from the depreciation fund and their directions of use.

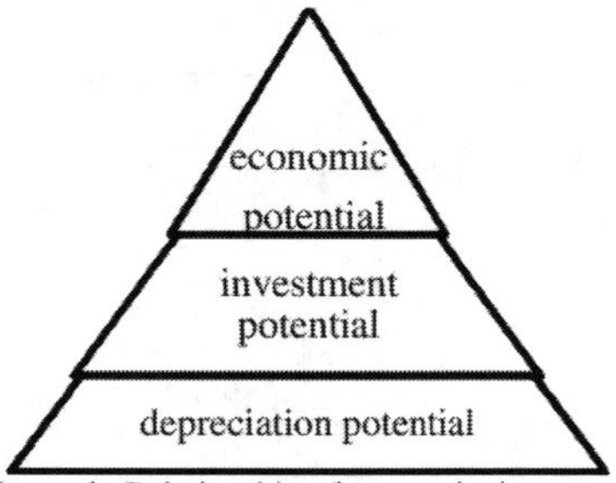

Figure 1. Relationship of economic, investment and depreciation potential

Depreciation charges as a result of the depreciation process are accumulated in the depreciation fund. Nowadays there is no interpretation of the concept of «depreciation fund» in the regulatory literature. According to R.G. Sabiryanova and Z.M. Bikmetova, the depreciation fund is a fund of funds, which is intended for the refund and renewal of fixed assets [12]. A.P. Grigoriev determines the depreciation fund as funds from depreciation that can be used to repair fixed assets [13]. Researchers highlight the use of the depreciation fund for the renewal of fixed assets. The nominal and real depreciation fund should be identified.

The accumulated depreciation for a particular period determines the maximum amount of accumulated funds that form the nominal depreciation fund of an enterprise. The accumulated depreciation forms the resources of the depreciation potential. Real depreciation fund will be considered as the amount of funds actually allocated for the purchase of new fixed assets. Modernization, renovation, repair, reconstruction, technical re-equipment, new construction and robotization are the directions of the use of the resources of depreciation potential. Table 4 shows the resources and directions of the use of the resources of amortization potential.

Table 4. Depreciation potential components

resources	*directions of resource use*
amount of accumulated depreciation	- new fixed assets;
	- modernization and renewal of fixed assets;
	- reconstruction and technical re-equipment;
	- robotization;
	- new construction

The index (coefficient) of the use of the amortization fund can be used as one of the indicators for the assessment of depreciation potential, which allows assessing its level of use. This index will act as a sub-index in relation to investment potential. The index (coefficient) of the use of the depreciation fund is determined by Formula 1.

$$K_{useDf} = \frac{Df_{real}}{Df_{nominal}} \quad (1)$$

Where Df_{real} – real depreciation fund, $Df_{nominal}$ – nominal depreciation fund.

In order to calculate the proposed index of the use of depreciation fund for an enterprise, it is necessary to collect information on the size of the nominal and real depreciation fund on the basis of financial statements and accounting data. To calculate the proposed index of the use of the depreciation fund for a region, it is necessary to use statistical data containing information on the accumulated depreciation charges and the amount of investments in real assets. According to the proposed formula in Table 5, this index was calculated on the basis of the statistical data of the Perm Territory.

According to the table, we see that the values of the utilization rate of depreciation fund have values exceeding one. This means that the entire amount of accumulated depreciation charges was invested in fixed assets. At the same time, the real amount of investments in 2013 and 2014 was twice higher than the nominal depreciation fund, which allows concluding that net profit is used as a source of investment or attracted funds. However, we note that there is a clear tendency to "approach" to one.

As an example, indices were calculated for several enterprises of the chemical industry of the Perm Territory, the results are shown in Table 6. Kamteks-Khimprom enterprise has significant fluctuations in the index during the studied period; the average value of the index is 1.1. This is a signal that for the entire studied period the created depreciation fund was used for its intended purpose.

The indices of Sorbent enterprise also have significant fluctuations: from 0.42 to 4.17 for the calculation period. Only in 2016 and 2017 the depreciation fund was not fully capitalized into fixed assets, while values exceeding one in subsequent years show that the unused balance of depreciation charges was directed to renewal, reconstruction, modernization and the acquisition of new fixed assets. The indices of Solikamsk Magnesium Plant allow revealing that in periods when the values did not exceed one, their level was quite high - 70 and 86% (in the corresponding years), which indicated 30% and 15% of unused resources of depreciation potential.

A constant excess of the index of one is not a prerequisite, since in this case the funds of the founders are diverted from the turnover or there is a financial and operational need to cover the difference between real investments and the created amortization fund.

If the index value approaches one within 5-7 years, this indicates the effectiveness of the developed depreciation policy of the enterprise and the intended use of depreciation funds, while one-time deviations of the coefficient are possible when the values are low (0.2-0.6). If the funds are not used in full according to their intended purpose in the current period, then they are accumulated (accumulated depreciation fund) and can be capitalized into fixed assets in the next period.

Table 5. Calculation of the index (coefficient) of the use of depreciation fund based on the statistical data of the Perm Territory

Indicator	2013	2014	2015	2016	2017	2018
Accumulated depreciation, mln. rub.	83498.77	84947.04	102909.42	118151.38	130974.58	127645.45
Fixed capital investments, mln. rub.	183030.30	171289.30	186384.60	189301.50	188233.30	184148.80
The use of depreciation fund	2.19	2.02	1.81	1.60	1.44	1.44

Table 6. Indices (ratios) of the use of depreciation fund based on the data of the accounting (financial) statements of chemical industry enterprises

Enterprise	2014	2015	2016	2017	2018	2019	2020
Kamteks-Khimprom	1.75	1.41	0.24	0.61	0.86	1.97	0.83
Sorbent	1.53	1.37	0.42	0.56	1.35	2.09	4.17
Solikamsk Magnesium Plant	1.02	0.86	1.83	0.70	1.64	1.13	1.22

However, in practice, the assets of the depreciation fund are not always used for their intended purpose. Firstly, this is explained by the lack of resources at enterprises to finance current activities and cash gaps. Secondly, the presence of inflation stimulates the preterm use of the depreciation fund for other purposes. Thirdly, it is connected with the risks associated with the storage of free funds and low deposit rates. In this case, the depreciation fund has a passive mechanism of action, since the funds are used for other purposes.

The calculation of the average index for the studied period will show how much the assets of depreciation fund were used for their intended purpose as a whole for the period, taking into account the redistribution of monetary resources over time.

4 Conclusion

The calculation of the index (coefficient) of the use of depreciation fund allows economic entities to understand where depreciation deductions are used. Since the use of depreciation funds is not reflected in accounting, the use of the proposed coefficient allows determining the degree of intended use of depreciation fund. Low values of the index of the use of depreciation fund indicate a weak depreciation potential of the subject (region, enterprise).

The practical relevance of this study is in the possibility to compare regions on the basis of the proposed index, identifying regions with a high depreciation potential. The calculation of the proposed coefficient at the enterprise level allows evaluating the effectiveness of depreciation policy, comparing enterprises on the basis of their depreciation potential in alternative investments.

References

1. T.Yu. Mazurina, *Finance and Credit.* **11,** 14–23, (2012).
2. *Investments in Russia* Retrieved from: https://rosstat.gov.ru/folder/210/document/13238
3. *Finance of Russia* Retrieved from: https://rosstat.gov.ru/folder/210/document/13237
4. N.A. Burakov, A. Ya. Rubinstein *Spatial Economics.* **16 (1)**, 20-50, (2020) DOI: 10.14530 / se.2020.1.024-050
5. I.M. Drapkin, E.O. Dubinina *Economy of the region,* 16 (1), 310-324, (2020). https://doi.org/10.17059/2020-1-23
6. E.G. Shcherban *Bulletin of the SSEU.* **2 (81),** 67-71, (2020)
7. L. D. Efanova, V. L. Polyakova *Bulletin of the University,* **5** , 141-144 (2019) DOI: 10.26425 / 1816-4277-2019-5-141-144
8. A.I. Golubeva, V.I. Dorokhova, A.V. Konovalov, Yu.V. Shumatbaeva, K.V. Pavlov *Bulletin of the Agroindustrial Complex of the Upper Volga Region.* **1 (49),** 59-67, (2020) DOI: 10.35694 / YARCX.2020.49.1.013
9. V.S. Antonyuk, A.V. Rezepin, D.V. Kremer *Bulletin of SUSU. Series "Economics and Management",* **14(2),** 18-27, (2020). DOI: 10.14529 / em200202
10. A.Yu. Vavilova, M.A. Letchikov, M.A. Sokolova, *Bulletin of the Udmurt University.* **2,** 31-43, (2006)
11. I.V. Grishina, A.O. Polynev *Modern productive force,.* 34–48, (2012)
12. R.G. Sabiryanova, Z.M. Bikmetova *Problems of modern economics.* **2,** 403-406, (2014)
13. A.P. Grigoriev *Economic analysis: theory and practice.* 17(8), 1584 – 1600, (2018) doi.org/10.24891/ea.17.8.1584

Modern approaches to Agile transformation of Russian financial and technological companies

Denis Vorontsov*

State University of Yaroslavl named after P.G. Demidov, 14 Sovetskaya Street, Yaroslavl, Russia, 150003

Abstract. The article considers the relevance of the Russian organizations Agile transformation operating in the financial technology industry. The current concepts of Agile-transformation are presented. The experience of implementing Agile transformations in large Russian banks is presented. The current processes occurring during the transformation of fintech companies are presented. The conditions and features of Agile transformation depending on the scale of a fintech company are considered. Changes in the corporate culture are analyzed from the point of view of project teams and the Agile companies as a whole. Approaches to the transformation of the company's organizational structure are presented, with the identification and description of new implemented structural elements and roles. The change in the project methodology of a Fintech company during Agile transformation is presented as a transition, implementation and adaptation of popular Agile frameworks and hybrid project approaches based on the project methodologies of Waterfall, Scrum and SAFe. The current and practical stages of Agile transformation of large fintech companies operating in the banking market are analyzed and described in detail. Assumptions are made about the potential problems caused by the escalating competition, restrictions and consequences caused by the Covid-19 pandemic. Medium-term challenges and prospects for the transition of companies to Agile in the Russian fintech market are presented.

1 Introduction

The active growth of Russian financial technology industry was provoked by the financial crisis in 2008. During this time, companies were forced to cut costs and improve efficiency and/or seek new technological solutions. The second significant factor in the development of the industry was the progress of mobile technologies, which allowed people to get round-the-clock and prompt access to the Internet. Today it is one of the fastest growing markets both in Russia and in the world. The number of users of financial services and the volume of their supply increase by 15–20% annually. The volume of payments, transfers, personal and corporate finance is growing by an average of 20%. Russia is in the top 3 countries in terms of the popularity of financial and technological services and penetration level reaches 82%. According to Deloitte estimates, in 2019 the volume of the Russian fintech market amounted to 60 billion rubles. (+ 11%), and in 2020 it grew to 65 billion rub (+ 8%).

The slowdown in growth is explained by the high level of uncertainty in new market and the impact of the Covid-19 pandemic. Despite this, analysts predict a twofold growth of the market by 2024 [1].

In order to be successful in this industry, companies must promptly deliver to customers a high-quality product increment, by which we mean a tangible result of one iteration/sprint (for example, an application prototype or an increment of new functionality to a product).

Moreover, it is extremely important to be flexible and be able to respond in a timely manner to feedback received from customers. Consequently, this leads to a transition from a hierarchical organization management structure and waterfall approaches to project management and financing, agile project management approaches and a new organizational structure. Therefore, more and more Russian companies choose agile development as an important financial and technological solution that is successfully integrated in international companies and helps companies win the competition.

Agile companies see significant improvements in key metrics as evidenced by research on productivity, time shifts, defect counts and development costs across 1,321 software development projects in the 10th State of Agile Report prepared by in 2016 by VersionOne. The parameters most subject to improvement are: the ability to manage changes (observed in 87% of cases); increased productivity (85%); increase in the transparency of projects (84%); reduction in the implementation time (79%); reduction of project risks (77%). The comparison of the effectiveness of Agile teams and teams working on all other methods of product / service development are: Productivity + 27%; Shift in terms of implementation -30%; the number of defects per request -70%; development cost -40% [2].

In this paper, the term Fintech (financial-technological) means an industry combining the sectors of financial services and modern IT products. The main players in this

* Corresponding author: vorontsovde@gmail.com

market include technology companies and classical financial institutions: banks, insurance and investment companies. All of them either create themselves or use third-party innovations to improve their work and make life of clients easier [3].

Under the term Agile development we mean a set of innovative modern ways of the development of IT products [4].

Today in Russia, agile development approaches are widely used in many fintech companies, but the greatest development is manifested by the industry leaders: Sberbank (state participation) and Alfa-Bank (private bank) [5].

Agile approaches are not a fintech solution itself, but a set of methods and practices designed to manage and create fintech products (mobile, virtual and augmented reality applications, front-end and backend solutions). The most commonly used agile development methodologies in Fintech (Scrum, Kanban, etc.) are aimed to minimize risks reducing development parts to a series of short cycles called iterations, which usually last from two to three weeks.

Accordingly, in this article, under the **Agile transformation** of a fintech company, we mean a deep reorganization of the **organizational structure**, a transition to an **agile methodology** of project management and a change in **company corporate** culture, as a result of which it will be able to work on Agile principles.

2 Results and discussion

2.1 Starting conditions for Agile transformation of a fintech company

The starting condition for launching the Agile transformation of fintech companies is the awareness of the position of a company at the current moment in time and the target place where it is necessary to move. The value attitudes of a fintech company should be taken into account in the context of the market and in the planning horizon of the strategy of a company. Most companies value the predictability of product delivery or the ability to adapt to changes and this has negative consequences. The more the system is tuned for predictability, the more difficult it is to change it. On the contrary, the more adaptability is attached to a system, the less predictable it is.

Moreover, it is necessary to take into account the essence of the value of a product for ta client (emergent ecosystem - an open ecosystem), or just focus on performance of assumed functions (convergent ecosystem - a closed ecosystem) [6].

The entire process of Agile transformation of a fintech company includes **the creation of teams, culture, structure and management methodology that are based on the principles of Agile**, as well as the definition and implementation of those metrics that are critical for a company as a starting point as they will be used to assess progress and track improvements made during transformation.

The agile transformation strategy will ultimately depend on the value system adopted by the team of a company and the needs and values of the customers.

2.2 Specifics of Agile transformation depending on the scale of a Fintech company

According to the size of a Fintech company, the Agile transformation of two types of companies should be considered separately: large companies such as classic banks, insurance and investment companies with more than 100 employees. The second type of companies is small, falling under the criteria of attribution to small businesses, as well as startups [7].

In most cases in small companies Agile transformation does not cause difficulties, since the change in structure and the introduction of agile methodologies is carried out by team building, training and adopting new development approaches. However, in large fintech companies, with a complex hierarchical structure and a heavy "legacy" in the form of an outdated information and technological architecture, changes need to be planned and organized in detail in order to eliminate the risks of loss of investment and effort invested in making changes that will create value for a client and therefore, for business, which will truly enable a fintech company to become flexible and adaptable to changes.

Contrary to established stereotypes the process of agile transformation of a large fintech company although follows the principles of the agile manifest, has little to do with so called management approaches of agile project. The agility of a solid company does not come from product separation, the implementation and ceremonies, but begins with a strategy for the management of internal investments and implementation of specific measures to develop a company culture. The main condition for the success of Agile transformation is not methods, not technology or experience, but company culture.

First of all the agile transformation of a large Fintech company is the work on the transition to a new structure and transformation of a company culture from a culture of control to a culture of development. Accordingly, this is the creation of effective self-organizing teams, the formation of a list of changes/backlog and the regular creation and setting of increments of a proven and workable software product. A fintech company-wide, this means to create chains of loosely coupled teams, measure team throughput, not team productivity, coordinate dependencies, work out negative impacts, continuously and rapidly improve products and processes through external and internal innovations, as well as quick product launch to the market [8].

2.3 Change in company culture of a Fintech company during Agile transformation

During the process of Agile transformation, a company culture is transformed according to the principles of Agile. This is a complex and risky process, as it involves a fundamental change in processes, as well as external and internal motivation of employees. In this case, Agile principles act as a transformation tool.

The culture change involves the creation of stable agile teams that are focused on a single project/service or business task. They do not switch between different approaches, as the culture of Agile does not involve switching to parallel or related projects and disintegration.

The advantages of cultural change appear in the long term and lead to access amplification of development teams, more opportunities for creativity, minimization of excessive control and micromanagement from a higher level. However, during the first period of company restructuring, productivity may decline as teams adapt to new principles and processes.

2.4 Change in the organizational structure of a Fintech company during the process of Agile transformation

As a part of the Agile transformation, a large-scale revision and change of the organizational structure of the management of a Fintech company is carried out. All elements are transformed by their one-time transition to a new state.

The organizational and functional structure of a company (functional mines/wells) is replaced by **Agile organization** with the following basic units: cross-functional teams, chapters (groups of specialists in one area of competence) and tribes (groups of interconnected teams formed for a specific product or business purpose). Overall responsibility for the business outcome rests on the tribe manager/board member of a fintech company involved in Agile.

Figure 1 shows the functional structure of the created Agile organization of a Fintech company

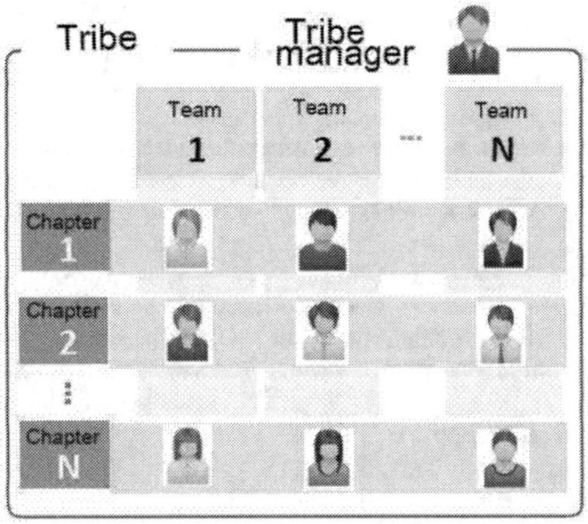

Figure 1. Agile organization structure

The size of the organizational units of an Agile organization must provide effective human communication and minimize bureaucracy. The number of tribes is not more than 150 people. The limitation is based on "Dunbar's number. In cases of exceeding the maximum tribe size or when close coordination of teams is required, the teams are combined into virtual clusters of up to 150 people, and then the tribe can be increased to 1500 people. [9].

The size of a chapter and a team should be no more than 9 people; the limit is based on the "Miller number".

The basic principle of tribe formation is self-sufficiency to meet a specific need.

According to the specifics and types of tasks there are two types of tribes and teams in a Fintech company: customer-oriented (they implement elements of business tasks and build an end-to-end customer experience) and platform, which create the basic modules of a product and ensure its architectural integrity and sustainability. A complex multi-level hierarchical management system is transformed into a flat structure with no more than three levels of the management hierarchy, excluding the board or CEO of a company.

Every level of management has the maximum level of independence in decomposition and determination of the approach and method of implementation of the task assigned to it. The setting of tasks is carried out according to the "What/How" principle, in which the higher level sets the task in the form of "what to do" or "what goal needs to be achieved" and the lower level itself determines "how" to do.

Figure 2 shows the final version of the Agile transformation of the hierarchical structure of a fintech company.

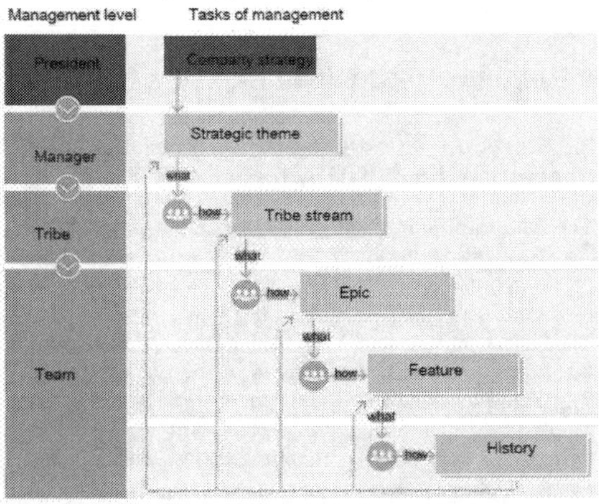

Figure 2. Hierarchical structure of Agile organization

The higher level does not participate in the tasks performed by lower level [10].

2.5 Change in the project methodology of a Fintech company during the process of Agile transformation

The process of agile transformation introduces a flexible methodology of project management both in the work of teams and at all levels of management of a Fintech company. It is necessary to note that there is no single Agile methodology. At the same time, various flexible methodologies or Frameworks are used. The most popular ones in the financial industry are Scrum, SAFe, eXtreme

Programming, Feature Driven Development, Crystal, as well as all possible hybrid approaches.

Accordingly, the waterfall approach of project implementation and major changes with a long preparation process is replaced by a flexible MVP (minimum viable product) development approach and quick feedback from a client for further improvements of this product. The principle of the creation of a product is being introduced iteratively without the desire to create everything at once. Respectively, the product is created incrementally, from simpler to more complex.

During the introduction of agile methodology, a number of changes take place. The bureaucratic process of start and management of projects is being replaced by a release process, in which changes become a routine process and the basis of a Fintech company. The transition to Agile releases is carried out in stages, depending on the stage of transformation and overcoming the limitations of existing production processes.

Partial involvement of management and staff in projects is transformed into full-time work in one team or on a specific task.

The adoption or approval of all decisions by the management is replaced by an open collaborative environment, implemented through the development of trust, support and encouragement of openness, professionalism and reliability, as well as the widest possible delegation of authority at all levels. Control over the performance of tasks is carried out at product demonstrations and a quarterly review of results. The integrity of approaches to work and architecture and compliance with standards is supported by the chapter mechanism and/or controlled by technologists.

2.6 Agile transformation stages of fintech companies

Successful transformations undergo through a large-scale and long path, consisting of conditional stages that can intersect with each other or be parallel.

A key step in the transformation is timely delivery of business value to a customer on a regular basis. The purpose is to reorganize the structure and approaches to the execution of project work. The most commonly used practice is to move from the "big bang" - large but infrequent releases/deliveries to small but regular deliveries or incremental product value. Transformation management is carried out in the same way.

Considering the experience of large organizations (Sberbank, Alfa-Bank, Diasoft and others), the Agile transformation of one company is different from others. At the same time, all considered companies went through the following stages:

The 1st stage. Formation of the final vision of Agile transformation and initiation of changes

At the beginning, a company in general is considered in connection with the definition of business goals for the upcoming changes. The final vision determines the approach to the formation of change leaders, the sequence of their initial actions, the type of project management model, metrics, tools and methodology for the identification and measurement of improvements. The understanding of where a company is approaching is formed before the movement begins. It is foreseen in advance that the plan will be subject to constant changes, since it includes working management hypotheses, structure and metrics that a company is using and will be continually refined as it transforms.

Large companies use well-established approaches to coordinate teams together in order to increase their productivity in the definition of final concept.

The 2nd stage. Formation of a transformation team

Agile transformation of a fintech company requires the implementation of changes in all key elements of organization structure and projects. Therefore, full management support is required. At this stage, the understanding and involvement of top managers in the course of what will happen is formed. The need for external consulting is explained.

The 3rd stage. Formation of the roadmap of Agile transformation

It is the creation of a detailed but flexible action plan with terms and results that a company will receive in return for investment. Groups or divisions are also appointed that lead and set the pattern for transformation.

The 4th stage. Creation of a pilot transformation plan

The process of Agile Transformation team constantly interacts with each other to plan, assess success and adjust the plan if it is necessary.

The main requirement for a team is to have a flexible 3-month plan and a creative view of the upcoming transformation. The 3 month plan is similar to flexible release planning. The plan lists all of the procedures and processes that will be affected by the changes over the next three months.

As a result, a team will prepare and launch a pilot project. The decision on the place and time of the pilot is made at the level of the president of a company. The results of a pilot will be a necessary basis for the identification and implementation of changes.

The 5th stage. Control over intermediate transformation results

This stage provides the collection, preparation and training of leaders and teams of the first stages of transformation, which are constantly progressing, moving from one state to other.

As the teams develop the final state of the pilot project and subsequent changes are reached. Basic and key metrics are determined, measurements are taken regularly and openly and information about the resulting improvements is spread. Similarly to the cyclicality and frequency of the iterations established by the methodology, a periodic assessment of the achievement of transformation metrics is carried out. Based on the control, retrospectives are carried out and changes to the process are made.

The 6th stage. Adaptation and training

At this stage, structure, management and metrics will begin to migrate into a coherent whole. An accessible internal base of best practices is being formed in a company. In the period of development and expansion of the scale of transformation, the centralized accumulation and use of the resulting best practices becomes a natural process.

A revision of the vision of the final state of changes is carried out based on the understanding of success in the Agile transformation process. Moreover, at this stage, new training and information materials, reminders for further actions and best practices of Agile transformation are created. In addition, the methods and approaches to adapt Agile coaches within a company are being developed, the level of development is regularly measured and work with attenuation and negative consequences are performed.

The 7th stage. Coordination of actions to the business results of a fintech company

The main reason to start the Agile transformation process is the desire to improve significantly business results, so the organization begins to justify the investment by making assumptions, experimenting, demonstrating and declaring results, changing strategy depending on the lessons learned.

The main purpose is to build the correct sequence of the required results, as well as to ensure that the transformation actions correspond to the inherent business results.

The 8th stage. Coordination of transformation results to business purposes

This stage includes tracking performance improvement towards desired business benefits and regular demonstration the positive dynamics of core business metrics to management.

The main purpose of this stage is to visualize a transparent and sustainable relationship between investment and achieved business results.

The 9th stage. Communication management and information support

This stage includes the transparent and ongoing communication in order to achieve success and overcome the difficulties faced by the transformation team. It also includes the creation of a positive atmosphere and energizes a team. The following formats are used: Roundtables of a transformation team with management, general meetings and other relevant communication channels. Due to the impact of the COVID-19 pandemic, face-to-face formats are being forcedly replaced by remote communication channels, video conferencing and webinars.

The 10th stage. Psychological safety of participants in the process of Agile transformation

The transparent and regular motivational meetings and conversations with the participants in the changes are being held, mainly about the benefits that they will receive and their place in the new structure of a company. The transparency and reliability of the measurement of the achievements of each participant is provided.

The implementation of these stages allowed major players in a Fintech market to launch or implement successfully the process of Agile transformation, preventing a decrease in the efficiency of a company when implementing changes. At the same time, we note the lack of uniform approaches to the stages of transformation and, as a consequence, the presence of variability in the results for similar companies.

2.7 Challenges and opportunities of Agile development in the Russian financial and technological segment

Financial and technological companies, being in a highly competitive, fast-growing environment, are clearly aware of the need for faster time to decision and time to market, as the main condition for survival in the modern rapidly changing world of financial services and the expansion of ecosystem.

The challenges faced by financial and technological companies require from management not only an instant response during decision making, but also a quick implementation of financial services and products. To provide speed, not only an appropriate structure to ensure the implementation of flexible and proven, fast-paced approaches are needed, but the adoption by employees of a culture of development based on the Agile principles.

The traditional hierarchical structures and the waterfall project approach used by the classic players of the Fintech market came under pressure of new players and, as a result, began to adapt to the requirements of the time, absorbing Agile approaches, which have recently been included and supplemented in the PMI, IPMA standards and PRINCE2. It is important that the use of Agile is causing a cultural transformation within large banks, changing the attitudes of people towards information technology, project work and ecosystem development for the better.

In 1-2 years, the competition between traditional banks and companies in the financial and technological sector

 https://doi.org/10.1051/shsconf/202111600066

will intensify even more without a banking license. New and fast IT companies will try to squeeze classic players out of the market, so the Fintech market, the success of which depends on the speed of response to external challenges and customer-centricity, is doomed to Agile transformation of the management structure and corporate culture of companies. Therefore, the demand for Agile transformation and specialists capable of carrying out these changes will grow at priority rates in the next 2-3 years.

3 Conclusion

The limitations and consequences of the COVID-19 pandemic, the rapid growth of technological solutions, the development of product and information platforms, as well as the construction of ecosystems of products and services, created high uncertainty and development risks for startups and existing large fintech players. Taking this into account, they will make more and more qualitative and quantitative requirements both to the process and the results of launched or planned Agile transformations of their companies.

References

1. *Fintech market development in Russia* Retrieved from: https://www.finam.ru/analysis/forecasts/razvitie-fintex-rynka-v-rossii-neobanki-i-startapy-20191211-142048/
2. The 10th report of the VersionOne company on the state of adaptive development methods Retrieved from: http://www.agile247.pl/wp-content/uploads/2016/04/VersionOne-10th-Annual-State-of-Agile-Report.pdf
3. *Russian fintech* Retrieved from: https://iq.hse.ru/news/334262759.html
4. Wolfson B. *Flexible project and product management*. (SPb.: Peter, 2016).
5. *Cases of agile transformation* Retrieved from: https://vc.ru/hr/100793-keysy-agile-transformacii-chast-pervaya-banki
6. *What determines the success of Agile-transformation* Retrieved from:: https://pro.rbc.ru/demo/602b21df9a7947f6aa338611
7. *Agile Research Report in Russia 2020* Retrieved from: https://agilesurvey.ru/
8. Foundations of Agile-transformation Retrieved from: https://habr.com/ru/company/rdtex/blog/452358/
9. *Experience in the implementation of flexible development methodology in the activities of retail and IT-blocks of the bank* Retrieved from: http://www.reglament.net/bank/raschet/2017_5/get_article.htm?id=5360
10. *Using the Agile-Approach in Business Practice* Retrieved from: https://alprex.ru/stati/ispolzovanie-agile-podhoda-v-praktike-biznesa

Citizen involvement in urban planning process in Russia

Elena Tselishcheva[1,*], *Elena* Kondyukova[2], *Aleksandr* Pobedin[1], and *Faina* Babanova[2]

[1]Ural Institute of Management, Russian Presidential Academy of National Economy and Public Administration, 66, 8 Marta st., Yekaterinburg, 620144 Russian Federation
[2]Ural Federal University named after the first President of Russia Boris Yeltsin, 19, Mira st., Yekaterinburg, 620002, Russian Federation

Abstract. The paper examines citizen involvement in the urban planning process. Although different forms of public control have been established in Russian legislation application of western approaches for urban planning process, organizations face low interest from local communities, on the one hand, and inability of local authorities to work with public initiative, on the other. The research is based on the multidisciplinary approach. The methodology includes deliberative democracy, advocacy planning conceptions, spatial economic theory, and urban studies. Methods of system, institutional, comparative analysis, focus groups and content analysis are used. The research has been conducted based on the materials of public space renovation projects in some major cities - regional capitals in Russia. The empirical base of the research includes acts regulating the urban planning sphere, mass media content about urban space renovation projects, materials from focus groups with representatives from local authorities and initiative citizen groups. The paper represent results of the international scientific research project "Legitimacy, urban planning and sustainability in Russia and Sweden (LUPSRUSS)" which was funded by the Swedish International Centre for Local Democracy (ICLD).

1 Introduction

If a country declares sustainable development goals in the constitution, it will have to design a communication mechanism between citizens and authorities. Western world countries spent more than hundred years to minimize conflicts of interest, and even after it, the problem does not entirely disappear.

While contemporary researchers apply to urban post-industrialism the conception with a global symbol of digital-cities and social innovations, discuss different aspects of innovative breakthrough, some Russian citizens still has not access to elementary communal facilities.

Social inequity has been growing up during the last two decades. Systematic economic, political and environmental turbulence takes away citizens' belief in a positive world picture of the future. In spite of the fact that turbulence can lead to creation of new structures, lack of stability in the transition period creates fear and rejection [1].

In this regard, cooperation between citizens and authorities depends on the level of development of social institutes and cover all aspects of life.

Participatory urban planning is one of the public cooperation forms. It indicates growing civil identity, which produce new models of communication between citizens and government. Authorities in western countries came a long way to develop trust relationship with citizens. Russia is at the start of this process. It will have to strengthen civil identity first through education, conscience and negotiation.

2 Conception of citizen engagement in urban planning process

Urban planning includes three aspects: 1) development of abandoned urban spaces; 2) manner of interaction between citizens and government in the process of urban plans actualization; 3) formation of civil identity in relation to public spaces development/redevelopment projects. The last aspect is the result of both previous ones, because the truth is born in a dialogue, communication process and even conflicts. Only through that, we can achieve a compromise – one of the alternative projects that satisfy all urban planning process stakeholders.

Participatory urban planning conception reflected in scientific publications by Davidoff P. (advocacy planning theory), Krumholz N. (equity planning theory), Lefebvre H., Harvey D., Lynch K. (right to the city conception), Arnstein S. (participatory decision making theory, the model «Ladder of citizen participation»), Sandercock L. (community planning and multiculturalism), Healy P. (collaborative planning conception) [2].

In 1968 Lefebvre H published the monograph «Le Droit à la ville». He was the French sociologist and philosopher, who introduced the concepts of the right to the city and the production of social space [3]. In his

* Corresponding author: *cefed@yandex.ru*

monograph he appealed to idea that citizens must be included in decision making process regarding to urban planning projects and formulated citizens' right to the city resources. Due to his publications, urban space became the form of social phenomena (space for social processes and social events) [4].

Although «the right to the city» conception appeared more than half of the century ago, Lefebvre H. became known in urban planning theory with some delay, that, in our opinion, was due to unavailability of society understand his ideas before achievement of higher level of civic engagement development. Tsunami wave of public initiative movement in urban planning covered the USA, Europe and Latin America at the end of the XX century and then inflamed Middle East in megacities Beirut and Istanbul [5].

Contemporary researchers criticize «the right to the city» conception for excessive radicalism and engagement of conflicting citizen groups in planning process because leads to intensification of social conflicts. For example, Pursell M. wrote: «I do not present the right to the city as the only or the best strategy for renewing urban democracy» [6]. On the other hand, citizens knew situation from the origin, and their well-reasoned arguments allow prevent «capture of the right to the city» by bureaucrats.

Nowadays on practice bureaucrats from administrative structures can still suppose that the scale of strategic urban plans (especially in mega-cities) is impossible for understanding by ordinary people without professional knowledge. Inhabitants cannot catch the essence of development project documents. That is why it is necessary to minimize their engagement in the decision-making process. But, even 50 years ago researchers warned that ignoring public interests at the times of growing civic conscience can lead to negative socio-economic consequences. Contemporary researchers, representatives of the new stage of urban theory, warn that citizens in mega-cities have got self-connection opportunity through social networks and other social innovations [7].

Economic geographer Harvey D. paid special attention to socio-psychological aspects of urbanization. Political struggle for «the right to the city» affects civic identity and changes not only living space but also a social relations system [8].

Authors of the advocacy planning theory [9] and equity planning conception [10] borrowed ideas from deliberative democracy theory, and found brilliant practical implementation in court cases in Europe and the United States in the 1970-90s. The court cases were mostly about protection of groups with a lower socio-economic status from negative consequences of urban space renovation projects or industrial environment pollution, which can lead to degradation of quality of life. Advocacy planning became both theory and practice of judicial and social support for vulnerable groups of citizens.

Healey P. developed a collaborative planning idea and put a special emphasis on necessity to involve all stakeholders in the urban planning process [11]. In his research he described reasons of urban conflicts and ways of its solution. Technical progress and innovations are considered as conflict reasons because it can lead to negative consequences and directly or implicitly affect all urban inhabitants. For example, conflicts about environment pollution reflect dissatisfaction of citizens with the loss of property benefits and their fear of the future. Healey P. suggested discuss consequences of urban projects in a form of mediation with the involvement of independent experts.

Canadian researcher L. Sandercock analyzed multicultural aspects of agglomerations' development and also was concerned about two negative tendencies: declination of the state's responsibility to the population and growing pressure from developers on decisions related to important urban development projects [12]. She wrote that big companies receive simplified access to redevelopment of public urban spaces.

In the middle of the 1960-s S. Arnstein, researcher and consultant of USA government on racial and social segregation, proposed the model «ladder of citizens participation». The model demonstrates different levels of participatory planning development and remains actual and applicable to characteristic of today public participation forms in different countries.

Unfortunately this scientific direction is poorly presented in works of Russian researchers.

3 Development of civic engagement in Russia

The Russian urban planning system for a long time reflected the Soviet Union's custom to plan urban development without asking habitants. The urban planning process was the privilege of authorities until the mid of the XX century, but after that some intention to consult with scientists and experts appeared. Risks of citizens' protest against some projects were smoothed out by active promotion of the picture of bright socialistic future and by call for total public welfare in the Soviet Union.

Initiative passed to developers in the post-Soviet period. They reshaped city maps as they wanted, built houses extremely tightly to each other, without taking into account the increasing burden on social and communal infrastructure. They could destroy parks and other green zones. This practice leads to significant declining of quality of life [13].

Participatory urban planning and development became popular and supported by authorities during the last few years. Now we can say that cooperation between authorities and citizens considered as a mechanism for monitoring the quality of construction projects and improvement projects, and as a tool to reduce social tension.

Intention to maintain social sustainability leads to understanding that civil city-management is impossible without deliberation. Habitants have right to the city – they can use existing spaces and transform them so they meet the needs of real people [3].

In spite of the fact that Russian patrician accept deliberation idea and know about public deliberation of urban planning documents with citizens, business, scientists, initiative projects were quite rarely implemented in real life.

During the research project we analyzed different forms of civic engagement in urban development planning by the example of largest Russian cities – regional capitals. In all cities, citizens can participate in the development of the following territorial planning documents:

- strategic plans of municipal socio-economic development (including a special chapter about spatial development);
- urban plans (general plans, land use and development plans);
- initiative budgeting projects (usually small projects in different spheres, only projects in the sphere related to territorial development were included in our research (for example, children play grounds construction, monuments renovation);
- projects of public spaces and semi-public spaces development (renovation) financed through the national project «Comfortable urban environment». The project includes projects in 40 Russian cities.

Public hearings, public deliberation, questionnaires, collection of signatures for support or against particular urban development projects became most popular forms in Russian cities for discussion of the above-mentioned documents.

Federal Law No. 212-FZ «About the basis foundations of public control system in the Russian Federation» (was signed and published in 2014) establishes different forms of civic engagement. The list of forms is opened so it is possible to develop the public control system further and implement new forms. Regional and municipal governments establish entire forms of public control in subnational legislation. Public control forms established in legislation and used in day-today practice are not the same (figure 1).

FORMS OF PUBLIC PARTICIPATION

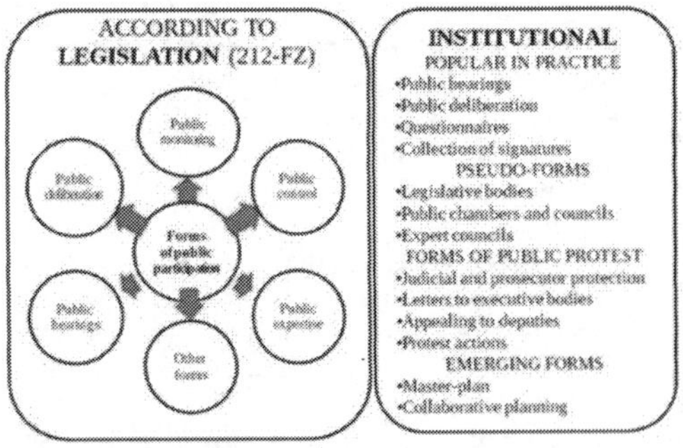

Fig. 1. Forms of public participation

Consideration of urban development documents with deputies (as they are elected representatives of citizens), experts and public council's officials equate to public participation forms.

From our point of view the total forced formation of public chambers and public councils in Russia is an attempt to impose from above convenient and acceptable to the authorities public participation forms. Unfortunately these forms don't reflect the interests of different citizens groups and have lack of feedback.

According to Federal Law No. 172-FZ «About strategic planning in Russian Federation» all long-term perspective documents or significant-volume documents (for example, strategies of socio-economic development, spatial development plans, complex development plans) are ought to be negotiated with:

- legislative bodies (deputies);
- scientific and professional community (expert councils and public councils created by executive bodies).

Both pseudo-forms of public participation identify opinion of professional groups, which does not always coincide with the opinion of citizens. We recommend using this forms in addition to the mentioned above – expert councils and negotiation with deputies allow one to examine in detail results of direct communication with habitants (public hearings, deliberation, questionnaires). Local authorities should recruit experts and deputies as mediators in the urban planning process.

The role of public chambers, public councils and expert councils in the participatory urban planning process worth a separate research work.

Municipal authorities often use public hearings and public deliberation for consideration of less-volume urban planning documents with habitants.

The concept of public hearings and public deliberation is related to the long-standing traditions of western sociology, but does not correspond to totalitarianism conception, which still exist in the post-Soviet space. Some researchers note that public hearings in Russian cities have formal character and their results can't affect decisions taken by authorities [14]. Shortcomings of public hearings include (see figure 2):

- late notification about the hearings' start;
- conducting of hearings in hardly accessible places in inappropriate time (working hours as usual);
- informing habitants about hearings in newspapers who never read;
- using special incomprehensible for ordinary people terminology during project presentation;
- ignoring remarks and questions from the audience.

Local governments and developers consider public hearings as excessive work that costs the time and money (a lot of noise without result).

Among public hearings participants can be a significant number of representatives from developers (have a clear financial interest), as well as public and political organizations (since any confrontational situation can be used to realize political ambitions). There were cases when in order to create a positive project image special «participants» were hired and paid for the «right» vote.

Municipal authorities consider public deliberation (according to Russian legislation means Internet discussions about urban development projects on a special Internet portal) as a better alternative to public hearings. It allow increasing availability for participants,

 https://doi.org/10.1051/shsconf/202111600067

informing citizens about urban development projects and simplifying the work of government.

Inhabitants had to visit city hall (or other specified location) to participate in public hearings and vote for city-planning changes (for example transfer of land from one category to other). They can ask questions and communicate directly with officials.

Public deliberation allows one to do all the same through the Internet. Loss of some active participants because of their computer illiteracy (mainly elderly population) is compensated with a vengeance by inclusion of a significant number of people who cannot attend public hearings because of their employment.

For participation in public deliberation the registration is required (you must enter personal data, including residence). That allows one to reduce risks of fraud when counting votes. On the other hand inhabitants without «permanent residence» as well as public organizations, groups of citizens (associations, clubs and etc.) are excluded from the communication.

Local authorities in the cities we studied promoting public deliberation instead of public hearings. Nonetheless, experts note that change of communication channel (virtual space instead of direct communication) does not deprive this form from inherent disadvantages of public hearings. Longstanding traditions administrate from above, personal interests of some officials and necessity spend organizational resources on «virtual» troubles cause local authorities' unwillingness to develop true participatory urban planning. Municipal authorities can make decisions contrary to the resolution of public discussion.

The situation is complicated by the fact that ordinary inhabitants don't have professional knowledge in the construction and urban development. Public discussion participants (both public hearings and public deliberation) only criticize urban development projects and have not practically applicable suggestions.

All the shortcomings as well as the advisory nature of public discussions' resolutions cause the loss of interest from inhabitants. Experts note a very low share of citizens involved in urban planning projects in Russia.

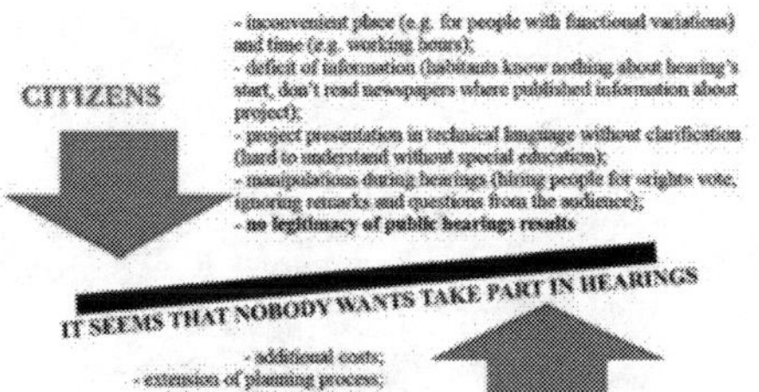

Fig. 2. Shortcomings of public hearings and public deliberation from different points of view

Bad examples of public discussions in big Russian cities can be classified as the «manipulation» and «therapy» stages of the model by S. Arnstein called «Ladder of citizens' participation» [15]. Public discussions in this cases «teach» and «treat» citizens and don't allow them take active part in urban development planning as well in projects implementation.

In practice habitants can express their opinion on different aspects of urban life very actively, have a strong position «for» or «against» some projects but they are not ready to make suggestions. That is why questionnaires and collecting signatures are very popular forms of public opinion expression in Russia. Questionnaires are conducted by municipal governments and as usual closely related to a place of inhabitants' residence (limited locality). Other form – collecting signatures – is usually initiated by active citizens (group of citizens) who want to promote some suggestion or to express their disagreement with some decision.

Civil society can affect decisions related to urban construction or development projects through deputies as public representatives or by applying (directly or in written form) to executive bodies, prosecutor's office. The mechanism of judicial protection is also can be used. Result of conducted focus-groups shows that people consider applying to deputies as ineffective and other specified above forms as very expensive, but possible in the extreme. All these forms indicate lack (or insufficient level) of communication between local authorities and citizens and used by citizens when local authorities make decisions contrary to their opinion.

Protest actions are even a more vivid form of expression citizens' disagreement in relation to urban development projects which implemented contrary to public opinion.

Sustainable development of cities creates a natural background for civic conscience growth and supported by a new generations «Y» and «Z» with IT-thinking [16]. This dilemma particularly affects megalopolises as dynamic business centers with implementation of fantastic ideas in both urban planning and business communications [17].

The new generation is opened to deliberation ideas, looks for comfortable living conditions, able to exchange information quickly. In this context society will develop along the way of strengthening civil self-organization.

Government which grabs «right to the city» and doesn't look for social compromise between all groups of stakeholders cannot be innovative and doesn't correspond to new needs of modern socium.

4 Results and discussion

We recommend increase effectiveness of public hearings and public deliberations. This allow moving from non-participation to tokenism stage (when citizens can at least be heard) according to the model of citizens' participation by S. Arnstein.

We recommend local governments pay more attention to small urban development projects, control procedures of their public discussion (for example, through interaction with deputies or developers). Local authorities should pay special attention to the following public discussion aspects:

- engage residents at all project cycle stages from the earliest;

- submit for public discussion clear and close to particular groups of citizen projects, divide scale projects into small, discuss separate aspects of whole project in focus-groups, make special surveys and consultations with experts (accessibility of renovated area for people with functional variations, bicycle roads, organization different zones for sport, promenade or play-grounds in parks and etc.);

- conduct regular public discussions with habitants at residential quarter, microdistrict level.

Local authorities should always keep in minds that legitimacy of public discussions depends rather on project changes under the pressure of public opinion, not on formal procedures [18].

New forms of public discussion (master plan development, participatory design, etc.) were introduced in some big Russian cities last years for public-space development projects. The Russian Agency of Strategic Initiatives (large non-profit organization in Russia responsible for stimulation of economic development) was developed and the Federal Ministry of construction, housing and communal services approved special methodological recommendations «The basic model for public participation process organization». This document details the national project «Comfortable urban environment». The positive experience of their approbation is described in the works of a number of authors [19, 20].

Participatory design conducted in small groups (in the form of project seminars) supposes organization series of meetings with residents throughout the whole project cycle (minimum three meetings), starting from the earliest stage of the project cycle where people formulate initial requirements (must be included in technical specification to the project development documents).

The master plan as a document is similar to a general city plan but is not included in the legally established list of urban planning documents. Master plan designers (usually consulting companies, architect bureaus) conduct sociological surveys of residents and other stakeholders (groups of active citizens, non-profit or public organizations), meetings or seminars. Public discussions are carried out at all stages of project cycle (development of the project concept, detailed planning, presentation of project results) in small groups. This form of participatory planning is applicable not only in urban development projects, but also in transport infrastructure development, ecology, stimulation of economic development projects. Public opinion forms base on the project improvement and citizens feel responsibility for positive urban development as if they are co-authors.

In spite of the fact that new forms of participatory urban planning became very popular in public space development project, it is still very hard to transfer their results into a general city plan. Other significant disadvantage of these forms is that they need a significant amount of time and resources.

5 Conclusion

Various aspects of participatory urban planning are described in papers of foreign researchers. This research direction has been developing for more than 50 years, but due to specifics of public administration, it was not actual in Russia for a long time.

Legally established forms of public participation with special accent on public chambers and councils don't correspond to institutional forms used in practice. Presentation of territorial planning documents to expert councils and in legislation bodies is widespread in practice and considered as public participation. Public hearings and public deliberation procedures, questionnaires are formal, their results don't affect the final decision-making.

Socio-political transformation in Russia requires further improvement of public participation forms. Municipal authorities receive significant benefit when organizing the social communication process correctly. First of all, residents have important for urban planners information and can help actualize it. Second, residents who take part in the project deliberation are often ready to make a personal contribution to its implementation.

Foreign scientists made significant contribution in this research direction development, but it still needs to be clarified:

- terminology (it is necessary to analyze the definitions of public participation forms used in scientific papers and legal acts);

- indicators and methodology for evaluation the degree and effectiveness of citizens' involvement in the processes of participatory urban planning.

Recommendations of western authors based on outdated 50 years old planning practice should be adapted to actual Russian conditions before application. For example the range of urban development stakeholders in Russia today includes not only residents.

First of all, functions of communication between state and citizens in Russia can belong to organized groups of people (public non-profit non-governmental organizations). Unfortunately they are currently excluded from deliberation because of gaps in legislation.

Second, significant attention must be paid to development of direct communication between citizens and big companies – urban development investors. Big companies in regional capital cities and also in mono-cities participate actively in housing building and public space development/renovation projects.

We should pay special attention to the new forms of public participation developing very fast.

References

1. L. Leonova, E. Alpatova, T. Rutkauskas, S. Kowalczyk, *International Conference on Sustainable Cities* (ICSC), **6,** (2016).
2. P. Akimov, *Urban research and practice, Pilot Issue* (2015).

3. H. Lefebvre, *Le droit à la ville. Collection Société et Urbanisme* (Paris, Anthropos, 1968).
4. S. Konchekov, 2018. *Architecture and modern information technologies*, **1(42)**, (2018).
5. J. Uitermark, W. Nicholls, M. Loopmans, Environment and Planning, **44 (11)** (2012).
6. M. Pursel, *GeoJournal*, **58 (2-3),** (2003).
7. F. Laloux, *Reinventing Organizations: A Guide to Creating Organizations Inspired by the Next Stage of Human Consciousness* (Nelson Parker, 2014).
8. D. Harvey, *Cosmopolitanism and the Geographies of freedom* (Columbia University Press. 2009).
9. P. Davidoff, *Journal of the American Institutes of Planners,* **31(4)** (1965).
10. N. Krumholz, *Journal of Architectural and Planning Research*, **4**, (1986).
11. P. Healey, *Collaborative Planning: Shaping Places in Fragmented Societies* (2nd Edition. Macmillan International Higher Education. 2006).
12. L. Sandercock, *Planning Theory and Practice*, **6(4)**, (2005).
13. R. Mukhatinova, *Youth scientific bulletin,* **12,** (2016).
14. I. Medvedev, *Zakon,* **3**, (2018).
15. S. Arnstein, *Journal of American Planning Association,* **35(4)**, (1969).
16. E. Shershneva, A. Gabitova, *International Conference on Continuing education for sustainable development: new challenge*, May 23-24, in Nur-Sultan, Kazakhstan, (2019).
17. E. Kondyukova, E. Shershneva, *WHATT,* **10**, 4. (2018).
18. A. Vasilache, *Case studies from small and mid-sized European cities*. (Belgrade, Serbia. 2010).
19. A. Lozhkin, *Moscow Urban Forum*, December 7, (Moscow, Russia, 2013).
20. A. Schukin, *The Expert,* **40 (1091)**, (2018).

The role of higher education institutions in the development of intellectual capital

Grigory Akopov[1], *Vazgen* Tiratsuyan[2,*], and *Yulia* Kurysheva[3]

[1] Rostov branch of Moscow Sate, Technical University of Civil Aviation , 262 B, Sholokhov street, Rostov-on-Don, 344009 Russian Federation
[2] *South-Russian Institute-branch of the Russian Academy of National Economy and Public Administration*, 70/54, Pushkinskaya Street, 344002, Rostov-on-Don, Russian Federation
[3] Petersburg State University, 7-9, University embankment, 199034, Saint Petersburg, Russian Federation

Abstract. This article deals with the questions of intellectual capital, summarizes the results of content analysis of the notion "intellectual capital", its essential characteristics and basic elements. Authors show the role of higher education system in generating and developing of intellectual capital of future professionals. A high level of intellectual capital is a key element of organization competitiveness and productivity and can be seen as a basis for a leading position in a particular industry. The study focuses on the development of intellectual capital of civil aviation professionals. It summarizes the results of analysis of Rostov branch of Moscow State Technical University of Civil Aviation's experience in the field of intellectual capital development during the education of future civil aviation engineers.

1 Introduction

Intellectual capital along with traditional factors of production is an essential element of competitiveness on macro- and microeconomic levels.

Intellectual capital can be defined as transferable knowledge that belongs to an organization and can be commercialized [1].

According to Stewart T., intellectual capital «is an intellectual material that includes knowledge, experience, information, intellectual property and contributes to creation wealth. It is a collective mental energy…». Another definition made by this author claims that intellectual capital is an aggregated knowledge of all company's employees, ensuring its competitiveness [2].

According to Brooking A., intellectual capital can be understood as a set of intangible assets which are vitally important for an organization [3].

There are other interesting researchers' views about the ration of intellectual and human capital.

Saint-Onge H. and Edvinsson L. distinguish human, structural and consumer capital in the structure of intellectual capital. Stewart T. agrees with them.

There is also an opinion among researchers that an implicit capital structure includes human, institutional and market capital [4].

According to the model of Brooking, intellectual capital consists of market, human, infrastructure assets and intellectual property [5].

It appears that intellectual capital has become or is becoming (depends on a particular socio-economic system and exact industry) a decisive factor stimulating efficiency of the company's activity and the most important one contributing to achieving high productivity.

2 Materials and methods

The role of higher education institutions in generation intellectual capital cannot be overemphasized. The development of intellectual capital is one of the fundamental tasks of any university as one of the most important components of intellectual capital is human capital. One of the most common approaches to assessing its quality is the length of the period of study [6]. By the early 70-s of the XX century the contribution of education and professional training to the overall structure of human capital in the US was estimated at about 85% [7].

Intellectual capital is a key factor for a successful innovative development of both an individual organization and a state as a whole. This allows one to produce high-tech goods which are competitive not only on the domestic but also on the external markets. For now, the Russian Federation is not among the key players of the global market of innovative manufacturing technologies. For example, according to some data, the share of the Russian Federation in the global exports of high-technologies has not exceeded 0.5% in the last decade. Moreover, there is a constant rick of lagging behind the leading nations [8].

One of the few competitive industries of the Russian Federation is the aviation industry (as a part of the aerospace industry). Therefore, it is vitally necessary to

* Corresponding author: amber.78@mail.ru

develop intellectual capital in the sphere of civil aviation more intensely in order to maintain and strengthen national positions on the global level. According to the international accountancy and business advisory network Baker Tilly Aeroflot-Russian Airlines is in the Top-10 ranking of Russian intellectual companies for public (8th position [9]. This allows the company to be one of the successful and efficient airlines in the world.

Strong intellectual capital makes it possible for leading companies of a particular industry to get a higher level of productivity in comparison with those who doesn't take in account the importance of this factor. For example, labor productivity among the leaders of the aviation industry in the Russian Federation is 6 times higher than in other enterprises of the same sector. High quality of intellectual capital and specific attention to its development are called among important factors of such indicators [6].

At the same time, there is generally a small share of highly qualified personnel among the outsiders of the aviation industry. The lack of highly skilled staff is one of the main factors of low productivity. Highly qualified personnel are assumed to have intellectual capital vitally important for a company. Consequently, it is possible to claim that about a significant role of intellectual capital in achieving high indicators if productivity, labor results and overall efficiency of a market player.

3 Results and discussion

The intellectual capital level from the starting point helps a company to estimate the potential of a candidate, its ability to work in modern circumstances characterized by variability of conditions, life-long learning, rapid changes of professional knowledge and skills. In our opinion, the further investments into the intellectual capital of a particular employee are more effective when the employee has already had a sufficient level of such a capital before starting his work. Consequently, a high level of intellectual capital, which started to be formed at the university, is a strong basement for the development of it during the whole professional life of a person. If this basis for intellectual capital isn't formed during the higher education period, it's the highest level to be achieved.

In recent years in Russian society there is an obvious problem linked with the deterioration in the cooperation of employers and universities and colleges. This problem gradually reduces higher education capacity as a main and the strongest source of intellectual capital builders. As a result, companies in different spheres have to become (and has already become) the main and only sources of intellectual capital formation, while higher education organizations play a secondary role or even don't participate at all. It is obvious that colleges and universities have to keep up to date, take into account constantly changing social and economic environment and start to take part in building intellectual capital during the education to fulfill their natural capacity. The establishment of sustainable long-term comprehensive links with leaders of the industry is also an important factor.

In contemporary economic reality the explosive development of information technologies and the development of the knowledge economy require constant updating of professional knowledge and skills of the graduates. In such circumstances the education system is searching for more effective approaches and forms of teaching future professionals. Intellectual capital becomes the main competitive advantage of a higher education and can be understood as a factor of additional educational services cost. Such "potentially commercialized" intellectual capital contributes to additional profit of the higher education institutions and reduces its dependence on budget funds. Therefore, intellectual capital of a higher educational organization, on the one hand, allows it to gain additional profit, on the other hand, additional profit can contribute to further development of intellectual capital. The rise of the knowledge economy forces universities to accumulate knowledge, skills, competencies and focus on their effective developing, framing a strong basic structure for collective intellectual capital.

Now in the context of postindustrial society when intangible assets become the most important assets of a company, elements of intellectual capital are acknowledged as main competitive advantages of any university. Key stakeholders with the decision-making power such as administration, influential representatives of teaching staff, honorable professors, founders realize this fact and strive to develop the ecosystem of higher education creating appropriate conditions for the development of intellectual capital and increasing its index.

The intellectual capital index provides more precise assessment of its quality. Among the relatively modern methods of valuation of intellectual capital is an IC-index method. It is a method of so-called second-generation and was developed in the 1990s. It differs from previous ones by an integrated approach combining several methods. It gives much more complex valuation of intellectual capital of an organization and helps to understand how it can bring profit.

The development of methods for measuring of intellectual capital has led researchers to creation of a holistic value approach (HVA). This approach takes into account the limitations of previous methods based on accounting approach that is ineffective for strategic management decision-making.

HVA method synthesizes the IC-index and inclusive valuation method (IVM). It assesses the contribution of financial and intangible assets as well as the shareholder value of the company to the final value. As a result, the method valuates the contribution of intangible assets to the cash flow.

In modern circumstances universities and colleges become the most influential creators of intellectual capital, but not all of them, only those with a certain potential, enabling them to implement several vectors of development simultaneously:

- formation and implementation of up-to-date educational programs and technologies that make possible to combine scientific and research activities during the process if education;

- project activities (organization, management, realization, group activities) in the sector of industry for which the organization prepare professionals;

- research activity, development of innovations and new approaches.

All these allow universities and colleges to decide the main task of all organizations of such type – to provide well-prepared high-skilled professionals [5].

Moreover, education and training of highly qualified specialists for the national aviation industry are the most important tasks facing the state.

In recent years the aviation industry has gained new impetus as part of national import substitution policy, the domestic civil aviation groups are developing promising components and aircraft vehicles (such as AP-14 family of aircraft engines and Irkut MC-21 airliner).

Despite the strong negative coronavirus pandemic impact on the aviation industry in 2020 the average rise of air travelers amounts during the 2019 – 2039 will be not less than 3.2 – 5.3% in the world an 8.9$ in Russia according to IATA (International Air Transport Association) survey [10].

This data predicts the long-term investment prospective of the aviation industry and the rise of request for a sufficient number of well-trained aviation engineers for the industry with high indicators intellectual capital. Of course, investors and major airline stakeholders, along with the intensification of fixed investment, should also develop intellectual capital. The peculiarity of intellectual capital is in that its systemic formation should take place only with help of specialized universities.

Such high-tech products can be operated and maintained only by highly qualified engineering specialists and only if there is no lack of such specialists in the industry it can grows steadily.

Thus, only in 2019 alone the State PJSC United Aircraft Corporation (UAC) recruited 1,187 persons from specialized educational institutions. (16.3% of the total number of employed), 813 specialists came from higher education institutions and 374 from professional colleges. A total amount of students with scholarship for education from aviation companies was 2,899 [11].

Thus, the main sources of the formation and generation of intellectual capital in the field of civil aviation are the educational system and the companies working in the sphere.

In our opinion, educational institutions play a key role in this process. Among other things this happens due to the fact that many aviation companies do not develop the intellectual capital of employees with help of their own resources, fearing to lose specialists because they gain more experience and become more in demand on the labor aviation market and would be able to change the work easily for better conditions.

So, we face emerging of a paradoxical situation: on the one hand, organizations are extremely interested in high level of intellectual capital; on the other, they deal with the issue with caution and fear. Moscow State Technical University of Civil Aviation has a special opinion about and practice of the formation of the intellectual capital in the aviation industry.

Not accidentally the university's mission specifically states that: "The mission of the University is to preserve and enhance the achievements of the engineering of mankind, to acquire and disseminate advanced knowledge and information, to advance the training of the intellectual elite of society through the integration of the educational process, fundamental research and innovation approaches".

It is important to note that the training of civil aviation engineers takes a special place in the national education system and has a great significant for the industry development.

It is well known that there are 28 thousand settlements in Russia that can be reached only by air transport.

The only and unique aviation university in the South of Russia, created directly for preparing and providing a civil aviation sphere with highly qualified engineers who are able to increase the intellectual capital of the industry, is situated in Rostov-on-Don. Its founder is the Federal Air Transport Agency of the Russian Federation.

The Aviation University was founded in Rostov-on-Don in June 16, 1969. The Higher Education Institute was established with the aim of improving the quality of training of aviation industry engineers in the south region of Russia, the countries of the South Caucasus and other nearby territories of 5 former republics of the USSR with the introduction of distance learning.

Thus, in modern language, Moscow State Technical University of Civil Aviation was created with the purpose of forming intellectual capital among aviation engineers. The university is fundamentally different from the majority of industrial colleges and schools as it realizes unique for the region educational programs.

The Rostov branch of Moscow State Technical University of Civil Aviation traditionally organizes international scientific and practical conference «AviaTrans». The decision to establish the conference was taken in the Rostov branch of institution in 2010, at the International Scientific and Practical Conference «Contemporary Issues of Air Transport Development in South Russia» and now it is held on a permanent basis since 2011 [12].

One of the key participants of the conference is the leading airline carrier of the Russian Federation and one of the best airlines in the world public joint stock company «Aeroflot-Russian Airlines», which is undoubtedly one of the leading domestic organizational structures at the same time forming and using intellectual capital of the aviation industry [13].

Every year, hundreds of experts, scientists, researchers, employees, young professionals and students of the aviation industry discuss current issues and exchange experience and innovations in the sphere of civil aviation at the scientific ground of the Rostov branch of Moscow State Technical University of Civil Aviation. This, in turn, contributes to the transfer of modern and up-to-date views, trends and knowledge to students studying within the walls of the Rostov branch of Moscow State University.

In 2021 the annual conference for the first time in the history of the Rostov branch of Moscow State Technical University of Civil Aviation and the main university of civil aviation was among the best scientific and practical measures of the industry in general. The university won a special government grant established by the national science funding body of the Russian government the Russian Foundation for Basic Research (RFBR) for the implementation of this project.

This made it possible to conduct this event more far-reaching with the publication of the scientific reports presented on the conference, conference papers were published in two volumes indexed in the Russian index of scientific citations and are available for professionals and students, which is also can be an indicator of the increase of intellectual capital.

Furthermore, the Conference is a way of development of a scientific discourse that also generates intellectual capital.

The main achievement of the conference is the implementation of the ideas discussed on panels of «AviaTrans» into the daily practice. Thus, according to the results of the international scientific-practical conference «AviaTrans» hold on September 27, 2017, the Rector of Moscow State Technical University of Civil Aviation declared the necessity of opening a face-to-face form of education in the Rostov branch of the university.

This strategic decision is directly related to the aspiration of Moscow State Technical University of Civil Aviation to become a leader and to further strengthen its positions as a global scientific and educational platform for the formation of intellectual capital in the sphere of civil aviation of Russia.

As a result of the joint efforts of the Head University and the Rostov department, on September 1, 2018 more than 20 students of a full-time form of education in the direction of «technical operation of aircraft and engines» have started to learn.

After opening of the full-time face-to-face education program in 2018 (20 budget students and a commercial one) the management of Rostov Branch of Moscow State Technical University of Civil Aviation set itself an ambitious goal of recruiting 50 full-time students in 2019. This figure was very symbolic because that year the Rostov branch of the university celebrated the 50th anniversary.

Since the opening of the face-to-face form of education in the Rostov branch of the Moscow State University the teaching and educational process has been substantially modernized. The shift from the usual distance learning modality has required from the administration and teaching staff the adoption new approaches in the learning process. Students has become more actively involved in various projects and events that contributes to the development of professional skills on theoretical and practical level as well.

In a relatively short period of time students of the Rostov branch took part in 14 scientific, scientific-cognitive, public-political events as a part of additional activity.

Our students take part in conferences and other competitive activities and become winners. By the recommendation of the Ministry of Transport of Russia a first-year student Dmitry Egorov reached the final stage of the International Olympiad on Aviation History, where he was the only representative of the South of Russia.

On the meeting of students and applicants with the Minister of Transport of the Rostov region took place in the Rostov branch of Moscow State University on the March 15th 2001, the head of the relevant department gave a high evaluation of the students' activities and their contribution to the potential development of the aviation industry in the region.

The Open-Door Days held by Rostov branch of Moscow State Technical University of Aviation have a significant meaning and become a decisive event for future applicants and aviation professionals. These events attract a great number of concerned teenagers seeking for professional education. The Open-Door Days is an important instrument of building a gradual personal intellectual capital trajectory, starting from school through higher education till working environment.

One of the most important criteria demonstrating the existence of high-level intellectual capital in the structure of a higher educational organization is not merely the formal achieved indicator of the demand for graduates in the labor market, but the actual state of affairs, when the vast majority of graduates get the work according to their education in the civil aviation industry. This indicator shows better that all skills, important professional ideas, mission and basic rules are shared and well comprehended by the students during the education, students are highly motivated and ready to apply all their knowledge for the development of the industry.

4 Conclusion

In order to be a powerful stakeholder in the industry, to gain public reputation and make its activity popular and open for society, the Rostov branch of Moscow State Technical University of Civil Aviation regularly promotes its brand via the mass media not just as a classical university, but as an advanced center of intellectual capital formation in the aviation industry.

Well-built relationship with mass media allows the university to regularly take place in regional news stimulating recognizability and interest of potential students and other stakeholders.

In the framework of the intellectual capital development the Rostov branch of Moscow State Technical University of Civil Aviation actively cooperates with various government and public organizations working in the field of civil aviation, including the Southern Interregional Territorial Air Transport Authority and the Federal Service for Supervision of Transport (Department of State Aviation

Supervision and Transport Security Supervision for the Southern Federal District).

Joint conferences, open-door day, organizations, involvement of employees from industry into the educational process make it possible for all participants of the formation of intellectual capital constantly receive practical and advanced experience.

The contemporary short-term goal of the University is to start the full-time form of education in another area of training, which is currently realized in distant format - «Technical operation of transport radio equipment».

Of course, this results and qualitative modernizations has become possible due to the acquisition of a new building for the Rostov brench of Moscow State Technical University of Civil Aviation, which is now being put into operation.

The new educational building will serve in the future for realization of innovative projects, deepening and broadening of cooperation links with aviation enterprises, raising the image of the university among the broad public, applicants and students. It will provide more comfortable conditions for fruitful work of professors and teaching staff. A new building will give a space for training students at an even higher and more modern level required by employers in the field of civil aviation, which will consequently lead to a qualitative increase in the intellectual capital of both the university and its graduates.

References

1. Ju. Duhnich, *Intellectual capital: basic elements, management, evaluation*. (2014) Retrieved from: https://www.cfin.ru/management/strategy/competit/Intellectual_Capital.shtml
2. T. A. Stewart, *Intellectual capital. New source of wealth of organization* (Moscow, 2007).
3. A. Brooking, *Intellectual capital* (St. Petersburg, Piter, 2001)
4. S.V. Baguzin, *Strategic management,* **4**, 249, (2014)
5. S.V. Panikarova, M. V.Vlasov, *Knowledge and intellectual capital management* (Ural. University, Ekaterinburg, 2015)
6. Ju.V. Simachev, *Economic questions,* **3**, 35, (2021_
7. T.A. Shtercer, *NGU Bulletin*, **6(2)**, 41, (2006)
8. *Experts have warned of the risk for Russia to "forever lag behind" in technology*, Retrieved from: https://www.rbc.ru/technology_and_media/13/04/2021/607478fc9a794731d03611ab
9. *Experts named the most intelligent companies in Russia,* Retrieved from: https://www.rbc.ru/business/19/09/2019/5d8359f49a7947371929d17f
10. *Why aviation is a promising area for long-term investors,* Retrieved from: https://quote.rbc.ru/news/article/5f6870669a794731cc4d4a8d
11. Annual report of Public Joint Stock Company «United Aircraft Corporation», 2019
12. G. L. Akopov, *Politics and society,* **4**, 126, (2011)
13. G. L. Akopov, *Innovative processes in Russia and world*, **1**, 5, (2015)

Project-based learning as an element of the university innovation ecosystem

Yulia Khodkovskaya[1,*], *Olga* Malykh[2]

[1]Ufa State Petroleum Technological University, 1 Kosmonavtov st., Ufa, 450076, Russian Federation

Abstract. The article presents an analysis of the development problems of project-based learning technology in Russian universities, which can make a significant contribution to achieving the goal of academic leadership. The issues of the formation of the competencies of a university graduate in the context of project-based learning are considered. It is shown that project-based learning should be included in a multifactor educational environment - the innovative ecosystem of the university. The relationship between the elements of the university's innovation ecosystem, such as education, individual learning path, project-based learning, digital content, polylingual content, non-formal education, partnership programs, supervising, mentoring, design and interdisciplinarity, has been determined. The article substantiates the criteria for the effectiveness of project-based learning, and also offers options for the formation of the innovative ecosystem of the university, as a new trend in the Russian educational environment, which allows students to be involved in research and project work.

1 Introduction

It has been proven many times in theory and practice that education is a key area of human economic activity. Over the past three decades, many problems associated with it have accumulated in Russia: organizational, financial, social, economic, etc. However, it cannot be said that society and the state do not care about these issues. So, in the 2020s, the implementation of the "5-100" program was completed, with the goal of entering the leading Russian universities to international ratings. In 2021, the program of strategic academic leadership "PRIORITET-2030" was launched, the result of which the government of the Russian Federation sees "training of personnel for priority areas of scientific and technological development of the Russian Federation, economic and social sectors, the development and implementation of breakthrough research and development, new creative and social and humanitarian projects, as well as the introduction of high technologies into the economy and social sphere".

There is a large amount of research in the field of education, among which, in the context of this article, it is of interest to analyze the possibilities of an interdisciplinary approach to project-based learning, which consistently connects such components as creative and critical thinking, accumulated experience, thus facilitating innovation [1]. An additional advantage of this approach is the professional stability of university graduates and their ability to work effectively in a team [2]. At the same time, many researchers note the complexity of organizing project-based learning, since not in all countries of the world it is possible to create such an innovative ecosystem today [3]. At the same time, it is important to understand the economic possibilities of the government in creating and supporting innovations in universities [4, 5]. Such elements of the university innovation ecosystem as digitalization of education and the involvement of participants in the educational process to the discussion of integrated curricula [6, 7], technology transfer algorithms and the commercialization of university research [8], as well as critical criteria for the success of business incubators, growth accelerators [9] are also actively being discussed.

2 RESULTS

A critical review of approaches to the definition and essence of project-based learning in Russian and foreign universities revealed its advantages in training university graduates, which is manifested in the flexibility of the educational technologies used, combining traditional forms with digital ones. It should be noted that many publications devoted to project-based learning in universities focus on the quality of training of specialists, activation of their zero gravity thinking, and, as a consequence, the formation and development of universal competencies of students in a digital educational environment.

Analysis of the results of project-based learning in leading universities in Russia and Europe showed that its application stimulates the research and development and educational activities of students, and the use of modern information and communication technologies in project-based learning significantly increases the effectiveness of training. In the surveyed universities in Western

*Corresponding author: khodkovskiy@bk.ru

Europe (Lancaster University Management School (UK), European School of Management and Technology (Germany), The University of Manchester (UK), Turku University of Applied Sciences (Finland), etc.) in the process of project-based learning students acquire interdisciplinary knowledge that determines the training of digital personnel. Thus, it can be concluded that in most European countries university education contains project-based learning as an indispensable component.

The results of studies of relevant practices of Russian universities (Moscow School of Management "Skolkovo", South Ural State University, Moscow Higher School of Social and Economic Sciences, etc.) made it possible to establish that the fundamental training of university students is based on the harmonious application of global educational trends, taking into account the national educational identity. In addition, the quality of university graduates is ensured through the integration of educational, research, economic and social processes. Moreover, this conclusion is the same for both Russian and foreign universities.

The competence of graduates in the modern labor market ensures their competitiveness. All competencies, taking into account the pace of development of scientific and technological progress, the changing demands of employers for the quality of education become obsolete, lose their importance, which manifests itself through the "half-life" of competencies, increasing the importance of improving the educational programs offered by universities.

Project-based learning contributes to the development of new, digital-economy-oriented competencies of graduates.

It should be noted that the competencies of university graduates in the innovation ecosystem of the university are formed as a synergistic result of "student consumption" of educational, research and development and business products offered by the university (Fig.1).

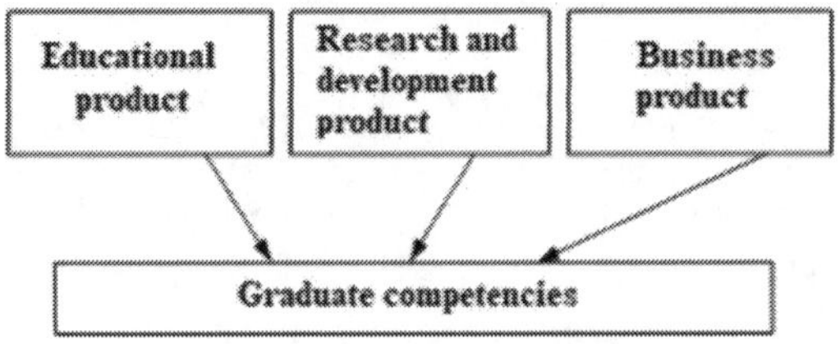

Fig. 1. Project-based learning: graduate competencies

For university graduates, supra-professional competencies aimed at developing social responsibility, safety management, and high-quality implementation of interaction interfaces are also important.

However, a comparison of the degree of involvement of domestic universities in the implementation of project-based learning technology leads to the need to solve the problem of taking into account the "multifactorial nature" of the existing educational environment when building the professional personal trajectory of students.

Despite the fact that the sectoral focus of Russian universities is very wide, the focus on consolidation, unification of universities, the creation of national educational and research and development centers, allows us to identify modern universities as global centers for the training of a polytechnic profile and science. Thus, the need for possessing universal competencies that increase the competitiveness of university graduates is actualized. In this aspect, project-based learning is an important tool of modern university education, an element of the university's innovation ecosystem.

Table 1. Effectiveness criteria of project-based learning

Consistency	Mastering knowledge, skills and abilities of professional activity based on the principle of "education throughout life": from theoretical foundations to improving the cognitive process and the development of universal competencies.
Agility	It manifests itself through changes in the multilevel training of university graduates, forms of advanced training, new specialties and specializations that meet the social and economic conditions of the development of society.
Polytechnology	Associated with project activities as part of the formation of a digital ecosystem of universities. It involves the use of a wide range of modern tools and teaching methods in the communication and information space.
Adaptability	Training of digital personnel based on project-based learning, taking into account the requests of stakeholders along the chain "science-university education-business community-APE".
Transparency	The openness and accessibility of project-based learning determine the opportunities to build a professional personal trajectory of a student in the organization of university education.
Flexibility	The advantage of project-based learning is the flexibility of the educational technologies used combining traditional forms with digital ones.

The peculiarities of the formation and development of the university education system in Russia have significantly changed the key principles, criteria for the effectiveness of higher education, focusing on the needs of the modern labor market. Such factors as the growth rate of digitalization of the economy, transfer of knowledge and technologies, integration processes, international cooperation in the field of education, a number of researchers are said to be the most significant for improving the system of university education.

Depending on the combination of factors and their significance, the university environment differs from mono- to polyuniversities, including additional professional education (APE). At the same time, there is a widespread opinion about the importance of vocational education, mastering practical skills, which actualizes the implementation of "double diploma" programs, a variety of specializations in the curricula of training programs for bachelors, specialists, masters, simultaneous mastering of the profession at production sites.

The effectiveness of university education in Russia is associated with project-based learning focused on the creative self-realization of students, increasing their competitiveness in the context of digitalization of business and society through the development of intellectual abilities, their creative application in the digital educational environment. Therefore, it becomes necessary to use project-based learning along the chain "science - university education - business community - additional professional education (APE)". The criteria for efficiency in this case can be considered consistency, dynamism, polytechnology, adaptability, transparency and flexibility (Tab. 1).

The presented criteria in general make it possible to assess the effectiveness of project-based learning, and, therefore, to identify it as an innovative approach to acquiring knowledge, mastering modern methods and skills.

Project-based learning as an element of the university's innovation ecosystem is implemented through a variety of communication links, both internal and external (Fig. 2). In this regard, the so-called "spillover effects" arise, taking into account the interactions of all interested parties along the chain "science - university education - business community - additional professional education (APE)". Effects of this kind can spread both within one country and in international cooperation.

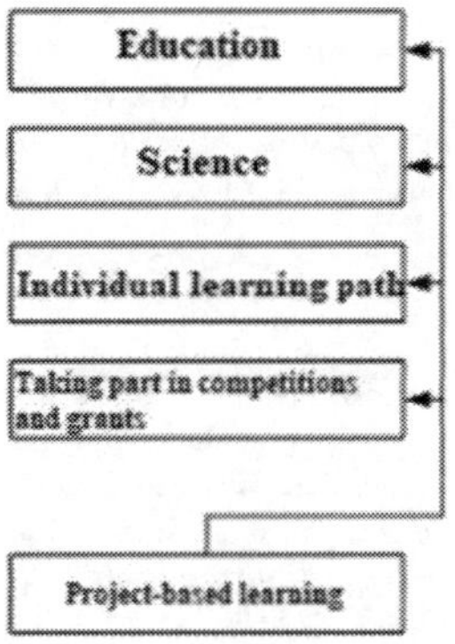

Fig. 2. Project-based learning as an element of the university innovation ecosystem

3 DISCUSSION

Many authors note that in a competitive market, the redistribution of economic resources occurs in favor of the ecosystem of the university that has the potential for growth, infrastructure resources that provide unhindered integration and interaction of material, financial, labor and information flows that form complex multi-level logistics chains [10], self-development and self-organization of all elements of the university innovation ecosystem.

The ecosystem of a modern university is a dynamically developing environment that creates opportunities for expanding educational, research activities, developing cooperation and partnerships both at the level of higher educational institutions and at the level of the business community.

The development of business cooperation and business partnerships in the university environment, for example, in the UK, Israel, Canada, Spain, Italy, South Korea, Finland, involves a significant revision and improvement of educational programs, taking into account a whole range of requirements for the quality of mastering universal competencies, fluency in technology digital modeling, management of complex technical or professional activities. In this aspect, digital learning is the university's tool that allows you to overcome the barriers to technology transfer, exchange effective methods, scale and replicate the results of scientific research in the business environment. In addition, project-based learning at the universities of aforementioned countries is actively developing due to the attraction of additional external funding, which means that research and development of universities are being introduced and adapted to the business community at an outstripping pace. Particularly popular are venture projects (in Russian practice, analogs are leadership projects, free projects), which allow you to quickly and easily use the resource capabilities of databases, cloud storages containing modern achievements of science and practice, solving applied problems of the business community.

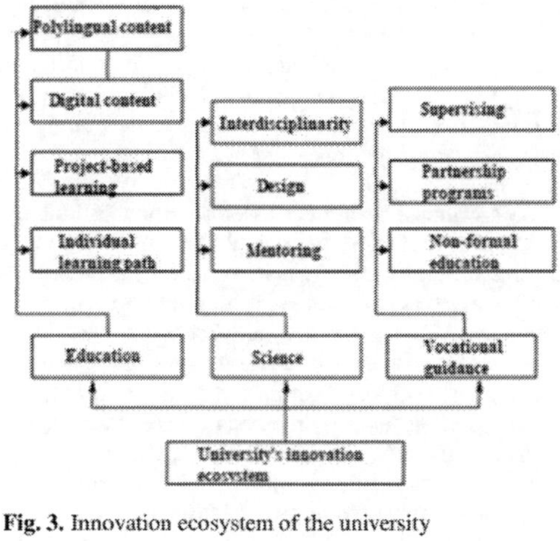

Fig. 3. Innovation ecosystem of the university

One of the elements of the innovation ecosystem of the university is polylingual content, which includes multilingual teaching of students, the importance of which has significantly increased in recent years (Fig. 3). This was facilitated by the processes of globalization, internationalization, integration interaction, the development of a system of international network, distance education format, the adoption by Russia of the provisions of the Bologna Declaration on a single European space in higher education [11].

The university's multilingual content connects the educational environment with the business community, ensures the building of business communications, using both the native language and "international languages".

For project training, the priority of poly-lingual content is confirmed by the fact that advanced foreign schools for training project managers in modern universities teach in English, which is recognized as the corporate language of business communication. Consequently, it becomes possible to get acquainted with primary sources, interact with representatives of science and business from other countries using multilingual knowledge. In addition, the consequences of the COVID situation have accelerated the development of distance educational programs, online courses, the availability of which increases as users learn "international languages".

Digital content as an element of the university's innovation ecosystem is inseparable from project-based learning. The educational programs of most modern universities in the era of digital transformations rely on the active use of digital technologies that provide higher speed, accuracy of calculations, variability of research results, which is important in project-based learning and at the same time attractive for business.

Project learning in a modern university environment is inextricably linked by accumulators of analytical information such as Big data, HDD and SSD drives, cloud storage, which requires in-depth digital knowledge and skills. As the analysis of the practice of Russian universities shows that joint R&D projects, on QR coding in recent years, which have become widespread in recent years, rely on digital decision-making models.

Also, in the innovation ecosystem of the university, an individual learning trajectory is important, its connection with project-based learning can be traced as a result of the development of critical thinking, creative, non-standard vision of solving a problem, searching for innovative ways to solve problems of science and business.

The advantage of project-based learning in building the individual trajectory of students is manifested in the fact that the formed information and communication space of universities makes it possible to choose a number of courses, taking into account the future professional activities of students (elective courses, workshops, free courses within the Student Academies, etc.). In addition, modern methods of cognition, online intensives, gamification, business trainings, master classes that develop new professional competencies in students are widely used in project-based learning.

Mentoring, design and interdisciplinarity are also elements of the university's innovation system. The importance of the above elements is explained by the fact that project-based learning is implemented along the chain "science - university education - business community - additional professional education (ADE)", increasing the importance of the elements describing the block "science" in the university ecosystem.

Mentoring is perceived in the innovation ecosystem of the university as an educational mechanism that takes into account different levels and needs for project-based learning. Mentoring, being one of the hallmarks of mentoring in project learning, is designed to ensure the continuous development of the learner's competencies throughout the entire educational/professional activity.

Design and interdisciplinarity are closely connected with one another, since in project training, not only knowledge in the field of modeling business processes, solving business problems are prioritized, but knowledge, skills and abilities acquired in interdisciplinary areas of professional activity are also important. It is design and interdisciplinarity that enhance the development of students' universal competencies that meet the needs of an innovative educational environment. The emphasis on mastering interdisciplinary competencies is determined by European network (ENAEE) [12], according to which graduates must have competencies that allow one:

- to think critically and solve environmental, economic, production and management problems, to understand their consequences for business and society;
- to build effective communication links;
- to develop innovative methods of analysis and problem solving;
- to communicate judgments that include reflections on relevant social and ethical issues.

4 Conclusion

Thus, the structural-competence approach as the basis of project-based learning at universities is not only associated with mastering universal competencies based on interdisciplinary knowledge, but, taking into account the needs of the business community, science, and the state, contributes to the transformation of professional competencies [13, 14].

Of course, the formation of a stable innovation ecosystem of the university is a new trend in the Russian educational environment, allowing students to be involved in research and project work at the university.

References

1. M.K. Habib, F. Nagata, K. Watanabe, *Educ.Sci.* **11**, 46. (2021) https://doi.org/10.3390/educsci11020046
2. T. Konrad, A. Wiek, M. Barth, IJSHE *International Journal of Sustainability in Higher Education* **22 (3)**, (2021)
3. S.I.H. Liyanage, F.G. Netswera, *Journal of the Knowledge Economy* (2021) https://doi.org/10.1007/s13132-021-00769-y

4. O.E. Malikh, I.K. Polyanskaya, M.E. Konovalova, O.Y. Kuzmina, O.V. Tarasyuk, I.V. Osipova, International Electronic Journal of Mathematics Education. **11(8)** 3104-3113 (2016).
5. A. Cahyadi, R. Magda. Economies. **9(32)** (2021) https://doi.org/doi:10.3390/economies9010032
6. K. Maaranen, H. Kynäslahti, *Technology, Pedagogy and Education* '(2021). DOI: 10.1080/1475939X.2021.1876756
7. M. Mielikäinen. *Industry and Higher Education.* (2021). https://doi.org/10.1177/0950422221994471
8. R. Ravi, M.D. Janodia, *Journal of the Knowledge Economy* (2021). doi.org/10.1007/s13132-021-00747-4
9. K.A. Siddiqui, M. Emad Al-Shaikh, I.A. Bajwa, A. Al-Subaie, *Entrepreneurship and Sustainability Issues.* **8(3),** 267-279. (2021). http://doi.org/10.9770/jesi.2021.8.3(15)
10. Luigi T. De Luca, *Propulsion physics* (EDP Sciences, Les Ulis, 2009)
11. D.R Peskova, Yu.V. Khodkovskaya, V.S. Charikov, R.B. Sharafutdinov. *GCPMED 2018. International Scientific Conference «Global Challenges and Prospects of the Modern Economic Development».* **122,** 1205–1212 (2019).
12. Bologna Declaration of 19 June 1999 (Wikisource: Documents of non-national or supranational organizations. Category: 1999 works)
13. *The European Network for Accreditation of Engineering Education (ENAEE)* Retrieved from: https://www.enaee.eu
14. N. Petch, Entrepreneur. (2016). Retrieved from: ttps://www.entrepreneur.com/article/271290
15. M. Lavčák, O. Hudec, Z. Sinčáková, *Journal of the Knowledge*, 1–19. (2019).

Management-developing of an algorithm for the design decisions in the innovative activity of an organization

Yuri Polyakov[1,*], *Andrey* Savchenko[1], *Mikhail* Savelyev [1], *Denis* Perevedentsev[1] and *Anna* Koscheeva[2]

[1]Mitra Association, 241, Pushkinskaya Str., Izhevsk, 426008, Russia
[2]Udmurt State University, 1, Universitetskaya Str., Izhevsk, 426034, Russia

Abstract. This article proposes a new algorithm for developing conceptual project decisions in terms of an organization or innovator's response to changing consumer demands through the development and investment in innovations. The theoretical basis for choosing the type of an algorithm was elaborated by studying the available referencies relevant to present issues of the applicability of types of algorithms for developing management decisions. A multistage multicriteria algorithm (hereinafter referred to as MMA) was developed on the basis of studied algorithms. On developing the stages of the MMA, a set of thinking methods (analysis – synthesis, deduction – induction), a decomposition method, and logical methods were used. MMA is considered as a set of step-by-step sequential descriptions of local processes, formulations and solutions of management tasks, used decision methods and decision-making criteria. The SMART(ER) method was used for managerial tasks at each stage to be set. When solving a management problem, a set of methods was used including an expert method, a factor analysis, brainstorm and methods for calculating economic efficiency. For choosing solutions at each stage to be made, a set of criteria was used. They are the weight of the negative impact on the result of the activity, the yes/no method, the maximum effect, costs minimizing, risks. The schematic design was defined for MMA to be developed. The schematic design was considered as a method of modeling and a method of graphical-analytical displaying of the stages of the process.

1 Introduction

The object of the research is the management and driving innovation. The algorithmic methods for making management decisions are under study.

The purpose of the research is to develop an algorithm for developing conceptual design decisions.

To solve the problems of conceptual design and development of complex self-organizing systems, large amounts of data are used today from heterogeneous and distributed sources, when working with which special highly loaded information systems for extracting data and knowledge about the characteristics of the project are created.

A high share of the uncertainty of this problem requires special methodological approaches to the analysis of the organization's response to a change in the structure of consumer needs, in the list of such approaches: expert assessment of knowledge about the behavior of stakeholders [1], system analysis of the links between the organizational structure and the workforce and information systems that support it [2, 9], planning the development of a project based on the theory of constraints and the critical chain method [3, 12], structural analysis of the social graph and the interconnections of project subsystems [13].

Highly specialized models of projects of organizational systems are being developed, so Takenberg S., Dukwitz S. and Shlik K. proposed a design and engineering approach including the analysis of process and actor-oriented behavior of system elements [5]. Bukhin E. and Rozens H. [6] presented a multipurpose optimization model of decision making in the implementation of a project based on the analysis of many alternative options. The authors of the article [7] describe examples of solving optimization problems of project development in various subject areas on the basis of their proposed approach "project tree", which is a tree-like structure of decision-making in time. The authors of [8, 10, 11] proposed graphical structures for generating knowledge about a project by situational and thematic analysis of its life cycle. The work [14] presents an optimization model for solving the problem of process planning and project support, taking into account the limitations of several executive modes and a managerial approach to the versatility of employees and prioritization of tasks.

The growing demand for special intelligent data processing systems is due to the need for digitalization of production processes, which is confirmed by new research in this area. For example, in [4], knowledge-based decision-making expert systems (KBDSS) are classified into three categories: technologies for modeling and knowledge representation, technologies for reasoning and inference, and web technologies, and recommendations for their integration into service

* Corresponding author: savaisai@gmail.com

production systems are developed. The authors of [15] proposed analytical approaches to the development and training of an autonomous algorithm for predicting the characteristics of projects, [16] presents a model method based on the registration and subsequent classification of facts about the operation of various organizational and engineering systems for data mining and estimation of project parameters. A study by the authors [17] describes a new approach to obtaining specific knowledge and making decisions to accelerate the development of expert systems, based on the generation of linguistic rules and replenishment of the base of heuristic expert knowledge.

Thus, recommender systems become an integral element of the processes of formation of expert knowledge and decision-making with collaborative filtering of alternatives, representing the modeling of the degree of confidence in the context of available incomplete data [19, 20]. Sakama S. and Inue K. In the article [18], they showed the importance of the abductive approach, including the analysis of the actual course of processes in organizational systems, and the search for explanatory hypotheses in the formation of new knowledge, since it is based on new relevant facts about the development of the project.

In our opinion, it is possible to develop an algorithm by introducing the concept of an innovative complex.

The innovative complex of the organization consists of the following structural elements:

- innovation and product complex which is a set of types and costs of work on the development of a product;
- innovation and technological complex which is a set of types and costs of work on the development of technologies;
- complex of innovative materials which is set of types and costs of work on the development of innovative materials.

2 Materials and Methods

The relevance of the study is due to the need to solve one of the priority management tasks aka the adaptation of the activities of managers during the transition to the digital economy, used for the processes of choosing the type and quality of a homogeneous product, usefulness for the consumer and manufacturer, and other consumer properties to an algorithm [21-24]. The main proposed approach to the management of the design of an innovative algorithm is the schematic design. Algorithm design Management-Developing of the algorithm is proposed to be carried out in the following sequence:

- analysis of the management situation in the organization;
- statistical observation;
- analysis of the management situation in innovation;
- putting forward/refuting hypotheses, structural elements of hypotheses that contribute to the appearance of adverse symptoms in the activities of the organization;
- identifying the reasons, methods for choosing and organizing an innovative system;
- formulation of problems, the discrepancy between the methods of selection and organization of the innovative system, certain criteria;
- putting forward/refuting alternatives, to solve the task of the innovative system.

3 Results and Discussion

Stage 1. Analysis of the management situation in the organization.

Using the method of factor analysis it is proposed to identify trends and factors that determine the tendencies in the results of the organization's activities, to assess the weight of the influence of a particular factor on the resulting indicator of the organization's performance, to identify the most weighty indications by the criterion of weight and to graphically simulate the results of the analysis in the form of a sketch. A sketch of the analysis results is shown in Figure 1.

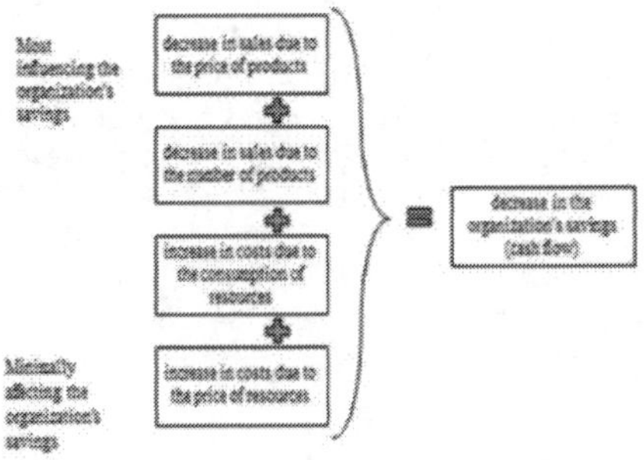

Fig. 1. The results of the analysis of the organization's activities

Stage 2. Statistical observation

It is proposed using the decomposition method to establish the relationship between the main results and factors that determine the trends of innovation with the results of the organization as a whole, etc. graphically with modeling as a sketch. A sketch of the decomposition results is shown in Figure 2.

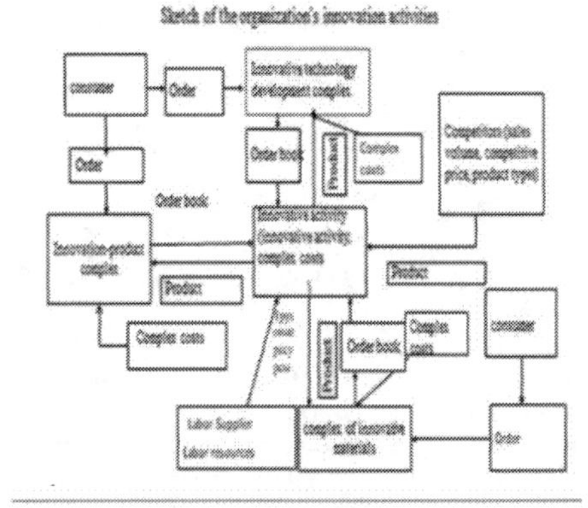

Fig. 2. Sketch of the organization's innovation activities

Stage 3. Analysis of the management situation in innovation

It is proposed, using the method of factor analysis, to identify trends and factors that determine the tendencies in the results of innovative activities of the organization, to assess the weight of the influence of a particular factor on the resulting indicator of the organization's innovative activity, to identify the most weighty indications by the criterion of weight and to graphically simulate the results of the analysis as a sketch. A sketch of the analysis results is shown in Figure 3.

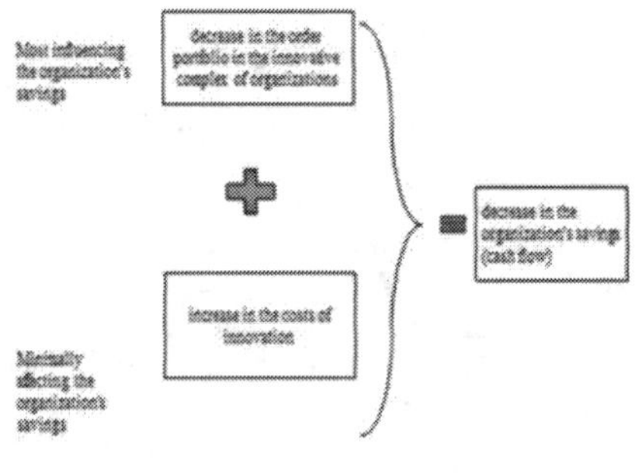

Fig. 3. An example of the result of the processes of identifying the set and selection of the most significant negative symptom of negative symptoms

Stage 4. Description of hypotheses

It is proposed, using logical methods and the decomposition method, to construct the structural elements and parameters of the hypothesis, using the factor analysis method to establish the correlation between possible changes in innovation and the appearance of the most significant negative trends in the organization's activities. One of the possible theoretical hypotheses is presented in Table 1.

Table 1. The relationship "changes in the innovative complex is the most significant negative symptom"

Types of innovative activities of the organization (hypothesis)	Organizational performance indicator (most significant negative indication)	Changing the structural elements of the hypothesis	Changing the parameters of structural elements
Innovation and product system	Decrease in sales volume due to product price	Decrease in the business acquisition	R&D, TESTS, PILOT PRODUCTION
Innovative and technological system		**Decrease in the business acquisition**	R&D, TESTS, PILOT PRODUCTION
Innovative materials		Decrease in the business acquisition	R&D, TESTS, PILOT PRODUCTION
Competitors		Increase of competitors' business acquisition	Production capacity

According to the criterion of weight, it is proposed to select the problematic structural element-parameter and graphically simulate the results of the analysis as a sketch. A sketch of the results of describing hypotheses is shown in Figure 4.

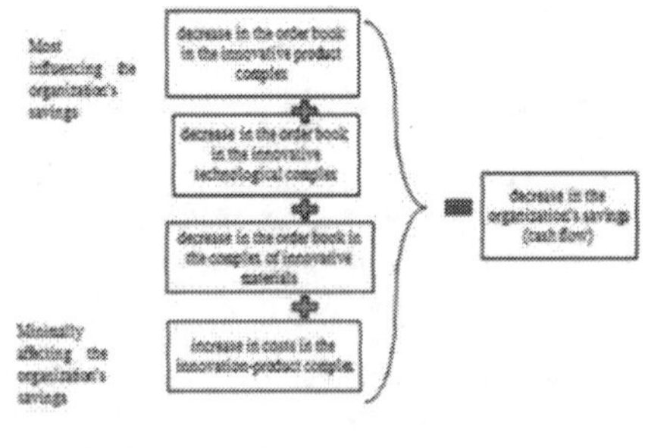

Fig. 4. Causal relationship "parameter - element - innovative activity - symptom"

Stage 5. Identifying the causes

It is proposed, using the grouping method, to group the causes into two groups: a set of causes influencing the appearance of a negative indication, due to the process of justifying the plan (investment plan and distribution of investments by types of innovative complexes localized in geographic markets) and a set of causes influencing the appearance of a negative symptom, due to the process of organizing work on the innovative complexes of the organization. By the interviewing specialists, by the method of compliance (yes/no), it is proposed to identify the essential factors. According to the criterion of significance, it is proposed to choose a significant factor and graphically simulate the results of identifying the cause factors in the form of a table. The results of identifying the causes are shown in Table 2.

Table 2. Results of identifying causes

Types of innovative activities of the organization (hypothesis)	Problematic structural element	Changing the parameters of structural elements	Causes	
			Justification	Organization
Innovation and product system	Decrease in the business acquisition of the innovative product complex	R&D, TESTS, PILOT PRODUCTION	No plan to develop new products	No work on the pilot production

6 stage. Formulating problems

It is proposed using the conformity method (yes / no) to establish a set of problems in the processes of justifying the plan and a set of problems in the processes of organizing work on the innovative complexes of the organization. According to the weight criterion proposed to select a significant issue and by graphical modeling we develop a table of the results of problem formulation. See Table 3.

Table 3. Problem Formulation Results

Types of innovative activities of the organization (hypothesis)	Problematic structural element	Changing the parameters of structural elements	Problems	
			Justification	Organization
Innovation and product system	Decrease in the business acquisition of the innovative product complex	R&D, TESTING, PILOT PRODUCTION	No prediction on the emergence of a new product of competitors	Lack of competencies in the modern organization of work in pilot production

Stage 7. Putting forward / refutation of alternatives for solving complex problems

It is proposed, using the brainstorming method and the scenario method, to form a set of alternatives for solving problems in the processes of substantiating the plan and in the processes of organizing work, to graphically model the formulation of a set of alternatives in the form of a table of results. The results of definition of alternatives are shown in Table 4.

Table 4. Set of alternatives

Problematic structural element	Changing the parameters of structural elements	A set of alternatives	
		Justification	Organization
Decrease in the business acquisition of the innovative product complex	R&D, TESTING, PILOT PRODUCTION …	Develop or upgrade a product	To develop or order a project to organize a pilot production of an innovative product system

It is proposed to form a sketch and graphically simulate in the form of a sketch the effectiveness of the innovation-product complex.

A sketch of the effectiveness of the innovation and product complex is shown in Figure 5.

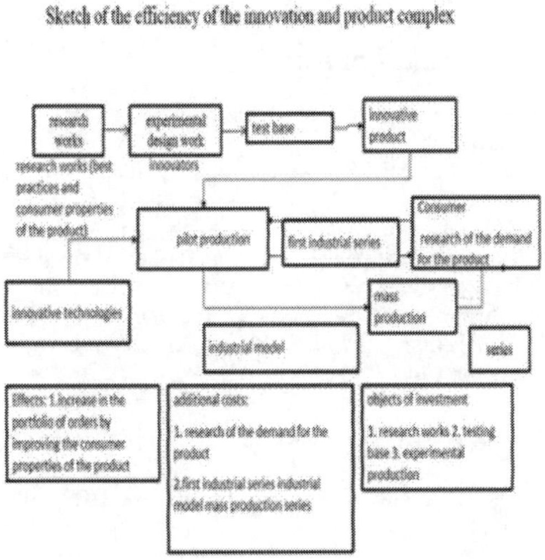

Fig. 5. Sketch of the efficiency of the innovation and product complex

According to the criterion of the maximum annual reduced effect, it is proposed to choose an alternative and graphically simulate the results of the choice of alternatives in the form of a table. The results of the choice of alternatives are shown in Table 5.

Table 5. Results of choosing alternatives

Main challenges	Alternatives to increasing the capabilities of the innovation and product complex	Calculation of the annual reduced economic effect (ΔPV) from the implementation of alternatives, million rubles.
No prediction of a new product developed by competitors	Developing a product	ΔPV p - due to product development
	Upgrade product	ΔPV m - due to product upgrades
The most significant negative symptom (decreased sales of the organization)		ΔPV p> ΔPV m - choose product development

7-stage MMA for the development process of conceptual design solutions in the innovation of the organization of the system consistently describes the operations (actions) of the manager in specific management situations. MMA enables a book of order and the organization development program to be formed.

4 Conclusions

Following from available referencies relevant to the study [25-37] it can be concluded that the proposed algorithm (MMA) is unique as compared to well known.

Novelty of MMA comprises the way of its embedding in logic of innovative system and two stages that are nomination or refutation of hypotheses and identification the causes of adverse effects. The MMA can be used by specialists in the development of management activities in a monocultural innovation organization.

Acknowledgements

The reported study was funded by RFBR, project number 20-010-00869.

References

1. F. Caron,. *Int. J. of Data Science*, **1- 1**, 42-57, (2015)
2. H. Kerzner, Project Governance. Project Management 2.0, (2015)
3. Q. Yuan, Y. Polychronakis, *Int. J. of Project Organisation and Management,* **4**, 339 – 367. 4 (2012)
4. P. Zarate, Sh. Liu, *International Journal of Information and Decision Sciences (IJIDS)*, **8**, 305 – 324, 3(2016)
5. S. Tackenberg, S. Duckwitz, Ch. Schlick, *Int. J. of Computer Aided Engineering and Technology,* **2**, 414 – 435, 4(2010)
6. Y. Bukchin, Sh. Rozenes,. *Int. J. of Project Organisation and Management,* **3**, 184 – 203, 2(2011)
7. A. Muhlemann, A. Lockett, A. Gear, *Decision Sciences,* **9**, 612-626, (1978)
8. M. Daradkeh, *Int. J. of Information and Decision Sciences,* **9**, 276 – 296, 3(2017)
9. R. Burke, S. Barron, *Project Organization Structures. Project Management Leadership,* (2015)
10. S. Lippe, J. Brocke, *Project Management Journal.* **47**, 76-96, (2016)
11. M. Gorog, *Project Management Journal,* **42**, 17-31 (2011)
12. D. McLain, *Project Management Journal*, **40**, 60-73, (2009)
13. P. Chinowsky, J. Taylor, *Engineering Project Organization Journal*, **2**, 15-26, 1-2(2012)
14. Q. Liu, X. Geng, M. Dong, Ch. Ye, *Engineering Optimization*, **49**, 896-914, 5 (2017)
15. L. Shi, L. Newnes, S. Culley, B. *Artificial Intelligence for Engineering Design, Analysis and Manufacturing,* **31**, 313-326, 3 (2017).
16. M. Quaglio, E. Fraga, E. Cao, A. Gavriilidis, F. Galvanin, *Chemometrics and Intelligent Laboratory Systems*, **172**, (2018).
17. H. Xing, S. Huang, J. Shi, *Artificial Intelligence for Engineering Design, Analysis and Manufacturing,* **17**, 221-234, 3 (2003)
18. C. Sakama, K. Inoue, *Theory and Practice of Logic Programming,* **31**, 671-715, 3 (2003)
19. F. Gohari, F. Aliee, H. Haghighi, *Information Sciences.* **422**, 21-50, (2018)
20. J. Herlocker, J. Konstan, L. Terveen, J. Riedl, *Journal ACM Transactions on Information System,* 22, 5-53, 1(2004)
21. Yu. Polyakov, M. Savelyev, D. Perevedentsev, O. Titova, M. Galiakhmetova, *SHS Web of Conf,* **89 01007** (2020)
22. Yu. Polyakov, A. Savchenko, N. Alekseeva, E. Panteleeva, M. Savelyev, *SHS Web of Conf,* **89 01006** (2020)
23. Yu. Polyakov, M. Savelyev, A. Savchenko, *Problems of regional economy,* **1-2**, 186 (2020)
24. M. Savelyev, *Problems of regional economy,* **3-4**, 133-140, (2020)
25. D. Meissner, M. Kotsemir,. *J Innov Entrep* **5**, 14 (2016)
26. F. Pleschak, H. Sabisch, *Innovationsmanagement.* (Stuttgart: Schäffer-Poeschl. 1996)
27. J. Witt, Produktinnovation: Entwicklung und Vermarktung neuer Produkte. (München: Vahlen, 1996)
28. G. Ebert, F. Pleschak, H. Sabisch, *Bericht - Fraunhofer-Institut für Systemtechnik und Innovationsforschung.* (Wiesbaden: Gabler, 1992)

29. W. Durfee, P. Iaizzo, *Engineering in Medicine, Academic Press*, 495-509, (2019)
30. G. Hughes, D. Chafin, *J Prod Innov Manag*, (1996)
31. R. Cooper, M. DeGroote, *Overhauling the new product process*. (Ind Mark Manage, 1996)
32. K. Ulrich, S. Eppinger, M. Yang, P*roduct Design and Development*, (Columbus: McGraw Hill, 2020)
33. C. Crawford, C. Di Benedetto, N*ew Products Management*. (Columbus: McGraw-Hill, 2021)
34. K. Brockhoff. *Forschung und Entwicklung - Planung und Kontrolle*. (Berlin: De Gruyter, 1999)
35. R. Rothwell, *International Marketing Review*, **11**, 7 – 31, 1(1994).
36. A. Berkhout, D. Hartmann, P. Duin, R. Ortt, *Int. J. Technology Management*, **34**. 390-404. 3/4 (2006)
37. S. Kline, N. Rosenberg, *An overview of innovation. The positive sum strategy: Harnessing technology for economic growth*. (Washington: National Academy Press, 1986)

Digitalization of communications with applicants as an innovative approach in modern higher education

Irina Baranova[1,*], *Marina* Gapon[1], and *Elena* Golova[1]

[1] FSBEI of HE Omsk State Agrarian University named after P.A. Stolypin, 1, Institutskaya square, Omsk, 644008, Russia

Abstract. In modern complex conditions, characterized by an unstable situation caused by a global pandemic, the acceleration of digitalization processes in all spheres of life, innovative software products that help users to receive complete and up-to-date information on emerging issues are becoming more and more important. Usually even such fateful decisions as the choice of a future profession and educational institution are made on the basis of data collection in mobile applications. In this regard, research in the field of digitalization of communications between universities, applicants and other stakeholders at this stage of development of Russian educational services is quite important. This paper is devoted to the study of the issue of the establishment of mutually beneficial links between applicants and Russian universities. The authors propose the implementation of an organizational project in order to create a multifunctional mobile application "Applicant" aimed to provide target groups of users with versatile, up-to-date information about the state of university education market. The project describes in detail and justifies such points as the formation of a team, the functions and competencies of its members, the main stages and environment of implementation, the implementation schedule, product and educational results.

1 Introduction

Modern conditions caused by the ongoing digitalization of economy, difficulties in the educational sphere that have appeared as a result of the Covid-19 pandemic, literally force educational organizations to look for new approaches to management. At the same time, the processes of the introduction of innovative solutions are observed in all areas: educational, scientific, pedagogic activities, etc.

A special place in the functioning of universities is occupied by work with applicants, which has undergone significant changes over the past year. Obviously, the further fate of educational institutions will depend on how effectively the enrollment campaign will take place. Therefore, the administration of universities is interested in the use of innovative methods in order to attract future students.

On the other hand, applicants also faced great difficulties. For many of them it was impossible to get offline information on the application. If earlier the use of online services was considered as one of the options to find information about universities, then in 2020 this state of affairs turned out to be the only possible one. At the same time, existing sites and mobile applications on this topic are characterized by scarcity and irrelevant information.

These circumstances indicate that there is a need to improve work with applicants both in their interests and in the interests of universities.

The studies of the digitalization of the attraction of students have been carried out over the past years quite actively by many Russian scientists. Thus, among the works highlighting these aspects of the education sector, it is possible to note the works of I.S. Zamulin, A.A.

Golubnichiy, N.D. Artyomov, P.G. Petrushi, D.K. Kozlova, V.Yu. Skaliukha, E.V. Nuzhnov and others [1-3]. In addition, it is necessary to note the activity of developers of various software products, one way or another related to the automation of work with applicants to higher educational institutions. At the same time, the overwhelming majority of studies are of a local nature, characterized by the reference of the research object to a specific university. A few works concern the digitalization of communications with applicants in terms of the creation of automated information programs for applicants and students.

In general, the analysis of the literature shows that there are gaps in the study of issues of digitalization of university activities in relation to those students who plan to get higher education.

In the context of the studied topic, special attention is paid to the review of sources of the use of innovative approaches in higher education. In recent years, Russian scientists have been studying innovations in higher education from various points of view, paying attention to the problems of distance learning, the integration of traditions and innovations, the use of innovative methods in the educational process, etc. Among the Russian researchers of this topic, we can mention O.A. Ivanova, F.U. Zairova, A.S. Gnennaya, S.V. Vivatenko, A.S. Veremchuk, E.A. Leushkina and others [4-7].

Special emphasis needs to be given to the authors specializing in the study of the "innovation-university-applicant" system. This list includes V.N. Kartashova,

* Corresponding author: iv.baranova@omgau.org

M.A. Lapina, M.P. Prokhorova, S.V. Bulganina, P.D. Zimnikova. Lektorova Yu.Yu. and others [8-11].

Thus, there is a high level of interest of Russian scientists in the research of the problems of the introduction of innovative approaches in the work with university applicants.

The purpose of the study is to deepen the theoretical and methodological approaches to the digitalization of communication processes between applicants and higher educational institutions, taking into account the use of modern innovative approaches.

The scientific novelty of the research is in the project proposed by the authors to improve communication with applicants in the modern digital environment.

The theoretical and practical importance of the study is in the improvement of the efficiency of higher education institutions with applicants, as well as expansion of the opportunities for applicants to search for information about the educational services market through the use of modern innovative approaches.

2 Materials and Methods

During the research on this topic, general scientific methods were used, such as analysis, synthesis, induction, deduction and comparison.

In the course of the study, the authors paid due attention to the search for information in the Internet that would help applicants to get a complete picture of the opportunities available on the educational services market for an informed choice of the future university (USE scores, education fees, university rating) [12].

The analysis of the data regarding the available mobile applications in modern search databases shows the presence of certain mobile applications, such as: «Russian universities », « Do the right thing », « Russian entrant », etc.

The disadvantages of these applications are their disunity, independence from each other, lack of communication with higher educational institutions and authorities (ministries, departments). Each of the applications provides fragmentary, incomplete and often irrelevant information for applicants.

That is, today there is no single database that would reflect all the data requested by applicants.

3 Results and Discussion

In order to fill the information niche that has appeared on the Russian market of university services in terms of the establishment of a digital environment for communication between applicants and universities, the authors propose to create an organizational project. The essence of the project (product result): development of a multifunctional mobile application for applicants, universities of the Russian Federation and other stakeholders (students, graduates of universities, organizations (employers)) in order to form a single information space between its participants.

Stakeholders who are interested in the use of the project result:

1. Project users: applicants, university students, university graduates, university staff, organizations (employers).
2. Customers of the project: ministries (departments), higher educational institutions.
3. Investors of the project: organizations interested in results (ministries, universities, employing organizations, other investors).

Table 1 shows the composition of the main tabs of a multifunctional mobile application, as well as the functions of these tabs.

Table 1. Main tabs of the multifunctional mobile application (MMA)

Application tabs	Functions performed by the application
user personal account	personal data and information about the user of the mobile application
higher educational institutions (university)	informing users about universities in various cities of the Russian Federation (university rating, areas of training, forms of study, number of budget places, education fees, exam passing score, university vacancies, etc.), university enrollment application
applicant (school student)	informing users about the number of registered applicants (school students) in the application database, places of their study, USEs, priority areas for enrollment, etc. university enrollment application
student	informing users about university students, forming a bank of vacancies for students, generating information about outstanding students, student schedule
university graduate	informing users about university graduates, including outstanding (successful) ones, the formation of monitoring the employment of graduates, a bank of vacancies for graduates in the areas of training
organizations (employers)	informing users about organizations, employers registered in the database, forming a bank of vacancies according to the availability of vacancies, informing graduate students about conditions of internship, practical training

Feedback from users of the developed multifunctional application is extremely important in the creation of this project. We need to figure out how the application works, what the inconvenience of tabs is, or what additional information may be needed. That is, it will be necessary to take into account the user experience as a result of the operation of this application.

Here we present the main stages of the project and the main tasks solved at each stage of the project.

The 1st stage: Problem statement

Tasks:

-To determine the existing problems (lack of relevant information about universities and their ratings, the number of budget places, education fees, data on areas of training, dormitories, applicants, students, the number of employed students, etc.).

-To conduct a survey, questioning schoolchildren, applicants, students, university staff, employers about what they want to see in the mobile application being developed.

-To monitor available search databases, applications (lack of a single database for the chain applicant (student) - university-student-graduate - university-employer).

The 2nd stage: Statement of a goal, hypothesis (to develop a multifunctional mobile application)

Tasks: What problems can MMA solve?

-To establish communication with ministries (universities) on the formation of the terms of reference for the project.

The 3rd stage: Determination of the main tasks

Tasks:

-To develop a work plan. (The project participants are divided into groups. Each group forms information in the context of its area of responsibility (by tabs), according to the identified problems, based on the results of the analysis of questionnaires and surveys.

The 4th stage: Creation of a mobile application

Tasks: Basic work according to plan.

- To learn to communicate in a team. Training of project participants to form the required competencies

The 5th stage: Project results

Tasks: -To test the developed mobile application

- To formulate conclusions

The 6th stage: Introduction

Tasks: -To distribute, to carry out an advertising campaign

The 7th stage: Opeartion

Tasks:-To check feedback from users of the mobile application

- To eliminate comments and take into account user experience

Table 2 presents the approximate timetable for the implementation of the project.

The important aspect of project development is the description of the environment for its implementation. Thus, the authors propose the following elements of the environment:

- The equipment for project implementation: separate classes, computers with technical equipment (printer, copying machine, scanner), tables, chairs, whiteboards (interactive).
- External participants: experts - technical consultants, programmers, administrative and management personnel of higher education institutions.

Table 2. Approximate timetable for the implementation of the project

Stage	Academic year (months from September to July)										
	09	10	11	12	01	02	03	04	05	06	07
I Problem statement	+	+									
II Statement of a goal, hypothesis (to develop a multifunctional mobile application)		+	+								
III Statement of main problems			+	+							
IV Determination of the main tasks					+	+					
V Project results							+				
VI Introduction								+			
VII Operation								+	+	+	+

A higher educational institution is the platform for the implementation of the project.

The requirements for the project participants: the number of participants is 30 people (approximately), 10 are school students - applicants, 10 students of secondary courses, 10 graduates (or graduate students).

Since school students and students are included in the team, the educational result obtained as a result of the project is predicted.

Stage I. Problem statement.

Educational results:

- ✓ Understanding of the problem field (lack of real information about universities and their ratings, insufficient information in the available applications about the availability of budget places, hostels, education fees, etc.) by applicants (school students) and students;
- ✓ Formation of skills when interacting with experts (the role of an expert is played by the administrative and managerial staff of universities, programmers, employers);
- ✓ Applicants and students at this stage studied various ways to search the data on the selected problem (monitoring of modern Internet portals, programs and search databases as Play Market, Google Play, Appstore, Yandex.Store, AndroidPIT, Getjar, etc.);

Values : sociability, initiative, independence.

Stage II. Purpose and tasks.

Educational results:

- ✓ Applicants and students learned to master some methods of work (goal setting, assignment of tasks);

Stage III. Project realization

Educational results:

- ✓ Development of communication links between applicants (school students) and students: the formation of a constructive dialogue, acceptance and respect of cultural and professional differences between each other

Stage IV. Presentation of results

Educational results:

- ✓ Skills and experience of public performance;
- ✓ Scientific article;
- ✓ Report;

Stage V. Reflection

Educational results:

- ✓ Understanding by applicants and students that they might have not analyzed all the search bases;
- ✓ Applicants and students can evaluate the results of their own work and form activities to improve the efficiency of the proposed product

Table 3 presents the characteristics of the project team. The composition of the project team, the main functions of the participants and competencies are determined.

Table 3. Project participants, their main functions and competencies

Project team composition	Roles (functions) of participants in the team	Knowledge, competence and skills	Work experience
Applicants (school students)	Analysis of available databases, problem solution, tab development	Knowledge of the main search bases, skills	Absent
First year students	Analysis of available databases, problem solution, tab development	Knowledge of the main search bases, skills	Part-time
Second students	Analysis of available databases, problem solution, tab development	Knowledge of the main search bases, skills; ability to pose and solve problems	Part-time
Graduate students	Analysis of available databases, problem solution, tab development	Knowledge of the main search bases, skills; ability to pose and solve problems	Part-time
Employees of higher education institutions (administrative and management personnel)	Compensation of the lack of knowledge and skills of the team, assistance in the narrow specialization of the issue	Scientific and professional knowledge and competence	Present
Employees of higher educational institutions (employees of the selection committee, employment department, teachers)	Compensation of the lack of knowledge and skills of the team, assistance in the narrow specialization of the issue	Scientific and professional knowledge and competence	Present
Representatives of organizations (employers)	Assistance in the narrow specialization of the issue	Scientific and professional knowledge and competence	Present
Laboratory assistant	Maintenance of the environment, provision of the safety of work and the availability of equipment, and operation of the equipment	Skills of working with equipment	Present
Experts (programmers)	Assistance in the narrow specialization of the issue, programming a mobile application	Skills of working with professional equipment and specialized software	Present
Representative of the ministry (department)	Is the customer of the project, pays for the project, prepares the terms of reference	Understanding the importance of the project in education	Present
Tutor	Is engaged in educational support, solves educational	General awareness and understanding of the topic	Present

	problems	of the project, the ability to build communications in a team, assistance in self-determination for project participants, pedagogical competencies, etc.	
Curator	Organizes all participants, ensures the product result of the project	Basic scientific and professional knowledge, organizational management, design, pedagogical competence, etc.	Present
Stakeholders (mobile app users, customers, investors)	Is interested in the result of the project, provide knowledge, information, money	Skills of the use of a mobile application; skills in project implementation and making a profit from it	Present

Thus, the proposed project includes the main particular elements. Every element is worked out in

detail and is ready for testing. Free MySQL DBMS, an open source relational database management system [13] can be chosen as a tool for the creation of a database of the system. The detailed development of this component of the project should be performed by programmers.

4 Conclusion

The implementation of the project results will solve the following problems:

- the lack of real information for applicants about the higher educational institutions of the Russian Federation according to the necessary search criteria (requests);
- the lack of information on the ratings of higher educational institutions;
-the impromptness, immobility when submitting documents for enrollment to a higher educational institution;
- the unemployment of students (graduates) of higher educational institutions (part-time / full-time employment), failure to provide a place for internship and practical training;
- the lack of information about outstanding graduates;
- the lack of information space for communication links in the chain «applicant (student) - university-student-graduate - university-employer».

Taking into that Internet commerce has such undoubted advantages as high speed, low cost, which allows increasing the efficiency of transactions [13], it is possible to predict the continued growth of demand for the purchase of goods and services through the worldwide network, including the demand for educational services, or at least, the search for information on these services.

The authors believe that the implementation of the project at the initial stage is possible in the form of a pilot based on one of the Russian regions. If positive results are obtained, it is possible to carry out certain adjustments and expand the scale to the level of the educational services market throughout the entire country.

References

1. I.S. Zamulin, A.A. Golubnichy, N. D. Artyomova, *Scientific and technical bulletin of the Volga region,* **9**, (2020)
2. P.G. Petrusha, D.K. Kozlova, *Economics and Entrepreneurship*, **10 (123)**, (2020)
3. V.Yu. Skaliukh, E.V. Nuzhnov, Proceedings of the International Scientific and Technical Congress Intelligent Systems and Information Technologies - 2019" (IS & IT-2019, IS & IT'19), *Department Intelligent Information Subsystem for Applicant Support*, (2019)
4. O.A. Ivanova, Scientific Research and Modern Education, (2018)
5. F.U. Zairova, Vectors of modern science, 1 (3), (2016)
6. A.S. Gnennaya, Priority models of social development in the era of modernization: economic, social, philosophical, political, legal aspects (2016)
7. S.V. Vivatenko, A.S. Veremchuk, Science and Education in the 21st Century: Theory, Methodology, Practice (2019)
8. V.N. Kartashova, Siberian teacher, **3 (118),** (2018)
9. M.A. Lapina, Energy saving and innovative technologies in the fuel and energy complex, (2018)
10. M.P. Prokhorova, S.V. Bulganina, Bulletin of Nizhnevartovsk State University, **3**, (2017)
11. P.D. Zimnikova, Yu. Yu. Lektorova, Translation, advertising and PR in modern communication, **1,** (2019)
12. Z. Uvaleyeva, D. Mukhiyayeva, I. Baranova, M. Valieva, O. Kopylova, *Journal of Entrepreneurship Education,* **22,** (2019)
13. 13.I.Kh. Gizzatov, A.A. Zlobina, M.A. Senilov, Society, **4 (19),** (2020)
14. E.E. Golova, I.V. Baranova, *Basic Research,* **11,** (2020)

Macroeconomic signs of an innovative economy by the case of Great Britain

Mikhail Savelyev [1*], *Mikhail* Kozyrev[1], *Andrey* Savchenko[1], *Vladimir* Koretsky[2], *Rail* Galiakhmetov[2]

[1]Mitra Association, Pushkinskaya Str. 241, 60, 426008 Izhevsk, Russia
[2]FSBEI HE M.T. Kalashnikov IzhSTU, Studencheskaya Str. 7, 426069Izhevsk, Russia

Abstract. By the case of the economic development of Great Britain, the hypothesis was verified that innovations at the macroeconomic level should accelerate economic growth and at the same time reduce development risks, stabilizing this growth, reducing its fluctuations under the influence of market factors. The economic development of Great Britain is investigated in 25 economic cycles for the period from 1830-2020. Economic development was investigated according to the parameters of economic growth and development risk in each of the considered cycles. Four types of economic development policy are theoretically described in terms of the dynamics of changes in growth and risk between the previous and subsequent cycles including progressive, regressive, aggressive and conservative. In relation to the identified periods of progressive development policy in Great Britain, the institutional innovations that led to this type of development were investigated. Among them was the great economic reform of the early Victorian era, the course of social or new liberalism and the popular budget before the First World War, the activities of the first Labor government immediately after this war, economic recovery after World War II in combination with the Marshall plan and nationalization, the era of the Conservatives and the politics of New Labor at the end of the 20th century. The study showed that the implementation of authentic national culture and institutions complementary to the existing authentic culture institutions of institutional innovations leads to a simultaneous decrease in the risk of development and acceleration of economic growth, which can be considered the most favorable policy of macroeconomic management of entrepreneurial activity in order to accelerate the application of technical and commercial innovations.

1 Introduction

The issue of managing innovative activity in the post-industrial world is acquiring basic, strategic importance for economic development, since innovations have become the main source of wealth, and not merely an additional factor of competitiveness, as it was peculiar for the industrial era. At the same time, economic science does not give an unambiguous answer with what tools, methods, means, institutions it is possible to effect on the subtle matters of the creativity of innovators and the activity of entrepreneurs.

Most researchers identify practices of influence at the microeconomic level: creating conditions for innovative activities of organizations. At the macroeconomic management level, research does not go beyond descriptive classifications. For example, one of the first to analyze the results of economic activity over long historical periods was A. Maddison [1], who widely used the methods of quantitative analysis (quantification) as a tool for historical analysis.

The work of W. Easterly and R. Levin shows the formation of modern empirical patterns of economic growth, such as, for example, the growing gap in the levels of GDP per capita between countries, national policies and their impact on long-term economic growth, etc. [2].

Conclusions on the importance of the influence of various factors, including cultural and institutional, on the country's macroeconomic indicators are presented in the works of other authors [3-5].

The descriptions of microeconomic models of innovation policy are more specific. For example, some the authors [6] identify three basic models of innovative development "Euro-Atlantic", which includes the British, "East Asian" and "alternative" and investigate their main features. The authors note that after the Second World War, European countries radically changed their research priorities, focusing on relatively cheap ways to obtain scientific and technical information. An illustrative example is Great Britain, which abandoned fundamental and costly research in the field of nuclear physics (with the exception of those directly related to the production of nuclear weapons). Radio astronomy and the study of the biological properties of high-molecular substances were chosen as the priority directions, where it achieved considerable success, laying the foundation for the creation of two

*Corresponding author: savaisai@gmail/.com

fundamental scientific disciplines - astrophysics and molecular biology.

In all other respects, the British innovation system, centered around a small number of top-class universities, copies the US innovation system, where universities are the basis of the national innovation system (hereinafter - NIS).

The newest model for the formation of the NIS, which was developed on the basis of the Euro-Atlantic model, is the triple helix model. In its finished form, it does not yet exist in any country. It received the greatest development in the USA, and its individual elements - in some developed countries of Western Europe, Brazil and Japan. The basis of this concept lies in the fact that in the chain "universities - business - government", it is universities that play an increasingly important role in a knowledge-oriented economy [7-10]. The triple helix idea is based on the synthesis of a number of sociological theories, the use of analogies from the biological sciences, as well as the similarity of the problem of the relative motion of three bodies, which does not have a general solution, but particular solutions are possible for some specific initial conditions. With regard to innovative development, the triple helix model describes the interaction of three institutions (science-state-business) at each stage of creating an innovative product. This is a dynamic model of interorganizational interactions that arises in the course of the evolution of the economy and society.

Moreover, to achieve synergy, network participants need to be in a mode of constant coordination (collaboration mode) [11].

Some researchers divide the stakeholders in the innovation process into groups. For example, H. Liu and S. Wyat distinguish primary (directly involved in innovation) and secondary (influencing the behavior of primary subjects) subjects, as well as institutions that affect the behavior of subjects of both groups [12].

At the same time, according to S. Botot and D. Satinsky [13], the triple helix model is a model of regional economic development, and not a model of national innovation policy in general.

At present, a complicated type of the triple helix model is being developed abroad - the four-link helix model described in 2009 by Y. Karayannis [14]. This model concerns interactive networking at the level of the entire national community, and not just between the three leading institutional sectors. Since other institutions, represented by various social strata, began to impact the innovation process, this circumstance found its theoretical embodiment in the addition of a fourth element to the triple helix, embracing representatives of civil society. It is believed that the fourth spiral characterizes the modern postindustrial economy better than the triple one, since in the 21st century. Civil society is acquiring a critical role in the creation and dissemination of new goods and values.

The principle of the four-link helix increases the likelihood of successful implementation by regional politicians of a locally-oriented entrepreneurial process, which stimulates intensive experimentation and invention and, at the same time, stimulates innovation [15].

Citizens not only participate in the actual development process, but can also propose new types of innovations; as a result, connections are established between users and other "actors" in industry, science or government [16].

This enables conclusion that for a number of countries the possibility of resorting to the Euro-Atlantic model of innovative development, despite all their advantages, is essentially closed to be made. Efforts to create fundamental science or a full high-tech cycle would not only require them to bear unbearable financial, time and organizational costs, but, most likely, would run into insurmountable obstacles in the form of peculiarities of national culture and psychology.

An alternative model, developed in accordance with the specifics of national culture, national psychology and traditions, enables these countries to join innovative development, turning national features into competitive advantages. Of the newly industrialized countries at the moment, only China, in addition to the simultaneous application of import substitution and export orientation strategies, is trying to implement a strategy of technological leadership.

This research is within the framework of this alternative model, which assumes that there are mechanisms for the interaction of culture with the institutional support of economic activity, and if these elements of macroeconomic management correspond to each other, they provide an opportunity for innovative development. At the same time, the set of successful institutional instruments is specific to each culture. We have presented the results of these studies in a number of works: both in the development of theoretical models and in the phenomenological description of various historical and cultural communities, civilizations and cultures [17].

This view of the search for the competitive advantages of economic development allows us to significantly expand the historical horizon of research on innovation management even beyond the industrial era. To the extent that the existing quantitative economic development database allows us to do so.

Research hypothesis. A decline in economic growth is traditionally associated with a decline in entrepreneurial activity and innovation. The increase in risk with the acceleration of economic growth speaks primarily of the opportunistic rather than innovative reason for the increase in entrepreneurial activity. This growth is usually unstable. Based on this, it is assumed that innovations at the macroeconomic level should accelerate economic growth and at the same time reduce development risks, stabilizing this growth, reducing its fluctuations under the influence of market factors. It is assumed that progressive economic development occurs due to institutional transformations that are authentic to culture and complementary to existing institutions of authentic culture.

To verify this hypothesis, an object is required that, over the longest possible period of time, has sufficiently reliable information about economic development in the absence of hostilities, large-scale social conflicts and revolutionary changes in the socio-economic system and political and legal structure on its territory. The United Kingdom best meets these criteria. In addition, this country has been a world economic leader for a long period of time. And after losing its leadership, the UK remains one of the largest national economies in the world. This position of the country also increases the likelihood of verifying the hypothesis, since it excludes distortions of economic indicators due to external influences or random phenomena.

2 Materials and Methods

The object of our research is the UK economy from 1830 to 2020. The periods are divided into 25 economic cycles. The economic cycle is defined as the period from the first year of growth or acceleration of economic growth to the last year of decline or slowdown in growth.

The subject is the search for changes in essential parameters of economic development between the previous and subsequent cycles, characterized by a simultaneous decrease in the risk of development and the acceleration of economic growth, as well as a phenomenological description of the institutional innovations that led to these changes.

The traditional use of only indicators of economic growth in macroeconomic studies is not informative enough. In this regard, by analogy with the technical analysis of the profitability of securities, the standard deviation of the average annual growth rates of real GDP was used as an indicator of stability.

The research method is a comparative statistical analysis of growth and risk indicators: the average annual growth rate of real GDP and the standard deviation of the same indicator, as a characteristic of its stability, are determined for each economic cycle.

If the subsequent period compared to the previous one had a lower risk indicator and a higher growth rate, then such changes in economic development can be considered progressive, the reverse change - regressive. If risks and growth are falling, then such development can be considered conservative, and if they grow, aggressive.

Initial data for analysis are data from Angus Maddison Historical Statistics database [18] up to 2008, and from 2009 up to 2020 from The Conference Board, Total Economy Database [19].

The periods for analysis are identified taking into account economic cycles. A total of 25 periods have been identified: 1830-1837, 1838-1842, 1843-1850. 1851-1855, 1856-1867, 1868-1879, 1880-1885, 1886-1892, 1893-1903, 1904-1908, 1909-1914, 1915- 1921, 1922-1926, 1927-1939, 1940-1947, 1948-1952, 1953-1958, 1959-1962, 1963-1969, 1970-1975, 1976-1981, 1982-1991, 1992-2001, 2002-2009, 2010-2020 and are grouped by known historical eras.

This research method is described in detail in [20] and applied to study the development of Russia in the imperial, Soviet and modern periods [20-21], in relation to German-speaking countries [20, 22], Portuguese-speaking countries [23], countries of the former French Empire [24] and newly industrialized countries [20, 25]. This method, combined with factor analysis and the method of production functions, was also used to study the effectiveness of the governments of post-Soviet Russia [26].

For meaningful interpretation of econometric results, the method of structural analysis of phenomenological data on institutional and innovative events that led to a progressive change in growth indicators and development risk is used. As a conceptual matrix for structural analysis, we used a matrix of cultures built on the scales of Individualism-Collectivism and Equality-Inequality [17].

According to this matrix, the culture of Great Britain belongs to an alienating culture dominated by the values of individualism and equality. Based on this assumption, it is suggested that institutional transformations in Great Britain during periods of progressive development correspond precisely to these principles. If, during the periods of progressive development policy, institutional transformations are found that are authentic to the alienating culture and complementary to the existing authentic culture institutions of economic development, then we can state that the hypothesis put forward in our previous works [19] about the connection between culture, institutions and economic development is confirmed.

3 Results and Discussion

In the Victorian era, the period of progressive economic policy was the period associated with the reforms of the Conservative government of Robert Peel, who was in power from 1841 to 1846. Under him, under pressure from free traders, foreign trade duties and indirect taxes were gradually reduced, which was balanced by an increase in income tax, and also was reduced working hours, the Bank of England received central status. These transformations are defined by historians as a great economic reform that led to a drop in development risks and the highest growth rates for the UK economy in the 19th century (see Figure 1).

The institutional competitive advantages achieved in this way in comparison with the rental economies of the main geopolitical competitors of Great Britain at that time were later used for continuous colonial conquests and elimination of competitors. Subsequent Victorian economic cycles can be characterized as extensive, non-innovative development, where risk indicators fluctuated significantly from cycle to cycle, and growth indicators gradually but steadily fell (see Figure 1).

Incomparably less progressive effect in the cycle of 1893-1903 was brought by the democratic reforms of

 https://doi.org/10.1051/shsconf/202111600072

local self-government in 1989 and budgetary support for the purchase of estates by tenants.

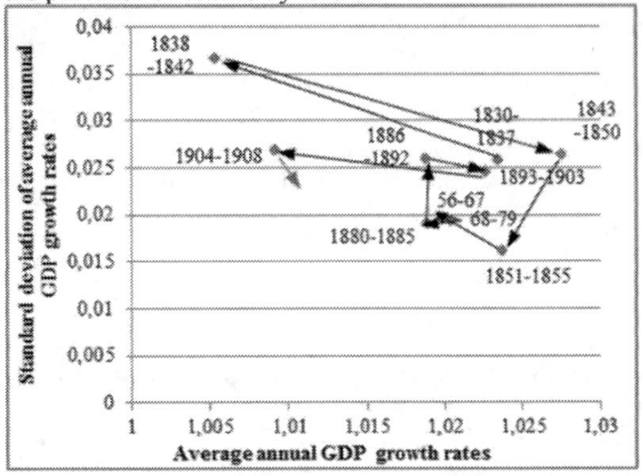

Fig 1. Economic development of the United Kingdom in the periods "The Regency" and "Victorian era" (1830-1837, 1838-1842, 1843-1850, 1851-1855, 1856-1867, 1868-1879, 1880-1885, 1886-1892, 1893-1903, 1904-1908).

The second historical period (see Figure 2) of progressive development policy is associated with the land reform in Ireland (since 1903) and its self-government (nominally since 1914, in fact since 1921), the suffragette movement and the activities of the Labor Party. Since 1906, liberal governments have adhered to the course of social or new liberalism: institutions for resolving labor conflicts and social security have been created, working hours have been reduced. The budget of 1910, known as "People's Budget", in addition to rallying the society with social reforms on the eve of the world war, financed military spending, which stimulated economic growth, and ensured the growth of budget revenues through taxes on the rich. Politically, this was achieved by narrowing the rights of the House of Lords and the introduction of professional parliamentarism in the House of Commons, which provided access to this chamber for low-income politicians.

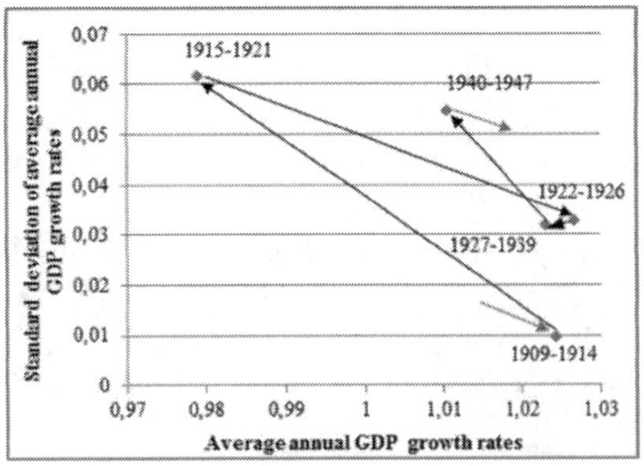

Fig. 2. Economic development of the United Kingdom in the periods "First World War" and "Second World War" (1909-1914, 1915-1921, 1922-1926, 1927-1939, 1940-1947).

During the period 1922-1926, the post-war recovery economy achieved the growth and risk indicators of the period 1843-1850. This happened against the background of universal suffrage introduced in 1918, the activities of the first Labor government and the continuation of social reforms.

These two economic cycles with progressive development policies have steadily ended in disastrous periods of regression during the two world wars and the Great Depression. The absence in this era of periods with politicians of conservative and aggressive development suggests that the economic system of the British Empire reached the limits of its extensive growth and did not bring the mother country a stabilizing or stimulating effect on internal economic development.

The period 1948-1958 is associated with the Marshall plan and the historical maximum average annual economic growth in the history of Great Britain, achieved in the next economic cycle (see Fig. 3). It began with the continuation of the social reforms of the Labor government and the nationalization of industrial, transport, information and financial organizations, which led to the creation of the "welfare state." This effect of accelerating growth lasted for a long time - 15 years - over 3 economic cycles and ended with the final collapse of the colonial empire and a significant loss of opportunities for political domination. A slight increase in risk in the 1959-1962 cycle did not exceed the risk indicator of 1948-1952, which retained the general progressive trend of 1948-1962.

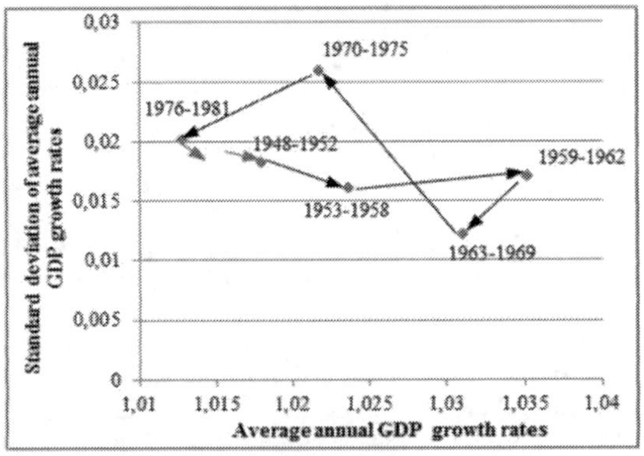

Fig. 3. Economic development of the United Kingdom in the periods "Postwar Britain" (1948-1952, 1953-1958, 1959-1962, 1963-1969, 1970-1975, 1976-1981)

The last, longest period of progressive politics was the period 1982-2001 (see Fig. 4), consisting of the "era of conservatives" and the period of "new Labor". Conservatives, relying on the ideas of monetarism and economic liberalism and using the boom in oil production in the North Sea, adopted a policy of stimulating business activity: suppressed inflation and labor movement, reduced social spending and direct taxes, stimulated the closure of inefficient enterprises and carried out large-scale privatization, thereby freeing up public resources for new investment spurt and formed the society of "People's Capitalism" and "Casino Capitalism". It should be noted that this period was characterized by a minimal drop in risks, possibly within the statistical error.

For the second consecutive year, the economic cycle of progressive development policy began with the devaluation of the national currency due to the actions of currency speculators. This has increased the competitiveness of the British economy. The progress was also influenced by the reallocation of resources in connection with the victory in the Cold War and the policy of loyalty to the United States. The New Labor policy combined a market economy, social justice for the most vulnerable strata of society, and the delegation of state powers to local authorities. Compared to previous cases of progressive development policy, this period was the least significant, and ended with a steady regression of the last two decades.

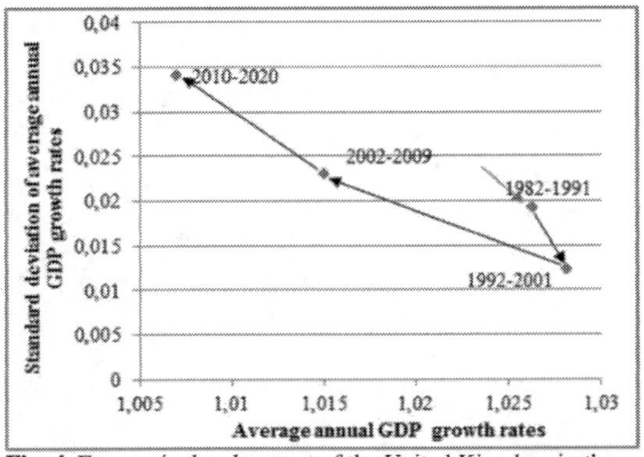

Fig. 4. Economic development of the United Kingdom in the period "European integration"(1982-1991, 1992-2001, 2002-2009, 2010-2020)

Thus, all known historical cases of simultaneous risk reduction and growth acceleration in the UK are associated with significant institutional innovation. It was they who created the opportunities for increased investment in technical and commercial innovation and the growth of entrepreneurial activity.

4 Conclusions

The study showed that the implementation of institutional innovations that are authentic and complementary to the national culture of Great Britain leads to a simultaneous decrease in development risk and acceleration of economic growth, which can be considered the most favorable policy of macroeconomic management of entrepreneurial activity in order to accelerate the use of technical and commercial innovations.

Within this logic, all the changes in institutions identified in the course of the study during the periods of progressive development of Great Britain are aimed at increasing the equality of people and removing barriers to their individual business activity. And the value principles of individualism and equality are fundamental signs of the alienating culture of Western civilization [19, 20, 22, 26].

The methodology used in this study to identify macroeconomic features of an innovative economy makes it possible to apply it to identify cases of progressive development policies in other countries and in different historical epochs, as well as to characterize and assess management impacts on innovation and entrepreneurial activity.

This research methodology makes it possible to identify institutional transformations in the largest countries of the main civilizations of the world during the period of progressive development, to assess what value characteristic of cultures they correspond to, and to classify their cultures according to the conceptual basis adopted here for the structural analysis of institutional transformations. In this case, for different cultures, progressive economic development should be generated by institutional changes that correspond to the value principles of their cultures.

Acknowledgements

The reported study was funded by RFBR, project number 20-010-00869

References

1. A. Maddison, *The World Economy: Volume 1: A Millennial Perspective, Volume 2: Historical Statistics* (2006)
2. W. Easterly, L. Ross, *Journal of Monetary Economics,* **50(1)**, 3 (2003)
3. D. Kaufmann, K. Aart, Z.-L. Pablo, *Aggregating Governance Indicators, World Bank Research Working Paper*, 2195 (1999)
4. D. Kaufmann, K. Aart, Z.-L. Pablo, *Governance Indicators, World Bank Research Working Paper*, **2196** (1999)
5. A. Helanterä, S.-E. Ollus, *Why they, why not we? An analysis of the competitiveness of Finland and Russia,* Retrieved from: https://iq.hse.ru/
6. V. Sergeev, E. Alekseenkova, V. Nechaev, *Journal of Political Philosophy and Sociology of Politics "Polity. Analysis. Chronicle. Forecast"*, **4**, 6-22, (2008)
7. H. Etzkowitz, L. Leydesdorff *EASST Review*, **1 (14)**, 14-19, (1995)
8. L. Leydesdor, Knowledge-Based Economy: Modelled, Measured, Simulated. (Boca Raton: Universal Publishers, 2006)
9. L. Leydesdorff, M. Meyer, *Research Policy*, **35**, 1441-1449, (2006)

10. L. Leydesdor, G. Zawdie *Technology Analysis & Strategic Management.* 22, 789-804, (2010)
11. L. Leydesdor, *Entropy,* 10, 391-410, (2008)
12. X. Liu, S. White *Research Policy,* 30, 1091-1113, (2001)
13. C. Botot, D.Satinsky, *Innovations.* (2011)
14. E. Carayannis, T. Barth, D. Campbell, *Journal of Innovation and Entrepreneurship,* 1 (1), 1-12, (2012)
15. S. Gackstatter, M. Kotsemir, D. Meissner *Foresight.* 16, 293-308, 4 (2014)
16. R. Arnkil, A. Järvensivu, P. Koski, T. Piirainen, *Final Report on Quadruple Helix Research for the CLIQ project,* Working Papers 85/2010. (Tampere: University of Tampere, 2010)
17. M. Saveliev, Multicultural institutionalism: General economic theory of civilizations. The political economy of traditionalism Network Society Project: [monograph] (Izhevsk: "Mitra", 2015)
18. A. Maddison *Historical Statistics,* Retrieved from: http://www.ggdc.net/
19. *Total economy database* (TED), Retrieved from: https://www.conference-board.org/
20. M. Savelyev, *Problemy regional'noj jekonomiki,* 3-4. 54-62, (2020)
21. M. Savelyev, N. Pushina, A. Savchenko, *E3S Web Conf,* 208 03052, (2020)
22. M. Savelyev, V. Ivanov, Yu. Polyakov,. *E3S Web of Conf.* 222 05009, (2020)
23. M. Savelyev, E. Kutyashova, A. Savchenko, V. Koretsky, Yu. Polyakov, *SHS Web Conf.* 101 02004, (2021)
24. M. Savelyev, T. Gruzdeva, A. Savchenko, V. Koretsky, N. Pushina, *SHS Web Conf.* 101 02005, (2021)
25. M. Savelyev, N. Sokolova, Yu. Polyakov. *IOP Conf. Ser.: Earth Environ. Sci.* 666, 062093. (2021).
26. M. Savelyev, N. Pushina, A. Bryndin, *IOP Conf. Ser.: Earth Environ. Sci.* 666, 062092, (2021).

Crowdfunding platforms as an example of innovation start-up's financial support

Julia Akhmetova*

Bauman State Technical University, 2-ya Baumanskaya st., 5/1, Moscow, Russian Federation

Abstract. Financial support is very important in the innovative sphere, because it creates the basis of the whole economic development. There are several ways how this support can be organized: from the family money to government subsidy. This paper focused on a relatively new way how a startup as a small innovative business can gain finance - crowdfunding platforms. As a preparation for the analysis theoretical concepts of innovative startup investment were considered and the unique characteristics of crowdfunding market were examined. The purpose of the study is to underline main characteristic of startups as a way of financial innovation performance, analyze the determinants of crowdfunding platform's effectiveness, define an optimal strategic behavior for a platform to accumulate greater amount of investment and increase the number of startups which achieve their financial goals. Analysis is based on the data from 105 crowdfunding platforms during last 3 year (from 2018 till 2020). As a result, based on the multiple linear regression an analysis was conducted which showed that age, platform type and scope are main characteristics of crowdfunding platform efficiency while commission type is not correlated with startup crowdfunding success rate.

1 Introduction

The development of innovations is a very important aspect of entrepreneurship as it is a determining factor in economic growth of the country. Innovation is a materialized result of scientific and technological activity, which is in demand from consumers through the introduction of new technologies or the improvement of existing products and services. The main characteristics of innovation are practical relevance, profitability, scientific novelty and compliance with market needs [1].

Innovative project is such a venture that aimed to solve the unique problem of creating, introducing, mastering innovations, displaying their results on consumers. Most common way to create an innovative project is by organizing an start-up, as it can be easily done by one person or a small group without significant control form others like government or mother company.

There are many different ways how innovative start-ups can be financed: starting from personal or family funds and ending with venture capital funding or crowdfunding platforms. Existence of multiple ways of innovative projects funding helps to overcome the initial stage of financing and increase the survival rate of the company. For example, according to Russian researches, small business does not fulfill the function of selecting the most promising innovations in mass production [2], as the procedure for gaining the state support can take more than a year, during which entrepreneurs will need to pass the control of a large number committees and departments and get their approval. Crowdfunding is one of the example how to solve this issue, being an innovative way to request online funding from a large audience (individuals and groups) on various projects, such as creative projects, business startups, charity events, etc., carried out through online platforms

The phenomenon of online crowdfunding comes from the United States, and crowdfunding platforms began to appear as early as 2006, initially collecting resources to support music groups and artists for a symbolic award. Nowadays there are many projects on such platforms starting from creating simple product to huge innovative start-ups that accumulate sufficient amount of funds. But despite the fact that crowdfunding platform is helping in raising funds for startups and getting an excellent informational occasion for publications in media, the innovative start-up funding fails much often than success.

This paper tries to understand the motivates behind investments in crowdfunding platforms and the factors that affect it efficiency.

2 Theoretic concepts

Startup is a new business that appeared on the market less than three years ago (the initial stage of entrepreneurship) that is trying to bring out a product or service different from the ones offered on the market, often creating a corresponding new niche in an environment of high uncertainty, aimed solving specific problems [3]. Thus, organization of start-ups is a prime application of innovation.

Organization of a small business through the creation of a new startup seems to be a rather attractive way of

corresponding author: julia.a.akhmetova@gmail.com

doing business in view of the possibility of implementing new, bold and innovative ideas, creating a business from scratch. However, the founders of an innovative startup project almost always face two problems: time constraints and insufficient funding.

Development of innovative startups is a new paradigm that has emerged as a result of the evolution of business over the past decade. Despite the existence of an opinion that the evolution of business models is natural and logical, there is an assumption that the startup boom (that has been going on since the early 2000s) shows the result of complex innovation processes in the world economy, technological progress and globalization. The volatility and complexity of the modern financial market creates new barriers for entrepreneurs in various fields, encourages them to explore new ways to attract funding, find opportunities to implement business ideas and try new ways of doing business [4]. In turn, startups aim not only technological improvement and optimization of production and business processes, but also provide companies with an increase in competitiveness in the market, beneficially contribute to the development of an existing business, opening up new horizons for development, investment and additional profit for the owners.

Government support of innovations can be carried out in different ways, including directly by the constituent entities of the Russian Federation, executive authorities or directly by the Russian Federation. At the same time the subjects can be not only higher educational institutions, state academies of sciences and scientific organizations, but also subjects of innovation infrastructure, including technological parks, incubators, information centers, innovation and venture funds.

However, because innovative start-up projects are characterized by a high level of risk with an extreme level of necessary funding, the usual ways of raising finance, such as a bank loan, seem to be unavailable to them, and personal finance and investments of friends and family members are insufficient. As a result, along with the development of startups, new ways of financial support for start-up projects and new financial intermediaries appeared, such as venture capital funds, private equity funds, as well as business angels, strategic investors and crowdfunding platforms.

3 Crowdfunding platforms

Crowdfunding platforms are an interesting way of startup financial support. The concept "crowdfunding" appeared in 2006 and in fact is a kind of collective microfinancing investment (what is close to peer-to-peer landing), the essence of which is the accumulation of funds from small investors using Internet platforms to fund risky projects. Many researchers who are working on the evaluation on crowdfunding platform effectiveness as a method of financing projects associate the success of project financing with the possibility of network interaction of participants - interpersonal communication between project creators and their investor-consumers. J. Birch [5] considered exchange in communities as reciprocal in cases where project creators also finance projects of other creators, accumulating mutual obligations and social capital within the community. He also noted that the projects of such creators are more successful.

The crowdfunding platforms showed the effectiveness of 68 percent [6] in predicting project success, and the European Commission says that crowdfunding can transform the financial system and become a key source of capital for SMEs in the long term [7]. The crowdfunding platforms can exit because they are charging a commission from funds attracted by the authors of projects to finance them, which varies from 2 to 10%.

The principle of crowdfunding is that a startup creates an Internet page (usually special sites are used, such as Kickstarter, Indiegogo, Planeta.ru, etc.), which tells about the idea and sets the amount of money needed to launch the product, and investors make a decision [8]: to invest or not, so it can be considered as an open appeal to the public. Having the ability to attract a large number of sponsors, developing a well-coordinated selection mechanism, optimizing commissions, advertising and penalties when using asymmetric pricing, allows crowdfunding platforms to apply different types of commissions for applicants and investors, thereby controlling the quality of both parties. The crowdfunding platform wants to gather as many successful entrepreneurs as possible and attract loyal sponsors who will invest on their platform constantly. To achieve this goal the crowdfunding platform must earn a reputation, i.e. become a reliable and convenient platform for both applicants and sponsors.

Today crowdfunding includes three areas: crowdlanding, lending by individuals to other individuals or legal entities through special Internet sites; crowdfunding - fundraising for the implementation of projects without further participation in the share capital; crowdfunding - fundraising for the implementation of projects with subsequent participation in equity capital. A feature of this method of funding lays in the absence of intermediaries, whether banks or venture capital funds, in the process of raising funds to create own project, as well as the lack of a legislative framework supporting the rights participants and guaranteeing the safety of investment and placement of business ideas. In some countries, such as the USA, there are several stimulation procedures as a source of capital for innovative start-ups. However, researchers know still not much about the dynamics of successful crowdfunding projects, factors of their success, as well as the functioning of various crowdfunding.

As main motivates behind investment in crowdfunding the researchers conclude in receiving a reward (as a material object: the product itself or a gift), helping others (close ones charity behavior), helping others, hoping to make a meaningful contribution to support the project and community membership (belonging to a select group, creators of a new one, the ability to interact with like-minded people) [8].

The biggest problem with crowdfunding platforms is that asymmetry of information about projects, lack of mediation and regulation of crowdfunding platform business and relationships between agents, ineffective selection of successful projects makes crowdfunding platform an extremely favorable environment for fraud and deceit.

At this moment there is only one law in Russia that regulates the activities of crowdfunding platforms, namely the federal law dated 02.08.2019 No. 259-FZ "On attracting investments using investment platforms and on amending certain legislative acts of the Russian Federation"[9]. This law introduced the concept of an "investment platform", explains the possibility of providing a loan for a share (in addition to the purchase of equity securities and the acquisition of utilitarian rights), limits the possibility of raising funds in the amount of up to 1 billion rubles, and the volume of investments of funds to 600 thousand rubles per year from an individual. The goals of the law are to increase the availability of the digital services market, protect investors, specify the legal status of crowdfunding platforms and create an effective tool for their legal regulation.

The main idea behind this work is to find the important factors that affect the performance of the platform, because there are few research papers that concentrate on this fact. Previous studies mostly cover that relationship between success of the innovative startup funding and certain features of the project founder as, for example [10], gender, ethnicity, entrepreneurial experience, positive psychological capital [11], founders education, investors motivation, media content about start-up or the project itself [12].

However, some researchers concentrate on the fact that innovative startup funding depend from country to country [13] and strictly based on cultural dimension [14], in this work I focused on 2 main types of crowdfunding platforms - AON and KIA, regardless of the location. The first group is characterized by a situation when the founder of a startup cannot save money in the event of a startup failure, that is, entrepreneurs bear more risk in this case, and entrepreneurs invest more capital helping in achieving significant and global goals. Thus, the choice of funding method is a signal theory, where a project that collects funding through the AON type is considered more reliable. Thus, the choice of the funding method refers to the signal theory, where a project that collects funding through the AON type is considered more reliable. However, this structure encourages the entrepreneur to provide the investor with more information, thereby increasing not only the cost of time and human resources, but also the cost of providing incorrect information if the startup is not implemented within the planning framework.

To other main characteristic that were considered during analysis I outline the type of commission based on the person who is paying it: investor or start-up founder. I found this determinant very important as I can affect the decision of investor or founder in fundraising launching as each rational individual should tends to decrease own expenses. Another determinant is the term of platform existence (in years) as the older platform is the greater reputation it can have, so it creates certain amount of trust not only between platform and investors, but also between platform and founders. Here we face interesting situation. Bigger and older platform, for sure, can help in raising bigger amount of money as they attract attention form many investors, have stronger reputation and bigger advertising budget. From the other side, bigger platform also contains bigger amount of project in it, making a new start-up that is looking for finance invisible in this crowd. This fact should be correlated to the last considered determinant - total number of projects launched on the platform during the year, which is interesting to considered in the field of correlation with the success rate of the platform.

3 Regression model

The regression model was constructed using the panel data from 105 crowdfunding platforms, which are usually presented in the form of the total number of projects and percentage of success, as well as news aggregators and sites describing the main parameters of startups. The initial analysis showed a high level of platform efficiency in the real estate market, where efficiency in 2018 - 2020 reached 62%, while other industries in the world were characterized by an average efficiency of about 32-35%.

As a result, the regression (1) was build:

$$SR = 0.24 + 0.18\,PT + 0.13\,CT + 0.03\,Y + 0.004\,PR \quad (1),$$

where

SR (Success rate) - percentage of successful projects completed by the platform during the year. By success I mean the achievement of the goal – collecting a certain amount of money, stated before the procedure started;

PT (platform type) - type of commission financing, where AON = 1 and KIA = 0;

CT (commission type) - type of commission divided by the situation where investors pay = 1 and 0, when the commission is paid by the startup founders;

Y (years) - number of years of platform existence;

PR (projects) - total number of projects launched on the platform during the year.

The correlation matrix demonstrates the following results (Table 1), which demonstrates a strong correlation between the type of commission and the type of platform, what is obvious because each platform is mainly associated with the certain type of business and as result a commission type that it represents. The other fact of existence in correlation between number of years of platform existence and total number of projects launched on the platform during the year was expected as the older platform should launching bigger amount of fundraising during the year.

Table 1. Correlation matrix

	PT	CT	Y	PR
PT	1	0.68	0.12	0.36
CT	0.68	1	0.07	0.08
Y	0.07	0.12	1	0.32
PR	0.36	0.08	0.32	1

4 Results

The analysis showed the significance of the "platform type", "number of years" and "number of projects" coefficients at the 5% significance level, as well as the significance of the "platform type" and "number of projects" coefficients at the 1% significance level as well. These results indicate that projects adhering to funding such as AON are more likely to be more successful, which was previously postulated by many studies because of the proper level of preparation by startup organizers before launching a project on crowdfunding platforms.

The significance of the "number of projects" coefficient seems to be interesting for research, since it can be considered as a reputation indicator of the platform. Investors and founders of startups often rely on the popularity of the platform when deciding whether to invest / launch fundraising, because it is easier for investors on such platforms to find projects that suit their financial interests, and gives founders the opportunity to compare their project with others, demonstrating significant differences or advantages over other projects.

The significance of the "number of years" coefficient only at the 5% significance level is explained by the relative novelty of the method of funding startups through crowdfunding platforms and, as a result, by the relative youth of all the platforms considered.

5 Conclusion

This paper examined the phenomenon of crowdfunding, an innovative way to request online funding from a large audience. Crowdfunding was viewed as an atypical form of the market, and the investment behavior of participants was viewed as involving risk and based on trust, in which there are both acts of altruism or reciprocity, and equivalent exchange relations. At this moment the effectiveness of crowdfunding platforms is not only obvious, but, as it was proved in the work, is directly related to a number of indicators that depend both on the decisions of the founders and investors themselves and on the reputation of the platform itself.

The main goal of the analyses was to underline the determinants of the innovative startup financial funding on crowdfunding platforms. The model shows strong relation between the level of success of the project in term of achievement the main goal – collecting the listed amount of money - and platform type and total number of projects launched on the platform. This bring to interesting point of consideration, as this factors together constitute crowdfunding platform reputation. The most important is that despite the fact that increase number of projects should decrease the total number of effective startups, it has contractional result in terms of startup effectiveness rate.

This leads us to the fact that this analysis needs detailed consideration of crowdfunding platforms in the future and the model itself suffers from lack of an in-depth analysis of the platform's reputational mechanism of influence and the construction of an appropriate index, which will include not only the lifetime of the platform in the market and the total number of launched projects, but also non-mathematical indicators of branding, reputation and additional services, which influence the attitude towards the platform both on the part of investors and on the part of startup founders. Despite this the work still has interesting and relevant conclusions that were not covered in previous research papers, so it constitutes its theoretical and practical importance.

References

1. L. Vnoučková, *Quality Innovation Prosperity,* **27**, 22 (2018)
2. N.V. Ketko, R.R. Zarafutdinov R.R. *Russian entrepreneurship*, 9, **23** (2011).
3. A. Damodaran, *Investment valuation: Tools and techniques for determining the value of any asset.* 3rd edn. (Hoboken, N.J.: Wiley, 2012)
4. S. Blank, B. Dorf, *The Startup Owner's Manual: The Step - By Step Guide for Building a Great Company (DIATEINO) Paperback.* (K & S Ranch; 1st edition, 2012)
5. G. Burtch, A. Ghos., S. Wattal *SSRN Electronic Journal,* **48**, 3 (2012).
6. M. D Greenberg, B. Pardo, K. Hariharan, E. Gerber, *ACM*, **1815** (2013)
7. European Commission. Commission StafWorking Document. *Crowdfunding in the EU Capital Markets Union. Brussels* (2016).
8. E. Gerber, J. Hui *ACM TCHI*, 20, **34** (2013)
9. Federal Law N 259-FZ "*On Digital Financial Assets, Digital Currency and Amendments to Certain Legislative Acts of the Russian Federation*" (2020)
10. Johnson, Stevenson & Letwin *Journal of Business Venturing,* **831** (2018)
11. Aaron, Jeremy, Regan, Stevenson& Allison *Journal of Business Venturing*, 33, **470** (2018).
12. Geigera, Oranburgb. *Journal of Business Venturing Insights,* **10** (2018)
13. M.K. Sanin, *NRU ITMO Economics and ecology management*, **57**, 4 (2015)
14. Mollick. *Journal of Business Venturing,* **29**, 16 (2014)
15. Cordova, Dolci & Gianfrate, *Geo-Informatics in Resource Management and Sustainable Ecosystem,* **167**, (2019).

Management of digital transformation of industrial enterprises based on maturity models

L. A. Malysheva [1*], *O. G. Kharlamova* [2]

[1]Bussiness School of Ural Federal University, professor, Doctor of Economics, Russia
[2]Bussiness School of Ural Federal University, senior lecturer, Russia

Abstract. This paper considers the issues of the implementation of the digital transformation of industrial enterprises: the analysis of the existing model (AS IS) and planning the desired model (TO BE), depending on different conditions. The authors present the existing business models of digital transformation, reveal different approaches to classification, as well as their drawbacks from the point of view of practical application. In general, the models can be applied to diagnostics, but not to planning the desired state. The principle of classification of transformation models based on the life cycle of the market is proposed: monopoly, oligopoly, competition and monopsony. Firstly, the life-cycle approach allowed applying the author's Dynamic Model of Changes in Corporate Strategies (Dynamics) to digital transformation. This model was proposed earlier for the classification of traditional business models. Secondly, the life-cycle approach allowed using the maturity models of the industry, strategies, product, processes, data etc. in order to build a planning algorithm for the desired business model. As a result of the lifecycle approach to the classification of business models, it was possible to develop an algorithm for diagnosing and planning the desired digital transformation model, taking into account the limitations of maturity levels and present it in the form of a Digital Dynamic Model of Corporate Strategy Changes (Dynamics).

1 Review of existing approaches to digital transformation

The process of digital transformation of companies is becoming more and more popular and relevant. Different researchers interpret this term differently. A. Prokhorov and L. Konik note the differences in understanding by business consultants, engineers and developers, personnel specialists [1]. There are clearly three approaches to digital transformation: 1) strategic, associated with changing business models; 2) technological, related to automation, data-based decision making; 3) cultural, associated with new thinking. The authors also highlight the areas where the changes take place. These are: strategy, technology, people and processes.

We believe that all three aspects have a cause-effect relation. In the course of digital transformation, strategies, business models, technologies, business processes, ways of storing data, people's thinking, and their behavior are changing. This is not a complete list of system elements subject to transformation.

In the report of NRU HSE it is noted that the proliferation of new business models significantly reduce the time of market launch of new products. These models are characterized by an iterative approach to meet customer needs [2]. The authors look at the process of transformation from the perspective of business models that are changing the way we make money. They distinguish the following categories of business models: 1) digital platforms; 2) service business models (SaaS - Software-as-a-Service); 3) product models (PaaS - Product-as-a-Service); 4) crowdsourcing, 5) monetization of personal data, etc.

A similar view of the transformation of business models is given in the article by P. Bilenko, Yu. Krasilnikova and S. Nochovny [3]. The authors consider 6 prerequisites for the emergence of new business models in the context of the transformation of industrial enterprises: 1) the increasing complexity and fragmentation of the supply network; 2) decentralization; 3) network effects of online platforms; 4) democratization of design and prototyping; 5) mass customization; 6) sustainable development trend.

The same authors identify 7 promising business models for digital transformation of enterprises and also emphasize the combination of models.

1. MaaS / XaaS (Manufacturing-as-a-Service / Anything-as-a-Service). The provision of value chain elements as services or service packages. Online platforms that aggregate requests from customers and place orders with partners and subcontractors
2. C2M-model (Customer-to-Manufacturer or Community-to-Manufacturer). Interaction of

factories with the end consumer. Personalization and customization of orders
3. Experimental crowdsourcing. Engagement of communities and experts to test and refine a product
4. Distributed factories. They are based on decentralization and suppose networking of e-commerce micro-shops
5. Technological hubs. Networking of project teams with experts, investors, educational organizations, factories and other ecosystem participants within the framework of high-tech accelerators
6. Selling software and monetizing data. Provision of value-added services based on data
7. Processing and reprocessing. Upgrading existing processes by modular upgrading of value chain elements.

In our opinion, the list of models characterizes the current situation, but it has a number of methodological shortcomings: it does not have completeness, mutual exclusion and clear grounds for classification. The reasoning at the level of examples of business models makes it possible to see development options, but does not provide an algorithm to choose an acceptable option and does not allow analyzing, planning and making decisions. We see two scientific challenges that will be discussed in this paper.

1. The classification of business models appropriate for making decisions on the preferred business model and suitable for diagnosing the current state and designing the desired one.
2. Algorithm for diagnostics and design of the desired state.

2 Classification of business models of digital transformation

Basic ideas:

1. The classification of business models should be based on the external conditions. The reasons for the change in the business model lie in the external conditions and a company behavior is a consequence of the influence of the external conditions and the result of the adaptation of a company
2. The classification of business models allows not only analyzing the existing state of a company, predicting the desired and ideal state of a company, but also developing a diagnostic and design algorithm

The purpose of the study is to propose a classification of the business models of a company, appropriate for both analysis and design of an ideal model

The proposed methodology is based on the Dynamic Model of Changing Corporate Strategies – Dynamics developed by the authors, which previously structured traditional business models [4]. This paper provides the extension of the DinAMIX model for digital business models.

2.1 Review of existing approaches to the classification of business models of digital transformation

One of the popular approaches to the classification of business models was proposed by Peter Wyle and Stephanie Warner [5]. The structuring is based on 2 principles: complete/incomplete knowledge of a client and the structure of business: linear, based on the value chain, or network, ecosystem.

According to the combination of factors, the authors determine 4 business models:

1. A supplier who sells his product through other companies, for example, marketplaces.
2. Omnichannel: a complex value chain that brings together various players in the market.
3. A modular manufacturer of finished products or services.
4. Ecosystem driver.

We see some limitations in the applicability of this classification, for example, in the absence of clear criteria to refer a company to one category. The second difficulty is in the fact that the basis of classification does not take into account external reasons for the division of companies, but internal factors that are a consequence.

Another classification is offered by Carsten Linz, Gunther Müller-Stevens and Alexander Zimmerman [6]. Their approach is based on the differences in transaction coverage and level of customization. Transaction Coverage: low, single or high, complex, integrative. Customization level is low, standardized, or highly customized.

Depending on the combination of factors, the authors determine 4 business models: product-oriented; platform-oriented; project-oriented and solution-oriented.

In our opinion, both the first and the second approaches are similar and quite correctly describe the behavior of companies. However, they do not give an answer to the question of the reasons for the changes, the limits of applicability, as well as clear criteria to refer a company to one type of business model.

The publications by A. Soolyatte [7], A. Kulizhsky [8], V. Kotelnikov [9], B. Garifullin and V. Zyablikov [10], Gassman O., Frankerberger K., Shik M. [11] and many other researchers are devoted to the classification of business-models.

In general, we are talking about the ways to monetize and analyze the behavior of companies. However the authors do not provide an answer to the question of the choice of an effective business model in each specific situation. Therefore, the task of the choice of a business model remains unresolved.

We believe that the choice of a business model should be based on external factors, for example, the type of market determined by the ratio of supply and demand.

In our previous works, we introduced a classification of traditional business models based on supply and demand in the author's business model DynaMIX. At the beginning of the cycle, demand exceeds supply, which corresponds to a small number of companies in the market under the conditions of monopoly. Then, as the market develops and the number of players increases, price fixing appears and an attempt to minimize costs due to transaction costs. Then the market becomes saturated and competitive. Finally, it degenerates into monopsony, the buyer market. Depending on this, 4 situations are determined that require different strategies and behavior of companies. These are: a monopoly characterized by growing demand; oligopoly with stable demand; competition and monopsony, with a significant excess of supply relative to demand.

Different types of markets allow talking about different strategies, types of products, business processes, structures, incentive systems for traditional models. Markets are also relevant when applied to digital transformation. We can use the same approach for digitalization conditions and consider the change in the market situation as a reason, and strategies, product type, business processes, information systems, data storage methods - as the areas of change. In the process of digitalization, data, methods of storing and processing them as well as the methods of decision-making become the most important. It is this area that significantly distinguishes digital transformation business models from traditional ones.

2.2 Classification of business models taking into account the market life cycle and maturity models

The lifecycle approach to classification allows developing discrete system states for each type of market.

At each stage, we recommend that companies choose different strategies, product types and, accordingly, behavior. It is a consequence of changes in the market and the adaptability of a company in certain areas: strategy, product, business processes, data collection and processing, decision-making algorithms. We classified business models by market type (Table 1).

Table 1. Classification of digital transformation business models by market type

Type of market	Monopoly	Oligopoly	Competition	Monopsony
Supply and demand ratio	Demand >> Supply	Demand> Supply	Demand <Supply	Demand << Supply
Strategy	Production growth	Reduced costs	Differentiation	Focusing
Type of product	Typical, standard	Cost optimized	Modular	Service on request
Model by P. Vail and S. Warner	Provider	Modular manufacturer	Omnichannel	Ecosystem driver
Model by K. Linz, G. Müller-Stevens, A. Zimmermann	Product	Platform	Project	Decision

Under the conditions of monopoly, it is profitable for companies to increase production volumes through product standardization. They act as a supplier of a typical product and can post information about it on marketplaces. The focus is on product standardization. With the transition to oligopoly, products with a lower price appear which necessitates a change in strategy in favor of cost reduction in order to maintain the same rate of return. Companies can offer modular solutions for placement on the platform and connect with their module to the value chain for a client, based on minimizing the cost of non-core functions.

In competitive conditions, it is useful for a company to place its offers on the platform and use seamless integration (omnichannel) with partners and suppliers for flexible customization of orders from a consumer. Moreover, every order will be a project, since it will differ in the set of modules, resources and costs. Finally, in a monopsony conditions, ecosystems consisting of many services and providing focus on consumer needs through a variety of services and the ability to fulfill a personalized request will be effective.

The Life Cycle concept is closely related to maturity models. Market maturity, business processes and structures are also present here. However, waiting for industrial enterprises, we will be interested in technological models of maturity to a greater extent, which become primary in the context of digitalization.

There are many maturity models and they are applied to different types of companies and functional areas, for example:

- data maturity model (DMM) - for companies engaged in data processing and commercializing data as a product [13];
- analytic maturity model of a company (A2CM2) - for companies specializing in forecasting and analytical data processing [14]
- model of technological readiness for commercialization (TRL) - for technology companies and those engaged in research and development (R&D - Research & Development) [15]
- capability maturity model integration (CMMI) - for companies with digital products [16].

We suppose that the abundance of maturity models is reasoned by the fact that at a given level of market development, different companies specialize in certain methods of data processing, analytics and research. In future, companies will expand their potential through the integration into ecosystem solutions.

Maturity models are used for diagnostics. They have categories and a scale for assessment, quantitatively specifying the ideal model and the current one.

The Data Maturity Model (DMM) has 6 segments: 1) Strategic dynamics; 2) Customer focus; 3) Digital culture, talents and skills; 4) Innovation and lean delivery; 5) Big data and artificial intelligence; 6) Technological leadership.

Each segment has its own subsections and weights, which depend on the type of company, in accordance with the classification of P. Vail and S. Warner. It is a supplier, modular manufacturer, omnichannel, ecosystem driver. We set the types of a company in accordance with the types of the market, taking into account the life cycle approach.

The subsections of the DMM are divided into categories and described. It is proposed to evaluate indicators on a 10-point scale, while the methodology provides a description of the best indicator value of 10 points. Depending on the final result, the following levels of maturity are distinguished: initial, controlled, definite, quantitatively controlled and optimal.

The Data Maturity Model (DMM) is good for diagnosis, but not good for planning, for the following reasons:

- intuitive choice of the type of company by the respondents, in accordance with the methodology of P. Vail and S. Warner;
- the multi-product nature of production enterprises, which does not allow a company to be classified as one type;
- the absence of a logical planning algorithm in the maturity model, which is not important in diagnostics, but critical in design.

In addition, as the market changes and the need to choose a more mature business model, a company will face the technological limitations of the existing data model.

In order to implement digital transformation, we propose to structure DynaMIX, taking into account maturity models, according to the following segments: market type, strategy, product type, type of business process, data model (DMM). For each product in the product portfolio, the desired model can be designed off-line.

The proposed classification of business models based on market types (Dynamics) has the following features:

1. It considers each product in the product portfolio separately. At the same time, with one product a company can act as a supplier, and with another it can be a modular manufacturer. Dividing products by market type makes it possible to choose more clearly the type of business model.
2. It matches the type of business model to the type of market, and also structures the levels of maturity of the potential of a company in relation to the life cycle of the market, product, business processes, data and digital culture.
3. It evaluates not the average level of maturity of the potential of a company, but detailed one for different products.

Thus, we obtain the structure of the maturity model, weight coefficients for structural elements and scales to assess the maturity level of factors of structural elements. According to the scale, we can determine the ideal state of the business model, ideal in each situational range, as well as the limits of applicability of the weights.

A business model based on maturity models is used for analysis (AS IS) and planning (TO BE), taking into account external factors, for example, the market situation. The implementation of the changes will require a transition from the existing business model to the desired one.

3 Classification results and algorithm to design a digital business model

In the course of digital transformation most companies use the services of vendors and consultants who offer their own solutions. The decisions are related to business models, however, companies still do not distinguish between traditional and digital models, relying on the popular business model of A. Osterwalder in the description [17].

Vendor solutions often do not provide a systemic vision and solve current problems without the compliance with a long-term strategy. Maturity models, in turn, do not cover all functional areas requiring changes. They do not provide a logical and systematic view of changes.

The use of DynaMIX as a logical foundation allows building logic of changes similar to changes in traditional models.

The changes in the market situation necessitate internal changes and adaptation of a company to environmental factors. The sequence of changes in the course of digital transformation is as follows:

1) Product strategy;
2) Product type;
3) Value proposition;
4) Type of behavior;
5) Business processes;
6) Data (DMM).

After the optimization of business processes, the organizational structure and employee motivation system are adjusted in traditional models.

In digital transformation, the Data Maturity Model (DMM) is important. The authors of the model do not provide a logical sequence for the analysis of the maturity level. During the process of planning, consistency is extremely important.

We propose the following sequence for the design of the desired DMM model:

1) Customer focus;
2) Technological leadership;
3) Innovation and Lean Manufacturing;
4) Big data and artificial intelligence;
5) Digital culture, talents and skills;
6) Strategic dynamics.

The logic is as follows. According to the type of market, a company chooses a different level of customer focus and type of a product. Each type of product requires a certain level of technological leadership and there is a certain connection to business processes. Innovation and lean production are also business processes, eliminating waste and inefficient functions. The processes generate data, thus the next step is to design a data model in the category of Big Data and Artificial Intelligence. For this, digital culture, talents and skills are important. The plan of development and training is drawn up. Finally, strategic dynamics are assessed and compared with the initially chosen strategy.

The Digital Dynamic Model of Changing Corporate Strategies was called DynaMIX - Digital.

We summarized DynaMIX in a table, while the planning algorithm was located in the left column of the table, the classifier of the model states was set by the types of markets, and, in fact, the business models themselves were arranged in columns, in relation to the types of markets. The calculations in the DMM model were made in accordance with the ODI - Open Data Institute methodology (Table 2) [18].

Table 2. Algorithm for the design of the desired business model based on DynaMIX

Type of market	**Monopoly**	**Oligopoly**	**Competition**	**Monopsony**
Demand/Supply	D>> S	D>> S	D>> S	D>> S
Strategy	Production growth	Reduced costs	Differentiation	Focusing
Type of product	Typical, standard	Modular, typical	Modular on request	Service on request
Value proposition according to the model of K. Linz, G. Müller-Stevens, A. Zimmermann	Product	Platform	Project	Decision
Behavior type: the model of P. Vail and S. Warner	Provider	Modular manufacturer	Omnichannel	Ecosystem driver
Business processes	BPM	BPR/BPI	BR	XE
DMM	**7,27**	**9,12**	**9,50**	**10**
1. Customer focus	5,54	7,79	9,91	10
2. Technological leadership	7,65	8,62	8,65	10
3. Innovation and Lean Manufacturing	6,23	10,00	9,06	10
4. Big data and artificial intelligence	7,78	9,43	9,47	10
5. Digital culture, talents and skills	8,72	9,87	9,82	10
6. Strategic dynamics	8,62	9,66	9,63	10

DDynaMIX and DynaMIX include traditional concepts of business process management. They do not change for digital models.

- BPM – Business Process Management ;
- BPR – Business Process Reengineering;
- BPI – Business Process Improvement;
- BR – Business Reengineering;
- XE – Cross Engineering – reengineering without limits (companies).

The data maturity model (DMM) calculates the ideal maturity level for each segment for each type of company, taking into account the weights of the questions and the recommended values.

The imposition of a data maturity model (DMM) on the market lifecycle allowed drawing the following conclusions.

1. Depending on the level of market maturity, the type of company becomes more complex, and with it the level of data maturity increases quantitatively.
2. Under the conditions of monopoly, the level of client orientation has the lowest status, as well as the values of all other categories. Indeed, the business of a company model in this situation is the simplest and it does not require special strategic dynamics.
3. Under the conditions of oligopoly, the maximum weight is assigned to the category "Innovation and Lean Manufacturing". This is explained by the strategy to reduce costs and optimize business processes.
4. Under the conditions of competition, a company chooses a differentiation strategy. It is customized. It provides a buyer with the opportunity to choose a product from different modules, based on a seamlessly assembled value chain. This situation corresponds to a high level of customer focus, a lower level of innovation and lean manufacturing due to a focus on product quality rather than costs. There is also a reduced level of strategic dynamics and digital culture, talent and skills due to possible integration with other suppliers and subcontractors.
5. Under the conditions of monopsony, the level of maturity in all categories is maximal.

4 Research results

During the transition to digital transformation, traditional business models are no longer valid. It is necessary to rely on best practices and descriptive business models. At the same time, maturity models are developed, which are actively used to assess the readiness for digital transformation. However, maturity models are poorly applicable to planning, since they are not related to the cause of changes, strategies, necessary elements of the system and also do not provide the algorithm for the design of the desired model.

Based on the research results, we propose the following:

1. To classify digital models by market type, continuing the logic of DynaMIX and ensuring the

continuity of business models. This approach became possible due to the underlying reason for any change - the market situation.

2. To supplement the Dynamic Model of Changing Corporate Strategies - DynaMIX with a data maturity model (DMM). It will make their combination appropriate for planning digital transformation, due to the interface with the external environment. We called it the generalized business model DynaMIX (Digital).
3. To develop a logical algorithm for the design of digital transformation based on the data maturity model (DMM). In this case, it is necessary to focus on the ideal correspondence of the model values to the types of companies.

5 Conclusion

The proposed Digital Dynamic Model of Changes in Corporate Strategies DynaMIX in terms of the data maturity model is focused on the entire company, and in terms of strategy, product type and business processes it is focused on each product separately.

The Data Maturity Level (DML) is a limit for the operation in different digital markets. A company can only choose the type of behavior based on the level of maturity of the data.

We propose to consider digital transformation as a sequential set of innovations: management, product and organizational ones. Today the digitalization of products and processes is popular. However, companies will not get a systemic effect without a strategy and organizational transformation.

Further studies of the digital transformation of industrial enterprises will reveal industry-specific features, implementation difficulties and classify successful practices and implementation algorithms.

References

1. A. Prokhorov, L. Konik *Digital transformation. Analysis, trends, world experience.* (Moscow: "ComNewsGroup", 2019).
2. G. I. Abdrakhmanova, K. O. Vishnevsky, L. M. Gokhberg *What is the digital economy? Trends, competencies, measurement of Ch-80,* XX Apr. International scientific conference on the problems of economic and social development, Moscow, April 9–12. (2019)
3. P. Bilenko, Y. Krasilnikova, S. Nochovny *Factories in containers, MaaS-platforms and local micro-shops: the main production business models of the 2020s.* Retrieved from: https://trends.rbc.ru/trends/innovation/6087d8629a7947e710a41556
4. L.A. Malysheva *Controlling Organizational Change: How Not to Drown in a Sea of Popular Concepts? Tutorial.* (Yekaterinburg: Educational and Methodological Center of USTU-UPI. 2010),
5. P. Vail, S. Warner, *Digital business transformation: Changing the business model for a new generation of organizations* (Moscow: Alpina Publisher, 2019).
6. C. Linz, G. Müller-Stephen, A Zimmerman, *Radical change in the business model: Adaptation and survival in a competitive environment* (Moscow: Alpina Publisher, 2019)
7. A. Soolyatte *Business models of companies: definition, evolution, classification.* Retrieved from: http://www.klubok.net/article2302.html/2009/30/01
8. Kulizhsky A. *Global competition of business models. How to choose and build your own?* Retrieved from: http://www.executive.ru/community/articles/7892220/2008/01/07
9. V. Yu. Kotelnikov *New business models for a new era of rapid change driven by innovation* (Moscow: Eksmo, 2007)
10. B.M. Garifullin, V.V. Zyabrikov *Creative Economy.* **13 (1),** 83-92 (2019). doi: 10.18334 / ce.13.1.39720
11. O. Gassman, K. Frankerberger, M. Schick, *Business models: 55 best templates.* (Moscow: Alpina Publisher, 2016).
12. E. Borisov, A. Minov, B. Slavin *Digital transformation of business models* Retrieved from: 4cio.ru ›content / 4CDTO / bori.pdf
13. *CMMI - Capability Maturity Model Integration* Retrieved from: https://www.tutorialspoint.com/cmmi/cmmi-overview.htm
14. *TRL - Technology Readiness Levels* Retrieved from: https://www.economy.gov.ru/material/file/535256e8193016b438e70338a9a0d42d/metodic_ta_mrg.pdf
15. *Model of maturity of analytical abilities of the company Advanced Analytics Capability Maturity Model (A2CM2)* Retrieved from: https://www.ssa.gov/data/data_governance_board/ACE_A2CM2_for%20DGB.pptx.pdf
16. *Model of maturity of data (DMM - Data Maturity Model).* Retrieved from: https://github.com/SuperNerb/Data-Maturity-Models
17. A. Osterwalder, I. Pigne. *Building business models* (Moscow: Alpina Publisher, Skolkovo Series, 2013).
18. *Methodology for calculating the level of maturity of data* Retrieved from: https://theodi.org/article/open-data-maturity-model-2/

Development of a matrix method for the assessment of information and economic security of an industrial enterprise

Anna Melnikova [1], *Elena* Mylnikova [1,*], *Irina* Moskvitina [1], *Nadezhda* Nagibina [2]

[1] Perm National Research Polytechnic University, 29, Komsomolsky Prospekt, Perm, 614045, Russia
[2] Perm State Institute of Culture, 18, street of the Newspaper «Zvezda», Perm, 614000, Russia

Abstract. The article focuses on the problem of the assessment of the state of industrial enterprise's economic and information security. The purpose of the research is to develop an scoring matrix for information and economic security of an industrial enterprise. The paper presents the necessary economic indicators and their normative values. It also presents the criteria for assessing information security, illustrates the developed matrix and defines development strategies for each element. More over, it describes the algorithm for the use of the matrix. The scientific novelty of the paper is in the development of a matrix that assesses both the economic activity and information of an industrial enterprise in order to determine the priority direction of the development. As a result of the work, the authors present a matrix of information and economic security of an industrial enterprise with individual development strategies for each element. The results of the study will be useful for industrial enterprises not only to analyze their activities, but also to determine the directions of further development.

1 Introduction

In modern conditions of repeating economic crises, an ever-growing number of competitors, deep penetration of the information environment into business, it is very important to monitor the state of information and economic security of an enterprise in order to change the development strategy in time [1]. The article focuses on the problem of the assessment of the state of industrial enterprise's economic and information security in order to determine ineffectively functioning elements of an enterprise and provide a development strategy on this basis. The main purpose is to develop the scoring matrix for information and economic security of an industrial enterprise. The fundamental research methods are the method of analysis and graphic interpretation of information.

The issue of differentiation of types of security was studied by such authors as A.A. Khachaturyan, V.N. Ageeva [2]. Ponomareva S. V. indicates the need to clarify the term digital security in the legislation, since there is no basic concept of it [3]. I.D. Kamskova proposed the stages of implementation of the information security system at enterprises and systematized the requirements for information security systems [4]. Some authors pay attention to such a topic as the information aspect of economic security [5, 6, 7, 8, 9].

There are many authors involved in the development of scoring matrices. For example, I. Ansoff developed Ansoff's strategic matrix designed to determine the tactics and strategy of product positioning on the market [10]. G. Schwartz and S. Davis presented a matrix of cultural risk assessment, revealing the level of risk of the use of a particular strategy in the context of the existing culture in a company [11].

A. Veretennikova, E. Popov and Zh. Omonovs have developed a matrix for evaluating the effectiveness of social innovations. This matrix helps to determine the features with which the development of social innovations is taking place. [12]. I. Lipsits offers a marketing assesment matrix as a marketing tool in the analysis and development of commercial policy [13]. There is also a BCG matrix developed by the Boston Consulting Group led by Bruce Henderson, which served as a model for the development of the matrix proposed by the authors [14].

2 Matrix of information and economic security of an industrial enterprise

The authors of this paper developed the scoring matrix of information and economic security of an industrial enterprise. The matrix presented in Figure 1 is based on assessments of the economic and information security of an enterprise.

In order to develop a matrix, it is necessary to designate two axes (from the origin): the vertical axis (economic security) and the horizontal axis (information security). Each axis is divided into two equal parts. One part corresponds to low indicators of information and economic security, while the other corresponds to high indicators.

Corresponding author: Elenamm152@yandex.ru

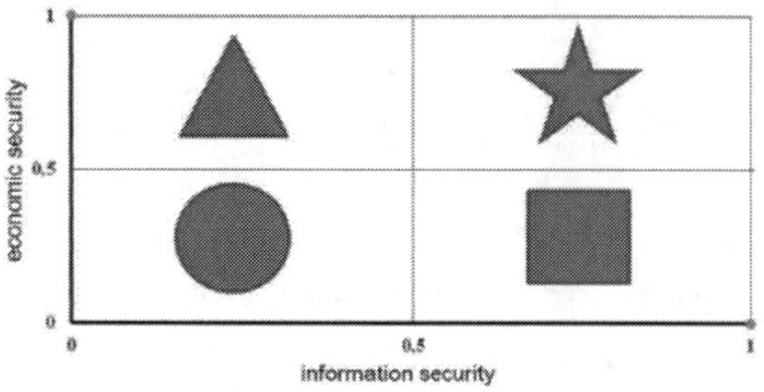

Figure 1. Matrix of information and economic security of an industrial enterprise

After the calculations and entering the results into the matrix, an industrial enterprise will be able to refer itself to one of the matrix elements: «star», «square», «circle», «triangle». Consequently, each element of the matrix has its own development strategies:

- «Star»– these are enterprises with a positive assessment of information and economic security, they need to hold their positions;
- «Square» – these are enterprises with an effective information security system, but economic security has a negative assessment. Here it is important to pay special attention to the economic indicators;
- «Triangle» – these are enterprises with good economic indicators, but information security is assessed critically, therefore, it is necessary to increase its level first;
- «Circle» – these are enterprises with a low level of economic and information security, thus it is necessary to completely reconsider their functioning

2 Indicators for the assessment of economic and information security of an industrial enterprise

Tables 1 – 4 present indicators, after the calculation of which, it is possible not only to determine the location of an industrial enterprise in the matrix, but also to draw a conclusion about the level of economic and information security at an enterprise, respectively.

Thus, if the ratio of economic security corresponds to an indicator equal to one (1) score, it means that the economic security of the enterprise is evaluated with high points.

If the economic security coefficient has a value in the range from 0.71 to 1.0 points, it means that the economic security of the enterprise is evaluated positively.

If the coefficient of economic security is in the range from 0.37 to 0.71 points or has a value equal to 0.5 points, it means that the assessment of the economic security of the enterprise is neutral.

Table 1. Assesment indicators of economic security. Financial component

Indicators of economic security	Assessment based on the degree of compliance with standard		
	1	0.5	0
Equity ratio	> 0.5	0.3 – 0.5	< 0.3
Working capital financed by equity to total assets ratio	> 0.1	0 – 0.1	< 0
Absolute liquidity ratio	> 0.2	0.1 – 0.2	< 0.1
Current liquidity ratio	1.5 – 3.0	1.0 – 1.5 or > 3.0	< 1.0
Altman Z-score	> 2.9	1.23 – 2.89	< 1.23
Financial stability index	> 0.5	0.3 – 0.5	< 0.3
Leverage ratio	< 1.0	1.0 – 2.3	> 2.3
Financial safety margin (%)	> 25	10 – 25	< 10
Determination of the type of financial strength	Ideal (1)	Norm (0.5)	Crisis (0)

Table 2. Assesment indicators of economic security. Production and marketing component

Indicators of economic security	Assessment based on the degree of compliance with standard		
	1	0.5	0
Return on Sales	> 0.2	0.1 – 0.2	< 0.1
Return on assets	> 0.1	0 – 0.1	< 0
The level of profitability of the main activity	Growth (1)	No changes (0.5)	Decline (0)
Ratio of receivables and payables (or)	0.9 – 0.1	0.5 – 0.9 or > 1.0	< 0.5

Table 3. Assesment indicators of economic security. Technical and technological component

Indicators of economic security	Assessment based on the degree of compliance with standard		
	1	0.5	0
Return on assets (growth, no changes, decline)	Growth (1)	No changes (0.5)	Decline (0)
Depreciation ratio of fixed assets	Decline (1)	No changes (0.5)	Growth (0)
Fixed asset usability ratio	> 0.5	0.3 – 0.5	< 0.3
Fixed asset renewal ratio	> 0.1	0 – 0.1	0

Table 4. Assesment indicators of economic security. Personnel component

Indicators of economic security	Assessment based on the degree of compliance with standard		
	1	0.5	0
Wage level ratio	> 1.0	0.5 – 1.0	< 0.5
Personnel stability ratio	> 1.0	0.5 – 1.0	< 0.5
Turnover ratio of personnel	0.03 – 0.07	0.08 – 0.12	> 0.12
Labor productivity	Growth (1)	No changes (0.5)	Decline (0)
Economic security ratio of an enterprise	0.71 – 1.0 High	0.37 – 0.7 Medium	> 0.37 Low

In the case when the ratio of economic security is less than 0.37 scores or is generally equal to 0, this indicates that a company has a critical assessment of economic security.

Next, it is necessary to analyze the information security of an enterprise, evaluate each indicator, calculate the average value and compare with the normative value, and then assign the following scores to information security: 1.0 (positive); 0.5 (neutral) or 0 (critical).

Table 5. Assesment indicators of information security. Environment and infrastructure

Assesment indicators of information security	Assessment based on the degree of compliance with standard		
	1	0.5	0
Access control system to other automated workstations	8 – 10	5 – 8	< 5
Video surveillance system in a company	8 – 10	5 – 8	< 5
Perimeter Security System	8 – 10	5 – 8	< 5
Confidential Disclosure Agreement	8 – 10	5 – 8	< 5
Clear regulation of employee responsibility	8 – 10	5 – 8	< 5
Hardware and software	8 – 10	5 – 8	< 5

Thus, with the average values of the information security coefficient from 8 to 10, a positive assessment of the level of information security of the enterprise is achieved.

If the average value of the information security coefficient from 5 to 8 points, then a neutral assessment of the level of information security of the enterprise can be obtained last.

However, if the average value of the information security ration is less than 5 scores, then this indicates that an enterprise has a critical assessment of information security.

Table 6. Assesment indicators of information security Hardware and Software

Assesment indicators of information security	Assessment based on the degree of compliance with standard		
	1	0.5	0
Access control arrangements	8 – 10	5 – 8	< 5
Security alarm	8 – 10	5 – 8	< 5
Authorized access to storage media	8 – 10	5 – 8	< 5
Clear distribution of responsibility for documents (carriers of classified information) between employees	8 – 10	5 – 8	< 5
Organization of classified document flow	8 – 10	5 – 8	< 5
Access control system for employees to copy and multiplying equipment	8 – 10	5 – 8	< 5
Information security ration of an enterprise	8 – 10 High	5 – 8 Medium	< 5 Low

3 Algorithm of operation with a matrix of information and economic security of an industrial enterprise

The authors of this work proposed the following algorithm to work with a matrix of information and economic security of an industrial enterprise:

1. To make calculations of economic safety indicators presented in Table 1.
2. To determine the average value of all ratios of economic safety and compare with the standard values.
3. To assign the resulting result the appropriate points (1.0, 0.5, or 0), which will depend on the degree of compliance with the regulatory values of the economic security coefficient.
4. To carry out the calculations of information security indicators, which are presented in Table 2.
5. To determine the average value of all information security coefficients and compare the result with the standard values.
6. To assign the obtained result the appropriate scores (1.0, 0.5 or 0) depending on the degree of

compliance with the normative values of the information security coefficient.

7. To mark the established coefficients of economic and information security on the corresponding axes of the matrix of information and economic security of an industrial enterprise.
8. To determine the location of an industrial enterprise in one of the matrix elements (star, square, circle or triangle).
9. To make a conclusion about what may be the priority direction of further development of an enterprise.

4 Conclusion

In modern market conditions, it is extremely important to analyze the economic component of the activities of an enterprise. Moreover, every enterprise needs to modernize its information security system in connection with the rapidly developing digital technologies [15].

The matrix proposed by the authors will be useful for industrial enterprises that want to analyze their activities and assess not only the economic state, but also the quality of information security and determine the development direction in accordance with the analysis.

The article consolidates all significant economic coefficients and regulatory and diagnostic indicators of the information and communication environment of economic entities of industry [16] what can be useful for the appropriate management and effective control of the level of risk and security of the entire information and economic infrastructure of the enterprise.

If top managers fix an imbalance in the information cycle or the financial and economic cycle of the company, then managers can turn to the proposed matrix model and identify the state of their management object by applying one of the author's strategies to eliminate negative trends in the security of the enterprise.

This research reveals four strategies for further adaptation of the enterprise, taking into account the identified negative determinants that have a destructive impact not only on the information security of the enterprise as a whole, but also on the entire socio-economic policy of the entire economic entity.

Working in matrix modeling, managers, performing the entire embedded algorithm, fall into one of the zones: a star, a square, a circle or a triangle. Each zone corresponds to a further way to solve the problems identified during the implementation of the entire sequence of actions developed by the authors of this article.

Taking into account the entire concept of work, the customers of the proposed matrix modeling can be both enterprises and public services, information and communication corporations, firms that work on outsourcing, but are engaged in auditing or ensuring the financial and economic security of enterprises, regulatory authorities, internal affairs bodies and the Prosecutor's office.

In future, it is planned to introduce this matrix and consider information and economic security through the example of high-tech industrial enterprises, taking into account the calculation of all the proposed coefficients and the entry of the obtained indicators of information and economic security in a single register, with the further development of an interactive map on the level of security of economic entities of the Perm Territory and other regions of the Russian Federation.

References

1. S. E. Doroshkin, N. V. Tarasova *Economics and business: theory and practice,* **2**, 128-133 (2020)
2. A. S. Melnikova, A. A. Khachaturian, V. N. Ageeva *Modern science: topical problems of theory and practice. Series: Economics and Law,* **3**, 107-112, (2020)
3. S. V. Ponomareva, H. H. Khachaturian, A. S. Melnikova *Lecture Notes in Networks and Systems,* **115,** 123-129 (2020)
4. I. D. Kamskova, *Prospects for the development of innovative technologie*, 170-172 (2014)
5. C. S. Teoh *Journal of Theoretical and Applied Information Technology,* **95 (23)**, 6512-22, (2017)
6. P. L. Yin and T. Bresnahan *Innovation Policy and the Economy,* **17 (1),** 95-124 (2017)
7. F. Kamal and W. Chen *Journal of International Business Studies,* **47 (5)**, 563-576 (2016)
8. V. V. Borchaninov, N. N. Kotova *Bulletin of the South Ural State University. Series: Economics and Management*, **1,** 20-27 (2017)
9. B. Carin *Economics,* **11 (19),** 3-19 (2017)
10. O. V. Kalinina, A. P. Maryukhta *Innovative Science,* **8,** 18-22 (2015).
11. A. V. Kolesnikov *Interactive Science,* **12,** 203-205 (2017).
12. E. V. Popov, A. Yu. Veretennikova *Regional economy: theory and practice,* **9,** 1752-1772 (2017).
13. I. V. Lipsits *Financial analysis and marketing assessment matrix* (2018)
14. E. S. Gridnev *Bulletin of Kamchatka State Technical University* 61-64 (2010)
15. Yu. A. Shcheglov *Bulletin of Penza State University*, **1**. 40-43 (2018)
16. E. M. Mylnikova, A. S. Melnikova, A. A. Kisareva *Issues of innovative economy* **10** (2), 941-950 (2020)

Formation of the approach to assess the effectiveness of project initiation for company development

Svetlana Belmas *, *Nikolay* Akatov, *Sergey* Kosyakin, and *Sergey* Belmas

Perm National Research Polytechnic University, 29 Komsomolsky Str., Perm, 614990, Russian Federation

Abstract. The paper presents the results of the assessment of the ability of a company to initiate and implement development projects. As a criterion for this ability, the authors propose a model for the assessment of the initiating impulse of changes. The authors propose the definition of this concept and present a model for its assessment based on a group of critical factors (willingness to act, responsibility, influence, faith and motives). A distinctive feature of the model is the proposal for a scheme for the formation of an impulse, which includes the stages of motivation, synthesis of activity and the formation of impulse itself. The paper presents a cognitive model and assesses the impulse formation depending on the value and significance of the measured critical factors.

1 Introduction

Many research works are devoted to the solution of the problems of the effectiveness improvement of strategic transformations of a company. It can be stated that this direction of research is a priority, since almost every research to some extent raises the issues of risks, feasibility, practicability and effectiveness of the implementation of strategies.

In this direction, the problem of the effectiveness of the initiation of development projects claims to act as the driving component of the problems of increasing the effectiveness of the strategic transformations of a company. Intuitively, each head of company agrees with the statement that the future fate of the project critically depends on the effectiveness of initiation. However, there are practically no rational criteria or models supporting this statement. At the same time, a detailed consideration of the initiation process itself, as the development of the conceived actions and the transition from the ideas to their realization, is manifested in the act of motivation or in the managerial impulse, which, both gains its strength and is transformed into ideas, or is lost, scattered, extinguished not reaching its goal.

The model of the initiating impulse of change (IIC) allows representing a wide range of possible results of the act of initiation and, accordingly, setting the requirements for management actions for its formation, taking into account the real complexity of this task, the level of uncertainty [1], the risk appetite of decision-makers and the project team. For example, for the case of a high level of IIC certainty, this is a tough scenario of the development of events (unambiguous comprehensibility of project idea), a given time (speed) mode and specificity of goals. For a scenario case and a wide range of future scenarios it is scenario analysis, scenarios and scenario behavior. On the other hand, for complete uncertainty, the following aspects will be important: the power of vision, value models of behavior, trust, risk appetite of a group of key participants and other characteristics typical of specific situation. The attempts to give a definition of such a complex concept as IIC are faced with the problem of ambiguity of its perception. Thus, in scientific and practical sources, the impulse is defined as the effective exit of the stage of initiation by the leaders of an innovative project, which characterizes the positive (negative) perception of the project idea by staff and their willingness to implement the project.

Summarizing the results of the study on the problem of IIC determination, the following conclusions can be drawn: it is necessary to highlight the essential characteristics of the final result of initiation, determine the factors (circumstances) that characterize this result; give an opinion on the possibility of their assessment and propose a model to assess IIC, which allows the management of a company to analyze the initiation and purposeful synthesis of solutions for the increase the feasibility of their initiatives.

Using the proposed algorithm, the attention should be paid to the distinct initiative result which refers to the project management area. And it is the project concept that comes as the final result of the initiation process when making the decision on the project validity. At the same time, it is worth noting that this statement is justified for the classical management option. When we are dealing with projects of high complexity and those implemented under the conditions of high uncertainty, a reasonable result of initiation goes to an active subject of management. So the most important thing becomes choosing a leader of changes, getting support of top-level leaders and creating conditions for the formation of an environment for team members' self-realization [2]. It is known from business practice that involved employees bring outstanding results for the company if they have

* Corresponding author: svetlana@rmc.edu.ru

the right for their initiatives, as well as support and trust of their leaders and feel confidence in justified assessment of the project results.

This is how the management logic in the Agile approach is characterized, which sets a new perception of the project life cycle: Iterative Life Cycle, an approach that allows using feedback to refine and clarify work in progress [2].

It is necessary to note that the result of initiation, during the transition from the classical model to the post-non-classical model of management, from actions under the conditions of "recognizable future" to actions under the conditions of "complete uncertainty", is increasingly focused on the activation of subjectivity and the creation of an environment for self-development and self-realization of active subjects of management, acting as the main guarantors of the feasibility of project. Thus, there are reasons to interpret the result of initiation as the creation of an organizational environment that provides the required level of strategic activity of management entities, manifested in their ability to form and maintain the required level of intensity of transformations in a company, set by the initiating impulse of changes.

The determination of the factors (circumstances) that characterize this result required the consideration of existing approaches to the solution of such problems. Thus, the study [3] presents the results of the assessment of the influence of a group of key aggregated factors (culture, organization, technology, information) on the readiness of a company for digital transformation. In the work [4], an attempt was made to assess the readiness of the IT department for digital business transformation by the development of a quantitative model to assess the level of the readiness of the IT department. There are existing approaches that use a comprehensive assessment of the readiness of a company, for example, using the EFQM 2020 model [5] in order to assess the ability of a company to implement development projects. In our opinion, the problem of the use of these approaches in order to determine the impulse characteristics of a company is in the fact that they focus more on the assessment of potential and to a lesser extent reflect the act of transforming initiatives and decision-makers into purposeful activities for the implementation of such initiatives.

For example, it is "a model for the assessment of the feasibility of a joint project in a network structure" [6] or "motivation to action" [7] with subsequent stages of its intensification. We made an attempt to form a model for the assessment of the impulse response, using a logical scheme for the formation of activity, which is based on "the five most important decision-making circumstances (readiness for action, responsibility, influence, faith and motives) and give the process either a powerful force or a paralyzing weakness" [7]. At the same time, the very act of impulse formation is considered in the form of three phases: motivation (managerial influence); the synthesis of activity (it includes the interaction of circumstances and the resolution of conflicts) and, in fact, the action, which is characterized by the resulting impulse.

2 Formation of a model for the assessment of the resulting impulse based on the factors of its formation and their relationship

Let us form a cognitive graph that takes into account the considered chains of cause-and-effect relationships leading to the result [8, 9, 10]. This graph in Figure 1 allows representing the formation of the result over time, depending on the level of influence of the leaders. The substantiation and calculation of the values of significant factors are presented below.

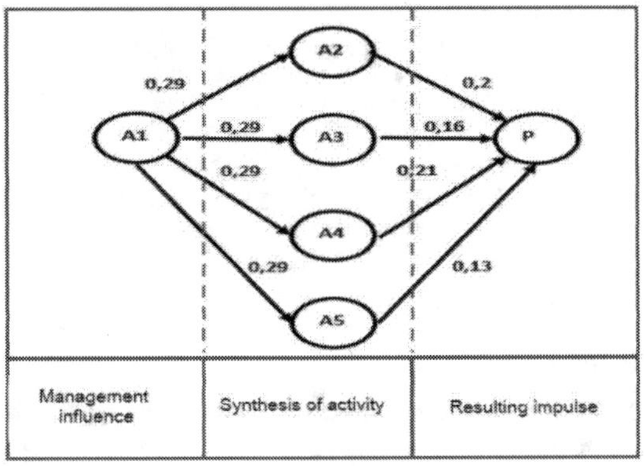

Figure 1. Graph showing the stages of the resulting impulse formation

The assessment of the actual value of the factors (circumstances) will be determined by a questionnaire survey of a group of experts of an enterprise.

Let us consider an approach related to the substantiation of the methodology, which is based on the results of processing field tests of the questionnaires presented below, with the purpose to find the value coefficients of one or another factor that determines the influence on the final result and analyze the cognitive model using these results.

An option of the questionnaire for interviewing experts is presented in Table. 1. The selected scale of expert assessments is from 1 to 5, where 1 is the minimum value of the factor and 5 is the maximum value.

Tables 2-5 present the results of processed field tests questionnaires. A summary of the field tests is given in Table 2 and shows the process of the key result formation. This process assumes accumulating consolidated sum (column Y) for each expert based on his answers to the questionnaire, which ensures the formation of a matrix with data suitable for statistical calculations and factor analysis.

Table 1. Questionnaire for expert survey

№	Influence factor and its meaning	
	The leaders' authority	A_1
1.1	Leaders take the position of a mentor and strive to help and teach. They are always in a dialogue with subordinates, they listen to them and consult them	
1.2	Leaders develop team spirit and dedication of employees	
1.3	Leaders support group activity and provide conditions for the generation and implementation of useful ideas and solutions	
	Level of trust	A_2
2.1	Established development goals of an enterprise are considered as realistic, achievable and credible	
2.2	Full confidence in the fairness of distribution of the results that will be obtained when the development goals of an enterprise are achieved	
2.3	The attitude of employees of a company is characterized by mutual trust, exchange of experience and friendly relations	
	Readiness to act	A_3
3.1	I have complete confidence in the feasibility of the set goals and their achievement.	
3.2	I have all the necessary competencies in full for productive activities, there is the possibility of their development	
3.3	I have the necessary authority and the necessary resources to work productively to achieve my goals	
	Willingness to take responsibility	A_4
4.1	There is a real involvement of personnel in the implementation of decisions for the development of an enterprise and a willingness to take responsibility for the consequences of decisions more than provided by the current instructions	
4.2	Assignment of responsibility for the implementation of projects of high complexity, as a rule, complies with the personal interests of the subject of management (project manager and team)	
4.3	Experience, values, positions of managers and mechanisms that form confidence in the objectivity and fairness of decisions in case of unfavorable outcomes	
	Level of motivation	A_5
5.1	A company has a common understanding and the same interpretation of the standards and regulations laid in the basis of motivation and performance	
5.2	The accepted norms and rates of remuneration are fully justified and fair	
5.3	A company has motivational provisions (incentives) that encourage the search for innovative solutions and their implementation	

To create a multiple regression model, the Regression program was used, which was connected to the numerical editor MS Excel. Filling out the "Regression" form with initial data assumes:
- the description of the input interval Y (indicating the range of cells in column Y of Table 2)
- the description of the input interval X (indicating the range of cells in columns A1-A5 of Table 2)
- the description of the report output parameters (indication of the cell from which the report starts).

Figure 2 shows the results of the regression analysis. For each of the factors, the coefficients A_ (i) were found, the Student and Fisher tests were carried out and the regression statistics were calculated [11].

Table 2. Formation of a key result

Y	A1	A2	A3	A4	A5	№
14	2	2	5	3	2	expert №1
15	2	2	2	4	5	expert №2
18	3	2	3	5	5	expert №3
15	1	5	2	3	4	expert №4
17	3	3	4	4	3	expert №5
18	5	4	5	2	2	expert №6
17	5	1	2	4	5	expert №7
18	5	2	2	5	4	expert №8
15	2	5	3	1	4	expert №9
19	5	3	5	3	3	expert №10
16	4	2	2	4	4	expert №11
11	3	1	1	3	3	expert №12
10	2	2	1	2	3	expert №13
20	5	3	3	5	4	expert №14
16	5	3	1	4	3	expert №15
12	3	1	4	3	1	expert №16
15	3	4	3	1	4	expert №17
14	3	2	4	4	1	expert №18
14	4	2	1	3	4	expert №19
20	4	5	3	4	4	expert №20
14	2	1	3	4	4	expert №21
19	5	3	3	5	3	expert №22
14	4	2	2	3	3	expert №23

From the data in Figure 2, it follows that, according to Student's test, all coefficients are significant (column p - the value contains data equal to 0, which is less than the 0.05 level). According to Fisher's test (the second table, column "Significance F"), we can conclude that the model is generally adequate and reliable with a probability of 0.95 (the significance of F is zero, which is less than the 0.05 level). The correlation analysis (the first table) shows that the static relationship between the key result and the explanatory factors are very strong (normalized $R^2 = 1$, which is greater than the level of 0.7).

Thus, we can conclude that the model is adequate and suitable for further calculations.

Consequently, the hypothesis about the correctness of the formation of the result $Y_i = \sum_{j=1}^{5} A_{i,j}$ has a solid confirmation, and the coefficients of the model are $A_i = 1, i = \overline{1-5},; A_0 = 0$ (column coefficients of the third table of Figure 2).

The next step as the method proceeds to the correlation analysis, the aim being to exclude collinear factors from the model and this principle goes as follows [11]:

- if two variables x_i, x_j are clearly collinear (correlation coefficient $R_{xi,xj} > 0.7$), then we exclude one of them;
- we include the factor that has the least tightness of connection with other factors;
- the closer to 1 the determinant of the cross-factorial correlation matrix, the less the multicollinearity of the factors.

Conclusion of the results

Regression statistics		
Multiple R	1	
R-square	1	
Normalized R-square	1	>0.7
Standard error	5.2384E-16	
Observations	23	

analysis of variance

	df	SS	MS	F	value F
Regression	5	162.869565	32.57391	8.37E+31	3.42E-266
Excess	17	6.616E-30	3.89E-31		<0.05
Total	22	162.869565			

	Coefficient	Standard error	t-stat	p-value	
A 0	-2.6645E-15	8.1941E-16	-3.25178	0.004694	
A 1	1	1.139E-16	8.78E+15	5E-262	
A 2	1	1.2079E-16	8.28E+15	1E-261	<0.05
A 3	1	1.2511E-16	7.99E+15	2E-261	
A 4	1	1.3924E-16	7.18E+15	2E-260	
A 5	1	1.4732E-16	6.79E+15	4E-260	

Figure 2. Regression Analysis Report

In order to calculate the cross-correlation matrix, in the numerical editor MS Excel, we should use the program "Correlation". Filling out the correlation form with initial data assumes:

- the description of the input data (indicating the range of cells of all columns, starting from 1, column Y to 24 of column A5 in Table 3),
- the description of the report output parameters (indication of the cell from which the report starts).

The report on the values of the coefficients of cross-correlation (R) is shown in Table 3

Table 3. Cross-correlation coefficient values (R)

	Y	*A1*	*A2*	*A3*	*A4*	*A5*
Y	1	0.62	0.4	0.34	0.48	0.32
A1	0.62	1	-0.04	0.03	0.37	-0.01
A2	0.43	-0.04	1	0.19	-0.34	0.11
A3	0.34	0.03	0.19	1	-0.08	-0.47
A4	0.48	0.37	-0.34	-0.08	1	0.21
A5	0.32	-0.01	0.11	-0.47	0.21	1

The determinant of the cross-factorial correlation matrix can be found using the built-in function = *MDETERM* (), argument being the range of cells of the explanatory factors. Its value in this case is 0.66, that is slightly less than the 0.7 level. The fact itself testifies the practical absence of collinear factors in the model and suggests ceasing the procedure to further selection of explanatory factors.

In order to determine the value influence of the explanatory factors on the result the following formula is applied as a problem solution:

$$\Delta_j = \frac{A_j S(AJ) R_{Y\,AJ}}{S(y)R} \quad (1),$$

where A_i – regression model coefficients (Figure 1 Table 3 $A_i = 1, i = \overline{1-5}$),

$S(x_i)$ - standard deviation of the factor x_i - can be found using the data in Table 3 and the function = *STDEV ()*,

R_{YXi} - the correlation coefficient of the factor x_i with the main result y (Table 4),

Δ_i –contribution (value) of explanatory factors to the key result

The results of the calculation of the values of the value coefficients of significant factors are shown in Tables 4. Table 5 shows the results of cognitive modeling.

Table 4. The values of the coefficients of significant factors

	Y	*A1*	*A2*	*A3*	*A4*	*A5*
A_j		1	1	1	1	1
R	1	0.62	0.43	0.34	0.48	0.32
S	2,72	1.27	1.26	1.27	1.16	1.11
Δ_j		0.29	0.20	0.16	0.20	0.13
					Total: $\sum_{j=1}^{5} \Delta_j$,	1

Table 5. The calculation results for a strong level of a leader's authority

	0	1	2	3	4	5
A 1	4.5	4.5	4.5	4.5	4.5	4.5
A 2	0	1.30	2.61	3.92	5	5
A 3	0	1.30	2.61	3.92	5	5
A 4	0	1.30	2.61	3.92	5	5
A 5	0	1.30	2.61	3.92	5	5
P	0	0	1.85	2.78	3.54	3.54

Similar calculations were performed for a low and medium level of the leader's authority, which allowed creating the dependencies shown in Figure 3.

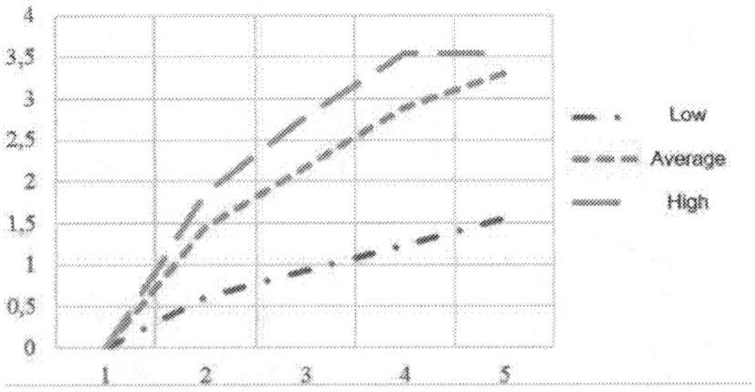

Figure 3. The behavior of the resulting impulse depending on the factor "the leaders' authority"

The results of the calculations made allow to illustrate the rate of change in the transient processes of the formation of stable results of the estimated resultant pulse (on the front edge of the pulse):

- the achieved level of the resulting impulse significantly depends on the assessment of the level and significance of the priority factors of the stimulation of the subjects of activity to proactive behavior and the implementation of purposeful actions;
- the sustainable results with a strong level of influence of managers are achieved 20% faster compared to the average level of influence and more than 2 times compared to a weak level;

- the formation of primary sustainable results with a strong level of influence of managers increases the level of results by 22% compared to the average level and 2.9 times higher compared to the low level.

3 Conclusion

This research confirms the practicability of using the initiating impulse of changes approach which proved its efficiency during assessment of initiating development projects in the company.

At the same time, focusing attention on the sequence of formation of the resulting impulse allows the management of a company to make targeted decisions to accelerate the transition from the formation of models and judgments on activities, directly to activities in order to achieve the result.

The presented method allows obtaining a visual representation of the rate of change of transient processes in the formation of the resulting impulse, depending on the value and significance of the key influencing factors. On the other hand, with a relatively simple procedure for the assessment of factors, it is possible to update quickly the assessment questionnaires, clarify the structure of key factors and the final assessments of impulse characteristics.

The proposed model for the evaluation of the initiating impulse of changes can be used as the basis for the assessment the willingness of a company to initiate development projects. The essence of which is to assess the organizational readiness of a company based on the identification of its ability to create and maintain an initiating impulse of changes necessary for the implementation of projects of various levels of complexity and uncertainty.

References

1. H. Courtney, J. Kirkland, P. Vigerio, *McKinsey Herald: Management Theory and Practice,* **1,** 68-81 (2002)
2. Project Management Institute, *Agile: practical guide,* Retrieved from: www.PMI.org
3. M. Gill, S. VanBoskirk, *The Digital Maturity Model 4.0. Benchmarks: Digital business transformation playbook* (2016),
4. E. A. Isaev, N.L. Korovkina, *Business Informatics*, **2 (44)** (2018)
5. F. A. Meyer *Radarise your business for success: EFQM* (Brussels Representative Office, 2005)
6. M.M. Gakashev, *Regional Economics and Management: Electronic Scientific Journal,* **4,** 20–31 (2017)
7. A. G. Teshnov, *Manage Decisions: How to Think, to Decide and to Act.* (Moscow: FLINT, 2021)
8. V. V. Davnis, V.I. Tinyakova, *Predictive Models of Expert Preferences,* (Voronezh: Publishing House of Voronezh state university6 2005)
9. O. N. Luchko, V.A. Marenko, *Educational Technologies,* **1**, 69–75 (2015)
10. Akatov N.B., Kosyakin S.I., *Bulletin of Perm National Research Polytechnic University: Socio-economic Sciences,* **3,** 109-124 (2019)

SHS Web of Conferences **116**, 00077 (2021) https://doi.org/10.1051/shsconf/202111600077
ICSR 2021

High-potential human resource management in the place marketing system

Lyudmila Simonova[1*] and *Anna* Zyryanova[2]

Tyumen State University, 6 Volodarskogo Street, 625003, Tyumen, Russia

Abstract. Serious competition for potentially valuable workers between the management teams of the regions is currently observed. The aim of the work is to form the concept of high-potential human resource management in the system of place marketing of the region (Tyumen region was taken as an example) . In the "war for talents", a region needs to compete for retaining and attracting primarily such target groups as youth and young families with children. Another significant target group is migrants - former residents of the region, who now live outside the region or country. It is this target group that is in the focus of the study, forming significant elements of the regional talent management as a system that are the ways of interaction between the talent management system and the external environment (region). The online survey, carried out for more than 100 ex-residents of Tyumen region, revealed the prospects of their participation in the regional economy and built the basis of the brand platform of Tyumen Region for this target audience. It is proposed to integrate such HR-marketing tools as EJM (Employee Journey Map), EX (Employee Experience), ERM (Employee Relationship Management) into the management system for this segment.

1 Introduction

Digital transformation provides great business opportunities and explosive economic growth. At the same time, technological and digital transformation is often not supported by the corresponding dynamics of labor productivity and qualitative changes in human resources, which is reflected in the well-known "productivity paradox" formulated by R. Solow based on the analysis of empirical data [1].

This gap between the technological and human factors of production is revealed in acute shortage of valuable personnel - 45% of employers in the world are not able to find a suitable employee, since there are simply no candidates on the market, who would possess the required qualifications, even in populous and rapidly developing China ("talent paradox") [2], in active headhunting of valuable personnel by competitors (executive search), in the growing imbalance between strategic jobs and the skill level and productivity of workers. According to Boston Consulting Group, more than 80% of Russia's working-age population does not have the skills and competencies to work in modern markets [3]. These resource constraints in relation to the quality and supply of human resources in Russia are aggravated by the consequences of the "demographic and qualification hole", long-term brain drain to foreign countries - according to RANEPA, 100,000 people leave Russia every year, 40% of which have higher education [2], an increase in the degree of differentiation of regions and an increase in interregional competition for investment and human resources.

Availability of qualified labor is becoming a systemic advantage that determines the interest of investors in the region. In the struggle for personnel, their training and retention, there is a serious competition between the managerial teams of the regions. Proof may be found in the results of the survey of the employees who work in American companies in Russia, conducted by the American Chamber of Commerce in partnership with the EY company. Among companies planning to expand their regional presence (30% of the total number of respondents), the main selection criterion is the availability of qualified personnel (25%) - this indicator outweighed the proximity to large sales markets (19%), the position of the region in the ranking of investment climate (17%), a developed transport system and personal support of the governor (14% each), tax conditions and benefits (8%). [4].

Both in the world and in Russia a real "war for talent" is unfolding. At the meso-level, this war acquires its own characteristics, but does not reduce its intensity. It is driven by digital transformation, the growing demand for talent and for the mobility of skills. In these circumstances, the task of regional governments is to develop mechanisms for the preservation, retention, accumulation, development and reproduction of human capital on their territory, which is especially important in relation to high-potential human resources.

* Corresponding author: lsim@utmn.ru

In this regard, the purpose of the work is to form the framework (concept) of high-potential human resources management in the regional system of territorial marketing.

3 Materials and Methods

Research in the field of organizational theory, strategy and economic development are focused on how the total human capital of a unit, a company, or the economy turns into a competitive advantage at the level of the firm and provides economic development [5; 6]. In this regard, the general scientific task at which the research is aimed is to study the mechanisms of how human capital influences the socio-economic development of regions. The main reason for this focus is the growing awareness of the fact that human resources and their interconnection are important for the activities of any actor, be it a firm, region, nation or economy. At the same time, the quality of human capital is of primary importance for the competitiveness of the regional economy. A fairly narrow group of people (economic agents) has a significant potential to influence it. Proceeding from the philosophy of human capital management, which is based on its strengths (Strengths-based philosophy), the concept of talent management will be approached as our task is to select "economic agents", groups of people with large human capital stock. The initial theoretical base is the resource theory [7; 8], which considers talent as unique and difficult-to-reproduce resources used to achieve competitive advantages. It is a unique combination of individual-personal and social characteristics that creates such resources and forms the degree of their demand (conjuncture) in the society.

The development of the problematics of talents, their specificity and special characteristics, classification and significance for the competitiveness of organizations has recently become widespread in foreign and domestic scientific literature and forms the modern agenda of global scientific research. Based on a generalization of the most authoritative sources in this area we will clarify the terminology of the basic concepts.[9; 10; 11; 12; 13; 14; 15; 16; 17; 18]

Talents are high-resource workers who can improve the performance of the region either right now, through their direct contribution, or in the long term, thanks to their high potential. In the world practice, such employees are called HiPo (High Potential) and in SHRM theory they are classified as strategic capital. Consequently, our basic concepts to operate are ‘talents’ as special resourceful groups and ‘strategic human capital’ - people who can influence society, the leaders of the intellectual and social development. [19].

Scientists consider talent in a narrow sense as the leaders; or as highly potential and/or highly productive employees [14]; or, in a broad interpretation, as all employees. The broad interpretation of the concept serves best the objectives of this study.

‘Talent management’ is a systematic identification, involvement, development, retention and the right placement of the employees who are valuable for the region due to either their high potential for the future or to what they are able to contribute to the regional economy at present - their specific competencies or the strategic positions they hold.

Increasing competition at the local and global level makes the tasks of attracting, developing, placing and retaining employees considered as talents challenging for company management and regional authorities [21; 11; 14; 20; 22; 23].

Traditionally, talent management has been viewed in relation to a firm or organization. But the research of the whole logical chain from forming individual KSAO («Knowledge, skills, abilities, and: other characteristics») of HiPo employees (nanolevel) to converting them into a competitive advantage of the firm, the industry and the region on the micro and meso levels is of interest [6]. At the same time, the return on human capital, its social performance, that is, the impact on the development and competitiveness of the region, is of fundamental importance.

For this study, the meso- and micro- levels are the most significant for the creation, reproduction and development of strategic human capital. Consequently, the objects of research are economic agents (firms, organizations, individuals), especially, HiPo-employees, and a region as a space of places for the implementation of dominant functions in the network society (according to M. Castells: financial markets, transnational production networks, information networks, network forms of management, social movements). It is the region that provides this space of places and creates an *environment for the* formation, accumulation and reproduction of strategic human capital. In connection with this study it is essential to take into account regional context, i.e. the combination of environmental factors affecting talent management [24]. This regional context acts as a prerequisite and the result of the strategic management of human capital [25].

This aspect, defining methodological specifics of the study, and reflecting the interaction of the regional environment and talent management, is implemented in the system of place marketing. The basic theoretical aspects of territorial marketing and branding are developed in the writings of scholars such as F. Kotler, S. Anholt, S. Raynisto, P.-P. Perre, M. Andersson, G. Trout, A. Rice et al. [26; 27; 28]

Recently, the era of employee power has entered the human resources market. The same applies to high-potential residents of regions, especially provincial ones. The main factor of choice is the region's prospects in terms of opportunities for self-realization and ensuring the required standard of living. Thus, the territory becomes, on the one hand, a product that, as marketers say, must be "well packaged" in order to become attractive to potential “customers”, and on the other hand, a brand with a well-thought-out communication system.

Interregional competition is unfolding for target groups of consumers of resources, who, in their turn, contribute to the development of the region, attracting new resources (people, funding, capital, information) to the region. Tightening of interregional competition - one of the threats already existing objectively – makes it necessary to integrate tools of place marketing into the system of the region's competitiveness management. In terms of managing such a segment of the target audience as human resources with high potential, HR marketing tools are most suitable, namely EJM (Employee Journey Map), EX (Employee Experience), ERM (Employee Relationship Management).

The most important area of competition in the region is the preservation and attraction of human resources. The regional context is studied using the example of the Tyumen region.

Former residents of the region who moved from Tyumen region to other regions of Russia and other countries were selected as the object for this study . The survey was conducted online by filling out a Google survey form, and 101 respondents participated in the survey. The purpose of the survey is to assess the possibility of interaction between Tyumen region and this target group, to explore the prospects for the participation of representatives of this segment of target group "Talents" in the regional economy and form the basis of the Tyumen region brand platform for this segment of the target audience. The objectives of the survey were as follows:

- to compile a socio-demographic portrait of the respondents (gender, age, marital status, educational level, the nature of professional activity, participation in ensuring subsistence)
- to track the geography of the outflow (from which settlement of Tyumen region the person moved, how many years ago, where he moved initially);
- to assess the level of satisfaction of respondents with their current life (quality of life, level of income);
- to identify the presence/absence of connections with Tyumen region at the moment (questions: Is your current occupation connected with Tyumen region?; Do you have any close relatives in the Tyumen region?; Do you keep in touch with the residents of Tyumen region? How often do you come to Tyumen region?)
- to find out the motives of migration;
- to assess the current opinion about Tyumen region;
- to analyze the factors of attractiveness of Tyumen region as a territory of permanent or temporary stay (questions: What was the main motive for your move? Would you recommend Tyumen region for visiting and living? How would you describe Tyumen region for those who have no idea about it?)

Although the data are still being analysed at the moment, the results obtained for each of the groups of questions allowed us to draw some conclusions.

4 Research Results

The bulk (71% of the respondents) were people aged 31 to 45 years. 24% - the age group 20-30 years old, 5% - people over 45 years old. Women accounted for 74% of the total number of respondents, men - 26%. This group of respondents can fully represent the target group "Talents", because 57% of the respondents have one higher education, 29% have more than one higher education, 8% have a candidate of sciences/PhD degree, 1% have a doctor of sciences degree, 5% have incomplete higher education (currently they are university students). According to their occupations, the respondents were divided into the following groups: 24% - office work, 19% - work in a non-governmental institution; 15% - work in a state institution; 13% - entrepreneurs, 9% - self-employed; 6% - creative industries; 2% - researchers; 8% do not work; 4% are studying. At the same time, 77% support their existence independently; 23% depend on spouses or parents.

Thus, we see that only 8% of respondents do not work, 92% are actively involved in various spheres of the economy, which means they acquire valuable experience and develop their human potential.

In order to understand the trends in the geography of migration, the sample should certainly be larger, however, and it allows us to draw some conclusions. In total, 64% were interregional migration (25% moved to Moscow, 9% - to St. Petersburg, 6% - to Kasnodar region, 6% - to Yekaterinburg; 4% - to the north (Khanty-Mansi Autonomous Okrug-Yugra and Yamalo-Nenets Autonomous Okrug); regions of the central Russia - 4%, regions of Russia to the east of Tyumen region - 3%, Tatarstan - 2%, Crimea - 1%). Among the countries attractive for relocation are the European countries (22%). 3% of respondents moved to North America (USA, Canada), 8% - to other countries of the world (Armenia, UAE, Egypt, Turkey, China, South Africa).

An important fact is that most people planning a move consider other regions of Russia, while former Tyumen residents are mainly attracted by Moscow, St. Petersburg, Krasnodar region, Yekaterinburg, and the north (Khanty-Mansi Autonomous Okrug and Yamalo-Nenets Autonomous Okrug). It is with these regions that it is necessary to build a competitive strategy for Tyumen region. It is quite obvious that Krasnodar region and warm countries attract Tyumen residents primarily for their climate.

Tyumen cannot compete with the capitals at the moment either in terms of the level of development of the creative environment (some respondents note that Tyumen is "boring", "nothing to do there", "nowhere to go"), or in terms of income. The same is with the countries of Europe and North America, where even a higher level of comfort and infrastructure development is observed. Accordingly, the only correct strategy of interaction with these regions and countries will not be competition, but partnership. In particular, the involvement of migrants in the activities of the region should be implemented remotely. In our opinion, this is an underestimated resource of Tyumen region.

Firstly, 100% of respondents keep in touch with friends and former colleagues - residents of Tyumen region, 97% have close relatives in Tyumen region, only 5% do not come to Tyumen region since they moved, the rest regularly come (38, 2% - more than 2 times a year, 38.2% - not more than once a year, 18.6% - once in several years).

Secondly, the majority of respondents demonstrated loyalty, describing their native region, and would recommend to visit Tyumen region to those, who has never been there, at least as tourists. And many say that if you want to work in the oil and gas industry, realize yourself in entrepreneurship, it is worth coming and living in Tyumen and the region, as it is “a rich, dynamic region with gorgeous nature”, “like Moscow, only smaller”, “a provincial cozy city that has everything for a happy life”, “a beautiful Siberian region with a good university”.

Thirdly, many called local residents “the main advantage” of Tyumen region. “Kind, easy-going, very adequate, modern, cool, talented, strong-willed, caring people who strive for development and will always come to the rescue” - such a collective image of the inhabitants of the Tyumen region was formed from the answers to the question: “How would you describe Tyumen region to those who have no idea about it?” In general, it should be noted that many of the answers were detailed and sincere, which reflected general positive attitude of the respondents and readiness to interact with the region.

Another unexpected discovery of our study is the proximity and reach of the representatives of the studied segment of the target group "Talents", despite the distance and years of residence in other places. 75 responses were received within the first 24 hours after the survey was published, the remaining 26 responses were received within the next three days. 40% expressed interest in receiving the survey results and left their email addresses, which in today's conditions of huge information flows and widespread spam, we consider a very good indicator. 15% of the respondents offered to be interviewed if necessary, although there was no such question in the survey.

Thus, we believe that systematic work using place marketing tools that can facilitate the involvement of talent in the economy of Tyumen region and make a tangible contribution to the further development of the region, might bring fruitful results.

5 Discussion

To build an effective system of interaction with consumers of the target group "Talents", it is proposed to introduce a CRM system (Customer Relationship Management) as a philosophy of building long-term productive relationships with representatives of this group at different stages of their interaction with the territory. In the work "Talent beyond Capitals"[28], the authors even propose to replace the familiar term CRM with "TRM" (Talent Relationship Management) and identify five main stages of interaction between talents and the territory (see Figure1).

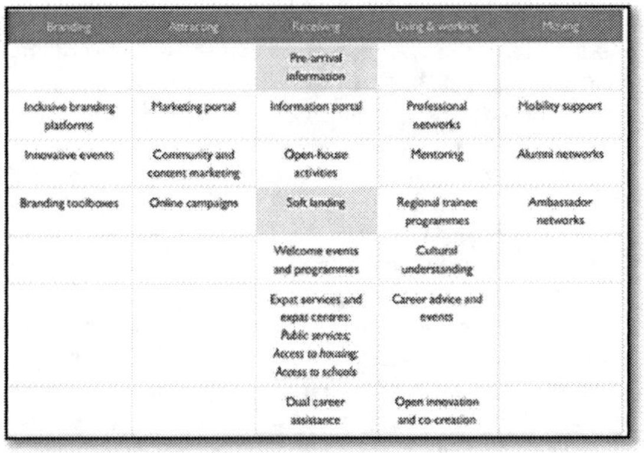

Branding	Attracting	Receiving	Living & working	Moving
		Pre-arrival information		
Inclusive branding platforms	Marketing portal	Information portal	Professional networks	Mobility support
Innovative events	Community and content marketing	Open-house activities	Mentoring	Alumni networks
Branding toolboxes	Online campaigns	Soft landing	Regional trainee programmes	Ambassador networks
		Welcome events and programmes	Cultural understanding	
		Expat services and expat centres; Public services; Access to housing; Access to schools	Career advice and events	
		Dual career assistance	Open innovation and co-creation	

Fig. 1. Stages of interaction of the territory with talents, proposed by the author of the concept of Talent Relationship Management [28]

Talent Relationship Management - a system of managing relationship with talent - is based on the principles, identical to CRM (Customer Relationship Management). This is not just a one-time "sale", it is a system of long-term relations aimed at mutually beneficial cooperation. Currently, recruiting companies have begun to actively develop technologies for attracting talent and use ERM (Employee Relationship Management) by analogy. In the Russian practice of regional management such targeted programs do not exist yet.

Customer Journey Map (CJM) – a roadmap of a customer in CRM – consists of three main "points" - stages of interaction between the company and the consumer: 1. Awareness stage; 2. Consideration Stage; 3. Decision Stage [29].

Similarly, a region should have a journey map of talent (or high-potential human resources) - the stages of interaction between territory and talent, which allows tracing the logic of "hiring" the place as a product of place marketing.

The overall strategy for creating a map of the talent journey from initial understanding to productive interaction with the territory, as well as the tactics of "guiding the talent along the route" must be adapted to the needs of each target group, combining the tools of territorial marketing in such a way that they best contribute to the achievement of marketing goals and objectives. allowed to form the desired image of the territory and led to the desired models of behavior of talents.

Key performance indicators of marketing a region in relation to the target group "Population" in general will be the following: accumulation of the population, migration inflow, growth in the wealth of citizens, a decrease in population differentiation by income level, a favorable demographic portfolio.

For the Talents segment, it is important to add such specific indicators as:

- inflow of highly qualified personnel,

- participation of highly qualified personnel from other regions in solving the problems of socio-economic development of the target region,

- duration of stay of the talent in the territory / amount of participation in hours

- economic, social, scientific, innovative or other result of the activity of a talent in the target region during the period of stay,

- the cost of attracting and maintaining talent in Tyumen region

- the economic effect of stay for the talent (income, expenses)

- in the long term, LTV - life-time value as an indicator of loyalty - the number of projects implemented with the participation of talent in the region.

6 Conclusion

The basic principles of talent management that characterize their interaction with territories may be formulated as follows:

- the influx of talent in the region, equivalent to an increase of its competitiveness;

- talents are mobile, so it is not reasonable to try and keep them forever;

- the relocation of talent (migration) does not mean its complete exclusion from the economic activity of the territory;

- the reciprocity of the hiring process "talent-region": we hire them, and they hire us to realize their potential, to earn money (possibly short-term), for a comfortable life, to be able to travel, etc. The "win-win" principle should become the basis of this relationship. Both parties must win.

To build an effective system of interaction with talents it is proposed to introduce a CRM-system (Customer Relationship Management) as a philosophy of building long-term productive relationship at different stages of their interaction with the region.

The competent analysis of the region's competitiveness factors and factors of choice of a region by target groups, the use of place marketing tools that meet the requirements of each stage of the "consumer's journey", as well as the integration of a kind of CRM systems into work with consumers of a territory- whether they are tourists, talents, the population, entrepreneurs, investors - will allow regional authorities to effectively manage the behavior of target groups based on the data collected at each stage.

In post-pandemic space the consumers' behavior and the consumption of territories as products will change. There will be a reassessment of the value orientations of target groups and the competitive advantages of a particular territory, and that may result in the ability of peripheral regions to receive an inflow of high-quality human capital.

It is not always advisable to fight for the physical presence of talent on the territory. Technically, the interaction of the territory with talents in the interests of the territory is also possible at a distance. This becomes especially relevant when the talent comes from this region. According to the research, such people are the most open to suggestions, they are much easier to engage. It is they who are able to increase the competitiveness of the region.

Acknowlegements

The reported study was funded by RFBR, project number 19-29-07131.

References

1. Solow R. *Rev. of Ec. and Stat.*, **39,** 3 (1957)
2. E. Vinogradova, *Vedomosti* (Dec. 20, 2019)
3. A. Auzan, *Novaya Gazeta,* **130** (Nov. 22, 2017)
4. N. Bleiman, *RBC plus,* **1-2 (154),** 122-127, (2020)
5. N. Gennaioli, R. L. Porta, F. L. de Silanes, and A. Shleifer, *The Quart. J. of Ec.*, **128**, 1 (2013)
6. C. E. Helfat, et al. *Dynamic capabilities: Understanding strategic change in organizations* (John Wiley & Sons, 2009)
7. E. T. Penrose *The Theory of the Growth of the Firm* (Basil Blackwell: London, 1959)
8. J. B. Barney *J. of Manag.*, **17**, 1 (1991)
9. P. Casse, *J. of Eur. Ind. Tr.*, **18,** 5 (1994)
10. G. K. Stahl, I. Bjorkman, E. Farndale, S. S. Morris, J. Paauwe, P. Stile, *Global Talent Management: How Leading Multinational Build and Sustain their Talent Pipeline* (INSEAD Fontainebleau: 2007)
11. P. Cappelli, *Talent on Demand: Managing Talent in an Age of Uncertainty* (Boston, 2008)
12. D. G. Collings, K. Mellahi *H. R. Manag. Rev.*, **19,** 4 (2009)
13. S. Swailes, *Bus. Ethics: A Eur. Rev.*, **22,** 1 (2013)
14. N. Dries, *H. R. Manag. Rev.*, **23,** 4 (2013)
15. A. R. Malik, P. Singh, *H. R. Manag. Rev.*, **24,** 4 (2014)
16. W. F. Cascio, J. W. Boudreau, *J. of W. Bus.*, **51**, 1 (2015)
17. C. Kontoghiorghes, *Int. J. of H. R. Manag.*, **27**, 16 (2016)
18. A. McDonnell, D. G. Collings, K. Melahi, R. Schuler, *Eur. J. of Int. Manag.*, **11**, 86-128 (2017)
19. L. Czaller, *Region. Stud.*, **51**, 3 (2017)
20. I. Bjorkman, M. Ehrnrooth, K. Makela, A. Smale, J. Sumelius, *H. R. Manag.*, **52**, 2 (2013)
21. J. W. Boudreau, P. M. Ramstad, P. J. Dowling, *CAHRS WP02-21* (2002)
22. P. R.Sparrow, H. Makram *H. R. Manag. Rev.*, **25,** 249-263 (2015)
23. J. H. Marler, J. W. Boudreau, *Int. J. of H. R. Manag.* **28,** 1 (2017)
24. V. Vaiman, H. Scullion, D. Collings, *Manag. Decis.* **50**, 5 (2012)
25. C. Brewster, *Manag. Int. Rev.* **39**, 45-64 (1999)
26. S. Anholt, *Destination Branding: Creating the Unique Destination Proposition* (Oxford, 2002)

27. S. K. Rainisto, *Success Factors of Place Marketing: A Study of Place Marketing Practices in Northern Europe and the United States* (Helsinki University of Technology, 2003)
28. P.-P. Pere, *Talent beyond capitals. Handbook.* (Stockholm: Future Place Leadership, 2018)
29. B. D. Temkin, *Mapping the customer journey* (Forrester Research Inc., 2010)

Business process model with "RandomForest" algorithm

Anatoly Kukartsev[1,2,*], *Alexandra* Fedorova[1], *Yuriy* Danilchenko[1], *Konstantin* Lobkov[2,3] and *Evgeniia* Korepanova[2]

[1]Siberian Federal University, 79, Svobodny Av., Krasnoyarsk, 660041, Russian Federation
[2]Reshetnev Siberian State University of Science and Technology, 31, Krasnoyarsky Rabochy Av., 660037 Krasnoyarsk, Russian Federation
[3]Krasnoyarsk State Agrarian University, 2 Kirenskogo, Krasnoyarsk, 660074, Russian Federation

Abstract. Currently, the areas of application of machine learning are multifaceted: artificial intelligence, financial applications, bioinformatics, intellectual games, speech and text recognition, computer language processing, medical diagnostics, technical diagnostics, text search and rubrication. Machine learning is an area of scientific knowledge associated with learning-capable algorithms. The use of machine learning methods is explained by the fact that for most intelligent complex tasks (for example, speech recognition, etc.) it is almost impossible to develop an obvious algorithm for their solution. However, you can teach a computer to learn how to solve such problems. In our article, we propose a model based on the machine learning algorithm "RandomForest", which allows one to recognize bots by HTTP sessions. The chosen algorithm provides many advantages: non-iterative learning; high quality of the resulting models (comparable to neural networks and ensembles of neural networks); a small number of adjustable parameters. It works well with missing data (retains good accuracy); internal assessment of the generalizability of the model; able to work with raw data without preprocessing. The algorithm was trained on a dataset of more than 5000 sessions. The prospects of this direction are obvious, since robotic traffic accounts for more than 40% of the total Internet traffic.

1 Introduction

The new digital world assumes a gradual transition of business to the online space. The 2020 quarantine measures accelerated this process. The concept of "transferring activity to the online" appeared, thus the business underwent a digital transformation [1-3]. It helped companies that had succeeded in the quick restructuring to stay afloat, to take the audience of less prepared competitors. But has the activity of people only increased in the online space? After reviewing data from traffic data, Imperva'sThreatResearchLab [1] produced a report for 2020 that shows an increase in network activity of automated means of interacting with content on websites. In this case, the share of human activity stands out - 62.8%; good bots - 13.1% and bad bots - 24.1% [4-6]. It should be noted that bad bot traffic has grown significantly and, at the moment, it is almost a quarter of all Internet traffic [7-9]. The damage is done to both businesses and individuals. In this regard, it is imperative to develop processes for filtering unwanted activity on websites [10, 11].

A bot is software designed to simulate the behavior of a real user on online social networks [2]. Many methods have been proposed for detecting and removing them. Research is carried out on the characteristics, detection and assessment of the impact of bots.

There are various techniques for detecting web bots in network traffic: limiting the frequency of requests to a host; checking blacklists of IP addresses; parsing the value of the User-Agent HTTP header; taking prints of the device; CAPTCHA implementation; behavioral analysis of network activity using machine learning [3].

Bot traffic analysis tools have become a must for any website.

2 Materials and Methods

The process of detecting bots by HTTP session. Session is a sequence of requests from one node (unique value of the IP address and the User-Agent field in the HTTP request) in a fixed time interval [3, 12]. The session will be evaluated offline using the data uploaded to the file, which will be divided into testset and trainingset, see Figure 1. This model uses key concepts:

-A random sampling of samples from a dataset when constructing trees.

-When dividing nodes, random sets of parameters are selected.

"RandomForest" optimization can be done through random search using the Randomized Search CV in Scikit-Learn. Optimization involves finding the best hyper parameters for the model on the current dataset (Figure 2) [14].

* Corresponding author: kempf@sibsau.ru

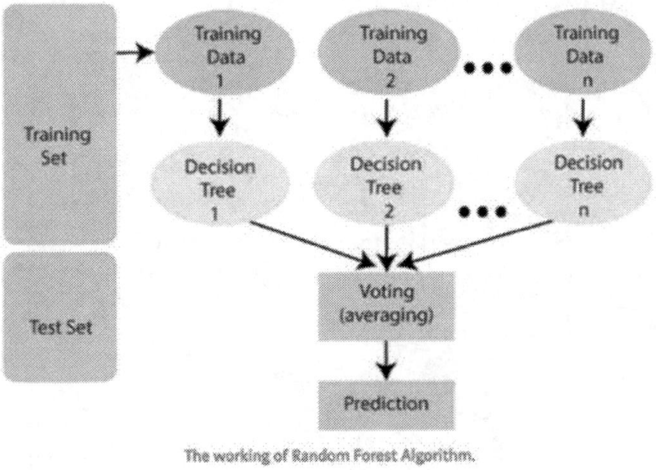

Fig. 1. Schematic diagram of the operation of the "RandomForest" algorithm.

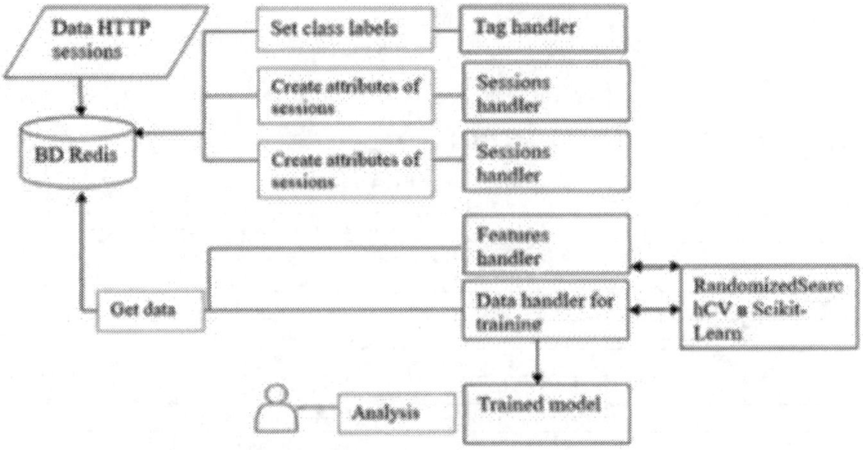

Fig. 2. Model of the process of working with data from HTTP sessions.

There is a significant difference from other models ; in this case, feature weighting is not used ; therefore any monotonous transformations will not affect the result (up to a numerical error) [4].

The algorithm for constructing a RandomForest consisting of N trees is as follows: For each n = 1, ..., N: Generate Xn fetch with Bootstrap.

Build the decision tree bn from the sample Xn:

According to a given criterion, we choose the best feature, make a partition in the tree according to it, and so on until the sample is exhausted.

The tree is built until each leaf contains no more than nmin objects or until we reach a certain tree height.

At each partition, firstly, m random features are selected from the initial ones and the optimal division of the sample is sought only among them.

3 Results and Discussion

The final classifier is a (x) = 1N∑i = 1Nbi (x), in simple words - for the classification problem we choose the solution by voting by the majority, and in the regression problem - by the average [9].

It is recommended to take m = n in classification problems, and m = n3 in regression problems, where n is the number of features. It is also recommended in classification problems to build each tree until each sheet contains one object, and in regression problems - until each sheet contains five objects.

Thus, a RandomForest is Boots trapaggregating over decision trees, during training of which, for each partition, features are selected from some random subset of features [5].

Let's create a simulation model for calculation with the main metrics of the unit-economy (Figure 3) [13-15].

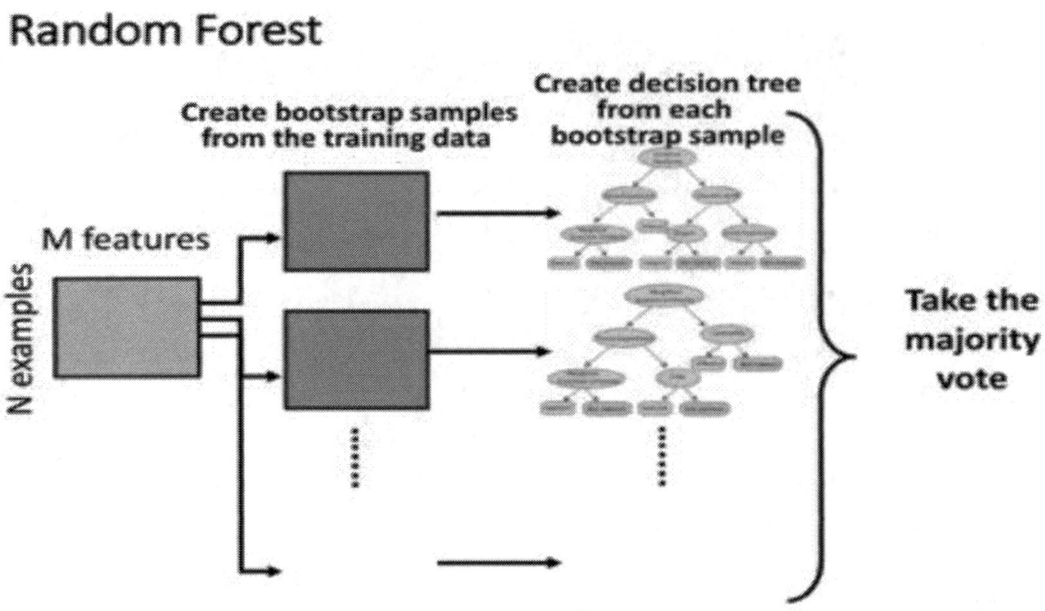

Fig. 3. Schematic diagram of the "RandomForest" algorithm.

We use session characteristics as parameters (Table 1).

Table 1. Session characteristics

n	Number of requests per session
n_pages	Number of requests per session in pages (URI ends with .htm, .html, .php, .asp, .aspx, .jsp)
n_static_request	Number of requests per session in static pages
n_sec	Session time in seconds
n_unique_uri	Number of requests in a session with a unique URI
headers_cnt	Number of headers per session
has_cookie	There is a cookie header
has_referer	There is a Referer header
(avg)mean_time_page	Average time per page per session
(avg)mean_time_request	Average time per request per session
(avg)mean_headers	Average number of headers per session

We will take 0 BOT and 1 PERSON as labels; that is, we will build a clear binary classification model: network activity is created by a web bot (bot label); network activity is created by a human (human label) [10].

In this case, we use quality metrics: accuracy, completeness and F-measure (Table 2, Figure 4).

The number of sessions is over 5000.

Data are divided into two subgroups: training and test. From the training data, we get the model.

To create and train the algorithm, we use:

```
fromsklearn.ensembleimportRandomForestClassifier
# Create a forest model from hundreds of trees
model = RandomForestClassifier(n_estimators=100,
bootstrap = True,
max_features = 'sqrt')
# Train on training data
model.fit(train, train_labels)
```

After a few minutes of training, the model is ready to make predictions for the test data:

```
# Current classification
rf_predictions = model.predict(test)
# Probabilities for each class
rf_probs = model.predict_proba(test)[:, 1]
```

Table 2. Training result

	Average accuracy	Average fullness	Average F-measure
bot	0.84	0.8	0.86
human	0.96	0.95	0.95

We will calculate the classification predictions (predict) along with the predictive probability (predict_proba).

```
from sklearn.metrics import roc_auc_score
# Calculate rocauc
roc_value = roc_auc_score(test_labels, rf_probs).
```

To test the effectiveness of the model, we consider the results..

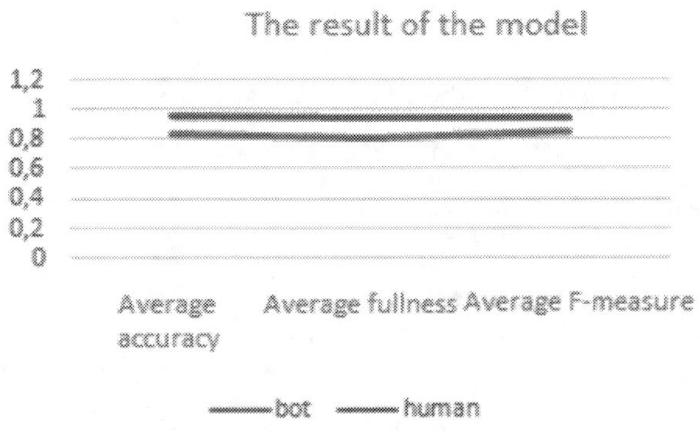

Fig. 4. Operation of the model with lazy fetch.

On lazy sampling, the values of the quality metrics are almost identical to the values of these metrics when the model requirements are met. It follows from this that the model based on these data can generalize the knowledge gained during training [7, 15].

Let's consider errors of the 1st kind. If these data are marked out expertly, then the error matrix will change significantly. This means that some errors were made when marking up the data for the model, but the model was still able to recognize such sessions correctly (Table 3, Figure 5) [15].

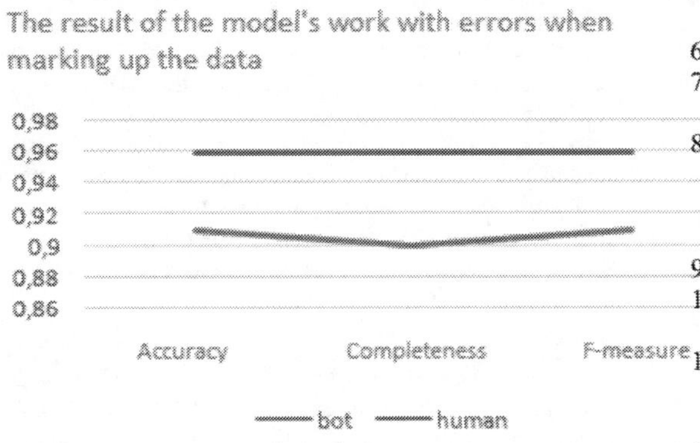

Fig. 5. Operation of the model with lazy fetch.

Table 3. Training result.

	Accuracy	Completeness	F-measure	Number
bot	0.91	0.9	0.91	1223
human	0.96	0.96	0.96	4220

5 Conclusions

As a result, an approach to the detection of web bots using a machine learning algorithm and the use of statistical features is described. The "RandomForest" machine learning algorithm is selected as the most appropriate. Implementation of the similarity of the model for recognizing bots by session in Python using the scikit-learn machine learning library [6]. A classical measure is used to assess the accuracy. 11 features were used to train the algorithm. The trained algorithm was tested on a random data set and the F-metric value was 0.98 [8].

References

1. E. Roberts, Imperva blog (2020)
2. A. Cabri, *IEEE 20th Int. Conf. on HPC. and Com.* (2018)
3. Z. Chu, S. Gianvecchio, H. Wang, *LNCS.* **6**, 11170 (2018)
4. A.A. Boyko, V.V. Kukartsev, E.S. Smolina, V.S. Tynchenko, Y.I. Shamlitskiy, N.V. Fedorova, *Jour. of Phys.: Conf. Ser.* **1353**, 012124 (2019)
5. P. De Meo, E. Ferrara, G. Fiumara, A. Provetti, *ACM.* **11**, 78 (2014)
6. N.D. Lyfenko, *Cyber security issues* **5**, 17 (2014)
7. F. Pedregosa, G. Varoquaux, A. Gramfort, *The J. of Machine Learning Research* **12**, 2825 (2011)
8. A.V. Kukartsev, A.A. Boyko, V.V. Kukartsev, V.S. Tynchenko, V.V. Bukhtoyarov, S.V. Tynchenko, *IOP Conf. Ser.: Mat. Sci. and Eng.* **537**, 042009 (2019)
9. L. Demidova, M. Ivkina, *SUMMA.* **359** (2020)
10. Y. Su, K. Weng, C. Lin, Z. Zheng, *IEEE Access* **9**, 9142 (2021)
11. A.A. Boyko, V.V. Kukartsev, D.V. Eremeev, V.S. Tynchenko, V.V. Bukhtoyarov, A.A. Stupina, *Jour. of Phys.: Conf. Ser.* **1353**, 012123 (2019)
12. I.R.H.T. Tangkawarow, J. Waworuntu, *IOP Conf. Ser.: Mat. Sci. and Eng.* **128**, 012010 (2016)
13. T. Zhu, *Jour. of Phys.: Conf. Ser.* **1607**, 012123 (2020)
14. A.A. Boyko, V.V. Kukartsev, V.S. Tynchenko, D.V. Eremeev, A.V. Kukartsev, S.V. Tynchenko, *Jour. of Phys.: Conf. Ser.* **1353**, 012138 (2019)
15. V. Khareva, O. Voronova, T. Khnykina, *IOP Conf. Ser.: Mat. Sci. and Eng.* **940**, 012057 (2020)

Modeling the cost of transferring funds taking into account the initial information

Dmitry Eremeev[1,*], *Galina* Zolotareva[1], *Natalia* Kazanskaya[2], *Natalia* Fedorova[1,3] and *Elena* Vaitekunene[1,3]

[1]Reshetnev Siberian State University of Science and Technology, 31, Krasnoyarsky Rabochy Av., 660037 Krasnoyarsk, Russian Federation
[2]Russian Academy of National Economy and Public Administration under the President of the Russian Federation, 57/43, Sredny prospect VO, 199178, St. Petersburg, Russian Federation
[3]Siberian Federal University, 79, Svobodny Av., Krasnoyarsk, 660041, Russian Federation

Abstract. The construction of a model of the optimal amount of money transfer, taking into account the uncertainty of the initial information, depends on the strategy practiced by the enterprise. The type of the strategy of the enterprise, first of all, depends on the choice of the decision-making criterion in conditions of uncertainty. For a machine-building enterprise, a priority strategy is such that the amount of money transferred should be equal to the amount of the cash gap. In this paper, using economic and mathematical methods for the model of forming an optimal portfolio of alternatives, a mathematical model for determining the amount of optimal transfer of funds, based on the construction of the function of transaction and opportunity costs, is developed. We also propose a method for calculating the absolute liquidity ratio, which is recommended for use not only in operational financial planning, but also for analyzing the current financial condition of an enterprise.

1 Introduction

The rise and development of mechanical engineering, the restoration of solvency, and the increase in investment attractiveness in order to ensure the competitiveness of products in the difficult economic conditions of the present period are impossible without improving the management system of the financial and economic activities of the enterprise [1-3]. Market transformations that have changed the internal and external environment of the functioning of economic entities has required more attention to the issues of managing the financial activities of the enterprise, which are the most important component of the entire management system [4-6]. The basis for improving the management of the company's financial activities is effective financial planning, which is present at all levels of planning. Along with the strategic and current levels of planning, more and more attention is paid to operational financial planning, which largely determines the success of the development and operation of the enterprise. The results of numerous studies [7-9] the lack of a system for planning financial flows and, above all, operational financial planning, as a planned receipt and expenditure of revenue from the sale of products [10], in many cases are the main reasons for the insolvency of Russian enterprises, which ultimately leads to bankruptcy. Thus, improving the efficiency of financial flow management, solving the problem of insolvency of the enterprise are impossible without the implementation of the process of operational financial planning. The solution to this problem is connected with the need to improve the approaches and methods of operational financial planning [11, 12], considering the industry specifics of Russian enterprises.

2 Materials and Methods

The results of scientific research of Russian and foreign scientists were used as the initial materials. Information from available sources describes the activities of machine-building enterprises in the Russian Federation.

The methodological basis of the research consists of the following methods: economic analysis; system analysis; mathematical statistics; economic and mathematical methods and models; methods of decision-making in conditions of uncertainty and risk.

3 Results

The data set, which shows the relationship between the amount of funds raised and the amount of transaction costs, is the input information for building a model of the optimal amount of money transfer.

To determine the functional relationship between the amount of funds raised and the amount of transaction costs, a graph is constructed using the available source information. The resulting point values on the graph must be approximated using one of the methods of mathematical analysis. In this case, the following methods can be used [5, 6]: uniform approximation. It is theoretically feasible to require for the best

* Corresponding author: eremeev.dmitriy@gmail

https://doi.org/10.1051/shsconf/202111600079

approximation, to maximize the value of $|f(x) - \phi(x)|$ on that interval $a \le x \le b$, where you want to get an approximate image $f(x)$, which would be the smallest (compared to the other choice $\phi(x)$). However, there is no method for efficiently obtaining such uniform approximations, except in certain special cases.

The second is the approximate least squares method. The most commonly used approximation is $\phi(x)$, for which the smallest value is value (1):

$$M = \int_a^b [f(x) - \varphi(x)]^2 dx \tag{1}$$

By requiring the reversal to zero of the partial derivatives of M in terms of the parameters, defining the function $\phi(x)$, we obtain an equation that allows us to find the best (in the specified sense) value of these parameters [13, 14].

The third is zoom in on individual points. In many cases, and especially if the function $f(x)$ is given by a graph or table to estimate the degree of approximation, the differences are considered $f(x) - \phi(x)$ not for all points of the interval (a, b) on which you want to approximate the function $f(x)$, but only for individual, pre-selected points x1, x2, …, xn. The function $\phi(x)$ is considered as the best approximation to $f(x)$ (using the least squares method), if (2):

$$S = \sum_{i=0}^{n} [f(x_i) - \phi(x_i)]^2 \tag{2}$$

It has the smallest value in comparison with other functions, among which the desired approximation is selected. If $\phi(x)$ is completely determined by the parameters k, l, m, …, then the best (in the specified sense) values of these parameters will be found by solving the system of equations (3):

$$\frac{\partial S}{\partial k} = 0, \frac{\partial S}{\partial l} = 0, \frac{\partial S}{\partial m} = 0, \tag{3}$$

The fourth is the approximation by the method of averages. Using the method of averages, the linear relationship between the "aligned" variables X and Y is first determined (obtained by the alignment method): Y=aX + b. To do this, the conditional equations Yi=aXi + b for available value pairs Yi and Xi are divided into two equal (or almost equal) groups in an ascending order of the variable Yi or Xi. Adding up the equations of each group, we get two equations, from which a and b are determined. Expressing X and Y in terms of the initial variables, we get the desired relationship between x and y. If at the same time not all the parameters are defined, then you should apply the same method again, aligning the other values of X and Y.

Note. For an approximate image of a given function $f(x)$ we choose the approximating function $\phi(x)$ from functions of a certain type.

The function obtained in this way is shown in Figure 1.

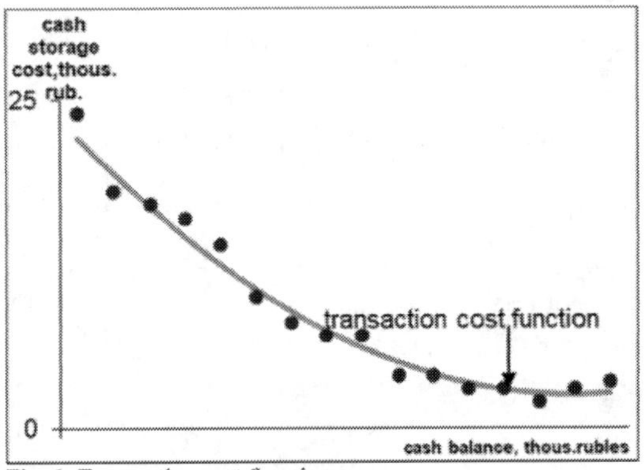

Fig. 1. Transaction cost function.

-To construct the opportunity cost function, it is necessary to obtain data on the relationship between the amount of money placed and the income received at the same time. In the dissertation research, the task of forming an optimal investment portfolio was not set, and therefore the initial data for constructing the function will be third-party.

A certain functional dependence is shown in Figure 2.

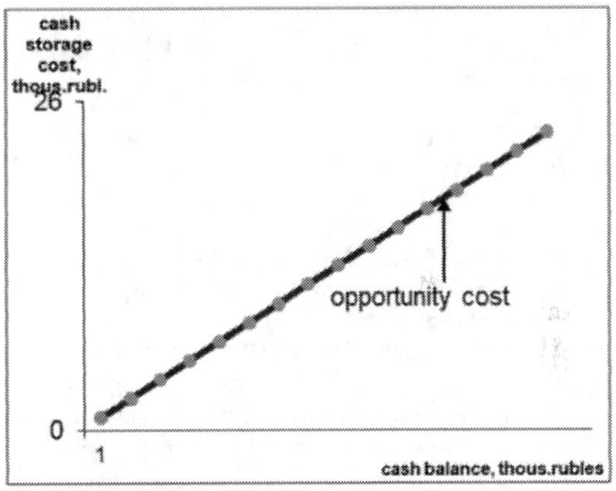

Fig. 2. Opportunity cost function.

-The optimal amount of money transfer can be determined by taking into account some competition between the desire for profit and security. With the increase in the liquidity reserve, the growing amount of opportunity costs is countered only by the costs of eliminating the shortage of means of payment – transaction costs. The optimal amount of money transfer is obtained when the sum of transaction and opportunity costs (a function of total costs) reaches its minimum. To find the minimum value of the function, you need to find the points where the first derivative is zero. Thus, the extremum points were determined. Then you can calculate the second derivative. If the found value is greater than zero, then the local minimum is located at this point. We can say that the point of correspondence to the amount of the optimal transfer of funds is not only the minimum value of the function, but also the point of intersection of the functions of alternative and transaction costs. As a result, it can be found by solving the system of equations (4):

$$\begin{cases} F=T(V) \\ F=A(V') \end{cases} \quad (4)$$

We obtained value of the optimal transfer of funds, determining the portfolio of alternatives by solving the dynamic programming problem [15]. The uncertain nature of the information about the receipt and expenditure of funds in the original payment calendar makes it necessary to take into account the uncertainty. For this purpose, a payment matrix is formed (Table 1).

Table 1. Payment matrix.

Alternative Portfolios	Total cash balance					
	Δ_1	Δ_2	Δ_3	Δ_4	…	Δ_m
1	Z_{11}	Z_{12}	Z_{13}	Z_{14}	…	Z_{1m}
2	Z_{21}	Z_{22}	Z_{23}	Z_{24}	…	Z_{2m}
…	…	…	…	…	…	…
n	Z_{n1}	Z_{n2}	Z_{n3}	Z_{n4}	…	Z_{nm}

Z_{nm} - value of total costs, according to the selected portfolio of alternatives for each total cash balance $\Delta_1<\Delta_2< \dots <\Delta_n$.

To use the constructed payment matrix, it is necessary to make a choice of the decision criterion, based on the propensity of the decision-maker to risk.

It is proposed to choose a solution based on one of four criteria [7-10].

The Wald decision criterion, or maximin, is a criterion of conservatism and an attempt to maximize the level of reliability. It presents the external conditions and makes the assumption that Murphy's law is fully confirmed. Therefore, by this criterion, it is necessary to determine the worst possible outcome of each strategy, and then choose the strategy that provides the best of the worst outcomes.

The Hurwitz decision alpha criterion involves determining the decision index d for each strategy, which is the weighted average of its extreme returns. The weighting factors are the coefficients of optimism α, which is applicable to the maximum return M, and its complement 1 -α, which is applicable to the minimum return, m. The cost of each strategy is thus equal to (5):

$$di= \alpha M+ (1-\alpha)mi. \quad (5)$$

The strategy with the highest cost for di is selected as the optimal one. The coefficient of optimism is in the range from 0 to 1, which allows the decision-maker to express their subjective attitude to risk with a certain degree of optimism.

The Savage decision criterion, sometimes called the mini-max loss criterion, examines losses that represent losses resulting from making a wrong decision. Loss is measured as the absolute difference between the return for a given strategy and the return for the most effective strategy within the same state.

There is a Beis postulate, which states that if the probabilities of a phenomenon are unknown, then they should be taken equal. In the Laplace decision criterion, this postulate is used to calculate the estimated cost of any strategy; therefore, the Laplace criterion is also called the Base criterion. The chosen strategy is the strategy with the highest estimated cost, assuming equal probabilities [16].

Based on the selected criteria, the payment matrix determines the final portfolio of alternatives and the amount of money transferred. The most difficult task for the decision-maker is to choose a specific criterion that is most suitable for solving the proposed problem. The choice of criteria should be logical under the circumstances. In addition, the selection criteria should take into account the philosophy, temperament and outlook of the company's management. This requires that decision-makers have sufficient knowledge about the possible conditions of the external and internal environment of the enterprise and the resulting benefits.

4 Discussion

In the study, as such a criterion for assessing solvency in a given period of time and for drawing up a payment calendar, the standard of absolute liquidity was proposed, showing the degree of coverage of current liabilities. The calculation of the absolute liquidity ratio is presented below:

$$\Psi = \frac{(A+P)}{C} \quad (6)$$

where A is the cash inflow; P - rolling cash balance at the beginning of the period; C - cash outflow.

The recommended value Ψ has the following limits: $1 \le \Psi \le b$, where b is the upper value of the absolute liquidity ratio, which is set by the company itself based on established practice.

The resulting standard of absolute liquidity can be used in operational financial planning and in the analysis of current financial activities, but it must be recalculated if there is a change in the operating conditions of the enterprise [17, 18].

Testing of the proposed methodology for planning highly liquid assets of the enterprise was carried out on the basis of JSC "DZNVA". The specifics of the activity of this enterprise is the production of products and goods of electrical engineering: low-voltage equipment, consumer goods. The proposed methodology for planning highly liquid assets at the enterprise was used by JSC "DZNVA" when developing an operational financial plan for 1 month.

At the initial stage, the company carried out work to determine the data on the basis of which the original payment calendar was compiled. The necessary data was taken from the information bases of the financial, planning department and accounting department [19]. The constructed payment calendar (table. 2) has a four-level method of specifying information due to a significant interval of uncertainty in the amount of the cash balance. Such a high share of uncertainty is due to the practice of selling products on the basis of prepayment at JSC "DZNVA" and as a result, it is impossible to determine the exact portfolio of orders for materials, components and services that need to be placed, received and brought to the enterprise. Taking into account the real possibilities in the implementation of alternatives allowed us to form an optimal portfolio of alternatives for the transfer of funds, which is presented in table 2.

Table 2. Structure of the optimal portfolio of alternatives for the transfer of funds

Alternatives / Portfolio amount (thous. rubles)	Amount of the sale of receivables (thous. rubles)	Amount of restructuring of accounts payable (thous. rubles)	Amount of securities sold (thous. rubles)	Amount of the loan received (thous. rubles)
428	428	0	0	0
512	500	0	0	12
590	500	0	0	90
685	500	140	0	45

The relationship between transaction costs and the amount of money transferred was established using the optimal portfolio of alternatives model and looks like this (7):

$$T = -0.0547V^6 + 3.2201V^5 - 70.396V^4 + 671.47V^3 - 1852.6V^2 - 13481V + 114499 \quad (6)$$

where V is the required amount of money transfer.

The function of opportunity costs due to the current practice of accumulating funds in the Savings Bank of the Russian Federation was set in the form of a linear relationship, which has the following form (8):

$$A = kV' \quad (7)$$

where k is the deposit rate of the Savings Bank of the Russian Federation for one month;

V' - the amount of the deposited funds.

A graphical interpretation of the optimal money transfer model is shown in Figure 3.

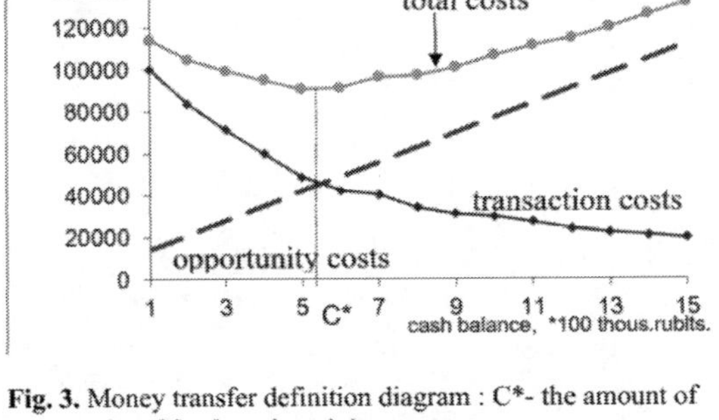

Fig. 3. Money transfer definition diagram : C*- the amount of the transfer of funds at the minimum cost.

The optimal transfer of funds, taking into account the uncertainty of the initial information on the inflow and outflow of funds in the original payment calendar, was 530,000 rubles. At the same time, the Savage criterion was used to make a decision in conditions of uncertainty. Based on the obtained optimal value of the transfer of funds, an optimal portfolio of alternatives was formed. And the final payment calendar was obtained, which allowed calculating the vector of standard coefficients of absolute liquidity Ψ (1,03; 1; 1,02; 1,01).

5 Conclusions

The algorithm for calculating the above methodology for planning highly liquid assets for enterprises and the vector of standard coefficients of absolute liquidity is proposed as the basis of the software product. This is necessary for further planning, accounting and control of the current financial and economic activities of the enterprise.

References

1. M.D.R. Vega Zavala, R.J. Santillán Salgado, *Contaduría y administración* **64**, 1 (2019)
2. D.V. Eremeev, A.A. Boyko, A.V. Kukartsev, A.V. Rozhkova, E.V. Mylnikovaand, L.N. Korpacheva *Jour. of Phys.: Conf. Ser.* **1582**, 012027 (2020)
3. N.V. Kolacheva, N.N. Bykova, *Bulletin of the NGIEI.* **44**, 29 (2015)
4. E.V. Antamanova, *RPPE.* **97**, 338 (2018)
5. A.Yu. Busheva, O.V. Eliseeva, *Bulletin of the Altai Ac. of Ec. and Law* **12**, 24 (2019)
6. Bursa, I.A., Ruga, V.O., Tahumova, O.V.: *Bulletin of the Altai Ac. of Ec. and Law* **11**, 25 (2019)
7. G.R. Vagazova, E.S. Luk'yanova, *Young scientist* **11.3**, 13 (2015).
8. E.A. Zapara, E.S. Almazova, *Alley of Science 2017.* **1**, 188 (2017)
9. E.P. Tetyuhina, *Sci. res.: prob. and pr.: Coll. of sci. pap. bas. on the mat. of the III In. sci. and pr. conf.* **1**, 32 (2019)
10. A.A. Boyko, V.V. Kukartsev, E.S. Smolina, V.S. Tynchenko, Y.I. Shamlitskiy, N.V. Fedorova, *Jour. of Phys.: Conf. Ser.* **1353**, 012124 (2019)

https://doi.org/10.1051/shsconf/202111600079

11. V.V. Kukartsev, A.A. Boyko, O.A. Antamoshkin, 2018 *Int. Russian Aut. Conf.* **2**, 1 (2018)
12. A.A. Boyko, V.V. Kukartsev, K.Y. Lobkov, A.A. Stupina, *Jour. of Phys.: Conf. Ser.* **1015**, 042006 (2018)
13. N.N. Kolodkina, N.I. Sutyagina, *KNJ* **4**, 157 (2014)
14. S.P. Zemtsov, D.V. Bukov, *Regional economy: theory and practice* **3**, 104 (2016)
15. I.Yu. Chuprova, *EGI.* **1**, 46 (2016)
16. M.A. Gorsky, L.G. Labsker, *Bulletin of the Altai Ac. of Ec. and Law* **4-2**, 179 (2020)
17. V.A. Semiglazov, *Innovation* **11**, 85 (2006)
18. A.A. Boyko, V.V. Kukartsev, V.S. Tynchenko, D.V. Eremeev, A.V. Kukartsev, S.V. Tynchenko, *Jour. of Phys.: Conf. Ser.* **1353**, 012138 (2019)
19. T.I. Ovchinnikova, I.N. Bulgakova, M.G. Khoreva, *Econ. analysis: th. and pr.* **8**, 22 (2006)

Using machine learning methods in problems with large amounts of data

Olga Kuimova[1,*], *Vladislav* Kukartsev[1,2], *Artem* Stupin[1], *Ekaterina* Markevich[2] and *Stanislav* Apanasenko[1]

[1]Reshetnev Siberian State University of Science and Technology, 31, Krasnoyarsky Rabochy Av., 660037 Krasnoyarsk, Russian Federation
[2]Siberian Federal University, 79, Svobodny Av., Krasnoyarsk, 660041, Russian Federation

Abstract. This article explores the use of artificial intelligence in medicine, in particular in radiology, pathology, drug development. The usefulness of robotic assistants in the medical field is revealed, including machine learning in medical science, as well as routing in hospitals. It also discusses such machine learning methods as classification methods, regression restoration methods, clustering methods. As a result, based on what is considered in this article, it is concluded that manual processing becomes more complicated and impossible with a large amount of data. There is a need for automatic processing that can transform modern medicine. And also, conclusions were made about how accurately the deep learning mechanisms can provide a more accurate result in the processing and classification of images compared to the results obtained at the human level. It became clear that deep learning not only aids in the selection and extraction of characteristics, but also has the potential to measure predictive target audiences and provide proactive predictions to help clinicians go a long way.

1 Introduction

In the modern world, great attention is paid to the development of computer technology. Every day, thousands or even millions of lines of code come out from under the fingers of programmers. Smart homes, smart cars, convenient electronic payment systems, online banking, entertainment services - all this would be impossible without writing code. However, the software capabilities are not limited to these examples only. While some technologies are designed to make human life easier in everyday life, others serve to preserve, and sometimes even save human life. The purpose of this article is to provide a brief excursion into the world of medicine and, in particular, into the world of technologies used in this area [1-3].

It's no secret that healthcare is one of the most important development niches at all times. And, like any other area, it undoubtedly needs modernization, and more importantly, optimization. Throughout the history of medicine, an unimaginable number of diseases have been studied, as well as ways to treat them. However, as practice shows, with the development of information technologies and over time, the database of this data is constantly being replenished. Moreover, more common issues often arise, for example, regarding the placement of patients or scheduling of appointments. Let's not forget the fact that the correct and timely diagnosis is constantly playing a leading role in the patient's recovery. After all, the process of training specialists is also extremely labor-intensive and resource-intensive [4-6].

From all of the above, an intermediate conclusion can be made that an exorbitant burden falls to the shoulders of a person, even the most successful in the field of medicine, which is almost impossible to cope with alone. This is where modern technologies come into play, designed to facilitate the work of professionals and help save lives. One of the most optimal solutions is the use of artificial intelligence technologies, in particular, the use of trained neural networks. The use of this technology makes it possible to facilitate the work of specialists in making even the most complex diagnoses, to increase the accuracy of detecting diseases, and also to solve a number of optimization issues that may arise in medical organizations [1]. So, neural networks are able to provide assistance not only in a quick and correct diagnosis, but also in solving more everyday issues related to the internal "kitchen" of medical institutions [7-9].

Further in the article we will give examples of the use of ML technologies directly in medicine, ML methods used to train neural networks, as well as the tasks solved by these neural networks [10-12].

2 Materials and Methods

2.1. The use of artificial intelligence in medicine

* Corresponding author: olga-barishnikova@yandex.ru

As mentioned earlier, the use of neural networks allows you to solve a number of medical problems that require immediate solution. One way or another, the most obvious and common issue solved by the use of neural network technologies is the diagnosis. So, today there are many examples of the use of this technology for the benefit of more accurate and effective medicine [5].

- AI in radiology.

Machines and algorithms have shown themselves to be effective in the classification of images, as well as in the recognition of objects in the photo. Hence, it naturally follows that they can effectively interpret imaging data, providing a wide range of possible applications of artificial intelligence algorithms, in particular in the field of radiology.

- AI in pathology.

Another interesting direction in the development of medicine is the development of an algorithm to assist a specialist.

Harvard Medical School is developing DeepLearning-based technology capable of detecting tissue pathologies. Combined with human experience, this technology can greatly enhance research efficiency by detecting pathological changes that the human eye may not be able to track. This technology in combination with humans has already shown excellent accuracy results (99.5%).

- Drug development.

The trend of using machine learning in medical research is quite popular. For example, the use of AI technologies can solve the following problem.

The experience of pharmaceutical companies shows that it takes approximately 12 years from the start of preclinical trials to the approval of a drug and treatment of patients. At the same time, only 0.1% of "candidate drugs" go to clinical tests. Approval is received by 20% of them.

A solution from San Francisco-based AtomWise, AtomNet uses all of the same deep learning technologies. A giant dataset of data on the interactions of molecules already known to science served as an excellent basis for training a neural network. The result was an algorithm that predicts the behavior of certain particles that interact. This technology can already be considered successful, as it helped to develop a cure for Ebola.

The first technology is surgical robot assistants.

Already today, many operations are carried out using machine vision technologies and human-controlled manipulators [6]. This combination is great for conventional surgeons. For example, this technology is able to "insure" the surgeon against accidents and inattention. Moreover, there is a clear advantage in improving the surgeon's visibility and reminding him of the order of the operation. In addition, the AI will assist in the selection of the tissue incision and suture to be applied, which will help alleviate the patient's pain during the recovery process.

The second is machine learning and medical science.

Departing a little from the topic of direct practical application of neural networks in diagnostics and operations, it is worth mentioning such an important problem of modern medicine as data structuring.

All medical facilities generate petabytes of various data about patients, their diseases and recovery processes. This data is unfortunately chaotic and unstructured. An additional complication is that, unlike, for example, business data, medical databases do not lend themselves well to traditional analytical methods.

A powerful AI-enabled platform is able to efficiently analyze information, find patterns and, therefore, create a certain structure that describes the available data. Thanks to this, it becomes possible to form more accurate data on the patient's health, which is an extremely important aspect in the field of medicine.

The third is routing in hospitals.

Another common, but no less important problem that can be solved by ML methods is routing in hospitals. This means the following. When a patient is admitted to the hospital, the medical staff carries out standard procedures: taking blood for analysis, taking anamnesis, etc. In 80% of cases, the patient is prescribed medication and sent home.

University College London Bloomsbury Clinic uses artificial intelligence systems to identify the very 20% of patients in urgent need of care. Based on standard analyzes, the system will give priority to one or another patient whose health is or will be under serious threat.

Scientists at the National Hospital of Neurology and Neurosurgery in the UK have developed a machine learning algorithm that analyzes information about appointments at a clinic and estimates the likelihood that a patient, for one reason or another, will miss an MRI scan. Their system considers such parameters as the age of a person, his address and distance to the clinic, weather conditions.

And scientists from the University are developing a system that monitors the "movement" of medical personnel and patients in the hospital: the approximate location according to the schedule, attendance of procedures, etc. Based on this data, the system will be able to determine situations or places where there is a potential shortage of equipment or the accumulation of queues.

This is not a complete list of the possibilities of applying machine learning technologies in medicine, but it shows the main branches of the development of this area, as well as the importance of developing and implementing software based on machine learning in medical organizations.

2.2 Machine learning methods

Machine learning is based on the idea that analytic systems can learn to identify patterns and make decisions with minimal human input. Machine learning is a class of artificial intelligence methods that are aimed not at solving a problem directly, but at learning in the process of solving similar problems. Machine learning focuses on developing computer programs that have ways to process data and learn on their own. This process is accompanied by the study of data (examples, instructions) in order to identify relationships or patterns in order to apply the acquired "knowledge" to make decisions or make predictions. The system "trains" in order to improve the accuracy of its forecasts and decisions [4].

The result directly depends on the choice of a machine learning method, since each method is most effective for a certain range of tasks. In addition, different methods use different resources, so some methods may be more cost-effective in a particular situation. In medicine, where no mistakes can be made, and performance should strive for the maximum level, the choice of teaching method is especially important.

Approaches to automatic learning can be divided into at least two types: with a teacher and without a teacher [7]. In the first case, the system receives training cases, each of which is represented by a set of input values along with the correct answer for it. The goal is to acquire the ability to respond appropriately to new cases. Supervised learning tasks include classification and regression reconstruction tasks.

2.3 Classification problem

The classification task is a formalized task where there are many objects (situations), divided into classes according to some rule. In the problem, a finite set of objects is given, for which it is known to which classes they belong. The set is commonly referred to as a sample. The class of the remaining objects is unknown. The classification problem is that it is required to build an algorithm capable of classifying an arbitrary object from the original set [8].

The phrase "classify an object" is usually understood to indicate the number or name of the class to which this object belongs. Classification is one of the sections of machine learning. The classification problem in machine learning falls under the heading of supervised learning.

There are various types of input data for the classification problem. Let's list some of them and reveal their essence:

• Feature description is one of the most common cases of classification problems. Each object in a class is described by a set of its specific characteristics, called attributes. Signs are both numeric and non-numeric.
• Matrix of distances between objects - each object in the class is described by the distance to all other objects of the training sample.
• Time series or signal is a sequence of measurements in time. Any dimension can be represented by a number or a vector, and sometimes by an indicative description of the object under study at a certain point in time.
• Video sequence or image, for example, X-rays of patients.
• Graphs, texts, results of database queries, etc. are perhaps the most difficult cases of processing input data. Usually they try to bring them to the first or second case by means of preliminary data processing and feature extraction.

There are also different types of classes that are included in the main typology of classification problems:

• Two-class classification,
• Non-overlapping classes,
• Multi-class classification,
• Fuzzy classes,
• Overlapping classes.

2.4 Classification methods

In the medical field, when constructing diagnostic systems, classification methods are often used, such as the naive Bayesian classifier (NaiveBayesClassifier), logistic regression (Logistic regression), decision trees (DecisionTrees), support vector machine (SupportVectorMachine) [9].

NaiveBayes consists of two types of probabilities that are calculated using the training data:
- probability of each class;
- conditional probability for each class for each value of x.

After calculating the probabilistic model, it can be used to make predictions with new data using Bayes' theorem. If you have real data, then, assuming a normal distribution, it is not too difficult to calculate these probabilities. Naive Bayes is called naive because the algorithm assumes that each input variable is independent. This is a strong assumption and does not match the actual data. Nevertheless, this algorithm is very effective for a number of complex tasks. For example, Naive Bayes is often used to classify text, filter spam, and generate recommendation systems [10].

SVM (Support Vector Machine) is a set of similar supervised learning algorithms used for classification and regression analysis problems. Belongs to the family of linear classifiers. A special feature of the support vector machine is that the empirical classification error continuously decreases and the gap increases, which is why the method is also known as the maximum gap classifier method.

The main idea of the method is to translate the initial vectors into a space of higher dimension and search for a separating hyperplane with a maximum gap in this space. Two parallel hyperplanes are constructed on both sides of the hyperplane separating the classes. The separating hyperplane is the hyperplane that maximizes the distance to two parallel hyperplanes [12]. The algorithm works under the assumption that the greater the difference or distance between these parallel hyperplanes, the smaller the average error of the classifier will be. This method is often used for face recognition, text categorization, handwriting recognition.

2.5 Regression reconstruction problem

The task of restoring regression is to find the approximate dependence of one group of values on another group of values. When solving the problem of restoring regression, the input data and the corresponding output values are analyzed in order to derive an approximate pattern, which will then be used to determine the output value with the new input data. This problem is encountered, for example, when assessing the value of real estate: by the size of the apartment, the height of the ceilings, the welfare of the area, the distance from the metro, etc. it is necessary to estimate the cost of housing. Also, this problem occurs, for example, when assessing mortality in certain injuries or when determining the healing time of an organ based on postoperative indicators [11].

2.6 Regression recovery methods

In the medical field, when constructing diagnostic systems, such regression methods as linear regression methods (Linear Regression), nonlinear regression methods (Nonlinear Regression), decision trees (DecisionTrees), and support vector method (SupportVectorRegression) are often used.

LinearRegression (linear regression) is one of the methods of regression analysis, which is used when there is a linear relationship between the set of independent variables (factors, regressors) and the dependent variable. In this case, we have a linear combination of the given basis functions, and the goal of the regression is to find the coefficients of this linear combination, while minimizing the loss function. This method is used in forecasting sales volumes, forecasting the value of securities, and analyzing the elasticity of demand.

DecisionTreeRegression is a method that builds a model using a tree structure. This method splits the dataset into smaller and smaller subsets while developing an associated decision tree. This method uses an algorithm that reduces the standard deviation [8].

2.7 Clustering problem

In unsupervised learning, the machine is not provided with outputs, only inputs, so it must try to find meaningful information on its own. An example of an unsupervised learning problem is the clustering problem. This task consists in the need to group a set of objects into subsets (clusters) in such a way that objects from one cluster are more similar to each other than to objects from other clusters according to some criterion. This task is often encountered in economic geography (dividing countries by indicators into groups with a similar economic situation), marketing (highlighting characteristic groups of consumers according to the degree of interest in a product), sociology (analyzing sociological surveys), medicine (identifying patterns of antibiotic resistance, highlighting various types of fabrics in the 3D image).

2.8 Clustering methods

The most popular clustering methods are the k-means method and DBSCAN.

K-means (k-means method) is the most popular clustering method. The operation of the algorithm is such that it seeks to minimize the total square deviation of cluster points from the centers of these clusters. This algorithm splits the set of elements into a predetermined number of clusters k. The main idea is that at each iteration, the center of mass is recalculated for each cluster obtained in the previous step. This method is used in marketing when dividing into groups by behavior (purchase history, activity in the application), identifying anomalies, categorizing inventory [1].

DBSCAN (Density-Based Spatial Clustering for Noisy Applications) is a density-based data clustering algorithm - if given a set of points in some space, then the algorithm groups together points that are closely spaced), marking as outliers the points that are lonely in areas with low density.

3 Results and Discussion

Artificial neural networks are an attempt to model the information retrieval capabilities of nervous systems. The main difference between neural networks and conventional computing systems lies in the massive parallelism and redundancy that they use to cope with the unreliability of individual computing units. At the moment, there are a large number of neural network architectures, but there are several main ones that are used most often.

RNN (Recurrent Neural Network) is one of the fundamental network architectures that other deep learning architectures are built on top of. RNNs can use their internal state (memory) to process variable-length input sequences [7]. Each processed information is intercepted, stored and used to calculate the final result. This makes them useful for speech recognition, for example. Moreover, a recurrent network can have connections that refer to previous layers (or even the same layer). This feedback allows them to retain a memory of past entries and solve problems over time. RNNs in medicine can be used to make a diagnosis based on data (symptoms), or they also compete with CNNs in the task of recognizing pathologies in images.

Artificial neural networks are an attempt to model the information retrieval capabilities of nervous systems. The main difference between neural networks and conventional computing systems lies in the massive parallelism and redundancy that they use to cope with the unreliability of individual computing units. At the moment, there are a large number of neural network architectures, but there are several main ones that are most often used.

RNN (Recurrent Neural Network) is one of the fundamental network architectures that other deep learning architectures are built on top of. RNNs can use their internal state (memory) to process variable-length input sequences. Each processed information is intercepted, stored and used to calculate the final result. This makes them useful for speech recognition, for example. Moreover, a recurrent network can have connections that refer to previous layers (or even the same layer). This feedback allows them to retain a memory of past entries and solve problems over time. RNNs in medicine can be used to make a diagnosis based on data (symptoms), or they also compete with CNNs in the task of recognizing pathologies in images [11].

According to the research conducted, the method of convolutional neural networks shows high efficiency when applied in the medical field. For medical image classification problems, this method shows higher accuracy than methods such as SupportVectorMachine or ArtificialNeuralNetworks. However, in other tasks, such as device design, methods such as SupportVectorMachine, RandomForest, or perceptron are used more often and show high efficiency. This method is often used to examine images in radiology.

A DBN (Deep Web of Trust) is a multi-layered network (usually deep, including many hidden layers) in which each pair of connected layers represents a restricted Boltzmann machine (Restricted Boltzmann Machine (RBM)). A DBN consists of many layers of hidden variables ("hidden units"), with connections between layers, but not between units within each layer. Unlike other models, each layer in the DBN network learns the entire input signal. In CNN, the first layers filter inputs based on basic characteristics only, while the second layers recombine all the simple patterns found by the previous layers. DBNs operate in a holistic manner and regulate each layer in order.

The operation of such a network is rather complicated and is rarely used in medicine. But it is still used in some cases, for example, to study the progression of a poorly understood disease.

DSN (Deep Stacking Network) architecture is different from others. DSN is also often referred to as DCN-DeepConvexNetwork. The DSN / DCN includes a deep learning network, but it is actually a collection of separate deep learning networks. Each network in a DSN has its own hidden layers that process data [3]. This architecture was designed to improve on the learning problem, which is quite complex when it comes to traditional deep learning models. Due to its many levels, DSN considers learning not the only problem to be solved, but a set of individual problems.

Networks with such an architecture are used as a rule to diagnose various diseases based on indirect indications. Thus, it is possible to determine, for example, the presence of diabetes mellitus in a person without taking a blood test.

5 Conclusions

The field of medical imaging is gaining importance with the requirement for accurate and efficient diagnosis of disease within a short period of time. As manual processing becomes more complex, stagnant and impossible with a large amount of data, there is a need for automatic processing that can transform modern medicine. Deep learning mechanisms can provide higher accuracy in image processing and classification compared to results obtained at the human level. In-depth learning not only assists in the selection and extraction of characteristics, but also has the potential to measure predictive target audiences and provide proactive predictions to help clinicians greatly. Machine learning cannot replace doctors, but it can help streamline routine tasks, increase forecast accuracy, and simplify solving atypical problems [12].

References

1. N. Stylianou, A. Akbarov, E. Kontopantelis, I. Buchan, K.W. Dunn, *Burns* **41**, 925 (2015)
2. Y. Xu, Y. Wang, J. Yuan, Q. Cheng, X. Wang, P.L. Carson, *Ultrasonics* **91**, 1 (2019)
3. T. Maruyama, N. Hayashi, Y. Sato, S. Hyuga, Y. Wakayama, H. Watanabe, T. Ogura, *J. of X-ray Sci. and Tech.* **26**, 885 (2018)
4. Y.C. Han, K.I. Wong, I. Murray, *IET Signal Proc.* **14**, 243 (2020)
5. I. Banerjee, Y. Ling, M.C. Chen, S.A. Hasan, C.P. Langlotz, N. Moradzadeh, M.P. Lungren, *Art. intell. in med.* **97**, 79 (2019)
6. Y. Acikmese, S.E. Alptekin, *Pro. Com. Sci.* **159**, 658 (2019)
7. Y. Gautam *ISA transactions* (to be published)
8. M. Nilashi, H. Ahmadi, A. Sheikhtaheri, R. Naemi, R. Alotaibi, A.A. Alarood, A. Munshi, T.A. Rashid, J. Zhao, *ESA*. **159**, 113562 (2020)
9. Q. Zhang, J. Zhou, B. Zhang, E. Wu, *Inf. Sci.* **547**, 945 (2021)
10. A.A. Boyko, V.V. Kukartsev, K.Y. Lobkov, A.A. Stupina, *Jour. of Phys.: Conf. Ser.* **1015**, 042006 (2018)
11. V.V. Bukhtoyarov, V.S. Tynchenko, E.A. Petrovsky, V.V. Kukartsev, A.I. Kuklina, *Jour. of Phys.: Conf. Ser.* **1118**, 012041 (2018)
12. A. Milov, V. Tynchenko, V. Bukhtoyarov, V. Tynchenko, V. Kukartsev, *Proc. of the Com. M.in Sys.and Soft.* **1295**, 480 (2020)

The impact of financial risks on the activities of an energy company

Tatyana Martynova[1,*], *Dmitry* Eremeev[1], *Natalia* Fedorova[1,2], Oleslav Antamoshkin[1,2] and *Konstantin* Lobkov[1,3]

[1]Reshetnev Siberian State University of Science and Technology, 31, Krasnoyarsky Rabochy Av., 660037 Krasnoyarsk, Russian Federation
[2]Siberian Federal University, 79, Svobodny Av., Krasnoyarsk, 660041, Russian Federation
[3]Krasnoyarsk State Agrarian University, 2 Kirenskogo, Krasnoyarsk, 660074, Russian Federation

Abstract. The activity of an enterprise in a competitive environment is always associated with one or another degree of risk. In order for the company to be able not only to survive in the market, but also to consolidate its position, the manager must be able to identify and assess risks in a timely manner, and also make effective management decisions to minimize them. This article assesses the risks, both in general terms and those that are specific to the energy sector enterprise, JSC "Lukoil". The authors revealed that the company under consideration (for the three-year analyzed period) is characterized by the following types of financial risks: credit risk, liquidity risk, currency risk, interest rate risk, risk in capital management. The article presents an assessment of the impact of financial risks on the financial results of a company.

1 Introduction

At present, financial risks have begun to play the most significant role in the overall "risk portfolio" of an enterprise. The financial risk of an enterprise is the likelihood of adverse financial consequences in the form of loss of income or capital under uncertainty in the conditions for carrying out its financial activities [1-3].

The current state of the energy industry is determined by the activities of a large number of Russian oil and gas companies. However, this paper examines the activities of a Russian company that is the leader in the industry on a global scale, JSC "Lukoil", a leading oil company.

Any field of activity is always accompanied by a certain type and level of risks. A specific risk classification cannot be identified. For each specific field of activity, its own risk factors are determined. This paper assesses the risks, both in general terms and those specific to the energy industry.

The choice of this company is explained by the fact that JSC "Lukoil" [4-6] is one of the leading and highly profitable oil companies, as well as information on production and final results are absolutely open and transparent. The company's activities can be divided into four main operating segments:

- Exploration and production,
- Processing,
- Trade and marketing,
- Petrochemistry,
- Energy.

Production in Russia accounts for 92% of the Company's existing reserves and 87% of marketable hydrocarbon production [7-9].

2 Materials and Methods

Financial risk management is a combination of techniques and methods that reduce the likelihood of these risks or localize their consequences. It is also one of the most important areas of activity of a financial manager, requiring deep knowledge of the basics of economics and finance of enterprises, mathematical and statistical methods, insurance, etc. The authors agree with the opinion of O.B. Braginsky [5] that in large foreign companies the financial manager can spend up to 45% of his working time on the risk management function.

In theory [8-11] and practice [12-16], there are approaches to assessing financial risks, which can be conditionally divided into three large groups:

- Assessment of the likelihood of occurrence. Financial risk as the likelihood of an adverse outcome, damage or loss.
- Assessment of possible losses in a particular scenario of the development of the situation. Financial risk as the absolute number of losses of a possible adverse event.
- Combined approach. Assessment of financial risk, both the probability of occurrence and the number of losses.

In the study, the third approach was used, since it shows not only the likelihood of risk, but also the possible damage to the financial and economic activities of the company, which is expressed in monetary terms.

The study formatted the process of financial risk management, which, from the author's point of view, includes several stages:

* Corresponding author: nvfed@mail.ru

- definition of the types of risks that the company may face;
- analysis and quantitative assessment of the degree of risk;
- choice of risk management methods.

At the last stage, these methods are applied in practice and the results of the measures taken are evaluated.

3 Results and Discussion

Based on the data [13], the authors made a conclusion about the prerequisites for the formation of financial risks in the activities of JSC "Lukoil". The main ones are the following:

- Liquidity. High volatility of prices for hydrocarbons and their products, exchange rates of world currencies, an increase in tariffs and prices of suppliers and other external factors can lead to an imbalance in the indicators of plans, budgets and investment programs of the "Lukoil" group and, as a rule, to a shortage of liquidity and sources of financing. Liquidity management of the "Lukoil" group is carried out centrally, the main tool is the rolling liquidity forecast. A global liquidity management system has been introduced and is functioning effectively. The liquidity indicators are monitored on a regular basis, in addition, the sensitivity of the indicators of plans, budgets and investment programs to changes in macroeconomic indicators is assessed.
- Volatility in the foreign exchange markets. The bulk of the company's revenue comes from the dollar-denominated proceeds from the sale of crude oil and petroleum products, while the bulk of operating and capital expenditures is denominated in rubles. In this regard, fluctuations in foreign exchange rates may have a significant impact on the company's financial results. The company manages foreign exchange risks using an integrated approach, also using natural hedging mechanisms, including managing the currency structure of monetary assets and liabilities and taking advantage of geographic diversification.
- Default by the counterparty and / or non-payment by the counterparty. Default events of the company's counterparties can lead to non-receipt or untimely receipt of proceeds for the delivered goods, also in relation to financial counterparties, non-return or partial non-return of funds placed on their accounts. This can negatively affect the financial results of the enterprise and entail the need to attract additional funds for fulfillment of the company's obligations. In order to minimize the risk of default and non-payment of counterparties, settlements are made with third parties outside the "Lukoil" group on a prepayment basis or are provided with letters of credit or bank guarantees from end customers.

Let us consider in detail each of the elements of financial risk inherent in "Lukoil". The first is credit risk.

The risk of non-fulfillment of obligations by counterparties to pay for delivered products is considered the most significant for the risk under consideration. In order to reduce this risk, the company focuses on cooperation with counterparties, which, in turn, have a high credit rating, uses guarantees from reliable banks and letters of credit. In most cases, it requires prepayment for the supplied products, in addition, it uses instruments to limit the concentration of credit risks that fall on one counter-party.

The second group of credit risks is the risks associated with the activities of counterparty banks and a decrease in their financial stability. To mitigate these risks, the company carries out centralized treasury operations, as well as operations that include raising and placing funds, including foreign exchange operations. The minimum exposure to credit risk is the carrying amount of financial assets.

Not past due receivables are not classified as high credit risk (Tables 1 and 2).

Table 1. Structure of the receivable's limitation period.

Indicators	Year			Growth rate, %	
	2016	2017	2018	2017	2018
Not expired	84.56	84.3	88.83	15.55	7.11
Overdue in 90 days	6.37	8.09	3.27	42.24	-52.71
Overdue in 90-180 days	4.99	1.09	3.36	-74.51	96.41
Overdue in 180-270 days	2.58	1.67	0.73	-24.89	-57.17
Overdue in 270-365 days	0.46	1.32	0.46	232.55	-62.48
More than 365 days overdue	1.04	3.51	3.36	290.41	19.12
Total trade and other receivables	100	100	100	15.89	2.79

Table 2. Structure of the reserve movement for doubtful debts.

Indicators	Year			Growth rate, %	
	2016	2017	2018	2017	2018
1st of January	93.72	91.94	104.89	6.70	44.43
Increase in provision charged to expense in 2018, the decrease	30.67	27.92	3.61	-1.00	-83.61
Write-off against the reserve	10.83	13.31	14.26	33.61	35.66
Exchange differences	12.95	2.64	9.50	-77.86	356.13
Other	0.60	3.91	3.48	604.09	64.18
Total	100	100	100	8.77	26.59

 https://doi.org/10.1051/shsconf/202111600081

The financial instruments used by the company and exposed to concentrations of credit risk consist primarily of cash equivalents, trade liabilities and over-the-counter derivatives. Money and cash equivalents are placed with a high rating.

The credit risk of forwards and swaps arises from the counterparties to the transaction from a leading international bank or a leading financial institution. The individual counterparty risk is managed within specified credit limits and includes the use of reverse sale requirements, which mitigates the risk of material default.

The company's liquidity is managed centrally. The global liquidity management system has been implemented in the company and is functioning effectively. It includes cash reallocation, an automatic concentration system, and rolling cash flow forecasts.

As the company operates in many countries, it is exposed to risks of adverse changes in foreign exchange rates. The exchange rate of the Russian ruble against the US dollar has the greatest impact on the results of operations, since export revenues are denominated in dollars, while the bulk of costs are incurred in Russia in rubles.

Within the framework of the centralized approach to the management of treasury operations and the company's liquidity, in general, the risks associated with unfavorable changes in exchange rates are considered, consolidated at the level of the corporate center. Currency risks, in a number of cases, are minimized on the trading floors due to operations with derivative financial instruments within the framework of corporate dealing. In order to reduce currency risks, loans are issued to companies in local currencies and within the framework of intra-group financing.

The structure of the assets book value and liabilities that form currency risk is presented in Table 3.

Table 3. Structure of the assets book value and liabilities that form currency risk.

Indicators	Total currency per year			Growth rate, %	
	2016	2017	2018	2017	2018
Financial assets					
Cash and cash equivalents	170 506	80 951	23 727	-52.52	-70.69
Trade and other receivables	126 242	168 519	160 523	33.49	-4.74
Loans issued	476 002	178 721	178 993	-62.45	0.15
Other financial assets	983	2 199	1 684	123.70	-23.42
Financial liabilities					
Loans and borrowings	128 977	(136 808)	(379 506)	6.07	177.40
Trade and other payables	37 450	(81 528)	(76 891)	117.69	-5.69
Net amount of risk	607 306	212 054	(91 470)	-65.08	-56.86

The net amount of risk in 2018 decreased by 56.86%, which caused a decrease in cash and cash equivalents - by 70.69%, trade and other receivables - by 4.74%, other financial assets - by 23.42%, and trade and other payables - by 5.69%.

Accordingly, the following exchange rates were applied (Table 4).

Table 4. Exchange rates.

Currency	Year		
	2016	2017	2018
US Dollar	60.66	57.60	69.47
Euro	63.81	68.87	79.46

The analysis of currency positions showed that the company conducts its activities, respectively, in rubles, US dollars and euros.

The net amount of risk in 2018 increased by 221.71%. We consider how the strengthening or weakening of currencies affects the valuation of financial assets and liabilities denominated in foreign currencies and the amount of profit (loss) before income tax (Table 5). All other variables remain the same.

Table 5. Strengthening (weakening) of currencies.

Currency	Year		
	2016	2017	2018
US Dollar (strengthening by 10%)	55080	22026	(7726)
Euro (strengthening by 10%)	3138	(249)	2566
Russian Ruble (strengthening by 10%)	(52445)	(19384)	4937

The weakening of the exchange rate of these currencies by 10% will have an equal and opposite effect on the company's profits (losses).

The Company is exposed to the risk of changes in interest rates both in the long term and in the short term.

Changes in interest rates may affect the value of the company's borrowed funds and the amount of its cash flows.

Monitoring the market situation on an ongoing basis allows us to take measures to improve the debt structure due to the optimal ratio of fixed and floating interest rates, control the need for additional financing and refinancing of existing debt, and increase the maturity of debt obligations.

The purpose of such measures is to reduce the impact of the company's considered risk.

* Corresponding author: nvfed@mail.ru

The structure of the financial instruments portfolio by interest rates is presented in Table 6.

The net amount of risk is also reduced by 16.68% due to the decline in all items, except for financial assets, which increased by 103.12%. Since the analysis showed a decline, the interest rate risk does not affect fixed-rate instruments, rather than variable-rate instruments.

The probable change in interest rates by 100 basis points as of 2017 and 2018 has been justified. This would be the amount of profit (loss) before tax for the amounts indicated below (Table 7).

Table 6. Structure of the financial instruments portfolio by interest rates.

Indicators	Growth rate, %	
	2016	2017
Fixed-rate instruments		
Financial assets	-43.27	103.12
Financial liabilities	-11.93	-3.53
Net amount of risk	-4.51	-18.54
Variable rate instruments		
Financial assets	59.47	-71.21
Financial liabilities	-11.53	-27.47
Net amount of risk	-20.28	-16.68

Table 7. Amount of profit (loss) before tax.

Indicators	Year			Growth rate, %	
	2016	2017	2018	2017	2018
Growth by 100 basis points					
Net financial liabilities	(2 504)	(1 996)	(1 663)	-20.29	-16.68
Decrease by 100 basis points					
Net financial liabilities	2 504	1996	1 663	-20.29	-16.68

Net financial liabilities decrease in 2017 and 2018, respectively. The analysis showed that the campaign has a loss before tax with an increase of 100 bps and a profit with a decrease of 100 bps. Therefore, it is profitable for the company to reduce interest

In order to ensure the continuation of the activities of all companies of JSC "Lukoil" in the future and to minimize profits for shareholders by optimizing the ratio of borrowed and own funds, the company performs capital management.

Capital includes long-term and short-term loans and borrowings, equity, and non-convertible interests. Net debt is not an indicator of IFRS and is calculated as the amount of loans and borrowings attracted, less cash and cash equivalents.

The ratio of net debt to equity allows you to assess the materiality of the amount of debt obligations to users of the financial statements (Table 8).

Table 8. Structure of the company's net debt to equity ratio.

Indicators	Year			Growth rate, %	
	2016	2017	2018	2017	2018
Net debt	437 223	285 970	42 397	-34.59	85.17
Capital	3 227 664	3 490 399	4 073 526	8.14	16.71
The ratio of net debt to capital	13.55%	8.19%	1.04%	-39.56	-87.30

The net debt to equity ratio decreased by 87.30% due to a decrease in net debt by 85.17%. The amount of capital is significantly greater than the net debt, which indicates a favorable trend in the company.

In 2018, net debt to equity accounted for 1.04%. The amount is not significant for the capital.

We will present a final assessment of the impact of financial risks on the company's financial performance. The structure of the revenue from sales is presented in Table 9.

Table 9. Structure of the revenue from sales.

Indicators	Share, %			Growth rate, %	
	2016	2017	2018	2017	2018
Total sales of products and services, rub.	19.53	18.58	16.68	8.04	21.51
Total exports of petroleum products, currency	80.47	81.42	83.32	14.92	38.52
Total sales revenue	100	100	100	13.58	35.36

Sales revenue showed an increase of 35.36% in 2018. Its increase was affected by all indicators, except for the sale of energy and related services in Russia, which decreased by 10.94%.

Other sales for export and other sales by foreign subsidiaries also decreased by 7.15%.

According to the data (Table 10) on financial income and expenses show that the company has a larger share of financial expenses than income for three years.

In comparison with 2017 the company experienced a 28.90% increase in revenue in 2018, and an increase in expenses by 40.13%. Thus, we see that in 2018, financial expenses exceed revenues by 96.1%. The company is subject to a budget deficit

Table 10. Structure of financial income and expenses.

Indicators	Share, %			Growth rate, %	
	2016	2017	2018	2017	2018
Income from interest on deposits	39.83	34.47	54.25	-11.16	102.89
Interest income on loans issued	49.51	44.32	33.20	-8.09	-3.44
Other financial income	10.65	21.21	12.55	104.45	-23.74
Total finance income	100	100	100	2.68	28.90
Interest expense	73.37	84.58	84.05	-42.62	39.26
The increment of the discount	4.94	9.90	7.82	16.44	10.68
Other financial expenses	9.41	5.52	11.39	-65.87	106.16
Total finance costs	100	100	100	-41.89	40.13

5 Conclusions

In general, we can say that the JSC "Lukoil" company is solvent and it has enough current assets to repay current liabilities. Accordingly, the equity capital is also sufficient to cover loans and borrowings. Also, cash and future proceeds from current activities cover the company's current debts. The company effectively uses its funds in order to make a profit.

The conducted research showed that the company finances its own activities with its own funds, it is financially independent and attractive to investors.

Long-term and short-term debt are particularly affected by market risk, debt risk, and liquidity risk. The company is also subject to a budget deficit, due to the excess of expenses over revenues. Fixed assets and cash are affected by liquidity risk, while inventory is affected by regional and market risk. The risk of the adequacy effects the share capital and retained earnings.

To reduce the impact of certain elements of the company's financial risks, it is necessary to increase various types of reserves. In the future, this will avoid financial losses.

References

1. E.J. Kwak, J.E. Grable, *Soc. Net. Analysis and Min.* **11**, 211 (2021)
2. K. Kiraci, Sci. *Annals of Ec. and Bus.* **66**, 335 (2019)
3. A.S. Avilkina, A.A. Tselykovskaya, *Ec. and eff. of prod. org.* **32**, 101 (2020)
4. L. Roa, A. Correa-Bahnsen, G. Suarez, F. Cortés-Tejada, M.A. Luque, C. Bravo, *Ex. Sys. with App.* **169**, 114486 (2021)
5. X. Wang, S. Hou, J. Shen, *Ec. Mod.* **96**, 196 (2021)
6. A.M.I. Lakshan, M. Low, C. de Villiers, SA. *Man. and Policy J.* **12**, 241 (2021)
7. J.F. de Guevara, J. Maudos, C. Salvador, *J. of Int. Money and Finance* **110**, 102288 (2021)
8. M.J. Garmaise, G.J. Natividad, o*f Fin. and Quan. An.* **56**, 249 (2021)
9. Z. Şenol, F. Zeren, *Avrasya Sosyal ve Ekonomi Araştırmaları Dergisi* **7(4)**, 1-16 (2020).
10. A. A. Boyko, V. V. Kukartsev, V. S. Tynchenko, D. V. Eremeev, A. V. Kukartsev, S. V. Tynchenko, *Journal of Physics: Conference Series* **1353(1)**, 012138, (2019)
11. Y. Cao, Y. Shao, H. Zhang, *ECR.* **17**, 1 (2021)
12. R. Kohli, S. Johnson, *MIS Qu. Exe.* **10**, 1 (2011)
13. A.A. Boyko, V.V. Kukartsev, V.S. Tynchenko, V.A. Kukartsev, E.A. Chzhan, A.S. Mikhalev, *Jour. of Phys.: Conf. Ser.* **1333**, 032009 (2019)
14. A. A. Boyko, et al. *Journal of Physics: Conference Series,* **1353 (1)**, 2019.
15. I. Ruble, The *J. of Eco. Asy.***20**, 132 (2019)
16. A. Yu, V. Rachok, *The sci. her.* **43-3**, 3 (2020)

Evaluation of the effectiveness of leveraging capital of a telecommunication company

Tatyana Martynova[1,*], Oleslav Antamoshkin[1,2], *Yuriy* Danilchenko[1], *Dmitry* Eremeev[1] and *Elena* Vaitekunene[1,2]

[1]Reshetnev Siberian State University of Science and Technology, 31, Krasnoyarsky Rabochy Av., 660037 Krasnoyarsk, Russian Federation
[2]Siberian Federal University, 79, Svobodny Av., Krasnoyarsk, 660041, Russian Federation

Abstract. Any company, and especially an actively developing one, constantly feels the need to attract financial resources. Debt capital plays an important role for companies as an additional means for expanding and carrying out economic activities. Accordingly, effective management of borrowed capital is economically beneficial for the company, and the wrong approach to the formation of borrowed sources of the company can adversely affect its financial condition. The object of research in this article was PJSC "MegaFon". A characteristic feature of this industry is the increased competition due to the aggressive policies of both telecom operators and companies that have entered this market from related industries. In the study, the process of forming borrowed funds by the company is presented in the form of five consecutive stages. The approbation of the proposed research scheme, using the example of PJSC "MegaFon", showed the ambiguity of the financial decisions made to attract borrowed capital. And the main recommendation is to attract borrowed funds only for the development of the company.

1 Introduction

Borrowed capital arises at a certain stage of production development and is conditioned by socio-economic relations, when production reaches a level of development at which the organization does not have enough equity capital to support production and expand it [1-3].

On the other hand, there are quite wealthy members of society and organizations who are willing not only to meet their personal needs, but also to lend their capital for the purpose of productive use. Of course, the lender does not care how his capital is used, it is in his interest to get the surplus value on it in the form of interest, but the borrower can get the surplus value from the borrowed capital received only if it is used productively.

Borrowed capital is a historical concept and, starting from the capitalist stage of historical development, has an objective nature. It is formed under the influence of economic and social factors. Having emerged in the earlier stages of capitalism in the form of loans, loans, and payables, with the development of its borrowed capital also takes the form of commercial credit-promissory notes, the emergence and growth of joint-stock companies encourages the emergence of bond loans. Resource provision of the enterprise is a necessary condition for its development. It is the availability of financial resources that determines the possibility of forming debt capital in the company. Leveraged capital is a catalyst for business processes, enabling businesses to increase the company's profit and value [4-6].

The essence of the company's borrowed capital is manifested in the implementation of operational, coordination, control and regulatory functions of the process of attracting external sources of financing. The company's management system must respond promptly to changes in internal and external factors, namely, changes in borrowing conditions, changes in borrowing methods, and the emergence of new ways to attract borrowed sources of financing.

Borrowed capital as a financial category expresses economic relations inherent only to it, mediated by a certain amount of funds transferred for temporary use by one economic entity to another. In this capacity, the borrowed capital represents the unity of the economic relationship and its form (monetary, tangible or intangible).

The use of borrowed capital to finance the company's activities has certain advantages. Borrowed capital has a wide range of opportunities to attract (with a high credit rating of the borrower), and with the help of financing from debt sources, it is possible to avoid attracting an additional number of shareholders and founders.

Accordingly, effective management of borrowed capital is economically beneficial for the company, and the wrong approach to the formation of borrowed sources of the company can adversely affect its financial condition [7-9].

2 Materials and Methods

* Corresponding author: ies_vel@mail.ru

Sources and instruments of debt financing are quite diverse. But it should be noted that the attraction of borrowed capital is based on the fundamental principles that determine its essence: repayment, payment and urgency.

In addition to the general advantages and disadvantages, each leveraged capital instrument may have its own advantages and disadvantages, which are derived from their specifics. In the scientific and educational literature, the following forms of attracting borrowed capital are distinguished: bank loans, bond issues, leasing.

When choosing the forms, methods of attracting debt capital, companies need to conduct a thorough analysis of their activities, as well as the conditions of the financing market, as well as the situation in the economic and financial sector of the country.

Borrowed capital has a wide range of opportunities to attract (with a high credit rating of the borrower), and with the help of financing from debt sources, it is possible to avoid attracting an additional number of shareholders and founders.

Accordingly, effective management of borrowed capital is economically beneficial for the company, and the wrong approach to the formation of borrowed sources of the company can adversely affect its financial condition.

The telecommunications services sector is one of the most important and promising areas of the Russian economy, the development of which is fundamentally necessary for the strategic development of all sectors. The Russian mobile market is represented by four main operators: MegaFon, MTS, VEON (VimpelCom) and Tele2.

PJSC "MegaFon" is an integrated digital communications company that occupies a leading position in the telecommunications market in Russia and the world. The company belongs to the three largest mobile operators in Russia.

"MegaFon" continues to actively expand its portfolio of digital products and services that create real value for customers in all business segments, which contributes to further growth in operational and financial performance and strengthens the company's digital leadership in the market.

Based on the available information [1-3] and the data of the analytical credit agency [6-9], we can say that too much borrowing reduces the financial stability of the enterprise, so the creation of an effective system of debt management is very important and should be based on the organization of planning, formation and use of debt capital.

In general, considering the opinions expressed in the theory and practice of financial management [10-14], I would like to note that the management of borrowing is a purposeful process of their formation from various sources and in different forms in accordance with the needs of the organization at various stages of its development.

The process [12] of forming borrowed funds by the company includes five main stages.

In the first stage, we study the dynamics of total borrowing in the period under review; rate these dynamics are compared with the growth rate of the sum of own financial resources of the volumes of operational and investment activities, the total assets of the organization.

At the second stage, the main forms of borrowing are determined, and the dynamics of the share of formed financial credit, commodity credit, and internal payables in the total amount of borrowed funds used by the company are analyzed.

At the third stage, the ratio of the volume of borrowed funds used by the organization by the period of their attraction is determined. To this end, a corresponding group used borrowed capital, we study the dynamics of the ratio of short- and long-term borrowings of the company and according to the amount of current and non-current assets.

At the fourth stage, the composition of specific creditors of the company and the conditions for their provision of various forms of financial and commodity (commercial) loans are studied.

At the fifth stage, the effectiveness of the use of borrowed funds in general and their individual forms in the organization is studied.

The information base of the study was the data of the open accounting statements according to the standards of IFRS "MegaFon" for a number of years.

3 Results and Discussion

Debt capital plays an important role for companies as an additional means for expanding and carrying out economic activities. But since the borrowed capital is raised on the terms of repayment with interest, companies must assess the current circumstances, considering their financial condition, before borrowing money to carry out economic activities.

In addition, the high proportion of borrowed funds, the high level of interest rate for the loan means that the attraction of new borrowed funds is an irrational decision, and it can worsen the financial condition and lead to bankruptcy.

Therefore, it is necessary to conduct systematic monitoring and a thorough analysis of borrowed funds before deciding on their additional attraction.

First of all, we will analyze the structure of the obligations of PJSC MegaFon. The results of the analysis are presented in Tables 1 and 2.

Table 1. Volumes of the liabilities structure of PJSC "MegaFon", million rubles.

Commitments	Years			
	2016	2017	2018	2019
Long-term loans and borrowings	195724	212097	288262	350066
Other non-current liabilities	33958	47891	39741	124848
Long-term liabilities	229682	259988	328003	474914
Short-term loans and borrowings	39389	52013	39232	25692
Other current liabilities	64397	84417	72371	87901
Current liabilities	107978	136430	111603	113593
Total liabilities	337660	396418	439606	588507

Table 2. Dynamics of the liabilities structure of PJSC "MegaFon", million rubles.

Commitments	Growth rate, %			Specific gravity, %			
	2017/2016	2018/2017	2019/2018	2016	2017	2018	2019
Long-term loans and borrowings	8.4	35.9	21.4	58.0	53.5	65.6	59.5
Other non-current liabilities	41.0	-17.0	214.2	10.1	12.1	9.0	21.2
Long-term liabilities	13.2	26.2	44.8	68.0	65.6	74.6	80.7
Short-term loans and borrowings	32.0	-24.6	-34.5	11.7	13.1	8.9	4.4
Other current liabilities	31.1	-14.3	21.5	19.1	21.3	16.5	14.9
Current liabilities	26.3	-18.2	1.8	32.0	34.4	25.4	19.3
Total liabilities	17.4	10.9	33.9	100	100	100	100

The structure of liabilities shows that PJSC MegaFon uses more long-term liabilities to finance its activities, namely, long-term loans and borrowings.

Let us consider in more detail the structure of borrowed funds of a telecommunications company.

The results of the analysis are presented in Tables 3 and 4, which show the dynamics of the increase in the company's borrowed funds.

Table 3. Volumes of the borrowed funds structure of PJSC "MegaFon", million rubles.

Elements of borrowed funds	Years			
	2016	2017	2018	2019
Loans in rubles-fixed rates	127203	161928	230855	236850
Loans in rubles-floating rates				38000
Loans in US dollars-floating rates	38661	34153	20598	
Loans in US dollars-fixed rates	7347	5082	6340	71
Loans in euros-floating rates	8103	9171	15575	14684
Total bank loans	181314	210334	273368	289605
bonds in rubles	55000	55000	55000	85000
Total long-term borrowings	236314	265334	328368	374605
Current portion of long-term loans and borrowings	(39385)	(50961)	(37909)	(23703)

Table 4. Dynamics of the borrowed funds structure of PJSC "MegaFon", million rubles.

Elements of borrowed funds	Growth rate, %			Specific gravity, %			
	2017/2016	2018/2017	2019/2018	2016	2017	2018	2019
Loans in rubles-fixed rates	27.3	42.6	2.6	53.8	61.0	70.3	63.2
Loans in rubles-floating rates			100.0		0.0		10.1
Loans in US dollars-floating rates	-11.7	-39.7	-100.0	16.4	12.9	6.3	0.0
Loans in US dollars-fixed rates	-30.8	24.8	-98.9	3.1	1.9	1.9	0.02
Loans in euros-floating rates	13.2	69.8	-5.7	3.4	3.5	4.7	3.9
Total bank loans	16.0	30.0	5.9	76.7	79.3	83.3	77.3
bonds in rubles	0.0	0.0	54.5	23.3	20.7	16.7	22.7
Total long-term borrowings	12.3	23.8	14.1	100	100	100	100
Current portion of long-term loans and borrowings	29.4	-25.6	-37.5	16.7	19.2	11.5	6.3

The analysis of the elements of the company's debt capital showed that the structure is dominated by long-term loans in rubles, the share of which increased from 53.8% to 70.3% in 2018.

At the same time, the company began to attract less loans in foreign currency, the share of which in the structure of borrowed capital for the analyzed period decreased from 22.9% to 12.9%.

It should be noted that in 2019, PJSC "MegaFon" prematurely repaid dollar loans with fixed and floating rates, the payment term for which was originally set for the end of 2021 and 2022.

As the analysis showed, during the analyzed period, the company did not attract short-term loans, loans to finance current activities. Thus, the short-term part of the borrowed capital is represented by long-term loans, the repayment period of which occurs within a year.

The main reason for raising debt capital is related to financing corporate needs, to finance the repurchase of ordinary shares in 2018-2019, as well as to restructure its loan portfolio.

Let us consider in more detail the basic information about the outstanding bonds of PJSC "MegaFon", presented in Table 5.

Table 5. Basic information on outstanding bonds of PJSC "MegaFon".

Bond Series	Volume, billion rubles	Coupon rate, %	Maturity date	Number of coupon periods
BO-001 R-02	10	9.9	29.05.2026	every six months
BO-001 R-03	15	7.85	03.10.2022	every six months
BO-001 R-04	20	7.2	15.02.2021	every six months
BO-001 R-05	20	8.55	31.01.2022	every six months
BO-001 R-06	5	8.9	11.03.2024	every six months
BO-002 R-01	5	8.9	11.03.2024	every six months
BO-002 R-02	10	8.9	06.04.2026	every six months

Since the company attracts loans in different currencies with a floating interest rate, it is necessary to determine the LIBOR and EURIBOR rates for the analyzed period, using additional information from the sources. The results of the calculations are presented in Table 6.

The data in Table 6 shows that the cheapest source of financing for the company is foreign currency loans. But attracting foreign currency loans increases the impact of currency risk.

Table 6. Interest rate on borrowed funds of PJSC "MegaFon", %.

Elements	Years			
	2016	2017	2018	2019
Loans in rubles with a fixed rate	9.6	9.6	9.3	11.13
Loans in rubles with a floating rate	-	-	-	6.9-7.32
Loans in US dollars with a floating rate	4.59	2.97	2.29	-
Fixed-rate US dollar loans	2.1	2.97	2.29	-
Floating-rate Euro loans	1.96	1.81	0.94	0.28
Bonds in rubles	9.8	8.9	8.6	8.6
Weighted average interest rate	8.3	8.4	8.3	9.7

The increase in the weighted average interest rate in 2019 to 9.7% is due to an increase in the interest rate on ruble-denominated loans with a fixed interest rate. Consequently, in 2019, the company's bonds were the cheapest borrowed capital.

Studying the borrowed funds, you must know the objective laws of the rate of change of the volume of attracted funds and the rate of change in working capital, production and product sales, net profit, etc.

In table 7 we will consider the dynamics of changes in key financial indicators of PJSC "MegaFon".

The results of Table 5 show that for the analyzed period, the excess of the rate of growth of borrowed capital in comparison with the growth of revenue may indicate a decrease in the efficiency of using borrowed funds.

With a decrease in net profit in 2017 and 2019, there is an increase in borrowed funds. At the same time, in 2018, the growth rate of net profit exceeded the growth rate of borrowed funds by many times.

It should be noted that in 2018, with a net profit growth of 517%, there is an increase in borrowed funds by 23.8%, a decrease in equity by 57.4%, as well as a decrease in the company's assets, which indicates a decrease in the financial stability of "MegaFon". Studying borrowed funds, it is necessary to determine their efficiency of use.

Table 7. Dynamics of changes in key financial indicators.

Indicators	Years			
	2016	2017	2018	2019
Growth rate of liabilities, %	-	17.4	10.9	33.9
Growth rate of borrowed funds, %	-	12.3	23.8	14.1
Growth rate of equity capital, %	-	33.2	-57.4	48.6
Growth rate of non-current assets, %	-	20.7	-8.6	36.9
Growth rate of current assets, %	-	17.0	-12.2	31.3
The growth rate of assets, %	-	21.6	-9.2	35.9
Revenue growth rate, %	-	1.7	4.3	4.0
The growth rate of net profit, %	-	-79.2	517.0	-75.1

In Table 8, we consider the indicators that characterize the effectiveness of the use of borrowed funds of PJSC "MegaFon".

Table 8. Indicators that characterize the effectiveness of the use of borrowed funds of PJSC "MegaFon".

Indicators	Years			
	2016	2017	2018	2019
The coefficient of concentration of the attracted capital	0.51	0.47	0.64	0.54
Operating profit interest coverage ratio	3.0	2.4	2.3	1.4
EBITDA interest coverage ratio	6.3	5.6	4.8	3.4
Debt load ratio	2.0	2.0	2.6	2.5
Financial leverage	1.9	1.6	4.7	3.6
Debt capital turnover ratio	1.34	1.21	1.02	0.93
Return on borrowed capital, %	10.8	2.0	10.0	2.2
Economic return on assets, %	12.4	10.3	11.5	9.0
Weighted average cost of borrowed capital, %	8.3	8.4	8.3	9.7
The effect of financial leverage, %	6.2	2.5	11.8	-1.9
Return on equity, %	16.1	10.8	21.0	5.3

The increase in the debt burden of PJSC "MegaFon" is due to the growth of borrowed funds, despite the increase in EBITDA. It should be noted that an adequate level of debt burden should be no more than two. Accordingly, a value greater than two means that PJSC "MegaFon" is heavily credited, therefore, has a low credit rating, which leads to unfavorable credit conditions.

Thus, the analysis of indicators that characterize the effectiveness of leveraging PJSC "MegaFon" showed that the attraction of borrowed funds to finance corporate needs, as well as to finance the repurchase of ordinary shares in 2018-2019 led to an increase in the debt burden, a decrease in the credit rating and return on equity.

5 Conclusions

Based on the conducted research, we can offer the following recommendations for improving the efficiency of the use of borrowed funds of PJSC "MegaFon":

- To achieve cost savings, increasing profits and profitability.
- To attract borrowed funds with a positive effect of financial leverage.
- To attract borrowed funds only for corporate purposes, namely, for the development of the company. Thus, the effective leverage will give the opportunity to PJSC "MegaFon" to flourish and to increase the competitiveness.

References

1. I.A. Kiseleva, M.S. Gasparian, E.N. Chernysheva, T.N. Voronkova, I.S. Androshina, *Int. J. of Rec. Tech. and Eng*. **8**, 6304 (2019)
2. L.D. Zubkova, S.M. D'yachkov, *Stud. on Rus. Ec. Dev*. **29**, 182 (2018)
3. M. Pendar, H. Tayar, S. Karimeh, *MSL*. **9**, 133 (2019)
4. Y. Yuniningsih, T.K. Pertiwi, E. Purwanto, *MLS*. **9**, 205 (2019)
5. M.D.R. Vega Zavala, R.J. Santillán Salgado, *Contaduría y administración* **64**, 1 (2019)
6. S. Dakua, *Int. J. of Fin. and Ec.* **24**, 427 (2019)

7. A.A. Boyko, V.V. Kukartsev, V.S. Tynchenko, V.A. Kukartsev, E.A. Chzhan, A.S. Mikhalev, *Jour. of Phys.: Conf. Ser.* **1333**, 032009 (2019)
8. A.H. Zadorozhaya, *Fin. An.: Sci. and exp.* **36**, 81 (2015)
9. T.A. Martynova, D.V. Eremeev, I.O. Knyazeva, *Fun. Res.* **11**, 100 (2019)
10. Boyko, A. A., Kukartsev, V. V., Tynchenko, V. S., Eremeev, D. V., Kukartsev, A. V., & Tynchenko, S. V. *Journal of Physics: Conference Series* **1353 (1)**, 012138 (2019)
11. D.A. Nikitin, *N. Cau. reg. Ser.: Soc. Sci.* **2**, 319 (2017)
12. G.A. Romitsyna, N.N. Romanovskaya, *Izvestiya TulGU. Ec. and legal sci.* **3**, 181 (2015)
13. A. A. Boyko, et al. *Journal of Physics: Conference Series.* **1353** (1), (2019).
14. I.V. Solntsev, *Ec. and Soc. Ch.: F., Tr., For.* **6**, 139 (2012)

SHS Web of Conferences **116**, 00083 (2021) https://doi.org/10.1051/shsconf/202111600083
ICSR 2021

Comparison of prices depending on factors in the secondary housing market

Vladislav Koktashev[1], *Vladimir* Makeev[1], *Pavel* Peresunko[1], *Anton* Mikhalev[1,*] and *Vadim* Tynchenko[1,2]

[1]Siberian Federal University, 79, Svobodny Av., Krasnoyarsk, 660041, Russian Federation
[2]Reshetnev Siberian State University of Science and Technology, 31, Krasnoyarsky Rabochy Av., 660037 Krasnoyarsk, Russian Federation

Abstract. The article considers the issue of pricing for the secondary real estate market regarding local causes (physical properties of housing). The aim of the study was to verify the following hypotheses: the influence of the pricing factors of residential real estate on its value is determined by its price segment and the influence of infrastructure on the value of apartments in different cities is the same. In the hypothesis test, data were used on the secondary housing market of the cities of Novosibirsk and Krasnoyarsk, taken from the site of «CIAN» apartment sale announcements and from various open data sources. During the study, non-parametric methods of machine learning, model-agnostic methods for the interpretation of predictive models, hierarchical clustering are involved. As a result of the work, the first hypothesis was confirmed and the second hypothesis was refuted, the high accuracy of forecasting the cost of an apartment was achieved, and the peculiarities of price formation for secondary housing objects were revealed and described.

1 Introduction

The real estate market is one of the most important indicators of the development of the city. In this regard, the question arises of the pricing of real estate. The reasons affecting price formation can be divided into two groups: local and global [1-3]. Local causes are those reasons that create the entire palette of real estate prices at a given time and depend on the physical properties of the housing. Global factors are understood as macroeconomic parameters: the level of development of the city, the level of income and life of the population. In this work, there is a study of the pricing of real estate with respect to local causes.

When choosing an apartment in the high price segment of real estate, higher priority will be given to factors that differ from the factors that guide the choice of comfort class real estate [2]. In this regard, there is a suspicion that the price segment determines the influence of the pricing factors of residential real estate on its value.

The global reasons for the pricing of apartments make it possible to compare the general price level in different cities and state that the price of a similar apartment in different cities will be proportional to the general price level of the city [1]. Accordingly, the question arises whether local causes will affect the cost of apartments in different cities equally?

Work [3] presents a study of the impact of the infrastructure area in which the apartment is located when predicting its value. Based on the obtained models, the authors came to the conclusion that the factors of the location of the apartment have a significant effect on its value. Also, due to the large errors of models based on training data, they put forward the assumption that there are special pricing laws for this type of apartment.

However, a massive assessment of real estate through the direct use of functional data approximation can lead to difficulties in interpreting the found dependencies, inconsistency in estimates of their coefficients [4-6]. The author notes the appropriateness of applying the approaches considered by him only in combination with other methods of processing pricing relationships. For example, one of such methods may be factor analysis and clustering of real estate, followed by setting the response function for each cluster [7].

Particularly noteworthy are the methods of forecasting the cost of an apartment, taking into account spatial heterogeneity and the interdependence of real estate objects. Instead of describing the process of pricing the real estate market of the city using one model, these methods provide for the construction of linear models with variable coefficients, the values of which are determined by the site to which the property being evaluated belongs [8, 9].

2 Materials and Methods

The object of this study is the secondary real estate market, the subject of the study is the influence of pricing factors on the value of real estate indicated in the announcement of its sale.

* Corresponding author: spaming1@yandex.ru

The aim of the study was to verify the following hypotheses:

- The influence of the pricing factors of residential real estate on its value is determined by its price segment.
- The impact of infrastructure on the cost of apartments in different cities is the same [10].

The data set used is structurally completely identical to the data set collected for the city of Krasnoyarsk in the work [3]. The main sources of information: a register of open data from the Fund for Assistance to the Reform of Housing and Communal Services, a database of announcements for renting and selling real estate of CIAN, as well as data from the Yandex.Directory service. In order to improve the quality of the data collected, the ads of real estate agencies and realtors were removed from the sample. For this, a restriction has been introduced on the number of repetitions of the telephone number in the received sample - no more than 3 announcements per one number. The final sample consists of the following parameters: number of rooms, total area, floor, maximum number of storeys of the house, type of walls and ceilings, repairs, number of balconies and loggias, type of bathroom, number of freight and passenger elevators, year of construction of the house, type of house heating, house breakdown as well as geolocation features. The number of organizations of various types located within a radius of 1000 meters from the house was used as geolocation features. The types of considered organizations are: bank, bar, kindergarten, gas station, cafe, cinema, metro station, clothing store, public transport stop, park and public garden, clinic, post office, dentistry, grocery store, moll, university, school. The total sample size was 2765 records [10].

During the study, nonparametric machine learning methods are involved:

- An insulating forest for detecting abnormal observations, which is the most suitable way to search for emissions compared to the probabilistic approach and metric methods [10].
- Random forest and gradient boosting of decision trees to predict the value of the property.

Quality metric of regression models - mean absolute percentage error (mean absolute percentage error - MAPE): for random forest models, its value was determined by the predictions on unselected samples, for gradient boosting models - according to the forecasts obtained in the process of moving control with 10 blocks and 5 repetitions.

The analysis and interpretation of regression models was carried out by model-agnostic methods Permutation Feature Importance, Accumulated Local Effects Plot, set out in [11].

Assessment of the degree of heterogeneity of the data was carried out according to Hopkins statistics: calculated statistics values exceeding 0.75 correspond to the rejection of the null hypothesis of a uniform distribution of data in favor of the alternative with a significance level of 10% [12]. The search for groups of nearby objects was carried out by the method of agglomerative hierarchical clustering for the matrix of Euclidean distances between observations, the estimation of inter-cluster distances was carried out according to the Ward's method [3].

3 Results and Discussion

Anomalies were detected in the source data using the isolation forest method [10]. For each observation, an anomaly estimate was obtained, after which 48 elements were excluded from the sample for which the value of this criterion exceeds 0.54.

A model of a random forest is constructed to predict the cost of apartments in Novosibirsk. The value of the MAPE model on the selected samples – out-of-bag error, estimating the magnitude of the error in the total population of objects - amounted to 11% [13]. Out-of-bag error of a similar model obtained on the initial sample without geospatial features was equal to 15%.

Figure 1 shows a diagram of the regression residuals, which clearly shows an increase in the spread of errors with an increase in the true value of the apartment.

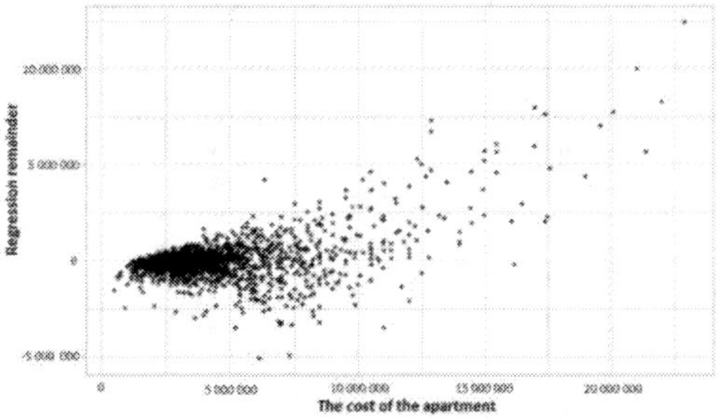

Fig. 1. Scatter chart of differences in forecasted and observed values.

In order to identify homogeneous groups of sample objects, the most important features in the model are identified and selected using Permutation Feature Importance algorithm (Figure 2, 10 most significant signs) [11]. A subsample has been generated for clustering of features that have a significance measure value greater than 1.5. The value of Hopkins statistics for this sample is 0.89, which confirms the heterogeneous nature of the data noted on the graph of the remnants of the regression model, and the presence of a tendency to group them.

Grouping of subsampling elements using the hierarchical clustering method was performed. As a result, two groups of residential real estate objects were obtained: the first includes 2085 objects, the second - 632. The resulting groups are shown in Figure 3. The average values of the most different from each other signs, rounded to hundredths, are presented in table 1 [9].

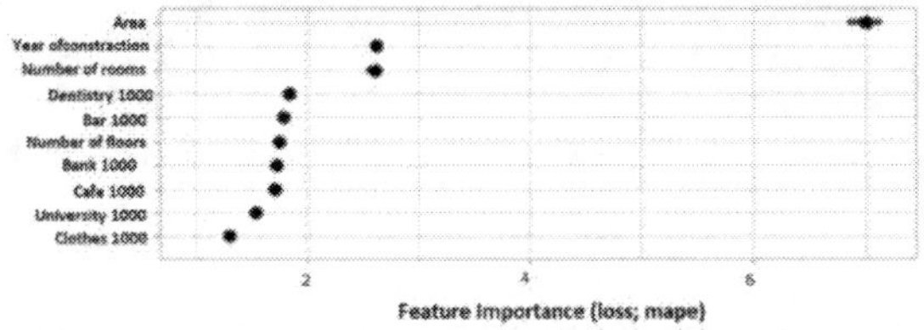

Fig. 2. The significance of the random forest model factors.

Table 1. The average values of the characteristics of the groups of real estate.

Characteristics	The average value of the characteristics	
	First group	Second group
Metrostation 1000	0.11	1.52
Cafe 1000	1.52	20.39
Bar 1000	1.94	15.24
University 1000	2.74	18.82
Cinema 1000	0.35	2.16
Clothes 1000	5.08	29.04
Bank 1000	5.31	30.17
Dentistry 1000	6.51	26.65
Park 1000	0.89	3.09
Apartmentboiler	0.01	0.03
The cost of the apartment	3563199.95	5368649.54

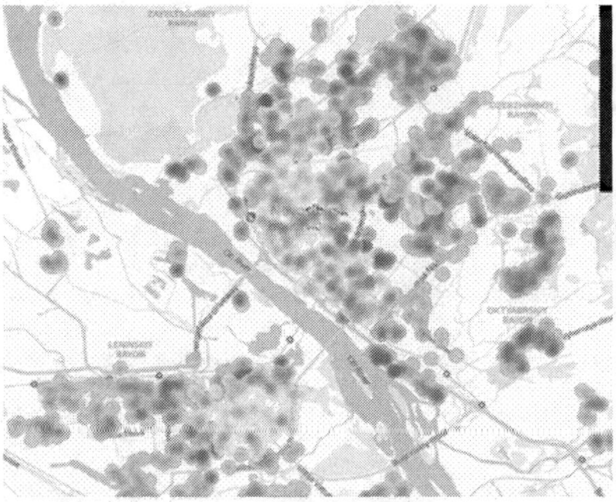

Fig. 3. Heat map of the cost per square meter of residential real estate.

For each of the selected groups of objects, a gradient boosting model is constructed that predicts the value of residential real estate (MAPE in both cases was 11%). The effect of the physical characteristics of the house and apartment, which is a subset of the most important distinguished features (see Figure 3), on the average output of the models is investigated. The effects were calculated using the Accumulated Local Effects Plot method; the corresponding graphs are presented in Figure 4 [4].

For the initial sample data on residential real estate in Novosibirsk and a similar sample of the city of Krasnoyarsk described in [3], common features are identified, random forest models are constructed to predict the cost of an apartment with the same hyperparameters. The accuracy of the models did not change: the MAPE values remained at the same level as was observed for models trained on data with a full set of features. The importance of each factor of both models is determined using the method previously used to select traits for clustering data. The obtained values of measures of the importance of features characterizing the infrastructure of the area around the apartment are displayed on the graph shown in Figure 5 [17].

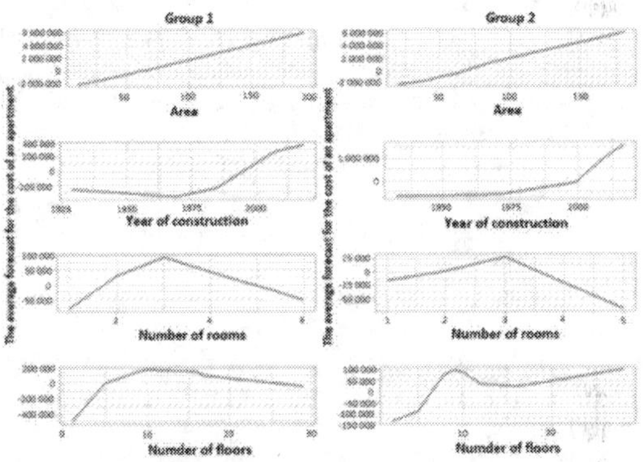

Fig. 4. The influence of signs on the average output of models.

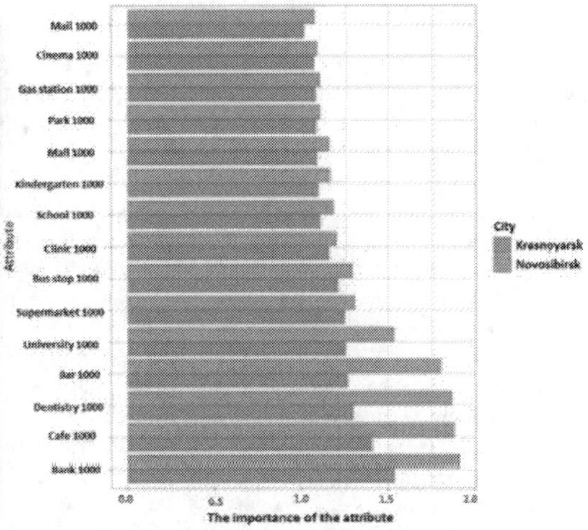

Fig. 5. The importance of geospatial features.

The models of a random forest predicting the cost of apartments in Novosibirsk have the following features:

- As in the case of Krasnoyarsk [3], for Novosibirsk there is also the problem of assessing the value of apartments of a high price category.
- The value of the MAPE model was 11%, which is less than the error characteristic of the models for mass valuation of real estate [14].
- The effect of improving the quality of the model due to the inclusion of infrastructure features observed in [3] for Krasnoyarsk is also present in Novosibirsk.

The growth of deviations between the predicted and the true cost of apartments in Novosibirsk is more directed towards underestimating the present values of the predicted model feature.

- Only a small number of features have a significant effect on model output. Among them, there are quite obvious pricing factors (apartment area, year of construction of the house, number of rooms, number of storeys of the house), and geospatial signs, which is explained by the fact that the signs that have the most significant influence on the output of the model are described by organizations more represented in areas with well-developed infrastructure, such as the city center. Groups of signs of wall materials, house ceilings, and the number of bathrooms practically do not contribute to the output of the model: the secondary housing market is regulated by demand, and not by the costs of construction production [15].

The following features are characteristic of the groups obtained by clustering the data of the Novosibirsk real estate market and the corresponding gradient boosting models.

- An increase in the average values of the area and cost of an apartment by groups is also accompanied by an increase in the average cost per square meter of housing: for the first group, it is 65,582 rubles, for the second - 84,160 rubles.
- Between the area and the predicted value of the apartment for both groups, a linear type of connection is noted, for other signs it is more complex.
- The influence of the year of construction on the average output of the model in the second group is not constant: it grows positively with an increase in the values of this attribute, which cannot be said about a similar dependence for the first group.
- The influence of the number of storeys on the average model output for apartments located in houses with more than 15 floors is opposite in each group: for the first group, it consists in decreasing the predicted value with an increase in the number of storeys in the house, for the second - in increasing it.
- The location of residential real estate significantly affects its value. Apartments from the second group, having a higher cost per square meter, are located mainly in the city center and surrounding areas, while apartments from the first group are located mainly on the outskirts of the city.

In general, the influence of infrastructure on the cost of an apartment in Novosibirsk is slightly higher: the largest difference between the significance is typical of those pairs of geospatial features of random forest models that describe the number of banks, cafes, dentistry, bars and universities in a radius of 1000 meters from the house in which the apartment is located [16].

5 Conclusions

The results obtained give reasons to confirm the first and refute the second hypothesis, namely:

- The influence of pricing factors on the value of residential real estate in different price groups may have a heterogeneous quantitative nature and an uneven trend.
- The influence of infrastructure factors on the formation of housing prices for different cities is different: perhaps one of the reasons for this is the variety of conditions for the formation of the urban environment. However, the results show that in both cases, the signs of infrastructure positively affect the quality of the forecast model for the cost of apartments.

Due to the rather high accuracy achieved in forecasting the cost of apartments and the visual clarity of the graphs given when interpreting machine learning models, the approaches used in this study can be applied to constructing models for mass valuation of real estate and analyzing the real estate market from a global perspective [17].

References

1. V.V. Koktashev, V.V. Makeev, E.G. Shchepin, P.V. Peresunko, *Jour. of Phys.: Conf. Ser.* **1353**, 012139 (2019)
2. F.T. Liu, K.M. Ting, Z.H. Zhou, *ACM - TKDD.* **6**, 1 (2012)
3. A.A. Boyko, V.V. Kukartsev, V.S. Tynchenko, V.A. Kukartsev, E.A. Chzhan, A.S. Mikhalev, *Jour. of Phys.: Conf. Ser.* **1333**, 032009 (2019)
4. N.V. Fedorova, V.V. Kukartsev, V.S. Tynchenko, Y.V. Danilchenko, S.N. Ezhemanskaya, N.V. Sokolovskiy, *IOP Conf. Ser.: Mat. Sci. and Eng.* **734**, 012084 (2020)
5. Z. Gongfu, *3rd Int. Conf. on Inf. Man., Inn. Man. and Ind. Eng.* **1**, 113 (2010)
6. N.V. Fedorova, V.V. Kukartsev, V.S. Tynchenko, Y.V. Danilchenko, S.N. Ezhemanskaya, N.V. Sokolovskiy, *IOP Conf. Ser.: Mat. Sci. and Eng.* **734**, 0120834 (2020)
7. S.Y. Kalashnikov, A.E. Godenko, Y.S. Kalashnikova, I.A. Tarasova, *IOP Conf. Ser.: Mat. Sci. and Eng.* **698**, 066003 (2019)
8. C. Zhou, *Jour. of Phys.: Conf. Ser.* **1802**, 032034 (2021)
9. M. Itoh, Y. Hagimori, K. Nonaka, K. Sekiguchi, *Jour. of Phys.: Conf. Ser.* **744**, 012222 (2016)
10. S.Y. Kalashnikov, A.E. Godenko, Y.S. Kalashnikova, I.A. Tarasova, *IOP Conf. Ser.: Mat. Sci. and Eng.* **698**, 066003 (2019)
11. L. Maksimenko, O. Korobova, O. Dudinova, X. Soskova, *IOP Conf. Ser.: Mat. Sci. and Eng.* **953**, 012043 (2020)
12. I.A. Panfilov, Y.A. Alekseeva, T.A. Panfilova, O.I. Karelin, D.V. Kustov, *Jour. of Phys.: Conf. Ser.* **1399**, 055092 (2019)

13. X. Liu, Q. Zhou, C.A. Wang, *Jour. of Phys.: Conf. Ser.* **1187**, 052104 (2019)
14. N. Lavrov, A. Druzhinin, N. Alekseeva, *IOP Conf. Ser.: Mat. Sci. and Eng*. **940 (1),** 012042, (2020)
15. O. Grineva, I. Shirochenskaya, *Jour. of Phys.: Conf. Ser.* **1425**, 012069 (2019)
16. F. Cui, *Jour. of Phys.: Conf. Ser.* **1629**, 012071 (2020)
17. Z. Tarmidi, N.H.A. Maimun, *IOP Conf. Ser.: Earth and Env. Sci.* **540,** 012047 (2020)

SHS Web of Conferences **116**, 00084 (2021) https://doi.org/10.1051/shsconf/202111600084
ICSR 2021

Fundamentals of effective activity of structural divisions of the company

Vladislav Kukartsev[1,*], *Valerii* Kernitskii[1], *Evgeniya* Semenova[1], *Olga* Kuimova[1] and *Vladimir* Khramkov[1]

[1]Reshetnev Siberian State University of Science and Technology, 31, Krasnoyarsky Rabochy Av., 660037 Krasnoyarsk, Russian Federation

Abstract. This article is devoted to the theoretical study of the foundations of the effective operation of the company's structural divisions. It examines the general directions of development of companies and the main strategies that are most relevant for companies that produce goods in modern market conditions. Also, in this article, attention is paid to specific areas of improvement of divisions in companies, the main of which are precisely the production departments. To understand the essence of unit improvement, leadership strategies were considered. Methods and factors influencing the improvement of the activity of a subdivision of a manufacturing enterprise are analyzed. All factors are considered in the context of the enterprise from two sides, both external and internal, and diagrams are built for them for visual presentation. On the basis of these factors, directions were developed to increase the economic efficiency of activities. The final stage was considered 8 ways to improve the efficiency of enterprise management. At the end of the work, conclusions were drawn about the reasons that necessitate the improvement of units in the modern market. The purpose of this work is to study the main directions for the development of companies and the activities of their structural divisions.

1 Introduction

The efficiency of the functioning of any company directly depends on the efficiency of its structural divisions, the possibility of rational interaction between adjacent structural divisions and labor resources in general. Evaluation of the effectiveness of the company as a whole is based on the degree of efficiency in the implementation of the working processes of its divisions; therefore, the improvement of the company as a whole is always local in nature, affecting certain departments and corresponding processes [1-3].

The dynamically changing nature of the external environment makes it necessary for enterprises to improve their activities on an ongoing basis in order to ensure all the greater efficiency of its functioning. This requires effective technologies, working methods and timely analysis in order to be able to adapt in time to changing market conditions [4-6].

The main issue that organizations face today is effective change management. In a dynamic market, organizations must be able to quickly adapt to changes in the external environment in order to survive. Often, the complexity and speed of change test the ability of managers and workers to adapt to a situation. This adaptation should consider not only external opportunities and emerging prospects, but also internal capabilities and the quality of the processes, because the growth of activities and a large market share do not say anything about the amount of profit that the company has at the end of the reporting year [7-9].

Changes in the company occur on the basis of management decisions that change one or more internal variables in the organization's goals, structure, tasks, technology and talent. In making such decisions, management must be proactive or reactive, i.e. or to be active on its own, or to respond to the demands of the situation. To be a market leader, you need to be among the first to notice trends and change, but in order to simply stay in the market and have good revenue indicators, it is enough to simply react to changes in time. If the management does not conduct adequate analysis and does not strive to improve its own activities, then such a company will soon lose in the competition based on price, quality or other factors that are important to consumers [10-12].

Because of the complexities of the state of some of the innovating and changing variables, the interdependencies of these variables, and the responses to innovation, effective change management is one of the most difficult but fundamental challenges for managers. Managers, as well as line personnel, must constantly notice and implement changes in the organization. There is synergy here - managers are focused more on the external sphere of the company's work, while the staff is focused on internal processes, because who else but they know better all the subtleties and nuances that can be changed for the better. Requirements for changes can come both from within the company in the form of the

* Corresponding author: vlad_saa_2000@mail.ru

needs and expectations of employees, and from the outside in the form of increasing competition, technological innovation and other factors [13-15].

2 Materials and Methods

Currently, there are many ways to improve the efficiency of the company [9, 10], each of which is aimed either at the head himself, or at any element of management activity. Work on improving work can begin with any element, improvement of which is most relevant for the organization. Relevance is expressed primarily by the degree of influence on the company's revenue and final profit. Therefore, the first and most important direction of improving the company's activities is to improve the functioning of its structural divisions, engaged in the main production:

A thorough diagnosis of organizational problems is absolutely essential as a starting point for carrying out planned organizational changes. In almost any organization, you can find patterns of behavior or procedures that have existed for a long time without changes, and no one can explain why and why this happens or find any meaning in these actions.

Diagnostics of changes in the organization should show the characteristic features of enterprise management and directions for its improvement, considering the stage of the life cycle, identify prospects and key competencies. Diagnostics should make it possible to formulate goals and key tasks, the consistent achievement of which will allow the organization to move in the desired direction. These goals may include: increasing profits; increasing production efficiency; the growth of the enterprise and the scale of its activities; building a business reputation; certain technological superiority and reduced risk of bankruptcy. The most common ways to improve the work of an organization, are [6]:

- improvement of the activities of the main production departments, which includes the improvement of both the main production processes and auxiliary ones;
- improvement of the management structure, its maximum simplification, decentralization of most functions, determination of the powers of managers, considering their qualifications and personal qualities;
- elaboration of a strategy for the development of the organization based on an analysis of its strengths and weaknesses, development of the philosophy and policy of the organization, covering all its functional areas;
- development of an organization's information system that ensures effective communication between employees and departments;
- development of a decision-making system, rules and management procedures, incentive systems;
- a system of continuous professional development of employees on the basis of training, retraining, development of initiative, creativity.

The diversity of the set of conditions in which companies operate, it is impossible to give a universal answer about how they should develop, because even for companies of the same size, operating in the same market, they can adhere to completely different goals, work for different target segments, and these differences are enough. The final development of their activities is radically different for them. In a dynamic environment, companies must independently decide on the choice of ways to improve management efficiency, depending on the specifics of the activity and the characteristics of the already established management system.

The extensive theoretical and methodological base of modern management allows one to constructively determine the directions of analysis and improvement of both the management system as a whole and its individual elements. At the same time, it is important for any leader to realize that the greatest efficiency of management activities will be achieved if various methods are used in a complex, focusing at the same time on the goals and strategies of the enterprise's development.

Only management that is adequate to market conditions can ensure the efficient use of resources (material, financial, labor, etc.) and solve the problem associated with increasing the competitiveness of the enterprise. This is a key factor in the success and prosperity of the enterprise [13].

A consistent analysis of the theoretical and methodological aspects of organization management allows us to form the following conclusion: management is defined as an integration process through which professionally trained specialists form organizations and manage them by setting goals and developing ways to achieve them.

The variety of purposes, goals and objectives of organizations determine their special differences; therefore, their management requires special knowledge and art, methods and techniques that consider the characteristic features of the organization. Management of the organization should be built considering these specific features, changing under the influence of external circumstances and as a result of the growth and development of the organization.

To ensure its long-term survival, a company must first of all produce a product that will consistently find buyers. This means that the product should be, firstly, interesting to the buyer so much that he is ready to give money for it. And, secondly, the buyer is interested in more than a similar product in terms of consumer qualities produced by other companies.

If a product has these two properties, then the product is said to have a competitive advantage. Consequently, an enterprise can conduct its production activities successfully only when their product has a competitive advantage.

Consideration of the creation and retention of competitive advantages involves the analysis of relations and, accordingly, the interaction of three subjects of the market environment [2]:

- the company-object of the analysis;
- its competitors who can sell a similar product that can satisfy the same need;
- the buyer.

The main thing in this market triangle is the buyer. Therefore, the competitive advantages of a product should consider those values that correspond to a specific consumer. Competitive advantage does not necessarily arise from product comparisons between different companies. It may be that there are no competitors on the market, but the product is not for sale. There is only one reason for this - the price. Economic theory says that a market exists when there are people willing to buy a product at a certain price. If the price set by the company does not correspond to the idea of an adequate consumer price, then there will be no sales. Therefore, the main thing to pay attention to is the cost of production activities, since the price makes the product more or less valuable in the eyes of the client.

Thus, the price is the main competitive advantage in the eyes of the buyer and very often he buys it precisely because it is cheaper than others. In this case, we can say that the company's production process is its competitive advantage, since its consequence is the characteristic of the product that is important to the consumer. Such purchases can occur even if the product has no consumer utility to the customer.

The second type of competitive advantage is product differentiation, i.e. how it differs from peers on the market. Differentiation is not necessarily related to the consumer qualities of the product (reliability, ease of use, good functional characteristics, etc.). It can be achieved due to such characteristics that have nothing to do with its consumer properties, for example, due to the brand, brand, and recognition [12].

Also, an enterprise can create competitive advantages in a product due to its market position. This is achieved by securing the buyer or, in other words, by monopolizing a part of the market, when the product is significantly different in some way from competitors. An example can be any multilateral market, when a group of users of a certain product creates a kind of ecosystem, and in order to have a relationship with them, other people are forced to also start using such a product. This, for example, can be any social network.

There are three strategies for creating competitive advantage that a company can focus on, and these are discussed in Table 1 [4].

All these strategies have significant differences among themselves and provide for the implementation of actions that are different from each other, so the company needs to determine which strategy to follow and work out the appropriate measures to gain a competitive advantage.

Just as the chosen strategy influences the choice of the vector of the company's development, so the factors influence the direction of the division's development. There are internal and external factors. External factors practically do not depend on the activities of the enterprise or unit itself and they include [8]:
- Business conditions.
- Demographic changes.
- Social changes.
- The level of demand and supply for labor.
- The level of prices for raw materials and energy resources.
- Access to finance.

Internal factors, on the contrary, affect the efficiency of the enterprise and completely depend on the management decisions taken. Internal factors are divided into [8]:
- Personnel management system.
- Application of technologies and know – how.
- Range of products.
- Price policy.
- Company image.
- Availability of capital and resources.
- Profit distribution system.

Table 1. Leadership strategies.

Strategy	Direction of the company
Price Leadership	Focused on its own costs. The goal is to minimize them as much as possible. Key sources of cost benefit: hands-on experience in relevant activities, economies of scale, reduced fixed costs, development of logistics, convenient geographic location.
Product differentiation	Endow your own product with distinctive features that can be relevant to the consumer and for which he is willing to pay. The main thing is to change the market, to shift the focus from some characteristics to others, where the company is the only one. The sales volume is growing due to the growth of the sold products in physical terms. To do this, you need to determine: who makes the decision to make a purchase, determine the consumer's criterion grid, how he chooses a product, find out where he gets information about the product from.
Spot satisfaction of the needs of the target segment	Create a product exclusively for the needs of a certain segment (for example, a phone with large buttons for the visually impaired). The goal of a strategy is to satisfy the unusual need of a specific group of people. This ensures that there is no competitive influence from similar products. With this approach, to attract a client, you can additionally use both the first and the second strategy.

A more detailed and expanded system of factors affecting the efficiency of the enterprise and the decisions made on the development of structural units is presented in Figure 1 for external factors, and in Figure 2 - for internal [6].

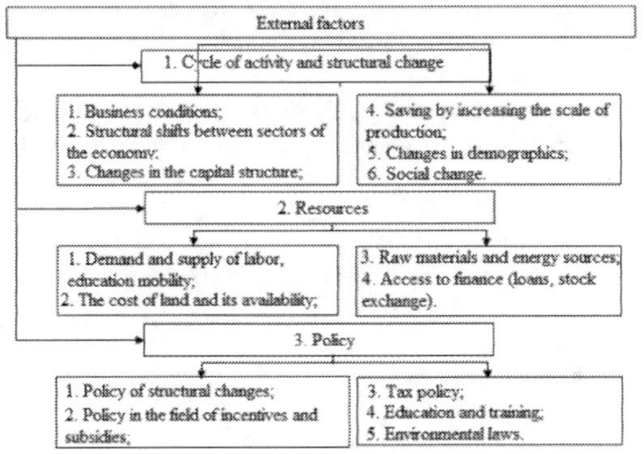

Fig. 1. External factors affecting the activities of enterprises.

Fig. 2. Internal factors affecting the activities of enterprises.

The above factors are the basis for the development of directions to increase the economic efficiency of enterprises. These areas include [4]:
- Increase in production volumes, assortment and quality of products.
- Increasing labor productivity.
- Reduction of costs and expenses.
- Expansion of sales markets.
- Improving the efficiency of using fixed assets.
- Inventory optimization.
- Development of relationships with suppliers.
- Introduction of the latest technologies.
- The use of modern technologies for managing production processes using appropriate programs.

3 Results and Discussion

It is also worth noting that in the modern conditions of a market economy, the basis for the successful operation of any company in our country is to improve management efficiency. There are many ways to improve management efficiency [14-17]. Their objects can be a leader, personal and professional qualities of managers, and elements of management activities. Work in the field of improving the efficiency of the organization's management can be started from any of its elements, that is, from the most relevant for the enterprise. It is only necessary to consider the relationship of the elements of the control system with each other. Ways to improve the efficiency of enterprise management are conventionally divided into eight groups [1,7]:
- improvement of the enterprise management structure is aimed at its maximum simplification, decentralization, distribution and determination of the powers of management employees, taking into account the requirements for qualifications and personal qualities;
- development of an enterprise development strategy based on an analysis of its activities, assessment of strengths and weaknesses, development of a philosophy, concept, mission and policy of the enterprise;
- development and implementation of the company's information system, which should ensure effective communication between employees and departments;
- introduction of a system for making and developing management decisions, rules and management procedures, incentive and motivation systems;
- development and implementation of a system of continuous professional development of employees based on training, retraining, development of initiative, creativity;
- development of an algorithm, regulations or systems for the selection, training, certification or assessment of the company's management personnel, career planning for managers, the use of a leadership style that is adequate to internal and external factors affecting the organization;
- the use of the most effective methods of personnel selection, its assessment, the formation of a workable, compatible composition of employees, the creation of the most favorable socio-psychological climate;
- work to create a culture of the organization, the development of shared values, recognized and approved by employees.

Thus, based on the study of the theoretical literature, the following logical conclusions can be drawn:
- the economic efficiency of a commercial organization is defined as the effectiveness of its functioning. Currently, there are at least three fundamental approaches to the concept and assessment of economic efficiency;
- analysis of the economic efficiency of the organization's activities, involves a comprehensive study of all aspects of the functioning of the research object using statistical, financial, economic and factor analysis methods;
- the result of assessing the effectiveness of the enterprise, in any case, is the search and identification of problems and weaknesses, which must subsequently be eliminated.

5 Conclusions

There are many reasons, internal and external ones that necessitate the improvement of divisions in today's market:
- The level of indicators of most processes tends to decrease over time, if not maintained. This means that just to maintain the current standards, it is necessary to

carry out a certain amount of work on the maintenance of production. If, in addition to this, the company wants to improve and renew itself; this will require not only maintenance, but also additional efforts.
- If some organization does not improve, then competitors are actively engaged in it. Therefore, the analysis and improvement of performance should be carried out by management on an ongoing basis.
- Modern consumers are becoming more and more demanding. The level of supplies and their quality are increasing all the time. This, in turn, leads to a rapid rise in consumer expectations. It is not always possible to exceed expectations, which is considered an ideal situation. You need to at least comply with them. If this is not done, then the company is guaranteed to lose a client [11].

References

1. A.V. Vyalov, *Lean production: a tutorial* (Komsomolsk-on-Amur, FGBOU VPO "KnAGTU", 2017)
2. A.A. Boyko, V.V. Kukartsev, E.S. Smolina, V.S. Tynchenko, Y.I. Shamlitskiy, N.V. Fedorova, *Jour. of Phys.: Conf. Ser.* **1353**, 012124 (2019)
3. T.Yu. Grubich, *Paradigms of modern sci.* **2**, 79 (2017)
4. N.A. Gukasyan, *Management of the enterprise. Just about the complicated* (St. Petersburg, BHV, 2020)
5. V.Zh. Dubrovsky, *Economy of the company* (Yekaterinburg, Ural State Publishing House. econom. University, 2019)
6. A.V. Kukartsev, A.A. Boyko, V.V. Kukartsev, V.S. Tynchenko, V.V. Bukhtoyarov, S.V. Tynchenko, *IOP Conf. Ser.: Mat. Sci. and Eng.* **537**, 042009 (2019)
7. A.V. Ilchenko, *Ac. prob. of econ. Sci.: mat. of the III Int. sci. conf.* **1**, 127 (2018)
8. I. Masaaki, *The way to reduce costs and improve quality* (Moscow, Alpina Publisher, 2017)
9. V.D. Markova, S.A. Kuznetsova, *Strategic management: concepts, concepts, decision-making tools: a reference guide* (Moscow, INFRA-M, 2019)
10. G. Mintzberg, *Strategic safari: an excursion through the wilds of strategic management* (Moscow, Alpina Publisher, 2019)
11. V.S. Tynchenko, V.V. Tynchenko, V.V. Bukhtoyarov, V.V. Kukartsev, V.A. Kukartsev, D.V. Eremeev, *IOP Conf. Ser.: Mat. Sci. and Eng.* **537**, 042010 (2019)
12. N.F. Revenko, *Organization of production and management at machine-building enterprises* (Moscow: Higher school, 2017)
13. E.A. Chzhan, V.S. Tynchenko, V.V. Kukartsev, N.V. Fedorova, A.S. Yamshchikov, D.A. Krivov, *IOP Conf. Ser.: Earth and Env. Sci.* **315**, 022106 (2019)
14. E. Volkova, *E3S Web of Conf.* **244**, 03023 (2021)
15. N. Havlovska, O. Illiashenko, O. Konoplina, I. Shevchuk, A. Hlynska, V. Prytys, *TEM J.* **9**, 1053 (2020).
16. O.V. Demyanova, A. Kireeva-Karimova, L.M. Zabirova, *Jour. of Eng. and App. Sci.* **12**, 4952 (2017)

Manufacturing enterprise management concept focused on maximum customer engagement

Vadim Tynchenko[1,2*], *Oksana* Kukartseva[1], *Aygul* Ogol[1], *Andrei* Yamshchikov[2] and *Aleksandr* Bondarev[1]

[1]Reshetnev Siberian State University of Science and Technology, 31, Krasnoyarsky Rabochy Av., 660037 Krasnoyarsk, Russian Federation
[2]Siberian Federal University, 79, Svobodny Av., Krasnoyarsk, 660041, Russian Federation

Abstract. This article examines one of the main directions for improving the activities of structural units - lean production, which is widely practiced abroad, but in Russian conditions it is mainly used by large manufacturing companies focused on the foreign market. The article discusses the methodology for the implementation of lean manufacturing. Seven types of losses were given, as well as a complete sequence of actions to eliminate them. We learned a value stream map, types of streams and their composition according to various factors. We examined the tools in the Lean Manufacturing system for the implementation of this technique. Also, enterprise management systems based on push and pull systems were described, diagrams for them were given that show the essence of the operation of these systems. At the end of the work, conclusions were drawn about the way to improve the enterprise, such as "lean manufacturing". The main objective of this article is to identify the essence and features of the research based on the principles of lean production in the structural unit of the research object.

1 Introduction

One of the most versatile ways to improve the performance of almost any enterprise is lean manufacturing. Its peculiarity lies in a thorough study of business processes and identification of all types of losses that may arise in the main production. Depending on the amount of losses and the ease of their measurement, a decision is made on specific measures to eliminate them [1-3].

The concept of "Lean Manufacturing" combines the optimization of production processes aimed at continuous improvement of product quality while constantly reducing costs, with the involvement of each employee in this process. The concept is maximally focused on market conditions of management. The main task of such production was the creation of competitive products in the required quantity, in the shortest possible time and with the least resource consumption [4-6].

The transition from conventional production to lean production indicates that the company has reconsidered its views on production relations within the enterprise, as well as on relations with suppliers and consumers, if necessary. Favorable conditions of the recent past - high consumer demand with a low supply of goods and services, contributed to the fact that the price of goods was set based on the costs of the enterprise for the production of products by adding often unjustified "cheating" [7-9].

However, over time, the market situation has changed. The market for goods and services gradually became saturated and became highly competitive. Offers from various manufacturers stimulate competition between domestic and foreign suppliers. A situation in which supply exceeds demand leads to a forced fall in the market price of goods and services, and, as a consequence, to a fall in the company's profits. The only way to avoid falling profits in such conditions is to reduce production costs, since prices for raw materials and resources are also determined by the market situation [10-12].

2 Materials and Methods

The possibility of reducing production costs is not always obvious to manufacturers: the consumption rates of raw materials and materials are being met, productivity is increasing. There is only one thing left - to return to the analysis of the structure of production costs. First of all, expenses should be classified into those that are really necessary and those that can be eliminated. In terms of lean manufacturing, the ability of a product to meet consumer expectations is called value. The consumer is ready to pay only for the conformity of the characteristics of the product to his expectations (for functionality, quality, lead time, price, etc.). If in the production process any activity is carried out for which the consumer does not intend to pay, i.e. activity that does not add value, such an activity is called loss. Within the framework of lean production, the following types of losses are distinguished [1, 4]:

* Corresponding author: vadimond@mail.ru

- losses due to overproduction - production of products that are not in demand; production of products in a larger volume earlier or faster than is required in the next stage of the process;
- wasted time due to waiting - interruptions in work associated with waiting for people, materials, equipment or information;
- losses due to excessive processing - additional processing of the product due to poor quality of the tool, design errors, etc.;
- loss due to unnecessary movement during operations - any movement of people, tools or equipment that does not add value to the final product;
- losses due to excess inventory - any excess flow of products into the production process, be it raw materials, semi-finished products or finished products;
- losses during transportation - unnecessary movement or movement over long distances of materials, parts, products;
- losses due to the release of defective products - products requiring inspection, sorting, disposal, replacement or revision.

Let's consider these losses in more detail. One of the most obvious ways to increase profits is to increase plant productivity. However, the pursuit of performance does not mean that there are enough consumers on the market who are willing to buy a particular product. A moment may come when a product that was previously lacking for everyone begins to accumulate in the warehouse. Then the management sets the task for sales managers to urgently sell everything, which is why they are forced to go out of their way to sell stale or spoiled goods. In this case, one cannot speak of any profit.

The main issue in this case is to compensate for the costs incurred, or even worse - to minimize losses. Ideally, they should be avoided and the solution to this problem here is quite simple. For this you need [1]:
- to produce only on time and only what the client wants (in fact, work only "to order");
- to produce goods in small batches, which are guaranteed to be sold (if demand falls, quickly switch to another type of product);
- to reduce losses during equipment changeovers, reduce changeover time to a minimum, making production in small batches profitable.

All this means that the profitability of production in large batches cannot be as efficient as it was previously thought. Today's customers are demanding. They need an assortment and variety right now, and losses due to storage of illiquid assets, due to excessive consumption of raw materials, due to problems with customers are not their concern.

Transportation losses. Any more or less complex production is a sequence of operations to transform raw materials, or semi-finished products into a final product. But between operations, all these materials need to be moved. Valuable transfer procedures are present even in conveyor production. After all, you need to deliver the raw materials to the conveyor or take the finished products to the warehouse. Of course, transportation is an integral part of production, only it does not create value at all, although it requires costs for fuel or electricity, for maintaining a transport fleet, for organizing transport infrastructure (roads, garages, overpasses, etc.). In addition, transportation is time consuming and risky because of product damage.

In order to reduce losses during transportation, it is necessary to create a map of the routes of vehicles and conduct a thorough analysis of the feasibility of one or another movement. After that, you should try to eliminate unnecessary transportation by redevelopment, redistribution of responsibility, elimination of removed stocks, etc. In the case of transportation, the system will not interfere: each movement of valuables must be justified by an appropriate regulatory document.

Losses due to waiting. This type of waste is associated with waiting for the start of material processing, and it indicates that the planning process and the production process are not coordinated with each other. This state of affairs is often found in the practice of domestic enterprises. The planning process itself is quite complex, since it requires the analysis of a large number of factors. These factors include: the structure of consumer orders, the state of the raw material market, equipment productivity, shift work schedule, etc. Truly optimal planning requires serious mathematical training and sophisticated interaction between sales, purchasing and manufacturing services. As a rule, the planning process is pseudo-optimal and is based on the subjective approach of people with some experience in production.

In addition to suboptimal scheduling, latency losses are significantly affected by uneven equipment throughput. In this case, the accumulation of products waiting to be processed can occur before the operation with the lowest throughput. The performance of such operations should be improved. If this is not possible, flexible equipment or redeployment of personnel between operations should be considered.

Losses from any inventory. Inventories reduce the free money supply of companies, which reduces the flow of profits and makes them dependent on lines of credit. But at present, the presence of stocks in production is considered quite common, and most importantly, a permissible phenomenon! Indeed, thanks to stocks, you can compensate for surges in consumer demand. Stocks allow the enterprise to produce products in case of interruptions in the supply of raw materials. Finally, inventory allows you to level the production flow. To answer this question, one should look at the problem of reserves from a different point of view [5]:
- Stocks are frozen circulating assets withdrawn from the circulating cycle.
- Stocks need maintenance (warehouse space, personnel, logistics, etc.).
- Inventories hide production problems: poor planning, strained relations with suppliers, uneven production flow, etc.

It turns out that stocks hide losses of other species, giving the impression of a safe working environment.

Losses due to manufacturing defects. The release of products that do not meet the requirements of the consumer entails obvious costs of raw materials, labor time, labor, processing costs and waste disposal. The traditional measure to reduce losses associated with the release of defective products is the organization of various controlling departments and services. It is believed that such units should take timely measures to prevent the release of scrap. But the fact that quality control services do not have the required leverage over production units is usually not taken into account. In any case, control is usually carried out only after the product has been manufactured. Consequently, the inspectors have no opportunity to influence the quality promptly. The only way out of this situation is to integrate quality management procedures into the production process.

Loss from over-processing. Earlier it was said that the consumer is ready to pay only for those properties of the product that are of value to him. If a consumer needs, for example, a smart phone, then he expects to receive a product of the appropriate quality, endowed with appropriate consumer properties for a certain price. Therefore, embedding a folding spoon in it, while doubling the price, is an unproductive activity that not only will not lead to more buyers, but, on the contrary, will push them away at its cost, because the additional functionality does not add any value to it.

The loss of over-processing should also include the maintenance of automatic equipment. For example, parts move along a conveyor that regularly stops due to skewing. A dedicated worker monitors the conveyor and corrects skewed parts. The labor of such a worker is also unnecessary processing and it is necessary to solve the root cause that lies in the conveyor. Its solution will lead to an increase in the speed of the conveyor itself and will reduce the cost of maintaining a special worker who corrects skewed parts.

Losses from unnecessary movements are typical of large enterprises. This type of waste can be described as meaningless fuss that does not add any value to the production process. From the outside, these movements may seem like hectic activity, but a closer look reveals that they, like the losses discussed earlier, do not contribute to the creation of value for the consumer. There is usually only one source of this type of loss - poor work organization. These include the lack of necessary instructions, poor training of personnel or low labor discipline. This type of loss is distinguished by the fact that they can be detected quite easily, and once discovered, take quite obvious measures to eliminate them.

At the first stage, Lean Manufacturing describes possible losses, and at the second stage it already identifies them in order to eliminate them [12]. The approach that is proposed within the framework of the lean manufacturing concept is based on a mandatory, comprehensive and detailed understanding of the mechanism for creating customer value of the manufactured product. To this end, the first step is to draw up a detailed description of the entire production process. If the production is complex, then the whole process can be broken down into processes that are described and analyzed separately. To describe production processes, a visual schematic representation is used, called a value stream map. The complete sequence of actions to eliminate losses includes the following steps [2]:

- Development of maps for creating value stream of processes.
- Development of checklists to help identify the causes of losses at each stage of the process.
- Collection of statistical information about the time of value creation and the time of losses, as well as any other information indicating the presence of losses, using the developed checklists.
- Building a future value stream map (no loss).
- Analysis of the causes of losses and elimination of procedures that do not create value.
- Standardization of working procedures and use in other processes.

Further, the process can be repeated. In addition, targeted efforts can be made to eliminate certain types of waste with suitable lean tools (more on this later).

A value stream map is a detailed description of a manufacturing activity. In order for the value stream map to be as accurate as possible, the work steps must be strictly followed. When drawing up a map, one should not miss even small and at first glance insignificant details and works. If the movement of material assets is controlled by the document circulation system, then the views and trajectories of the documents being drawn up should be displayed on the map. Often it is the irrationality of the workflow that is the reason for the loss of time or the accumulation of stocks.

In production, the following types of flows are distinguished:

- Material flow (describes the movement of materials within production).
- Information flow (tells each process what to produce or what to do next).
- The flow of people or processes.

When analyzing the current state, the assessment is carried out according to the main factors. Resources:

- Determination of the range of work performed.
- Determination of the number of involved personnel.
- Determination of the amount of equipment involved.

Distance:

- Determination of all movements.
- Determination of the sequence of operations.
- Measuring the distance of each movement.

Time:

- Timing of operations.
- Timing of movements.
- Timing of the entire value stream.

Lean has its own tools. One of these is checklists, which are used to monitor and record results. They serve as primary documentary evidence reflecting the results of monitoring the performance of a particular production operation. Completing the checklists should be left to independent observers to eliminate the subjective component of the observations. Moreover, at the most critical points of observation, several independent monitors should be assigned. This will allow obtaining a consistent statistical sample of the observation results [13].

In the course of observations, inspectors should make records characterizing the peculiarities of the performance of a particular operation, as well as records indicating the possible presence of one of the types of losses. Based on the analysis of checklists, a decision is made on the presence or absence of losses in operations and a map of the value stream is drawn up, considering the desired state of the flow, after which they begin to eliminate the identified losses.

To eliminate losses, as well as to prevent their occurrence within the framework of lean production, many methods have been developed for the rational organization of production activities. Let's consider the most important ones, the use of which is a prerequisite for the successful implementation of a program for the implementation of lean manufacturing [11].

The first of these is a pull, which is a cascading production system in which the supplier (the employee of the subsequent operation of those processes), who is upstream, does nothing until the consumer, who is below, informs him about it (buyer's order - assembly - release of parts - order to the supplier). Pulling allows you to reduce your inventory to almost zero.

Another approach to improving the production process is the flow of single items - a method of work in which a machine or process processes no more than one item at a time (creating a one-item flow) [3]. Advantages of one-piece flow [7]:

- Built-in quality. The flow of one-off items greatly simplifies the integration of quality. Each operator is at the same time a controller and one tries to solve the problem on the spot, without passing it on to the next stage. Even if he missed the defects, and they passed on, they will be detected very quickly and the problem will be identified;
- Genuine flexibility. If equipment becomes part of a production line, the ability to use it for other purposes will be reduced. The lead time is reduced to the limit, which means that you can more flexibly respond to the needs of the consumer, making what he really needs. Various 3D printers are an example of this approach. The flexibility of production increases, the transition to a new range of products, which requires a change in consumer demand, is carried out at the same time more quickly;
- Productivity increase. When the work is divided into departments, then the maximum productivity is estimated by the load of people and equipment. In fact, it is difficult to determine how many people it takes to make a given number of items in high volume production because productivity is not measured in terms of value-adding work. If there is a one-piece flow cell, then non-value-added work such as moving materials is minimized and you can see who is overwhelmed and who is idle at a glance.
- Freeing up areas in the shop. When equipment is distributed to sites, significant areas in between are lost and are often occupied by reserves;
- Increased security. One-piece flow will automatically lead to increased safety by reducing the amount of material that needs to be moved around the plant;
- Improving morale. The flow of one-off products means that most of the time people are busy creating added value and can quickly see the fruits of their labor, and when they see their successes, they feel satisfaction [10].

One of the losses discussed earlier was called waiting. Waiting loss occurs when parts exit from a previous operation but cannot be processed in the next. If the previous operation continues to produce parts, then another loss is formed - stock. If stocks grow too quickly, then it becomes necessary to transport them to the warehouse (another loss). As you can see, it is very important to submit parts for the subsequent operation only when the need arises. This method of work is called Just-In-Time, which involves working on order for the next operation. As with any custom work, this approach requires clear planning to achieve flexible equipment utilization. The main factor for successful planning is to reduce the amount of resources processed and transferred to the next operation and the use of methods to control the production flow.

3 Results and Discussion

Improvement of the main production should be based on the optimization of all types of flows discussed earlier. Materials management is carried out mainly in two ways: with a push-out control system and with a pull-out control system.

A push-out (pushing) control system is a material flow control system in which objects of labor are fed from a previous technological operation to a subsequent operation in accordance with a pre-formed schedule.

A production plan is developed for each shop and the implementation is monitored, and the system is shown in Figure 1 [14, 15].

A common drawback of a push system is insufficient tracking of demand with the mandatory creation of safety stocks [9]. In contrast to the pushing system, there is a pulling (pulling) control system in which objects of labor are fed from the previous stage of the production process to the next only when there is a need for them [16]. With this method of organizing production, the central system does not interfere with the exchange of material flows between sections of the enterprise, does not establish current production targets for them [8].

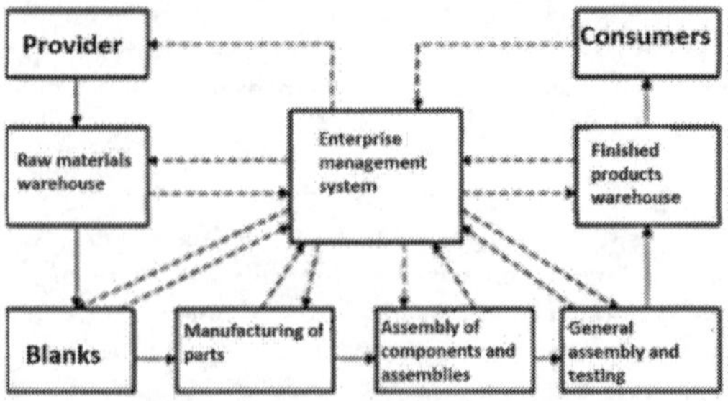

Fig. 1. Enterprise management system based on a push system.

5 Conclusions

One of the most versatile ways to improve the performance of almost any enterprise is lean manufacturing. Its peculiarity lies in a thorough study of business processes and identification of all types of losses that may arise in the main production. Depending on the amount of losses and the ease of their measurement, a decision is made on specific measures to eliminate them.

The production systems of "lean" enterprises are focused on the complete elimination of losses and continuous improvement of all processes. All employees of the company, from workers to senior management, are involved in the daily work of preventing all possible types of losses and continuous improvement. The company's management makes decisions that consider the prospect of further development, while momentary financial interests should not be decisive. The main thing on which the activities of not only the company's management, but also of each employee, should be concentrated is the timely detection, solution and prevention of problems and losses [6].

References

1. J. P. Wumek, D. Jone, *Lean Manufacturing: How to get rid of losses and make your company prosper* (Moscow, Alpina Publisher, 2018)
2. A.V. Vyalov, *Lean production: a tutorial* (Komsomolsk-on-Amur, FGBOU VPO "KnAGTU", 2017)
3. V.D. Markova, S.A. Kuznetsova, *Strategic management: concepts, concepts, decision-making tools: a reference guide* (Moscow, INFRA-M, 2019)
4. I. Masaaki, *The way to reduce costs and improve quality* (Moscow, Alpina Publisher, 2017)
5. A.A. Boyko, V.V. Kukartsev, E.S. Smolina, V.S. Tynchenko, Y.I. Shamlitskiy, N.V. Fedorova, *Jour. of Phys.: Conf. Ser.* **1353**, 012124 (2019)
6. N.V. Fedorova, V.V. Kukartsev, V.S. Tynchenko, Y.V. Danilchenko, S.N. Ezhemanskaya, N.V. Sokolovskiy, *IOP Conf. Ser.: Mat. Sci. and Eng.* **734**, 012084 (2020)
7. S. Indrawati, A. Azzam, A.C. Ramdani, *IOP Conf. Ser.: Mat. Sci. and Eng.* **598**, 012062 (2019)
8. V. Salnikov, D. Galimov, O. Mikheeva, A. Gnidchenko, A. Rybalka, *Rus. J. of Econ.* **3**, 240 (2017)
9. N.F. Revenko, *Organization of production and management at machine-building enterprises* (Moscow, Higher school, 2017).
10. E.A. Chzhan, V.S. Tynchenko, V.V. Kukartsev, N.V. Fedorova, A.S. Yamshchikov, D.A. Krivov, *IOP Conf. Ser.: Earth and Env. Sci.* **315**, 022106 (2019)
11. N.N. Pushina, N.G. Sokolova, V.P. Koretskiy, *IOP Conf. Ser.: Mat. Sci. and Eng.* **753**, 062034 (2020)
12. E. Volkova, *E3S Web of Conf.* **244**, 03023 (2021)
13. N. Havlovska, O. Illiashenko, O. Konoplina, I. Shevchuk, A. Hlynska, V. Prytys, *TEM J.* **9**, 1053 (2020)
14. O.V. Demyanova, A. Kireeva-Karimova, L.M. Zabirova, *J. of Eng. and App. Sci.* **12**, 4952 (2017)
15. E.L. Moskvicheva, A.M. Mukhametshina, A.N. Erofeyev, K.V. Savelyev, *IOP Conf. Ser.: Mat. Sci. and Eng.* **862** 042051 (2020)
16. A.A. Boyko, V.V. Kukartsev, K.Y. Lobkov, A.A. Stupina, *Jour. of Phys.: Conf. Ser.* **1015**, 042006 (2018)

AUTHOR INDEX

AUTHOR INDEX

AUTHOR INDEX